Methods of Analysis for Functional Foods and Nutraceuticals

Second Edition

FUNCTIONAL FOODS AND NUTRACEUTICALS SERIES

Series Editor
G. Mazza, Ph.D.
Senior Research Scientist and Head
Food Research Program
Pacific Agri-Food Research Centre
Agriculture and Agri-Food Canada
Summerland, British Columbia

Functional Food Carbohydrates (2007)
Costas G. Biliaderis, Ph.D. and Marta S. Izydorczyk, Ph.D.

Functional Food Ingredients and Nutraceuticals: Processing Technologies (2007)
John Shi, Ph.D.

Dictionary of Nutraceuticals and Functional Foods (2006)
N. A. Michael Eskin, Ph.D. and Snait Tamir, Ph.D.

Handbook of Functional Lipids (2006)
Edited by Casimir C. Akoh, Ph.D.

Handbook of Functional Dairy Products (2004)
Edited by Collete Short and John O'Brien

Handbook of Fermented Functional Foods (2003)
Edited by Edward R. Farnworth, Ph.D.

Herbs, Botanicals, and Teas (2002)
Edited by G. Mazza, Ph.D. and B.D. Oomah, Ph.D.

Functional Foods: Biochemical and Processing Aspects Volume 2 (2002)
Edited by John Shi, Ph.D., G. Mazza, Ph.D., and Marc Le Maguer, Ph.D.

Methods of Analysis for Functional Foods and Nutraceuticals, Second Edition (2008)
Edited by W. Jeffrey Hurst, Ph.D.

Functional Foods: Biochemical and Processing Aspects Volume 1 (1998)
Edited by G. Mazza, Ph.D.

Methods of Analysis for Functional Foods and Nutraceuticals

Second Edition

EDITED BY

W. Jeffrey Hurst

CRC Press
Taylor & Francis Group
Boca Raton London New York

CRC Press is an imprint of the
Taylor & Francis Group, an **informa** business

CRC Press
Taylor & Francis Group
6000 Broken Sound Parkway NW, Suite 300
Boca Raton, FL 33487-2742

Printed in the United States of America on acid-free paper
10 9 8 7 6 5 4 3 2 1

International Standard Book Number-13: 978-0-8493-7314-5 (Hardcover)

Library of Congress Cataloging-in-Publication Data

Methods of analysis for functional foods and nutraceuticals / editor, W. Jeffrey Hurst. -- 2nd ed.
p. ; cm. -- (Functional foods and nutraceuticals series)
Includes bibliographical references and index.
ISBN 978-0-8493-7314-5 (hardback : alk. paper) 1. Functional foods--Analysis. I. Hurst, W. Jeffrey (William Jeffrey), 1948- II. Series: Functional foods & nutraceuticals series.
[DNLM: 1. Food, Fortified--analysis. 2. Dietary Supplements--analysis. 3. Plant Extracts--chemistry. 4. Vitamins--analysis. QU 145.5 M592 2007]

QP144.F85M48 2007
613.2--dc22 2007049891

Visit the Taylor & Francis Web site at
http://www.taylorandfrancis.com

and the CRC Press Web site at
http://www.crcpress.com

Contents

Series Editor's Preface

The Functional Foods and Nutraceuticals book series, launched in 1998, was developed to provide a timely and comprehensive treatment of the emerging science and technology of functional foods and nutraceuticals, which are shown to play a role in preventing or delaying the onset of diseases, especially chronic diseases. The first ten volumes in the series, *Functional Foods: Biochemical and Processing Aspects.* Volumes 1 and 2; *Herbs Botanicals and Teas*; Handbook of Fermented Functional Foods; *Methods of Analysis for Functional Foods and Nutraceuticals; Handbook of Fermented Functional Foods; Handbook of Functional Dairy Products; Handbook of Functional Lipids; Dictionary of Functional Foods and Nutraceuticals; Processing Technologies for Functional Foods and Nutraceuticals*; and *Functional Food Carbohydrates*, have received broad acceptance by food, nutrition, and health professionals.

The latest volume, *Methods of Analysis for Functional Foods and Nutraceuticals*, second edition, edited by Dr. W. Jeffery Hurst (The Hershey Company Technical Center, Hershey, Pennsylvania) is organized into 10 chapters contributed by 22 leading scientists. This edition provides an updated, in-depth treatment of the peer-reviewed literature on methods of analysis for phytoestrogens, ω3 fatty acids, conjugated linoleic acid, flavonoids, anthocyanins, carotenoids, chlorophylls, water soluble vitamins, amino acids, and carbohydrates.

Dr. Hurst has assembled a group of outstanding international contributors in the forefront of analysis for functional foods and nutraceuticals, and together they have produced an outstanding reference book, which is expected to be a valuable resource for researchers; teachers; students; food, nutrition and health practitioners; and all those working in the functional food and nutraceutical industry.

It is hoped that this book will serve to further stimulate the development of functional foods and nutraceuticals and provide consumers worldwide with products that prevent diseases and enhance the quality of life of people everywhere.

To Dr. Hurst and all the contributors, I extend my sincere gratitude, and I hope that the reader will find this book informative and stimulating.

G. Mazza

Preface

Functional foods and nutraceuticals encompasses a wide variety of topics ranging from foods with positive health attributes to even dietary supplements. The analysis of this diverse category can create its own set of challenges since sometimes it is not known what is the active agent. Hence an analyst looks for a marker compound, but the compound of interest might occur at an extremely low level requiring alternative detection methods to be developed or the matrix could offer unique opportunities to the analysts. The first edition of this book addressed some of these issues with its focus on differing classes of compounds based on their biological significance such as vitamins and carbohydrates. Since the first edition, the area has continued to evolve with many of the topics that play important roles and these topics are included in this revision. For example, antioxidants are increasingly in the news with not only the scientific community interested but also the general public. This interest by all parties is reflected in this book with three chapters included on polyphenolic compounds. As with all chapters in this second edition, international experts have contributed their time and energies to provide the readers of this volume with timely and accurate information. Although chapters have been added on polyphenolic compounds, the majority of the remaining chapters have been revised to include updated information and references. Furthermore, those topics that were not revised have been thoroughly reviewed to ensure information is appropriate with over 85% of the volume being new or containing revised information. Those involved in this second edition feel that it makes a useful contribution to the literature on methods of analysis of functional foods and nutraceuticals.

Editor

W. Jeffrey Hurst, PhD, is a senior staff scientist for the Hershey Company and a member of the Hershey Center for Health and Nutrition, and has been active in the analysis of foods for over three decades. During this period, he has written or edited numerous books on topics ranging from lab robotics to lab automation and food analysis. Additionally, he has published over 200 papers and given presentations on an eclectic series of subjects in food analysis and analysis of biological samples. He is active in the AOAC (Association of Official Analytical Chemists) International and serves as study director on selected topics related to chocolate and cocoa. Furthermore, he serves on the AOAC task force on food allergens and on the working group on methods of analysis of dietary supplements. He is a member of the American Society for Mass Spectrometry (ASMS), a member of the Institute of Food Technology (IFT), the American Chemical Society (ACS), the Association for Laboratory Automation, and Groupe Polyphenol. He is also a member of the editorial board of the *Journal of Liquid Chromatography and Related Techniques* and is a reviewer for a variety of food and analytically related journals.

At Hershey, his emphases are on the applications of separations and spectroscopy to food analysis and the evaluation of new and unique analytical technologies in the food arena. Recently, Dr. Hurst has also focused on the chemistry of polyphenolic compounds.

Contributors

Laura Bravo
Department of Metabolism and Nutrition
Instituto del Frio (CSIC)
Madrid, Spain

Lan Dao
Western Regional Research Center, Agricultural Research Service
U.S. Department of Agriculture
Albany, California

Gary Dobson
SCRI
Dundee, Scotland, United Kingdom

Miguel Ángel Fernández-Muiño
Área de Nutrición y Bromatología
Departamento de Biotechnología y
Ciencia de los Alimentos
Universidad de Burgos
Burgos, Spain

Lourdes Gallardo-Guerrero
Department of Food Biotechnology
Instituto de la Grasa (CSIC)
Sevilla, Spain

Beatriz Gandul-Rojas
Department of Food Biotechnology
Instituto de la Grasa (CSIC)
Sevilla, Spain

Dámaso Hornero-Méndez
Department of Food Biotechnology
Instituto de la Grasa (CSIC)
Sevilla, Spain

Farah S. Hosseinian
Department of Food Science
University of Manitoba
Winnipeg, Manitoba
Canada

Rafael Japón-Luján
Department of Analytical Chemistry
University of Córdoba
Córdoba, Spain

Manuel Jarén-Galán
Department of Food Biotechnology
Instituto de la Grasa (CSIC)
Sevilla, Spain

William R. LaCourse
Department of Chemistry and Biochemistry
University of Maryland
Baltimore, Maryland

María D. Luque de Castro
Department of Analytical Chemistry
University of Córdoba
Córdoba, Spain

José M. Luque-Rodríguez
Department of Analytical Chemistry
University of Córdoba
Córdoba, Spain

Larry R. Massom
Technical Development
Lovance Laboratories, Inc
Madison, Wisconsin

Raquel Mateos
Department of Metabolism and Nutrition
Instituto del Frio (CSIC)
Madrid, Spain

María Isabel Mínguez-Mosquera
Department of Food Biotechnology
Instituto de la Grasa (CSIC)
Sevilla, Spain

B. Dave Oomah
Agriculture and Agri-Food Canada
Pacific Agri-Food Research Centre
Summerland, British Columbia, Canada

Antonio Pérez-Gálvez
Department of Food Biotechnology
Instituto de la Grasa (CSIC)
Sevilla, Spain

María Roca
Department of Food Biotechnology
Instituto de la Grasa (CSIC)
Sevilla, Spain

María Teresa Sancho-Ortiz
Área de Nutrición y Bromatología
Departamento de Biotechnología y
Ciencia de los Alimentos
Universidad de Burgos
Burgos, Spain

Gary Takeoka
Western Regional Research Center,
Agricultural Research Service
U.S. Department of Agriculture
Albany, California

Felicidad Valls-García
Área de Nutrición y Bromatología
Departamento de Biotechnología y
Ciencia de los Alimentos
Universidad de Burgos
Burgos, Spain

1 Phytoestrogens

B. Dave Oomah and Farah S. Hosseinian

CONTENTS

1.1 INTRODUCTION

Several interesting events have taken place since the publication of the original chapter on phytoestrogen (Oomah, 2002). Most important of all was the early termination of one of the trial arms of the Women's Health Initiative (WHI) that was evaluating hormone replacement therapy (HRT) (Writing Group for the Women's Health Initiative Investigators, 2002), although information has been updated recently (Stefanick et al., 2006). This led to uncertainty in the risks and benefits of nonhormonal therapies including diets high in phytoestrogens for alleviating hot flashes after menopause and their avoidance as far as possible. Dietary exposure of phytoestrogens thus became a priority in human health resulting in the establishment and compilation of a comprehensive database for lignan contents of Dutch plant foods (Milder et al., 2005), for isoflavones in diets in the United Kingdom (Clarke and Lloyd, 2004) followed by further emphasis and focus on exposure and toxicology of phytoestrogens regarding human health (NTP-CERHR Expert Panel Report on the reproductive and developmental toxicity of genistein, 2006). In this regard, the United Kingdom Committee on Toxicity of Chemicals in Food, Consumer Products and the Environment (COT) has advanced a definition of phytoestrogen as "any plant substance or metabolite that induces biological responses in vertebrates and can mimic or modulate the actions of endogenous estrogens usually by binding to estrogen receptors."

The general concern with environmental issues has led to the development of eco-friendly methods of phytoestrogen preparation (Cacace and Mazza, 2006; Downing et al., 2007; Yu et al., 2007). New sources of phytoestrogen are continually being identified, isolated, and characterized and their effects on animal and human health are also being evaluated. These developments have been important as quantification of phytoestrogens in foodstuffs as well as in pharmacological and toxicological studies are dependent on accurate and precise analytical methodology. As a result, the information available on phytoestrogen concentrations in foodstuffs and biological matrices has increased significantly in recent years. The analysis of phytoestrogen content in 2000 foods, for instance, is expected to be the cornerstone of the nutritional prospective database of diet and cancer started in 1993 of the European Prospective Investigation of Cancer (EPIC).

Phytoestrogens are diphenolic compounds that structurally resemble the human sex hormone, estrogen. This shining similarity of these compounds at the molecular level provides phytoestrogens the ability to mildly mimic, and in some cases act as an antagonist to estrogen. Phytoestrogens can be grouped into three main classes: coumestans, isoflavones, and lignans. Coumestans, of which coumestrol is the most common, are found in red clover and alfalfa, with smaller amounts in split peas and bean sprouts. Isoflavones are found primarily in soybeans and soy products and are probably the best known and intensely investigated for their substantial health benefits among the three classes. Lignans are found in many cereals and grains, the highest amounts occurring in flaxseed. Despite their more widespread occurrence in foods, and greater consumption in Western populations, the lignans have received comparatively little attention. Although the number of studies supporting the health benefits of lignans is growing, dietary consumption of lignans is plummeting. According to 2002 data from the Framingham Heart Study, daily intake of lignans by American women has dropped to less than 1 mg/day (Schatzman, 2007).

Evidence is beginning to accrue that phytoestrogens may offer protection against a wide range of human conditions including breast, bowel, prostrate and other cancers, cardiovascular disease, brain function, alcohol abuse, osteoporosis, and menopausal symptoms (Bingham et al., 1998). The reported health benefits of the naturally occurring isoflavones, genistein, and daidzein include relief of menopausal symptoms (Murkies et al., 1998; Nestel et al., 1999), reduction of osteoporosis (Gennari et al., 1998), improvement in blood cholesterol levels (Sharma, 1979), and lowering the risk of certain hormone-related cancers (Peterson and Barnes, 1991) and coronary heart disease (Federal Register, 1999). The basis for these effects has not been established, but the weak estrogenic activity of isoflavones, sometimes referred to as phytoestrogens, may be a factor in conferring these properties. As a result, many food manufactures are striving to provide products containing soy or isoflavones to consumers.

The interest in phytoestrogens as functional foods stems from research suggesting that the intake of dietary phytoestrogens plays an important role in prevention of menopause symptoms, osteoporosis, cancer and heart disease (Brandi, 1999), and lung cancer (Schabath et al., 2005). Phytoestrogen supplementation with flaxseed or soy flour has been reported to increase vaginal cell maturation, an indication of estrogen activity in postmenopausal women (Wilcox et al., 1990), and significantly reduce menopause symptom scores, particularly hot flash and vaginal dryness (Brzezinski and Debi, 1999). Dietary studies indicate substantial reduction in breast cancer risk among women with high urinary excretion of phytoestrogens, particularly the isoflavone equol and the lignan enterolactone (Ingram et al., 1997). The lower incidence of prostate cancer in Asian men compared to men from North America and Europe has also been speculated to be due to the higher dietary intake of isoflavones and lignans (Adlercreutz, 1990; Morton et al., 1997). A recent preliminary study indicates that high isoflavone exposure is associated with low-plasma estradiol in postmenopausal women suggesting the involvement of diet–gene interactions (Low et al., 2005). Phytoestrogens may alter sex hormone and modulate their metabolism at the cellular level by biochemical mechanisms. They influence intracellular enzymes, protein synthesis, growth factor action, malignant cell proliferation and differentiation, angiogenesis, calcium transport, Na^+/K^+ ATPase, vascular smooth muscle cells, and lipid oxidation (Adlercreutz, 1995; Adlercreutz and Mazur, 1997). Lignans from the medicinal plant *Schisandra*, namely gomisin M_1, even exhibit potent anti-HIV activity (Chen et al., 2006).

Isoflavonoid and lignan compounds exhibit a range of mammalian health-promoting activities that are currently the focus of intense study (Table 1.1). Dietary genistein reduces susceptibility to mammary cancer in rats (Fritz et al., 1998) and

TABLE 1.1
Selected Animal Health-Promoting Effects of Isoflavonoid and Lignan Compounds

Compound	Effect	Species
Genistein	Prevention of mammary cancer	Rat
Genistein	Inhibition of prostate tumor growth	Human (cell lines)
Genistein; biochanin A	Inhibition of stomach tumor growth	Human (cell lines)
Genistein, daidzein	Activation of natural killer cells	Human
Genistein	Reduction of serum triglycerides and cholesterols	Rat
Genistein, daidzein	Protection against low-density lipoprotein oxidation	Human
Genistein	Prevention of bone loss	Mouse
Genistein	Antiangiogenesis	Human (cell lines)
Daidzin	Suppression of alcohol consumption	Hamster
7-Hydroxymatairesinol	Inhibition of mammary cancer	Rat
7-Hydroxymatairesinol	Inhibition of uterine cancer	Rat
7-Hydroxymatairesinol	Inhibition of prostate cancer	Rat
Secoisolariciresinol	Inhibition of mammary cancer	Rat

Sources: From Dixon, R.A. and Steele, C.L., *Trends Plant Sci.*, 4, 394–400, 1999; Unkila, M., *HMRlignan Webinar*, October 24, 2005. With permission.

helps to prevent bone loss caused by estrogen deficiency in female mice (Ishimi et al., 1999). These effects are mirrored by epidemiological evidence that implicates dietary isoflavones as major factors associated with reduced cancer risk in populations consuming a high soy diet (Setchell and Cassidy, 1999). Lignans are chemopreventive in animal models since they inhibit mammary, uterine, and prostate cancers in rats (Unkila, 2005). Potential harmful effects have also been shown for isoflavones from forage crops in causing specific infertility problems of sheep feeding on clover (Shutt and Cox, 1972). Their negative role cannot be excluded in the early stage of carcinogenesis (Messina et al., 1994).

These and other studies have provided the impetus for the manufacture of functional foods and nutraceutical products to meet the phytoestrogenic needs for women seeking alternatives to hormone replacement medications. For example, manufacturers in Australia promote the use of commercially produced bread, Burgen Soy-Lin that combines soy and flaxseed (Payne, 2000) and claims to contain 220 mg of isoflavones and lignans per four slices (Jorgensen et al., 1998). An example of a neutraceutical in this regard is an isoflavone extract of soybean, 17-β-estradiol (Estrace, Roberts Pharmaceuticals Canada, Inc.), that is being marketed as a prescriptive oral estrogen for effective management of menopause (Oomah and Mazza, 2000). A number of isoflavone extracts are available in pill form as extracts of soy (i.e., Estroven, Healthy Woman, and others) or red clover (Promensil) in doses ranging from 40 to 55 mg/day (Carusi, 2000). Novasoy daily soya isoflavone concentrates (20%, 40%, and 70%) developed using controlled delivery technology for supplements by ADM and Bonistein and a pure synthetic genistein developed by DSM nutritional products for bone health are two new isoflavone products (van Nieuwenhuyzen, 2007). In France, more than 30 food supplements, mostly containing soy and a few with other plants such as clover, alfalfa, yam, hops, or cimicifuga, for menopausal women have been commercialized (Bennetau-Pelissero et al., 2003). HMRLignan, a source of 7-hydroxymatairesinol isolated and purified from the Norway spruce (*Picea abeis*), released by Switzerland-based Linnea may offer cardiovascular protection, since it effectively reduced inflammation and inhibited the production of reactive oxygen species from white blood cells (Nutraingredients, 2007). The rapid growth of functional foods and interest in health benefits of plant foods have created the need for standardized methodology for the analysis of phytochemicals, phytoestrogens in particular, in raw and processed products as well as their metabolites in mammalian tissues and fluids. However, investigation of the possible benefits of phytoestrogens is hampered by lack of analytical standards and, hence, inadequate methods for the measurement of low levels in most foods. This problem may prove to be a major dilemma for regulatory authorities, clinicians, and others wishing to advise the general public on whether these compounds really do have the health benefits attributed to them. The complexity of phytoestrogens, with at least 15 different chemical forms of the isoflavones alone occurring in foods, is a major limitation in the study of metabolism of these substances in human subjects (Bingham et al., 1998).

1.2 GENERAL ASPECTS

1.2.1 Sources

More than 300 plants are known to possess estrogenic activity (Farnsworth et al., 1975) with isoflavones and coumestans (occurring) as the most common estrogenic compounds. The dietary sources of phytoestrogens are only partially known. Oilseeds, such as flaxseed, have the highest detected contents of lignans. Seaweeds, legumes, cereal brans, whole cereals, vegetables, and fruits contain lesser amounts (Thompson et al., 1991). Rye has a high content of plant lignans that has been hypothesized as one of the factors in rye bran responsible for inhibiting prostate cancer growth (Bylund et al., 2000). Wheat, barley, corn, amaranth, millet, and oat bran are common dietary sources of the lignan 7-hydroxymateiresinol, abbreviated as 7-HMR. Considerable amounts of 7-HMR (about 60% of total lignans) mainly in the unconjugated free form are found in coniferous trees, especially in the heartwood of spruce (Unkila, 2005). Mammalian isoflavonoids are derived mostly from soybeans and fermented soy products. Some alcoholic beverages, such as beer and bourbon, and several varieties of legumes, such as peas and beans, also contain isoflavonoid precursors, but at lower levels (Franke et al., 1994; Gavaler et al., 1995; Lapčík et al., 1998). It has been suggested that resveratrol, a weak phytoestrogen found in grapes and wine, may contribute to the beneficial effects of wine consumption (the "French Paradox") (Kopp, 1998). Resveratrol also exhibits antiestrogenic activity and inhibits the growth of human breast cancer cells (Lu and Serrero, 1999). Similarly, the identification of the potent phytoestrogen, 8-prenylnaringinin in hops (*Humulus lupulus*) and beer has been suggested to be a contributing factor in the beneficial health effects of moderate beer consumption (Milligan et al., 1999).

Tea contains high levels of lignans but only low levels of isoflavones (Mazur et al., 1998). The precise phytoestrogen content of many individual foods has only recently been documented (Nakamura et al., 2001; Clarke and Lloyd, 2004; Hubert et al., 2005; Milder et al., 2005; Smeds et al., 2007) and their effect on human health needs further studies to unravel their complexity.

The isoflavonoids, genistein (4′,5,7-trihydroxyisoflavone), biochanin A (5,7-dihydroxy-4′-methoxyisoflavone), daidzein (4′,7-dihydroxyisoflavone), and formononetin (7-hydroxy-4′-methoxyisoflavone), are supposed to be health-promoting dietary factors of plant origin. They are particularly abundant in seeds and other parts of many plant species belonging to Leguminosae. In the human diet, they are represented by the genera *Glycine*, *Phaseolus*, *Cicer*, *Lens*, and *Pueraria*. In legumes, the isoflavonoids are present mainly as 6″-*O*-malonylglycosides and β-glycosides and only minimal amounts are present in the aglycone form (Lapčík et al., 1998). The genus *Trifolium* that include red clover and lucerne rich in genistein, daidzein, biochanin A, and formononetin, and important sources of phytoestrogens for domestic animals may be effective in reducing bone loss induced by ovariectomy of animals (Occhiuto et al., 2007). It has been previously reported that phytoestrogens such as daidzein and genistein are present in crude drugs including the root of *Puerarie lobata* or the root of *Glycyrrhizae glabla* (Shiizaki et al., 1999). In fact, deoxymiroestrol from *Pueraria mirifica*, a rejuvenating folk medicine from Thailand, is considered to be the compound with the highest estrogenic potency among the known phytoestrogens (Chansakaow et al., 2000).

Only relatively few of the many isoflavones identified in plants possess oestrogenic activity. Examples of such substances include daidzein and genistein, their glucosides daidzin and genistin, and their methyl ether derivatives formononetin and biochanin A, respectively (Price and Fenwick, 1985). Similarly, only a few of the plant coumestans exhibit oestrogenic activity, the most important of which are coumestrol and 4′-methoxy-coumesterol (Figure 1.1). Lignans, as the precursors of lignin formation in cell walls, are present as minor constituents in many plants. They contain a 2,3-dibenzylbutane structure (Table 1.2) and were first identified in humans in the late 1970s following studies in monkeys in which compounds were detected in the urine with apparent similarities to urinary steroid hormone metabolites (Setchell and Adlercreutz, 1988).

In plants, isoflavones play major roles in the defense response to pathogen attack (Rivera-Vargas et al., 1993), and in nodulation and nitrogen fixation (Pueppke, 1996). Isoflavones also impart negative taste components to foods (Okubo et al., 1992) and the reduction of isoflavone concentrations would be of value for other products. A benefit of manipulating isoflavone concentrations is that increased levels of isoflavones may increase resistance to various pathogens (Padmavati and Reddy, 1999). Developing other grain crops that can synthesize isoflavone would provide food manufacturers with alternatives to soy for use in their products.

1.2.2 Chemistry

The basic structural feature of isoflavonoids is two benzene rings linked through a heterocyclic pyrane ring at position 3. The most abundant isoflavonoids are genistein

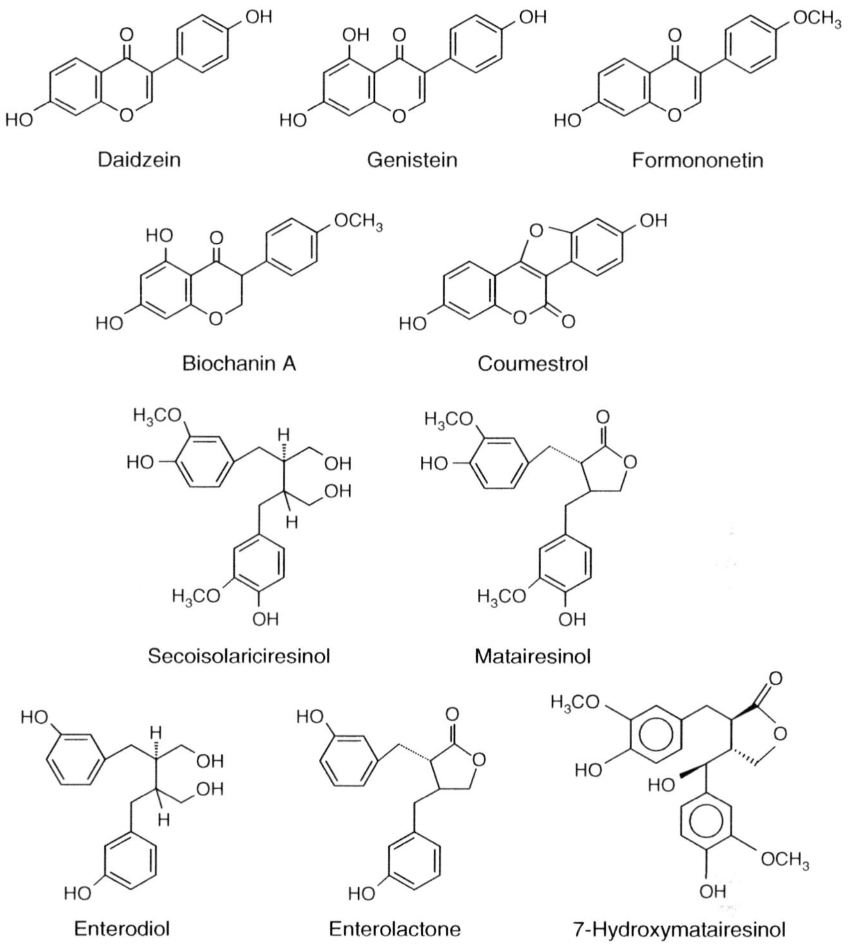

FIGURE 1.1 Structure of common phytoestrogens.

and daidzein. Other compounds in this group include formononetin and biochanin A, which in humans can be metabolized to daidzein and genistein, respectively. Daidzein can be further converted to *O*-desmethylangolensin (*O*-DMA) and equol, and genistein to *p*-ethyl phenol. Lignans are a group of diphenolic compounds with dibenzylbutane skeleton structures. The mammalian lignans enterolactone and enterodiol are produced by the bacterial flora in the colon from the plant lignans matairesinol and secoisolariciresinol, respectively (Kardinaal et al., 1997).

1.2.2.1 Ultraviolet Absorption Spectrum

Isoflavonoids have distinct ultraviolet (UV) absorption spectra (Table 1.3); it is therefore possible to enhance detection selectivity using detector wavelengths at 260 and 280 nm, as well as to provide a positive confirmation of identity.

TABLE 1.2
Nomenclature of Lignans

Compound	Lignan Class	Nomenclature
Secoisolariciresinol (SDG)	Oxydiarylbutane	([*R*-(*R**′,*R**)]-2,3-bis[(4-hydroxy-3-methoxyphenyl) methyl]-1,4-butanediol)
Matairesinol	Diarylbutyrolactone	([3*R*-*trans*]-dihydro-3,4-bis[(4-hydroxy-3-methoxyphenyl)methyl]-2(3*H*)-furanone)
7-Hydroxymatairesinol (7-HMR)		(4-[hydroxy(4-hydroxy-3-methoxyphenyl)methyl]-3-(4-hydroxy-3-methoxybenzyl)-dihydrofuran-2-one)
Lariciresinol	Furanoid	([2*S*-(2α,3β,4β)]-tetrahydro-20(4-hydroxy-3-methoxyphenyl)-4-[(4-hydroxy-3-methoxyphenyl) methyl]-3-furanmethanol)
Isolariciresinol	Tetrahydronaphthalene	([1*S*-(1α,2β,3α)]-1,2,3,4-tetrahydro-7-hydroxy-1-(4-hydroxy-3-methoxy-phenyl)-6-methoxy-2, 3-napthalenedimethanol)
Arctigenin	Diarylbutyrolactone	((3*S*,4*S*)-3-[(3-methoxy-4-hydroxyphenyl)methyl]-4-[(3,4-dimethoxyphenyl)methyl]dihydro-2(3*H*)-furanone)
Pinoresinol	Furofuran	([1*S*-(1α,3α,4β,6α)]-4,4′-(tetrahydro-1*H*,3*H*-furo(3,4-)furan-1,4-diyl)bis(2-methoxyphenol))
Nordihydroguaiaretic acid (NDGA)		(1,4-bis[3,4-dihydrooxyphenyl]-2,3-dimethylbutane)
Hinokinin	Diarylbutyrolactone	(3*S*,4*S*-3,4-bis(1,3-benzodioxol-5-ylmethyl)dihydro-2-(3*H*)-furanone)

The spectrum of isoflavones consists of two absorption maxima at 245–275 and 300–330 nm with low relative intensities in the latter range. The UV absorption spectra of daidzein, formononetin, and genistein in methanol (λ_{max}) are 238 sh, 249, 260 sh, 303 sh; 240 sh, 248, 259 sh, 311; and 261, 328 sh nm, respectively (Markham, 1982). The concentration of stock solutions of isoflavonoids can be determined by absorbance readings at the wavelength with maximum absorption (λ_{max}) using molar extinction coefficients(ε) after proper dilution with ethanol (acetonitrile for coumestrol) using the following values: daidzein ($\lambda_{max} = 250$ nm, $\varepsilon = 20{,}893$), genistein ($\lambda_{max} = 263$ nm, $\varepsilon = 37{,}134$), formononetin ($\lambda_{max} = 256$ nm, $\varepsilon = 29{,}512$), biochanin A ($\lambda_{max} = 263$ nm, $\varepsilon = 27{,}542$), coumestrol ($\lambda_{max} = 339$, $\varepsilon = 22{,}330$), and equol ($\lambda_{max} = 281$ nm, $\varepsilon = 6761$) (Franke et al., 1994). Therefore, monitoring carried out at 260, 280, and 342 nm (for coumestrol) provides sensitive detection at or very near the absorption maximum of isoflavonoids (Figure 1.2) (Franke et al., 1995). Absorption spectra of the isoflavones, daidzein, genistein, formononetin, and biochanin A, in water show strong absorption between 250 and 270 nm and a rather weak band or shoulder in the 300–350 nm region. Coumestrol has maxima at 240 and 355 nm (Beekman et al., 1999). In electrospray (ES) mass spectrometry, positive ionization of genistein yields two prominent ions at $m/z = 271$ and 312 at a ratio of 4.5:1. These correspond to the singly protonated ion of genistein, $[MH]^+$, and the commonly observed acetonitrile solvent adduct $[M\text{-}CH_3CN\text{-}H]^+$. With daidzein a single

TABLE 1.3
Nomenclature, $[M+H]^+$, Fragment Ions, Ultraviolet (UV) λ_{max}, and Structural Characteristics of Isoflavones

Compound	$[M+H]^+$ (m/z)	Fragment Ion (m/z)	λ_{max} (nm)	Isoflavone Part					
				R_1	R_2	R_3	R_4	R_5	MW
Daidzin (daidzein-7-*O*-β-D-glucoside)	417	255	248, 302 sh	H	H	H	glc	H	416
Glycitin (glycitein-7-*O*-β-D-glucoside)	447	285	257, 320	H	H	OMe	glc	H	446
Genistin (genistein-7-*O*-β-D-glucoside)	433	271	258, 330 sh	H	OH	H	H	H	270
Glycitein	285		257, 320	H	H	OMe	H	H	284
Daidzein	255		248, 302	H	H	H	H	H	254
Genistein	271		260, 330 sh	H	OH	H	H	H	270
Formononetin	269		250, 301 sh	Me	H	H	H	H	268
Biochanin A	285		262, 325 sh	Me	OH	H	H	H	284

prominent ion is observed at $m/z = 255$, corresponding to $[MH]^+$ (Jung et al., 2000). Infrared (IR) spectral characteristics of malonylated soybean isoflavones 6″-*O*-malonyldaidzin, 6″-*O*-malonylglycitin, and 6″-*O*-malonylgenistin from soybean seeds are similar (1735 and 1695 cm^{-1}) (Kudou et al., 1991).

1.2.2.2 Fluorescence Spectrum

The fluorescence properties of the isoflavones daidzein, genistein, formononetin, biochanin A, and coumestrol are markedly different despite the closely analogous molecular structures. Under aqueous alkaline conditions, daidzein, formononetin, and coumestrol exhibit strong fluorescence, whereas isoflavones with an additional hydroxy group at position 5, genistein and biochanin A, do not show any emission at all. Both daidzein and formononetin show a broadband without vibrational details, with maximum at 478 nm because of their molecular structures. The spectrum of coumestrol emerges at distinctly shorter wavelengths with a maximum at 441 nm although a broadband is also observed (Beekman et al., 1999). Formononetin exhibits large Stokes' shift dependent on pH, with less-pronounced shift in emission from 495 nm for neutral molecule to 470 nm for the anion at high pH (De Rijke et al., 2006).

1.2.3 Biosynthesis

1.2.3.1 Isoflavonoids

The biosynthesis of isoflavonoids has recently been reviewed by Dixon (1999). Isoflavones may be either constitutively synthesized under the control of developmental programs or induced in response to environmental stress such as pathogen

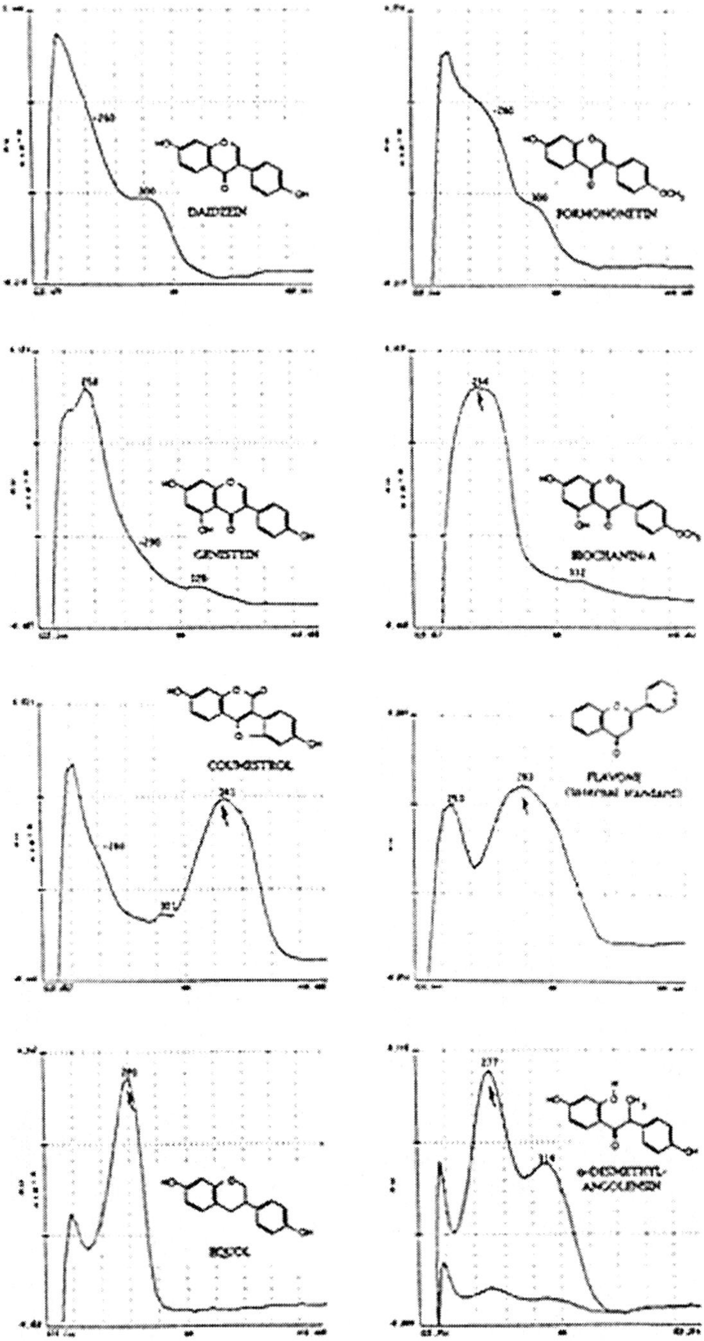

FIGURE 1.2 Structure and ultraviolet (UV) scan of phytoestrogens. (Reprinted with permission from Franke, A.A., et al., *Proc. Soc. Exp. Biol. Med.*, 208, 18–26, 1995.)

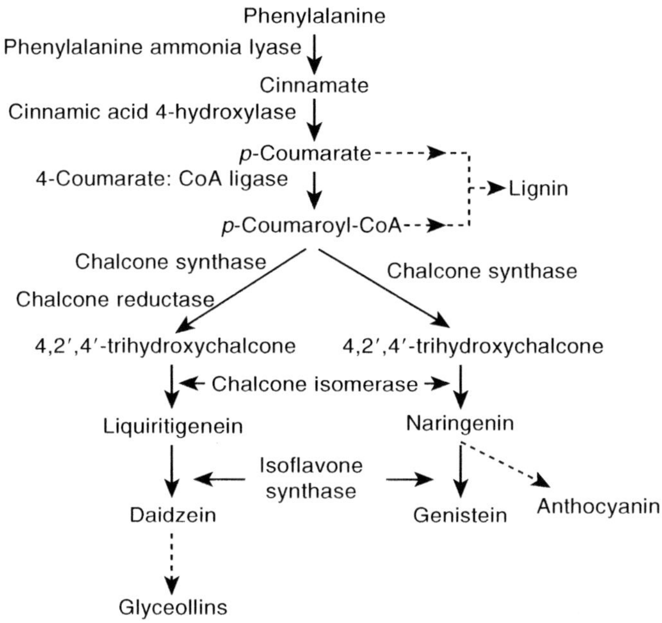

FIGURE 1.3 Isoflavonoid biosynthesis via the phenylpropanoid pathway. (Reprinted with permission from Jung, W., Yu, O., Lau, S.-M.C., O'Keefe, D.P., Odell, J., Fader, G., and McGonigle, B., *Nat. Biotechnol.*, 18, 208–212, 2000.)

infection. Isoflavones are synthesized by a branch of the phenylpropanoid pathway of secondary metabolism by specific enzymes, L-phenylalanine ammonia-lyase (PAL), cinnamate 4-hydroxylase (C4H), and 4-coumarate ligase (4CL), as well as the acetyl-CoA carboxylase (ACCase) (Dixon and Pavia, 1995) (Figure 1.3). The B-ring of the C_{15} of isoflavonoids originates from the phenylpropane unit of 4-coumaroyl-CoA, whereas the A-ring is derived from acetyl-CoA via the condensation of malonyl CoA (Dixon, 1999). The isoflavone synthase catalyzes the first step of isoflavone biosynthesis, the oxidation of 7,4′-dihydroxyflavanone (liquiritigenin) or 5,7,4′-trihydroxyflavanone (naringenin) to daidzein or genistein, respectively. Conversion of naringenin to genistein is carried out via a protein with isoflavone synthase activity. The presence of isoflavone synthases in other legumes and sugar beet and the potential for producing isoflavones in non-isoflavone-producing crops have been demonstrated (Jung et al., 2000).

1.2.3.2 Lignan

The biosynthesis of lignan has recently been revised based on the discovery of the dirigent proteins that guide phenolic radical coupling (Davin et al., 1997; Davin and Lewis, 2000). Lignans are derived mainly via differential partitioning of the monolignol, coniferyl alcohol to yield the lignan (+)-pinoresinol which in turn serves as the precursor of both (−)-secoisolariciresinol and (−)-matairesinol (Figure 1.4).

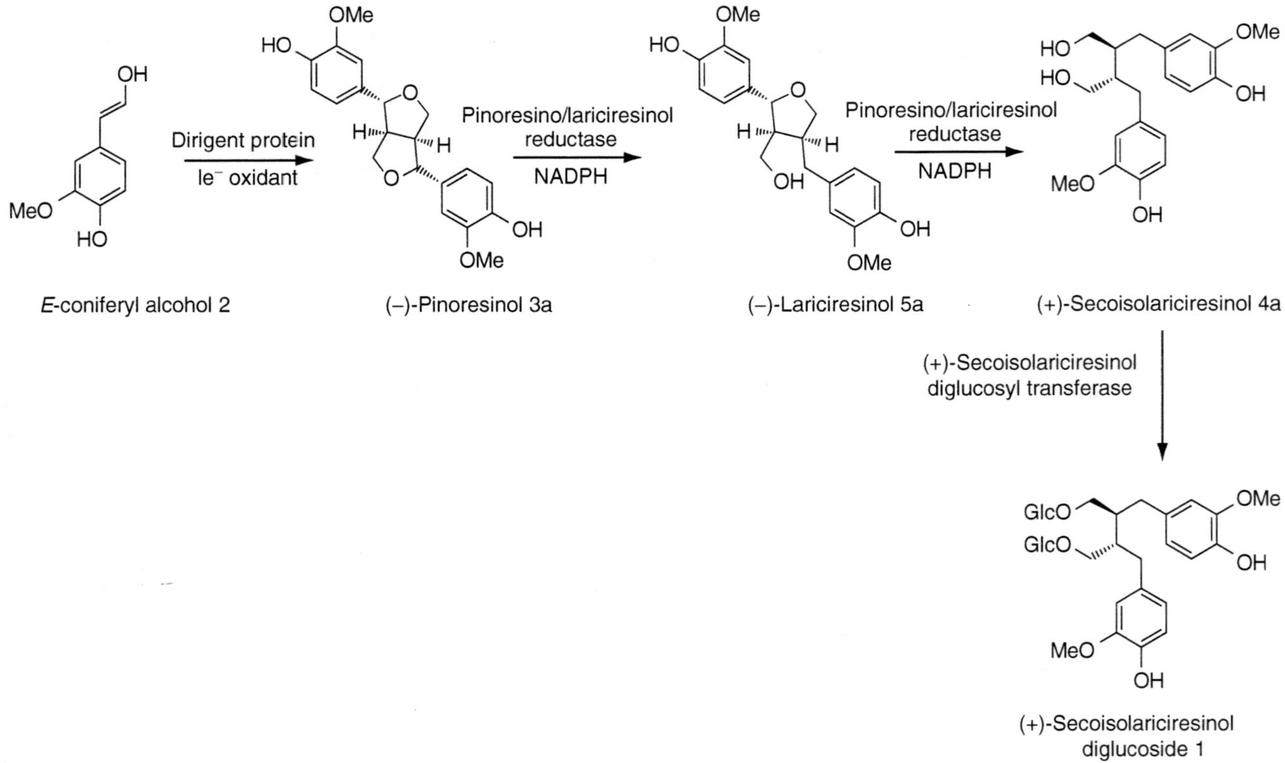

FIGURE 1.4 Biosynthesis of the lignan secoisolariciresinol diglucoside (SDG) in flaxseed. (From Ford, T.D., Davin, L.B., and Lewis, N.G., *Proc. 58th Flax Inst. United States*, March 23–25, Fargo, ND, 39–48, 2000.)

Two genes encoding the corresponding protein involved in the formation of pinoresinol and lariciresinol have been obtained from developing flaxseed (Ford et al., 2000). Elucidation of the pathway leading to the cancer chemopreventive agent (−)-secoisolariciresinol diglucoside (SDG) was first accomplished using *Forsythia intermedia*, a plant that both produces and further metabolizes enantiometrically pure (+)-pinoresinol, a dimeric lignan formed from (*E*)-coniferyl alcohol. It involves the regio- and stereo-selective intermolecular phenoxy radical coupling of two molecules of (*E*)-coniferyl alcohol by a pinoresinol dirigent protein to yield pinoresinol. Sequential enantiospecific reduction of this intermediate then occurs by a reductase to consequently generate lariciresinol and then secoisolariciresinol. Glycosylation of secoisolarisiresinol is accomplished by secoisclariciresinol diglucosyl transferase that appears to be localized mainly in the seed coat. Elucidation of the lignan biosynthetic pathway is leading to the development of strategies for enhancing the levels of SDG in staple dietary foodstuffs and maximizes yields of lignans used to treat or protect against human disease (Jung et al., 2000).

1.2.4 Metabolism

The common dietary phytoestrogens, isoflavones and lignans, show similar patterns of metabolism in animals and humans. Both isoflavones and lignans undergo significant metabolism by bacteria in the gastrointestinal tract (Setchell and Adlercreutz, 1988). In foods, the isoflavones are conjugated with glucose to form glycosides and in soybean, the major glycosides are daidzin, genistin, and glycitin. These glycosides are deconjugated to their aglycones daidzein, genistein, and glycitein by acid hydrolysis or bacterial action in the rumen of cows and sheep. Daidzein and genistein are further converted to equol and *p*-ethylphenol in the rumen of sheep (Cox and Davies, 1988). In humans, daidzein and genistein are metabolized to dihydrodaidzein and dihydrogenistein, respectively that are further metabolized to equol and *O*-DMA and 6′-hydroxy-*O*-DMA in the case of genistein (Setchell and Adlercreutz, 1988).

The lignans also occur in plants as secoisolariciresinol and matairesinol with glucose residues in the diglucosides attached to the phenolic or side-chain OH groups. The glucose and methyl groups are removed by colonic bacterial flora during fermentation to form the diphenols enterodiols that is further oxidized to enterolactone. Matairesinol is converted to enterolactone through dehydroxylation and demethylation (Kurzer and Xu, 1997). 7-Hydroxymatairesinol, the major lignan manufactured from Norwegian spruce tree, is not attached to sugar molecules and therefore need not be cleaved in the intestinal tract resulting in sustained increase in circulating plasma enterolactone concentration (Unkila, 2006).

Absorption of the dietary phytoestrogens is facilitated by hydrolysis of the sugar moiety by human gut bacterial β-glucosidases, gastric hydrochloric acid, and by β-glucosidases in foods (Kelly et al., 1993). After absorption in the small intestine, isoflavones and lignans are conjugated with glucuronic acid and sulfate, these conjugates are excreted through urine and bile and undergo enterohepatic circulation. The conjugated phytoestrogens, after excretion in the bile, are deconjugated once again by gut bacteria (Kurzer and Xu, 1997). As dietary phytoestrogen metabolism is predominantly determined by the gastrointestinal flora, antibiotic use or bowel disease and gender will modify metabolism. Concurrent dietary intake, in particular high dietary fiber, vegetable, and fruit intake and duration of exposure, exert a major influence on lignan and isoflavone metabolism (Murkies et al., 1998). Recent studies suggest that soy consumption in humans can influence thyroid function by disrupting the effects of estradiol and may have profound effects in human infants raised on soy formula because of the high daily intake when adjusted for body weight (Patisaul, 2005).

1.3 METHOD OF ANALYSIS

1.3.1 Sample Preparation and Extraction

1.3.1.1 Sample Preparation

Sample preparation and extraction are important in the analysis of phytoestrogens, since large losses can occur depending on the technique used and the isoflavone or lignan investigated (Bingham et al., 1998). Before extraction, samples are sometimes spiked with various volumes of stock solution containing 20% of standard diadzein, genistein, and biochanin A (Hutabarat et al., 1998). The spiking solution (1.25%, w/v *tert*-butylhydroquinone in methanol) can also be combined with the extractant (1 M HCL) (Wang et al., 1990). In serum, sodium acetate buffer (0.1 M, pH 5.0) and an internal standard containing d_4-equol, d_4-daidzein, d_4-genistein, d_4-enterodiol, and d_4-enterolactone are added, equilibrated for 1 h at room temperature before hydrolysis that is allowed to proceed overnight at 37°C after the addition of β-glucuronidase laryl sulfatase (Morton et al., 1999). The choice of flavone as the internal standard is based on its excellent similarity with phytoestrogens, its elution in an empty and late part of the chromatogram avoiding interference with the analytes, and its stability against heat and acids (Franke et al., 1995). The choice of internal standard is important especially when modifying gradient high-performance liquid chromatography (HPLC) for rapid separation. For example, internal standards such as fluorescein and robigenin overlap with genistein when acetonitrile–water gradient is used to separate 12 isoflavones within 20 min, whereas formononetin (5 μg/mL) is most appropriate under the same conditions (Hsieh et al., 2004). A synthetic compound, 2,4,4′-trihydroxydeoxybenzoin (THB, 1-[2,4 Dihydoxy-phenyl]-[3-hydroxyphenyl]-ethanone), has been used as an internal standard for soy isoflavone analysis (Downing et al., 2007), as a reference peak for retention time adjustments, and as an external standard (Song et al., 1998), but the use of this compound is limited by lack of its commercial availability. Internal standards such as *p*-nitrophenol (30 mg/L) have also been used with solvents (66% acetronitrile, 0.4% NaCl in water) during extraction (Vänttinen and Moravcova, 1999). Several other internal standards have been proposed for quantitation of isoflavones by HPLC including apigenin (an analogue of genistein), acetophenone, 2′-methoxy-flavone, and 6-methoxy-flavone. Use of internal standards is infrequent although it allows for the precise determination of analyte recoveries and can account for losses during the extraction process. These standards range from deuterated (2*H*) or carbon-13 (^{13}C)-labeled stable phytoestrogens isotopes with prevailing use in gas chromatography–mass spectrometry (GC-MS) methods, or enantiomeric compounds not naturally present in the sample (Wang et al., 2002). However, the extent of hydrolysis of lignan conjugates cannot be determined when adding unconjugated labeled standards and this aspect has to be optimized by maximizing the yield of lignans from the food matrix. For this purpose, several internal standards, $^{13}C_3$-pinoresinol, $^{13}C_3$-lariciresinol, $^{13}C_3$-medioresinol, $^{13}C_3$-syringaresinol, $^{13}C_3$-secoisolariciresinol, and $^{13}C_3$-matairesinol, have been synthesized for quantification of lignans in foods using isotope dilution (ID)-GC-MS (Penalvo et al., 2005). Glycoside, β-glucuronide, and sulfate esters of suitable analogous compounds have been used in liquid chromatography (LC)-based techniques to correct extraction and incomplete hydrolysis. Thus, only 7 out of 22 studies on phytoestrogens content of foods used an internal standard (Reinli and Block, 1996). Previously, most investigators relied on urine and blood levels as markers of intake of isoflavones and lignan precursors in foods (Bingham et al., 1998). Availability of newer standards or their rapid synthesis has sparked increased interest in their use in protocols aimed to serve as reference analytical methods for building accurate food databases (Penalvo et al., 2005).

1.3.1.2 Extraction

Extraction methods vary widely depending on the sample and the phytoestrogen of interest. Several different solvents are in use including ethanol, methanol, acetonitrile, diethyl ether, and acetone for extraction, although ethanol and methanol appear to be the most common. Samples are extracted with 80% methanol (1:6 meal to methanol ratio) for 30 min, with recovery rates for genistein and daidzein >90% with intraday and interday coefficients of variance below 2% (Wang et al., 2000a); aqueous methanol at elevated temperature (80%, 60°C for 2 h with occasional shaking) (Morton et al., 1999); methanol/water (3:1, v/v, refluxing for 1 h under nitrogen) (Rong et al., 1998); acidified ethanol (2 M HCl containing 0.05% butylated hydroxytoluene (BHT) and 20 ppm flavone and refluxing at 80°C for 1 h) (Hutabarat et al., 1998); acidified methanol (1 M HCl with gentle stirring over a steam bath for 2 h) (Wang et al., 1990); aqueous ethanol (80%, refluxing for 2 h) (Setchell et al., 1987); acidified ethanol (3 M trifluoroacetic acid (TFA) in 96% ethanol, refluxing at 100°C for 1 h) (Beekman et al., 1999); and acetonitrile (66% with 0.4% NaCl and 30 mg/L *p*-nitrophenol as an internal standard in water for 1 h) (Vänttinen and Moravcova, 1999). Best extraction efficiencies of soy phytoestrogens are obtained by refluxing 2-3 h with 77% ethanol containing 2 M HCl (Franke et al., 1995). Reversed-phase (RP) HPLC analysis shows that in most foods, the β-glucosides predominate when isoflavones are extracted with hot aqueous methanol or ethanol (80°C, 4 h). Extraction at room temperature results in a different chromatogram with predominance of 6″-*O*-malonylglucoside (Barnes et al., 1994). Lipid removal by hexane partitioning results in concentration-dependent loss of isoflavone upon washing with hexane, therefore lipid removal is deemed not essential for isoflavone quantitation. Omission of the lipid-removal step also reduces sample preparation time substantially. A simple method to extract isoflavones from an aqueous solution is to partition it with an organic solvent immiscible in water; ethyl acetate is safer to handle and more efficient than diethyl ether commonly used for this purpose. The recoveries of the internal standards for this method are in the range of 70%–90%. Within these limitations, the method is accurate to at least 100 ng of daidzein or genistein per gram of freeze-dried food (Liggins et al., 1998). Isoflavones from food matrices (e.g., meat products) are extracted using *N*,*N*-dimelthylformamide (DMF)/water (2/1, v/v) or with acetonitrile/water (2/1, v/v). The acetonitrile/water solvent system has the advantage of extracting low levels of fat that can interfere with the measurements by affecting the electroosmotic flow (Mellenthin and Galensa, 1999).

Genistein and its glycosidic conjugates have been extracted from soybean and soy foods using pressurized solvent extraction (PSE). The PSE method conducted at 80°C, 138 bar (2000 psi) for 20 min and at 5 mL/min flow rate is rapid, requires no filtration and compares favorably with reference method (60 min procedure) for the stirring extraction of isoflavones (Downing et al., 2007). Lignans from flaxseed have also been extracted using pressurized low-polarity water with the highest concentration obtained at 140°C, 5.2 MPa, total volume of 30–40 ML/g of flaxseed, and flow rate of 0.5 mL/min (Cacace and Mazza, 2006). Supercritical fluid extraction (SFE) is used as a rapid, selective, and convenient method for sample preparation before analysis of compounds in natural product matrices (Slanina and Glatz, 2004). The most recent advance in this technology has been the online coupling of SFE with supercritical fluid chromatography (SFC) for extraction and separation of neolignans, magnolol, and honokiol in Magnolia cortex. The extraction is performed for 1 min with carbon dioxide containing 5% methanol and the extract is concentrated by NH2 SFC column. After SFE, the concentration of methanol is increased to 15% and the trapped extracts are analyzed. The combination of extraction, concentration, and analysis is completed in 5 min (Slanina and Glatz, 2004). Other methods for extracting lignans have been described elsewhere (Oomah, 2003).

1.3.1.3 Hydrolysis

Since the phytoestrogen exists in plants mainly as glycoside conjugates, in biological fluids from humans and animals as glucuronide or sulfate conjugates, hydrolysis of the conjugate moiety is required before HPLC analysis. Direct quantitation of soy isoflavones without any hydrolysis step but including calibration curves of the acetyl and malonyl derivatives of daidzin, genistin, and glycitin have been described (Griffith and Collison, 2001). Basic hydrolysis breaks the ester bond of the manolyl and acetyl groups esterified to the alcoholic function of the sugar moieties of isoflavones, while acidic hydrolysis breaks the carbon–oxygen bond between the isoflavone ring and the sugar. The hydrolysis step should be considered essential when analyzing animal tissues or fluids but unnecessary for soy protein preparations (Setchell et al., 1987). Although most of the analytes are heat and acid stable, enzymatic rather than acidic hydrolysis for urine analysis is recommended to be able to include the metabolite equol in the assay which degrades under acidic conditions (Franke et al., 1995). Enzymatic hydrolysis is also more effective than acid hydrolysis, especially if the latter is performed at elevated temperature leading to partial destruction of isoflavones (Table 1.4). Hydrolysis of glycosidic bond is best

TABLE 1.4
Effect of Hydrolysis on Concentration of Isoflavones in Soy Food

Compound	Genistein ($mg \cdot kg^{-1}$ DW)			Daidzein ($mg \cdot kg^{-1}$ DW)		
	Average	SD	RSD (%)	Average	SD	RSD (%)
Soy flour[a]	53.0	4.9	9.33	68.0	8.0	11.76
Soy flour[b]	406.1	58.9	14.51	516.1	80.9	15.68
Tofu[a]	76.4	2.3	3.03	74.0	1.6	2.14
Tofu[b]	726.9	38.4	5.28	644.5	2.4	0.37

Source: From Vänttinen, K. and Moravcova, J., *Czech. J. Food Sci.*, 17, 61–67, 1999.

[a] Without hydrolysis.
[b] With enzymatic (β-glucuronidase) hydrolysis.

accomplished overnight at 37°C with enzyme using β-glucosidase preparation in 0.1 M acetate buffer (pH 5.0) (Setchell et al., 1987), in borate buffer pH 4.0 (Vänttinen and Moravcova, 1999), or a preparation of *Aspergillus niger* since it produces good consistent recoveries from different isoflavone glycoside sources (Liggins et al., 1998). Crude preparations of *Helix pomatia*, β-glucosidase from almonds, and acid hydrolysis produce reasonable recoveries of isoflavones from soy flour but are not applicable to different sources of isoflavone glycosides (Liggins et al., 1998). A second hydrolysis may be performed with a combined β-glucuronidase and sulfatase preparation in 0.5 M acetate buffer (pH 4.5) and extraction and recovery of the isoflavones repeated (Setchell et al., 1987). The native β-glucosidases released during the extraction of isoflavones can hydrolyze the isoflavone derivatives to their aglycones and TRIS (hydroxymethyl) aminomethane buffer has been added to the extraction solvent as the enzyme inhibitor (Delmonte and Rader, 2006). For acid hydrolysis, the sample is incubated in triflouroacetic acid at 95°C for 1 h. After incubation, the mixture is neutralized with CaO then filtered before high-performance capillary electrophoretic (HPCE) analysis (Vänttinen and Moravcova, 1999).

1.3.1.4 Lignan Extraction

Lignan extraction with aqueous alcohol is superior to other solvent and results in the extraction of ester-linked complex that can be chromatographed on RP-HPLC columns, although quantitation is not possible. SDG is generally isolated by treating an aqueous alcoholic extract with sodium methoxide (1.6% NaMeOH in anhydrous MeOH). The resulting extract is acidfied (4 N H_2SO_4 to pH 3.0) before purification on silica gel prepared in $CHCl_3$, and eluted with $CHCl_3$/MeOH/H_2O (65:35:50) to obtain SDG. HPLC analysis of the SDG is possible although several other contaminating compounds elute very close to SDG. An alternative to the isolation of the ester complex first is the enzymatic conversion of the lignan glycosides to the corresponding aglycone reported by Obermeyer et al. (1995). Although β-glucuronidase releases the aglycone secoisolariciresinol from the complex, hydrolysis is incomplete. Apparently, base hydrolysis produces the highest yield of SDG because of complete hydrolysis (Westcott et al., 1998). Combinations of enzymatic and acid hydrolyses have been proposed to release the aglycones that can be derivatized and analyzed by GC. However, an artifact, anhydrosecoisolariciresinol is produced as a result of acid hydrolysis thereby producing an unreliable estimate of the lignan content of flax and its products. This artifact is now considered as a naturally occurring lignan chemically identical to shonanin, and as such should be quantified alongside the other phytoestrogens (Liggins et al., 2000). The aqueous base hydrolysis is robust and produces SDG as the principal lignans with cinnamic acid glycosides in trace amounts. The glycosides can then be quantitatively converted to aglycones that are analyzed by HPLC or derivatized and analyzed by GC (Muir et al., 2000).

1.3.1.5 Hydrolysis of Glycosidic Bonds of Lignans

Two basic approaches are used to hydrolyze the glycosidic bonds between plant lignans and carbohydrate and both methods have been used to produce analytical results presenting the lignan concentrations in various foods (Thompson et al., 1991; Mazur et al., 1996). The Thompson method employs anaerobic in vitro fermentation of food containing plant lignans with gut bacteria producing enterolactone and enterodiol, the concentration of which is subsequently measured by GC and flame ion detection. The second method described by Mazur et al. (1996) employs hot acid to break the glycoside bond and is a multistepped process using both enzyme and hot acid to hydrolytically remove the carbohydrate component of lignan and isoflavone phytoestrogens in food. In this procedure, the food extract is first subjected to enzyme hydrolysis, liberating the isoflavones from their respective glycosides.

Since the lignans are not completely hydrolyzed by enzymes, hot acid is used to liberate the aglycones that are partitioned off with organic solvents. Both organic fractions of isoflavone and lignan aglycones are then combined, further purified, derivatized, and analyzed by GC-MS. However, the stability of products from acid hydrolysis has recently been questioned (Liggins et al., 2000). It appears that a naturally occurring lignan called shonanin (3,4-divanillyltetrahydrofuran) is also liberated from its glycosides alongside secoisolariciresinol as a result of acid hydrolysis. To account for the presence of shonanin and its hydrolytic product enterofuran in foods, Liggins et al. (2000) modified the procedure of Mazur et al. (1996) and simplified it for the quantification of lignans, secoisolariciresinol, mateiresinol, and shonanin in food after hydrolytic removal of any conjugated carbohydrate. The modification includes acid hydrolysis of food samples for 1–3 h, followed by neutralization with sodium hydroxide, separation of the aglycones from aqueous to organic phase, and quantitating the aglycones by GC-MS after the formation of trimethylsilyl derivatives of the lignans. Contents of lignans in reference foods are reported as the combined concentrations of secoisolariciresinol and shonanin, alongside the concentration of mateiresinol after 1–3 h of hydrolysis since optimum time of hydrolysis for maximum yield of lignans varies between foodstuffs. The source of variation because of the degradation of certain lignan by acid hydrolysis has prevented routine analysis of large batches of samples. This is circumvented by the recently proposed alkaline hydrolysis-based method (Penalvo et al., 2005) that has similar efficiency in the extraction of lignans with only minimal degradation of matairesinol that can be accurately determined using internal standards. The method combines methanolic extraction and alkaline hydrolysis as the step before enzymatic (*H. pomatia*) hydrolysis, followed by C_{18}-cartridge extraction of released aglycones and purification by DEAE (diethylaminoethyl)-Sephadex before GC-MS. This method is highly sensitive and can be used for determination of six most common and abundant lignans in the diet.

Enzymatic cleavage is a relatively mild method compared to acid hydrolysis. However, the commonly used enzymes β-glucuronidase, sulfatase, and cellulase release lignan aglycones only from glycosidic, but not from additionally esterified forms thereby requiring an additional base solubilization step before enzyme hydrolysis. This limitation in lignan analysis has now been overcome by the development of a nondestructive two-step method consisting of methanolysis followed by enzymatic cleavage in sol-gel cellulose columns (Schwartz and Cichna-Markl, 2005). The enzyme columns are produced by entrapping and immobilizing commercially available cellulose eliminating enzyme-born impurities in the process. Quantitative hydrolysis of both lignan and isoflavone glycosides is achieved within 5 min and are as effective as acid hydrolysis (Schwartz and Cichna-Markl, 2005).

1.3.2 Sample Treatment

1.3.2.1 Solid-Phase Extraction

Klejdus et al. (1999) used different solid-phase extraction (SPE) methods for isolation of daidzein, genistein, formononetin, and biochanin A to compare two red clover varieties. Filtered ethanolic (80%) extract of isoflavone is diluted with water (1:3, extract to water) before SPE. The cartridges are conditioned with either water or methanol, samples are then passed through, washed with methanol (5% or 10%) to remove impurities, and the retained isoflavones are eluted with 80% methanol. The extract is evaporated to dryness, dissolved in mobile phase, and directly injected onto an RP-HPLC column.

C_{18} and C_8 sorbents have low recoveries because of the extremely hydrophobic surfaces. The best recoveries $\approx$100% are obtained with conditioning on a Speed ABN cartridge (among the cartridges tested are C_{18}, C_8, amide 2, RP 105, ABN, and HLB). Conditioning is not a prerequisite since high recoveries of over 90% are obtained with ABN cartridges. SPE provides for high reproducibility (low relative standard deviation [RSD]) and low consumption of plant materials. SPE with polyamide or alternatively RP_{18} cartridges is required for samples with low isoflavone content or with a matrix interfering with the experiment. The RP_{18} columns should be conditioned with two-column volumes of methanol followed by the same volume of water before eluting the polyphenols with methanol. Only two-column volumes of water are required to pretreat the polyamide column (Mellenthin and Galensa, 1999). Isoflavones can be isolated from aqueous sample (human urine) by a simple procedure based on liquid–liquid extraction using a ChemElut column (containing diatomaceous material that absorbs and retains water from the matrix) connected online with a Florosil cartridge for effective purification. The high surface area of the cartridge enables efficient interaction between sample and extracting solvent. This procedure that does not require vacuum has been used in the isolation of 8-prenylnaringenin (a hop-derived phytoestrogen) from beer, by methanol elution from the SPE column and diethyl ether–ethyl acetate (1:1, v/v) desorption from the Florisil cartridge (Wang et al., 2002).

SPE cartridge containing 500 mg of C_{18} sorbent (AccuBond II SPE ODS-C_{18}) has also been used for concentrating and purifying lignans subsequent to methanolysis and enzymatic hydrolysis (Schwartz and Sontag, 2006). The hydrolysate is applied to a preconditioned (5 mL of methanol, 5 mL of deionized water) SPE cartridge, the analytes are eluted with 5 mL of methanol/water (80:20, v/v) after a washing step with 2 mL of methanol/water (20:80, v/v), and the SPE-eluate is either diluted or concentrated before HPLC analysis (Schwartz and Sontag, 2006). Silica-based C_8 SPE cartridge (Speed Scan ABN) suitable for extraction of polar compounds provides consistent recoveries of genistein and daidzein from human plasma only when diluted two times with aqueous 0.5% formic acid solution. Direct loading of the plasma sample into the cartridge without dilution reduces recoveries of the analytes due partially to their dissociation and to protein interaction (Yang et al., 2005). A dual-SPE method has been used for analysis of a red clover extract using C_{18} sorbent with an acid wash followed by an alkaline second step for complete elution of all analytes (De Rijke et al., 2006). Analysis of seven SPE columns shows that maximum yields of the isoflavone daidzein, genistein, coumestrol, apigenin, biochanin A, and kaempferol are achieved with Oasis HLB SPE column supplied by waters. This is probably due to the composition of the sorbent with the hydrophilic *N*-vinylpyrrolidone enhancing the wettability of the copolymer and the lipophilic divenylbenzene increasing phase retention necessary for trapping the phytoestrogens (Bajer et al., 2007). Divenylbenzene-based cartridge, especially Strata X (Phenomenex) and HLB Oasis, provides extremely accurate, reproducibly high recoveries, and the best sample cleanup/concentration (6.25:1) in less than 10 min for isoflavone analysis from soy extracts (Rostagno et al., 2005).

1.3.2.2 Matrix Solid-Phase Dispersion

Analytes in matrix solid-phase dispersion (MSPD) are extracted from samples homogeneously dispersed in a solid support, generally C_{18}- or C_8-bonded silica, thereby enabling simultaneous sample extraction and cleanup with good recoveries and precision. MSPD has been compared to Soxhlet and ultrasonic extraction for the LC-MS determination of isoflavone aglycones and glycosides in *Radix astragalis* widely used in Chinese medicine (Xiao et al., 2004). SPE cartridge (C_{18}) placed at the bottom of the MSPD column provides sequential cleanup especially if the MSPD procedure is optimized in regard to elution profile and layering of the column with the analyte. MSPD provides high-extraction efficiency for the isoflavones formononetin and calycosin, but low efficiency for glycosides compared to Soxhlet and ultrasonic extractions (Xiao et al., 2004). A similar MSPD sample preparation procedure for analysis of red clover leaves provided sufficiently high concentrations of seven isoflavones to enable their identification with only 500 mg of sample (de Rijke et al., 2004). The extraction efficiency in this particular instance is lower compared to those of solvent extraction (SE). Recent comparison of various extraction techniques indicates that maximum yields of daidzein and genistein are obtained by Soxhlet extraction, ultrasonic extraction, and ultrasonic homogenization techniques, while maximum yields of apigenin and biochanin A from herbs are obtained by SFE. However, MSPD provides the highest yield of daidzein from soy flour (Bajer et al., 2007).

1.3.2.3 Solid-Phase Microextraction

Solid-phase microextraction (SPME) is generally combined with GC analysis for analyte extraction from liquid or gaseous sample, or from the headspace above a sample using a fused-silica fiber coated with polyacrylate or polydimethylsiloxane as a stationary phase. For HPLC applications, the SPME fiber is placed in the mobile phase of the HPLC instrument and desorption of analytes occurs by flushing with the mobile phase followed by detection. SPME-LC-MS can improve detection and throughput by decreasing sample manipulation and eliminating derivatization of the analytes. This method has been applied for extraction of genistein and daidzein from human urine after 3 h consumption of 35 g soy protein with a Carbowax-templated polydivenylbenzene resin (CW-TPR) providing the best performance of four evaluated SPME fibers (Satterfield et al., 2001). Another preconcentration technique using an open tubular fused-silica capillary with an inner surface coating as the SPME device, known as in-tube SPME, coupled with an HPLC photodiode array detection system has been used to determine daidzein and genistein in soy foods (Mitani et al., 2003). High analyte recoveries (above 97%) can be achieved from spiked foods with low-detection limits (0.5 ng/mL) and without interfering peaks when microextraction and desorption are optimized.

1.3.3 Chromatographic Separation

1.3.3.1 Paper Chromatography

Two-dimensional paper chromatography has been used to demonstrate the presence of the phytoestrogen coumestrol in alfalfa (Bickoff et al., 1967). This method was later used by Knuckles et al. (1976) to ascertain the coumestrol content of vegetable products. According to this method, coumestrol is extracted by soaking the products in 75% ethanol for 16 h. The extract is applied on the chromatogram that is first developed in acetic acid/water (1:1, v/v), then in isopropyl alcohol/ammonium hydroxide (2:1, v/v). The intensity of the coumestrol spots is measured fluorometrically without elution from the paper. The detection limit of coumestrol is 0.1 μg/g. Thirteen of the 16 vegetable products analyzed by two-dimensional paper chromatography contained coumestrol (Knuckles et al., 1976). Paper chromatography is no longer in use for separation or detection of phytoestrogens.

1.3.3.2 Thin Layer Chromatography

Thin layer chromatography (TLC) has been used in the identification of isoflavonoids. Precoated polyamide 6 TLC plates have been used for the initial fractionation of isoflavones and phenolics from soybeans and soy products (Pratt and Birac, 1979).

The methanolic extracts are spotted on 20 × 20 cm plates and developed using methanol/acetic acid/water (90:5:5). Bands are eluted with methanol and rechromatographed on polyamide using chloroform/methanol/methyl ethyl ketone (12:2:1). For aglycones, the plates are developed in ethyl acetate/petroleum ether (3:1) and then in ethanol/chloroform (1:1). Bands are detected with appropriate chromogenic sprays or with UV lamp at 366 nm (Pratt and Birac, 1979). Diethyl ether extract chromatographed on silica plate developed three times in dichloromethane/methanol (95:5, v/v) gave the following R_f: Daidzein—0.20, genistein—0.30, formononetin—0.55, and biochanin A—0.75, respectively (Table 1.5) (Lapčík et al., 1998). TLC carried out on a Kieselgel 60 F_{254} plate using chloroform/methanol/2% acetic acid (7:3:1, v/v, lower layer) and visualized by heating at 120°C for 10 min after spraying with 10% H_2SO_4 gave the following R_f values of 0.41, 0.45, 0.43, 0.08, 0.12, 0.10, 0.56, 0.61, and 0.59 for daidzin, glycitin, genistin, 6″-*O*-malonyldaidzin, 6″-*O*-malonylglycitin, 6″-*O*-malonylgenistin, 6″-*O*-acetyldaidzin, 6″-*O*-acetylglycitin, and 6″-*O*-acetylgenistin, respectively (Kudou et al., 1991). Isoflavones obtained by SFE methods have also been evaluated by TLC on glass plates precoated with silica gel with fluorescent indicator (Chandra and Nair, 1996). The supercritical carbon dioxide extracts analyzed using chloroform/methanol (10:1) produce the following R_f values of 2.5, 3.1, 5.0, and 5.7 for daidzein, genistein, formononetin, and biochanin A, respectively when visualized under a UV lamp at 254 nm (Chandra and Nair, 1996).

TLC-silica (straight phase) and rechromatography on RP octadecylsilica column HPLC have been used to fractionate isoflavonoids from diethyl ether extracts of beer (Lapčík et al., 1998). The fractions are analyzed by radioimmunoassay specific for daidzein/formononetin and genistein/biochanin A. The diethyl ether extract is chromatographed on a Merck PSC silica plate along with standards of daidzein, formononetin, genistein, and biochanin A in separate tracks. The plate is developed

TABLE 1.5
Chromatographic Properties of Isoflavonoids

Compound	TLC System 1[a] R_f	TLC System 2[b] R_f	HPLC System 1[c] Retention Time (min)
Daidzein	0.20	0.29	6.5
Genistein	0.30	0.44	8.5
Formononetin	0.55	0.69	12.8
Biochanin A	0.75	0.81	21.7

Source: From Lapčík, O., Hill, M., Hampl, R., Wähälä, K., and Adlercreutz, H., *Steriods*, 63, 14–20, 1998a.

R_f, the chromatographic mobility as the ratio of the distance of the center of the spot from the start to the distance between the start and the solvent front.

[a] Merck PSC silica plate (Art. 13794) developed three times in dichloromethane/methanol (95:5, v/v).
[b] Merck DC-Alufolien (Art. 5583) developed twice in dichloromethane/methanol (95:5, v/v).
[c] Column ET 250/4 Nucleosil 100-5 C18, mobile phase—methanol/water (60:40, v/v); flow rate, 0.8 mL/min; temperature, 35°C.

three times in dichloromethane/methanol (45:5, v/v), viewed under UV light, scraped, eluted with methanol, and analyzed by radioimmunoassay (RIA). The fractions containing isoflavonoids are rechromatographed on RP-HPLC, eluted, collected, and assayed for immunoreactivity with two antisera (daidzein/formononetin and genistein/biochanin A) (Lapčík et al., 1998). The presence of lignans and their precursors in flaxseed has been monitored by TLC and quantified by an HPLC method developed by Harris et al. (1994). TLC carried out on HP Kieselgel 60 F_{254} plate using chloroform/methanol/acetic acid (90:10:1, v/v/v) gave the following R_f values of 0.01, 0.17, 0.57, 0.59, 0.70, and 0.71 for chlorogenic acid, gallic acid, 4-hydroxybenzoic acid, coumaric acid, ferulic acid, and SDG, respectively. Because of the success of the TLC and HPLC methods, Harris et al. (1994) suggest that high-performance TLC (HPTLC) may allow for the analysis of SDG and its metabolites in a large number of flaxseed samples in a short time. One of the major advantages of HPTLC is the potential of analyzing samples, especially oilseeds, without employing a defatting step. Recent development in HPTLC analysis of SDG includes optimization of TLC conditions by eluting HPTLC silica gel plates in a horizontal developing chamber, quantitation in single-beam reflectance mode and application of a five-point calibration (1–10 μg/spot range) using an external standard (Coran et al., 2004). The proposed method is precise, reproducible, and accurate and can be used in place of HPLC for determination of SDG in complex matrices.

The intracellular accumulation and mechanism of genistein transport in breast cancer cells and the effective reversal of drug resistance have been demonstrated by silica gel TLC of secreted [^{3}H] genistein and its corresponding metabolites (Imai et al., 2004). [^{3}H]-labeled genistein has also been used in two-dimensional RP TLC to demonstrate accumulation of genistein derivatives in lipoproteins for increased oxidation resistance of low-density lipoprotein (LDL) during intake of soybean isoflavones (Kaamanen et al., 2003). In this method, chromatographic (Sephadex LH-20) fractions are applied to RP TLC plates (20 × 20 cm, RP-8 F_{254} s), developed in the first dimension with methanol/acetonitrile/H_2O (90:5:5, v/v), and the second with *n*-heptane/diisopropylether/acetic acid/methanol (60:40:4:2, v/v). The location of the standards (nonradioactive genistein, genistein 4′-*O*-oleate, genistein 7-*O*-oleate, and genistein 4′,7-*O*-dioleate) is visualized under UV light after rhodamine staining and [^{3}H]-label is determined by scintillation counting after scraping the spots from the plates (Kaamanen et al., 2003). The strong fluorescence of formononetin permits its detection and quantitation in black cohosh root using TLC on silica gel 60. The mobile phase for formononetin separation consists of ethyl acetate/toluene/glacial acetic acid (17.5:80:2.5, v/v), the silica gel layer (0.25-mm thick) is heated at 100°C for 30 min and scanned using a densitometer in the fluorescence mode at 450 nm with excitation at 250 nm. This method is accurate and precise with 0.08 ppm limit of detection and content of formononetin in rhizomes ranging from 3.1 to 3.5 μg/g (Panossian et al., 2004). However, this same TLC method using methanol as the extracting solvent failed to provide definitive formononetin quantitation in black cohosh from United States (Jiang et al., 2006).

Genistein as a metabolite in transgenic plants overexpressing soybean isoflavone has been purified by TLC before quantitation by HPLC (Liu et al., 2007).

This procedure involves grinding the plant tissues in liquid N_2 to fine powders, extracting with 80% ethanol overnight at room temperature with brief agitation, filtration, and concentration by evaporation followed by acid hydrolysis. Hydrolysates are extracted with ethyl acetate, evaporated to dryness, resuspended in 80% methanol, loaded beside authentic genistein dissolved in methanol on silical gel plates, and placed in a mixture of chloroform and methanol (15:1) for migration to separate genistein. After migration, the plates are viewed under UV lamp and the fluorescent materials with the same migration rate as the authentic genistein are scraped into glass tubes, extracted with 80% methanol and the extract resolved, detected, and quantified by HPLC and electrospray ionization (ESI)-MS (Liu et al., 2007). TLC continues to be an inexpensive method mainly for initial examination of plant material and for monitoring various stages of analyte purification because of its simplicity and relatively high throughput. Some recent examples of TLC procedures applicable to lignan analysis have been summarized elsewhere (Slanina and Glatz, 2004).

1.3.3.3 Gas Chromatography

Isoflavonoids and lignans are normal constituents of body fluids and have been identified in most animal and human body fluids by GC-MS (Setchell and Adlercreutz, 1988). GC-MS has been successfully used in the analysis of phytoestrogens in urine (Adlercreutz et al., 1986, 1991a; Kelly et al., 1993; Cassidy et al., 1994), plasma (Adlercreutz et al., 1993; Morton et al., 1994; Joannou et al., 1995), saliva (Finlay et al., 1991), semen (Dehennin et al., 1982), prostatic fluid (Morton et al., 1997), and feces (Adlercreutz et al., 1995; Kurzer et al., 1995). The presence of at least one hydroxyl group in the structure of common dietary phytoestrogens makes them difficult for direct GC-MS analyses. The hydroxyl group is generally derivatized with *N,O*-bis(trimethylsilyl)-trifluoroacetamide (BSTFA) before GC-MS, making the analysis more time consuming (Barnes et al., 1998). However, GC-MS is accepted as the technique sufficiently sensitive for low-level analysis of lignans and isoflavones (Mazur et al., 1996). In a more recent study (Zeleniuch-Jacquotte et al., 1998), isoflavonoid assay by ion-exchange chromatography and GC-MS showed high coefficients of variation (CV $\leq$ 10%) unlike those of lignans, indicating that it is not well suited to examine the relation of serum isoflavonoid with disease risk. GC-Ion mobility, an atmospheric pressure detection system for GC, has been used to detect the presence of mammalian lignans in urine samples (Atkinson et al., 1993).

Morton et al. (1999) describe a simple, robust method employing ID-GC-MS for the determination of phytoestrogens in biological samples and food matrices. Briefly, samples are hydrolyzed with β-glucuronidase, the aglycones extracted, and the phytoestrogen fraction isolated by chromatography on Sephadex LH 20. This fraction is then derivatized for GC-MS by reaction with BSTFA to form trimethylsilyl derivatives. For GC-MS, the column (silica capillary column 10-μ phase thickness) is programed from 190°C to 245°C at 49.9°C/min under a helium pressure of 25 kPa. Calibration standards and samples are injected in BSTFA reagent in a splitless injector maintained at 230°C. The interface and source are operated at 250°C and 230°C, respectively. ID-GC-MS is carried out in the selected ion recording (SIR) mode where the intensity of selected ions for each analyte and deuterated internal standard is monitored and expressed as peak ratio. A calibration curve of peak height ratio against concentration is used to calculate the concentration of the analyte (Morton et al., 1999). The levels of daidzein and genistein produced from soybean hulls, soybean hypocotyls, dehulled soybeans, and other soy-based foods as well as the levels of phytoestrogens in the sera from Japanese men have been evaluated by this method.

Isoflavones, together with other flavonoid aglycones, have been separated into well-resolved, sharp, and symmetrical peaks by a rapid and sensitive capillary column gas chromatographic method (Koupai-Abyazani et al., 1992). The tailing factor or asymmetric ratio of the trimethylsilyl derivatives of flavonoid aglycones is, on the average, 50% greater in packed column GC than in capillary column GC. As a general rule, the retention times of the flavonoid derivatives increase with the number of ether groups, substitution at different positions of the flavonoid nucleus affects the retention characteristic differently. For example, the retention times for 4,5,7-triydroxyisoflavone and 4′,6,7-trihydroxyisoflavone are 27.59 and 37.07 min, respectively, because of the ether group in position 6 resulting in a longer retention time. However, ether group added to position 5, as in the case of daidzein to genistein, produces small differences in retention times (25.92 to 27.59 min). The number of hydroxyl group of the isoflavones (daidzein and genistein, two and three hydroxyl groups, respectively) has very little effect on retention time.

The main phytoestrogens, matairesinol, secoislariciresinol, anhydrosecoisolariciresinol, biochanin A, genistein, formononetin, daidzein, and coumestrol, have been determined in dry food samples by a very sensitive, but complicated isotope dilution gas chromatography–mass spectometry in the selected ion monitoring (ID-GC-MS-SIM) method (Mazur et al., 1996). This is the only method, at present, by which all these lignans and isoflavonoids may be simultaneously determined in foods. These compounds are measured by ID-GC-MS-SIM mode using synthesized deuterated internal standards for the correction of losses during the procedure. Significant limitations of lignan analysis by GC-MS reside in the insufficient volatility of the parent compound without derivatization and the conversion of the conjugates to their corresponding aglycones. These limitations can be overcome by devising a series of hydrolysis and purification procedures (Adlercreutz et al., 1993). A description of the main steps of the method is provided in the flow diagram (Figure 1.5).The diphenolic glycosides are converted to their respective aglycones by a three-step hydrolysis involving a rehydration with distilled water, followed by enzymatic and acid hydrolysis. Enzymatic hydrolyses are carried out with *H. pomatia* juice which hydrolyzes the isoflavonoid glycosides effectively and the lignan glycosides minimally. Thus, the ether extract contains mainly lignans that are hydrolyzed completely with acid without affecting the isoflavones. Purification and separation are carried out in two ion-exchange chromatographic steps followed by derivatization and GC-MS. Neutral steroids and other contaminants are removed by DEAE- and QAE-Sephadex chromatography before separation and identification by GC-MS. The method is highly specific because of the use of deuterated internal standards, the retention time of which is very similar to those of the nondeuterated standards (Table 1.6). The within- and between-assay imprecision values of the method vary

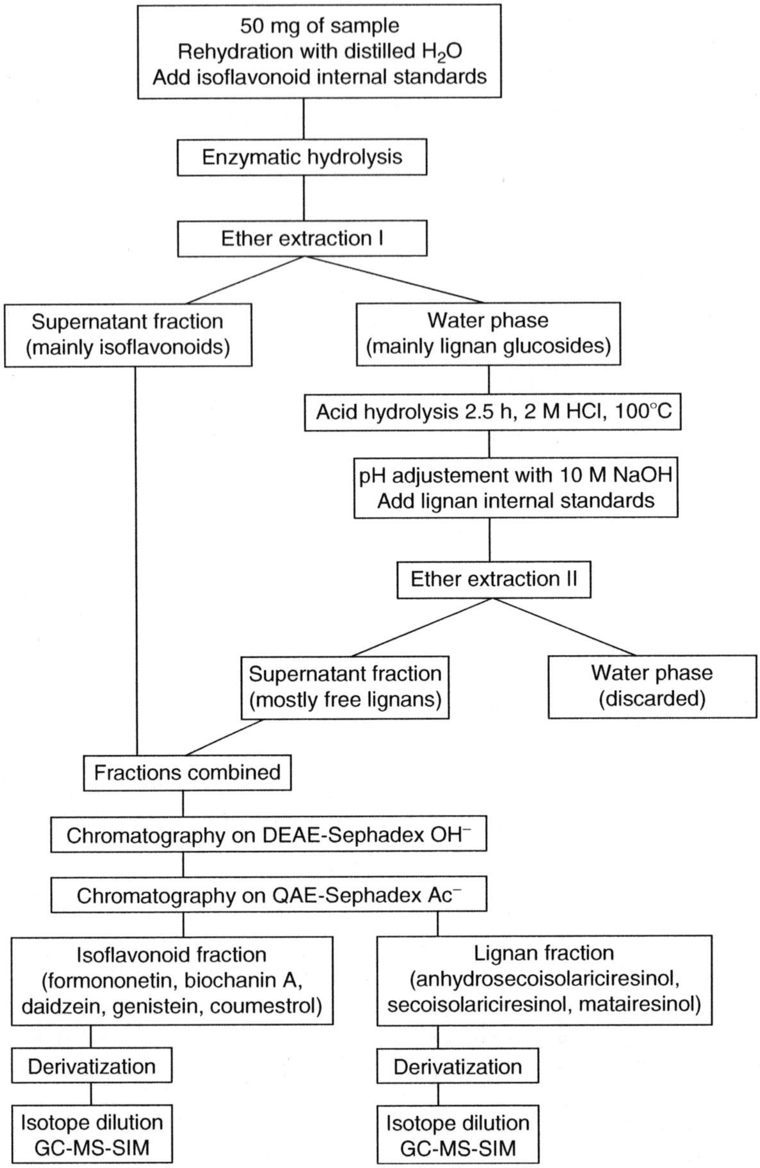

FIGURE 1.5 Flow diagram of the isotope dilution gas chromatography–mass spectometry in the selected ion monitoring (ID-GC-MS-SIM) method for the determination of the isoflavonoid and lignan phytoestrogens in food. (From Mazur, W.M., Wähälä, K., Ojala, S., Makkonen, A., Hase, T., and Adlercreutz, H., *Br. J. Nutr.*, 79, 37–45, 1998.)

from 3.1% to 9.6% and 7.0% to 21.2%, respectively. The mean recovery of authentic standards processed through the whole procedure varies from 95.5% to 105.5%. The method gives four to five times higher secoisolariciresinol values in flaxseed

TABLE 1.6
Relative Retention Times in Isotope Dilution Gas Chromatography–Mass Spectometry in the Selected Ion Monitoring (ID-GC-MS-SIM) of Phytoestrogens and Their Deuterated Homologs

Compound	Time (min)	Relative Retention Time
Phytoestrogen fraction		
Formononetin		0.802
d_3-Formononetin		0.800
Biochanin A		0.841
d_4-Biochanin A		0.839
Daidzein		1.003
d_4-Daidzein[a]	8.821	1.000
Genistein tri-TMS		1.029
di-TMS		1.050
d_4-Genistein tri-TMS		1.028
di-TMS		1.049
Coumestrol		1.168
d_4-Coumestrol		1.165
Lignan fraction		
ANHSEC		0.916
d_8-ANHSEC		0.914
SECO		0.946
d_6-SECO		0.945
Matairesinol		1.001
d_6-Matairesinol[a]	8.732	1.000

Source: From Mazur, W., Fotsis, T., Wähälä, K., Ojala, S., Salakka, A., and Adlercreutz, H., *Anal. Biochem.*, 233, 169–180, 1996.

[a] Reference compound.

compared to published results (Aldercreutz and Mazur, 1997) probably because of the higher accuracy of the methodology in the determination of secoisolariciresinol and its hydrolytic product, anhydrosecoisolariciresinol. Although the methodology is complicated, it simultaneously determines eight phytoestrogens even at very low levels. However, one disadvantage is that the concentration of the various compounds is so different that analysis has to be carried out twice, once for the abundant compounds and once for the minor components. The sensitivity of the method is approximately 2–3 mg/100 g. The method is suitable for the measurements of isoflavonoids and lignans in as little as 50 mg of foodstuffs. This method has been used to measure phytoestrogens in many foods including 52 leguminous seeds, teas and coffees (Mazur et al., 1998), vegetables, fruits and berries, oilseeds, nuts, and beer (Lapčík et al., 1998). Modifications involving a combination of methanolic extraction and alkaline hydrolysis followed by enzymatic hydrolysis with stable $^{13}C_3$-labeled standards added to samples before pretreatment allow for routine, sensitive analysis of the most common food lignans (Penalvo et al., 2005). The ID-GC-MS-SIM mode has been used in measuring and reporting quantitative results for plant estrogens in grains and cereals, oilseeds and nuts, berries, fruits, vegetables, leafy vegetables, and beverages (Mazur, 1998).

Liggins et al. (1998) also used GC-MS-SIM mode to develop a simple analytical method for routine quantification of the isoflavones daidzein and genistein in food using synthetic glucosides daidzein and genistein as internal standards, combined with the food before extraction. In this method, daidzein and genistein are liberated from their respective glycosides by hydrolytic enzymes from *A. niger* in aqueous buffer. The aglycone isoflavones are partitioned into ethyl acetate, which is evaporated under nitrogen, derivatized with *N-tert*-(butyldimethylsilyl)-*N*-methyltrifluoroacetamide (TBDMS) and quantified using GC-MS-SIM mode. Derivatization of isoflavones with TBDMS improves stability, volatility, and chromatographic resolution of the ethers. The recoveries of the internal standards for this method are in range of 70%–90%. Within these limitations, the method is accurate to at least 100 ng of daidzein or genistein per gram of freeze-dried food (Liggins et al., 1998).

GC-MS in the positive ion mode has been used to confirm the identity of lignans by retention times and mass spectra of their trimethylsilyl ether (TMS) derivatives after separation and purification by HPLC-DAD (diode array detector) (Meagher et al., 1999). Apparently, lignans with different structures such as isolariciresinol and pinoresinol are detectable only in the positive ion and total ion modes but not if SIM is utilized. Although the total ion current profile of lignan is complex, lignan peaks are well resolved on the nonpolar column (Figure 1.6). Isolariciresinol has a shorter retention time on GC than other lignans although it is unresolved on HPLC because of coeluting sterol peaks (Meagher et al., 1999).

GC-MS continues to be the technique of choice for phytroestrogen analysis in human biological fluids partly because of its reproducibility, sensitivity, and limits of detection (0.04–10 nmol/L). Representative examples of recent phytoestrogens analysis, mostly isoflavones using GC are provided in Table 1.7. GC-MS followed by the GC-MS/MS method has been used to measure phytoestrogens for in vivo human study (Grace et al., 2003a, 2003b, 2004). The phytoestrogen conjugates are enzymatically (β-glucuronidase) hydrolyzed to their aglycones, extracted using SPE cartridges, and derivatized to trimethylsilyl derivatives for analysis using ID-GC-MS. The assay is rapid, sensitive with low limits of detection (1–5.3 ng/mL), and consists of one SPE stage before derivatization for GC-MS analysis (Grace et al., 2003b; Low et al., 2005). Hydrolysis followed by solvent extraction is recommended when GC-MS is used for analysis because it is easier to form volatile di- or trimethylsilyl derivatives from the aglycones than from the glycosides (Prasain et al., 2004).

The ID-GC-MS-SIM method of Adlecreutz and coworkers (1993) is perhaps the most frequently used in clinical and medical studies, although intra- and total-assay variabilities of all analytes are considered to be high, probably reflecting low-analyte concentrations in biological fluids (Wilkinson et al., 2002). Other GC-MS methods report lower inter-assay variation of 6%–11% isoflavones in serum. The latest method of measuring urinary excretion of isoflavones (daidzein, genistein, equol,

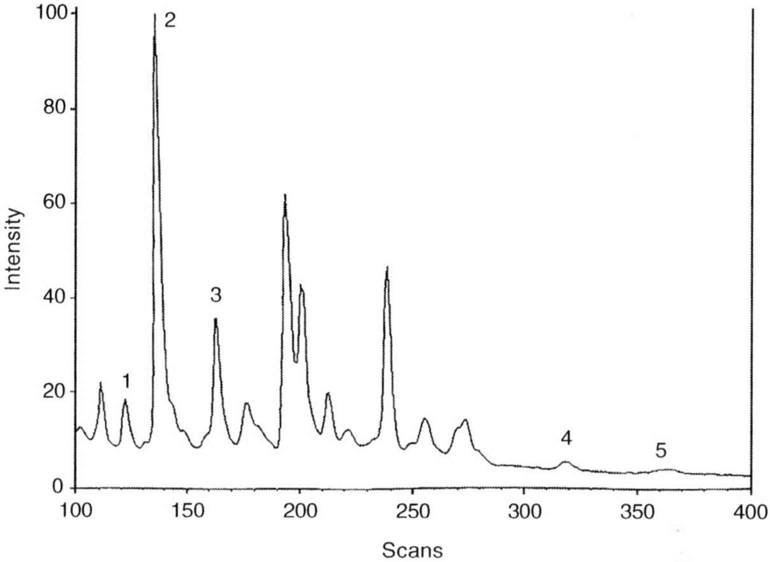

FIGURE 1.6 Gas chromatography–mass spectometry (GC-MS) chromatogram of an acid-hydrolyzed methanolic flaxseed meal extract. Peaks 1–4 are isolariciresinol, secoisolariciresinol, divanillyltetrahydrofuran, and matairesinol, respectively. (Reprinted with permission from Meagher, L.P., Beecher, G.R., Flanagan, V.P., and Li, B.W., *J. Agric. Food Chem.*, 47, 3173–3180, 1999.)

and *O*-DMA) and lignans (enterolactone and enterodiol) involves GC-MS in the SIM mode with deuterated internal standards. The intra-run CV for duplicate quality-control urine aliquots is less than 5% for all analytes and the inter-run variations range from 6.2% to 22.4% depending on analytes (Atkinson et al., 2006). GC-MS in the SIM mode has also been applied for its reproducibility in quantitative analysis of 17 phytoestrogens in 22 traditional medicinal herbs. The purification procedure is simplified by using SPE (Oasis HLB cartridge) following enzymatic and acid hydrolyses and forming TMS derivatives of phytoestrogens with a mixture of MSTFA/NH4I/DTE (1000:4:5, v/w/w) (Lee et al., 2004). The recovery for this procedure exceeds 84% for all analytes with a 0.2 μg/g limit of detection and within-day and day-to-day variations of 0.2%–16% (Lee et al., 2004).

1.3.3.4 High-Performance Liquid Chromatography

Although GC-MS detection method applied to the analysis of phytoestrogens is sensitive, it is also time consuming because of the cumbersome sample preparation steps. Separation by LC obviates the need for derivatization. Hence HPLC has proven to be the method of choice since the 1980s. Generally, phytoestrogens can be separated with mixtures of methanol or acetonitrile and aqueous acids or buffers as modifiers using RP C_{18} stationary matrices. Using acidified solvents has been recommended by compilers of the U.S. Department of Agriculture-Iowa State University Database. In the latest review on the measurement of food flavonoids,

TABLE 1.7
Recent Gas Chromatography (GC) Analysis of Phytoestrogens

Analytes	Hydrolysis/Extraction	Internal Standard	GC, GC-MS Conditions	Recovery	Result	References
Human urine (premenopausal women); daidzein, genistein, and equol; and *O*-desmethylangolensin enterolactone, and enterodiol	Enzymatic hydrolysis by β-glucuronidase/sulfatase; SPE (C-18) cartridges; and isolation of conjugated fractions by acetate form of DEAE-Sephadex chromatography	Deuterated standards of corresponding analytes	Silica capillary column (12.5 m × 0.22 mm ID, 0.25 μm), 100°C–280°C at a rate of 30°C/min, ESI-MS-SIM	Recovery: 97%–106%, CV: 0.8%–15.2%	Two fractions obtained: fraction 1 contained equol, enterolactone, enterodiol, and matairesinol; fraction 2 contained *O*-desmethylangolensin, daidzein, and genistein	Adlercreutz et al. (1991)
Human urine (women); daidzein, genistein, glycitein, and equol; and *O*-desmethylangolensin, enterolactone, and enterodiol	Enzymatic hydrolysis of conjugates to the aglycones; extraction on SPE (C-18) cartridges; and derivatization to trimethylsilyl derivatives	^{13}C-labeled of corresponding analytes	Dimethyl-polysiloxane column (15 m × 0.25 mm ID, 0.25 m), 160°C–300°C at a rate of 10°C/min, ESI-MS-SIM	Recovery: 93%–104 %; CV—(within-assay): 2.0%–6.5%, (between-assay): 3.2%–26.5%, LOQ: 1.2–5.0 ng/mL	Large range in amount and type of excreted phytoestrogens	Grace et al. (2003b)
Animal urine (rats and rats associated with human faecal bacteria); daidzein, genistein, and equol; and *O*-desmethylangolensin, matairesinol, enterolactone, and enterodiol	Enzymatic hydrolysis, ether extraction, and ion-exchange chromatography	Deuterated standards of corresponding analytes	Silica capillary column (12.5 m × 0.22 mm ID, 0.25 μm), 100°C–280°C at a rate of 30°C/min, ESI-MS-SIM		Evidence for the role of the gut microflora in absorption and metabolism of phytoestrogens; demonstrated inability of some rats to produce equol	Bowey et al. (2003)

Human plasma (women); daidzein and genistein	Hydrolysis by ß-glucuronidase/sulfatase (*Helix pomatia*) pH 4.5 at 37°C; clean up with SPE C18-bond elut cartridge; and separation through a small Sephadex LH-20 column (7 × 0.4 cm)	^{13}C-labeled of corresponding analytes	Silica capillary column (12.5 m × 0.25 mm ID, 0.25 μm), 260°C–310°C at a rate of 10°C/min, ESI-MS-SIM		Plasma genistein concentrations are higher than daidzein; bioavailability of isoflavones is greater when ingested as β-glycosides rather than aglycones	Setchell et al. (2001)
Human urine and plasma (premenopausal women); Ingestion of ^{13}C-labeled isoflavone daidzein and genistein; 7-day dietary diaries; 24-h pooled urine and collection of blood sample (10 mL) from (2–72 h)	Hydrolysis of conjugates to aglycones; extraction on SPE (C-18); blood samples centrifuged (3000 × *g*, 10 min, 4°C) and serum is removed, hydrolyzed enzymatically; and urine and plasma samples are separated on Sephadex-LH20 chromatography	^{13}C-labeled of corresponding analytes	Similar to Setchell et al. (2001)	CV (within-assay): 0.5% for daidzein and 1.0% for genistein, (between-assay): 5.0% for daidzein, 7.0% for genistein, and 10% for equol (5–7 ng/mL)	Serum concentration of ^{13}C-genistein are significantly greater than those of ^{13}C-daidzein	Setchell et al. (2003)
Human urine and plasma (women aged 41–76 years); 7-day dietary diaries; daidzein, genistein, glycitein, *O*-desmethylangolensin, equol, enterodiol, and enterolactone	For urine, hydrolysis of conjugates to the aglycones; extraction on Strata C18-E SPE cartridges; derivatization to trimethylsilyl derivatives for plasma, all steps are similar to urine analysis except, blood samples are centrifuged (3000 × *g*, 10 min, 4°C) and serum is removed before hydrolysis	Triply ^{13}C-labeled standards	Similar to Grace et al. (2003)	Pearson correlation coefficients > 0.8	Significant relationships ($P < .02$) between both urinary and serum concentrations of isoflavones and dietary intakes; spot urine correlated strongly with that in serum	Grace et al. (2004)

(*Continued*)

TABLE 1.7 (Continued)
Recent Gas Chromatography (GC) Analysis of Phytoestrogens

Analytes	Hydrolysis/Extraction	Internal Standard	GC, GC-MS Conditions	Recovery	Result	References
Female Sprague–Dawley rats for 10 days, sesamin (15 mg/kg body weight) and a 10% sesame seed diet	In vitro fermentation with human fecal inoculum		Silica capillary column (12.5 m × 0.25 mm ID, 0.25 μm), 260°C–310°C at a rate of 10°C/min, ESI-MS-SIM		Conversion of sesamin to the mammalian lignan, although at a lower rate (1.1%) compared with that of SDG (57.2%); lignan concentration in sesame seed (2180 μmol/100 g) is higher than that in flaxseed (820 μmol/100 g)	Liu et al. (2006)

TABLE 1.8
Gradient High-Performance Liquid Chromatography (HPLC) Analysis of Isoflavones in Soybean Products

HPLC Condition

Column: Brownlee aquapore C_8 (300 × 4.5 mm ID, 5 μm)
Mobile phase: Eluent (A)—acetonitrile; eluent (B)—water–trifluoroacetic acid (0.1%)
Use gradient of 0%–46.4% acetonitrile (2.25%/min) in 0.1% aqueous TFA
Flow rate: 1.5 mL/min
Detection: UV 262 nm

Sample Preparation

Extract soybean foods with 80% aqueous MeOH (10 mL/g), containing 1.25 mg fluorescein (internal standard), by stirring 1 h at 60°C, and centrifuging 10 min at 2500 *g*. Dry the supernatants, redissolve in 50% MeOH (5 mL), and defat by partitioning with hexane (4 × 20 mL). Evaporate the aqueous methanol phase to dryness, suspend in 80% MeOH (10 mL), and centrifuge 2 min at 14,000 *g*.

Source: From Coward, L., Barnes, N.C., Setchell, K.D.R., and Barnes, S., *J. Agric. Food Chem.*, 41, 1961–1967, 1993.

Merken and Beecher (2000) point to the diverse and ever increasing use of HPLC in the analysis of isoflavonoids. Several HPLC methods have been tested using various mobile phases and columns (Table 1.8). The most commonly used HPLC method applies a linear gradient for the separation and quantitative analysis of isoflavonoids from soybean and soybean-based products (Ohta et al., 1979; Eldridge, 1982a, b; Seo and Morr, 1984; Coward et al., 1993; Barnes et al., 1994; Wang and Murphy, 1994). Nonlinear gradient elution HPLC has also been used for the determination of isoflavonoid and coumestan phytoestrogens from many legumes and legume-based products (Murphy, 1981; Franke et al., 1994, 1995). Most investigators prefer gradient conditions for HPLC to maximize the separation of complex mixtures of phytoestrogens as they exist in plant and food products. Methods for the separation of phytoestrogens in plant extracts by HPLC have been described (Dziedzic and Dick, 1982; Eldridge, 1982a, b; Murphy, 1982; Nicollier and Thompson, 1982; Farmakalidis and Murphy, 1984; Pettersson and Kiessling, 1984; Sachse, 1984; Seo and Morr, 1984) using gradient elution systems. The gradient HPLC analysis of isoflavones in soybean foods is summarized in Table 1.8. The concentration of acetonitrile increases by 2.25% per minute enabling the separation of both the isoflavone β-glucoside conjugates and the aglycones in a single chromatographic run (Lee, 2000). Runs are generally an hour maximum with equilibration between runs. A striking exception is the 340-min run used for the HPLC of isoflavones in soy sauces for pattern recognition analysis (Kinoshita et al., 1997, 1998). Quality control measures for routine analysis of soy isoflavones by HPLC-DAD (Murphy et al., 2002) have been promoted as a reference method and form the basis of current protocol for analysis of isoflavones (Zhang and Schwartz, 2005).

Isocratic elution systems are also commonly encountered in the analysis of phytoestrogens. Thus, West et al. (1978) used HPLC analysis to separate the isomeric trihydroxyisoflavones and the genistein from soybean with isocratic conditions. However, this method was not applicable to phytoestrogens other than genistein. A simple isocratic RP system, with methanol 0.1 M ammonium acetate, pH 4.6 (60:40, v/v) as mobile phase on an Hypersil ODS column has been reported by Setchell et al. (1987) for the rapid and effective separation of the phytoestrogens daidzein, genistein, coumesterol, formononetin, and biochanin A. In this system, phytoestrogens on Hypersil ODS separate and elute in the following order: daidzein, genistein, coumestrol, formononetin, and biochanin A. Methanol is a better modifier than acetonitrile and is essential for the separation of genistein and coumestrol. The pH and buffer concentration of the mobile phase affects the retention but not the resolution. Increasing the pH or the buffer concentration decreases the retention of all compounds while maintaining the resolution (Setchell et al., 1987). Hutabarat et al. (1998) succeeded in obtaining an iosocratic HPLC method to separate daidzein, genistein, and biochanin A in soybeans. The isocratic separation of all analytes is achieved in less than 24 min with acetonitrile (33%) and water–acetic acid (67%) (99:1, v/v). Daidzein elutes at 3 min (capacity factor $k' = 0.92$), genistein at 5 min ($k' = 2.24$), biochanin A at 21 min ($k' = 12.76$), and flavone an internal standard at 23 min ($k' = 14.16$). RSD values for six injections of phytoestrogens for this method are 1.26%–1.51%. The mean recoveries of the compounds ranged from 100% to 127% for daidzein, 101% to 114% for genistein, and 97% to 135% for biochanin A (Hutabarat et al., 1998).

RP-HPLC has frequently been described for the determination of the isoflavones, daidzein, and genistein (Carlson and Dolphin, 1980; Murphy, 1981; Kitada et al., 1986; Lundh et al., 1988; Barbuch et al., 1989; Wang et al., 1990; Franke and Custer, 1994, 1996; Coward et al., 1996; Franke et al., 1998; Gamache and Acworth, 1998). Up to 12 isoflavones including daidzin, glycitin, genistin, malonyl daidzin, malonyl glycitin, malonyl genistein, daidzein, acetyl genistein, acetyl daidzin, and genistein can be detected in soybean and soy-derived products (Plewa et al., 1999; Wang et al., 2000). In the mature soybean, only the glucosides and the malonyl glucosides are found. The acetylglucosides and the aglycones are degradation products of the malonyl glucosides and glucosides, respectively. These generally appear during commercial processing after acid- or heat-mediated hydrolysis of the malonyl glucosides and glucosides to form acetyl glucosides and the aglycones. The soy isoflavones are separated by HPLC according to their retention times (Figure 1.7, Table 1.9). Generally, the glycosides elute quantitatively in the early part of the HPLC chromatogram (between 10 and 25 min) and aglycones eluting much later (t_R between 34 and 45 min) (Rong et al., 1998).

HPLC has been used to determine the levels of genistein, daidzein, biochanin A, formononetin, and equol in foods, formulas, and drinks consumed in Australia. Samples are diluted with water (1:10, v/v), hydrolyzed enzymatically (glucosidase 0.4 μ/mL, 24 h at 37°C), and the aglycone form extracted in absolute ethanol, filtered and injected onto the HPLC. The lower limits of detection with this procedure for daidzein, genistein, and equol are 54, 48, and 94 ng/mL, respectively, within batch precision of 0.7%–3.0% CV (Knight et al., 1998).

Franke et al. (1995) describe a simple and fast procedure to extract and simultaneously hydrolyze phytoestrogens and their conjugates from foods, and present a fast and selective HPLC method for precise determinations of the most common

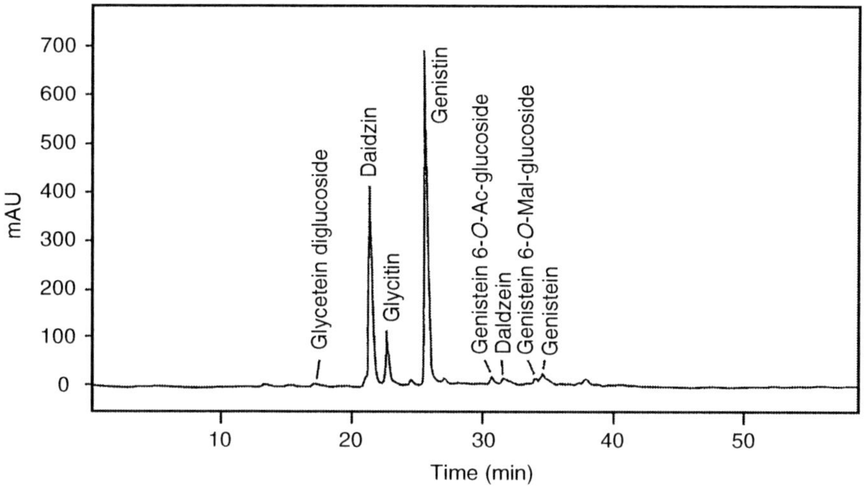

FIGURE 1.7 High-performance liquid chromatography (HPLC) of isoflavonoids extracted with 70% methanol from a soybean by-product. (Reprinted with permission from Plewa, M.J., Wagner, E.D., Berhow, M.A., Conway, A., Rayburn, A.L., and Anderson, D., *Teratog. Carcinog. Mutagen.*, 19, 121–135, 1999.)

TABLE 1.9
Retention Times of Isoflavonoids from Soybean By-Product Separated by High-Performance Liquid Chromatography (HPLC)

Isoflavones	Retention Time (min)	R_f
Daidzin	21.0	0.41
Glycitin	22.3	0.45
Genistin	25.2	0.43
Glycitein (6-*O*-acetylglucoside)	26.8	0.61
Glycitein (6-*O*-malonylglucoside)	27.4	0.12
Daidzein (6-*O*-acetylglucoside)	27.7	0.56
Genistein (acetylglucoside)	29.2	0.59
Daidzein (6-*O*-malonylglucoside)	30.1	0.08
Genistein (acetylglucoside)	31.0	nr
Daidzein	32.0	nr
Glycitein (6-*O*-malonylglucoside)	32.5	nr
Genistein (6-*O*-malonylglucoside)	33.0	0.10
Glycitein	34.0	nr
Genistein	37.4	nr

Source: Adapted from Plewa, M.J., Wagner, E.D., Berhow, M.A., Conway, A., Rayburn, A.L., and Anderson, D., *Teratog. Carcinog. Mutagen.*, 19, 121–135, 1999.

Note: R_f values from Kudou, S., Fleury, Y., Welti, D., Magnolato, D., Uchida, T., Kitamura, K., and Okubo, K., *Agric. Biol. Chem.*, 55, 2227–2233, 1991.

nr, not reported.

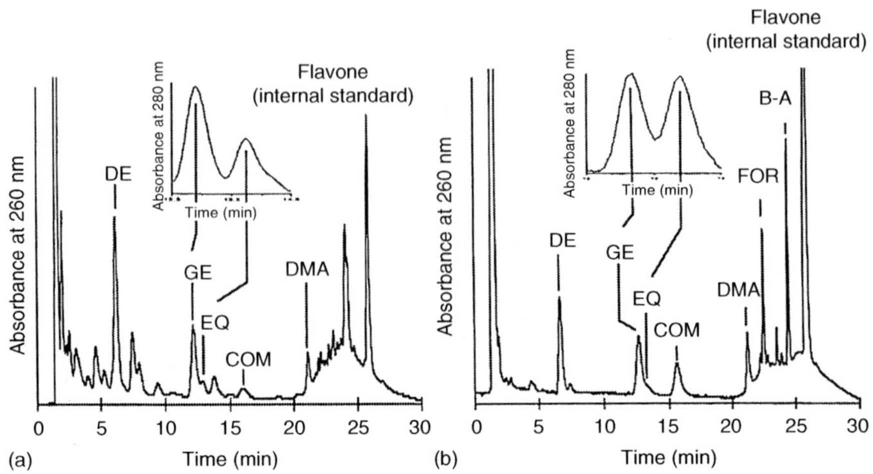

FIGURE 1.8 High-performance liquid chromatography (HPLC) separation of urinary phytoestrogens (a) and standards (b) monitored at 260 and 280 nm (insert). Peak identification: DE, daidzein; GE, genistein; COM, coumestrol; FOR, formononetin; B-A, biochanin A; flavone, internal standard. (Reprinted with permission from Franke, A.A., Custer, L.J., Cerna, C.M., and Narala, K., *Proc. Soc. Exp. Biol. Med.*, 208, 18–26, 1995.)

phytoestrogen genistein, biochanin A, daidzein, formononetin, and coumesterol using flavone as internal standard. A Nova-Pak C_{18} column in combination with an acetonitrile/acetic acid (10% in water) elution system showed best selectivity, recovery, and peak shape for all analytes of interest among several HPLC columns and solvent system tested using authentic phytoestrogen standards (Figure 1.8). Monitoring at 260, 280, and 342 nm at or very near the absorption maximum of the analytes provided sensitive detection. Coumestrol is additionally monitored with fluorescence detection to increase selectivity. Detection limits obtained from authentic standards range between 1.3 and 4.2 ng/mL and allow sensitive phytoestrogen quantitations (Barnes et al., 1998). The procedure of Barnes et al. (1998) has been further modified with improved selectivity, sensitivity, and precision for the quantitative detection of phytoestrogens in human urine and serum. The method employed by the National Health and Nutrition Examination Survey (NHANES) uses enzymatic deconjugation of the phytoestrogens followed by SPE and RP-HPLC to resolve the analytes. The phytoestrogens are detected with a Sciex API heated nebulizer–atmospheric pressure chemical ionization (HN-APCI) interface coupled with MS-MS. This method allows for rapid detection of the major isoflavones and lignans in human serum and urine with limits of detection in the low ppb range. Selectivity of the method is insured by the combination of tandem MS and chromatographic resolution and precision is improved with internal standards for each analytes.

1.3.4 Detection

Traditionally, RP-HPLC analysis combined with UV detection has been used to measure isoflavones in soy products because of the high concentration of isoflavones from these sources (Eldridge, 1982a; Coward et al., 1993; Barnes et al, 1994; Wang and Murphy, 1994; Franke et al., 1995). However, foods with much lower levels of isoflavones, or those with other phytoestrogens such as coumestrol or the lignans, present a greater technical challenge. The maximum absorption of daidzein is at 249.8 nm with a shoulder at 301.9 nm, while those of genistein and biochanin A are at 259.2 nm (Hutabarat et al., 1998). Isoflavones are generally detected at 236 nm (Graham, 1991a, b), 260 nm (Garrett et al., 1999), 262 nm (Barnes et al., 1994), and 280 nm (Kinoshita et al., 1997, 1998). Wang used UV detection at 254 nm and fluorescence detection at 365 and 418 nm for excitation and emission, respectively (Wang et al., 1990). DAD and coulometric detection were used simultaneously for detection of isoflavones in soy foods (Franke et al., 1998).

Fluorescence detection (Pettersson and Kiessling, 1984; Wang et al., 1990; Franke et al., 1994), amperometric detection (Kitada et al., 1986; Setchell et al., 1987; Dewald et al., 1991), and thermospray mass spectrometry (TSP-MS) with SIM (Kitada et al., 1986; Setchell et al., 1987; Dewald et al., 1991) have been very useful in increasing the sensitivity and selectivity of commonly used UV detection (Dziedzic and Dick, 1982; Eldridge, 1982a; Van de Casteele et al., 1982; Pettersson and Kiessling, 1984; Jones et al., 1989; Mellenthin and Galensa, 1997). Müllner and Sontag (1999) describe a reliable and sensitive method for quantification of soybean phytoestrogens, daidzein and genistein, by HPLC with coulometric electrode array detection using bisphenol A as internal standard. Samples are acid hydrolyzed (10 M HCl with ethanol) into aglycones, purified in a C_{18} cartridge, and the phytoestrogen eluted with methanol (80%). The phytoestrogens are separated on an RP column (Hypersil Elite C_{18}), eluted with methanol/acetonitrile/50 mm sodium acetate pH 4.8 (40:20:40, v/v/v), and detected in a coulometric electrode array detector using an oxidative detection mode (+390 to +810 mV in increments of 60 mv versus palladium reference electrode). Daidzein, genistein, and biochanin A have similar recoveries at 95%–98% and detection limits of 0.9, 1.0, and 1.4 mg/kg, respectively using this method.

1.3.4.1 Electrochemical Detection

The phytoestrogens can be detected by UV absorption at 260–280 nm with a detection limit of about 5 ng injected (signal-to-noise ratio of 3 at 0.002 aufs). The phytoestrogens are also electroactive because of the presence of phenolic groups, and can therefore be detected with electrochemical detection (ED). Coumestrol is the most electroactive compound, followed by genistein and daidzein. The optimum potential for the simultaneous sensitive detection of all three compounds is +0.75 V, and at a sensitivity of 3 nA, the detection limits of coumestrol, genistein, and daidzein are 5, 10, and 15 pg injected, respectively. ED is more sensitive than the UV detector for coumestrol, genistein, and daidzein. Formononetin and biochanin A require high operating potential ($>\pm 1.2$ V) resulting in baseline instability in ED, for the detection of a wide range of phytoestrogens. However, in preparations containing only daidzein and genistein, ED is the obvious detection system of choice since it allows for a much smaller sample size to be used, resulting in a simpler and cleaner matrix. For the specific detection of coumestrol, a lower operating potential

(+0.45 to 0.5 V) may be used since few compounds are electroactive at these low potentials (Setchell et al., 1987).

1.3.4.2 Coulometric Array Detection

A less-expensive technique is the combination of HPLC with a coulometric array detector (Gamache et al., 1993). The advantages of this technique are the simplicity of extraction, only hydrolysis or simple alcohol extraction is needed, high selectivity of the detector excluding most contaminant, and ability to determine conjugated phytoestrogens. Comparisons with GC-MS and immunoassay results using both biological fluids and food samples have shown that this technique in most instances yields specific results and can separate 13 different phytoestrogens and their metabolites in one HPLC run (Aldercreutz, 1999).

Nurmi and Adlercreutz (1999) developed an HPLC method for profiling 13 phytoestrogens and their metabolites particularly for nonsupplemented plasma samples using coulometric electrode array detection. The method is less sensitive compared to GC-MS, but higher than other HPLC methods using DAD or UV detection. Detection limits range from 13.4 (secoisolariciresinol) to 40.3 pg (genistein) on column. The intra- and inter-assay ranges from retention time variations are negligible and frequent calibration eliminates detector response variation. The accuracy of the method for six analytes ranges from 69% for enterodiol to 118% for genistein. The intra- and inter-assay precision CV range from 1.5% to 14% and 9.9% to 44%, respectively.

Separation of the phytoestrogens is carried out by using gradient elution with sodium acetate buffer (50 mM, pH 5.0)/MeOH (80:20), and sodium acetate buffer (50 mM, pH 5.0)/MeOH/ACN (40:40:20), as eluents A and B, respectively. A low (0.3 mL/min) flow rate results in a long analysis time of 85 min. A pretreatment with β-glucuronidase and sulfatase is required to hydrolyze the conjugated compounds of plasma by incubation at 37°C for 16 h. The sensitivity of coulometric electrode array detection enables the analysis of low-level plasma phytoestrogens. Phytoestrogen metabolism partly explains the difficulties in determining the precision for all compounds in the same sample. This method is useful for monitoring plasma phytoestrogen profiles in nonsupplemented samples. The higher the concentration of the compound, the better the method precision, even if concentrations close to detection limit determination are satisfactory, when compared to other methods with similar sensitivity.

Gamache and Acworth (1998) applied coulometric array detection, which uses a series of flow-through electrochemical sensors, each providing 100% electrolytic efficiency for measurement of phytoestrogens in complex matrices. The binary gradient RP C_{18} chromatography consisted of 50 mm sodium acetate buffer (pH 4.8) acetic acid/methanol/acetonitrile solvent system (20:20:0 and 40:40:20, v/v/v). Eight coulometric sensors are set at 260, 320, 380, 440, 500, 560, 620, and 680 mV (versus Pd reference). Compounds are resolved in 30 min via both their oxidation/reduction characteristics and chromatographic behavior. Maximal oxidation potentials are 380 mv for coumestrol, 500 mv for daidzein and genestein, 560 mv for estradiol, equol, and estriol, 440 mv for diethylstilbestrol, 620 mv for enterodial, enterolactone, and daidzin with 5–50 pg limits for detection. Ethanol extracts of urine samples from human subjects are evaporated and reconstituted in 20% methanol before HPLC analysis. Daidzein, enterodiol, enterolactone, equol, and genistein are determined in urine with less than 5% (RSD) inter-assay imprecision and 85%–102% recovery.

This technique is based on the use of multiple electrochemical detectors placed in series after the analytical column and maintained at different potentials. Analytes are detectable at 10 pg per injection and the response is linear over three orders of magnitude for all analytes. Coulometric array detection provides low picogram sensitivity and a wide linear response range, suitable for the measurement of phytoestrogens, with a high degree of selectivity. This technique is easy to use with both isocratic and gradient elution chromatography and may provide an alternative to chromatographic methods that use MS as a means of detection. HPLC with coularray detection has been suggested as a highly sensitive method for measurement of multiple analytes with high resolution despite its high cost and other disadvantages (Wilkinson et al., 2002). This method potentially does not require the deconjugation of glucosides (in food matrices) or glucuronides and sulfates in human biological fluids.

1.3.5 Liquid Chromatography–Mass Spectrometry

1.3.5.1 Isoflavones

LC-MS is a powerful technique for analysis of complex mixtures, where additional analytical information is required to positively confirm the identity of the separated compounds or few separations are obtained. Among the interfacing systems used to couple LC with MS, the newly developed electrospray/ionspray (ES/ISP) mass spectrometric liquid interface allows undoubted advantages in terms of sensitivity and capability to analyze large, thermally labile, and highly polar compounds; in addition, tandem MS techniques are useful for structural elucidation studies.

The use of the hyphenated technique LC-MS provides important advantages because of the combination of the separation capabilities of LC and the power of MS as an identification and conformation method. The combination of LC with various types of MS systems (Coward et al., 1996; Barnes et al., 1998) culminates in advantages of these techniques that need less purification and no derivatization compared to the GC-MS determinations and are relatively specific, particularly the MS-MS techniques. Other LC-MS combinations may not be as specific as the GC-MS method because of steroids interference when biological samples are analyzed after short purification. The inability of GC-MS to analyze nonvolatile, high polar, or thermally unstable substances has led to a growing interest in the development of LC-MS as a valuable analytical technique. In this context, LC-MS is a viable system for analyzing naturally occurring compounds such as phytoestrogens in foods and biological fluids. LC-MS and LC-MS/MS technologies are sensitive methods that have become the preferred methods for direct analysis of phytoestrogens. The accuracy of these methods has been improved by highly automated sample preparations, low-detection limit (10 pg/mL), and use of triply ^{13}C-labeled standards (Setchell et al., 2003; Muir, 2006).

LC-MS/MS coupled with triply labeled ^{13}C-NMR has been used to measure isoflavones (daidzein, *O*-DMA, equol, genistein, and glycitein) and two mammalian lignans, enterodiol and enterolactone in urine and plasma (Valentin-Blasini et al., 2000; Taylor et al., 2005). Enzymatic hydrolysis by β-glucuronidase (*H. pomatia*) under optimal conditions (pH 5, 37°C) is required to ensure complete hydrolysis of all conjugates (2 and 16 h for urine and plasma, respectively). Seven phytoestrogens (daidzein, genistein, glycitein, *O*-DMA, equol, enterodiol, and enterolactone) have been measured by LC-MS and LC-MS/MS in spot urines and serum samples of women (aged 45–75 years, 7-day food diaries) who later developed breast cancer (Grace et al., 2004). Results compared with urine and serum samples from healthy controls showed significant correlation ($P < .02$) between both urinary and serum concentrations of isoflavones and lignans and dietary intakes (Grace et al., 2004). Stable isotope-labeled internal standards helped the precision of the assay, resulting in interday CV values of <10% for most compounds studied (Grace et al., 2003). The limit of detection is <50 ng/mL and precision is generally <10% CV for conjugates. Representative examples of recent phytoestrogens analysis using LC-MS and LC-MS/MS are provided in Table 1.10. A high throughput quantification of phytoestrogens in human urine and serum using LC-MS/MS is the most recent development described by Grace et al. (2007). The technique uses SPE in 96-well plate format for the analysis of nine phytoestrogens (genistein, daidzein, equol, *O*-DMA, glycitein, enterodiol, enterolactone, naringenin, and secoisolariciresinol) with LC-MS/MS analysis based on a shallow gradient requiring only 10 min analysis time per sample. The method is highly reproducible (5.5% mean inter-assay CV across all analytes) and repeatabe (3% CV) and has been used in the analysis of 200 urine samples from women diagnosed with invasive breast cancer 1 year earlier.

There are different approaches to the hyphenation of HPLC with MS and some commercial interfaces are available. Atmospheric pressure ionization (API), a soft and highly efficient method suitable for analysis of polar, ionic, high-molecular-mass, and thermally labile compounds amenable to LC has greatly increased the popularity of LC-MS. API-based interfacing systems are ES and ISP, which are liquid-based interfaces.

Analytical methods utilizing thermospray (TSP)-LC-MS have been useful in analyzing phytoestrogens in various soy protein preparations (Setchell et al., 1987). Best separations of the phytoestrogens daidzein, genistein, coumestrol, formononetin, and biochanin A are obtained in RP mode with methanol/ammonium acetate buffer (0.1M, pH 4.6; 3:2v/v) containing 0.25 mM EDTA (ethylenediamine tetraacetic acid) as mobile phase. Extracts containing isoflavones, their polar conjugates, and related substances obtained after aqueous 80% ethanol extraction of soybean products are submitted to enzymatic hydrolysis before LC analysis. The compounds are positively identified in the chromatographic effluent by performing MS analysis in the m/z 100–300 range; daidzein and genistein are detected in high concentrations in soybean milk formulas, soybean flakes, and textured soybean protein. Optimum ionization of phytoestrogens can be achieved by operating the mass spectrometer in continuous scanning mode over a mass range of 100–300 Da with vaporizer and jet-block temperatures of 135°C and 215°C, respectively (Setchell et al., 1987). In another study of the same group (Barbuch et al., 1989), TSP-MS/MS

TABLE 1.10
Selected Examples of Isoflavone and Lignan Analysis by High-Performance Liquid Chromatography (HPLC) and Liquid Chromatography–Mass Spectrometry (LC-MS)

Sample/Analyte	Sample Preparation	Internal or External Standard	HPLC Conditions	Result	References
Human urine and plasma (premenopausal women); Ingestion of ^{13}C-labeled isoflavone daidzein and genistein; 7-day dietary diaries; and 24-h pooled urine and collection of blood sample (10 mL) from (2–72 h)	Serum recovered from centrifuge (3000 × *g*, 10 min, 4°C); blood samples are enzymatically hydrolyzed; urine and plasma samples are separated on Sephadex-LH20 chromatography	Dihydro-flavone	RP-C_{18} column (250 mm × 4.6 mm ID) with PDA detector; A: 10 mmol $AcONH_4$/1 L H_2O (0.1% TFA); B: ACN; gradient: 0% B for 2 min; 0%–50% B, 2–24 min; hold 50% B for 5 min; flow rate: 1.0 mL/min	The bioavailability of both isoflavones is linear at lower intakes (0.4 mg/kg body weight) and nonlinear at higher intakes (0.8 mg/kg body weight)	Setchell et al. (2001)
Rat plasma; daidzein, genistein, glycitein, equol, 4-ethyl phenol, and biochanin	Hydrolysis by glucuronidase and sulfatase (*Helix pomatia*); liquid–liquid extraction method using ethyl acetate	Phenolphthalein β-glucuronide, 4-methylumbelliferyl sulfate, and apigenin	RP-Zorbax SB-CN column (75 × 4.6 mm × 3.5 μm), PDA-detecotor; 0.1% formic acid in water–acetonitrile (85 + 15, v/v)	Recovery of the reported isoflavones range from 86% to 100%	Sepehr et al. (2006)

(*Continued*)

TABLE 1.10 (Continued)

Selected Examples of Isoflavone and Lignan Analysis by High-Performance Liquid Chromatography (HPLC) and Liquid Chromatography–Mass Spectrometry (LC-MS)

Sample/Analyte	Sample Preparation	Internal or External Standard	HPLC Conditions	Result	References
Plant organism, genistein, quercetin, and kaemferol	Extraction with acetone and acetone/MeOH (1:1) (for conjugates); acid hydrolysis, extraction with ethyl acetate (for aglycones)		RP-ODS column (250 mm × 4.6 mm ID, 5 μm), A: 1% H_3PO_4 in H_2O; B: ACN; flow rate: 1.0 mL/min. For conjugates: gradient, 5% B for first 5 min, 5%–10% B, 5–10 min; 5%–15% B, 5–15 min; hold 15% B for 5 min; 15%–17% B, 20–25 min; 17%–23% B, 25–30 min; 23%–50% B, 30–65 min. For aglycone: gradient, 27% B for first 5 min; 27%–30% B, 5–45 min; 30%–100% B, 45–46 min		Liu et al. (2002)
Human breast cancer MCF-7 cells, genistein, and biochanin	MCF-7 cell line in six-well plates is grown to 70% confluence; media are collected, sonicated, and incubated with proteinase K (0.1 mg/mL) for 4 h at 37°C; extracted by SPE (Sep-Pak C18 cartridges) to collect the ^{14}C isoflavones and their metabolites	^{14}C-genistein	C_8 column (30 cm × 4.6 mm ID), linear gradient of 0%–45% CAN (at 4.5%/min) in 0.1% (v/v) TFA; flow rate: 1 mL/min	Genistein inhibits breast cancer cell growth, the major metabolites are sulfate esters of isoflavones	Peterson et al. (1998)
Human inflammatory cells (leukemia HL-60 cells),	HL-60 cell is induced with DMSO-supplemented 10% FBS/RPMI-1640 media for 7 days; cell medium exchanged with K-H buffer;	3′-Chlorinated genistein	RP-C_8 (10 cm × 4.6 mm ID), 6 min isocratic run, isocratically with 40% ACN in 10 mmol/L NH_4OAc; flow rate: 1 mL/min	Genistein and daidzein both undergo chlorination and nitration when added to the	D'Alessandro et al. (2003)

genistein, and daidzein	activated with PMA (10 μmol/L) in the presence of genistein (10 μmol/L); treated with catalase (5 U/mL); centrifuged for 5 min at 800 × *g* at 4°C; the supernatant is removed and extracted twice using diethyl ether (2 mL); ether layers are evaporated under air and redissolved with 80% methanol before analysis			medium of stimulated inflammatory cells	
Food (white, whole wheat, flax, and multigrain breads) for SDG	Defatted sample is extracted with aqueous MeOH (70%), base hydrolysis		Symmetry C_{18} column (250 mm × 4.6 mm ID, 5 μm), A: 1% AcOH/H_2O; B: MeOH; gradient: 5%–60% B, 0–44 min; held 60% B for 4 min; 50%–60% B, 48–55 min; UV-PDA at 280 nm	Recovery is 73%–75%; SDG is detected only in products containing flax (0.06–1.98 μM/g of DW or 19–602 μM/loaf)	Muir and Westcott (2000)
Separation of major oxidation products of SDG and SECO	Oxidation of SDG (0.01M) and SECO (0.01M) with AAPH (0.1M) in a shaking bath at 60°C for 5 h; aliquots (20 μL) are removed through the septum via syringe (to minimize the introduction of air to the system) at 5 min intervals for 1 h and 30 min intervals to the end of the reaction	SDG and SECO	Analytical HPLC: Symmetry C_{18} column (150 mm × 4.6 mm, 5 μm); A: 0.05% TFA in H_2O; B: 0.05% TFA in acetonitrile; 55 min run; gradient: 10%–60% B; 0–9 min, 10% B; 20 min, 40% B; 40–50 min, 60% B; 51–55 min, 10% B; flow rate: 0.4 mL/min; PDA detector (200–400 nm). Prep-HPLC: Symmetry C_{18} (25 cm × 10 cm × 6 μm); flow rate: 20 mL/min; other parameters are similar to analytical HPLC except the use of methanol instead of can	SECO generates five and SDG generates four oxidation products; demonstrate SDG and SECO are chain-breaking antioxidants	Hosseinian (2006)

(Continued)

TABLE 1.10 (Continued)
Selected Examples of Isoflavone and Lignan Analysis by High-Performance Liquid Chromatography (HPLC) and Liquid Chromatography–Mass Spectrometry (LC-MS)

Sample/Analyte	Sample Preparation	Internal or External Standard	HPLC Conditions	Result	References
Secoisolariciresinol, matairesinol, lariciresinol, and pinoresinol in different foods (broccoli, bread, flaxseed, and tea)	Alkaline (0.3 M NaOH) methanolic (30% MeOH) extraction, followed by enzymatic hydrolysis (overnight at 37°C) using β-glucuronidase/sulfatase from *Helix pomatia* (0.83 mg in 1 mL 0.05 M sodium acetate buffer, pH 5.0); samples extracted twice with 2 mL diethyl ether, diethyl ether evaporated, and dried samples are redissolved in 0.3 mL methanol	50 ng of SDG-d8 and 50 ng of matairesinol-d6	Symmetry C_{18} column (150 mm× 3.0 mm ID, 5 μm); flow rate: 0.4 mL/min at 40°C; A and B consisted of water and methanol, respectively; gradient: 0–0.5 min, 30% B; 0.5–12 min, linear gradient from 30% to 50% B; 12–15 min, isocratic at 50% B; 15–15.2 min, linear return to 30% B; 15.2–19 min, isocratic at 30% to equilibrate; run time was 19 min	Yield of alkaline-methanolic extraction increases up to 80%; Recovery is 73%–123%, except for matairesinol added to bread (51%–55%)	Milder et al. (2004)
Human urine (men); muffins containing sesamin (180 mg) and sesamolin (71 mg)	Urine is collected for four 12-h intervals; deconjugated by β-glucuronidase/sulfatase; extraction on SFE containing silica, eluted with chloroform/methanol (4:1, v/v), dried and dissolved in 80% ethanol before separation	Naringenin	RP-C_{18}-DAD ODS column (5 μm, 250 × 4.6 mm); A: 0.01 M phosphate buffer (pH 2.8), 5% ACN; B: ACN; 0–5 min, 15% B; 30 min, 30% B; 40–50 min, 70% B; flow rate: 1.0 mL/min; detection at $\lambda_{max} = 234, 285$ nm	30% of sesamin ingested is excreted in urine in the form of a major catechol metabolite	Moazzami et al. (2007)

Oil-soluble lignans (sesamin and sesamolin)	Extraction with hexane/isopropanol (3:1, v/v); oil-soluble lignans present in hexane fraction	Sesamin and sesamolin	Silica column (4.6 × 250 mm) equipped with a fluorescence detector (excitation and emission wavelengths, 296 and 324 nm), hexane/1,4-dioxane (94:4, v/v) as mobile phase	Sesamin: 7–712 mg/100 g seeds and sesamolin: 21–297 mg/100 g seeds	Moazzami and Kamal-Eldin (2006)
Lignan di- and triglucosides including sesaminol di- and triglucosides and sesamolinol diglucoside	Extraction of defatted flour with 70%–85% ethanol; TLC plates developed by chloroform/methanol (1:1, v/v)	Naringenin	ODS column (5 μm × 250 × 4.6 mm), PDA detector; A: 0.01 M phosphate buffer (pH 2.8) containing 5% ACN and B: ACN. The elution conditions: 0–5 min,15% B; 30 min, 30% B; 40%–50%, 70% B; flow rate: 1.0 mL/min	Increased recovery overtime (from 15 to 72 h) when defatted sesame flour is extracted with 85% ethanol; sesaminol (16–720 mg/g seeds) and sesamolinol content range (0–124 mg/100 g seeds)	Moazzami et al. (2006b)
A single dose of sesame oil muffins (94 mg sesamin, 42 mg sesamolin) with a capsule containing deuterium-labeled α-tocopherol (d6-α-T) and γ-tocopherol (d2-γ-T)	Blood and urine samples collected over 72 h are analyzed for labeled (d6 and d2) and unlabeled (d0) tocopherols and their metabolites	Deuterated α- and γ-tocopherols	HPLC with fluorescence detection	Sesamin and sesamolin decrease urinary excretion of γ-tocopherol in humans and increase human plasma γ-tocopherol concentrations	Frank et al. (2004)
Measurement of intact sulfate and glucuronide phytoestrogen conjugates in human urine	Sample pretreatment not necessary, other than addition of internal standards and pH adjustment; urine is injected directly onto the analytical column	[$^{13}C_3$]daidzein and[$^{13}C_3$] genistein	ODS column (250 mm × 3.2 mm ID, 5 μm), A: 0.1% AcOH/H_2O; B: 0.1% AcOH/ACN; gradient: B,10%, hold for 5 min; 10%–22% B, 5–20 min; 22%–30% B, 20–35 min; 30%–46% B, 35–36 min; 46%–80% B, 36–56 min; 80%–85% B, 56–60 min; 85%–100% B, 60–65 min; held 10% B for 10 min; flow rate: 0.5 mL/min	80% glucuronide, 13% monosulfates, 7% free daidzein, 0.9% sulfoglucuronides, 0.4% diglucuronide, and <0.1%disulfate; CV: <10%; accuracy: <12%; LOD: <50 ng/mL	Clarke et al. (2002)

(Continued)

TABLE 1.10 (Continued)
Selected Examples of Isoflavone and Lignan Analysis by High-Performance Liquid Chromatography (HPLC) and Liquid Chromatography–Mass Spectrometry (LC-MS)

Sample/Analyte	Sample Preparation	Internal or External Standard	HPLC Conditions	Result	References
Rat urine/deuterated ENL	Feeding rats with rye or wheat bran (15% diet) for 5 days; urine samples are collected for 24 h, extracted twice with ethyl acetate, and centrifuged at 2400 × *g* for 10 min. The supernatant fluid is dried under nitrogen, redissolved in 500 μL of 25% aqueous MeOH and filtered	Genistein	C_{18} column (150 mm × 2.1 mm ID, 5 μm), A: 5% ACN/0.1% AcOH; B: 40% ACN/0.1% AcOH; gradient: 0%–100% B, 0–3 min; held 100% B for 12 min; flow rate: 0.2 mL/min	Dietary lignins are precursors of mammalian lignans in rats; ENL urinary excretion is reduced from 18.6 to 5.3 nmol/d when lignans are removed from rye bran and from 30.5 to 6.2 nmol/d when removed from wheat bran	Begum et al. (2004)

Note: SPE, solid-phase extraction; CV, coefficient variation; LOD, limit of detection; TFA, trifluoroacidic acid; ACN, acetonitrile; ENL, enterolactone; END, elactrodiol; MAT, matairesinol; SDG, secoisolariciresinol diglucoside; SECO, secoisolariciresinol.

proved to be a useful technique yielding valuable structural information absent from the mass spectra of the phytoestrogens recorded under TSP-MS conditions. The resultant product-ion spectra contained fragment characteristics of each phytoestrogen subclass, allowing the use of tandem MS to confirm identification and propose structures for unknown compounds. TSP-MS/MS analysis of these substances has also been carried out in the neutral loss operating mode by monitoring the neutral loss of ion (because of consecutive releases of CO), common to all members of this family. Several soy protein preparations have been investigated to confirm the presence or absence of phytoestrogens; in one soybean product analyzed, daidzein, genistein, and an unknown phytoestrogen of the biochanin A subclass were found, this unknown was tentatively identified as 6,7-dihydroxy-4′-methoxyisoflavone using its product spectrum.

HPLC-TSP-MS has been investigated for its potential to identify individual phytoestrogens in the HPLC effluent. When TSP ionization is interfaced with HPLC-MS, ions are formed by spraying the column eluate at high enough heat to evaporate the solvent. The mass spectra generated in the TSP ionization process are characterized (Figure 1.9) by intense protonated molecular ions $[MH^+]$ with no significant fragmentation of the molecule. Since most of the ionization resides in a single ion, SIM of the $[MH^+]$ for each phytoestrogen affords a more specific method of detecting these compounds with a 100-fold improvement in sensitivity over the scanning mode or UV detection alone. It is also suggested that these compounds would be ideally suited to HPLC-MS/MS detection, where after focusing the $[MH^+]$ ion, collision-induced dissociation (CID) would yield fragmentation specific for each compound, thereby assisting structural elucidation of these and unknown phytoestrogens or metabolites separated by HPLC (Setchell et al., 1987).

HPLC with APCI-MS and ISP-MS has been investigated for the first time for the analysis of isoflavones and their conjugates in soy food products (Barnes et al., 1994). The analytes are detected in both PI and NI mode (positive ion and negative ion) using the HN-APCI and the ISP interfaces. The more valuable spectral data about each isoflavone conjugate are obtained using the HN-APCI system in the PI mode. Enhanced sensitivity has also been observed for the positive isoflavone aglucone ions produced in the HN-APCI interface, these ions are about 1.5–3 times more intense than the protonated molecules generated in the ISP interface. The effect of mobile phases containing 0.1% acetic acid or 10 mM ammonium acetate is also described as for the major ions detected in the HN-APCI NI mass spectra of the isoflavone conjugates (Table 1.11).

LC-MS of two isoflavonoids in a licorice root extract powder using particle beam (PB) interface has been described (Weinberg et al., 1992). The LC-PB-MS approach identified formononetin and isoliquiritigenin in the LC eluates on the basis of the corresponding PB-EI mass spectra. In the case of the GC-MS method, the corresponding monolated or disilated derivatives are required since the involatility of the analytes is not amenable to direct analysis by GC-MS (Figure 1.10). High-performance liquid chromatography (HPLC) has also been coupled to photodiode array (PDA) and to mass spectroscopy (MS) using atmospheric pressure chemical ionization (APCI) or electrospray ionization (ESI) in combination with collision-activated decomposition (CAD) (HPLC-APCI-CAD-MS or HPLC-ESI-CAD-MS)

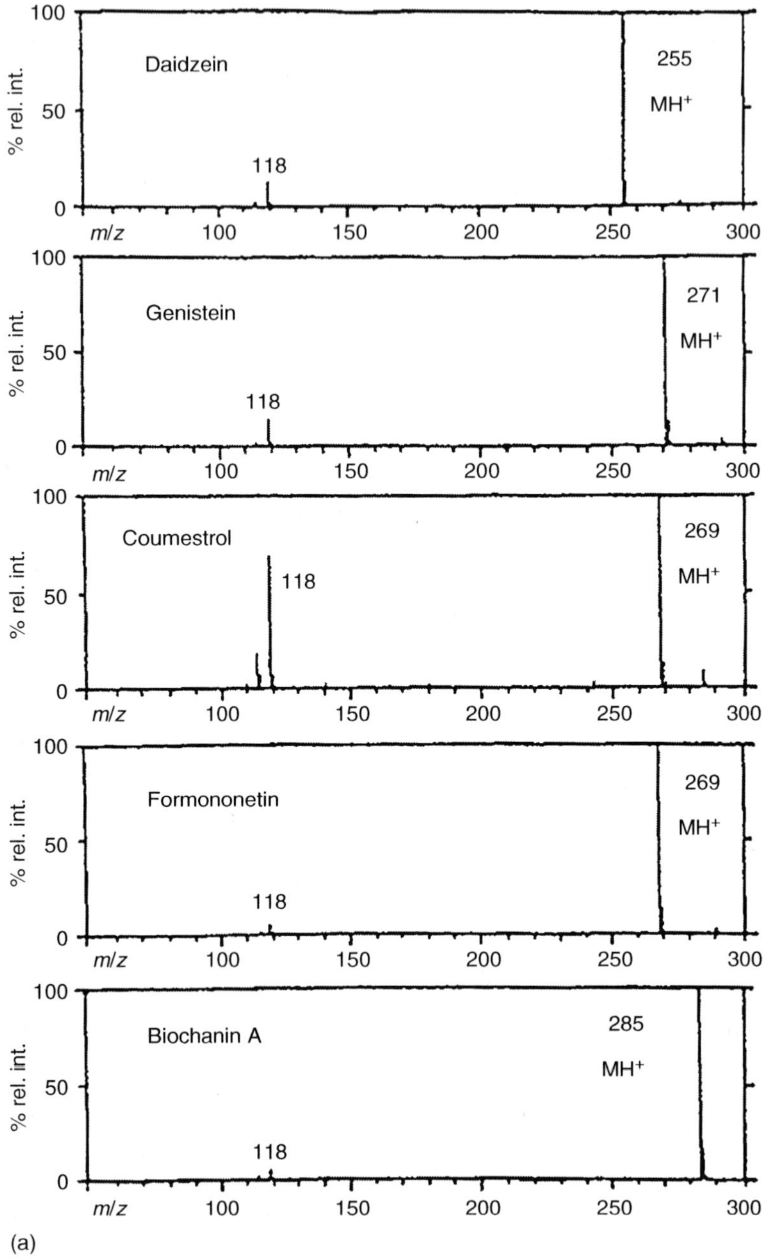

FIGURE 1.9 Mass spectra (a) and total-ion current chromatogram (b) obtained for the high-performance liquid chromatography (HPLC)–thermospray mass spectometry (MS) analysis of a mixture of phytoestrogen standards daidzein (D), genistein (G), coumestrol (C), formononetin (F), and biochanin A (B). (Reprinted with permission from Setchell, K.D.R, Welsh, M.B., and Lim, C.K., *J. Chromatogr.*, 386, 315–323, 1987.)

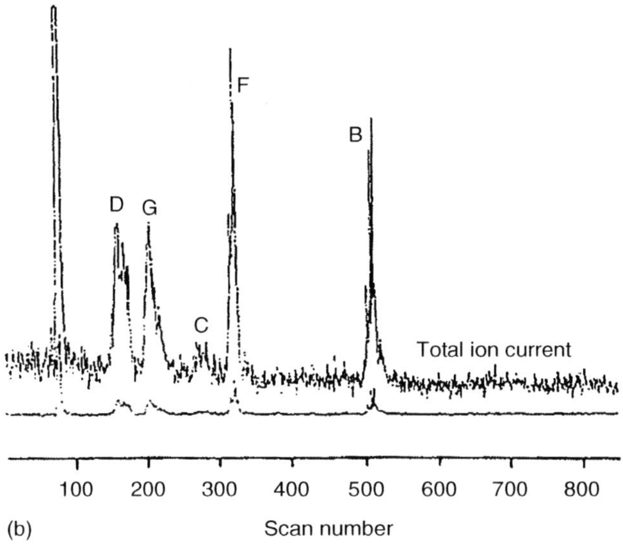

FIGURE 1.9 (Continued)

for identification of glycosides and HPLC-APCI-CAD-MS for identification of aglycones in investigating the isoflavones of the roots of *Pueraria lobata* (Willd.) Ohwi (*Pueraria radix*) (Rong et al., 1998).

TSP as an HPLC-MS interface has been largely superseded by atmospheric pressure ionization (API) because of its recognized limitations in robustness and stability of the ion beam. The API techniques of APCI and ESI are highly sensitive, show greater ionization stability, and are more universally applicable than other HPLC-MS techniques. Both APCI and ESI involve the formation and ionization of an aerosol at atmospheric pressure. The suitability of APCI technique, where spray formed by pneumatic nebulization from a capillary in a heated probe is ionized by a high-voltage corona discharge, has been demonstrated in the determination of isoflavones in plasma (Coward et al., 1996). Aramendia et al. (1995b) have examined synthetic mixtures of isoflavones by capillary electrophoresis/ESI-MS using the negative ionization mode.

Using RP-HPLC interfaced with ESI to an MS, it is possible to obtain a mass/intensity map of all isoflavone metabolites in a single 20 min analysis. Analysis of isoflavonoid conjugate in serum/plasma samples requires initial extraction, but the conjugates can be measured intact either by capillary RP-HPLC-ESI-MS or on regular RP columns by HPLC-HN-APCI-MS. When it is only necessary to measure the total isoflavonoids and their metabolites in blood, hydrolysis can be performed directly in serum/plasma samples and flavonoids recovered by ether or ethyl acetate solvent extraction (Barnes et al., 1998).

In the HN-APCI interface, the mobile phase is passed down a quartz tube heated to 400°C–500°C. Ionization is accomplished by first ionizing the air within the interface by a corona discharge needle. In this type of interface, sensitivity is

TABLE 1.11
Extraction Systems for Lignan Isolation from Flaxseed and Flaxseed-Containing Foods

Extraction	Hydrolysis	Purification	Level SDG (μmol/g seed)	References
MeOH/Dioxane (1:1) 24 h	Ba methoxide	Cellulose column	3.15	Bakke and Klosterman (1956)
In vitro fermentation	Na methoxide	Silical gel ($CHCl_3$:MeOH:H_2O)	0.96–3.15	Thompson et al. (1997)
β-glucuronidase	β-glucuronidase	C-18 SPE	1.19–1.97	Obermeyer et al. (1995)
β-glucuronidase	β-glucuronidase + 2 M HCl 2.5 h, 100°C	Ether extraction/DEAE-Sephadex OH^- QAE-Sephadex Ac^-	9.05–10.21	Mazur and Adlercreutz (1998)
Reflux, 80% MeOH, 2 h	β-glucuronidase	C-18 SPE + lipophillic gel chromatography	0.22–3.41	Setchell et al. (1999)
70% aqueous alcohol	NaOH	C-18 SPE	5.24–15.74	Westcott and Muir (1998)
95% EtOH/Dioxane (1:1) 8 h	nr	nr	0.001–0.004	Harris et al. (1994)
SCO_2 + THF/H_2O (1:1)	nr	nr	7.15	Wilson et al. (1993)
Shaker, 80% MeOH, 4 h, 55°C	1 M HCl 1 h, 100°C	EtOAC/Hexane (1:1)	nr	Meagher et al. (1999)

Source: Adapted from Muir, A.D., Westcott, H.D., Ballantyne, K., and Northrup, S., *Proc. 58th Flax Inst. United States*, Flax Institute of the United States, Fargo, ND, 23–32, 2000.

Note: nr, not reported.

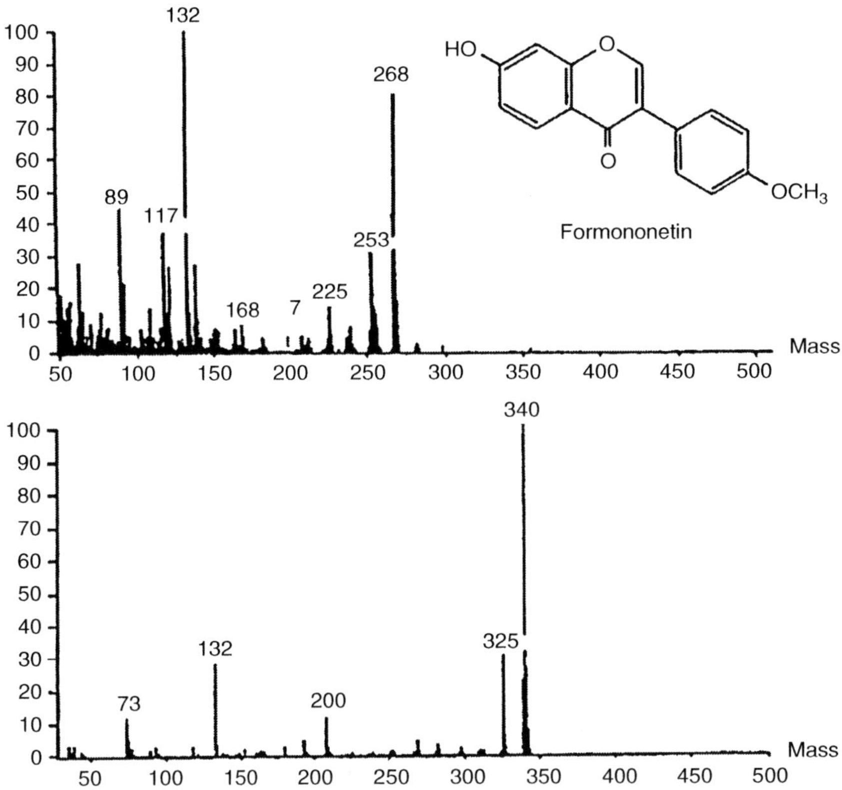

FIGURE 1.10 Mass spectra of formononetin obtained by liquid chromatography–mass spectometry (LC-MS) (top) and monosilylated formononetin obtained by gas chromatography–mass spectometry (GC-MS) (bottom). (Reprinted with permission from Weinberg, D.S., Manier, M.L., Richardson, M.D., and Haibach, F.G., *J. High Resol. Chromatogr.*, 15, 641–654, 1992.)

determined by the total mass of solutes present, rather than their concentration as in the case of ESI-MS. Solutes obtained from the fractionation of isoflavonoids and other phytoestrogens generally extracted with 80% aqueous and separated on Aquapore C_8 RP-HPLC column can be introduced directly into the mass spectrometer via the HN-APCI interface operating in both the positive and the negative modes. In the scan mode, ions entering the mass spectrometer are analyzed over an m/z range from 50 to 800. MS-MS daughter ion spectra are obtained by passing molecular ions selected by the first quadrupole into an argon gas collision cell and analyzing the fragment ions in a third quadrupole (Barnes et al., 1998).

Measurement of the isoflavonoids and their metabolites in urine from those consuming a heavy soy diet can be carried out by HPLC-UV analysis (Franke and Custer, 1994) as well as HPLC-MS. In the latter method, multiple reaction ion monitoring (MRM) is used, combining each parent molecular ion with a specific daughter ion produced by CID. The data from the two methods are significantly correlated ($n > 0.92$, $P \leq .0001$). However, the concentrations of daidzein detected by the HPLC-HN-APCI-MRM method are 40% lower than those by HPLC-UV. Thus at low urinary isoflavone concentration found in those consuming a regular American diet, HPLC-UV is inadequate, and HPLC-MS becomes the method of choice. In plasma and serum, where isoflavone concentrations are low (<1 μM), high-pressure collision cell is used for extensive fragmentation of the ions with a 10-fold increase in sensitivity. Because of the high selectivity of the parent–daughter ion combination, the chromotographic separation time is reduced from 25–30 min to 6 min and reproducibility is improved under isocratic conditions.

HPLC-MS using ESI and HN-APCI interfaces is highly applicable to the analysis of isoflavones and their metabolites in foods and biological fluids. The interfaces permit the detection of diagnostic ions without the need for derivatization and multistep extraction procedures. Analysis of isoflavones has also been carried out using the HN-APCI interface (Aramendia et al., 1994; Barnes et al., 1994; Coward et al., 1996). For this procedure, the instrument is initially calibrated in the positive APCI mode using a mixture of polyethylene glycols over the mass range m/z 80–1100. Tuning is then optimized in ESI on the m/z 42 background ion followed by optimization on the protonated molecule $[M+H]^+$ ions of daidzein and genistein at m/z 255 and 271. Typical operating conditions are as follows: capillary voltage 2.4 kV, cone voltage offset 5 V, high-voltage lens 0 kV, and source temperature 150°C. Scanned acquisitions are made over the range m/z 80–620 with a scan time of 2 s (Figure 1.11). At low voltage (20 V), daidzin and genistin loss of glucose molecule occur readily to yield the corresponding aglycone masses at m/z 255 and 271, respectively. Detection is based on a single ion and the retention time because fragmentation is precluded because of the highly conjugated nature of the molecule. This interface is well suited to simply applying previous HPLC methods in which isoflavonoids are detected by their UV absorbance. Identification of physiologic conjugates (β-glucuronides, sulfates) using the HN-APCI interface can only be based on chromatographic mobility.

Barnes et al. (1998) developed a microbore HPLC-ESI-MS method for the determination of the phytoestrogens daidzein and genistein in soy flours and baby foods. In this procedure, the samples are hydrolyzed and extracted with acetonitrile–water before analysis. LC is performed in a microbore Primesphere $5C_8$ using a water/acetonitrile/acetic acid mobile phase at a flow rate of 60 μL/min and detection is by API in the form of pneumatically assisted ESI-MS. The limit of detection for daidzein and genistein in the flour and food samples is 0.2 and 0.7 mg/kg, respectively. The method is very robust and reliable when operated over a long time period, generating precision data with 4%–15% CV. HPLC coupled with MS, such as ESI of treated nebulizer APCI, can directly access the intact molecular weight of isoflavones, both conjugated and unconjugated (Barnes et al., 1994, 1998; Aramendia et al., 1995).

In LC-ESI-MS, the quadrupole detector can be replaced by an IT (ion traps) or a time-of-flight (TOF) detector. The IT has the advantage due to carrying out first, fragmentation of the parent molecular ion and second, the daughter ions (Wu et al., 2003). This is important in the analysis of glycosides of isomeric form of compounds (Barnes, 2006). For example, the glycosides of genistein (in soybean) and apigenin (in parsley) have the same molecular weight (m/z 431 for their $[M-H]^-$ molecular ions)

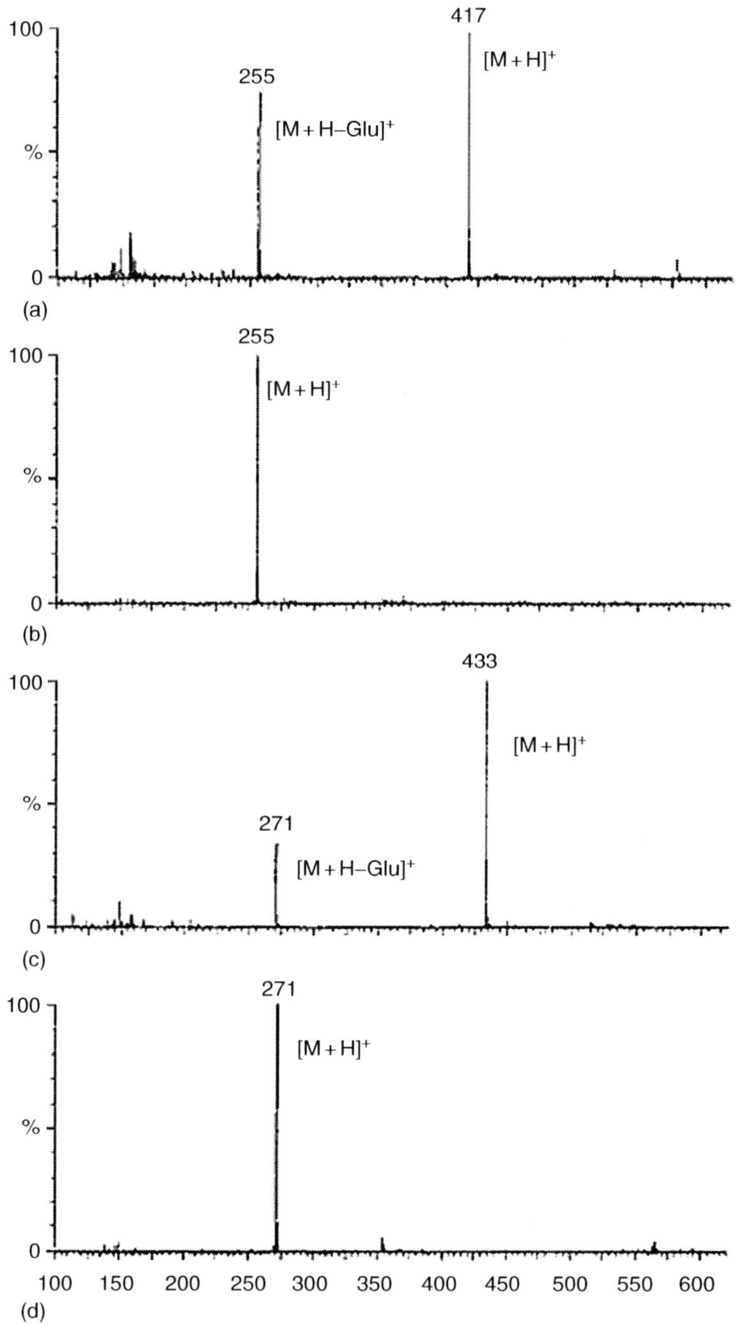

FIGURE 1.11 High-performance liquid chromatography electrospray ionization–mass spectrometry (HPLC-ESI-MS) of (a) daidzin, (b) daidzein, (c) genistin, and (d) genistein obtained at low-cone voltages. Reprinted from Barnes, S., Coward, L., Kirk, M., and Sfakianos, J., *Proc. Soc. Exp. Biol. Med.*, 217, 254–262, 1998b.

with very close retention time in HPLC. Fragmentation of these glucosides in a TQ (triple quadrupole) leads to an aglycone daughter ion (*m/z* 269), making them indistinguishable. However, IT shows the fragmentation of the *m/z* 269 ion that leads to unique daughter ions (*m/z* 133 for genistein and *m/z* 149 for apigenin), allowing their structural measurements.

Although HPLC is currently the most widely used analytical technique to quantify isoflavones (Murphy et al., 1997; Song et al., 1998), it is time consuming with tedious sample preparation. To overcome these problems, Wang and Sporns (2000) developed a simple protocol for the uses of matrix-assisted laser desorption/ionization time-of-flight mass spectrometry (MALDI-TOF-MS) to analyze isoflavones from food samples with minimal sample preparation. MALDI-TOF-MS advantages over other methodologies include speed of analysis (spectrum is obtained in minutes), high sensitivity, wide applicability combined with good tolerance toward contaminants, and the ability to analyze complex mixtures (Karas, 1996).

For the MALDI-TOF-MS procedure, acetonitrile crude extract of isoflavone is diluted 10-fold, loaded on a Sep-Pak C_{18} cartridges where the isoflavones are retained and later eluted with 70% methanol. The eluent is applied directly to MALDI-TOF-MS to acquire the isoflavone spectrum of interest. In MALDI-TOF-MS, isoflavones exhibit only fragmentation corresponding to loss of their carbohydrate residues. Daidzin and genistin fragments produce $[M\text{-}162+H]^+$ ions at *m/z* 255 and 271, respectively similar to the fragmentation pattern observed in ESI-MS (Barnes et al., 1998). The response of daidzin in MALDI-TOF-MS is more than twice that of genistin. MALDI-TOF-MS identification of isoflavones is similar in retention time and peak positions to those obtained by HPLC methods (Wang and Sporns, 2000). MALDI-TOF-MS can easily study changes in isoflavones because of processing or can serve as a rapid analytical tool for authenticity.

1.3.5.2 Lignans

At least six different methods have been proposed for the analysis of lignans in flaxseed and products containing flaxseed, and three different methods for analysis of lignan metabolites in mammalian tissues and fluids since the lignan SDG was first isolated from flax by Bakke and Klosterman (1956). In addition, several variations exist for many of these methods. Variations in extraction of lignans from flaxseed and products containing flaxseed further complicate its analysis. Results from these studies have been inconsistent mainly because of differences in methodology.

Lignans in flaxseed, similar to isoflavones in soybeans, mostly occur in the conjugated forms as ester-linked complex, hence extraction conditions require conversion to aglycones before analysis. Several of the commonly used extraction procedures are inefficient in extracting lignans from plant materials (Table 1.12). Most extraction methods involve the use of a combination of alcohol and other solvent. β-Glucuronidase enzyme is also used to extract lignans as a first step in a multistep hydrolysis process. A comparison of enzymatic hydrolysis of lignan versus extraction and chemical hydrolysis of lignans has shown that more lignan can be detected with extraction and chemical hydrolysis than with enzymatic release (Westcott et al., 1998).

TABLE 1.12
Comparison of Retention Times (Minutes) for Lignan Aglycons by High-Performance Liquid Chromatography (HPLC) Diode Array Detector (DAD) with Those of the Lignan Trimethylsilyl Ether (TMS) Derivatives by Gas Chromatography–Mass Spectrometry (GC-MS)

	GC-MS		HPLC-UV	
Compound	**Isolated**	**Standard**	**Isolated**	**Standard**
NDGA[a]		8.01		44.3
Isolariciresinol[b]	8.32			
Secoisolariciresinol	8.44	8.41	16.8	16.9
Anhydroseco[a]	9.04		36.0	36.2
Matairesinol	10.54	10.48	31.6	31.5
Lariciresinol		10.01		18.6
Hinokinin		10.55		17.9
Arctigenin		11.13		40.5
Pinoresinol	11.19	11.17	26.8	25.9

Source: From Meagher, L.P., Beecher, G.R., Flanagan, V.P., and Li, B.W., *J. Agric. Food Chem.*, 47, 3173–3180, 1999.

[a] Nordihydroguaiaretic acid (NDGA), anhydroseco, divanillyltetrahydrofuran.
[b] Isolariciresinol was not resolved on HPLC because of a coeluting sterol peak.

In mammalian tissues, lignans have been commonly analyzed by GC (Fotsis et al., 1982) and later GC-MS (Adlercreutz et al., 1995), although these compounds can be more appropriately assessed by HPLC. A significant limitation of lignan analysis by GC-MS resides in the insufficient volatility of the parent compound without derivatization and the conversion of the conjugates to their corresponding aglycones. These limitations have been overcome by devising a series of hydrolysis and purification procedures (Adlercreutz et al., 1993). However, quantitation becomes increasingly difficult especially when complicated purification procedures are employed in GC-MS methods, unless internal standards are used. The preferred method is the use of 2*H*-labeled aglycones and SIM to increase sensitivity (Adlercreutz et al., 1993). However, only compounds for which internal standard is available such as enterodiol, enterolactone, and deuterated matairesinol have been reported in many studies.

Analytical systems as an alternative to the costly mass spectrometers or improvement to the sensitivity of HPLC systems have been proposed. GC-Ion mobility, an atmospheric pressure detection system for GC, has been used to detect the presence of mammalian lignans in urine samples (Atkinson et al., 1993). Recently, electrochemical detection has been used to determine enterolactone and enterodiol at *p*-mole concentrations only after the conversion of the conjugates to their corresponding aglycones (Gamache et al., 1999).

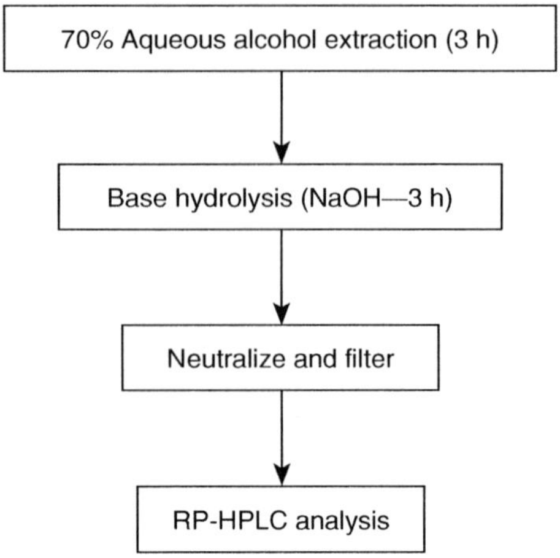

FIGURE 1.12 Flowchart for extraction of lignans from flax containing samples. (Reprinted with permission from Muir, A.D., Westcott, H.D., Ballantyne, K., and Northrup, S., *Proc. 58th Flax Inst. United States*, Fargo, 23, 2000.)

A new standard method for analysis of lignans in flaxseed and products containing flaxseed has been proposed (Muir et al., 2000) and is outlined in Figure 1.12. Extraction is carried out with aqueous alcohol, either ethanol or methanol, with a liquid-to-solid ratio of 20:1. The isolated ester complex is hydrolyzed with dilute base for 3 h at room temperature, neutralized (0.5 mL 0.5 N HAC), filtered, concentrated if necessary, and subjected to HPLC analysis. Quantitation is achieved by using SDG as external standard. When higher sensitivity is required, samples can be analyzed by LC-MS using soft ionization techniques such as ESI or APCI. Results are expressed as mg SDG/g samples or preferably in μmol/g of sample.

LC-MS techniques with soft ionization such as ESI and APCI overcome the time consuming and large sample requirements of GC-MS for analysis of lignans in urine and plasma. The urine sample (in MeOH, 1:1, v/v) can be analyzed directly by ESI-MS in the NI mode without chromatographic separation (Figure 1.13). Four principal ions, *m/z* 477, 473, 381, and 377 are detected and further fragmented by collision with argon producing daughter ions consistent with derivatives of enterodiol (MW = 302 from *m/z* 477 and 381 fragments) and enterolactone (MW = 297 from *m/z* 473 and 377), respectively. The mass loss suggests that *m/z* 477 and 473 are the glucuronides and *m/z* 381 and 377 are the monosulfates of enterodiol and enterolactone, respectively.

While ESI is probably the most widely used soft ionization technique, APCI can also be used to quantitate lignan metabolites. The sample is chromatographed on C_{18} column and the effluent subjected to APCI in the NI mode producing fragmentation pattern similar to that obtained by ESI. This mammalian lignan conjugates in biological fluids can be identified simply using LC-MS (ESI or APCI) either by direct injection of the sample or after chromatographic separation before mass spectroscopic analysis.

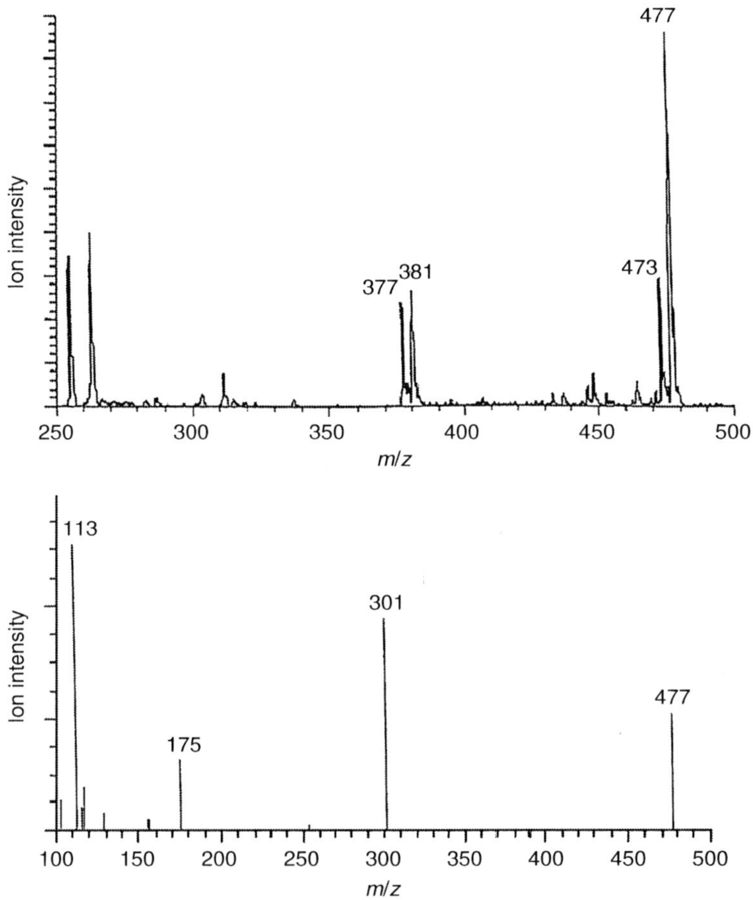

FIGURE 1.13 Direct electrospray ionization–mass spectrometry (ESI-MS) of urine sample of rabbit maintained on a diet containing 15% flaxseed. Negative ion electrospray of rabbit urine (top) for ions corresponding to the glucuronides of enterodiol ($m/z = 477$) and enterolactone ($m/z = 473$) and their corresponding sulfates ($m/z = 381$) and ($m/z = 377$). Daughter ion scan of m/z 477 (bottom). (Reprinted with permission from Muir, A.D., Westcott, H.D., Ballantyne, K., and Northrup, S., *Proc. 58th Flax Inst. United States*, Flax Institute of the United States, Fargo, ND, 23–32, 2000.)

The isolation and characterization of potent phytoestrogens, the lignans isolariciresinol, pinoresinol, secoisolariciresinol, and matairesinol, from flaxseed meal have been described (Meagher et al., 1999). The extraction method combined the removal of the lignan glycosides from the plant matrix with an alcoholic solvent system (methanol/water, 80:20 for 4 h at 55°C), followed by acid (1 M HCl) hydrolysis (1 h at 100°C) to release the aglycones. The acid-hydrolyzed methanolic extract is separated by RP-HPLC and lignans detected at 280 nm with DAD. GC-MS is used to characterize the lignans after derivatization. Lignan aglycones isolated

TABLE 1.13
Performance and Noise Determination for High-Performance Capillary Electrophoretic (HPCE) Analysis of a Standard Mixture (5 mg/L) of Daidzein and Genistein

	Daidzein	Genistein	*p*-Nitrophenol
Noise (6 × SD)[a] (MAU)	0.1646	0.1648	0.2122
Drift (mAU/h)	3.394	2.947	1.5119
Migration time (min)	13.52	13.87	20.19
Theoretical plate number	93.672	88.241	73.011
Resolution	1.72	1.61	—
S/*N* (stability/noise ratio)	47.20	57.40	23.1
Limit of detection (mg/dm^3)	0.32	0.26	—
Limit of quantification (mg/dm^3)	1.00	0.90	—

Source: From Vänttinen, K. and Moravcova, J., *Czech. J. Food Sci.*, 17, 61–67, 1999.

[a] Time range from 15 to 20 min.

from flaxseed have shorter retention time on GC than on HPLC (Table 1.13) (Meagher et al., 1999).

Acid hydrolysis of methanolic lignan extract leads to the isolation of divanillyltetrahydrofuran (anhydrosecoisolariciresinol), previously reported (Mazur et al., 1996), as an artifact of acid hydrolysis resulting from the dehydration of secoisolariciresinol but is now proven to be a naturally occurring lignan (Meagher et al., 1999). Enzyme hydrolysis of defatted flaxseed meal is not efficient at removing lignan from the plant matrix compared to extraction with an organic solvent system. Combination of aqueous methanolic extraction and subsequent acid hydrolysis renders a lignan extract that can be further purified by partitioning with ethyl acetate and hexanes. Purification of the HPLC fractions before GC-MS analysis is not required since the crude lignan extract can be directly applied to GC-MS without the need for a C_{18} preparative column or an ion-exchange column used in other reported methods (Setchell et al., 1983; Mazur et al., 1996) as the lignan peaks of interest are well resolved on the nonpolar column.

The most recent analytical method as described by Johnson et al. (2000) involves extraction of defatted flaxseed flour with dioxane/ethanol, aqueous base hydrolysis, solid-phase purification of an SDG-containing fraction, and quantitative analysis by HPLC. After base hydrolysis and release of SDG from its polymer with aqueous NaOH, the extracts are acidified to pH 3.0 to prevent ionization of the carboxylic and phenolic groups. The acidification leads to the formation of salts that are removed by SPE in a C_{18} reversed-SPE column with almost quantitative recovery (>99.5%) of SDG from the SPE column.

The gradient consisting of solvent A (5% acetonitrile in 0.01 M phosphate buffer, pH 2.8) and solvent B (acetonitrile) is mixed together (A/B, v/v) to give the following: 0 min (100:0), 30 min (70:30), and 32 min (30:70) at 1 mL/min to provide good separation of SDG and requires less time than other HPLC gradients

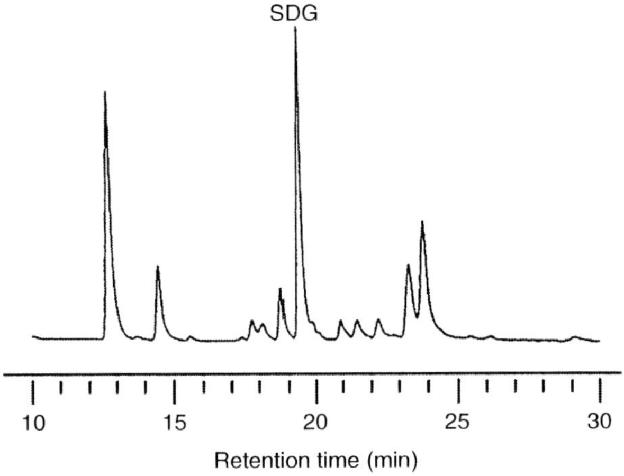

FIGURE 1.14 High-performance liquid chromatography (HPLC) of flaxseed hydrolyzate recorded at 280 nm with a diode array detector (DAD). (Reprinted with permission from Johnson, P., Kamal-Eldin, A., Lundgren, L.N., and Aman, P., *J. Agric. Food Chem.*, 48, 5216–5219, 2000.)

(Figure 1.14). In this system, SDG elutes at 19.5 min with *p*-coumaric acid and ferulic acid glucoside eluting at 12 and 15 min, respectively. The yield of SDG is independent of the solvent (methanol or water) used for base hydrolysis. This study is the first to provide data on the variation in SDG content in whole flaxseeds (6–13.3 mg/g). The nature and utility of the SDG polymer in animal or human nutrition have not been reported.

1.3.5.3 Ionspray Mass Spectrometry of Lignans

ISP-MS and tandem MS (MS/MS) have been used to rapidly detect and unambiguously identify secolarisiresinol in flaxseed (Bambagiotti-Alberti et al., 1994a). According to the authors, SDG is present only in the methanolyzed extract suggesting that it is complexed and embedded in flaxseed. Detection of SDG is possible within the products arising from methanolysis bypassing any chromatographic separation. The only drawback is the lack of structural information, this being typical of soft ionization techniques such as ISP.

In this procedure, air-dried hexane-defatted flaxseed meal is extracted with methanol/dioxane mixture (1:1) and analyzed by MS flow injection analysis (FIA). The methanol–dioxane extract is methanolyzed with methanol and sodium methoxide, neutralized with acid (1 N H_2SO_4), filtered and analyzed with FIA-MS and MS/MS. The molecular weight of SDG is 686 Da.

Later Bambagiotti-Alberti et al. (1994b) set out to test the comparative performance of LC-ISI-MS (liquid chromatography ionspray ionization mass spectrometry) and LC-CF-FAB-MS (liquid chromatography continuous-flow fast-atom bombardment mass spectrometry) of the two isomeric forms of the lignan SDG. The LC separation is performed using a gradient of acetonitrile and water containing trifluoroacetic acid (0%–30%). The eluate is introduced into the ISI interface by splitting after the UV detector. The ISI total ion current (TIC, mass range *m/z* 600–800) and mass chromatograms of methanolyzed dioxane–methanol extract of defatted flaxseed meal show two peaks corresponding to the two isomers of SDG present in flaxseed.

A comparison of performance achieved in the analysis of lignans by ISI and CF-FAB reveals that in the PI mode, ISI provides stable quasi-molecular ion species which, by MS/MS, give fully diagnostic product-ion spectra. ISI is seen to be almost useless in the NI mode since it fails to produce CID-prone ion species. CF-FAB, on the other hand, gives rise to spontaneous diagnostic fragmentation along with the expected quasi-molecular species, allowing positive identification without MS/MS intervention. However, in the PI mode, a dense spectral background can hamper recognition of the diagnostic peaks, especially at low-analytic concentrations. In the NI mode, the peaks of interest lie on smooth background.

1.3.6 Capillary Electrophoresis

Capillary electrophoresis (CE) is a powerful technique that affords rapid, high-resolution separations (10^4–10^6 theoretical plates) while requiring only a few femtomoles of sample. The technique is applicable to a wide range of analytes present in buffered aqueous solution as charged species. The utility of CE, however, is greatly enhanced by MS detection, particularly with a soft ionization technique such as ESI to produce ions even from thermally labile, nonvolatile, polar compounds.

The mass spectrometer equipped with an ESI source is operated in the NI mode at a probe tip voltage of 1–3.5 kV. The extraction cone voltage serving primarily to focus ions into the mass analyzer varies from −25 to −75 V. As a general rule, compounds with a methoxy group (biochanin A and formononetin) exhibit one fragment ion at *m/z* [M-H-CH_3] when high-extraction voltages are used. The mass spectrometer is scanned from *m/z* 90–325 at 25s per scan. A coaxial sheath liquid consisting of water/propanol (80:20) at a flow rate of 10 μL/min is the fluid for CE-ESI-MS. Nitrogen is used as both a drying gas (50 L/h) and the ES nebulizing gas (10 L/h). Mass spectral data are acquired with SIR mode (0.2 s dwell time, 0.2 mass unit span) for the [M–H] ion. For CE, samples are electrophoresed at 305 V/cm resulting in very stable currents of 35 μamp and detected by UV at 260 nm with scanning between 190 and 400 nm.

Analysis with CE is complete within 4 min (Figure 1.15) for an isoflavone mixture (1 mM) containing about 3 fmol of each compound after 20 cm of capillary using UV detection at 260 nm. The signal reproducibility is very good with less than 10% peak variation for the same sample. The success of the CE-ESI-MS analysis of isoflavones relies on many factors, including the ESI interface and buffer composition. Analyte detection and sensitivity vary widely with buffer concentration, the optimum signal is obtained at the lowest possible concentration (10–25 mM). CE with its exceptionally low flow rate (nanoliters per minute) is easier to interface with MS than LC, since no flow splitting is required.

Dietary phytoestrogens have been determined by a fast and selective CE method. In this procedure, isoflavones, daidzein, and genistein are separated on an uncoated

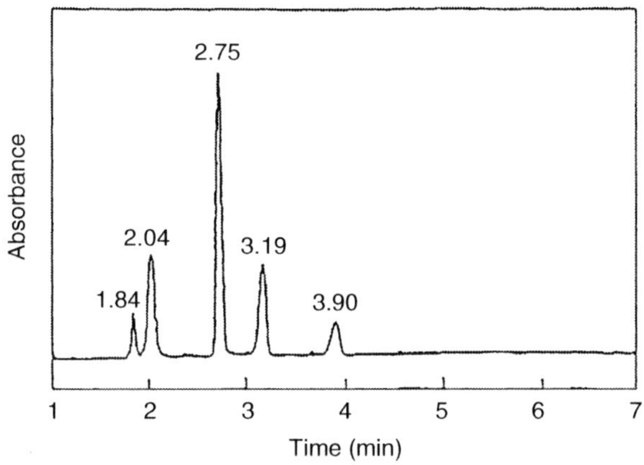

FIGURE 1.15 Electropherogram of isoflavone mixture with ultraviolet (UV) detection at 260 nm. Peaks: genistein (1.89 min), daidzein (2.04 min), pseudobaptigenin, formononetin, and biochanin A (2.76 min), isoliquirtigenin (3.19 min), and biochanin A 7-glucoside (3.90 min). (Reprinted with permission from Vänttinen, K. and Moravcova, J., *Czech. J. Food Sci.*, 17, 61–67, 1999.)

fused-silica column using borate buffer and DAD at 254 and 268 nm, respectively. The linear response ranges from 5 to 100 mg/dm^3 with minimum detectable limit at 0.4 mg/dm^3 for both analytes. The relative response factors for daidzein and genistein are 0.519 and 0.755, respectively when *p*-nitrophenol is used as a standard (Vänttinen and Moravcova, 1999). Aramendia et al. (1995) separated isoflavones on an uncoated fused-silica CE column (110 cm $\times$ 75 μm ID) using ammonium acetate buffer (25 mL) and UV and ESI-MS detection. ESI-MS enables the determination of the molecular mass of isoflavones and also the presence of various functional groups according to observed losses from the [M–H] ion during CID by adjusting some MS parameters. Peng and Ye (2006) used a fused-silica capillary (75 cm $\times$ 25 μm ID $\times$ 360 μm OD) with 50 mM borate buffer (pH 9.5) and a 14-kV voltage as optimal parameters for the determination of isoflavones in red clover by CE with electrochemical detection. The analytes are separated within 25 min with linear response over three orders of magnitude and detection limit ranging from 2 to 5 $\times$ 10^{-5} mg/mL.

For quantitative investigations using capillary zone electrophoresis (CZE), an addition of 3-isobutyl-1-methylxanthin as an internal standard is recommended. The CZE separations are performed on fused-silica capillary of 50 or 70 μm ID with boric acid adjusted to pH 8.6 as separation buffer. Before each run, the capillary is conditioned with NaOH followed by boric acid. Detection is at 260 nm by DAD with scans from 200 to 400 nm (Mellenthin and Galensa, 1999).

Soy isoflavones have been separated by CZE and HPLC (Figure 1.16) (Mellenthin and Galensa, 1999). The main advantage of the CZE is the short analysis time (8 min) compared to the 50 min by HPLC. During the CZE separation, the glycosides daidzin and genistin with their higher molecular weight are detected earlier than their aglycones because the negatively charged isoflavone molecules move toward the anode. Daidzin and daidzein are detected before genistin and

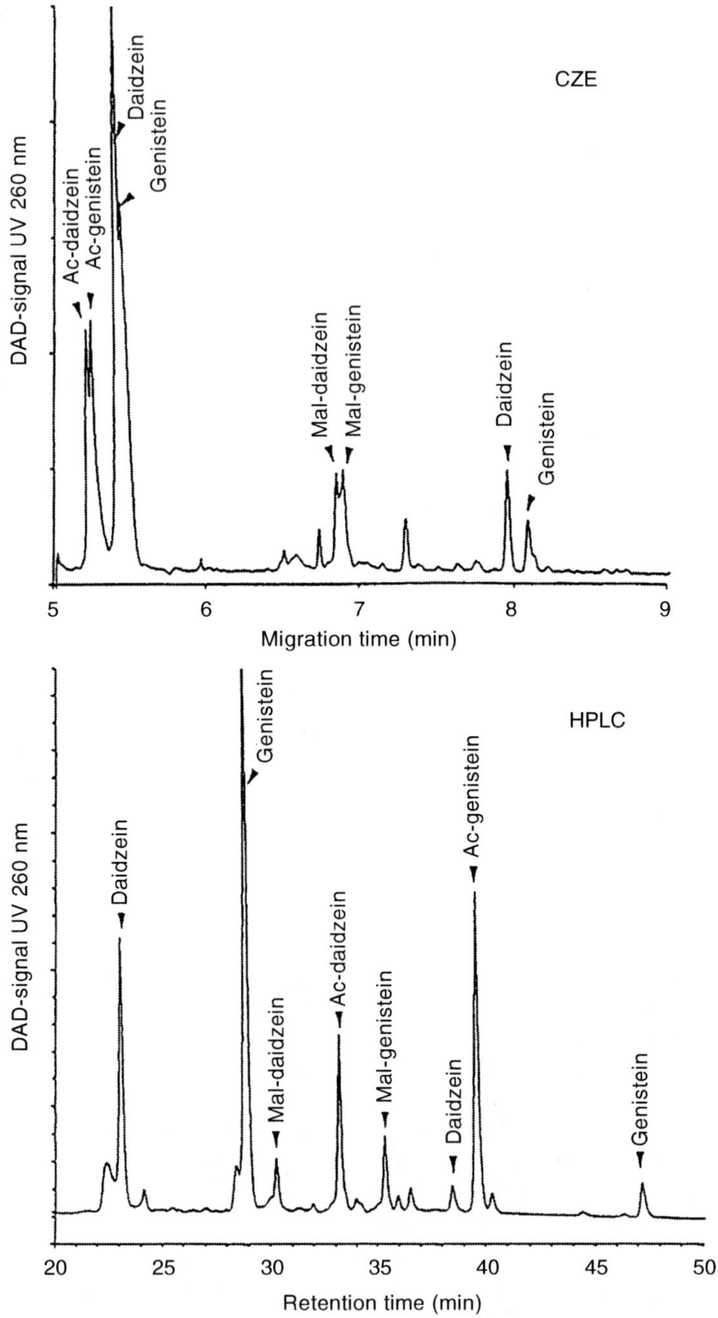

FIGURE 1.16 Separation of isoflavones from toasted soy flour by capillary zone electrophoresis (CZE) and high-performance liquid chromatography (HPLC). (Reprinted with permission from Mellenthin, O. and Galensa, R., *J. Agric. Food Chem.*, 47, 594–602, 1999.)

genistein, respectively because of charge effects (the extra hydroxy group of genistein tends to lose a proton). However, reproducibility of migration time in CZE is poor. In HPLC, daidzein derivatives elute earlier than those of genistein. The detection limit in HPLC is about 0.01–0.03 mg/L compared with 0.1–0.5 mg/L in CZE. The HPLC method has superior repeatability, is less dependent on matrix effects, and more sensitive, and therefore more suitable than CZE for the detection of small amounts of analytes. In CZE, the daidzein peak and in particular both glycoside peaks in the front part of the electrophoregram can be overlapped by other interfering peaks from the matrix. In addition, the moderate reproducibility of migration times in CZE and the narrow detection range of 4 min for eight isoflavones lead to large variation of the results.

1.3.7 High-Performance Capillary Electrophoresis

HPCE is now the most rapidly expanding analytical technique because of its speed and very high separation efficiency, and interest in performing single-step analyses with simple sample preparation technique in HPCE is considerable. HPCE has been introduced into analytical methods of daidzein and genistein (Shihabi et al., 1994; Aramendia et al., 1995). Separation by HPCE generally requires a buffer preferably sodium borate (200 mM) adjusted to pH 8.6, an uncoated capillary (150 mm ID × 50 cm, 56 cm length), sample injection performed in pressure mode (150 mbar.s), and UV detection set at 254, 268, and 400 nm for daidzein, genistein, and *p*-nitrophenol, respectively. The separation potential at 10 kV corresponding to an electric current of 60 mA, column temperature of 25°C, and run time of 30 min are used for the extracts. The migration times under these conditions for daidzein and genistein are 13.52 and 13.80 min, respectively (Figure 1.17). The standard deviation of

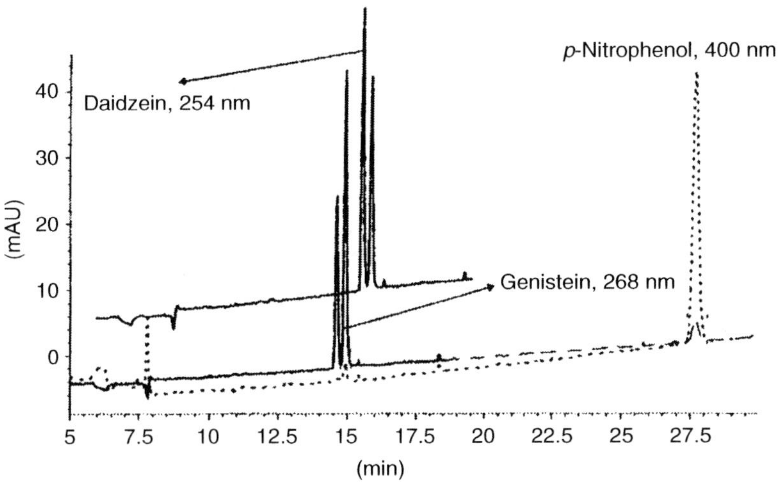

FIGURE 1.17 Electropherogram of a standard mixture of daidzein, genistein, and the internal standard *p*-nitrophenol. (Reprinted with permission from Vänttinen, K. and Moravcova, J., *Czech. J. Food Sci.*, 17, 61–67, 1999.)

TABLE 1.14
Specificity of Radioimmunoassays (RIAs)

Compound	Daidzein/Formononetin RIA (% Cross-Reactivity)	Genistein/Biochanin A RIA (% Cross-Reactivity)
Daidzein	100	5.5
Formononetin	59.7	3.9
Genistein	1.3	100
Biochanin A	1.5	173
Dihydrodaidzein	2.4	0.36
Equol	1.1	0

Source: From Lapčík, O., Hill, M., Hampl, R., Wähälä, K., and Adlercreutz, H., *Steriods*, 63, 14–20, 1998a.

Note: Cross-reactivities of selected isoflavonoids and flavonoids in RIA were calculated as the ratio of 50% intercepts of analyte and of potential cross-reactant.

fluctuation of migration time is low 0.34 min with RSD of 2.50 and 2.48 for daidzein and genistein, respectively. The resolution and other performance indicators for HPCE of daidzein and genistein are presented in Table 1.14. A major concern of the HPCE method is the instability of the baseline during analysis resulting in decreasing migration times (Vänttinen and Moravcova, 1999).

1.3.8 Micellar Electrokinetic Chromatography

High-separation power such as electrodriven separation methods is quite attractive for the determination of phytoestrogens in biological samples. However, poor concentration detection limits and the influence of interferences from sample constituents are major disadvantages of the method. For these reasons, laser-induced fluorescence (LIF), a technique known for its favorable concentration detection limits, is combined with online recording of the fluorescence emission spectra. Micellar electrokinetic chromatography (MEKC) is combined with a special mode of LIF detection for the determination of isoflavonoids and especially formononetin in red clover. In this LIF mode, deep-UV excitation is performed and fluorescence spectra are recorded online. The native-fluorescent isoflavonoids are excited at 275 nm, by a modified large-frame argon-ion laser and DAD mounted on a spectrograph. Online recording of the fluorescence emission spectra simplifies identification of peaks in the electropherogram. Compared to conventional UV absorption detection, interferences are drastically reduced since only sample constituents exhibiting native fluorescence are detected. The identification and quantification of formononetin in a clover extract at a concentration of 3×10^{-5} M or 17 μg formononetin per gram in red clover demonstrate the potential of this approach.

The electrodriven separation of the isoflavonoids is optimized using UV absorbance detection, and complete separation is achieved in 12 min using a micellar system of 50 mM aqueous borate buffers at pH 8.7 containing 50 mM SDS (Figure 1.18). Severe conditions are necessary to hydrolyze and extract all phytoestrogens

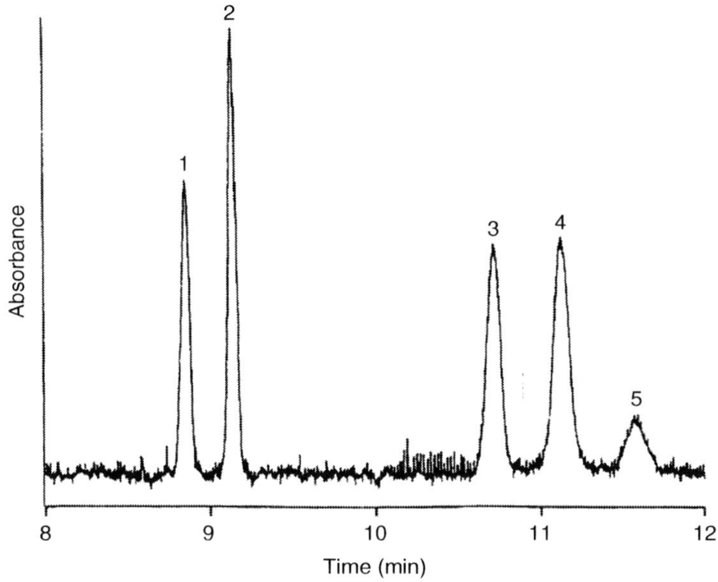

FIGURE 1.18 Electropherogram of an aqueous solution of 5 nm of daidzein (1), genistein (2), formononetin (3), biochanin A (4), and coumestrol (5) using micellar electrokinetic chromatography–ultraviolet (MEKC-UV). (From Beekman, M.C., Lingeman, H., Brinkman, U.A.Th., and Gooijer, C., *J. Microcolumn Sep.*, 11, 347–350, 1999.)

and their conjugates (primarily water-soluble glycosides). After SPE, the sample is diluted fourfold and subsequently analyzed using MEKC. Figure 1.19 shows part of a typical electropherogram—recorded using a new capillary—with prominent peaks at 8.53, 8.97, and 9.62 min (denoted as *X*, *Y*, and Z, respectively) in addition to some smaller ones. The repeatability of migration times of a mixture of standards at 5 μm concentration is 1%–2%. The detection limits for the three fluorescent analytes, daidzein, formononetin, and coumestrol, in the standard solution at 275 nm LIF detection are between 0.1 and 0.4 μm (signal-to-noise ratio (S/N) 3, where N is peak-to-peak noise). At analyte concentrations of 1 μM, the relative standard deviations of the peak areas are less than 9% ($n = 5$) (Beekman et al., 1999).

1.3.9 Immunoassay

Immunoassay is one of the most sensitive analytical techniques described for phytoestrogen measurements in human biological fluids. Detection limits for target analytes are below 1 nmol/L for most methods with an average recovery of 92% and inter-/intra-assay variation less than 10% for most methods (Wilkinson et al., 2002). Immunoassays have been validated against other analytical techniques and excellent correlation has been demonstrated. Immunoassay can be a powerful alternative to GC-MS- or HPLC-based methods for studying the bioavailability and metabolic fate of dietary isoflavonoids because of its ease, relatively low cost, availability for

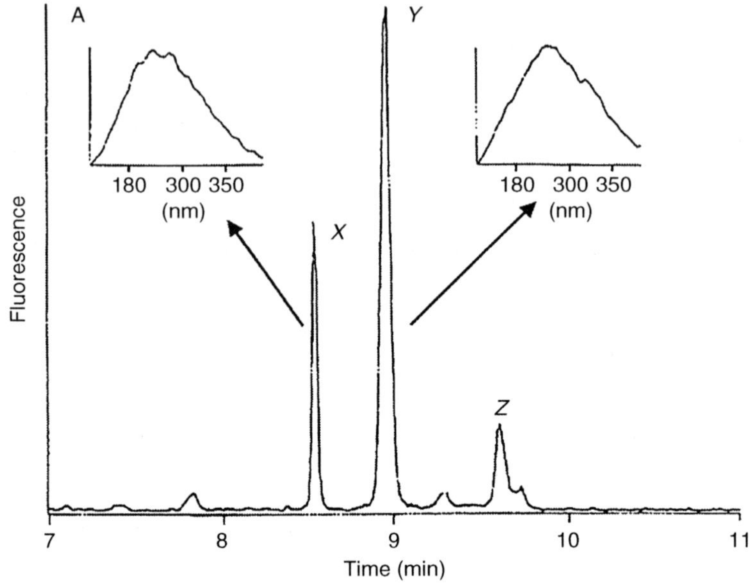

FIGURE 1.19 Electropherogram of clover extract with main peaks indicated as formononetin (*X*), daidzein (*Y*), and coumestrol (*Z*). (From Beekman, M.C., Lingeman, H., Brinkman, U.A.Th., and Gooijer, C., *J. Microcolumn Sep.*, 11, 347–350, 1999.)

automation, and suitability for epidemiological studies. GC-MS and HPLC methods are neither suitable for screening purposes in large populations, nor sensitive enough for assay of unconjugated phytoestrogens in plasma. Antibodies to some phytoestrogens including daidzein have been raised in rabbits as early as 1969 to be used in sheep for passive immunization as protection from an excess of phytoestrogens (Bauminger et al., 1969). The development of sensitive RIA for daidzein, genistein, and formononetin was attempted by Cox et al. (1972), but was never used for quantitative assays. Immunoassays specific for daidzein and its 4′-derivatives (formononetin 4′-sulfate and 4′-glucuronide) and for genistein and its 4′-derivatives (biochanin A, 4′-sulfate, and 4′-glucuronide of genistein) have recently been developed (Lapčík et al., 1998).

Lapčík et al. (1997) established a RIA for daidzein based on polyclonal antibodies against daidzein-4′-*O*-carboxymethyl-ether coupled to bovine serum albumin (BSA). Daidzein is measured in serum, plasma, or urine by either direct RIA or RIA following an extraction step with diethyl ether. The dry residue after ether evaporation is dissolved in assay buffer (20 mM sodium phosphate in saline) containing sodium azide and BSA and an aliquot is measured by RIA. The radioligand and the antibody are dissolved together with the biological fluid in the assay buffer, incubated after vortex mixing, the bound and free portions separated by dextran-charcoal absorption, and radioactivity measured in a gamma counter. In the case of direct

RIA, the samples of serum or urine are appropriately diluted with the assay buffer to obtain the signal within the range of the calibration curve.

The intra- and inter-assay CV, depending on the method (direct or extraction) and concentration of daidzein in the sample, range from 4.1% to 11.5% and from 5.6% to 21.7%, respectively. The assay has low cross-reactivity with other chemically related compounds, 2%–4% for dihydrodaidzein, 1%–3% for genistein, 1%–5% for biochanin A, and 1%–6% for equol. However, the daidzein values obtained following diethyl ether extraction of human sera are only 8% of that obtained by direct RIA probably because of the cross-reactivity of daidzein 4′-glucuronides with sulfates present in serum. The RIA method for daidzein is more sensitive than previous methods used for quantitation of phytoestrogens in biological fluids. The direct RIA method is well suited for screening and kinetic studies because the signal obtained is proportional to the free analyte after a soy load.

A RIA method has been used to quantitate formononetin in murine plasma and mammary glandular tissue for animal model studies (Wang, 1998). This assay utilizes an antiserum raised in rabbits following immunization with formononetin-7-*O*-carboxymethyl-ether coupled to BSA. The bound and free forms of formononetin are separated by adding dextran-coated charcoal. The RIA procedure enables the quantification of 4 ng/mL of plasma or 50 pg/mg of mammary tissue. The reliability and reproducibility of the assay, demonstrated by intra- and inter-assay variation, are 6.5% and 11.9%, respectively. The RIA correlates well with an HPLC method ($r = 0.980$) in the determination of formotonetin extracts in red clover. However, the RIA values are on average 5% higher than those obtained by HPLC, probably because of cross-reactions from other compounds. The mean recovery of formononetin addition in plasma and mammary tissue homogenates is about 87% and 98%, respectively. In practical terms, the assay is efficient since the procedure can readily run 100 samples in a normal working day. Therefore, this RIA, which avoids any sample extraction and preparation, should be suitable for routine monitoring of formononetin in biological fluids or in plants.

The fractions from diethyl ether extracts of beer have been analyzed by RIA specific for daidzein/formononetin and for genistein/biochanin A (Lapčík et al., 1998). For this procedure, polyclonal antibodies against conjugates of daidzein 4′-carboxymethyl ether and genistein 4′-carboxymethyl ether with BSA are raised in rabbits. The conjugates of daidzein and genistein are labeled with radioiodinated tyrosine methylester for use as radioligands in competitive RIA. The cross-reactivities of the antisera are summarized in Table 1.14. The first immunoassay is highly specific for daidzein and formononetin, the specificity of the genistein/biochanin A immunoassy is lower with 4%–5% cross-reaction of daidzein and formononetin. The sensitivity of the assay is 0.08 nmol/L for daidzein/formononetin and 0.15 nmol/L for genistein/biochanin A. The intra- and inter-assay of this procedure are 5.7% and 9.8% for the daidzein/formononetin and 7.3% and 13.2% for genistein/biochanin A RIAs, respectively (Lapčík et al., 1998).

RIAs for formononetin in forage, and free conjugated daidzein and genistein in plasma and free isoflavones in beer have also been developed (Wang et al., 1994; Lapčík et al., 1997, 1998a, b). However, the short shelf-life of the labeled compounds has led to the development of time-resolved fluoroimmunoassay (TR-FIA) that combine the advantages of other nonradioisotopic assays with a 10- to 100-fold increase in sensitivity and assay range in comparison with conventional enzyme immunoassay (EIA) and fluoroimmunoassay methods. The first TR-FIA for daidzein in urine and enterolactone in plasma has recently been published (Adlercreutz et al., 1998; Kohen et al., 1998). The latest TR-FIA method uses a europium chelate as label for the determination of the phytoestrogens daidzein and genistein in plasma for screening and kinetic studies (Wang et al., 2000). In this procedure, 4′-*O*-carboxymethyl derivatives of daidzein and genistein are synthesized, coupled to BSA, and used as antigens to immunize rabbits. Immunoassay is carried out after enzymatic hydrolysis and ether extraction. The method is highly sensitive, with a detection limit of 1.8 and 3.1 pg/20 μL for daidzein and genistein, respectively and is precise, with intra- and inter-assay CV between 3.2% and 6.3%. Although the antisera cross-reacts with some isoflavonoids, values obtained by the TR-FIA method correlates highly ($r \geq 0.95$) with those based on ID-GC-MS. The advantages of this technology have been described as being highly sensitive, low background interference, wide dynamic range, reliable, relatively inexpensive, and practical with short incubation time and completion of 96 samples in 4 h (Wang et al., 2000). These advantages enable the TR-FIA method to be well established in routine immunodiagnostic work and have been extensively used for large studies with methods developed for the analysis of enterolactone, genistein, daidzein, *O*-DMA, and equol. RIA is considered to be the most sensitive (0.002 pmol/assay for daidzein) technique for detecting phytoestrogens in foods and human biological fluids, based on an analysis of 90 reports (Wilkinson et al., 2002).

The disadvantage related to the use of radioactive material associated with immunological assays has led to the development of more convenient ELISA (enzyme-linked immunosorbent assay) technique based on the use of specific antibodies directed against genistein, daidzein, and equol. The ELISA method has been used to measure isoflavones in plasma and urine of women in a kinetic study and in a survey performed on healthy menopausal women and also to assay soy food and food supplements based on soy (Bennetau-Pelissero et al., 2003). The method involves enzyme (*H. pomatia*) hydrolysis monitored using external standard (genistin or daidzin), followed by liquid–liquid extraction with acid–ethyl acetate, evaporation of the organic phase, and dissolution in a PBS (phosphate-buffered saline) assay buffer. Serial dilutions of the analytes are required before addition in a coated well where specific antibodies are added, incubated, and the plates washed before measurements. The inter-assay variation is about 13% for the same sample on 10 different plates and the intra-assay variation is about 5% measured on the same sample assayed 12 times on the same plate. The sensitivity of the assay is 40, 15.6, and 10 ng/mL and the limit of detection is 10, 3.9, and 2.5 ng/mL for daidzein, genistein, and equol, respectively. This technique cannot be used for clover or alfalfa exposure because of high cross-reactivity with formononetin (51% anti-diadzein) and biochanin A (53% anti-genistein). The extraction and hydrolysis procedures have been validated and compared with HPLC methods performed in 23 participating laboratories. The method is described as cheap, rapid, sensitive, and simple requiring only basic extraction steps.

1.3.10 Bioassay

In the age of genomic bioassays, using reporter genes for the detection or quantitative evaluation of estrogen is becoming commonplace. Phytoestrogens through their estrogenic effect are known to regulate transcription of specific genes via estrogen receptors (Shiizaki et al., 1999). A highly sensitive bioassay system has been developed by placing estrogen responsive elements upstream to the reporter gene and this assay has been used to determine the estrogen activity in Chinese herbal prescription used for postmenopausal disorder (Shiizaki et al., 1999). For this procedure, plasmids are constructed by inserting sequences of estrogen responsive elements upstream to the luciferase gene, stable transfection of these plasmids into MCF-7 human cancer cells is obtained, and cell strains with luciferase activity are then used to evaluate the herbal prescriptions. In contrast to immunoassay, the bioassay is independent of the structures of active substances, more sensitive, and is considered highly suitable for assessing phytoestrogens.

1.4 CONCLUSIONS

A wide range of analytical techniques are available for the detection, quantitation, identification, evaluation, and monitoring of phytoestrogens in biological materials. The methods vary from simple to complex depending on extraction, separation, fractionation, identification, and detection of the analytes. HPLC has become the method of choice for the evaluation and determination of phytoestrogens, especially from plant sources because of its widespread availability, affordability, and adequate sensitivity. The development of API sources, such as the ESI and APCI, has added LC-MS to the list of bioanalytical assay providing higher specificity, precision, accuracy, sensitivity, applicability, and wider dynamic range than HPLC.

Not withstanding the sudden surge in the analytical capability of the separation and identification of phytoestrogens on the basis of their molecular structure, HPLC-MS method is not quite the standard yet. However, the ability of HPLC-MS to characterize and quantify the full range of phytoestrogens can help in accurately assessing dietary intake of phytoestrogens that may interfere with epidemiological attempts to elicit relationship between phytoestrogens and health. To date only the complicated ID-GC-MS-SIM method can determine phytoestrogens from the three main groups: isoflavones, lignans, and coumestans. Only one method for the quantitation of isoflavones has undergone collaborative study for approval as an official method. Methods such as CE with LIF detection have the ability to detect the distribution, localization, and diversity of phytoestrogens in individual cells and organelles and may lead to early diagnosis of phytoestrogen-deficient diseases. It is therefore necessary to rethink and reevaluate new analytical techniques so that all phytoestrogens can be detected and identified simply and reliably. There is a serious and urgent need of accurate methodologies for the determination of phytoestrogens in foods and biological fluids to assess their bioavailability and distribution in tissues for evaluating metabolic effects. Recent development in quick assay (10 min) for nine phytoestrogens in urine samples of women diagnosed with breast cancer is a clear indication of progress in the area of metabolomic fingerprinting related to phytoestrogens.

REFERENCES

Adlercreutz, H., Western diet and Western diseases: Some hormonal and biochemical mechanisms and associations, *Scand. J. Clin. Invest. Suppl.*, 201, 3, 1990.

Adlercreutz, H., Fotsis, T., Banwart, C., Wähälä, K., Brunow, G., and Hase, T., Isotope dilution gas chromatographic–mass spectrometric method for the determination of lignans and isoflavonoids in human urine including identification of genistein, *Clin. Chim. Acta*, 199, 263–278, 1991a.

Adlercreutz, H., Fotsis, T., Bannwart, C., Wähälä, K., Mäkelä, T., Brunow, G., and Hase, T., Determination of urinary lignans and phytoestrogen metabolities, potential antiestrogens, and anticarcinogens, in urine of women on various habitual diets, *J. Steroid Biochem. Mol. Biol.*, 25, 791–797, 1986.

Adlercreutz, H., Fotsis, T., Lampke, J., Wähälä, K., Mäkelä, T., Brunow, G., and Hase, T., Quantitative determination of lignans and isoflavonoids in plasma of omnivorous and vegetarian women by isotope dilution gas chromatography–mass spectrometry, *Scand. J. Clin. Lab. Invest.*, 53, 5–18, 1993.

Adlercreutz, H., Fotsis, T., Watanabe, S., Lampe, J., Wähälä, K., Mäkelä, T., and Hase, T., Determination of lignans and isoflavonoids in plasma by isotope dilution gas chromatography–mass spectrometry, *Cancer Prev.*, 18, 259–271, 1994.

Adlercreutz, H., Honjo, H., Higashi, A., Fotsis, T., Hämäläinen, E., Hasegawa, T., and Okada, H., Urinary excretion of lignans and isoflavonoid phytoestrogens in Japanese men and women consuming a tradional Japanese Diet, *Am. J. Clin. Nutr.*, 54, 1093–1100, 1991b.

Adlercreutz, H., van der Wildt, J., Attalla, H., Wähälä, K., Mäkelä, T., Hase, T., and Fotsis, T., Lignan and isoflavonoid conjugates in human urine, *J. Steroid Biochem. Mol. Biol.*, 52, 97–103, 1995.

Adlercreutz, H. and Mazur, W., Phyto-oestrogens and Western diseases, *Ann. Med.*, 29, 95–120, 1997.

Adlercreutz, H., Fotsis, T., Watanabe, S., Lampe, J., Wähälä, K., Mäkelä, T., and Hase, T., Determinations of lignans and isoflavonoids in plasma by isotope dilution gas chromatography-mass spectrometry, *Cancer Prev.*, 18, 259–271, 1994.

Adlercreutz, H., Wang, G.J., Lapčík, O., Hampl, R., Wähälä, K., Mäkelä, T., Lusa, K., Talme, M., and Mikola, H., Time-resolved fluoroimmunoassay for plasma enterolactone, *Anal. Biochem.*, 265, 208–215, 1998.

Adlercreutz, H., Phytoestrogens. State of the art, *Environ. Toxicol. Pharmacol.*, 7, 201–207, 1999.

Aramendia, M.A., Borau, V., Garcia, I., Jimenez, C., Lafont, F., Martinas, J.M., Porras, A., and Urbano, F.J., Determination of isoflavones by capillary electrocphoresis/electrospray ionization mass spectrometry, *Rapid Commn. Mass Spectrom.*, S153–S157, 1995a.

Aramendia, M.A., Garcia, I., Lafont, F., and Marinas, J.M., Determination of isoflavones using high-performance liquid chromatography with atmospheric-pressure chemical ionization mass spectrometry, *Rapid Commn. Mass Spectrom.*, 9, 503–508, 1994.

Aramendia, M.A., Garcia, I., Lafont, F., and Marinas, J.M., Determination of isoflavones using capillary electrophoresis in combination with electrospray mass spectrometry, *J. Chromatogr.*, 707, 327–333, 1995b.

Atkinson, D.A., Hill, H.H., and Shultz, T.D., Quantification of mammalian lignans in biological fluids using gas chromatography with ion mobility detection, *J. Chromtogr. Biomed. Appl.*, 617, 173–179, 1993.

Atkinson, C., Lampe, J.W., Scholes, D., Chen, C., Wähälä, K., and Schwartz, S.M., Lignan and isoflavone excretion in relation to uterine fibroids: A case-control study of young to middle-aged women in the United States, *Am. J. Clin. Nutr.*, 84, 587–593, 2006.

Bajer, T., Adam, M., Galla, L., and Ventura, K., Comparison of various extraction techniques for isolation and determination of isoflavonoids in plants, *J. Sep. Sci.*, 30, 122–127, 2007.

Bambagiotti-Alberti, M., Coron, S.A., Ghiara, C., Giannellini, V., and Raffaelli, A., Revealing the mammalian lignan precursor secoisolariciresinol diglucoside in flaxseed by ionspray mass spectrometry, *Rapid Commn. Mass Specrom.*, 8, 595–598, 1994a.

Bambagiotti-Alberti, M., Coron, S.A., Ghiara, C., Moneti, G., and Raffaelli, A., Investigation of mammalian lignan precursors in flaxseed: First evidence of secoisolariciresinol diglucoside in two isomeric forms of liquid chromatography/mass spectrometry, *Rapid Commn. Mass Spectrom.*, 8, 929–932, 1994b.

Barbuch, R.J., Coutant, J.E., Welsh, M.B., and Setchell, K.D.R., The use of thermospray liquid chromatography/tandem mass spectrometry for the class identification and structural verification of phytoestrogens in soy protein preparations, *Biomed. Environ. Mass Spectrom.*, 18, 973–977, 1989.

Barnes, K.A., Smith, R.A., Williams, K., Damant, A.P., and Shepherd, M.J., A microbore high performance liquid chromatography/electrospray ionization mass spectrometry method for the determination of the phytoestrogens genistein and daidzein in comminuted baby foods and soya flour, *Rapid Commn. Mass Spectrom.*, 12, 130–138, 1998a.

Barnes, S., Coward, L., Kirk, M., and Sfakianos, J., HPLC–mass spectrometry analysis of isoflavones, *Proc. Soc. Exp. Biol. Med.*, 217, 254–262, 1998b.

Barnes, S., Kirk, M., and Coward, L., Isoflavones and their conjugates in soy foods: Extraction and analysis by HPLC–mass spectrometry, *J. Agric. Food Chem.*, 42, 2466–2474, 1994.

Barnes, S., Basics of mass spectrometry, Retrieved 2006, from http://www.uab.edu/botanicals/workshop/pdf/2006_d1/6BarnesBotanicalsLC-MS.pdf.

Bauminger, S., Lindner, H.R., Perel, E., and Arnon, R., Antibodies to a phytooestrogen: Antigenicity of genistein-coupled to a synthetic polypeptide, *J. Endocrinol.*, 44, 567–578, 1969.

Beekman, M.C., Lingeman, H., Brinkman, U.A.Th., and Gooijer, C., Determination of the isoflavone formononetin in red clover (*Trifolium pratense* L.) by micellar electrokinetic chromatography combined with deep-UV laser-induced wavelength-resolved fluorescence detection, *J. Microcolumn Sep.*, 11, 347–350, 1999.

Begum, A.N., Nicolle, C., Mila, I., Lapierre, C., Nagano, K., Fukushima, K., Heinonen, S.-M., Adlercreutz, H., Rémésy, C., and Scalbert, A., Dietary lignins are precursors of mammalian lignans in rats, *J. Nutr.*, 134, 120–127, 2004.

Bennetau-Pelissero, C., Arnal-Schnebelen, B., Lamothe, V., Sauvant, P., Sagne, J.L., Verbruggen, M.A., Mathey, J., and Lavialle, O., ELISA as a new method to measure genistein and daidzein in food and human fluids, *Food Chem.*, 82, 645–658, 2003.

Bickoff, E.M., Loper, G.M., Hanson, C.H., Graham, J.H., Witt, S.C., and Spencer, R.R., Effect of common leafspot on coumestans and flavones in alfalfa, *Crop Sci.*, 7, 259–261, 1967.

Bingham, S.A., Atkinson, C., Liggins, J., Bluck, L., and Coward, A., Phyto-oestrogens: Where are we now? *Br. J. Nutr.*, 79, 393–406, 1998.

Bowey, E., Adlercreutz, H., and Rowland, I., Metabolism of isoflavones and lignans by the gut microflora: A study in germ-free and human flora associated rats, *Food Chem. Toxicol.*, 41, 631–636, 2003.

Brandi, M.L., Phytoestrogens and menopause, *Environ. Toxicol. Pharmacol.*, 7, 213–216, 1999.

Brzezinski, A. and Debi, A., Phytoestrogens: The "natural" selective estrogens receptor modulators? *Eur. J. Obstet. Gynecol.*, 85, 47–51, 1999.

Bylund, A., Zhang, J.-X., Bergh, A., Damber, J.-E., Widmark, A., Johansson, A., Adlercreutz, H., Aman, P., Shepherd, M.J., and Hallmans, G., Rye bran and soy protein delay growth and increase apoptosis of human LNCaP prostate adenocarcinoma in nude mice, *The Prostate*, 42, 304–314, 2000.

Cacace, J.E. and Mazza, G., Pressurized low polarity water extraction of lignans from whole flaxseed, *J. Food Eng.*, 77, 1087–1095, 2006.

Careri, M., Mangia, A., and Musci, M., Overview of the applications of liquid chromatography–mass spectrometry interfacing systems in food analysis: Naturally occurring substances in food. *J. Chromatogr. A*, 794, 263–297, 1998.

Carlson, R.E. and Dolphin, D., High-performance liquid chromatographic method for the analysis of isoflavones, *J. Chromatogr.*, 198, 193–197, 1980.

Carusi, D., Phytoestrogens as hormone replacement therapy: An evidence-based approach, *Primary Care Update OB/GYNS*, 7, 253–259, 2000.

Chandra, A. and Nair, M.G., Supercritical carbon dioxide extraction of daidzein and genistein from soybean products, *Phytochem. Anal.*, 7, 259–262, 1996.

Chansakaow, S., Ishikawa, T., Seki, H., Sekine, K., Okada, M., and Chaichantipyuth, C., Identification of deoxymiroestrol as the actual rejuvenating principle of "Kwao Keur," *Pueraria mirifica*. The known mirosterol may be an artifact, *J. Nat. Prod.*, 63, 173–175, 2000.

Chen, M., Kilgore, N., Lee, K.-H., and Chen, D.-F., Rubrisandrins A and B, lignans and related anti-HIV compounds from *Schisandra rubriflora*, *J. Nat. Prod.*, 69, 1697–1701, 2006.

Clarke, D.B., Lloyd, A.S., Botting, N.P., Oldfield, M.F., Needs, P.W., and Wiseman, H., Measurement of intact sulfate and glucuronide phytoestrogen conjugates in human urine using isotope dilution liquid chromatography–tandem mass spectrometry with [$^{13}C_3$] isoflavone internal standards, *Anal. Biochem.*, 309, 158–172, 2002.

Clarke, D.D. and Lloyd, A.S., Dietary exposure estimates of isoflavones from the 1998 UK total diet study, *Food Add. Contam.*, 21, 305–316, 2004.

Coran., S.A., Gianenellini, V., and Bambagiotti-Albert, M., High-performance thin-layer chromatographic–densitometric determination of secoisolariciresinol diglucoside in flaxseed, *J. Chromatogr. A*, 1045, 217–222, 2004.

Coward, L., Barnes, N.C., Setchell, K.D.R., and Barnes, S., The antitumor isoflavones, genistein and daidzein, in soybean foods of American and Asian diets, *J. Agric. Food Chem.*, 41,1961–1967, 1993.

Coward, L., Kirk, M., Albin, N., and Barnes, S., Analysis of plasma isoflavones in the reversed-phase HPLC-multiple reaction ion monitoring–mass spectrometry, *Clin. Chim. Acta*, 247, 121–142, 1996.

Cox, R.I., Wong, M.S., Braden, A.W., Trikojus, V.M., and Lindner, H.R., The formation and specificity of antibodies to phyto-oestrogens in the sheep, *J. Reprod. Fertil.*, 28, 157–158, 1972.

Cox, R.I. and Davies, L.H., Modification of pasture oestrogens in the gastrointestinal tract of ruminants, *Proc. Nutr. Soc. Aust.*, 13, 61–67, 1988.

D'Alessandro, T., Prasain, J., Benton, M.R., Botting, N., Moore, R., Darley-Usmar, V., Patel, R., and Barnes, S., Polyphenols, inflammatory response, and cancer prevention: Chlorination of isoflavones by human neutrophils, *J. Nutr.*, 133, 3773S–3777S, 2003.

Davin, B.L. and Lewis, N.G., Dirigent proteins and dirigent sites explain the mystery of specificity of radical precursor coupling in lignan and lignin biosynthesis, *Plant Physiol.*, 123, 453–461, 2000.

Davin, B.L., Wang, H.B., Crowell, A.L., Bedgar, D.L., Martin, D.M., Sarkanen, S., and Lewis, N.G., Stereoselective biomolecular phenoxy radical coupling by an auxiliary (dirigent) protein without an active center, *Science*, 275, 362–366, 1997.

De Rijke, E., De Kanter, F., Arise, F., Brinkman, U.A.Th., and Gooijer, C., Liquid chromatography coupled to nuclear magnetic resonance spectroscopy for the identification of isoflavone glucoside malonates in *T. pratense* L. leaves, *J. Sep. Sci.*, 27, 1061–1070, 2004.

De Rijke, E., Out, P., Niessen, W.M.A., Arise, F., Gooijer, C., and Brinkman, U.A.Th., Analytical separation and detection methods for flavonoids, *J. Chromatogr. A*, 112, 31–63, 2006.

Dehennin, L., Reiffsteck, A., Jondet, M., and Thibier, M., Identification and quantitative estimation of a lignan in human and bovine semen, *J. Reprod. Fertil.*, 66(1), 305–309, 1982.

Delmonte, P. and Rader, J.I., Analysis of isoflavones in foods and dietary supplements, *J. AOAC Int.*, 89, 1138–1146, 2006.

Dewald, H.D., Worst, S.A., Butcher, J.A., and Saulinskas, E.F., *Electrocanalysis*, 3, 777–782, 1991.

Dixon, R.A., Isoflavonoids: Biochemistry, molecular biology, and biological functions, in *Comprehensive Natural Products Chemistry*, Vol. 1, Sankawa, U. (Ed.), 773–823, Elsevier, Amsterdam, 1999.

Dixon, R.A. and Pavia, N.L., Stress-induced phenylpropanoid metabolism, *Plant Cell*, 7, 1085–1097, 1995.

Dixon, R.A. and Steele, C.L., Flavonoids and isoflavonoids—a gold mine for metabolic engineering, *Trends Plant Sci.*, 4, 394–400, 1999.

Downing, J.M., Chung, O.K., Seib, P.A., and Hubbard, J.D., Pressurized solvent extraction of genistein and its β-glucoside conjugates from soybean flours and soy-based foods, *Cereal Chem.*, 84, 44–47, 2007.

Dziedzic, S.Z. and Dick, J., Analysis of isoflavones in bengalgram by high-performance liquid chromatography, *J. Chromatogr.*, 234, 497–499, 1982.

Eldridge, A.C., Determination of isoflavones in soybean flours, protein concentrates, and isolates, *J. Agric. Food Chem.*, 30, 353–355, 1982a.

Eldridge, A.C., High-performance liquid chromatography separation of soybean isoflavones and their glucosides, *J. Chromatogr.*, 234, 494–496, 1982b.

Farmakalidis, E. and Murphy, P.A., Oestrogenic response of the CD-1 mouse to the soya-bean isoflavones genistein, genistin and daidzin, *Food Chem. Toxicol.*, 22, 237–239, 1984.

Farnsworth, N.R., Bingel, A.S., Cordell, G.A., Crane, F.A., and Fond, H.S.S., Potential value of plants as source of new antifertility agents, *J. Pharmacol. Sci.*, 64, 57700–57733, 1999.

Federal Register, Food labeling: Health claims; soy protein and coronary heart disease, Food and Drug Administration, HHS, Final rule, *Fed. Regist.*, 64, 57700–57733, 1999.

Finlay, E.M.H., Wilson, D.W., Adlercreutz, H., and Griffiths, K., The identification and measurement of phyto-estrogens in human saliva, plasma, breast aspirate or cyst fluid, and prostatic fluid using gas chromatography–mass spectrometry, *J. Endocrinol.*, 129 (Suppl): 49, 1991.

Ford, T.D., Davin, L.B., and Lewis, N.G., Chemopreventive flaxseed lignans: Toward bio-engineering the metabolic pathways, *Proc. 58th Flax Inst. United States*, March 23–25, Fargo, ND, 39–48, 2000.

Fotsis, T., Heikkinen, R., Adlercreutz, H., Axelson, M., and Setchell, K.D., Capillary gas chromatographic method for the analysis of lignans in human urine, *Clin. Chim. Acta.*, 121, 361–371, 1982.

Frank, J., Kamal-Eldin, A., and Traber, M.G., Consumption of sesame oil muffins decreases the urinary excretion of γ-tocopherol metabolites in humans, *Ann. N.Y. Acad. Sci.*, 1031, 365–367, 2004.

Franke, A.A. and Custer, L.J., Daidzein and genistein concentrations in human milk after soy consumption, *Clin. Chem.*, 42, 955–964, 1996.

Franke, A.A., Custer, L.J., Cerna, C.M., and Narala, K.K., Quantitation of phytoestrogens in legumes by HPLC, *J. Agric. Food Chem.*, 42, 1905–1913, 1994.

Franke, A.A., Custer, L.J., Cerna, C.M., and Narala, K., Rapid HPLC analysis of dietary phytoestrogens from legumes and from human urine, *Proc. Soc. Exp. Biol. Med.*, 208, 18–26, 1995.

Franke, A.A., Custer, L.J., Wang, W., and Shi, C.Y., HPLC analysis of isoflavonoids and other phenolic agents from foods and from human fluids, *Proc. Soc. Exp. Biol. Med.*, 217, 263–273, 1998.

Franke, A.A. and Custer, L.J., High-performance liquid chromatographic assay of isoflavonoids and coumestrol from human urine, *J. Chromatogr. B*, 662, 47–60, 1994.

Fritz, W.A., Coward, L., Wang, J., and Lamartiniere, C.A., Dietary genistein: Perinatal mammary cancer prevention, bioavailability and toxicity testing in the rat, *Carcinogenesis*, 19, 2151–2158, 1998.

Gamache, P., Ryan, E., and Acworth, I.N., Analysis of phenolic and flavonoid compounds in juice beverages using high-performance liquid chromatography with coulometric array detection. *J. Chromatogr.*, 635, 143–150, 1993.

Gamache, P.H. and Acworth, I.N., Analysis of phytoestrogens and polyphenols in plasma, tissue, and urine using HPLC with coulometric array detection, *Proc. Soc. Exp. Biol. Med.*, 217, 274–280, 1998.

Gamache, P.H., Freeto, S.M., and Acworth, I.N., HPLC-electrochemical detection for phytoestrogen analysis: Isoflavones and lignans in human serum and animal tissue, *J. Med. Food*, 2, 125–129, 1999.

Garrett, S.D., Lee, H.A., Friar, P.M.K., and Morgan, M.R.A., Validation of a novel estrogen receptor-based microtitration plate assay for the determination of phytoestrogens in soy-based foods, *J. Agric. Food Chem.*, 47, 4106–4111, 1999.

Gavaler, J.S., Rosenblum, E.R., Deal, S.R., and Bowie, B.T., The phytoestrogen congeners of alcoholic beverages: Current status, *Proc. Soc. Exp. Biol. Med.*, 208, 98–102, 1995.

Gennari, C., Agnusdei, D., Crepaldi, G., Isaia, G., Mazzuoli, G., Ortolani, S., Bufalino, L., and Passeri, M., Effect of ipriflavone—a synthetic derivative of natural isoflavones—on bone mass loss in the early years after menopause, *Menopause*, 5, 9–15, 1998.

Grace, P.B., Mistry, N.S., Carter, M.H., Leathem, A.J.C., and Teale, P., High throughput quantification of phytoestrogens in human urine and serum using liquid chromatography/tandem mass spectrometry (LC-MS/MS), *J. Chromatogr. B*, 853, 138–146, 2007.

Grace, P.B., Taylor, J.I., Botting, N.P., Fryatt, T., Oldfield, M.F., Al-Maharik, N., and Bingham, S.A., Quantification of isoflavones and lignans in serum using isotope dilution liquid chromatography/tandem mass spectrometry, *Rapid Commn. Mass Spectrom.*, 17, 1350–1357, 2003a.

Grace, P.B., Taylor, J.I., Botting, N.P., Fryatt, T., Oldfield, M.F., and Bingham, S.A., Quantification of isoflavones and lignans in urine using gas chromatography/mass spectrometry, *Anal. Biochem.*, 315, 114–121, 2003b.

Grace, P.B., Taylor, J.I., Low, Y.L., Luben, R.N., Mulligan, A.A., Botting, N.P., Dowsett, M., Welch, A.A., Khaw, K.T., Wareham, N.J., Day, N.E., and Bingham, S.A., Phytoestrogen concentrations in serum and spot urine as biomarkers for dietary phytoestrogen intake and their relation to breast cancer risk in European prospective investigation of cancer and nutrition-norfolk, *Cancer Epidemiol. Biomarkers Prev.*, 13, 698–708, 2004.

Graham, T.L., A rapid high-resolution high performance liquid chromatography profiling procedure for plant and microbial aromatic secondary metabolites, *Plant Physiol.*, 95, 584–593, 1991a.

Graham, T.L., Flavonoid and isoflavonoid distribution in developing soybean seedling tissue and in seed and root exudates, *Plant Physiol.*, 95, 594–603, 1991b.

Griffith, A.P. and Collison, M.W., Improved methods for the extraction and analysis of isoflavones from soy-containing foods and nutritional supplements by reverse-phase high-performance liquid chromatography and liquid chromatography–mass spectrometry, *J. Chromatogr. A*, 913, 397–413, 2001.

Hampl, R., Lapčík, O., Wähalä, K., Starka, L., and Adlercreutz, H., Radioimunoanalýza fytoestrogenù isoflavonoidovérady, *Chem. Listy*, 92, 44–50, 1998.

Harris, R.K., Greaves, J., Alexander, D., Wilson, T., and Haggerty, W.J., Development of stability-indicating analytical methods for flaxseed lignans and their precursors, in *Food Phytochemicals II: Teas, Spices, and Herbs*, Osawa, C.-T., Huang, M.-T., and Rosen, R.T. (Eds), American Chemical Society, 547, 295–305, Washington DC, WA, 1994.

Hosseinian, F., Antioxidant properties of flaxseed lignans using in vitro model systems, *Ph.D. Thesis*, College of Pharmacy and Nutrition, University of Saskatchewan, Saskatoon, Saskatchewan, Canada, 2006, 234.

Hsieh, H.C., Kao, T.H., and Chen, B.H., A fast HPLC method for analysis of isoflavones in soybean, *J. Liq. Chromatogr. Relat. Technol.*, 27, 315–324, 2004.

Hubert, J., Berger, M., and Daydé, J., Use of a simplified HPLC-UV analysis for soyasaponin B determination: Study of saponin and isoflavone variability in soybean cultivars and soy-based health food products, *J. Agric. Food Chem.*, 53, 3923–3930, 2005.

Hutabarat, L.S., Mulholland, M., and Greenfield, H., Development and validation of an isocratic high-performance liquid chromatographic method for quantitative determination of phytoestrogens in soya bean, *J. Chromatogr.*, 795, 377–382, 1998.

Hutchins, A.M., Lampe, J.W., Martini, M.C., Campbell, D.R., and Slavin, J.L., Vegetables, fruits and legumes: Effect on urinary isoflavonoid phytoestrogen and lignan excretion. *J. Am. Diet Assoc.*, 95, 769–774, 1995a.

Hutchins A.M., Slavin, J.L., and Lampe, J.W., Urinary isoflavonoid phytoestrogen and lignan excretion after consumption of fermented and unfermented soy products, *J. Am. Diet Assoc.*, 95, 545–551, 1995b.

Imai, Y., Tsukahara, S., Asada, S., and Sugimoto, Y., Phytoestrogens/flavonoids reverse breast cancer resistance protein/ABCG2-mediated multidrug resistance, *Cancer Res.*, 64, 4346–4352, 2004.

Ingram, D., Sanders, K., Kolybaba, M., and Lopez, D., Case-control study of phytoestrogens and breast cancer, *Lancet*, 350, 990–994, 1997.

Ishimi, Y., Miyaura, C., Ohmura, M., Onoe, Y., Sato, T., Uchiyama, Y., Ito, M., Wang, X., Suda, T., and Ikegami, S., Selective effects of genistein, a soybean isoflavone, on β-lymphopoiesis and bone loss caused by estrogen deficiency, *Endocrinology*, 140, 1893–1900, 1999.

Jiang, B., Kronenberg, F., Balick, M.J., and Kennelly, E.J., Analysis of formononetin from black cohosh (*Actaea racemosa*), *Phytomedicine*, 13, 477–486, 2006.

Joannou, G.E., Kelly, G.E., Reeder, A.Y., Waring, M., and Nelson, C., A urinary profile study of dietary phytoestrogens: The identification and mode of metabolism of new isoflavonoids, *J. Steroid Biochem. Mol. Biol.*, 54, 167–184, 1995.

Johnson, P., Kamal-Eldin, A., Lundgren, L.N., and Aman, P., HPLC method for analysis of secoisolariciresinol diglucoside in flaxseeds, *J. Agric. Food Chem.*, 48, 5216–5219, 2000.

Jones, A.E., Price, K.R., and Fenwick, G.R., Development and application of a high-performance liquid chromatographic method for the analysis of phytoestrogens, *J. Sci. Food Agric.*, 46, 357–364, 1989.

Jorgensen, K., Suter, D.A.I., Thompson, W.K., Dalais, F.S., and Rice, G.E., Wahlqvist., Burgen® Soy-Lin™. Development of an innovative functional staple food, *Food Aust.*, 50, 297–299, 1998.

Jung, W., Yu, O., Lau, S.-M.C., O'Keefe, D.P., Odell, J., Fader, G., and McGonigle, B., Identification and expression of isoflavone synthase, the key enzyme for biosynthesis in legumes, *Nat. Biotechnol.*, 18, 208–212, 2000.

Kaamanen, M., Adlercreutz, H., Jauhiainen, M., and Tikkanen, M.J., Accumulation of genistein and lipophilic genistein derivatives in lipoproteins during incubation with human plasma in vitro, *Biochim. Biophys. Acta*, 1631, 147–152, 2003.

Karas, M., Matric-assisted laser desorption ionization MS: A progress report, *Biochem. Mass Spectrom.*, 24, 897–900, 1996.

Kardinaal, A.F.M., Waalkens-Berendsen, D.H., and Arts, C.J.M., Pseudo-oestrogens in the diet: Health benefits and safety concerns, *Trends Food Sci. Technol.*, 8, 327–333, 1997.

Kelly, G.E., Nelson, C., Waring, M.A., Joannou, G.E., and Reeder, A.Y., Metabolites of dietary (soya) isoflavones in human urine, *Clin. Chim. Acta*, 223, 9–22, 1993.

Kinoshita, E., Ozawa, Y., and Aishima, T., Novel tartaric acid isoflavone derivatives that play key roles in differentiating Japanese soy sauces, *J. Agric. Food Chem.*, 45, 3753–3759, 1997.

Kinoshita, E., Sugimoto, T., Ozawa, Y., and Aishima, T., Differentiation of soy sauce produced from whole soybeans and defatted soybeans by pattern recognition analysis of HPLC profiles, *J. Agric. Food Chem.*, 46, 877–883, 1998.

Kirkman, L.M., Lampe, J.W., Campbell, D.R., Martini, M.C., and Slavin, J.L., Urinary lignan and isoflavonoid excretion in men and women consuming vegetable and soy diets, *Nutr. Cancer*, 24, 1–12, 1995.

Kitada, Y., Ueda, Y., Yamamota, M., Ishikawa, M., Nakazawa, H., and Fujita, M., Determination of isoflavones in soy bean by high-performance liquid chromatography with amperometic detection, *J. Chromatogr.*, 366, 403–406, 1986.

Klejdus, B., Vitamvasova, D., and Kuban, V., Reversed-phase high performance liquid chromatographic determination of isoflavones in plant materials after isolation by solid-phase extraction, *J. Chromatogr. A*, 839, 261–263, 1999.

Knight, D.C., Eden, J.A., Huang, J.L., and Waring, M.A., Isoflavone content of infant foods and formulas, *J. Pediatr. Child Health*, 34, 135–138, 1998.

Knuckles, B.E., deFremery, D., and Kohler, G.O., Coumestrol content of fractions obtained during wet processing of alfalfa, *J. Agric. Food Chem.*, 24, 1177–1180, 1976.

Kohen, F., Lichter, S., Gayer, B., Deboever, J., and Lu, L.J.W., The measurement of the isoflavone daidzein by time resolved flourescent immunoassay: A method of assessment of dietary soya exposure, *J. Steroid Biochem. Mol. Biol.*, 64, 315–320, 1998.

Kopp, P., Resveratrol, a phytoestrogen found in red wine. A possible explanation for the conundrum of the 'French paradox'? *Eur. J. Endocrinol.*, 138, 619–620, 1998.

Koupai-Abyazani, M.R., Creaser, C.S., and Stephenson, G.R., Separation and identification of flavone, flavonol, isoflavone and flavanone aglycones by capillary column gas chromatography, *Phytochem. Anal.*, 3, 80–84, 1992.

Kudou, S., Fleury, Y., Welti, D., Magnolato, D., Uchida, T., Kitamura, K., and Okubo, K., Malonyl isoflavone glycosides in soybean seeds (*Glycine max* Merrill), *Agric. Biol. Chem.*, 55, 2227–2233, 1991.

Kurzer, M.S., Lampe, J.W., Martini, M.C., and Adlercreutz, H., Fecal lignan and isoflavonoid excretion in premenopausal women consuming flaxseed powder, *Cancer Epidemiol. Biomarkers Prev.*, 4, 353–358, 1995.

Kurzer, M.S. and Xu, X., Dietary Phytoestrogens, *Annu. Rev. Nutr.*, 17, 353–381, 1997.

Lapčík, O., Hampl, R., Al-Maharik, N., Salakka, A., Wahala, K., and Adlercruetz, H., A novel radioimmunoassay for daidzein, *Steroids*, 62, 315–320, 1997.

Lapčík, O., Hampl, R., Hill, M., Wahala, K., Al-Maharik, N., and Adlercruetz, H., Radioimmunassay of free genestein in human serum, *J. Steroid Biochem. Mol. Biol.*, 64, 261–268, 1998.

Lapčík, O., Hill, M., Hampl, R., Wähälä, K., and Adlercreutz, H., Identification of isoflavonoids in beer, *Steroids*, 63, 14–20, 1998.

Lee, H.S., HPLC analysis of phenolic compounds, in *Food Analysis by HPLC*, Nollet, L.M.L. (Ed.), Marcel Dekker, Inc., New York, 775–824, 2000.

Lee, H.S., Jung, B.H., Kim, S.Y., and Chung, B.C., Determination of phytoestrogens in traditional medicinal herbs using gas chromatography–mass spectrometry, *J. Nutr. Biochem.*, 15, 452–460, 2004.

Liggins, J., Bluck, L.J.C., Coward, W.A., and Bingham, S.A., Extraction and quantification of daidzein and genistein in food, *Anal. Biochem.*, 264, 1–7, 1998.

Liggins, J., Grimwood, R., and Bingham, S.A., Extraction and quantification of lignan phytoestrogens in food and human samples, *Anal. Biochem.*, 287, 102–109, 2000.

Liu, C.J., Blount, J.W., Steele, C.L., and Dixon, R.A., Bottlenecks for metabolic engineering of isoflavone glycoconjugates in *Arabidopsis, Proc. Natl. Acad. Sci. USA.*, 99, 14578–14583, 2002.

Liu, Z., Saarinen, N.M., and Thompson, L.U., Sesamin is one of the major precursors of mammalian lignans in sesame seed (*Sesamum indicum*) as observed in vitro and in rats, *J. Nutr.*, 136(4), 906–912, 2006.

Liu, R., Hu, Y., Li, J., and Lin, Z., Production of soybean isoflavone genistein in non-legume plants via genetically modified secondary metabolism pathway, *Metab. Eng.*, 9, 1–7, 2007.

Low, Y.-L., Taylor, J.I., Grace, P.B., Dowsett, M., Scollen, S., Dunning, A.M., Mulligan, A.A., Welch, A.A., Luben, R.N., Khaw, K.-T., Day, N.E., Wareham, N.J., and Bingham, S.A., Phytoestrogen exposure correlation with plasma estradiol in postmenopausal women in European prospective investigation of cancer and nutrition—Norfolk may involve diet–gene interactions, *Cancer Epidemiol. Biomarkers Prev.*, 14, 213–220, 2005.

Lu, R. and Serrero, G., Resveratrol, a natural product derived from grape, exhibits antiestrogenic activity and inhibits the growth of human breast cancer cells, *J. Cell. Physiol.*, 179, 297–304, 1999.

Lundh, T.J.O., Pettersson, J., and Kiessling, K.H., Liquid chromatographic determination of the phytoestrogens daidzein, formononetin, coumestrol, and equol in bovine blood plasma and urine, *J. Assoc. Off. Anal. Chem.*, 71, 938–941, 1988.

Markham, K.R., *Techniques of Flavonoid Identification*, Academic Press, New York, 113, 1982.

Mazur, W., Phytoestrogen content in foods. *Baillière's Clin. Endocrinol. Metabol.*, 12, 729–742, 1998.

Mazur, W., Fotsis, T., Wähälä, K., Ojala, S., Salakka, A., and Adlercreutz, H., Isotope dilution gas chromatographic–mass spectrometric method for the determination of isoflavonoids, coumestrol, and lignans in food samples, *Anal. Biochem.*, 233, 169–180, 1996.

Mazur, W.M., Wähälä, K., Ojala, S., Makkonen, A., Hase, T., and Adlercreutz, H., Lignan and isoflavonoid polyphenols in tea and coffee, *Br. J. Nutr.*, 79, 37–45, 1998.

Mazur, W.M. and Adlercreutz, H., Naturally occuring oestrogens in food, *J. Pure Appl. Chem.*, 70, 1759–1176, 1998.

Meagher, L.P., Beecher, G.R., Flanagan, V.P., and Li, B.W., Isolation and characterization of the lignans, isolariciresinol and pinoresinol in flaxseed meal, *J. Agric. Food Chem.*, 47, 3173–3180, 1999.

Mellenthin, O. and Galensa, R., Sojanachweis in fleischprodukten mittels kapillarzonenelektrophorese und diodenarraydetektion. *Dtsch. Lebensm. Rdsch.*, 92, 171–175, 1996.

Mellenthin, O. and Galensa, R., Nachweis von soja in fleischprodukten durch diodenarraydetektion ausgewählter isoflavonoide: Vergleich von CZE und HPLC, *Lebensmittelchemie*, 51, 23, 1997.

Mellenthin, O. and Galensa, R., Analysis of polyphenols using capillary zone eletrophoresis and HPLC: Detection of soy, lupin and pea protein in meat products, *J. Agric. Food Chem.*, 47, 594–602, 1999.

Merken, H.M. and Beecher, G.R., Measurement of food flavonoids by high-performance liquid chromatography: A review, *J. Agric. Food Chem.*, 48, 577–599, 2000.

Messina, M.J., Persky, V., Setchell, K.D.R., and Barnes, S., Soy intake and cancer risk: A review of the in vitro and in vivo data, *Nutr. Cancer*, 21, 113–131, 1994.

Milder, I.E.J., Arts, I.C.W., Van de Putte, B., Venema, D.P., and Hollman, P.C.H., Lignan contents of Dutch plant foods: A database including lariciresinol, pinoresinol, secoisolariciresinol and matairesinol, *J. Nutr.*, 93, 393–402, 2005.

Milder, I.E.J., Arts, I.C.W., Venema, D.P., Lasaroms, J.J.P., Wähälä, K., and Hollman, P.C.H., Optimization of a liquid chromatography–tandem mass spectrometry method for quantification of the plant lignans secoisolariciresinol, matairesinol, lariciresinol, and pinoresinol in foods, *J. Agric. Food Chem.*, 52, 4643–4651, 2004.

Milligan, S.R., Kalita, J.C., Heyerick, A., Rong, H., De Cooman, L., and De Keukeleire, D., Identification of a potent phytoestrogen in hops (*Humulus lupulus* L.) and beer, *J. Clin. Endocrinol. Metab.*, 83, 2249–2252, 1999.

Mitani, K., Narimatsu, S., and Kataoka, H., Determination of daidzein and genistein in soybean foods by automated on-line in-tube solid-phase microextraction coupled to high-performance liquid chromatography, *J. Chromatogr. A*, 986, 169–177, 2003.

Moazzami, A. and Kamal-Eldin, A., Sesame seed is a rich source of dietary lignans, *J. Am. Oil Chem. Soc.*, 88, 719–723, 2006.

Moazzami, A.A., Andersson, R.E., and Kamal-Eldin, A., Characterization and analysis of sesamolinol diglucoside in sesame seeds, *Biosci. Biotechnol. Biochem.*, 70, 1478–1481, 2006a.

Moazzami, A.A., Andersson, R.E., and Kamal-Eldin, A., HPLC analysis of sesaminol glucosides in sesame seeds, *J. Agric. Food Chem.*, 54, 633–638, 2006b.

Moazzami, A.A., Andersson, R.E., and Kamal-Eldin, A., Quantitative NMR analysis of a seasamin catechol metabolite in human urine, *J. Nutr.*, 137, 940–944, 2007.

Morton, M.S., Arisaka, O., Miyake, A., and Evans, B., Analysis of phyto-estrogens by gas-chromatography–mass spectrometry, *Environ. Toxicol. Pharmacol.*, 7, 221–225, 1999.

Morton, M.S., Matos-Ferreira, A., Abrances-Monteiro, L., Coreia, R., Blacklock, N., Chan, P.S. F., Cheng, C., Lloyds, S., Chieh-ping, W., and Griffiths, K., Measurement and metabolism of isoflavonoids and lignans in the human male, *Cancer Lett.*, 114, 145–151, 1997.

Morton, M.S., Wilcox, G., Wahlqvist, M.L., and Griffiths, K., Determination of lignans and isoflavonoids in human female plasma following dietary supplementation, *J. Endocrinol.*, 142, 251–259, 1994.

Muir, A.D., Flax lignans—analytical methods and how they influence our understanding of biological activity, *J. Assoc. Off. Anal. Chem. Int.*, 89, 1147–1157, 2006.

Muir, A.D., Westcott, H.D., Ballantyne, K., and Northrup, S., Flax lignans—recent developments in the analysis of lignans in plant and animal tissue, *Proc. 58th Flax Inst. United States*, Flax Institute of the United States, Fargo, ND, 23–32, 2000.

Muir, A.D. and Westcott, N.D., Quantitation of the lignan secoisolariciresinol diglucoside in baked goods containing flax seed or flax meal, *J. Agric. Food Chem.*, 48, 4048–4052, 2000.

Müllner, C. and Sontag, G., Determination of some phytoestrogens in soybeans and their processed products with HPLC and coulometric electrode array detection, *Fresenius J. Anal. Chem.*, 364, 261–265, 1999.

Murkies, A.L., Wilcox, G., and Davis, S.R., Phytoestrogens, *J. Clin. Endocrinol. Metab.*, 83, 297–303, 1998.

Murphy, P.A., Separation of genistin, daidzein and their aglucones, and coumesterol by gradient high-performance liquid chromatography, *J. Chromatogr.*, 211, 166–169, 1981.

Murphy, P.A., Phytoestrogen content of processed soybean products, *Food Technol.*, 36, 62–64, 1982.

Murphy, P.A., Song, T., Buseman, G., and Barua, K., Isoflavones in soy-based infant formulas, *J. Sci. Food Agric.*, 45, 4635–4638, 1997.

Murphy, P.A., Barua, K., and Hauck, C.C., Solvent extraction selection in the determination of isoflavones in soy foods, *J. Chromatogr. B*, 777, 129–138, 2002.

Nakamura, Y., Kaihara, A., Yoshii, K., Tsumura, Y., Ishimitsu, S., and Tonogai, Y., Content and composition of isoflavonoids in mature or immature beans and bean sprouts consumed in Japan, *J. Health Sci.*, 47, 394–406, 2001.

Nestel, P.J., Pomeroy, S., Kay, S., Komesaroff, P., Behrsing, J., Cameron, J.D., and West, L., Isoflavones from red clover improve systemic arterial compliance but not plasma lipids in menopausal women, *J. Clin. Endocrinol. Metab.*, 84, 895–898, 1999.

Nicollier, G.F., Separation and quantitation of estrogenic isoflavones from clovers by high-performance liquid chromatography, *J. Chromatogr.*, 249, 399–402, 1982.

Nicollier, G.F. and Thompson, A.C., Phytotoxic compounds from *Melilotus alba* (white sweet clover) and isolation and identification of two new flavonoids, *J. Agric. Food Chem.*, 30, 760–764, 1982.

NTP-CERHR Expert Panel Report on the reproductive and developmental toxicity of genistein, April, NTP-CERHR-Genistein-06, http://cerhr.niehs.nih.gov, 2006.

Nutraingredients, HMRLignan may offer cardiovascular protection, http://www.nutraingredients.com/news/ng.asp?n = 74987-hmrlignan-may-offer, 2007.

Nurmi, T. and Adlercreutz, H., Sensitive HPLC method for profiling phytoestrogens using coulometric electrode array detection: Application to plasma analysis, *Anal. Biochem.*, 274, 110–117, 1999.

Obermeyer, W.R., Musser, S.M., Betz, J.M., Casey, R.E., Pohland, A.E., and Page, S.W., Chemical studies of phytoestrogens and related compounds in dietary supplements: Flax and chaparral, *Proc. Soc. Exp. Biol. Med.*, 208, 6–12, 1995.

Occhiuto, F., De Pasquale, R., Guglielmo, G., Palumbo, D.R., Zangla, G., Samperi, S., Renzo, A., and Circosta, C., Effects of phytoestrogenic isoflavones from red clover (*Trifolium pratense* L.) on experimental osteoporosis, *Phytother. Res.*, 21, 130–134, 2007.

Ohta, N., Kuwata, G., Akahori, H., and Watanabe, T., Isoflavonoid constituents of soybeans and isolation of a new acetyl daidzin, *Agric. Biol. Chem.*, 43, 1415–1419, 1979.

Okubo, K., Fijima, M., Kobayashi, Y., Yoshikoshi, M., Uchida, T., and Kudou, S., Components responsible for the undesirable taste of soybean seeds, *Biosci. Biotechnol. Biochem.*, 56, 99–103, 1992.

Oomah, B.D., Phytoestrogens, in *Methods of Analysis for Functional Foods and Nutraceuticals*, Hurst, W.J. (Ed.), CRC Press, Boca Raton, FL, 1–63, 2002.

Oomah, B.D., Processing of flaxseed fiber, oil, protein, and lignan, in *Flaxseed in Human Nutrition*, 2nd edition, Thompson, L.U. and Cunnane, S.C. (Eds), AOCS Press, Champaign, IL, 363–386, 2003.

Oomah, B.D. and Mazza, G., Functional foods, in *The Wiley Encyclopedia of Science and Technology*, 2nd edition, Vol. 2, Francis, F.J. (Ed.), Wiley, New York, 1176–1182, 2000.

Padmavati, M. and Reddy, A.R., Flavonoid biosynthetic pathway and cereal defence response: An emerging trend in crop biotechnology, *Plant Biochem. Biotechnol.*, 8, 15–20, 1999.

Panossian, A., Danielyan, A., Mamikonyan, G., and Wikman, G., Methods of phytochemical standardization of rhizome *Cimicifugae racemosae*, *Phytochem. Anal.*, 15, 100–108, 2004.

Patisaul, H.B., Young investigator perspectives phytoestrogen action in the adult and developing brain, *J. Neuroendocrinol.*, 17, 57–64, 2005.

Payne, T.J., Promoting better health with flaxseed in bread, *Cereal Food World*, 45, 102–104, 2000.

Penalvo, J.L., Haajanen, K.M., Botting, N., and Adlercreutz, H., Quantification of lignans in food using isotope dilution gas chromatography/mass spectrometry, *J. Agric. Food Chem.*, 53, 9342–9347, 2005.

Peng, Y.Y. and Ye, J.N., Determination of isoflavones in red clover by capillary electrophoresis with electrochemical detection, *Fitoterapia*, 77, 171–178, 2006.

Peterson, G. and Barnes, S., Genistein inhibition of the growth of human breast cancer cells—independence from estrogen receptors and multidrug resistance gene, *Biochem. Biophys. Res. Commn.*, 179, 661–667, 1991.

Peterson, T.G., Ji, G.P., Kirk, M., Coward, L., Falany, C.N., and Barnes, S., Metabolism of the isoflavones genistein and biochanin A in human breast cancer cell lines, *Am. J. Clin. Nutr.*, 68, 1505S–1511S, 1998.

Pettersson, H. and Kiessling, K.H., Liquid chromatography determination of the plant estrogens coumestrol and isoflavones in animal feed, *J. Assoc. Off. Chem.*, 67, 503–506, 1984.

Plewa, M.J., Wagner, E.D., Berhow, M.A., Conway, A., Rayburn, A.L., and Anderson, D., Antimutagenic activity of chemical fractions isolated from a commercial soybean processing by-product, *Teratog. Carcinog. Mutagen.*, 19, 121–135, 1999.

Prasain, J.K., Wang, C.C., and Barnes, S., Mass spectrometric methods for the determination of flavonoids in biological samples, *Free Radic. Biol. Med.*, 37, 1324–1350, 2004.

Pratt, D.E. and Birac, P.M., Source of antioxidant activity of soybeans and soy products, *J. Food Sci.*, 44, 1720–1722, 1979.

Price, K.R. and Fenwick, G.R., Naturally occurring oestrogens in food—a review, *Food Addit. Contam.*, 2, 73–106, 1985.

Pueppke, J.L., The genetics and biochemical basis for nodulation of legumes by rhyzobia, *Crit. Rev. Biotechnol.*, 16, 1–51, 1996.

Reinli, K. and Block, G., Phytoestrogen content of foods—a compendium of literature values, *Nutr. Cancer*, 26, 123–148, 1996.

Rivera-Vargas, L.I., Schmitthenner, A.F., and Granham, T.L., Soybean flavonoid effects on metabolism by *Phytophthora sojae*, *Phytochemistry*, 32, 851–857, 1993.

Rong, H., Stevens, J.F., Deinzer, M.L., De Cooman, L., and Keukeleire, D., Identification of isoflavones in the roots of *Pueraria lobata. Planta Medica*, 64, 620–627, 1998.

Rostagno, M.A., Palma, M., and Barroso, C.G., Solid-phase extraction of soy isoflavones, *J. Chromatogr. A*, 1076, 110–117, 2005.

Sachse, J., Quantative hochdruckflüssigchromatographie von isoflavonen in rotklee (*Trifolium pratense* L.), *J. Chromatogr.*, 298, 175–182, 1984.

Satterfield, M., Black, D.M., and Brodbelt, J.S., Detection of the isoflavone aglycones genistein and deidzein in urine using solid-phase microextraction–high-performance liquid chromatograophy–electrospray ionization mass spectrometry, *J. Chromatogr. B*, 759, 33–41, 2001.

Schabath, M.B., Hernandez, L.M., Wu, X., Pillow, P.C., and Spitz, M.R., Dietary phytoestrogens and lung cancer risk, *J. Am. Med. Assoc.*, 294, 1493–1504, 2005.

Schatzman, D., Ingredients to watch, *Nutr. Outlook*, 10, 24–31, 2007.

Schwartz, H. and Cichna-Markl, M., Analysis of phytoestrogens in foods using sol-gel enzyme columns for sample preparation, *J. Sol-Gel Sci. Technol.*, 35, 211–220, 2005.

Schwartz, H. and Sontag, G., Determination of secoisolarisiresinol, lariciresinol and isolariciresinol in plant foods by high performance liquid chromatography coupled with coulometric electrode array detection, *J. Chromatogr. B*, 838, 78–85, 2006.

Seo, A. and Morr, C.V., Improved high-performance liquid chromatographic analysis of phenolic acids and isoflavonoids from soybean protein products, *J. Agric. Food Chem.*, 32, 530, 1984.

Sepehr, E., Robertson, P., Gilani, G.S., Cooke, G., and Lau, B.P.-Y., An accurate and reproducible method for the quantitative analysis of isoflavones and their metabolites in rat plasma using liquid chromatography/mass spectrometry combined with photodiode array detection, *J. AOAC Int.*, 89, 1158–1167, 2006.

Setchell, K.D.R., Childress, C., Zimmer-Nechemias, L., and Cai, J., Method for measurement of dietary secoisolariciresinol using HPLC with multichannel electrochemical detection, *J. Med. Food*, 2, 193–198, 1999.

Setchell, K.D.R., Faughnan, M.S., Avades, T., Zimmer-Nechemias, L., Brown, N.M., Wolfe, B.E., Brashear, W.T., Desai, P., Oldfield, M.F., Botting, N.P., and Cassidy, A., Comparing the pharmacokinetics of daidzein and genistein with the use of 13C-labeled tracers in premenopausal women, *Am. J. Clin. Nutr.*, 77, 411–419, 2003.

Setchell, K.D.R. and Adlercreutz, H., Mammalian lignans and pytoestrogens: Recent studies on their formation, metabolism and biological role in health and disease, in *Role of the Gut Flora in Toxicity and Cancer*, Rowland, I.R. (Ed.), Academic Press, San Diego, CA, 315–345, 1988.

Setchell, K.D.R., Borriello, S.P, Hulme, P, Kirk, D.N., and Axelson M., Nonsteroidal estrogens of dietary origin: Possible roles in hormone-dependent disease, *Am. J. Clin. Nutr.*, 40, 569–569, 1984.

Setchell, K.D.R., Brown, N.M., Desai, P., Zimmer-Nechemias, L., Wolfe, B.E., Brashear, W.T., Kirschner, A.S., Cassidy, A., and Heubi, J.E., Bioavailability of pure isoflavones in healthy humans and analysis of commercial soy isoflavone supplements, *J. Nutr.*, 131, 1362S–1375S, 2001.

Setchell, K.D.R. and Cassidy, A., Dietary isoflavones: Biological effects and relevance to human health, *J. Nutr.*, 129, 758S–767S, 1999.

Setchell, K.D.R., Lawson, A.M., McLaughlin, L.M., Patel, S., Kirk, D.H., and Axelson, M., Measurement of enterolactone and enterodiol, the first mammalian lignans, using stable isotope dilution and gas-chromatography-mass spectrometry, *Biomed. Mass Spectrom.*, 10, 227–235, 1983.

Setchell, K.D.R, Welsh, M.B., and Lim, C.K., High-performance liquid chromatographic analysis of phyoestrogens in soy protein preparations with ultra-violet, electrochemical and thermospray mass spectrometric detection, *J. Chromatogr.*, 386, 315–323, 1987.

Sharma, R.D., Isoflavonoids and hypercholesterolemia in rats, *Lipids*, 14, 535–540, 1979.

Shihabi, Z.K., Kute, T., Garcia, L.L., and Hindsdale, M., Analysis of isoflavones by capillary electrophoresis, *J. Chromatogr.*, 680, 181–185, 1994.

Shiizaki, K., Goto, K., Ishige, A., and Komatsu, Y., Bioassay of phytoestrogen in herbal medicine used for postmenapausal disorder using transformed MCF-7 cells, *Phytother. Res.*, 13, 498–503, 1999.

Shutt, D.A. and Cox, R.I., Steroid and phytoestrogen binding to sheep uterine receptors in vivo, *J. Endocrinol.*, 52, 299–310, 1972.

Slanina, J. and Glatz, Z., Separation procedures applicable to lignan analysis, *J. Chromatogr. B*, 812, 215–229, 2004.

Smeds, A.I., Eklund, P.C., Sjöholm, R.E., Willför, S.M., Nishibe, S., Deyama, T., and Holmbom, B.R., Quantification of broad spectrum of lignans in cereals, oilseeds, and nuts, *J. Agric. Food Chem.*, 55, 1337–1346, 2007.

Song, T., Barua, K., Buseman, G., and Murphy, P.A., Soy isoflavone analysis: Quality control and a new internal standard, *Am. J. Clin. Nutr.*, 68 (Suppl.), 1474s–1479s, 1998.

Stefanick, M.L., Anderson, G.L., Margolis, K.L., Hendrix, S.L., Rodabough, R.J., Pasket, E.D., Lane, D.S., Hubbell, F.A., Assaf, A.R., Sarto, G.F., Schenken, R.S., Yasmeen, S., Lessin, L., and Chlebowski, R.T., Effects of conjugated equine estrogens on breast cancer and mammography screening in postmenopausal women with hysterectomy, *J. Am. Med. Assoc.*, 295, 1647–1657, 2006.

Taylor, J.I., Grace, P.B., and Bingham, S.A., Optimization of conditions for the enzymatic hydrolysis of phytoestrogen conjugates in urine and plasma, *Anal. Biochem.*, 341, 220–229, 2005.

Thompson, L.U., Robb, P., Serraino, M., and Cheng, F., Mammalian lignan production from various foods, *Nutr. Cancer*, 16, 43–52, 1991.

Thompson, L.U., Rickard, S.E., Cheung, F., Kenaschuk, E.O., and Obermeyer, W.R., Variability in anticancer lignan levels in flaxseed, *Nutr. Cancer*, 27, 26–30, 1997.

UK Government COT working group document available at http://www.food.gov.uk/multimedia/webpage/phytoreportworddocs.

Unkila, M., Optimizing the health benefits of plant lignans, *HMRlignan Webinar*, October 24, 2005.

Unkila, M., The use of a lignan for the manufacture of a composition for preventing or alleviating of symptoms relating to estrogen deficiency, *World Patent* WO 072647 A1, 2006.

Valentin-Blasini, L., Blount, B.C., Rogers, H.S., and Needham, L.L., HPLC-MS/MS method for the measurement of seven phytoestrogens in human serum and urine, *J. Expo. Anal. Environ. Epidemiol.*, 10(6, Pt 2), 799–807, 2000.

Van de Casteele, K., Geiger, H., and Van Sumere, C.F., Separation of flavonoids by reversed-phase, high-performance liquid chromatography, *J. Chromatogr.*, 240, 81–94, 1982.

Van Nieuwenhuyzen, W., Health ingredients Europe 2006 exhibition matures, *Int. News Fats Oils Relat. Mater.*, 18(2), 142–144, 2007.

Vänttinen, K. and Moravcova, J., Phytoestrogens in soy foods: Determination of daidzein and genistein by capillary electrophoresis, *Czech. J. Food Sci.*, 17, 61–67, 1999.

Wang, C., Sherrard, M., Pagadala, S., Wixon, R., and Scott, R.A., Isoflavone content among maturity group 0 to II soybeans, *J. Am. Oil Chem. Soc.*, 77, 483–487, 2000a.

Wang, C.-C., Prasain, J.K., and Barnes, S., Review of the methods used in the determination of phytoestrogens, *J. Chromatogr. B*, 777, 3–28, 2002.

Wang, G., Kuan, S.S., and Francis, O.J., A simplified HPLC method for the determination of phytoestrogens in soya bean and its processed products, *J. Agric. Food. Chem.*, 38, 185–190, 1990.

Wang, G.J., Lapčík, O., Hampl, R., Uehara, M., Al-Maharik, N., Stumpf, K., Mikola, H., Wahala, K., and Adlercruetz, H., Time-resolved fluorimmunoassay of plasma daidzein and genestein, *Steroids*, 65, 339–348, 2000b.

Wang, H.J. and Murphy, P.A., Isoflavone content in commercial soybean foods, *J. Agric. Food Chem.*, 42, 1666–1673, 1994.

Wang, J. and Sporns, P., MALDI-TOF MS analysis of isoflavones in soy products, *J. Agric. Food Chem.*, 48, 5887–5892, 2000.

Wang, W., Radioimmunoassay determination of formononetin in murine plasma and mammary glandular tissue, *Proceedings of the Society for Experimental Biology and Medicine*, 217, 281–287, 1998.

Weinberg, D.S., Manier, M.L., Richardson, M.D., and Haibach, F.G., Identification and quantification of anticarcinogens in garlic extract and licorice [*Glycyrrhiza sp.*] root extract powder, *J. High Resol. Chromatogr.*, 15, 641–654, 1992.

West, L.G., Birac, P.M., and Pratt, D.E., Separation of the isomeric isoflavones from soybeans by high-performance liquid chromatography, *J. Chromatogr.*, 150, 266, 1978.

Westcott, N.D. and Muir, A.D., Overview of flax lignans, *Int. News Fats Oils Relat. Mater.*, 11, 118–121, 2000.

Westcott, N.D., Muir, A.D., and Northrup, S.F., Analysis of flax lignans after enzymatic and chemical extraction, *Proc. 57th Flax Inst. United States*, Fargo, ND, 27–30, 1998.

Wilcox, G., Wahlquist, M.L., Burger, H.G., and Medley, G., Oestrogenic effects of plant foods in postmenopausal women, *Br. Med. J.*, 301, 905–906, 1990.

Wilkinson, A.P., Wähälä, K., and Williamson, G., Identification and quantification of polyphenol phytoestrogens in foods and human biological fluids, *J. Chromatogr. B*, 777, 93–109, 2002.

Wang, W.G., Tanaka, Y., Han, Z.K., and Cheng, J., Radioimmunoassay for quantitative analysis of formononetin in blood plasma and rumen fluid of wethers fed red clover, *J. Agric. Food Chem.*, 42, 1584–1587, 1994.

Wilson, T.C., Harris, R.K., Winters, T.R., Greaves, J.G., and Haggerty, W.T., Supercritical fluid extraction of polar materials from animal feed as and plant matrices, 205th ACS National Meeting, Denver, *CO. Am. Chem. Soc.*, 33, 391–394, 1993.

Writing Group for the Women's Health Initiative Investigators, Risks and benefits of estrogen plus progestin in healthy postmenopausal women: Principal results from the Women's Health Initiative randomized controlled trial, *J. Am. Med. Assoc.*, 288, 321–333, 2002.

Wu, Q., Wang, M., and Simon, J.E., Determination of isoflavones in red clover and related species by high-performance liquid chromatography combined with ultraviolet and mass spectrometric detection, *J. Chromatogr. A.*, 1016, 195–209, 2003.

Xiao, H.B., Krucker, M., Albert, K., and Liang, X.M., Determination and identification of isoflavonoids in *Radix astragali* by matrix solid-phase dispersion extraction and high-performance liquid chromatography with photodiode array and mass spectrometric detection, *J. Chromatogr. A*, 1032, 117–124, 2004.

Yang, S., Zhou, X., and Xu, Y., Determination of genistein and daidzein in human plasma by liquid chromatography and tandem mass spectrometry, *J. Liq. Chromatogr. Relat. Technol.*, 27, 481–499, 2005.

Yu, J., Liu, Y.-F., Qui, A.-Y., and Wang, X.-G., Preparation of isoflavones enriched soy protein isolate from defatted soy hypocotyls by supercritical CO_2, *Lebensm. Wiss. u. Technol.*, 40, 800–806, 2007.

Zeleniuch-Jacquotte, A., Adlercreutz, H., Akhmedhanov, A., and Tonido, P., Reliability of serum measurements of lignans and isoflavonoid phytoestrogens over a two year period, *Cancer Epidemiol. Biomarkers Prev.*, 7, 885–889, 1998.

Zhang, Y.C. and Schwartz, S.J., Analysis of isoflavones in soy foods, in *Handbook of Food Analytical Chemistry*, Vol. 2, Wrolstad, R.E., Acree, T.E., Decker, E.A., Penner, M.H., Reid, D.S., Schwartz, S.J., Shoemaker, C.F., Smith, D., and Sporns, P. (Eds.), John Wiley & Sons, New Jersey, 519–535, 2005.

2 Analysis of Fatty Acids in Functional Foods with Emphasis on ω3 Fatty Acids and Conjugated Linoleic Acid

Gary Dobson

CONTENTS

2.1 ROLE OF FATTY ACIDS IN HEALTH

2.1.1 INTRODUCTION

Fatty acids are ubiquitous molecules in biological systems. They occur as components of lipids, notably phospholipids and glycolipids in membranes, and triacylglycerols in seed oils of plants, oily fish, and adipose tissue (fat) in animals. They are present in appreciable quantities in many foodstuffs and although there has been a great deal of concern over the consumption of too much fat, it has to be remembered that fatty acids are essential to our well being. There is a range of different types of fatty acids, varying in chain length and number of double bonds (degree of unsaturation), and any one source normally contains several different types. Fatty acids with even numbers of carbon atoms predominate and there are short-chain (SCFA, C_4–C_{10}), medium-chain (MCFA, C_{12}, C_{14}), long-chain (LCFA, C_{16}–C_{22}), and very long-chain (LCFA > C_{22}); sometimes the C_{20}–C_{24} fatty acids are referred to as LCFA. Common saturated fatty acids (SFA) are palmitic (16:0) and stearic (18:0) acids. Unsaturated fatty acids (UFA) are divided into monounsaturated fatty acids (MUFA) with one double bond and polyunsaturated fatty acids (PUFA) with two or more double bonds; in this review we will define LC-PUFA as containing 20 and 22 carbons. Important UFA in nutrition include the MUFA oleic acid (OA, 18:1ω9; the double bond is nine carbons from the terminal methyl end of the molecule), diunsaturated linoleic acid (LA, 18:2ω6; containing two methylene-interrupted double bonds, the one nearest the terminal methyl is six carbons down from it), triunsaturated α-linolenic acid (ALA, 18:3ω3), tetraunsaturated arachidonic acid (AA, 20:4ω6), pentaunsaturated eicosapentaenoic acid (EPA, 20:5ω3), and hexaunsaturated docosahexaenoic acid (DHA, 22:6ω3). It can be seen from this list that some PUFA are of the ω3 (or *n*-3) type whereas others are ω6 (or *n*-6), the fatty acids in each group being biosynthetically related. LA and ALA are termed essential fatty acids because they cannot be biosynthesized by humans and they must be provided in the diet from vegetable or animal sources. The more unsaturated and longer chain ω6 and ω3 acids may be biosynthesized from LA and ALA (Figure 2.1), respectively, or they may be obtained from the diet. The double bonds in UFA can exist in two geometrical forms. The *cis* form overwhelmingly predominates in nature although *trans* fatty acids occur naturally in milk and ruminant tissues, and are formed in partially hydrogenated vegetable oils (PHVO) used in foods.

The simple health message has been to eat less saturated fat and more unsaturated fat. SFA, in general, increase the risk of cardiovascular disease (CVD) by raising low-density lipoprotein (LDL) cholesterol. However, not all SFA are equal in this respect; lauric (12:0), myristic (14:0), and palmitic acids are more harmful than stearic acid which appears to be neutral in its effect on LDL cholesterol

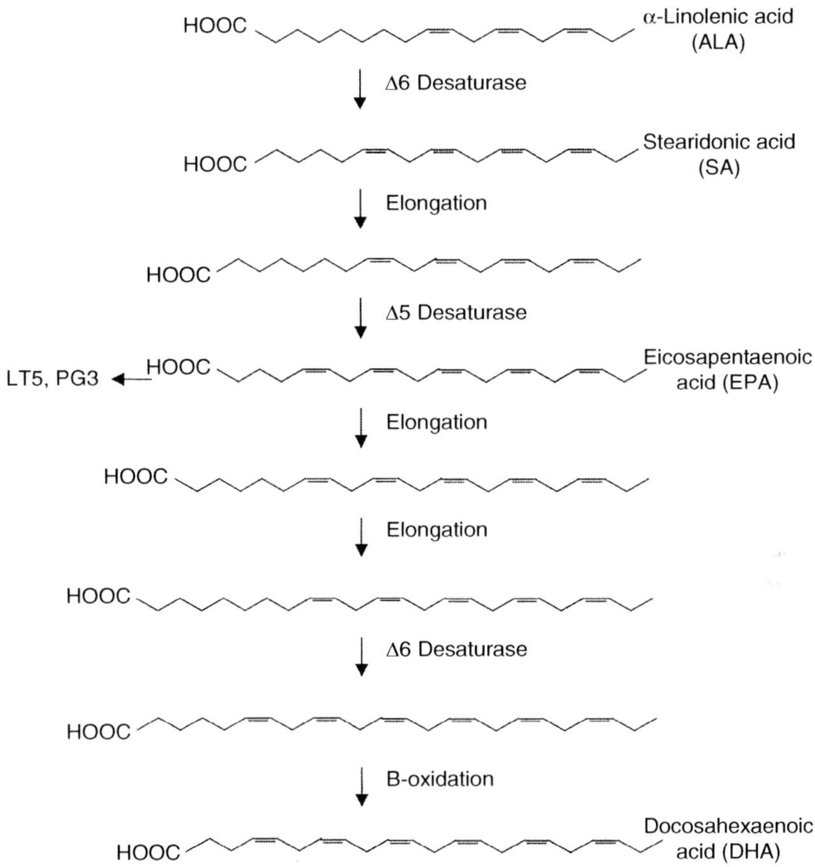

FIGURE 2.1 Biosynthetic pathway of ω3 fatty acids (LT5, series-5 leukotrienes; PG3, series-3 prostaglandins).

(Molkentin, 1999). It is generally considered that OA decreases LDL cholesterol levels. It is interesting that lauric acid has been implicated as having anticancer properties (Enig, 1994) and infants require MCFA as a rapid energy source (Willis, Lencki, and Marangoni, 1998), thus highlighting the over-simplification of whether a fatty acid is either good or bad. Butyric acid (4:0), present in milk, is also thought to have anticarcinogenic properties and SCFA in general have been shown to reduce serum cholesterol and triglyceride levels (Molkentin, 1999). The role of *trans* fatty acids in nutrition is controversial but there is considerable evidence that they raise triglycerides and LDL cholesterol, and lower high-density lipoprotein (HDL) cholesterol, that is, they have a negative impact on risk of CVD.

2.1.2 ω3 Fatty Acids

Over the past few years, interest in the role of fatty acids in health has focussed on LC-PUFA, particularly the ω6 fatty acid AA, and more on the ω3 fatty acids EPA and DHA (Horrocks and Yeo., 1999; Ruxton et al., 2004; Schmidl, 1996, Simopoulos, 1994, Shahidi and Miraliakbari, 2005; Uauy-Dagach and Valenzuela, 1996). There has been increasing evidence for the role of ω3 fatty acids in promoting health effects and protecting against certain diseases. DHA and AA, present in human milk, are essential for normal visual and cerebral function in infants. On the basis of several randomized trials, DHA has benefits for preterm infants at least in the first 4 months of life, but there were no consistent significant benefits in visual acuity or general development in term infants. There is also some evidence that maternal ω3 LC-PUFA consumption affected later development in terms of better visual acuity, higher developmental scores, and higher IQ. ω3 LC-PUFA may protect against diseases of the retina (SanGiovanni and Chew, 2005).

DHA and EPA cause a number of effects that are considered to protect against CVD (Bucher et al., 2002; Harper and Jacobson, 2005; Ramaa et al., 2006; Ruxton et al., 2004). These effects include lowering of triglyceride levels by decreasing VLDL (very low-density lipoprotein) synthesis, antithrombotic activity by decreasing platelet aggregation, and lowering of blood pressure and anti-atherogenic activity. Epidemiological evidence indicates that fish consumption is related to a lower incidence of CVD, but hard evidence can only come from randomized intervention studies. An evaluation of several clinical trials concluded that fish oil reduces overall mortality, mortality due to myocardial infarction (MI), and sudden death in patients with CVD but did not affect nonfatal MI. It is considered that ω3 LC-PUFA are most beneficial in secondary prevention of CVD. In two major studies of subjects who had survived an acute MI, fish consumption reduced overall mortality and fish oil supplementation additionally resulted in decrease in sudden deaths and cardiovascular deaths.

There is evidence for immunomodulatory effects of ω3 LC-PUFA, and there have been many studies on the efficacy of treating various inflammatory conditions (Beckles Willson, Elliott, and Everard, 2003; Curtis et al., 2004; Mickleborough and Rundell, 2005; Ruxton et al., 2004; Shahidi and Miraliakbari, 2005; Woods, Thien, and Abramson, 2003). Rheumatoid arthritis is the only inflammatory condition for which there is substantial evidence that ω3 LC-PUFA can have beneficial effects (Curtis et al., 2004; Ruxton et al., 2004; Shahidi and Miraliakbari, 2005). It is likely that ω3 LC-PUFA exert their antiarthritic effect through modulation of inflammatory cytokine production. Although treatment is effective in alleviating symptoms such as duration of stiffness, the required dose is high and progression of the disease may not be affected. There appears to be benefits for cystic fibrosis (Beckles Willson, Elliott, and Everard, 2003), but conflicting evidence for the effect of ω3 LC-PUFA in treating inflammatory bowel disease (Ruxton et al., 2004; Shahidi and Miraliakbari, 2005) and asthma (Mickleborough and Rundell, 2005; Shahidi and Miraliakbari, 2005; Woods, Thien, and Abramson, 2003), although positive effects in treating athletes with exercise-induced bronchoconstriction are more conclusive (Mickleborough and Rundell, 2005).

An assessment of several studies concluded that there was no significant effect of ω3 LC-PUFA on reducing risk in various types of cancer, and that dietary supplementation with ω3 LC-PUFA was unlikely to prevent cancer (MacLean et al., 2006). On the other hand, beneficial effects of ω3 LC-PUFA on bone metabolism and on bone and joint diseases have been noted (Watkins et al., 2001).

Mental health is an area that has received considerable attention. Imbalances in fatty acid status may be linked to childhood behavioral and learning disorders such as attention deficit hyperactivity disorder (ADHD), dyslexia, dyspraxia, and autism, but limited evidence from intervention trials did not demonstrate a consistent benefit from EPA/DHA intake, and large-scale studies are needed (Richardson, 2004; Ruxton et al., 2004). Animal and human studies have shown that deficiency in dietary ω3 LC-PUFA is associated with memory loss and diminished cognitive function. Biological, observational, and epidemiological studies have identified fish consumption and ω3 LC-PUFA intake as potential protective factors in protecting against dementia, but to date no full-randomized trials have been completed (Issa et al., 2006; Lim et al., 2006; Ruxton et al., 2004). Depression is associated with overproduction of inflammatory cytokines and schizophrenia is related to oxidative stress (Shahidi and Miraliakbari, 2005). There is some evidence for the therapeutic benefit of ω3 LC-PUFA treatment in both these conditions but the evidence is inconclusive and preliminary (Peet and Stokes, 2005; Shahidi and Miraliakbari, 2005). ω3 Fatty acids may act by modulating eicosanoid production and decreasing oxidative stress, respectively.

EPA and AA (and dihomo-γ-linolenic acid, 20:ω6, DGLA), present in membrane phospholipids, are precursors of eicosanoids (comprising prostaglandins, leukotrienes, and thromboxanes) that exhibit a variety of physiological functions such as regulating the inflammatory response and platelet aggregation. The structures of the eicosanoids from the two fatty acids are different and they have separate and competitive functions, those from AA are more prothrombotic and pro-inflammatory. The balance of the different types of eicosanoids, determined by the ratio of AA and EPA, influence the physiological outcome, those from EPA resulting in a more favorable response (antithrombotic, anti-inflammatory, and anti-vasoconstrictive). It is now recognized that the ratio of ω6 to ω3 acids in the modern diet is generally too high (10:1 to 25:1) for optimal health and therefore it is recommended to consume more ω3 fatty acids (present in high concentrations in oily fish) and less ω6 to redress this balance to a ratio of 2:1 to 4:1 (Simopoulos, 2000).

The role of ALA in increasing levels of EPA and DHA has also received attention. Competition between ALA and LA for the same enzymes that are involved in both the ω3 and the ω6 pathways influences the levels of EPA and DHA. Earlier studies suggested that conversion of ALA to EPA and DHA was low because of the rate-limiting effect of the Δ6-desaturase, but there is now evidence that dietary ALA can increase EPA, but less so DHA, in tissue, and can affect eicosanoid production. However, the effects of supplementation with ALA are not the same as with EPA and DHA. Although ALA can reduce blood pressure, it does not decrease plasma triglycerides. Flaxseed, containing high levels of ALA and other phytochemicals with a potentially health-promoting effect, can positively affect some CVD risk factors (e.g., lowers total and LDL cholesterol, decreases some markers of inflammation, raises levels of ALA and EPA), but the effect on others (e.g., hypotensive effect) are inconclusive (Bloedon and Szapary, 2004). Flaxseed oil can also enhance factors, such as plasma, platelet, and tissue EPA levels, that are protective against CVD, but serum triglyceride levels, for example, are not affected (Oomah, 2001). It would appear that the role of ALA in health is more tentative than that of EPA and DHA.

2.1.3 Conjugated Linoleic Acid

Conjugated linoleic acid (CLA) is another fatty acid that is currently receiving considerable attention because of a range of properties that may make a positive contribution to health (Banni and Martin, 1998; Fritsche and Steinhart, 1998; Nagao and Yanagita, 2005; Pariza, 1999; Roche et al., 2001; Salas-Salvado, Marquez-Sandoval, and Bullo, 2006; Wahle, Heys, and Rotondo, 2004). CLA is the collective name for a range of conjugated octadecadienoic geometrical and positional isomers. Positional isomers may range from 6,8- to 13,15-positions and *cis-trans*, *trans-cis*, *cis-cis*, and *trans-trans* geometrical isomers may occur. CLA is formed in ruminants during microbial biohydrogenation of LA, and moreover from Δ9-desaturation of *trans*-vaccenic acid (VA, 11*t*-18:1; also a product of biohydrogenation) in tissues, particularly the mammary gland and in adipose cells; it is often referred to as natural CLA. It is also produced on an industrial scale by alkaline isomerization of LA (from sunflower or safflower oil) and is then referred to as commercial CLA. Low levels are also produced during catalytic hydrogenation and during heating processes (Ledoux and Laloux, 2006), although selective hydrogenation processes can produce oils of high CLA content (up to 24% of the oil) and low levels of other *trans* fatty acids (Jang, Jung, and Min, 2005).

In natural CLA, *cis*-9,*trans*-11-octadecadienoate (9*c*,11*t*-18:2, rumenic acid; Figure 2.2) is always the major isomer, although a variety of other isomers can occur as minor components. Noteworthy are the 7*t*,9*c* isomer, occurring at a level of about 10% of the major isomer, and the 10*t*,12*c* and 11*t*,13*c* isomers that can occur at significant levels under certain conditions (Lock and Bauman, 2004). In industrial preparations, the 9*c*,11*t* and 10*t*,12*c* isomers are the major components, but there are varying amounts of particularly the 8*t*,10*c* and 11*c*,13*t* isomers, and sometimes other positional isomers. *Cis-cis* and *trans-trans* isomers are also produced, usually as minor components.

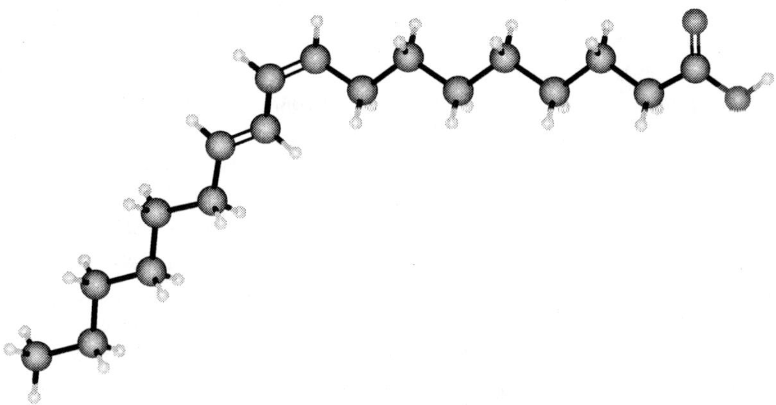

FIGURE 2.2 Structure of *cis*-9,*trans*-11-octadecadienoate (9*c*,11*t*-18:2), the major natural conjugated linoleic acid (CLA).

Since the publication of the first edition of this chapter (Dobson, 2002) the potential health-promoting effects of CLA have continued to be a topic that is being extensively studied. It has become apparent that the considerable potential suggested by animal studies has not been demonstrated in humans, some positive effects either being diminished in humans or remaining inconclusive. The physiological properties of CLA have been extensively reviewed (Banni and Martin, 1998; Fritsche and Steinhart, 1998; Nagao and Yanagita, 2005; Pariza, 1999; Roche et al., 2001; Salas-Salvado, Marquez-Sandoval, and Bullo, 2006; Wahle, Heys, and Rotondo, 2004) and include inhibition of carcinogenesis and atherosclerosis, hypotensive, antidiabetic and anti-inflammatory effects, enhancement of immunological function, and effects on body composition by reducing body fat while enhancing lean body mass.

The anticarcinogenic effects have been demonstrated in animals and human cell lines (Banni, Heys, and Wahle, 2003; Nagao and Yanagita, 2005) and both the 9*c*,11*t* and 10*t*,12*c* isomers appear to be equally effective. Inhibition against skin, stomach, mammary, and colon tumors in mice and rat (Banni, Heys, and Wahle, 2003; Kritchevsky, 2000), and against multistage carcinogenesis (Lee et al., 2005), is evident. However, there is some evidence that particularly the 10*t*,12*c* isomer can elicit pro-carcinogenic effects in animal models (Wahle, Heys, and Rotondo, 2004). There is limited indirect evidence for an anticarcinogenic effect in humans, but no intervention trials have been published (Nagao and Yanagita, 2005). It was concluded that CLA supplementation might be effective against breast cancer in humans but further evidence is required. It is interesting that dietary VA can decrease tumor formation in the rat mammary gland because of its conversion to rumenic acid (Corl et al., 2003).

An anti-atherogenic effect of CLA, demonstrated by reducing aortic fatty streak area, has been reported in studies with rabbits and hamsters (Kritchevsky, 2003; Nagao and Yanagita, 2005). In a variety of animals, CLA reduces plasma total and LDL cholesterol, but in some studies on humans, CLA had an adverse effect on the lipid profile by decreasing HDL cholesterol (Salas-Salvado, Marquez-Sandoval, and Bullo, 2006).

The hypotensive and antidiabetic effects (improves glucose tolerance, alleviates hyperinsulinemia, and delays onset of diabetes), possibly attributable to the specific action of the 10*t*,12*c* isomer (Belury, 2003; Zulet et al., 2005), have been shown in animals. However, CLA supplementation has been reported to have adverse effects on insulin sensitivity and glucose metabolism in patients with type 2 diabetes mellitus (Zulet et al., 2005). In obese men, the two major CLA isomers can each induce adverse effects on insulin sensitivity and can increase markers of oxidative stress and inflammation (Zulet et al., 2005).

The enhancement of immunologic function has been demonstrated in a variety of models including pigs and humans (Bassaganya-Riera, King, and Hontecillas, 2004; Cook et al., 2003). It has been suggested that the anti-inflammatory properties of CLA, by decreasing pro-inflammatory eicosanoids and increasing anti-inflammatory cytokines (Salas-Salvado, Marquez-Sandoval, and Bullo, 2006; Troegeler-Meynadier and Enjalbert, 2005; Zulet et al., 2005), may be responsible for its anticarcinogenic and anti-atherosclerosis effects (Zulet et al., 2005). In rats, CLA also has anti-inflammatory activity in bone and can increase the rate of bone formation (Watkins and Seifert, 2000).

The most attractive properties of CLA that might be utilized in a functional food are perhaps the fat-to-lean partitioning effect and body weight reduction. In rats, CLA, probably the 10*t*,12*c* isomer, may exert its hypolipidemic effect by suppressing fatty acid synthesis and enhancing Β-oxidation and energy consumption (Nagao and Yanagita, 2005). However, although the effects are considerable in mice and rats, and less in pigs, there is not enough evidence in humans to indicate a significant effect (Kovacs and Mela, 2006; Larsen, Toubro, and Astrup, 2003; Salas-Salvado, Marquez-Sandoval, and Bullo, 2006; Wang and Jones, 2004). It has been highlighted that, to some extent, results in humans have been conflicting because of differences in isomer mixtures, dose levels, and subject populations (Evans, Brown, and McIntosh, 2002). It seems that in humans CLA may have an effect on body composition rather than on body weight (Kovacs and Mela, 2006). Long-term randomized trials are needed to determine the efficacy and indeed the safety of CLA supplementation in humans.

2.1.4 *Trans* Fatty Acids

Trans fatty acids occur as complex mixtures of geometrical and positional isomers in milk fat and in PHVO, and at low levels in deodorized oils (Ratnayake, 2004). *Trans*-18:1 isomers are the major *trans* acids in PHVO and milk fats, but occur at very low levels in deodorized oils. They range from 5*t*-18:1 to 16*t*-18:1 and in PHVO the distribution is approximately Gaussian, centered around 9*t*- and 10*t*-18:1, whereas in milk fat VA usually accounts for about 70% of the total *trans* monenes. The distribution of isomers can vary considerably between different samples. For example, 10*t*-18:1 is the major *trans* monoene in the milk of cows fed fish meal (Kramer et al., 2004). *Trans*-18:2 and -18:3 fatty acids are also present and again they exist in a range of geometrical and positional isomeric forms, the distribution of which differs between the different sources. *Trans* polyenes are the major *trans* acids (but at low levels) in deodorized oils, but occur at lower levels than *trans* monoenes in milk fats and PHVO, the levels in PHVO depending on the degree of hydrogenation. The presence of *trans*-18:3 isomers in PHVO and deodorized oils depend on whether ALA is present in the unprocessed oils. The mixtures are so complex that not all 18:2 and 18:3 isomers have been identified.

Intake of *trans* fatty acids is associated with a higher risk of CVD (Aro, 1998). Several countries have introduced mandatory labeling of total *trans* fatty acids in food products. However, the specific *trans* fatty acids responsible for the detrimental effects have not been identified nor is it known whether PHVO and ruminant fats present the same risk. Most work has been done using PHVO but, as noted above, ruminant fats have a different distribution.

2.1.5 Other Fatty Acids

Another nutritionally important fatty acid is γ-linolenic acid (GLA, 18:3ω6). This ω6 acid is converted to DGLA resulting in increased prostaglandin production and decreased inflammation. It has been used to treat rheumatoid arthritis, premenstrual syndrome, atopic eczema, and other inflammatory skin conditions, and has anticancer activity (McDonald and Fitzpatrick, 1998). Stearidonic acid (SA, 18:4ω3) is a precursor to EPA, and an increase in its consumption may be an efficient way to increase the longer chain ω3 acids, by bypassing the possible rate-limiting step involving the Δ6 desaturase conversion of ALA to SA. Limited evidence has shown that SA might have anticancer and anti-inflammatory properties (Yang, Gunstone, and Kallio, 2003). SA-containing oils such as blackcurrant seed oil can decrease plasma total and LDL cholesterol levels and increase HDL cholesterol levels, relieve the symptoms of atopic eczema, and reduce blood pressure. Some of these effects were shown in animals, others in humans, and require confirmation. Moreover, it remains to be determined whether the effects were due to SA as blackcurrant oil contains other active fatty acids, notably ALA and GLA, that are at higher levels than SA; the hypotensive activity was, however, greater than other GLA-containing oils (borage, evening primrose).

The health benefits of OA have been implicated in olive oil as part of the Mediterranean diet (Wahle, Heys, and Rotondo, 2004). LDL particles rich in MUFA are less susceptible to oxidation than those containing PUFA. Some of the health benefits resulting from a higher intake of MUFA in olive oil are probably due to a decreased intake of SFA and PUFA such as LA which may result in greater desaturation of ALA to beneficial ω3 LC-PUFA, therefore decreasing CVD risk. There is also evidence to indicate that OA is responsible for lowering LDL cholesterol and has beneficial effects on immune function.

Palmitoleic acid is another fatty acid that deserves a brief mention. There have been reports concerning a correlation between palmitoleic acid levels in plasma and tissues and CVD risk factors. There are claims for hypocholesterolemic and hypoglyceridemic activities, enhancement of HDL cholesterol and depression of LDL cholesterol levels, reduction of stroke incidence, and protection against certain cancers in animal models (Yang, Gunstone, and Kallio, 2003). However, the amount of data is limited and further investigations of these effects are required.

2.1.6 Recommended Intake

Weighing up the beneficial/harmful effects of the various fatty acids, it has been suggested that the ideal dietary fats for adults should be low in cholesterol and *trans* fatty acids, high in OA, EPA, and DHA, low in myristic acid, have an LA/ALA ratio of 4:1, and obtain <2% energy from LA (Willis, Lencki, and Marangoni, 1998). Fats for infants should be similar to human milk in containing low amounts of *trans* acids, high AA and DHA but not EPA, have an LA/ALA ratio of 7:1, 4%–8% MCFA, and have palmitic acid mainly in position 2 of triacylglycerol for optimum absorption. It has been recommended that pregnant and lactating women, especially, should include DHA in the diet because of increased ω3 LC-PUFA demand (Koletzko et al., 2001). Although breast feeding is preferable, it was suggested that infant formula should contain at least 0.2% of total fatty acids as DHA and 0.35% as AA, or even more for preterm infants.

The recommended level of ω3 LC-PUFA intake has changed over the past few years and is a subject of debate with varying levels recommended by different bodies. For the general population (healthy adults), a range of 200–650 mg/day is generally recommended, for people at risk of CVD 500–1000 mg/day are advised, and to reduce triglyceride levels 2–4 g/day has been suggested (Garg et al., 2006). It has been estimated that in the United States the amount of ω3 LC-PUFA consumed at 100–200 mg/day falls short of the recommendations, whereas in Japan the amounts are much higher at 1–2 g/day.

2.2 FUNCTIONAL FOODS ENRICHED WITH FATTY ACIDS AND THE FACTORS AFFECTING THEIR ENRICHMENT

2.2.1 Foods Enriched with ω3 Fatty Acids

Although humans can synthesize ω3 LC-PUFA, it is accepted that dietary intake of the preformed acids is required to attain the levels necessary for health-promoting effects. The obvious way to increase ω3 LC-PUFA intake is to eat more fish, but many people do not like the taste of fish, and marine fish supplies are declining. Taking a fish oil supplement such as cod liver oil is another possibility, but a more attractive proposition is to be able to obtain ω3 acids from a variety of foods in which these fatty acids are either normally absent or present at low levels. Consequently, a number of functional foods to which ω3 acids have been added have been developed. Such foods are more established in Japan than Europe and the United States (Garcia, 1998). It is interesting that in Japan, where ω3 enhanced foods range from bread to meat products (e.g., sausages) and soft drinks, the emphasis is often on the benefits of DHA for the brain rather than for the cardiovascular system (Hilliam, 1996).

The ω3 acids can be added in a variety of different forms. The simplest form is fish oil with a typical ω3 content of about 15%–25% and variable EPA to DHA ratio depending on the origin. For example, EPA predominates over DHA in herring and menhaden but the reverse is true for salmon and tuna (Padley, Gunstone, and Harwood, 1994). Therefore, there is a range of oils that can be used for different applications. For example, low EPA oils are used for infant formulae because EPA can produce negative effects such as reducing body weight (Willis, Lencki, and Marangoni, 1998). Oils containing about 40% DHA but no EPA, developed from the alga *Crypthecodinium cohnii* are particularly useful for this purpose (Becker and Kyle, 1998). Fungal oils have also been developed for incorporation into infant formulae and food (Kolanowski and Laufenberg, 2006). ω3 Concentrates, prepared by winterization or fractional distillation, contain about 30% ω3 acids but, if the triacylglycerols are converted to free fatty acids (FFA) or alkyl esters, higher concentrations can be achieved by using urea fractionation (Shahidi, 1998).

Incorporating the fish oil (essentially all triglyceride) directly into the food is a possibility, but highly UFA are prone to oxidation resulting in off-flavors, short shelf life, and reduction in ω3 content. Foods most suitable for incorporation are those that are frequently consumed, not strongly heated or stored for long periods, and packed in absence of light and oxygen (Kolanowski and Laufenberg, 2006). In one study (Kolanowski, Swiderski, and Berger, 1999), in which a variety of foods were enriched with fish oils, up to 0.5% EPA/DHA could be incorporated into fat spreads and other products of high sweetness and flavor without detection of fish off-flavors, but even low levels (0.05% EPA/DHA) could not be tolerated in more bland products such as milk and orange juice. However, low-fat milk products are a common vehicle for ω3 fortification with fish oil (and vegetable oils, flaxseed oil) up to about 0.08% (Sharma, 2005), which even at this level (200 mg in a 250-mL serve) contributes a significant amount of ω3 intake. Some products, such as salad dressing had long (4 months) shelf-lives, whereas in others (e.g., low pH juices) the fish oil was unstable.

Highly refined and stabilized (with antioxidants) fish oils, to remove off-flavors, inhibit oxidation, and extend shelf life, have been successfully incorporated into products such as low-fat spreads (containing about 1%–2% EPA/DHA) and bread (0.2%–0.5% EPA/DHA) (Madsen, 1998; Newton and Snyder, 1997). Antioxidants such as vitamin E may also be added into the spread (Hilliam, 1996). A low-fat spread into which a stabilized fish oil had been incorporated at a level of 3% (0.8% EPA/DHA) could be stored for up to 3 months without changes in stability parameters (e.g., peroxide value, sensory quality) (Kolanowski et al., 2004).

Microencapsulated marine oils were developed to overcome the problems of oxidation and to extend shelf life of foods (Andersen, 1995; Lauritzen, 1994; Shahidi, 1998). These are powders (typically containing 10% ω3 acids) in which the oil and antioxidants are entrapped within a suitable coating such as gelatin or sucrose, and starch is added to prevent clumping. They allow a high level of addition, are stable during manufacture and storage of the food, and do not give rise to a fish taste in the product. Microencapsulated powders are used mainly in dry goods such as cereal products and powder-based products (Michelsen, Madsen and Gotfredsen, 2001; Muggli, 1997). They have been incorporated into a host of products including low-fat spreads, bread, breakfast cereals, biscuits, fruit/cereal bars, diet powder, low-fat cakes, fruit juice, pasta, salad dressings, milk drinks, soups, baby follow-on food, and infant formula (Garg et al., 2006; Lauritzen, 1994; Muggli, 1997). Infant formula can have a shelf life of up to 2 years in this form (Andersen, 1995). They can be added to products that have a high water content (e.g., dairy products, fruit beverages) but for economical reasons, fish oils are usually used for such products. In any case, when the powders dissolve in water the stability becomes similar to that of an unencapsulated oil (Michelsen, Madsen, and Gotfredsen, 2001). Microencapsulated powders are not recommended for addition to products of high fat/oil content.

ALA in flaxseed has been incorporated into breads, baked goods, cereals, and energy bars, with up to 1 g ALA per bar/cereal serving (Haumann, 1998; Oomah and Mazza, 1998). Although ALA is less susceptible to oxidation than EPA and DHA, it is still a potential problem in flaxseed oil in which the ALA content is high. Microencapsulation of flaxseed oil has been developed to expand the range of potential foods (Oomah and Mazza, 1998).

ω3 PUFA can be increased in animal products such as meat, milk, and eggs by including sources of ω3 acids into the feed, but the degree of incorporation varies according to the type of animal. Incorporation is greatest in monogastrics, especially birds, where the fatty acid composition of the meat resembles that of the diet. In polygastrics, in which PUFA are modified to some extent by hydrogenating bacteria in the rumen, incorporation is less efficient (Bourre, 2005; Lock and Bauman, 2004). Additionally, at least in milk, fatty acids are not transported effectively in plasma which is a major source of mammary fatty acids (Lock and Bauman, 2004). It has been estimated that feeding with linseed or rapeseed can increase ALA in eggs by twenty- to forty-fold, in chickens ten-fold, in pork six-fold, and in beef two-fold, and feeding algae or fish extracts can increase DHA in fish by twenty-fold, in chickens seven-fold, eggs three- to six-fold, and in beef two-fold. Including ALA in fish feed is more effective, in terms of EPA/DHA incorporation, in vegetarian than carnivorous fish, but most fish consumed are carnivorous. In attempting to increase the ω3 acids in meat (beef, lamb, and pork), diets rich in ALA (linseed, forages) tend to increase ALA and EPA but not DHA; DHA can be increased directly with fish oils (Raes, De Smet and Demeyer, 2004). Feeding cows fish meal compared to fish oil, and oilseeds compared to oils, tends to result in greater incorporation in milk because there is greater protection of the fatty acids from rumen biohydrogenation (Lock and Bauman, 2004).

Increasing ω3 acids in the diet is usually accompanied by decreasing the dietary ω6 acid content, resulting in less ω6 acids in the meat and a more favorable ω6/ω3 ratio (PUFA to SFA ratio is less affected). Also, grass-fed cattle have been shown to have lower levels of total *trans*-18:1 and SFA than long-term grain-fed cattle (Ponnampalam, Mann, and Sinclair, 2006). However, increasing the PUFA content in meat increases the potential for oxidation, resulting in undesirable sensory and health effects. Including vitamin E in the diet of cattle, pigs, and poultry can inhibit oxidation and prolong the shelf life of the meat (Jimenez-Colmenero, Carballo, and Cofrades, 2001).

Eggs normally contain high levels of ω6 and are a poor source of ω3 PUFA. They have been fortified with DHA or ALA by feeding hens with fish oils, seal blubber, flaxseed oil, or micro-algae rich in oil (Haumann, 1998; Pacetti et al., 2005; Sim, 1998; Van Elswyk, 1997). 175 mg of DHA and 350 mg total ω3 acids can be incorporated into each egg by feeding a mixture of micro-algae and flaxseed, and up to 700 mg ω3 acids (mainly ALA) by feeding flaxseed alone (Haumann, 1998). It is interesting that EPA incorporation was always low, even with fish oils. It has been shown, in eggs produced using a diet containing refined seal blubber, that there is more incorporation of ω3 acids into egg phospholipids compared to triacylglycerols, and into the *sn*-2 position rather than *sn*-1,3 positions (Pacetti et al., 2005). DHA-rich egg lecithin has been used in infant formula in an attempt to increase DHA incorporation into phospholipids rather than triglyceride (Haumann, 1998).

There are conflicting reports as to whether fishy off-flavors, from fish oils and flaxseed, in eggs can be eliminated by incorporating tocopherols into the feed (Sim, 1998; Surai and Sparks, 2001), although off-flavors are apparently absent if the level of fish oil (preferably stabilized) in the diet is 1.5% or less, and for flaxseed 5% or less (Surai and Sparks, 2001). Off-flavors have been reported to be absent if the feed contains micro-algae (Haumann, 1998). The cooking characteristics of ω3 eggs are reported to be normal (Surai and Sparks, 2001).

Studies in humans on the beneficial effects of ω3 eggs have indicated improvements in CVD risk factors such as decreased plasma triglyceride, enriched ω3 PUFA in plasma lipids, and lower blood pressure (Surai and Sparks, 2001). The effect on cholesterol levels, by increasing egg consumption, might be of concern to consumers, but there is evidence that total plasma cholesterol does not change significantly and is sometimes reduced.

There have been a number of studies that have examined the effect of foods enriched with ω3 PUFA on CVD risk factors. For example, in a recent study, a liquid egg product, containing fish oil and providing 1.3 g of EPA + DHA, when given to subjects with moderately elevated triglyceride levels over a 3 week period, resulted in significant improvements in triglyceride levels and EPA and DHA levels were enhanced in serum phospholipids (Rose and Holub, 2006). Other studies have shown that ω3-enriched eggs, milk, and processed foods can improve plasma and blood levels of ω3 PUFA, reduce plasma triglyceride levels, and increase plasma HDL cholesterol levels (Li et al., 2003).

2.2.2 Foods Enriched with Conjugated Linoleic Acid

Although humans may have a limited capacity to synthesize CLA from VA, the main source is dietary. CLA can occur naturally at low levels in a range of products but is highest in ruminants, both in meat (0.1%–1.9% of total fat in lamb and beef) and in dairy products (0.3%–1.8% of total fat) (Fritsche and Steinhart, 1998; Parodi, 2003; Schmid et al., 2006). It is very low in nonruminants (<0.1% of total fat in pigs and chicken), although slightly higher in turkey (0.2%). For a detailed account of the CLA content of various foods, the reader is referred to Parodi (2003). It has been estimated that the daily human intake of CLA is between 100 and 400 mg (Schmid et al., 2006). However, considering the low levels of CLA in food, the dietary CLA intake falls far short of the high levels that have been estimated (1.5–3.5 g/day; Banni and Martin, 1998) to have a potential effect on health.

CLA is involved in the production of two distinct types of functional food production. Firstly, it is included in the diet of animals to alter body composition and produce leaner meat products or low-fat milk, but the CLA content in the product is irrelevant to its functionality (Parrish et al., 2003). Secondly, and of more concern to this review, is the development of products with health-promoting effects owing to the enrichment of the CLA content. CLA, mainly as the FFA, is available as a supplement in capsule form. In a study of the relative absorption and palatability of different forms of CLA in a yogurt-based functional food, the ethyl ester form was absorbed into chylomicrons the least whereas the free acid had the poorest taste; the triacylglycerol was suggested to be the most suitable form overall (Fernie et al., 2004). Indeed, it has been noted that the FFA form is restricted to capsules because of the unpleasant taste, and the triacylglycerol form is the most expensive but the best form to add to foods (Kovacs and Mela, 2006). A structured lipid incorporating CLA into the triacylglycerols of extra virgin olive oil has been developed for possible inclusion in functional foods but stability was a problem (Lee et al., 2006). The low oxidative stability of CLA can be improved by microencapsulation (Andersen, 1995; Kovacs and Mela, 2006).

Although there are products, such as milk-fruit juice drinks, in which enrichment has been achieved through addition of CLA (Sharma, 2005), much effort has gone into increasing natural levels particularly in milk. Levels of up to about 30 mg/g of milk fat can be attained, compared to typical values of 3–12 mg/g (Fritsche and Steinhart, 1998; Parodi, 2003). Many factors including diet (grass/silage quantity and grass lushness, plant-oil supplements), breed, age of animal, and processing conditions (heating, dry fractionation, protein quality, and starter cultures) affect the CLA content (Collomb et al., 2006; Fritsche and Steinhart, 1998; Stanton et al., 2003). CLA content of milk seems to vary more between individual cows than different breeds, and is related to rumen biohydrogenation and Δ9-desaturase activity in the mammary gland (Lock and Bauman, 2004; Peterson, Kelsey, and Bauman, 2002). In cheese, aging, processing, type of package, origin of milk, and lipid and protein content all affect the CLA content (Lin et al., 1998; Werner, Luedecke, and Schultz, 1992).

Although some bacteria, especially propionibacteria, can form CLA during cheese ripening, the contribution from presently used starter cultures is thought to be minor (Sieber et al., 2004). On the other hand, the level of CLA in fermented milk products, such as yogurt, can be enhanced up to about twofold by using, under selected conditions, an appropriate lactic acid bacterial strain as a starter culture, and including LA as a substrate (e.g., sunflower oil) (Kim and Liu, 2002). The levels can be increased further by fermenting with an probiotic bacterium in coculture with a traditional yogurt culture in the presence of an LA substrate (Xu, Boylston, and Glatz, 2005), thus producing a functional food with dual functionality.

Considering that the majority of CLA in milk and ruminant meat comes from VA, a major strategy for maximizing CLA production is to maximize VA output through biohydrogenation of LA and ALA (Collomb et al., 2006; Lock and Bauman, 2004). This can be achieved by feeding grass compared to concentrate, but supplementation of feed concentrate with plant oils high in LA (e.g., sunflower, safflower, soybean), and sometimes ALA (e.g., linseed), is particularly effective at least for milk (Bessa et al., 2000; Fritsche and Steinhart, 1998; Stanton et al., 2003). Oilseeds rather than oils tend to be used because they are cheaper, and oils can disturb the rumen environment and oxidize more readily, although oilseeds are more protected against biohydrogenation. Also, feeding oils at high levels can result in depressing the total milk fat levels. The effects of high LA oils on CLA levels appear to be more pronounced in milk fat than in beef (Raes, De Smet, and Demeyer, 2004; Schmid et al., 2006). In beef, it has been shown that sunflower has a similar effect to grass on enhancing CLA levels and is more effective than rapeseed or rapeseed oil (Schmid et al., 2006); conflicting results have been obtained for linseed and soybean oil. Limiting hydrogenation of VA to stearic acid is another possibility for enhancing VA levels, and it would appear that this mechanism is responsible for the increase in CLA levels in milk by including dietary fish oils (that do not supply high levels of 18 carbon PUFA precursors for VA) (Lock and Bauman, 2004). It is interesting that including fish meal in the diet increased the levels of 7*t*,9*c* and 10*t*,12*c* isomers, but not the 9*c*,11*t* isomer in milk (Cruz-Hernandez et al., 2004).

CLA and VA levels in cows are always highly correlated (Molkentin, 1999) and therefore it is difficult to increase CLA levels in ruminant products without increasing VA levels. Since there may be limited conversion of VA to CLA by humans, VA in this regard may be beneficial. However, considering the possible detrimental effects of *trans* fatty acids on health, it may be preferable to minimize the levels of these fatty acids for nutritional reasons, and also for labeling purposes. However, as already mentioned, it is not certain whether *trans* monoenes from ruminants are harmful.

In monogastrics because the fat in the diet is unmodified before digestion and absorption, enhancing CLA levels in tissue must be achieved by either including CLA or VA as a substrate for endogenous CLA synthesis (Schmid et al., 2006). Significant increases in CLA content of tissue in pigs and chickens has been achieved by including CLA in the diet (Raes, De Smet, and Demeyer, 2004; Schmid et al., 2006). The CLA isomer composition of the meat approximately reflects the isomer distribution in the diet, usually a roughly equal proportion of the 9*c*,11*t* and 10*t*,12*c* isomers, although there may be some preferential incorporation of the former isomer. Including CLA in the diet of pigs and chickens increases the levels of SFA and decreases MUFA and to some extent PUFA, by down-regulating Δ9-desaturase activity. Although, nutritionally, this may be regarded as a negative effect, in pigs it can be regarded as a positive attribute as it increases the oxidative stability of the meat and makes it firmer and therefore easier to carve (Dugan et al., 1997). Including hydrogenated rapeseed (high in *trans* fatty acids) in the diet of pigs was successful in enhancing CLA levels (Schmid et al., 2006).

Attempts at enriching eggs with CLA have been less successful than with ω3 acids. The percentage of CLA of the total fatty acids increases with increasing percentage of CLA in the feed; CLA at 0.5%–5% in the feed results in 2%–19% CLA in the egg (Aydin, 2005; Shang et al., 2005; Szymczyk and Pisulewski, 2003, 2005). However, the major drawback is that with increasing dietary CLA there is increasing deposition of SFA (16:0, 18:0) and decreasing deposition of MUFA (16:1, 18:1) and more variable effects on PUFA (LA, ALA, AA, DHA), resulting in a nutritionally less favorable profile, although cholesterol was reduced (Szymczyk and Pisulewski, 2005). Other negative quality effects include changes in pH and color (Aydin, 2006), but only when eggs are stored at 4°C or 15°C and not at 21°C–24°C, and rubbery and hard yolks when the eggs are hard-cooked (Watkins et al., 2003).

The CLA-induced changes in SFA/UFA and egg quality can be prevented by also including high levels of tallow (9.5%; Aydin and Cook, 2005), olive oil (10%; Aydin, Pariza, and Cook, 2001), and canola oil (5% and 10%; Aydin, 2005) in the feed. However, the effect was only demonstrated at a level of 0.5% CLA in the feed, and the added oil or fat caused a reduction in the CLA content of the egg, resulting in about 2% CLA in the egg. Inclusion of vitamin E in the feed had a negligible effect on preventing the CLA-induced SFA/UFA alterations (Szymczyk and Pisulewski, 2005).

CLA was found to be higher in the yolk than the albumen and was incorporated into neutral and polar lipids, although in contrast to incorporation of ω3 acids (Pacetti et al., 2005), the greatest concentration was in neutral lipids (Watkins et al., 2003). The incorporation of CLA isomers is partially selective with *trans-trans* isomers, 9*c*,11*t* and 11*c*,13*t* being preferentially incorporated, compared with 10*t*,12*c* and 8*t*,10*c* (Szymczyk and Pisulewski, 2003; Yang et al., 2002).

2.2.3 Foods Enriched with Other Fatty Acids

Although GLA is used mainly in an encapsulated form as a supplement, there is some limited use in functional foods including infant formula, bread, drinks, and honey (Yang, Gunstone, and Kallio, 2003). The GLA is sometimes added in the form of evening primrose oil.

There are a number of products, particularly spreads, that incorporate high levels of OA often using olive oil. The low levels or absence of *trans* fatty acids is apparent in a whole range of products through reducing the use of PHVO.

2.2.4 Foods with Optimal Balance of Different Fatty Acids

Sometimes, in developing a functional food, the aim is to alter more than one aspect of the fatty acid profile. For example, in developing the spread "Pact" (MD Foods Viby, Denmark), the aim was to encompass an overall healthy fatty acid profile (Madsen, 1998). As well as increasing the ω3 content, the total fat content was reduced, total SFA and ω6 acids were lowered, and *trans* fatty acids were virtually excluded. Another example is Appetize (Bunge Foods USA, Bradley, IL), a shortening from blends of vegetable oils and animal fats, formulated to minimize the disadvantages of animal fats and *trans* fatty acids in raising plasma and LDL cholesterol levels (McDonald and Fitzpatrick, 1998). Cholesterol was removed from the animal fats, *trans* fatty acids were avoided, and the balance of LA and myristic acid was optimized. Indeed in validating the potential health benefits of a fat, the overall fatty acid composition must be examined, and not just single components in isolation, because some fatty acids (e.g., ω3 acids, CLA) have potentially positive health attributes whereas others (e.g., some SFA, *trans* acids) may have a negative impact.

2.3 EXTRACTION AND ANALYSIS OF FATTY ACIDS

2.3.1 Introduction

A fatty acid analysis of a food may be required for a variety of reasons, which in turn may influence the type of analysis to be performed. Information may be required in-house on the composition of the food/ingredients or for monitoring a process, or it may be required for regulatory reasons, that is for labeling or for registration of a product. Consequently, the type and complexity of the information required will vary, and will also depend on the regulations in a particular country. Information may be required only on those fatty acids (e.g., ω3 acids) of functional significance, on the ratio of such acids to other fatty acids (e.g., ω6/ω3) or on all fatty acids. In fact, a single method, usually using gas chromatography (GC), often gives information on all the fatty acids present. A fatty acid profile may be obtained in which all fatty acid components are expressed as a weight percent of the total fatty acids. Such information may be all that is required in analyzing, for example, the composition of a fish oil to be used for inclusion in a food. On the other hand, quantification of each fatty acid in terms of milligram per gram of food may often be required. Labeling requirements may be for total fat and saturated fat only, or additionally for total monounsaturated and total polyunsaturated fat. The polyunsaturated content may be broken down further into total ω6 and ω3, and the ω3 may be further broken down into individual fatty acids, DHA, EPA, and ALA. Irrespective of the requirements, all these labels would require a fatty acid profile but, with increasing detail, increasing discriminatory power between individual components would be necessary.

In most respects, the standard approach taken to analyze the fatty acids of functional foods is similar to that for conventional foods. The steps are to extract the total lipids or fatty acids, convert the fatty acids to a suitable derivative (often to fatty acid methyl esters (FAME)), and analyze the derivatized fatty acids by a suitable chromatographic technique, usually GC with flame ionization detection (FID), Other chromatographic techniques, including gas chromatography–mass spectrometry (GC-MS) and high-performance liquid chromatography (HPLC), may be required. Nonchromatographic techniques such as infrared (IR) spectroscopy may be used in some situations, perhaps because of the speed of analysis.

The choice of extraction method will be influenced by the type of food matrix and the types of lipids present. The analysis method will depend on the types and complexities of the fatty acid mixtures as well as the nature and degree of information required. In conventional foods, the fatty acids will normally be mainly in an esterified form as triacylglycerols and phospholipids. In functional foods, the fatty acids may also be in these forms, but it is possible that they may have been incorporated in other less natural forms such as alkyl (e.g., ethyl) esters, free fatty acids (FFA), or as microencapsulated oils. The analyst should bear in mind that the form in which the fatty acids occur may influence the analytical approach.

2.3.2 Extraction

Care should be taken to minimize changes in the fatty acid composition of the food before analysis. Lipids are susceptible to enzymic lipolysis and UFA are susceptible to enzymic and nonenzymic oxidation. Therefore, food samples should be kept at low temperatures (−20°C or lower), preferably under an atmosphere of nitrogen or in an oxygen-impermeable wrap, before extraction.

Two basic approaches can be taken during extraction. If only the fatty acid composition of the sample is required, as is our concern in this chapter, then the sample can be directly hydrolyzed to recover the total FFA (or even transesterified directly to produce FAME, as described in Section 2.3.3.1.4). The other approach is to extract the total lipids and then in a second step either transesterity the lipids to FAME, or hydrolyze them and then methylate the FFA.

2.3.2.1 Hydrolysis Followed by Recovery of Free Fatty Acids

In the United States, the total fat is defined as the total fatty acids expressed as triacylglycerols. Therefore, procedures for releasing fatty acids for total fat determination may be applicable to analysis of fatty acids *per se*. In the Association of Official Analytical Chemists (AOAC) official method 996.06 (AOAC, 1995), the sample is digested either with hydrochloric acid, to hydrolyze lipids and also carbohydrate and protein with which lipid material may be associated with, or with ammonia for dairy products, in the presence of an internal standard (triundecanoin). The fatty acids are then extracted with a petroleum ether–diethyl ether solvent mixture. As pointed out by Ackman (1999), the potential drawback of using hydrochloric acid is that not all lipids may be hydrolyzed totally to FFA. Presumably, this does not matter if an appropriate fatty acid derivatization procedure for FFA and esterified fatty acids (EFA) is subsequently employed (see Section 2.3.3.1), as long as all lipids are extracted, but this may be uncertain by employing a relatively nonpolar solvent mixture. Additionally, in view of the problems of artifact formation under acidic conditions during methylation with some fatty acids, including CLA (see Section 2.3.3.1.2), the method may not be the first choice for some applications.

A number of gravimetric or volumetric methods are commonly used for determining the total fat and oil content of a range of different samples. The Gerber method (suitable for milk, milk products, and cheese), the Werner–Schmid method (suitable for foods of low-sugar content e.g., cheese, meat, mayonnaise, and cereal products), and the Weilbull Stoldt method (suitable for a wide range of foods including those of high-sugar content such as chocolate products) all use a strong acidic hydrolysis procedure (McLean and Drake, 2002) and are therefore not ideally suited for obtaining a sample for subsequent fatty acid analysis.

A more universal hydrolysis method is to treat the sample with alkali. Typically, the sample is refluxed with 0.1–1 M ethanolic potassium hydroxide for 1 h. A butanolic alkali method (Gertz and Fiebeg, 2000) has been adopted as a German standard method. Following hydrolysis, nonfatty acid components, if required, can be recovered by extracting several times with an appropriate solvent (e.g., isohexane-diethyl ether, 1:1 by volume). After acidification with dilute hydrochloric acid, the FFA can be extracted with solvents such as diethyl ether. Multiple extractions with solvent will ensure complete recovery of fatty acids especially if shorter chain fatty acids (C_{12} and below), that have a greater affinity for aqueous mixtures than longer chain acids, are present. Alkaline hydrolysis may result in the formation of small amounts of fatty acid ethyl esters, particularly if the ratio of ethanol to water is too high, but again these will be subsequently converted to FAME along with the FFA if an appropriate derivatization procedure is employed. The ethyl esters, however, will be lost if the sample is extracted before acidification to recover nonsaponifiables such as sterols. So this step should be avoided, if possible.

In the experience of our laboratory, compared to methods involving extraction of total lipids, alkaline hydrolysis can give better recovery of FFA, at least in certain types of samples particularly powders (baby formula, soup, drinks), probably because of the strong association of lipid with other components and possibly the presence of lipid in a microencapsulated form. If the form (free or esterified) in which the fatty acids occur in the food is unknown, then it may be advisable to hydrolyze the sample so that all fatty acids are in the free form, and then use an appropriate derivatization method (see Section 2.3.3), although hydrolysis could also be done after lipid extraction.

2.3.2.2 Extraction of Intact Total Lipids

The total lipids may be extracted from a sample, perhaps because, in addition to the total fatty acids, the lipid composition is to be analyzed. There are many extraction methods available and further information may be obtained from reviews (Ackman, 1999; Christie, 1989, 2003; Shahidi and Wanasundara, 1997). The reader is advised to seek out further methods in the literature relevant to the type of sample under investigation. Some general guidelines will be given below. Lipids are distinguished from other macronutrients by being soluble in organic solvents and insoluble in water. However, lipids are diverse in structure and cover a wide range of polarities and therefore a single extraction procedure may not extract all lipids. Neutral lipids, such as triacylglycerols, are bound hydrophobically and can be extracted with nonpolar solvents such as hexane or diethyl ether, but most foods contain large amounts of water which reduces the effectiveness of diethyl ether, which is hygroscopic and becomes saturated with water resulting in incomplete extraction. Polar lipids such as phospholipids, the major lipids of egg yolk for example, are held by stronger bonds and more polar solvents are required for extraction.

Chloroform/methanol in a ratio of 2:1 (by volume) as in the procedure of Folch, Lees, and Stanley (1957) is generally recognized as an appropriate solvent mixture for extraction of most lipids. In a comparison of various methods for extracting total lipids from different food products, a chloroform–methanol procedure was found to be the most effective (Hubbard et al., 1977). In practice, the extraction mixture also often contains water depending on the amount in the material to be extracted. The Bligh and Dyer procedure (1959) was developed for tissue samples that contain considerable amounts of water and takes into account the amount of water in the sample. We have found that this procedure is suitable for extracting aqueous samples such as drinking yogurts.

In the Folch method (Folch, Lees, and Stanley, 1957), the sample is homogenized with chloroform/methanol (2:1, v/v) and the extraction is usually repeated with fresh solvent to ensure complete extraction of lipids. The ratio of chloroform to methanol may have to be altered in the initial extraction, depending on the amount of water in the sample, to obtain a monophasic system. After extraction, nonlipid contaminants (e.g., sugars, amino acids) are removed by adjusting the ratio of chloroform/methanol/water (0.88% potassium chloride) to 8:4:3 (by volume) to give a biphasic system. After vigorous shaking and allowing the layers to settle, the upper aqueous layer is discarded and the lower organic layer, containing the lipids, is evaporated to dryness. When other extraction methods are used, the Folch wash is often applied to remove nonlipid contaminants; the extracting solvent is evaporated off and the residue is suspended in chloroform/methanol/water, 8:4:3.

Even the Folch (Folch, Lees, and Stanley, 1957) and Bligh and Dyer (1959) methods may not extract all lipids. For example, polar-acidic phospholipids may not be totally recovered in the Folch washing procedure, but in the Bligh and Dyer method, where a more polar organic phase is employed, only very polar lipids such as phosphoinositides may be lost but incomplete recovery of triacylglycerols may result. So it is apparent that no single extraction method can recover all possible lipids. Other extraction procedures include the use of diethyl ether and petroleum ether for the extraction of dairy products (Sehat et al., 1998a), in which by for triacylglycerols are the major lipid class, and water-saturated *n*-butanol for the extraction of materials high in starch (Morrison, Tan, and Hargin, 1980).

There are a range of more modern methods for extracting lipids, and although they may offer certain advantages over more traditional methods, they have not been subjected to the same degree of testing over the range of sample types. Supercritical fluid extraction (SFE) uses supercritical carbon dioxide as extracting fluid at low temperatures and is therefore a mild method (Carrapiso and Garcia, 2000). Cost of equipment is an obvious drawback but there are other potential problems.

In materials in which there are strong interactions between the lipids and the matrix, methods involving alkaline or acid pretreatment may be required. Extractability decreases with samples of high water content and therefore freeze-drying may be required before extraction. Incomplete extraction particularly of polar lipids can be overcome to some extent by including polar co-solvents such as ethanol. Extraction of nonfat material, including water, can also be a problem. In spite of the potential problems, it has been shown for several foods, including meat and some bakery and dairy products, that extraction is as efficient as conventional methods.

There are a number of accelerated solvent extraction procedures which offer a more rapid extraction and the use of less solvent than conventional extraction methods. Microwave-assisted extraction is applicable to solid foods and has been shown to give comparable efficiencies to traditional hydrolysis/extraction methods when applied to meat, milk, and egg powders (Carrapiso and Garcia, 2000). A focused microwave Soxhlet extraction method was used to extract fat from bakery products (Priego-Capote, Ruiz-Jimenez, and Luque de Castro, 2007) and hydrolyze/extract fat from cheese (Garcia-Ayuso et al., 1999), for subsequent analysis of the fatty acids, and gave results that were comparable or even better than Folch and Weinbull-Berntrop methods, respectively. Dynamic ultrasound-assisted extraction of bakery products also gave comparable results to the Folch method (Ruiz-Jimenez, Priego-Capote, and Luque de Castro, 2004). Pressurized solvent extractions are possible at higher temperatures than normal, resulting in rapid extractions. The method has been applied successfully to fish (Dodds et al., 2004), and when applied to poultry meat it gave comparable recoveries and data quality (determined by fatty acid profile) to the Folch method, although the recoveries were slightly less than an acid hydrolysis method (Toschi et al., 2003).

Once a lipid or fatty acid fraction has been obtained, care is needed to minimize oxidation. Samples should be kept at low temperatures (−20°C in freezer) in the presence of an inert atmosphere (usually by flushing tubes with nitrogen) and dissolved in nonpolar solvents (e.g., isohexane) containing an antioxidant (e.g., butylated hydroxytoluene) at an appropriate concentration (about 0.005%).

2.3.3 Derivatization of Fatty Acids

2.3.3.1 Methyl Esters

2.3.3.1.1 General Procedures

Fatty acids are normally analyzed by GC as FAME. These volatile derivatives have favorable chromatographic properties and are easily prepared (Brondz, 2002; Christie, 2003; Rosenfeld, 2002). Broadly, there are two types of common derivatizing reagent: acid and alkaline. Acid reagents, including sulfuric acid, hydrochloric acid, boron trifluoride, and acetyl chloride in methanol, will catalyze the formation of methyl esters of both EFA (e.g., triacylglycerols and phospholipids) and FFA (normally only present in small amounts, but formed when samples are subjected to hydrolysis treatments). On the other hand, alkaline reagents, including potassium hydroxide/methanol and sodium methoxide, transesterify EFA only and do not react with FFA, *N*-acyl lipids such as sphingomyelin or alk-1-enyl ethers as in plasmalogens. Most fatty acids including LC-PUFA can be treated with either reagents with no adverse effects. However, some fatty acids, including CLA (see Section 2.3.3.1.2), form artifacts to varying degrees, depending on the type of reagent and the reaction conditions, with acid reagents.

For samples in which it is known that only EFA are present, sodium methoxide treatment can be used. The sample, in toluene (not necessary if only phospholipids are present), is heated at 50°C for 10 min with 0.5 M sodium methoxide in anhydrous methanol. If cholesterol esters are present then a longer reaction time of about 30 min is required. The reaction is stopped with acetic acid, saturated sodium chloride is added, and the FAME are extracted twice with isohexane. The isohexane extracts are dried over anhydrous sodium sulfate and the solvent is removed under nitrogen or on a rotary evaporator.

Samples containing FFA, or when the form of the fatty acids is unknown, can be methylated with 1% sulfuric acid in dry methanol by leaving overnight at 50°C. Again, toluene is included in the reaction mixture when nonpolar lipids, such as triacylglycerols, are present, but is not necessary for FFA or phospholipids. Sodium chloride solution (5%) is added and the FAME are extracted twice with isohexane. The isohexane extract is then washed with sodium hydrogen carbonate solution (2%) to remove traces of acid, and then dried and evaporated as above.

Boron trifluoride/methanol (12%–14%, w/v) is a widely used acid catalyst, but it should be noted that the use of old reagent or solutions that are too concentrated can result in production of artifacts and loss of PUFA (Christie, 1989, 2003) The official methods, American Oil Chemists' Society (AOCS) method Ce 1b-89 (American Oil Chemists' Society, 2005) and AOAC method 991.39 (AOAC, 1995), for analysis of marine oils use 0.5 M sodium hydroxide to hydrolyze the sample to release FFA that are then reacted with boron trifluoride/methanol.

2.3.3.1.2 Special Approaches for Conjugated Linoleic Acid

Methylation procedures for use with CLA have been the subject of extensive investigation (Aldai et al., 2005; Cruz-Hernandez et al., 2004; Yurawecz, Kramer, and Ku, 1999; Kramer and Zhou, 2001). CLA, if totally present in an esterified form, can be converted to FAME with alkaline reagents such as sodium methoxide (e.g., 0.5 M, 50°C, 15 min) or potassium hydroxide/methanol (e.g., 0.2 M, 50°C, 15 min) without causing any changes. However, under acidic conditions, the major *cis-trans* isomers may isomerize to *trans-trans* isomers and allylic methoxy artifacts may also be formed (Kramer et al., 1997; Ostrowska et al., 2000; Shantha, Decker, and Hennig, 1993; Werner, Luedecke and Schultz, 1992; Yamasaki et al., 1999; Yurawecz, Kramer, and Ku, 1999). Acid-catalyzed methylation can also form additional CLA by reacting with allylic hydroxy oleates, oxidation products of oleate, present in foods (Yurawecz et al., 1994).

It has been reported that isomerization/artifact formation was unacceptable with HCl/methanol at 80°C–100°C for up to 60 min (Kramer et al., 1997; Ostrowska et al., 2000; Yamasaki et al., 1999). Milder conditions (e.g., 60°C for 20 min) produce minimal isomerization/artifacts (Park and Pariza, 1998), but phospholipids were not totally methylated (Kramer et al., 1997). Boron trifluoride/methanol has been used under a variety of conditions but isomerization/artifact formation was apparent, certainly at temperatures of 70°C and above after only a few minutes (Kramer et al., 1997; Ostrowska et al., 2000; Shantha, Decker, and Hennig, 1993; Werner, Luedecke, and Schultz, 1992; Yamasaki et al., 1999). Increasing, harsh acid conditions are required to release fatty acids as methyl esters from FFA, triacylglycerols, phospholipids, alk-1-enyl ethers in plasmalogens (released as aldehydes), and *N*-acyl lipids such as sphingomyelin. It seems that with the exception of FFA it is not possible to completely methylate the other forms without causing some degree of isomerization/artifact formation. Sulfuric acid/methanol under mild conditions (1 N H_2SO_4, 55°C, 5 min) was reported to completely methylate free CLA without problems (Park et al., 2002). In our laboratory, we use 1% or 0.36 N sulfuric acid (50°C, up to 60 min) without problems. Commercial CLA is often in the FFA form, and can be methylated using these conditions, as can FFA (containing CLA) that has been extracted by alkaline hydrolysis.

CLA probably exists mainly in an esterified form in most foods. However, there may be some free CLA and in meat products some may be incorporated into alk-1-enyl ethers in plasmalogens and *N*-acyl lipids. The levels of free CLA are likely to be low unless the product has degraded or it has not been extracted appropriately. It is possible that foods already containing CLA in an esterified form may be enriched with the readily available free from. However, this is unlikely considering the poor organoleptic properties of this form and also the availability of the triacylglycerol form (Fernie et al., 2004; Kovacs and Mela, 2006). There is a lack of knowledge on the amount of CLA incorporated into alk-1-enyl ethers in plasmalogens and *N*-acyl lipids. However, there may be a requirement to ensure that CLA, in all possible forms, is methylated, while minimizing alterations in the CLA. This is a difficult proposition. A number of alternative reagents have been looked at for methylating CLA and it is worth briefly considering their suitability.

Tetramethylguanidine (TMG) in methanol has been reported to be the only base-catalyzed reagent that can react with both FFA and EFA (Schuhardt and Lopes, 1988). However, although isomerization has not been reported, there have been problems with methylating FFA (Shantha, Decker and Hennig, 1993), incomplete derivatization of esterified forms (Kramer et al., 1997), losses of *trans-trans* isomers (Ostrowska et al., 2000), and formation of compounds that interfered with the GC analysis of low molecular weight FAME present in milk (Kramer et al., 1997). It seems that strict anhydrous conditions are essential for methylating FFA and, as an alternative to other alkali reagents, TMG is only possibly suitable for transesterifying esterified CLA. Park et al. (2002) favor the use of TMG (20% TMG/methanol, 100°C, 5 min) either on its own for samples containing only triacylglycerols or in combination with sulfuric acid/methanol (1 N sulfuric acid, 55°C, 5 min) for samples containing phospholipids with or without triacylglycerols (e.g., egg yolk).

Trimethylsilyldiazomethane (TMS-DM) will form methyl esters of FFA only, and has been used in several studies for analysis of commercial CLA (Sehat et al., 1998b; Yurawecz et al., 1998). However, although isomerization was minimal, several artifacts of unknown origin and identity were produced (Kramer et al., 1997; Park et al., 2001). It seems that artifact formation is minimal if sufficient methanol is present (Aldai et al., 2005). A procedure involving alkali hydrolysis (of extracted lipid or tissue) followed by TMS-DG treatment has been suggested (Aldai et al., 2005), but concerns with TMS-DG over quantification, possibly because of solubility problems, is a distinct disadvantage (Aldai et al., 2005; Park et al., 2001).

Different approaches are possible for methylating esterified CLA with CLA in other forms. For esterified CLA with free CLA, the extracted lipid could be treated with a mild alkaline hydrolysis procedure (e.g., using 0.1 M potassium hydroxide in 90% aqueous ethanol and leaving overnight at room temperature) to release the FFA, followed by a mild acid methylation procedure using sulfuric acid/methanol as described above. FFA extracts released directly from the food material by alkaline hydrolysis could be used, but care should be taken not to use conditions that are too harsh as some isomerization of the CLA might result. Alternatively, the lipids could be treated with an alkaline transesterification procedure, such as sodium methoxide followed by mild sulfuric acid/methanol treatment. To additionally release CLA as the methyl esters from *N*-acyl lipids such as sphingomyelin, stronger acid conditions such as hydrochloric acid/methanol at 80°C for 30 or 60 min have been recommended but the CLA will be substantially or even totally converted to *trans-trans* isomers (Cruz-Hernandez et al., 2004; Kramer and Zhou, 2001). Presumably, assuming methoxy artifacts are not formed under such conditions, this would give an estimate of the total CLA present. It was also recommended to treat another sample at the same time with sodium methoxide to obtain the amount of esterified CLA, together with the isomer distribution. Hydrolysis of sphingomyelin to produce FFA will be incomplete under mild alkaline hydrolysis conditions, but is almost complete with prolonged alkaline treatment (1 M potassium hydroxide for 10 h), and under acidic conditions (e.g., 1.5% hydrochloric acid at 70°C for 4 h) (Christie, 1989). It would be worthwhile to determine the extent of isomerization/artifact formation under such conditions.

2.3.3.1.3 Samples Containing Short-Chain Fatty Acids

Samples such as dairy products that contain SCFA (C_4–C_{10}) need special treatment during derivatization because of the high volatility and solubility in aqueous solutions (Christie, 2003). Derivatized samples should be extracted with the required volume of solvent for subsequent analysis by GC, so that volume reduction by evaporation of solvent is avoided to reduce losses of SCFA esters. Solvents that are more polar than hexane or isohexane, such as diethyl ether, may be used for extraction, and multiple extraction of the aqueous phase (which may contain a high proportion of salt) may be performed, to minimize losses. Methods are available in which there are no aqueous extraction or solvent removal steps, and in which the reaction is carried out at room temperature.

Derivatives, other than methyl esters, namely isopropyl esters (Wolff, Bayard, and Fabien, 1995) and butyl esters (Iverson and Sheppard, 1986), have been used for dairy samples. Isopropyl esters have other advantages for quantitation of SCFA (see Section 2.3.4.2.1). These esters can be prepared by using procedures where methanol is replaced by isopropanol or butanol, for example using sulfuric acid or sodium isopropoxide/butoxide as catalyst. Isopropyl esters have been prepared from cheese fat using isopropanol/sulfuric acid to determine the CLA content (Lavillonniere et al., 1998). However, presumably there are the same potential problems as with other acidic methods for making CLA methyl esters.

2.3.3.1.4 Combined Extraction and Methylation

As an alternative to either hydrolyzing the sample or extracting the total lipids before forming FAME, rapid one-step *in situ* extraction/methylation methods have been developed. The success of such methods depends on the choice of solvent mixed with methanol to give effective solubilization of all lipids, on the choice of catalyst, and on the amount of water in the sample (Carrapiso and Garcia, 2000). Samples must either contain low levels of water or the water must be removed to ensure that methylation is complete. One such procedure was developed for examining the fatty acid composition of processed foods (Ulberth and Henninger, 1992). The method involves freeze-drying the sample and treating with methanol-HCl/toluene to release the FAME. The method gave comparable results to those involving extraction with chloroform/methanol and acid hydrolysis-Soxhlet extraction.

In our laboratory, a similar approach is used for examining the fatty acids of egg yolk, and as an alternative to freeze-drying, water can be removed by mixing with anhydrous sodium sulfate. Freeze-drying is also a convenient way to treat heterogeneous samples such as meat before direct transesterification, as it is an effective way to homogenize the sample and therefore obtain a representative subsample. A similar method to that of Ulberth and Henninger (1992) but without removing the water from the samples was successfully used to transesterify pig adipose tissue (Carrapiso et al., 2000) and various fish tissues (Meier et al., 2006). In both cases, the recoveries were higher than Folch extraction, and the two methods gave similar fatty acid profiles. It is notable that fish samples contain a high proportion of water, but it is claimed that there are no problems as long as the water is less than about 10% of the methanol volume (Meier et al., 2006).

In a study of CLA in ovine muscle, a direct transesterification method using potassium hydroxide/methanol (0.2 M, 50°C,15 min) gave superior quantitative results to direct methods using either sodium methoxide or acid catalysts (hydrochloric acid, boron trifluoride) (Murrieta, Hess, and Rule, 2003). In a study in beef, aimed at analyzing all fatty acids, including CLA and ω3 and ω6 fatty acids, by GC, the samples were directly hydrolyzed (5 M potassium hydroxide in methanol/water, 1:1, v/v, 60°C, 60 min), followed by methylation with TMS-DM (Aldai et al., 2006).

2.3.3.2 Dimethyloxazoline Derivatives

Fatty acids are often examined by GC-MS as 4,4-dimethlyoxazoline (DMOX) derivatives (see Section 2.3.4.2.3) and are easily prepared by reacting the FFA or FAME/intact lipids with an excess of 2-amino-2-methyl-1-propanol to form an *N*-acyl intermediate which cyclizes at 190°C after 6 and 18 h, respectively. Addition of water is followed by extraction of the derivatives with diethyl ether/isohexane (1:1) and the extract is washed with water (Christie, Sebedio, and Juaneda, 2001). The products should be dried over anhydrous sodium sulfate to prevent ring opening. Although no adverse effects of using high temperatures for derivatizing CLA have been reported, a milder procedure, involving formation of the ethanolamide of the fatty acid followed by treatment with trifluoroacetic anhydride to cyclize the derivative, is available (Kuklev and Smith, 2003).

2.3.4 Qualitative and Quantitative Analysis of Fatty Acids

As already indicated, the analyst may not be interested in analyzing only the functional fatty acids, but may also require information on the other fatty acids for a variety of reasons. The complexity of the fatty acid mixture will depend not only on the presence of the functional component, for example fish oil, but also on other fatty acid components of the food. The fatty acid profile of fish oil is reasonably complex because of the range of chain lengths and degrees of unsaturation, and presence of positional isomers. The additional presence of PHVO (containing *trans* fatty acids), coconut or palm kernel oil (containing SCFA), or butterfat (containing *trans* fatty acids, SCFA, and CLA), for instance, would complicate the mixture further. The analyst may be faced with samples in which not all of the component parts are known.

2.3.4.1 Pretreatment

Following derivatization the sample often requires no further treatment before analysis. A common problem in animal FAME samples is the presence of cholesterol, which may interfere with a later run when the sample is analyzed by GC. Cholesterol and other relatively polar components can be removed by putting the sample on a small column or solid-phase extraction (SPE) cartridge of silica or Florisil, and eluting the FAME with hexane-diethyl ether (95:5, v/v). In some cases, a prefractionation step is included to concentrate minor FAME or separate FAME, that are poorly resolved by GC, in a complex mixture. This may involve reversed-phase (RP)-HPLC where fatty acids are separated according to chain length and degree of unsaturation, or silver-ion thin-layer chromatography (TLC) and HPLC where separation is according to number and geometry of double bonds. Although often the fatty acid profile of the total lipids is required, sometimes the fatty acids of individual lipid classes such as triacylglycerols or phospholipids are to be determined and in this case, before derivatization, the total lipids are fractionated by adsorption TLC or HPLC or column chromatography either on a large-scale or small-scale using SPE columns. The reader is referred to *Lipid Analysis* by Christie (2003) for details of fractionation procedures.

The standard approach is to analyze the FAME by GC and, fortunately, the majority of fatty acids can be separated using modern capillary columns. Although different types of columns [varying in type of chemical phase, length, internal diameter (i.d.), and phase thickness] may be required for different types of mixtures, in practice usually two different columns will suffice: a moderately polar polyethylene glycol or Carbowax type column (e.g., Carbowax-20M or CP-Wax 52*c*B) of medium length (about 25 m) and a long (100 m) polar cyanopropyl column (e.g., CP-Sil 88 or SP-2560).

2.3.4.2 ω-3 and Other Fatty Acids

2.3.4.2.1 Qualitative Analysis by Gas Chromatography of FAME

EPA and DHA are the major ω3 acids from fish oil, but ALA is a significant minor component and may also be derived from some vegetable oils, such as flaxseed, rapeseed, and soybean oils. Other ω3 PUFA that are present in fish oils at low but significant levels include 16:3, 16:4, SA, 20:4, and 22:5, and possibly 20:3. The ω6 acids in fish oils are at relatively low abundance. LA is the major ω6 acid, and may be derived from many other sources including vegetable oils, followed by AA. Other ω6 acids from fish oil include GLA, 20:2 and 20:3. The ω6 acids 22:3, 22:4, and 22:5 are negligible in fish oils but may be present in some foods, for example in some meat products. Other PUFA, such as 20 and 22 carbon ω9 acids, may be encountered at low levels. As well as PUFA, a range of SFA, and MUFA will be present. Palmitic and stearic acids are usually the most abundant SFA, and OA is the major MUFA, but a range of chain lengths and double-bond positional isomers (ω11, ω9, and ω7) may be encountered.

Separations of complex FAME mixtures, with chain lengths up to 24 carbons and up to six double bonds, including all ω3 and ω6 acids, can be achieved on capillary columns of polyethylene glycol or Carbowax type columns (usually of dimensions 25–30 m long, 0.25 mm i.d., and a film thickness of 0.2 μm) as recommended in the AOCS official method Ce 1b-89 (American Oil Chemists' Society, 2005). A GC profile of FAME from a fish oil/sunflower oil mixture on a column of this type is shown in Figure 2.3. These columns are preferred because of the stability, and the predictable order in which all compounds elute. An analysis on this type of column would normally be the first step for a sample of unknown composition. Typically in our laboratory, the following conditions are used: an

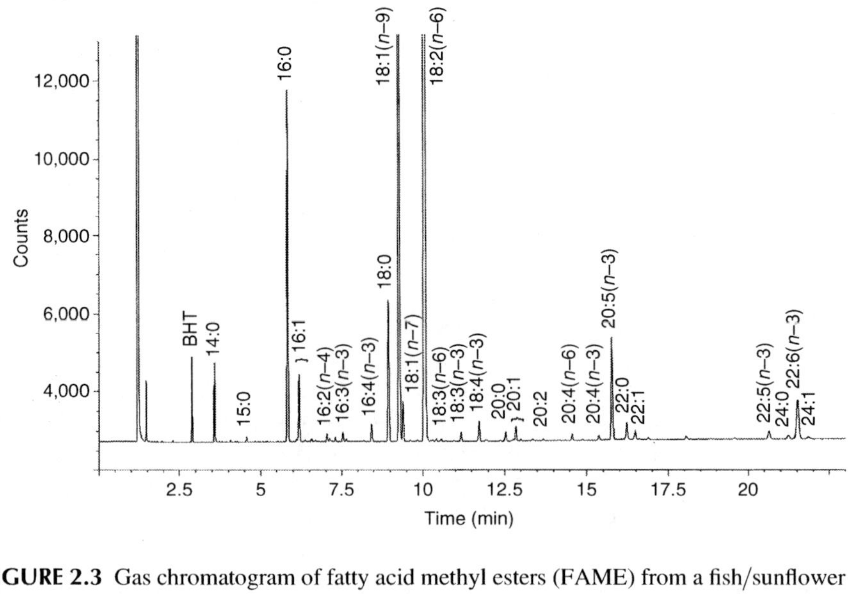

FIGURE 2.3 Gas chromatogram of fatty acid methyl esters (FAME) from a fish/sunflower oil mixture, separated on a CP-Wax 52*c*B capillary column (25 m × 0.25 mm i.d. × 0.2 μm film thickness; Chrompack BV, Middleburg, the Netherlands). The oven temperature was held at 170°C for 3 min, then was programed at 4°C/min to 220°C and held at 220°C for 8 min. Hydrogen was the carrier gas at a flow rate of 1 mL/min and a split ratio of 50:1 was used.

initial oven temperature of 170°C for 3 min, followed by an increase to 220°C at a rate of 4°C/min and then holding this temperature for 15 min, hydrogen as carrier gas at an initial flow rate of 1 mL/min, and a split ratio 50:1. Typically, 1 μL of sample at a concentration of about 3 mg/mL in isohexane (containing 0.005%, w/v, butylated hydroxytoluene as an antioxidant) is injected.

Shorter chain esters elute before longer chain esters and, for a given chain length, retention time increases with increasing degree of unsaturation. All FAME of a given chain length elute before all those of chain length two carbons longer, the only exception is that DHA elutes between 24:0 and 24:1ω9 (as the column ages, 24:0 and DHA can overlap). Double-bond positional isomers are also separated. The isomer with double bonds nearer the ester group elutes first, that is 20:4ω6 has a shorter retention time than 20:4ω3. Therefore, the order of elution for C_{20} esters would be SA $<$ 20:0 $<$ 20:1ω11 $<$ 20:1ω9 $<$ 20:1ω7 $<$ 20:2ω6 $<$ 20:3ω6 $<$ 20:3ω3 $<$ AA $<$ 20:4ω3 $<$ EPA $<$ 22:0. There is the possibility of overlap between 21:5ω3 (present in some marine oils) and 23:0 (often used as internal standard (IS), see Section 2.3.4.2.2) which would lead to inaccuracies, and separation of these two fatty acids is a requirement of the AOCS method. By using appropriate standards the identification of peaks can be verified by comparison of retention times. With some exceptions (e.g., *trans* [see Section 2.3.4.3] and SCFA [see below]), a fish oil FAME standard will contain the majority of components likely to be encountered in food samples. Synthetic FAME standards, that mimic the composition of fish oils, are readily available.

It is interesting that in an analysis of the C_{20} and C_{22} PUFA in milk from cows fed a fish meal diet, the pairs EPA/22:0 and 22:5ω3/24:0 co-eluted and DHA eluted after 24:1ω9 on a 60-m Supelcowax 10 column, although there had previously been no problems on a 30-m column of the same type (Kramer, Blackadar, and Zhou, 2002). Normally, separations would be expected to improve on longer columns.

From the GC analysis, each FAME can be expressed as a percentage of the total FAME. The areas under the GC peaks are approximately equivalent to the masses of the FAME they represent. Therefore, the area percent of each peak (as a percentage of the total areas of all FAME peaks) is approximately equal to the weight percent of each FAME. However, using this approach, the longer chain FAME are overestimated and shorter chain FAME are underestimated. The area percent values can be multiplied by a theoretical correction factor (TRF) (tables are published; e.g., see Christie, 1989) and values renormalized to give the true weight percent values. The corrections will only make small differences over chain lengths C_{14}–C_{24}, but will make significant differences for shorter chain lengths and may even give inaccurate results for C_4–C_8 FAME.

If SCFA are present (indicated by a series of peaks eluting just after the solvent front), as in dairy fats, then GC conditions should be adjusted to include a low-temperature step (often 70°C or lower), before programing up to the normal starting temperature. In any case, dairy fats contain *trans* fatty acids and CLA and long cyanopropyl columns are preferable for their analysis (see Section 2.3.4.3). Instead of FAME, the use of other esters, particularly isopropyl esters, should be considered for analyzing SCFA, particularly if the samples contain dairy fats. Details on analysis of fatty acids from dairy fats using isopropyl esters can be found in the publications of Wolff and colleagues (Wolff, 1995; Wolff, Bayard, and Fabien, 1995). One of the reasons for analyzing such samples as isopropyl esters is that, irrespective of chain length, areas are equivalent to masses without the need for correction factors (Wolff, Bayard, and Fabien, 1995). The analyst need only use specific methods for SCFA, and adhere to any necessary precautions, if (1) any (SCFA or otherwise) or all fatty acids are to be expressed as a percentage of the total fatty acids or (2) the concentrations, in mg/g of fat or sample, are required for SCFA. When information is required only on the concentration of specific fatty acids (e.g., ω3 acids), other than SCFA, then the SCFA can be ignored and special treatment is not required.

Alternative GC approaches are being developed to improve the identification, resolution, and quantification of FAME, including ω3 PUFA, in complex lipid samples. Taking advantage of the fact that the polarity of cyanopropyl columns, and therefore the elution characteristics of FAME, can vary with temperature, the shifts in equivalent chain length (ECL) values of FAME on a 70-m BPX70 (70% cyanopropyl) column under five different analytical conditions were used to identify, with high accuracy using multivariate methods, the PUFA in fish oils and geometrical isomers of 18:1, 18:2, and 18:3 (Mjos, 2003).

The 2D GC (two-dimensional GC, GC × GC) is an emerging technique in which analytes are separated on the first column and then cryogenically trapped before they are desorbed on to a second column of different polarity. The first column is usually of similar dimensions (e.g., 20–100 m long and 0.22–0.25 mm i.d.) to conventional 1D (one-dimensional) columns, resulting in retention times of normal values. The second column is considerably shorter (about 1 m) and narrower (narrow bore, 0.05–0.1 mm) and analytes are only retained for seconds. This results in greater resolving power of analytes and increased sensitivity. The method is especially applicable to complex mixtures in which some analytes overlap when analyzed by 1D GC.

The chosen arrangement for separating complex FAME mixtures has involved a nonpolar column, where separations are primarily due to volatility, followed by a more polar column, where separations occur according to polar interactions (Hyotylainen et al., 2004; Western et al., 2002). In one study (Western et al., 2002), a marine oil (blue mussel) was analyzed on a 25-m BP1 (100% polydimethylsiloxane) column followed by a 1-m BPX70 column and there was greater resolution and sensitivity, and improved quantitation, compared to a 1D system. Identifications were based on retention times compared to standards, although it was noted that the 2D retention map was a useful aid to tentative identifications in the absence of appropriate standards. The second study (Hyotylainen et al., 2004) involved the analysis of milk FAME of a certified reference material on a 30-m HP-1 column followed by a 0.35-m DB-Wax column. Again improved separations were achieved but only some compounds were identified due to lack of suitable standards. With the exception of low molecular weight FAME, good quantitative agreement was obtained for the few compounds that had certified levels.

Fatty acids have been analyzed as TMS esters, using GC on nonpolar columns, in a variety of sample types, including edible oils (Eras et al., 2001). The separations are not as good as FAME on polar columns, and TMS esters are susceptible to hydrolysis if low levels of moisture are present.

2.3.4.2.2 Quantitative Analysis by Gas Chromatography of FAME

To determine the concentration of a fatty acid in terms of milligram per gram of sample a known mass of an IS is added to the sample. The IS should be added at the beginning of the extraction procedure (but may be added to a proportion of the extracted fat if the fat content has been determined accurately, gravimetrically), preferably in a form (e.g., triacylglycerol) similar to that in which the bulk of fatty acids occur in the sample, although sometimes a FAME IS is used. It must be a fatty acid that is either not present in the sample or is at negligible levels, and it should be added at a level so that it represents an average peak size in the GC analysis. Usually an IS containing an odd-chain fatty acid is selected, but the analyst should be aware that odd-chain fatty acids are not always absent, and some can be significant components of dairy products for example. The chain length of the IS should be representative of the fatty acids of interest within a sample; triundecanoin (containing 11:0) may be suitable for smaller chain lengths in a milk fat, but 23:0 methyl ester is more suitable for determining EPA and DHA. It is advisable to carry out the analysis in triplicate, and add the IS to two of the samples, but not to the third which can serve as a check that the fatty acid component of the IS is not present in the sample (or if it is at low levels then it can be allowed for). It is worth noting that even when an IS is added at the beginning of extraction, it will not allow for an inefficient extraction process, because only the lipid in the sample (and not the IS) will be bound in the food matrix. However, the IS may allow for losses in subsequent steps (e.g., washing of organic phase, methylation), certainly if it is structurally similar to the lipids in the sample.

The amount of a FAME in a sample analyzed by GC as FAME can be calculated from the following equation:

$$\text{FAME (mg/g)} = \frac{(A_x)(W_{\text{IS}})(\text{CF}_x)}{(A_{\text{IS}})(W_{\text{S}})} \times 1000$$

where

A_x = Area for FAME x
A_{IS} = Area for internal standard
W_{IS} = Weight of internal standard added to the sample, in mg
W_{S} = Weight of sample, in mg
CF_x = TRF of FAME x relative to that of the IS, for example, a value of 0.98 for EPA relative to 23:0 as IS

The above equation assumes that the IS was added as a FAME. If it were added as a triglyceride then the equation would also give the concentration of x as triglyceride. Indeed in the United States, total fat is expressed as sum of triglyceride equivalents. However, if the IS was added as a FAME but the concentration is required in triglyceride equivalents, then the equation would include another correction factor on the denominator. The correction factor would be calculated by dividing three times the molecular weight of the IS as FAME by the molecular weight of the IS as triglyceride. In practice, the correction will be small, for example, the value is 1.004 for EPA, with 23:0 methyl ester as IS.

Performing the calculation for every fatty acid will then allow the amount of any chosen group of fatty acids to be calculated. For example, the total PUFA or total ω3 can be calculated simply by summing the relevant values for individual FAME. In addition, the ratio of ω6/ω3 can be determined.

The AOCS official method Ce 1b-89 for quantifying fatty acids in marine oils (American Oil Chemists' Society, 2005) has a mandatory requirement for a 23:0 IS, a split injection at 50:1 is recommended, and application of TRF is suggested for optimum accuracy. Schreiner (2005) has evaluated the accuracy of measuring ω3 LC-PUFA using the official method and concluded that, although the use of a 23:0 IS and application of TRF gave accurate results with an on-column injection system, the results were significantly less accurate using a split injection because of discrimination in the injector. Accuracy was improved considerably when 19:0 and 21:0 were used as IS for C_{20} and C_{22} PUFA, respectively. Although 19:0 and 21:0 are usually avoided because they are in crowded parts of the chromatogram, it was found that by optimizing analytical conditions 19:0 could elute between GLA and ALA, and 21:0 between 20:3ω6 and AA. Because odd-chain fatty acids may be present at low levels in some foods, the use of a range of ethyl ester and d_3-methyl esters has been suggested as alternative IS, for use with FID (ethyl ester only) and MS (including selected-ion monitoring [SIM] mode) detectors, for quantifying fatty acids in foods (Thurnhofer and Vetter, 2006).

2.3.4.2.3 Complementary and Alternative Approaches

Identification of fatty acids by GC, using appropriate standards, is often adequate for most purposes. Sometimes, GC-MS in electron impact (EI) mode is used to unambiguously verify the identity of a fatty acid. Low-bleed Carbowax columns (e.g., Supelcowax 10) can be used for this purpose. Methyl esters will normally only give molecular weight information, so that the chain length and degree of unsaturation can be verified. To determine the exact structure of the fatty acid, including position of the double bonds, other derivatives are used. DMOX derivatives are often used for this purpose (Dobson and Christie, 1996). DMOX derivatives have similar chromatographic properties to methyl esters, eluting slightly later, and resolution of peaks is usually retained. In the mass spectra, simple radical-induced cleavage occurs at every carbon in the chain, resulting in gaps of 14 amu between adjacent methylene groups in a saturated chain, and a gap of 12 amu when a double bond is encountered. For example, in the mass spectrum of the DMOX derivative of EPA (Figure 2.4), gaps of 12 amu between m/z 300 and 312, 260 and 272, 220 and 232, and 180 and 192 establish double bonds at Δ17, Δ14, Δ11, and Δ8. The odd-numbered mass at m/z 153 is characteristic for the remaining double bond at Δ5.

By selecting specific ions, GC-MS can be useful for detecting co-eluting analytes that might be identified as a single component by GC-FID. A GC-MS method using SIM was developed for determining the percent fatty acid profile, including SFA, MUFA, PUFA, and branched FA, in foods (Thurnhofer and Vetter, 2005). Although the method was generally inferior to GC-FID, it was superior for detecting minor fatty acids and co-eluting branched FA and MUFA (a problem in milk samples) which would be difficult to determine by GC-FID without a silver-ion TLC or HPLC prefractionation step.

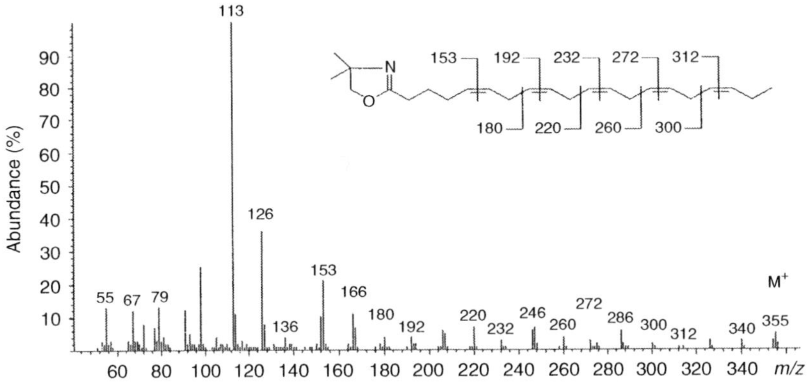

FIGURE 2.4 Gas chromatography–electron impact (GC-EI) mass spectrum of the 4,4-dimethyloxazoline (DMOX) derivative of eicosapentaenoic acid (EPA).

A Fourier transform infrared (FTIR) spectrometer can also be used as a GC detector (Mossoba et al., 2001) and is particularly useful for distinguishing between *trans* and *cis* double bonds. IR spectroscopy, encompassing mid-IR, near-IR (NIR), and Raman spectroscopy, on its own is also a useful tool in lipid analysis (Van de Voort, Sedman, and Russin, 2001). Application to analyzing *trans* fatty acids is mentioned in Section 2.3.4.3.2. More recent methods using IR, NIR, or Raman spectroscopy have used multivariate techniques, particularly partial least squares (PLS) regression, to develop calibration models for measuring a variety of fatty acid quality parameters. A method using FT-NIR for determining iodine value (a measure of the degree of unsaturation in the fatty acids of oils and fats) is now an AOCS official method.

Recently, attempts have been made to use NIR, coupled with multivariate approaches, in rapid methods for determining the fatty acid composition of different materials. To date, only materials with simple fatty acid compositions, such as oilseeds (Kim, Park, and Choung, 2007) and forages (Foster, Clapham, and Fedders, 2006) have been studied, and it seems that at the moment only the levels of major fatty acids can be predicted with reasonable accuracy, compared to GC determinations. Such approaches might have applications as rapid screens in plant breeding, for example, where a high degree of accuracy is not required. However, with further refinement, there may be applications for rapid analysis of more complex fatty acid compositions in food, for example. One possible drawback of the method would be the effort that would have to be put into establishing product-specific calibration models.

HPLC methods have been developed for analyzing EFA and FFA, but the main application is in sensitive analysis of fatty acids in biological materials such as plasma (Brondz, 2002; Chen and Chuang, 2002; Lima and Abdalla, 2002; Miwa, 2002). The carboxylic acid group can be derivatized with a variety of groups to give sensitive UV or fluorescence detection, and the response is often linear over a wide range. FFA can be detected using the universal evaporative light-scattering detector

(ELSD), but in this case, careful calibration is required for quantification. A variety of MS detectors (quadrupole, ion trap, time-of-flight) using different ionization techniques (electrospray [ESI], atmospheric pressure chemical ionization [APCI]) have also been used for analyzing fatty acids (Lima and Abdulla, 2002). With suitable derivatives, detection by chemiluminescence (Ohba, Kuroda, and Nakashima, 2002) and under appropriate conditions electrochemical detection (Kotani, Kusu, and Takamura, 2002) are also possible.

RP-HPLC is the commonest form of HPLC for analyzing fatty acids, and separations differ from GC in that unsaturation have a much greater effect on elution time (each double bond decreasing it by the equivalent of about two carbons, e.g., 16:0 and 18:1 elute closely), so that the method may be used as a complementary technique to GC. Separations of SFA, MUFA, and PUFA of equal quality to GC have been claimed with the use of 2-nitrophenylhydrazide derivatives (Chen and Chuang, 2002). Indeed, GC has limitations in the size of chain length that can be analyzed and highly unsaturated FA may be subjected to thermal degradation, but HPLC does not have these drawbacks, and uncommon very LCFA of around 30 carbons or more can be analyzed. However, the applications to food are limited.

Silver-ion HPLC is a powerful technique for separating fatty acids according to number, position, and geometry (*cis* double bonds are held stronger than *trans*) of double bonds (Dobson, Christie, and Nikolova-Damyanova, 1995) and has found application as a fractionation method for determining *trans* monoenes in PHVO and milk fats (see Section 2.3.4.3.1), and for simplifying complex fatty acid mixtures, such as those from milk fats, before GC analysis. It is also used to separate CLA isomers (see Section 2.3.4.4). Supercritical fluid chromatography (SFC), often used with an FID, is another technique that has been applied to the analysis of FAME and FFA, in fish and vegetable oils for example (Senorans and Ibanez, 2002).

^{1}H nuclear magnetic resonance (NMR) spectroscopy has been applied to the analysis of lipid extracts (Diehl, 2001) and, using integrals of signals derived from protons at or near to double bonds (and from the methylene group α to the carbonyl for calibration), the percent of the total fatty acids of saturates, monoenes, dienes, and so on can be determined. ^{13}C NMR is more powerful and, using the signals from the terminal methyl region of an ultra high-resolved spectrum, total $\omega3$, $\omega6$, $\omega7$, $\omega9$, and SFA can be distinguished and quantified. Hence the total $\omega3$ as a percent of the total fatty acids and the ratio $\omega6$ to $\omega3$, for instance, can be calculated. Using the carbonyl signals, which are influenced by the distance of the double bonds from it, the different fatty acids at both the *sn*-1 and *sn*-2 positions in triacylglycerols and phospholipids can be distinguished. These ^{13}C NMR approaches have been used to analyze the fatty acids in lipid extracts of DHA-enriched eggs.

2.3.4.3 *Trans* Fatty Acids

A range of methods has been developed to identify and quantify *trans* isomers present in PHVO and milk fats, that have complex fatty acid compositions. Detailed descriptions of the methods involved are beyond the scope of this chapter (because *trans* fatty acids may be detrimental to health and therefore do not fall under the umbrella of a functional food fatty acid), although the absence or low levels of *trans* acids in a food can be considered as a positive nutritional attribute. Their analysis has been covered adequately elsewhere (Ackman, 1999; Cruz-Hernandez et al., 2004; Ledoux, Laloux, and Wolff, 2000; McDonald and Mossoba, 1996; Ratnayake, 1998, 2004; Wolff, Bayard and Fabien, 1995; Wolff, Precht and Molkentin, 1998), and an overview will be presented here with emphasis on recent developments.

2.3.4.3.1 *Analysis by Gas Chromatography*

Trans fatty acids are commonly analyzed by GC and on most columns the *trans*-18:1 esters (see Section 2.1.4) are poorly resolved from each other and from *cis*-18:1 esters, which occur as a range of positional isomers from 5*c*-18:1 to 16*c*-18:1, centered around OA in PHVO, and mainly OA with less 11*c*-18:1 in milk fat. Separations on long highly polar columns coated with phases containing either 100% cyanopropyl (e.g., CP-Sil 88 or SP-2560) or a high proportion of these groups (e.g., BPX-70 containing 70% cyanopropyl) are required. The 50-m columns do not give optimum separation of all components and therefore the 100-m columns are preferred. Using PHVO, Ratnayake et al. (2002) found that a column temperature of 180°C was optimum for separation of the earlier eluting *trans*-18:1 from *cis*-18:1 isomers on the 100-m columns of both CP-Sil 88 and SP-2560, and separation was slightly better on the SP-2560 column. All *trans*-18:1 isomers were separated with the exception that the 13*t*-18:1 and 14*t*-18:1 isomer pair overlapped with 6*c*-18:1, 7*c*-18:1, and 8*c*-18:1, and 15*t*-18:1 overlapped with the major *cis*-18:1 peak (mainly OA). *Trans* monoenes of other chain lengths (e.g., 16:1) occur as minor components, and mixtures of several *trans*-18:2 and -18:3 isomers eluting before LA and ALA respectively, can be separated and included in the total *trans* acids.

On cyanopropyl columns, in contrast to Carbowax columns, although retention time increases with increasing chain length and degree of unsaturation, there is considerable chain-length overlap. ALA and its *trans* isomers can overlap with 11*c*-20:1 and the separation is also affected by column temperature: 180°C was optimum for SP-2560 and 170°C for CP-Sil 88 (Ratnayake et al., 2002). Recently, conditions were optimized for both columns for AOCS official method Ce 1h-05 (American Oil Chemists' Society, 2005; Ratnayake, Hansen, and Kennedy, 2006). Similar columns, often under different conditions, have been used for analyzing *trans* fatty acids in milk fats (Ledoux, Laloux, and Wolff, 2000), and the methodology can be applied to lipid samples isolated from various foods. For example, the total *trans* fatty acid contents (and other fatty acids) in dairy powders and infant formulae were determined by GC on a 100-m CP-Sil88 column using total peak areas of *trans*-18:1 and -18:2 peaks relative to a methyl undecanoate (11:0) IS (Golay et al., 2006). Infant formulae may contain milk fat and deodorized vegetable oils. GC analysis on a 100-m CP-Sil 88 column indicated only a single *trans*-18:3 isomer (9*c*,12*c*,15*t*-18:3) as a marker of deodorized vegetable oils in such a matrix (Dionisi, Golay, and Fay, 2002).

Kramer and colleagues (Kramer, Blackadar, and Zhou, 2002) compared a 100-m CP Sil88 column and a 60-m Supelcowax 10 column for analysis of milk fat FAME, and the former column gave better overall separation. There was better separation of *trans*-18:1 and *cis*-18:1 positional isomers, *trans*- from *cis*-18:1 (and *trans*- from *cis*-14:1, 15:1, and 16:1), -18:2, and -18:3 isomers, CLA isomers, and PUFA. The drawbacks were poor separation of some SCFA, SFA, and MUFA (e.g., 10:1 and 11:0 co-eluted) and overlap of 18:3 and 20:1 isomers, all achievable on the Supelcowax 10 column. Both columns had the problem of branched fatty acids overlapping with monoenes. Fractionation by silver-ion TLC or HPLC would overcome all the co-elution problems.

A shorter (60 m) column of TC-70 (similar to BPX-70) has been used to determine the total *trans* content (*trans*-16:1, -18:1, 20:1, -18:2, and -18:3) in unhydrogenated oils and PHVO with acceptable results compared to the AOCS method Ce 1h-05 (American Oil Chemists' Society, 2005; Shirasawa et al., 2007). The analysis time was reduced from 74 to 40 min and although the same degree of information concerning the composition of individual isomers would not be obtained by this method, it is not important when only total *trans* fatty acids are required.

GC analyses are usually carried out using an FID, but sometimes GC-MS in EI mode has been used to aid identification of FAME. GC-MS has also been used for quantification, although MS detectors are generally regarded as not as quantitative as FID. Such an approach was used to analyze the *trans* fatty acids in bakery products (Priego-Capote, Ruiz-Jimenez, and Luque de Castro, 2007; Ruiz-Jimenez, Priego-Capote, and Luque de Castro, 2004).

GC analysis of a *trans*-monoene fraction, isolated by either silver-ion TLC or HPLC, provides an accurate assessment of the *trans*-monoene isomer distribution. A combined silver-ion TLC/HPLC and GC approach is considered to be the most accurate way for determining the total *trans*-18:1 content in PHVO and is also applicable to milk fat (Buchgraber and Ulberth, 2001). A *trans* plus saturated FAME fraction is isolated and, together with the unfractionated FAME sample, is analyzed by GC. Using 18:0 in both samples as an IS, the proportion of *trans* of the total monoenes, and hence the percent *trans* of total fatty acids, can be determined. A similar approach was used to determine the *trans*-20:1 and -22:1 content of partially hydrogenated fish oils (Wilson et al., 2000). As an alternative to silver-ion HPLC, RP-HPLC using two columns was used to isolate a trans-18:1 plus 16:0 fraction and the *trans*-18:1 content was subsequently determined by GC using 16:0 as an IS (Juaneda, 2002).

2.3.4.3.2 Alternative Methods for Determining Total Trans Fatty Acids

Alternative methods using IR and FTIR spectroscopy, based around the absorption of *trans* double bonds at 967 cm^{-1} are also used for determining total *trans* content. A long-established IR method for quantifying total *trans* (in PHVO, for example) is an official method (Cd 14–61) of the AOCS (American Oil Chemists' Society, 2005). The method has been improved and an FTIR method was adopted as the official AOCS method Cd 14d-99 (American Oil Chemists' Society, 2005). One drawback of the method was inaccuracies in determining the total *trans* content in biological matrices of low *trans* fat or low total fat content. There were interferences, particularly from CLA in dairy and ruminant meat products (Kramer et al., 1997). The method has been further improved by applying standard addition techniques that eliminate interferences from CLA, for example (Mossoba et al., 2001). This method improves the accuracy of measuring total *trans* in products low in total fat (<5%), low in *trans* fat (<10% of total fat) when interferences are observed, and in dairy and ruminant meat products which contain CLA. The authors (Mossoba et al., 2001) indicated that the method could be used for most foods to determine compliance with FDA (Food and Drug Administration) regulations that state that labeling as "saturated fat-free" can be claimed if foods contain less than 0.5 g saturated fat and less than 0.5 g *trans* fat per serving.

FT-NIR methods, using multivariate calibration models with statistical approaches such as PLS, have been developed to analyze the total *trans* fatty acid content of oils and fats (Azizian and Kramer, 2005; Li et al., 2000). Calibration models were developed using total *trans* determinations by the official AOCS FTIR method Cd 14d-99 (American Oil Chemists' Society, 2005; Li et al., 2000) and by GC (Azizian and Kramer, 2005). In one study (Li et al., 2000), models were developed for both a broad range of oil types and for restricted oil types, whereas in the other study (Azizian and Kramer, 2005), models were developed for different levels of *trans* fatty acids. Both studies produced results that agreed well with the official method. The use of product-specific calibrations was highlighted as a requirement for accurate predictions (Li et al., 2000). It remains to be seen whether such methodology can have application in accurately analyzing the *trans* fatty acid content of foods. GC can also be coupled to IR to give information on the *trans* content at each chain length (Mjos and Pettersen, 2003).

2.3.4.4 Conjugated Linoleic Acid

CLA is another type of fatty acid that cannot be fully evaluated by standard methodology, and substantial effort has gone into developing suitable approaches. A complete qualitative and quantitative determination affords a considerable challenge because of the complexity of geometrical and positional isomers. In functional foods, CLA could potentially come from either natural or commercial sources or possibly a mixture of the two, and the isomer distribution will vary between and within these two sources. There are a variety of different approaches that can be taken, and a complete characterization can usually only be obtained by using a combination of methods. In the following paragraphs, the analytical methodology used for characterizing commercial and natural CLA samples will be considered, followed by some recommendations on how the CLA content of a functional food might be determined.

2.3.4.4.1 Isomer Distribution in Commercial CLA

Initial separations of commercial CLA were achieved by GC. It was recognized that long polar columns, similar to those used for separation of *trans* isomers, were required to give optimum separations (Eulitz et al., 1999; Kramer et al., 1997, 1999; Yurawecz et al., 1999). A 100-m column coated with 100% cyanopropyl (e.g., CP-Sil 88 or SP-2560) is normally used. Typical conditions involved holding the temperature at 70°C for 4 min, temperature programing at 13°C/min to 175°C, holding at 175°C for 27 min, then temperature programing at 4°C/min to 215°C and holding at 215°C for 31 min, a total run time of 80 min (Eulitz et al., 1999), Similar conditions have been used for separating natural CLA isomers, for example in milk

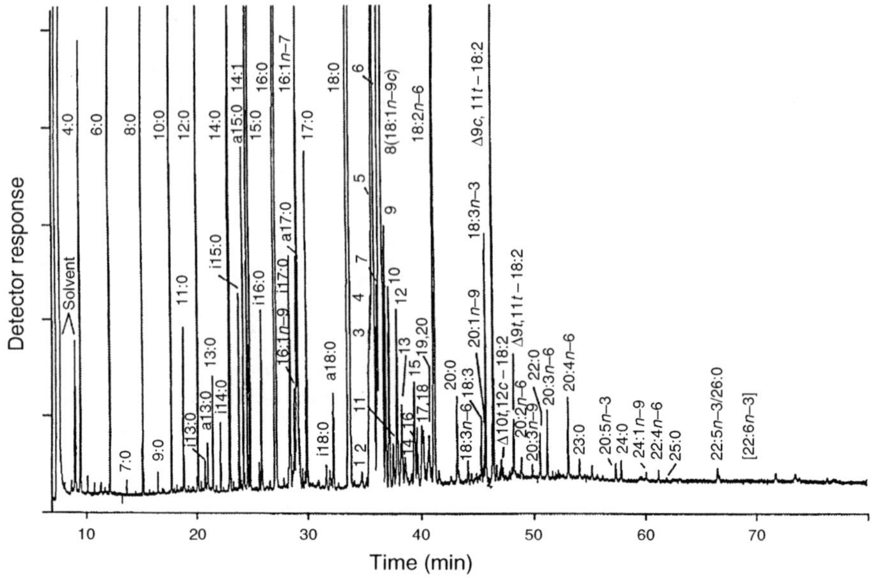

FIGURE 2.5 Gas chromatogram of bovine milk fatty acid methyl esters (FAME), separated on a SP-2560 capillary column (100 m × 0.25 mm i.d. × 0.2 μm film thickness; Supelco Inc., Bellefonte, PA). The oven temperature was held at 70°C for 4 min, then was programed at 13°C/min to 175°C, held at 175°C for 27 min, programed at 4°C/min to 215°C, and finally held there at 215°C for 31 min. Hydrogen was the carrier gas and a split ratio of 15:1 was used (i, *iso*; a, *anteiso*; 1–20 are *cis* and *trans* monounsaturated and diunsaturated FAME). (Reproduced from Kramer, J.K.G., Fellner, V., Dugan, M.E.R., Sauer, F.D., Mossoba, M.M., and Yurawecz, M.P., *Lipids*, 32, 1219–1228, 1997.)

(Figure 2.5) (Kramer et al., 1997). The program was developed to separate all fatty acids in a range of sample matrices. Under these conditions, all CLA isomers elute just after ALA and 20:1ω9 and before 20:2ω6, but under suboptimal conditions these peaks may interfere with CLA.

The order of elution of isomers in commercial CLA was 9*c*,11*t*, followed by other *cis-trans* isomers 11*c*,13*t* and 10*t*,12*c*, then the *cis-cis* isomers (8*c*,10*c*, 9*c*,10*c*, 10*c*,12*c*, 11*c*,13*c*, in that order), and finally the *trans-trans* isomers, the 11*t*,13*t* isomer eluting first followed by 8*t*,10*t*, 9*t*,11*t*, and 10*t*,12*t* all of which overlapped (Figure 2.6). The 8*t*,10*c* isomer had a slightly longer retention time than the 9*c*,11*t* isomer, but at best occurred as a shoulder on the 9*c*,11*t* peak, and could not be quantified separately. These two isomers could only be resolved on very long (120 m) polar (BPX70) columns (Christie, Sebedio, and Juaneda, 2001).

Further information on the isomer distribution could be obtained using silver-ion HPLC. The first silver-ion HPLC separation was achieved with a commercial CLA free acid converted to the methyl esters (Sehat et al., 1998b). A ChromSpher 5 (4.6 mm i.d. × 250 mm) column was employed with 0.1% acetonitrile in hexane as mobile phase at a flow rate of 1 mL/min and UV detection at 233 nm. Isomers were separated into three groups of peaks corresponding to *trans-trans*, *cis-trans*,

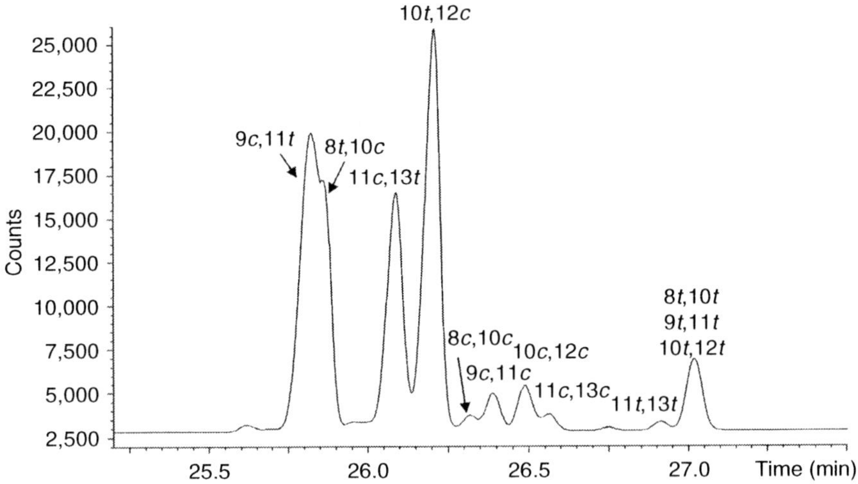

FIGURE 2.6 Partial gas chromatogram of a commercial conjugated linoleic acid (CLA) methyl ester mixture, separated on a CP-Sil 88 capillary column (100 m × 0.25 mm i.d. × 0.2 μm film thickness; Chrompack UK Ltd, London). The oven temperature was held at 160°C for 3 min, then was programed at 2°C/min to 220°C and held at 220°C for 20 min. Hydrogen was the carrier gas at a constant flow rate of 1 mL/min and a split ratio of 50:1 was used.

and *cis-cis* and eluting in that order. Within each group, separation of the 11,13, 10,12, 9,11, and 8,10 positional isomers was achieved to varying degrees, with the isomers always eluting in that order. Compared to GC, the most notable aspects of the separation were the improved separation of the trans/trans isomers and the separation of 9*c*,11*t* from 8*t*,10*c*-18:2. Continuous improvement in separation was achieved by increasing the number of silver-ion columns in series up to a maximum of six columns, but three (Eulitz et al., 1999; Sehat et al., 1999) or two (Rickert et al., 1999) columns was suggested as a good compromise, and all 12 isomers were well resolved (Figure 2.7A).

An alternative approach to improving resolution has been to use alternative derivatives to methyl esters. Utilizing only a single silver-ion HPLC column, the separation of a commercial CLA mixture was substantially improved as the *p*-methoxyphenacyl esters (Nikolova-Damyanova, Momchilova, and Christie, 2000). Recognizing that methods for forming methyl esters of CLA may cause geometrical isomerization or the formation of artifacts, another group (Ostrowska et al., 2000) separated CLA mixtures as free acids by silver-ion HPLC by incorporating acetic acid into the hexane-acetonitrile mobile phase. Separation was slightly improved compared to methyl esters on a single column. Even RP-HPLC has been utilized in a highly sensitive method using 9-anthryldiazomethane derivatives for analyzing the CLA composition of dairy products (Nishimura et al., 2005).

GC-MS of the DMOX derivatives readily allows the location of double bonds in CLA isomers (Spitzer, Marx, and Pfeilsticker, 1994) and has been used in confirming

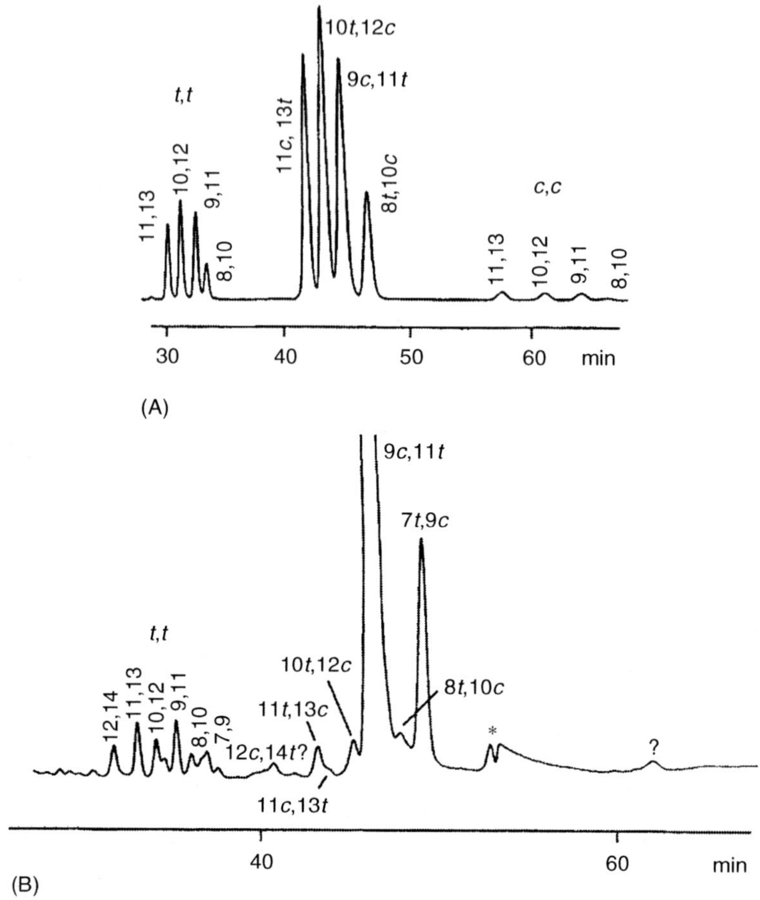

FIGURE 2.7 Silver-ion high-performance (CLA) chromatogram of the conjugated linoleic acid methyl ester liquid region of (A) a commercial CLA-methyl ester mixture and (B) cheese total fatty acid methyl esters FAME, using three ChromSpher 5 Lipids analytical silver-impregnated columns (each 250 mm × 4.6 mm i.d., 5 μm particle size; Chrompack, Bridgewater, NJ) and ultraviolet (UV) detection at either 234 nm (A) or 233 nm (B). The mobile phase was 0.1% acetonitrile in hexane and operated isocratically at a flow rate of 1 mL/min. (Reproduced from (A) Eulitz, K., Yurawecz, M.P., Sehat, N., Fritsche, J., Roach, J.A.G., Mossoba, M.M., Kramer, J.K.G., Adlof, R.O., and Ku Y., *Lipids*, 34, 873–877, 1999; (B) Sehat, N., Rickert, R., Mossoba, M.M., Kramer, J.K.G., Yurawecz, M.P., Roach, J.A.G., Adlof, R.O., Morehouse, K.M., Fritsche, J., Eulitz, K.D., Steinhart, H., and Ku, Y., *Lipids*, 34, 407–413, 1999.)

the isomers in commercial CLA (Eulitz et al., 1999; Rickert et al., 1999; Sehat et al., 1998b). For example, in the mass spectrum of the 8,10-18:2 derivative, gaps of 12 amu between m/z 182 and 294, and 208 and 220 confirmed the position of the double bonds, and similar gaps between m/z 224 and 236, and 250 and 262 confirmed the position of the double bonds in 11,13-18:2 (Figure 2.8) (Sehat et al., 1998b). Intense allylic ions at m/z 248 and 290 in 8,10-18:2 and 11,13-18:2,

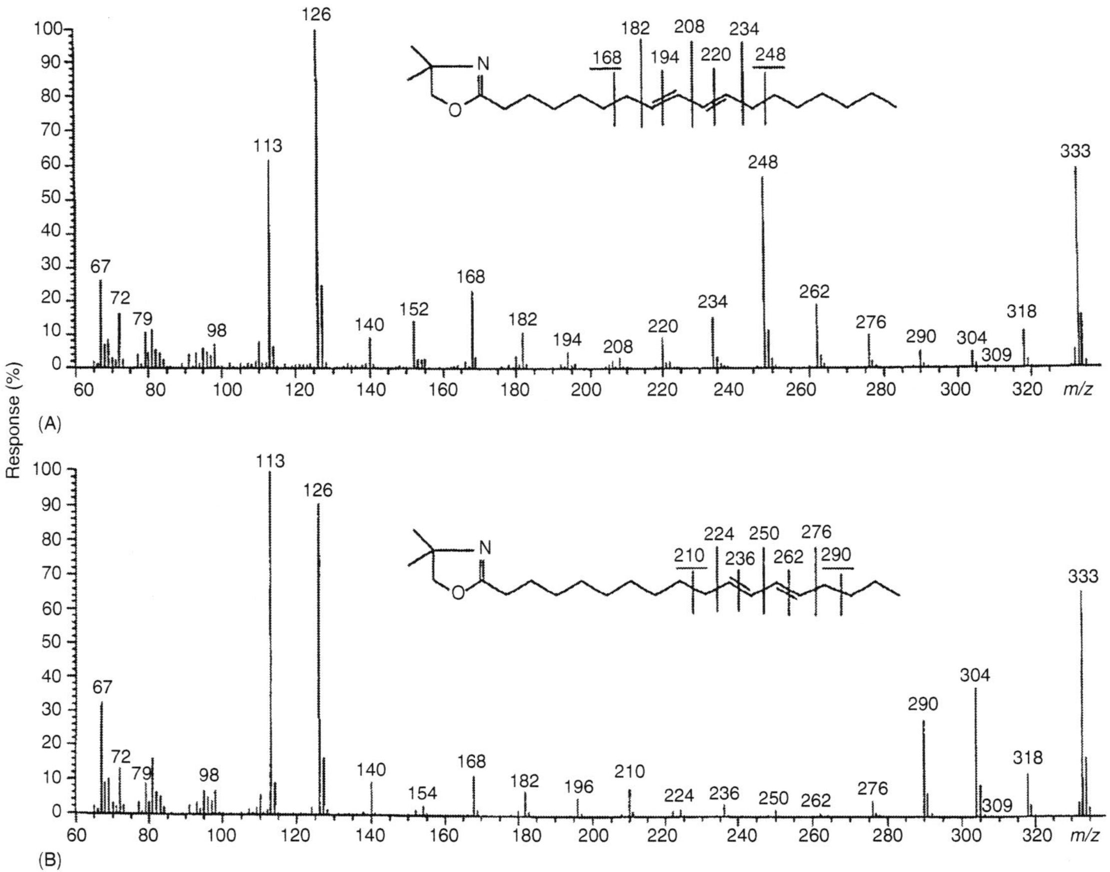

FIGURE 2.8 Gas chromatography–electron-impact (GC-EI) mass spectra of the 4,4-dimethyloxazoline (DMOX) derivatives of (A) *trans*-8,*cis*-10-18:2 and (B) *cis*-11,*trans*-13-18:2 from a commercial conjugated linoleic acid (CLA) mixture. (Reproduced from Sehat, N., Yurawecz., M.P., Roach, J.A.G., Mossoba, M.M., Kramer, J.K.G., and Ku, Y., *Lipids*, 33, 217–221, 1998b.)

respectively, were also characteristic. GC-MS is particularly useful for detecting the *8t*,10*c* isomer in the 9*c*,11*t* isomer peak.

Another approach, that can be applied to commercial CLA, utilized the reactivity of conjugated systems with the dienophile, 4-methyl-1,2,4-triazoline-3,5-dione (MTAD) to form Diels-Alder cycloaddition products (Dobson, 1998). The products were analyzed by GC-MS and the position of the original double bonds could be deduced from the mass spectra. Quantitative information on the distribution of positional isomers could be obtained but the method did not distinguish between geometrical isomers. The method is useful for complementing other approaches, for example in estimating the amount of 8*t*,10*c* isomer that is difficult to separate from 9*c*,11*t*-18:2 by GC. However, it may be possible to utilize the claim that *cis-trans*, and *trans-trans* isomers give MTAD adducts of different stereochemistry (*trans* and *cis* isomers of the chains about the ring, respectively) that can be resolved by GC-MS (Reaney, Liu, and Taylor, 2001). Analyzing tissue or food samples by this technique is likely to be problematic because it has been observed that MTAD may react to some extent with methylene-interrupted unsaturated fatty acids (Dobson, 1998).

Information on the isomer distribution in CLA mixtures can be obtained by other approaches. A powerful 2D ^{13}C NMR spectroscopy technique has been used to quantify all 20 isomers (all four geometrical forms of 7,9–11,13 isomers) in a complex commercial CLA mixture (Davis, McNeill, and Caswell, 1999a, b). The method was only applied to pure CLA mixtures and, together with the need for relatively large sample sizes, it would appear that such an approach would be restricted to analyzing commercial CLA mixtures before incorporation into foods.

2.3.4.4.2 *Isomer Distribution in Natural CLA from Ruminants*

In commercial CLA samples, usually only one of the *cis-trans* forms of each positional isomer (e.g., 9*c*,11*t* but not 9*t*,11*c*) occurs (Eulitz et al., 1999). In natural samples, such as cheese, both geometrical isomers may occur (Eulitz et al., 1999; Lavillonniere et al., 1998; Sehat et al., 1998a) and their complete separation is a challenge. All the isomers present in commercial CLA have been found in cheese, the food that has been studied in most detail (Lavillonniere et al., 1998; Rickert et al., 1999; Sehat et al., 1998a, 1999; Yurawecz et al., 1998). The 9*c*,11*t* isomer accounted for about 80%–90% of the total CLA. In addition, application of silver-ion HPLC (using three columns; Figure 2.7B), GC (Figure 2.9A), and GC-MS (Figure 2.9B) (Lavillonniere et al., 1998; Sehat et al., 1998a, 1999; Yurawecz et al., 1998), using similar conditions to those used for commercial CLA, revealed all four possible geometrical isomers of each of the positional isomers from 7,9–12,14, with the exception of 8*c*,10*t* and 12*t*,14*c*. The 6*t*,8*t* and 13*t*,15*t* isomers were tentatively identified (Rickert et al., 1999). The 7*t*,9*c*-isomer was often the second major isomer (3%–16% of the total CLA) and occurred as a shoulder (not quantifiable by GC) at the leading edge of the 9*c*,11*t* peak (Sehat et al., 1998a; Yurawecz et al., 1998), but was resolved as a single component (eluting just after 8*t*,10*c*) by HPLC (Rickert et al., 1999; Sehat et al., 1999).

By GC, none of the *cis/trans* isomers occur as pure peaks; 7*t*,9*c*, 8*t*,10*c*, and 9*c*,11*t* isomers overlap, as do the 9*t*,11*c* and 10*c*,12*t*, and 10*t*,12*c* and 12*c*,14*t* isomers (Figure 2.9A). The 11*c*,13*t* isomer can coincide with 21:0 and 11*t*,13*c*-18:2

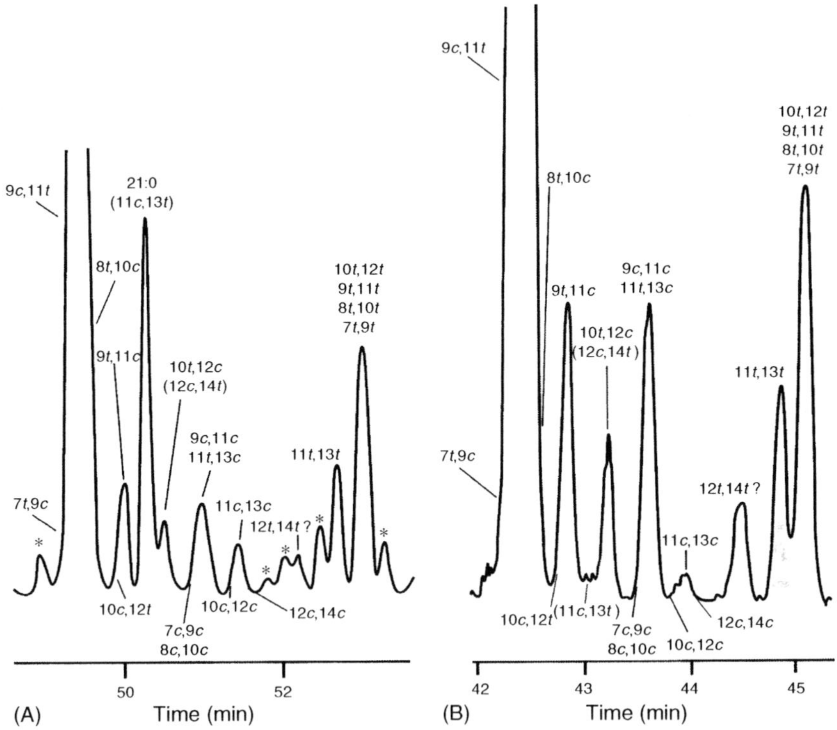

FIGURE 2.9 Comparison of partial gas chromatograms of the conjugated linoleic acid (CLA) methyl ester region from cheese total fatty acid methyl esters with flame-ionization detection (FID) (A) and the same region with high-resolution selected-ion recording (SIR) at m/z 294.2559 (molecular weight of CLA methyl ester) (B). Separations were performed using a CP-Sil 88 capillary column (100 m × 0.25 mm i.d. × 0.2 μm film thickness; Chrompack Inc., Raritan, NJ) using hydrogen as carrier gas. For gas chromatography (GC), the oven temperature was held at 75°C for 2 min, programed at 5°C/min to 180°C, held at 180°C for 33 min, programed at 4°C/min to 225°C, and finally held at 225°C for 43.8 min. For gas chromatography–mass spectrometry (GC-MS), the oven temperature was held at 75°C for 2 min, programed at 5°C/min to 170°C, held at 170°C for 40 min, programed at 5°C/min to 220°C, and finally held at 220°C for 20 min. Peaks indicated by an asterisk were not found in the SIR chromatogram and are not CLA isomers. CLA isomers in brackets were detected at low levels. (Redrawn from Roach, J.A.G., Yurawecz, M.P., Kramer, J.K.G., Mossoba, M.M., Eulitz, K., and Ku, Y., *Lipids*, 35, 797–802, 2000. Some identities have been annotated using information from Delmonte, P., Roach, J.A.G., Mossoba, M.M., Losi, G., and Yurawecz, M.P., *Lipids*, 39, 185–191, 2004; Sehat, N., Kramer, J.K.G., Mossoba, M.M., Yurawecz, M.P., Roach, J.A.G., Eulitz, K., Morehouse, K.M., and Ku, Y., *Lipids*, 33, 963–971, 1998a.)

overlaps with the first cis,cis peak. Within the cis/cis isomers, presumably because of the wide range in the relative amounts of the different positional isomers, only two peaks are apparent, the first comprising the 7,9–9,11 isomers and the second the 10,12–12,14 isomers. The 12*t*,14*t* isomer elutes first followed by the 11*t*,13*t* isomer and then the major trans/trans peak, comprising the 7*t*,9*t*–10*t*,12*t* isomers.

By silver-ion HPLC, all the *trans-trans* isomers separate well, as do all *cis*/*trans* isomers (including the 7*t*,9*c* and 8*t*,10*c* isomers that do not separate from the major 9*c*,11*t* peak by GC), except the 10,12*c*/*t* and 9,11*c*/*t* isomers. The minor 10*c*,12*t* and 9*t*,11*c* isomers (not shown in Figure 2.7B), overlap with the relatively more abundant 10*t*,12*c* and 9*c*,11*t* peaks, respectively. The *cis-cis* isomers, present at low levels, are often obscured by the 18:1 peak (Sehat et al., 1999).

Because of the resolution problems, particularly by GC, of CLA isomers in natural samples, reconstructed ion chromatograms (RIC) of specific ions have been used to aid identification of the isomers by GC-MS in the form of DMOX derivatives. Roach (1999) has highlighted seven diagnostic ions per isomer, giving a total of 28 ions (56 ions were highlighted but some occurred more than once) for isomers ranging from 6,8-18:2 to 13,15-18:2, which are useful for identifying positional isomers (Table 2.1). However, the ions are not necessarily specific to any one positional isomer and therefore some degree of chromatographic separation, which is often the case, is useful. The use of selected ions enabled the previously undetected 7*t*,9*c* isomer to be found in a variety of samples (Yurawecz et al., 1998).

The presence of non-CLA peaks, particularly 21:0 and also 20:2 isomers, in the GC profile of the CLA methyl ester region from cheese made identification of CLA isomers difficult without using high-resolution selected-ion recording (SIR) GC-MS of the molecular ion of CLA (Figure 2.9B) (Eulitz et al., 1999; Roach et al., 2000). The 21:0 is often at similar concentrations to the minor CLA isomers and can elute anywhere between 11*c*,13*t*-18:2 and 10*c*,12*c*-18:2 depending on the GC conditions

TABLE 2.1
Diagnostic Ions for 4,4-Dimethlyoxazoline (DMOX) Derivatives of Conjugated Linoleic Acid (CLA) Positional Isomers Ranging from 6, 8-18:2 to 13,15-18:2

Isomer	*n*−2 Ion (*m*/*z*)	*n*−1 Ion (*m*/*z*)	*n* Ion (*m*/*z*)	*m*−1 Ion (*m*/*z*)	*m* Ion (*m*/*z*)	*m*+2 Ion (*m*/*z*)	*m*+3 Ion (*m*/*z*)
6,8[a]	140	154	166	180	192	220	234
7,9[b]	154	168	180	194	206	234	248
8,10[c]	168	182	194	208	220	248	262
9,11	182	196	208	222	234	262	276
10,12	196	210	222	236	248	276	290
11,13	210	224	236	250	262	290	304
12,14	224	238	250	264	276	304	318
13,15	238	252	264	278	290	318	332

Source: Adapted from Roach, J.A.G., *Advances in Conjugated Linoleic Acid Research*, AOCS Press, Champaign, IL, 1999, pp. 126–140.

The double bonds are at carbons n and m, respectively.
The molecular ion is $m/z = 333$.

[a] m/z 154 is weak and m/z 220 is more abundant than m/z 234.
[b] m/z 154 is weak and m/z 234 is more abundant than m/z 248.
[c] m/z 248 is more abundant than m/z 262.

(Cruz-Hernandez et al., 2004). SA, if present, would also be expected to elute in the CLA region. By silver-ion HPLC, interference in the CLA region from residual absorption of MUFA and PUFA has been observed and it was suggested that the absorption at 205 nm (maximum for nonconjugated double bonds) should also be recorded to detect the presence of these fatty acids (Cruz-Hernandez et al., 2004). RP-HPLC has been used to isolate a total CLA fraction, also containing 14:0, 16:1, and LA, from cheese total FAME, before analysis by GC (Figure 2.10) (Lavillonniere et al., 1998). This served not only to concentrate the CLA isomers but

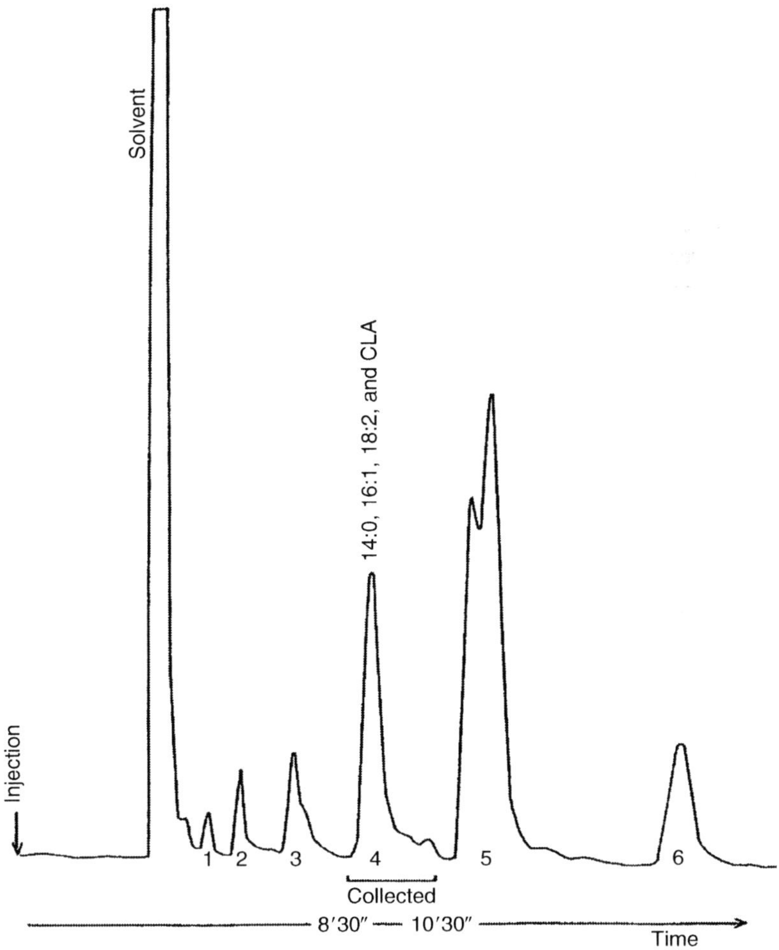

FIGURE 2.10 Preparative reversed-phase high-performance liquid chromatography (RP-HPLC) separation of fatty acid methyl esters (FAME) (5–10 mg) from cheese on a Nucleosil C18 column (250 mm × 10 mm i.d.) with acetonitrile as mobile phase at a flow rate of 4 mL/min. Detection of FAME was by refractometry and the fraction containing conjugated linoleic acid (CLA) (also containing 14:0, 16:1, and 18:2) was monitored by ultraviolet (UV) detection at 234 nm. (Reproduced from Lavillonniere, F., Martin, J.C., Bougnoux, P., and Sebedio, J.-L., *J. Am. Oil Chem. Soc.*, 75, 343–352, 1998.)

also appeared to eliminate other contaminating peaks in the CLA region of the GC profile. It would also eliminate the 18:1 peak that can overlap with the *cis-cis* isomers by HPLC (Sehat et al., 1999).

A major problem with silver-ion HPLC of CLA isomers is drift in elution times or retention volumes (RV) with time. This is a problem when attempting to identify the isomer distribution in natural CLA samples by comparison of RV values with those of an appropriate standard. Attempts to overcome drifts were only partially successful when 0.5% diethyl ether was incorporated into the mobile phase (Delmonte et al., 2003), but a hexane/2% acetic acid system was more effective (Delmonte et al., 2005); the latter system also altered the elution behavior of some *cis-trans* isomers, and there was better separation of the 10,12 *c*/*t* isomer pair in a standard. By substituting propionitrile for acetonitrile as modifier in the mobile phase, problems of RV stability were also greatly improved, with negligible loss in resolution, enabling isomers to be identified by comparison to adjacent runs of a standard mixture (Muller et al., 2006).

A CLA standard mixture for use in GC or HPLC is essential for identifying CLA isomers in natural samples. A commercial CLA mixture is useful but it does not contain all the isomers found in natural samples. Other standard mixtures can be prepared. A mixture of all 16 possible isomers ranging from 8,10-18:2 to 11,13-18:2 can be obtained by isomerization of commercial CLA with either iodine or *p*-toluenesulfinic acid as catalyst (Eulitz et al., 1999). Another containing all 32 possible isomers from 6,8-18:2 to 13,15-18:2 has been synthesized (Delmonte et al., 2004). The relative retention times (RRT), relative to GLA, of all 32 isomers in this mixture have been determined by GC on a 100-m CP-Sil 88 (Table 2.2). The relative retention volumes (RRV), relative to 9*c*,11*t*-18:2, of the same mixture was determined by silver-ion HPLC using three columns with two different mobile phases (0.1% acetonitrile/0.5% diethyl ether/hexane and 2% acetic acid/hexane, which gives more stable RV) (Delmonte et al., 2005). Use of RRV eliminated problems with RV drift, enabling isomers to be identified with greater confidence. The method was applied to cow milk, plasma, and rumen fluids.

A powerful tandem mass spectrometry (MS/MS) technique using acetonitrile chemical ionization was capable of determining double-bond position and geometry in CLA as the methyl esters (Michaud et al., 2003, 2005). The acetonitrile generated a (1-methyleneimino)-1-ethenylium ion which reacted with the CLA isomers to form a $[M+54]^+$ ion, which upon collisionally activated dissociation produced two ions as a result of carbon–carbon cleavage at a position vinylic to either side of the conjugated diene system. The two ions were characteristic of the position of the double bonds, one (α) containing the ester group, and the other (ω) the terminal methyl group. The ion resulting from vinylic cleavage to a *trans* double bond was more abundant than that from a *cis* double bond. Therefore, the ratio of the abundance of α to that of ω was highest (>4.8) for a *cis-trans* arrangement, lowest (<0.5) for *trans-cis*, and intermediate (0.7–3.2) for *cis-cis* and *trans-trans*, thus allowing differentiation of the geometrical isomers. When used in conjunction with GC retention time data, the majority of CLA isomers in milk samples were identified.

GC-MS does not distinguish between geometrical isomers, but this can be achieved by examining the IR spectra obtained by GC-FTIR of the methyl esters

TABLE 2.2
Relative Retention Times (RRT) of Conjugated Linoleic Acid (CLA) Methyl Ester Isomers on a 100-m CP-Sil 88 Column[a]

Isomer	RRT	Isomer	RRT
7*c*,9*t*	1.083	9*c*,11*c*	1.137
8*c*,10*t*	1.085	10*c*,12*c*	1.145
6*c*,8*t*	1.086	13*c*,15*t*	1.147
6*t*,8*c*	1.086	12*t*,14*c*	1.149
7*t*,9*c*	1.091	11*c*,13*c*	1.151
9*c*,11*t*	1.091	12*c*,14*c*	1.159
8*t*,10*c*	1.097	13*t*,15*c*	1.166
10*c*,12*t*	1.104	13*c*,15*c*	1.166
9*t*,11*c*	1.109	12*t*,14*t*	1.171
11*c*,13*t*	1.114	11*t*,13*t*	1.184
12*c*,14*t*	1.118	13*t*,15*t*	1.184
10*t*,12*c*	1.122	9*t*,11*t*	1.191
6*c*,8*c*	1.128	8*t*,10*t*	1.191
8*c*,10*c*	1.131	10*t*,12*t*	1.192
7*c*,9*c*	1.131	6*t*,8*t*	1.192
11*t*,13*c*	1.136	7*t*,9*t*	1.193

Source: Adapted from Delmonte, P., Roach, J.A.G., Mossoba, M.M., Losi, G., and Yurawecz, M.P., *Lipids*, 39, 185–191, 2004.

[a] RRT relative to γ-linolenic acid (GLA) methyl ester calculated using $(RT_{isomer} - RT_{solvent})/(RT_{GLA} - RT_{solvent})$. Oven temperature programe: 75°C for 2 min, then 5°C/min to 175°C, holding at 175°C for 33 min, then at 5°C/min to 225°C, and finally holding at 225°C for 8 min.

(Lavillonniere et al., 1998; Mossoba et al., 1999b; Rickert et al., 1999; Sehat et al., 1998a, 1999; Yurawecz et al., 1998). Two peaks at 978–988 cm^{-1} and 946–949 cm^{-1} were characteristic of *cis*/*trans* isomers, but there was only one peak at 986–993 cm^{-1} for *trans-trans* isomers, and *cis-cis* isomers lacked peaks in this region of the spectrum. Unsaturated carbon C–H stretch bands at 3002/3020 cm^{-1}, 3017 cm^{-1}, and 3005/3037 cm^{-1} were highly characteristic of *cis-trans*, *trans-trans*, and *cis-cis* isomers, respectively. GC-FTIR does not distinguish between *cis-trans* and *trans-cis* isomers. One way to achieve this was to examine the *cis* and *trans* monoenes, derived from partial hydrazine reduction, by GC-MS of the DMOX derivatives (Lavillonniere et al., 1998).

2.3.4.4.3 *Measuring Total CLA*

In a functional food, at the most basic level, the total CLA content may be required. In this case, the total FAME may be analyzed by GC on a long polar column such as a 100-m CP-Sil 88 using conditions as detailed above (Eulitz et al., 1999; Kramer et al., 1997, 1999; Yurawecz et al., 1999). The CLA peaks, eluting between ALA and 20:2ω6 (Figure 2.5), would be identified with an appropriate standard mixture, and summed and quantified using an appropriate IS (e.g., 23:0) (Kramer et al., 1997; Mossoba et al., 1999a). Quantitation, in terms of milligram of CLA per gram of sample, would be carried out by including an appropriate internal standard during extraction similar to the approaches outlined in Section 2.3.4.2.2. Sometimes in ruminant-derived CLA samples, the 9*c*,11*t* isomer is the only peak that is measured as an estimate of the CLA content in natural samples (Fritsche and Steinhart,1998; Fritsche et al., 1999; Lin et al., 1998), and in any case, this peak includes the second most abundant peak, the 7*t*,9*c* isomer and also the 8*t*,10*c* isomer. On a Carbowax type column, the CLA region also comes after ALA, but unless only the major isomer(s) is required, it is advisable not to use this type of column as it has been observed that the *trans-trans* isomers overlap with the 20:1 isomers (Kramer, Blackadar, and Zhou, 2002). In ruminant products, other CLA peaks may represent a significant minority of the CLA and should be included in the total. Moreover, in products such as eggs, where commercial CLA is the source, the 10*t*,12*c* isomer will be of similar abundance to the 9*c*,11*t* isomer, and the 11*c*,13*t* and 8*t*,10*c* isomers may be significant, as may *cis-cis* and *trans-trans* isomers.

A possible problem with using GC only would be the presence of non-CLA peaks, particularly 21:0, in the CLA region of the chromatogram (Eulitz et al., 1999; Roach et al., 2000; Sehat et al., 1999) that may be mistakenly included as CLA. SIR GC-MS would help to distinguish CLA from non-CLA peaks (Eulitz et al., 1999; Sehat et al., 1999). Alternatively, before GC, a CLA fraction could be isolated by RP-HPLC thus minimizing the chance of interfering non-CLA peaks by GC while serving to concentrate the CLA peaks (Figure 2.10) (Lavillonniere et al., 1998). In this case, quantitation would be achieved by relating the total CLA peak areas to that of LA (isolated in the CLA fraction by RP-HPLC), which in turn would be quantified from the GC trace of the total FAME by reference to the IS.

Other approaches to quantitation of total CLA have included SFC with FID (Yurawecz, Kramer, and Ku, 1999) and RP-HPLC with UV detection at 234 nm (e.g., Banni et al., 1998). In both methods, CLA isomers elute as a single or poorly resolved peak. Disadvantages are that the former requires nonroutine equipment, and the latter requires an external calibration procedure. On the other hand, UV detection has the advantage over GC and SFC in that the method has a high degree of specificity, thus minimizing problems of misidentification and interference from non-CLA compounds, although there is minor interference from isolated double bonds (Banni and Martin, 1998). A more accurate RP-HPLC method, where second derivative spectra are used to overcome this problem, is recommended by Banni and Martin (1998). UV spectroscopy of total FAME samples, without chromatographic separation, would not be recommended because of the absorption of other compounds, particularly oxidation products of CLA, which also absorb at 234 nm (Banni and Martin, 1998).

FTIR spectroscopy has been used with PLS approaches to develop calibration models for simultaneous analysis of isolated *trans* fatty acid content and total CLA content in beef fat with reasonable accuracy (Christy, Egeberg, and Ostensen, 2003). The method used a sample to which CLA was added, but was not validated with samples containing endogenous levels of CLA against an independent method.

Raman spectroscopy was also used with multivariate statistics to analyze total CLA in milk samples and there was generally good agreement with the values obtained from a GC method (Meurens et al., 2005).

2.3.4.4.4 Recommended Approaches for Analyzing CLA

The above sections confirm that CLA can be a complex mixture and that it can be analyzed by a variety of techniques. The degree of sophistication of the adopted analytical approach will depend on the type and detail of information required. With the use of appropriate standards, quantitation of all the CLA isomers can be achieved with a combination of GC (Figure 2.9) and silver-ion HPLC (Figure 2.7B) alone. GC would be performed on a 100-m CP-SIl 88 or SP-2560 column in the presence of an IS (e.g., 23:0) using conditions as detailed above (Eulitz et al., 1999; Kramer et al., 1997, 1999; Yurawecz et al., 1999). HPLC would be performed using three ChromSpher 5 Lipids columns (each 250 mm × 4.6 mm i.d., 5 μm particle size) using the more stable mobile phase of 0.1% acetonitrile/0.5% diethyl ether/hexane and a flow rate of 1 mL/min and detection at 233 nm (Delmonte et al., 2003). The use of a standard either of a commercial CLA, containing four rather than two positional isomers, or for ruminant samples one containing all geometrical isomers from 6,8-18:2 to 13,15-18:2 (Delmonte et al., 2004) is recommended to aid identifications. The use of RRT (Table 2.2) (Delmonte et al., 2004) and RRV (Delmonte et al., 2005) for GC and HPLC, respectively, is also advisable. The analyst must then decide on whether other techniques (e.g., GC-MS of DMOX derivatives, GC-FTIR) are warranted, depending on the required degree of certainty of identifications of the isomers.

Total CLA could be carried out by GC only. For higher accuracy, in view of the possible overlapping problems with 21:0 and other minor components, that might be more of a problem with a complex food matrix, it may be worth isolating a CLA fraction by RP-HPLC (Figure 2.10) before GC analysis and quantifying relative to LA as described in Section 2.3.4.4.3 (Lavillonniere et al., 1998).

To determine the amounts of all individual isomers, for all samples (apart from pure commercial CLA) it is recommended to isolate a CLA fraction by RP-HPLC, which would also eliminate any oleate that might interfere with silver-ion HPLC measurements of the *cis-cis* isomers (Lavillonniere et al., 1998), before GC and silver-ion HPLC analysis. If a fractionation step is not included then analyses of ruminant CLA should be performed at high sample loading because the level of CLA is only 0.5%–4% of the total fatty acids. The approach would be to quantify single or groups of isomers by GC and then determine the proportion of the isomers within the groups by HPLC.

For commercial CLA, foods to which commercial CLA has been added and foods derived from nonruminant animals fed on commercial CLA, the task is relatively straightforward. From the GC profile (Figure 2.6), all four *cis-cis* isomers (8*c*,10*c*–11*c*,13*c* isomers) and 11*t*,13*t*, 10*t*,12*c*, and 11*c*,13*t* can be individually quantified. The 8*t*,10*c*/9*c*,11*t* and 8*t*,10*t*/9*t*,11*t*/10*t*,12*t* peaks can also be quantified by GC. The proportions of the individual isomers within each of these peaks then can be determined by silver-ion HPLC.

For ruminant CLA, the following groups of isomers can be quantified by GC (Figure 2.9): (1) 7*t*,9*c*, 8*t*,10*c*, 9*c*,11*t*, 9*t*,11*c*, 10*c*,12*t*, 11*c*,13*t*, 10*t*,12*c*, and 12*c*,14*t*;

(2) 11*t*,13*c*, 7*c*,9*c*, 8*c*,10*c*, 9*c*,11*c*, 10*c*,12*c*, 11*c*,13*c*, and 12*c*,14*c*; and (3) 12*t*,14*t*, 11*t*,13*t*, 10*t*,12*t*, 9*t*,11*t*, 8*t*,10*t*, and 7*t*,9*t*. For groups 2 and 3, the proportions of the individual isomers within the same groups can be determined by HPLC (Figure 2.7) and therefore the amounts of these isomers can be calculated. Similarly, within group 1, the proportions of 7*t*,9*c*, 8*t*,10*c*, 11*c*,13*t*, and 12*c*,14*t* isomers can be determined by HPLC, and the amounts can be determined. The amounts of 7*t*,9*c* + 8*t*,10*c* + 9*c*,11*t* and 10*t*,12*c* + 12*c*,14*t* can be deduced from the GC profile and therefore the levels of 9*c*,11*t* and 10*t*,12*c* can be calculated. The proportions of 9*c*,11*t* + 9*t*,11*c* and 10*t*,12*c* + 10*c*,12*t* within group 1 can also be determined by HPLC, and therefore the amounts of the remaining two isomers, 9*t*,11*c* and 10*c*,12*t* can be calculated. The values for the individual isomers will not be very accurate. Some minor isomers and some *cis-cis* isomers, which appear as broader peaks by HPLC because of their later elution time, may be difficult to quantify or even detect.

2.4 CONCLUDING REMARKS

Currently, ω3 fatty acids (EPA, DHA, and ALA) are the major types of fatty acids incorporated into functional foods because of their potential benefits against CVD. The major sources of EPA and DHA are fish oils and algae, and flaxseed for ALA. ω3 fatty acids have also been incorporated into eggs. The analyst who is particularly interested in the ω3 content, may also require a qualitative and quantitative analysis of some (e.g., *trans*, total SFA, or total PUFA) or all the fatty acids in the food, for labeling requirements or for evaluating the total health potential of the food.

CLA is another fatty acid that might have beneficial effects such as helping to control obesity, but the effects in humans do not appear to be as great as in animals. There are CLA-containing products, such as milk-based drinks. CLA is composed of a mixture of various positional and geometrical isomers whose distribution varies between the two main sources: commercial CLA or ruminant products with enhanced levels of CLA. Different isomers can produce different physiological effects, so the different sources may have different activities. Notably, the 10*t*,12*c* isomer may be responsible for effects on body composition and is high in commercial CLA, but normally at low levels in dairy products.

To obtain a total fatty acid profile, to include ω3 fatty acids, and other potentially beneficial fatty acids such as GLA, the approach is normally straightforward and involves extraction of lipids in the presence of an appropriate, derivatization of fatty acids to methyl esters, and analysis by GC. Appropriate extraction methods, depending on the type of food and lipids presents, must be employed to extract the total lipids efficiently. An alternative is to hydrolyze the sample directly to release all the fatty acids in the free form followed by derivatisation. An appropriate methylation step is required. Alkaline methods are milder, but will only derivatize EFA. Acidic methods are used to derivatize samples containing FFA with or without EFA. Separation of FAME can normally be achieved on Carbowax type columns of medium length, but samples such as milk fats or PHVO, containing complex mixtures of *trans* fatty acids, require longer polar columns and may require the use of additional techniques. Milk fats also contain SCFA and specific precautions or adaptations of protocols are necessary for their analysis.

Special derivatization procedures are required for forming methyl esters of CLA because isomerization and artifact formation can occur under acidic conditions. In choosing a procedure, the form of the CLA should be taken into account. A prefractionation step, using RP-HPLC, is recommended to concentrate the CLA and to remove other components that can interfere with the separation of the isomers. The total CLA content can be measured by GC, and a complementary GC and silver-ion HPLC approach, with the use of a comprehensive standard, can be used to analyze all isomers.

There is a range of alternative and complementary analytical techniques available for analyzing fatty acids. Increased sensitivity can be achieved by analysing specific derivatives by RP-HPLC. GC-MS of specific derivatives are useful for determining position of double bonds. FTIR, including GC-FTIR, is used for analyzing trans fatty acids. NMR, particularly ^{13}C NMR, can give considerable structural details, although samples size (for ^{13}C NMR) may be a limitation. MS/MS has been applied to CLA to give detailed structural information.

The last few years have seen the emergence of techniques that on the one hand can examine samples of increasing complexity, and on the other hand can afford rapid analyses. The former includes 2D GC which has been applied to complex fatty acid mixtures and has potential for examining complex food matrices. The latter includes FTIR and NIR methods in conjunction with multivariate statistical approaches to measure total *trans* fatty acids and CLA, for example, but there are even applications for measuring simple fatty acid profiles.

ACKNOWLEDGMENTS

The author is supported by the Rural and Environment Research and Analysis Directorate of the Scottish Government. The assistance of Dr. W.W. Christie and Dr. C. Scrimgeour in preparing some of the figures is greatly appreciated.

REFERENCES

Ackman, R.G., Application of gas-liquid chromatography to lipid separation and analysis: Qualitative and quantitative analysis, in *Fatty Acids in Foods and their Health Implications*, Chow, C.K. (Ed.), Marcel Dekker, New York, 1999, pp. 47–65.

Aldai, N., Murray, B.E., Najera, A.I., Troy, D.J., and Osoro, K., Derivatization of fatty acids and its application for conjugated linoleic acid studies in ruminant meat lipids, *J. Sci. Food Agric.*, 85, 1073–1083, 2005.

Aldai, N., Osoro, K., Barron, L.J.R., and Najera, A.I., Gas-liquid chromatographic method for analysing complex mixtures of fatty acids including conjugated linoleic acids (*cis*9*trans*11 and *trans*10*cis*12 isomers) and long-chain (*n*-3 or *n*-6) polyunsaturated fatty acids—Application to the intramuscular fat of beef meat, *J. Chromatogr. A*, 1110, 133–139, 2006.

American Oil Chemists' Society, *Official Methods and Recommended Practices of the American Oil Chemists' Society*, 5th edition, Firestone, D. (Ed.), AOCS Press, Champaign, IL, 2005, Official Methods Cd 14–61, Cd 14d-99, Ce 1b-89, and Ce 1h-05.

Andersen, S. Microencapsulated omega-3 fatty acids from marine sources, *Lipid Technol.*, July, 81–85, 1995.

AOAC, Official Method 996.06 and 991.39, *Official Methods of Analysis of AOAC International*, 16*th* edition, AOAC International, Gaithersburg, MD, 1995.

Aro, A., Epidemiological studies of *trans* fatty acids and coronary heart disease, in *Trans Fatty Acids in Human Nutrition*, Sebedio, J.L. and Christie, W.W. (Eds.), The Oily Press Ltd, Dundee, Scotland, 1998, pp. 235–260.

Aydin, R., The effects of conjugated linoleic acid (CLA) and canola oil on the fatty acid composition and quality of eggs from laying hens, *S. Afri. J. Anim. Sci.*, 35, 172–179, 2005.

Aydin, R., Effect of storage temperature on the quality of eggs from conjugated linoleic acid-fed laying hens, *S. Afr. J. Anim. Sci.*, 36, 13–19, 2006.

Aydin, R. and Cook, M.E., The influence of conjugated linoleic acid enriched tallow on egg hatchability and quality in chicken, *Int. J. Poult. Sci.*, 4, 536–542, 2005.

Aydin, R., Pariza, M.W., and Cook, M.E., Olive oil prevents the adverse effects of dietary conjugated linoleic acid on chick hatchability and egg quality, *J. Nutr.*, 131, 800–806, 2001.

Azizian, H. and Kramer, J.K.G., A rapid method for the quantification of fatty acids in fats and oils with emphasis on *trans* fatty acids using Fourier transform near infrared spectroscopy (FT-NIR), *Lipids*, 40, 855–867, 2005.

Banni, S. and Martin, J.-C., Conjugated linoleic acid and metabolites, in *Trans Fatty Acids in Human Nutrition*, Sebedio, J.L. and Christie, W.W. (Eds.), The Oily Press Ltd, Dundee, Scotland, 1998, pp. 261–302.

Banni, S., Angioni, E., Contini, M.S., Carta, G., Casu, V., Lengo, G.A., Melis, M.P., Deiana, M., Dessi, M.A., and Corongiu, F.P., Conjugated linoleic acid and oxidative stress, *J. Am. Oil Chem. Soc.*, 75, 261–267, 1998.

Banni, S., Heys, S.D., and Wahle, K.W.J., Conjugated linoleic acids as anticancer nutrients: Studies in vivo and cellular mechanisms, in *Advances in Conjugated Linoleic Acid Research*, Vol. 2, Sebedio, J.-L., Christie, W.W., and Adlof, R. (Eds.), AOCS Press, Champaign, IL, 2003, pp. 267–282.

Bassaganya-Riera, J., King, J., and Hontecillas, R., Health benefits of CLA—lessons from pig models in biomedical research, *Eur. J. Lipid Sci. Technol.*, 106, 856–861, 2004.

Becker, C.C. and Kyle, D.J., Developing functional foods containing algal docosahexaenoic acid, *Food Technol.*, 52, 68–71, 1998.

Beckles Willson, N., Elliott, T.M., and Everard, M.L., Omega-3 fatty acids (from fish oils) for cystic fibrosis (Cochrane Review), in *The Cochrane Library*, Issue 2, Update Software, Oxford, United Kingdom, 2003.

Belury, M.A., Conjugated linoleic acids in type 2 diabetes mellitus: Implications and potential mechanisms, in *Advances in Conjugated Linoleic Acid Research*, Vol. 2, Sebedio, J.-L., Christie, W.W., and Adlof, R. (Eds.), AOCS Press, Champaign, IL, 2003, pp. 302–315.

Bessa, R.J.B., Santos-Silva, J., Ribeiro, J.M.R., and Portugal, A.V., Reticulo-rumen biohydrogenation and the enrichment of ruminant edible products with linoleic acid conjugated isomers, *Livestock Prod. Sci.*, 63, 201–211, 2000.

Bligh, E.G. and Dyer, W.J., A rapid method of total lipid extraction and purification, *Can. J. Biochem. Physiol.*, 37, 911–917, 1959.

Bloedon, L.T. and Szapary, P.O., Flaxseed and cardiovascular risk, *Nutr. Rev.*, 62, 18–27, 2004.

Bourre, J.M., Effect of increasing the omega-3 fatty acid in the diets of animals on the animal products consumed by humans, *M S-Med. Sci.*, 21, 773–779, 2005.

Brondz, I., Development of fatty acid analysis by high-performance liquid chromatography, gas chromatography, and related techniques, *Analytica Chim. Acta*, 465, 1–37, 2002.

Bucher, H.C., Hengstler, P., Schindler, C., and Meier, G., *N*-3 polyunsaturated fatty acids in coronary heart disease: A meta-analysis of randomized controlled trials, *Am. J. Med.*, 112, 298–304, 2002.

Buchgraber, M. and Ulberth, F., Determination of *trans* octadecenoic acids by silver-ion chromatography–gas liquid chromatography: An intercomparison of methods, *J. Assoc. Off. Anal. Chemists Int.*, 84, 1490–1498, 2001.

Carrapiso, A.I. and Garcia, C., Development in lipid analysis: Some new extraction techniques and in situ transesterification, *Lipids*, 35, 1167–1177, 2000.

Carrapiso, A.I., Timon, M.L., Petron, M.J., Tejeda, J.F., and Garcia, C., In situ transesterification of fatty acids from Iberian pig subcutaneous adipose tissue, *Meat Sci.*, 56, 159–164, 2000.

Chen, S.-H. and Chuang, Y.-J., Analysis of fatty acids by column chromatography, *Analytica Chim. Acta*, 465, 145–155, 2002.

Christie, W.W., *Gas Chromatography and Lipids*, The Oily Press Ltd, Dundee, Scotland, 1989.

Christie, W.W., *Lipid Analysis. Isolation, Separation, Identification and Structural Analysis of Lipids*, The Oily Press Ltd, Bridgwater, England, 2003.

Christie, W.W., Sebedio, J.L., and Juaneda. P., A practical guide to the analysis of conjugated linoleic acid (CLA), *Inform*, 12, 147–152, 2001.

Christy, A.A., Egeberg, P.K., and Ostensen, E.T., Simultaneous quantitative determination of isolated *trans* fatty acids and conjugated linoleic acids in oils and fats by chemometric analysis of the infrared profiles, *Vib. Spectrosc.*, 33, 37–48, 2003.

Collomb, M., Schmid, A., Sieber, R., Wechsler, D., and Ryhanen, E.-L., Conjugated linoleic acids in milk fat: Variation and physiological effects, *Int. Dairy J.*, 1347–1361, 2006.

Cook, M.E., Butz, D., Li, G., Pariza, M., Whigham, L., and Yang, M., Conjugated linoleic acid enhances immune responses but protects against the collateral damage of immune events, in *Advances in Conjugated Linoleic Acid Research*, Vol. 2, Sebedio, J.-L., Christie, W.W., and Adlof, R. (Eds.), AOCS Press, Champaign, IL, 2003, pp. 283–291.

Corl, B.A., Barbano, D.M., Bauman, D.E., and Ip, C., The *cis*-9, *trans*-11 CLA derived endogenously from *trans*-1 18:1 reduces cancer risk in rats, *J. Nutr.*, 133, 2893–2900, 2003.

Cruz-Hernandez, C., Deng, Z., Zhou, J., Hill, A.R., Yurawecz, M.P., Delmonte, P., Mossoba, M.M., Dugan, M.E.R., and Kramer, J.K.G., Methods for the analysis of conjugated linoleic acids and *trans*-18:1 isomers in dairy fats by using a combination of gas chromatography, silver-ion thin-layer chromatography/gas chromatography, and silver-ion liquid chromatography, *J. AOAC Int.*, 87, 545–561, 2004.

Curtis, C.L., Harwood, J.L., Dent, C.M., and Caterson, B., Biological basis for the benefit of nutraceutical supplementation in arthritis, *Drug Discovery Today*, 9, 165–172, 2004.

Davis, A.L., McNeill, G.P., and Caswell. D.C., Analysis of conjugated linoleic acid isomers by ^{13}C NMR spectroscopy, *Chem. Phys. Lipids*, 97, 155–165, 1999a.

Davis, A.L., McNeill, G.P., and Caswell. D.C., Identification and quantification of conjugated linoleic acid isomers in fatty acid mixtures by ^{13}C NMR spectroscopy, in *Advances in Conjugated Linoleic Acid Research*, Vol. 1, Yurawecz, M.P., Mossoba, M.M., Kramer, J.K.G., Pariza, M.W., and Nelson, G.J. (Eds.), AOCS Press, Champaign, IL, 1999b, pp. 164–179.

Delmonte, P., Kataoka, A., Corl, B.A., Bauman, D.E., and Yurawecz, M.P., Relative retention order of all isomers of *cis*/*trans* conjugated linoleic acid FAME from the 6,8- to 13,15-positions using silver ion HPLC with two elution systems, *Lipids*, 40, 509–514, 2005.

Delmonte, P., Roach, J.A.G., Mossoba, M.M., Losi, G., and Yurawecz, M.P., Synthesis, isolation, and GC analysis of all the 6,8- to 13,15-*cis*/*trans* conjugated linoleic acid isomers, *Lipids*, 39, 185–191, 2004.

Delmonte, P., Roach, J.A.G., Mossoba, M.M., Morehouse, K.M., Lehmann, L., and Yurawecz, M.P., Synthesis and isolation of *trans*-7,*cis*-9-octadecadienoic acid and other CLA isomers by base conjugation of partially hydrogenated gamma-linolenic acid, *Lipids*, 38, 579–583, 2003.

Diehl, B.W.K., High resolution NMR spectroscopy, *Eur. J. Lipid Sci. Technol.*, 103, 830–834, 2001.

Dionisi, F., Golay, P.A., and Fay, L.B., Influence of milk fat presence on the determination of *trans* fatty acids in fats used for infant formula, *Analytica Chim. Acta*, 465, 395–407, 2002.

Dobson, G., Identification of conjugated fatty acids by gas chromatography–mass spectrometry of 4-methyl-1,2,4-triazoline-3,5-dione adducts, *J. Am. Oil Chem. Soc.*, 75, 137–142, 1998.

Dobson, G., Analysis of fatty acids in functional foods with emphasis on ω3 fatty acids and conjugated linoleic acid, in *Methods of Analysis for Functional Foods and Nutraceuticals*, Hurst, W.J. (Ed.), CRC Press LLC, Boca Raton, FL, 2002, pp. 65–99.

Dobson, G. and Christie, W.W., Structural analysis of fatty acids by mass spectrometry of picolinyl esters and dimethyloxazoline derivatives, *Trends Anal. Chem.*, 15, 130–136, 1996.

Dobson, G., Christie, W.W., and Nikolova-Damyanova, B., Silver-ion chromatography of lipids and fatty acids, *J. Chromatogr. B*, 671, 197–222, 1995.

Dodds, E.D., McCoy, M.R., Geldenhuys, A., Rea, L.D., and Kennish, J.M., Microscale recovery of total lipids from fish tissue by accelerated solvent extraction, *J. Am. Oil Chem. Soc.*, 81, 835–840, 2004.

Dugan, M.E.R., Aalhus, J.L., Schaefer, J.A.L., and Kramer, J.K.G., The effects of conjugated linoleic acid on fat to lean repartitioning and feed conversion in pigs, *Can. J. Anim. Sci.*, 77, 723–725, 1997.

Enig, M.G., Rationale for lauric oils as physiologically functional foods for human immunodeficiency virus-infected individuals, *Inform*, 5, 471F (Abstr.), 1994.

Eras, J., Montanes, F., Ferran, J., and Canela, R., Chlorotrimethylsilane as a reagent for gas chromatographic analysis of fats and oils, *J. Chromatogr. A*, 918, 227–232, 2001.

Eulitz, K., Yurawecz, M.P., Sehat, N., Fritsche, J., Roach, J.A.G., Mossoba, M.M., Kramer, J.K.G., Adlof, R.O., and Ku Y., Preparation, separation, and confirmation of the eight geometrical *cis/trans* conjugated linoleic acid isomers 8,10- through 11,13-18:2, *Lipids*, 34, 873–877, 1999.

Evans, M.E., Brown, J.M., and McIntosh, M.K., Isomer-specific effects of conjugated linoleic acid (CLA) on adiposity and lipid metabolism, *J. Nutr. Biochem.*, 13, 508–516, 2002.

Fernie, C.E., Dupont, I.E., Scruel, O., Carpentier, Y.A., Sebedio, J.L., and Scrimgeour, C.M., Relative absorption of conjugated linoleic acid as triacylglycerol, free fatty acid and ethyl ester in a functional food matrix, *Eur. J Lipid Sci. Technol.*, 106, 347–354, 2004.

Folch, J., Lees, M., and Stanley, G.H.S., A simple method for the isolation and purification of lipids from animal tissues, *J. Biol. Chem.*, 226, 497–509, 1957.

Foster, J.G., Clapham, W.M., and Fedders, J.M., Quantification of fatty acids in forages by near-infrared reflectance spectroscopy, *J. Agric. Food Chem.*, 54, 3186–3192, 2006.

Fritsche, J. and Steinhart, H., Analysis, occurrence, and physiological properties of *trans* fatty acids (TFA) with particular emphasis on conjugated linoleic acid isomers (CLA)—A review, *Fett/Lipid*, 100, 190–210, 1998.

Fritsche, J., Rickert, R., Steinhart, H., Yurawecz., M.P., Mossoba, M.M., Sehat, N., Roach, J.A.G., Kramer, J.K.G., and Ku, Y., Conjugated linoleic acid (CLA) isomers: Formation, analysis, amounts in foods and dietary intake, *Fett/Lipid*, 101, 272–276, 1999.

Garcia, D.J., Omega-3 long-chain PUFA nutraceuticals, *Food Technol.*, 52, 44–49, 1998.

Garcia-Ayuso, L.E., Velasco, J., Dobarganes, M.C., and Luque de Castro, M.D., Accelerated extraction of the fat content in cheese using a focused microwave-assisted Soxhlet device, *J. Agric. Food Chem.*, 47, 2308–2315, 1999.

Garg, M.L., Wood, L.G., Singh, H., and Moughan, P.J., Means of delivering recommended levels of long chain *n*-3 polyunsaturated fatty acids in human diets, *J. Food Sci.*, 71, R66-R71, 2006.

Gertz, Ch. and Fiebeg, H.-J., Determination of fat content by the Caviezel method (rapid method), *Eur. J. Lipid Technol.*, 102, 154–158, 2000.

Golay, P.-A., Dionisi, F., Hug, B., Giuffrida, F., and Destaillats, F., Direct quantification of fatty acids in dairy powders with special emphasis on *trans* fatty acid content, *Food Chem.*, 101, 1115–1120, 2006.

Harper, C.R. and Jacobson, T.A., Usefulness of omega-3 fatty acids and the prevention of coronary heart disease, *Am. J. Cardiol.*, 96, 1521–1529, 2005.

Haumann, B.F., Alternative sources for *n*-3 fatty acids, *Inform*, 9, 1108–1119, 1998.

Hilliam, M., Functional foods: The western consumer viewpoint, *Nutr. Rev.*, 54, S189–S194, 1996.

Horrocks, L.A. and Yeo, Y.K., Health benefits of docosahexaenoic acid (DHA), *Pharmacol. Res.*, 40, 211–225, 1999.

Hubbard, W.D., Sheppard, A.J., Newkirk, D.R., Prosser, A.R., and Osgood, T., Comparison of various methods for the extraction of total lipids, fatty acids, cholesterol, and other sterols from food products, *J. Am. Oil Chem. Soc.*, 54, 81–83, 1977.

Hyotylainen, T., Kallio, M., Lehtonen, M., Lintonen, S., Perajoki, P., Jussila, M., and Riekkola, M.-J., Comprehensive two-dimensional gas chromatography in the analysis of dietary fatty acids, *J. Sep. Sci.*, 27, 459–467, 2004.

Issa, A.M., Mojica, W.A., Morton, S.C., Traina, S., Newberry, S.J., Hilton, L.G., Garland, R.H., and MacLean, C.H., The efficacy of omega-3 fatty acids on cognitive function in aging and dementia: A systematic review, *Dementia Geriatr. Cognit. Disor*ders, 21, 88–96, 2006.

Iverson, J.L. and Sheppard A.J., Determination of fatty acids in butter fat, using temperature-programmed gas chromatography of the butyl esters, *Food Chem.*, 21, 223–234, 1986.

Jang, E.S., Jung, M.Y., and Min, D.B., Hydrogenation for low *trans* and high conjugated fatty acids, *Comp. Rev. Food Sci. Food Saf.*, 1, 22–30, 2005.

Jimenez-Colmenero, F., Carballo, J., and Cofrades, S., Healthier meat and meat products: Their role as functional foods, *Meat Sci.*, 59, 5–13, 2001.

Juaneda, P., Utilisation of reversed-phase high-performance liquid chromatography as an alternative to silver-ion chromatography for the separation of *cis*- and *trans*-C18:1 fatty acid, isomers, *J. Chromatogr. A*, 954, 285–289, 2002.

Kim, K.S., Park, S.H., and Choung, M.G., Nondestructive determination of oil content and fatty acid composition in perilla seeds by near-infrared spectroscopy, *J. Agric. Food Chem.*, 55, 1679–1685, 2007.

Kim,Y.J. and Liu, R.H., Increase of conjugated linoleic acid content in milk by fermentation with lactic acid bacteria, *J. Food Sci.*, 67, 1731–1737, 2002.

Kolanowski, W. and Laufenberg, G., Enrichment of food products with polyunsaturated fatty acids by fish addition, *Eur. Food Res. Technol.*, 222, 472–477, 2006.

Kolanowski, W., Swiderski, F., and Berger, S., Possibilities of fish oil application for food products enrichment with ω-3 PUFA, *Int. J. Food Sci. Nutr.*, 50, 39–49, 1999.

Kolanowski, W., Swiderski, F., Jaworska, D., and Berger, S., Stability, sensory quality, texture properties and nutritional value of fish-oil enriched spreadable fat, *J. Sci. Food Agric.*, 84, 2135–2141, 2004.

Koletzko, B., Agostoni, C., Carlson, S.E., Clandinin, T., Hornstra, G., Neuringer, M., Uauy, R., Yamashiro, Y., and Willatts, P., Long chain polyunsaturated fatty acids (LC-PUFA) and perinatal development, *Acta Paediatrica*, 90, 460–464, 2001.

Kotani, A., Kusu, F., and Takamura, K., New electrochemical detection method in high-performance liquid chromatography for determining free fatty acids, *Analytica Chim. Acta*, 465, 199–206, 2002.

Kovacs, E.M.R. and Mela, D.J., Metabolically active functional food ingredients for weight control, *Obesity Rev.*, 7, 59–78, 2006.

Kramer, J.K.G. and Zhou, J., Conjugated linoleic acid and octadecenoic acids: Extraction and isolation of lipids, *Eur. J. Lipid Sci. Technol.*, 103, 594–600, 2001.

Kramer, J.K.G, Blackadar, C.B., and Zhou, J., Evaluation of two GC columns (60-m SUPELCOWAX 10 and 100-m CP Sil 88) for analysis of milkfat with emphasis on CLA, 18:1, 18:2 and 18:3 isomers and short- and long-chain FA, *Lipids*, 37, 823–835, 2002.

Kramer, J.K.G., Cruz-Hernandez, C., Deng, Z., Zhou, J., Jahreis, G., and Dugan, M.E.R., Analysis of conjugated linoleic acid and *trans* 18:1 isomers in synthetic and animal products, *Am. J. Clin. Nutr.*, 79, 1137S–1145S, 2004.

Kramer, J.K.G., Fellner, V., Dugan, M.E.R., Sauer, F.D., Mossoba, M.M., and Yurawecz, M.P., Evaluating acid and base catalysts in the methylation of milk and rumen fatty acids with special emphasis on conjugated dienes and total *trans* fatty acids, *Lipids*, 32, 1219–1228, 1997.

Kramer, J.K.G., Sehat, N., Fritsche, J., Mossoba, M.M., Eulitz, K., Yurawecz, M.P., and Ku, Y., Separation of conjugated fatty acid isomers, in *Advances in Conjugated Linoleic Acid Research*, Vol. 1, Yurawecz, M.P., Mossoba, M.M., Kramer, J.K.G., Pariza, M.W., and Nelson, G.J. (Eds.), AOCS Press, Champaign IL, 1999, pp. 83–109.

Kritchevsky, D., Antimutagenic and some other effects of conjugated linoleic acid, *Br. J. Nutr.*, 83, 459–465, 2000.

Kritchevsky, D., Conjugated linoleic acid in experimental atheroscerosis, in *Advances in Conjugated Linoleic Acid Research*, Vol. 2, Sebedio, J.-L., Christie, W.W., and Adlof, R. (Eds.), AOCS Press, Champaign, IL, 2003, pp. 292–301.

Kuklev, D.V. and Smith, W.L., A procedure for preparing oxazolines of highly unsaturated fatty acids to determine double bond positions by mass spectrometry, *J. Lipid Res.*, 44, 1060–1066, 2003.

Larsen, T.M., Toubro, S., and Astrup, A., Efficacy and safety of dietary supplements containing CLA for the treatment of obesity: Evidence from animal and human studies, *J. Lipid Res.*, 44, 2234–2241, 2003.

Lauritzen, D., Food enrichment with marine omega-3 fatty acids, *Food Ingredients*, 1–2, 41–44, 1994.

Lavillonniere, F., Martin, J.C., Bougnoux, P., and Sebedio, J.-L., Analysis of conjugated linoleic acid isomers and content in French cheeses, *J. Am. Oil Chem. Soc.*, 75, 343–352, 1998.

Ledoux, M. and Laloux, L., Conjugated linoleic acids: Occurrence in food and physiological properties, *Sci. Aliment.*, 26, 291–314, 2006.

Ledoux, M., Laloux, L., and Wolff, R.L., Analytical methods for determination of *trans*-C18 fatty acid isomers in milk fat. A review, *Analysis*, 28, 402–412, 2000.

Lee, J.H., Lee, K.-T., Akoh, C.C., Chung, S.K., and Kim, M.R., Antioxidant evaluation and oxidative stability of structured lipids from extravirgin olive oil and conjugated linoleic acid, *J. Agric. Food Chem.*, 54, 5416–5421, 2006.

Lee, K.W., Lee., H.J., Cho, H.Y., and Kim, Y.J., Role of the conjugated linoleic acid in the prevention of cancer, *Crit. Rev. Food Sci. Nutr.*, 45, 135–144, 2005.

Li, D., Bode, O., Drummond, H., and Sinclair, A.J., Omega-3 (n-3) fatty acids, in *Lipids for Functional Foods and Nutraceuticals*, Gunstone, F.D. (Ed.), The Oily Press Ltd, Bridgwater, England, 2003, pp. 225–262.

Li, H., Van de Voort, F.R., Ismail, A.A., Sedman, J., and Cox, R., *Trans* determination of edible oils by Fourier transform near-infrared spectroscopy, *J. Am. Oil Chem. Soc.*, 77, 1061–1067, 2000.

Lim, W.S., Gammack, J.K., Van Niekerk, J., and Dangour, A.D., Omega 3 fatty acid for the prevention of dementia, *Cochrane Database Syst. Rev.*, Issue 1, Art. No. CD005379, 2006.

Lima, E.S. and Abdalla, D.S.P., High-performance liquid chromatography of fatty acids in biological samples, *Analytica Chim. Acta*, 465, 81–91, 2002.

Lin, H., Boylston, T.D., Luedecke, L.O., and Schultz, T.D., Factors affecting the conjugated linoleic acid content of cheddar cheese, *J. Agric. Food Chem.*, 46, 801–807, 1998.

Lock, A.L. and Bauman, D.E., Modifying milk fat composition of dairy cows to enhance fatty acids beneficial to human health, *Lipids*, 39, 1197–1206, 2004.

MacLean, C.H., Newberry, S.J., Mojica, W.A., Khanna, P., Issa, A.M., Suttorp, M.J., Lim, Y.W., Traina, S.B., Hlton, L., Garland, R., and Morton, S.C., Effects of omega-3 fatty acids on cancer risk—A systematic review, *J. Am. Med. Assoc.*, 295, 403–415, 2006.

Madsen, S., Development of a reduced-fat spread enriched with long-chain n-3 fatty acids, *Lipid Technol.*, November, 129–132, 1998.

McDonald, B.E. and Fitzpatrick, K., Designer vegetable oils, in *Functional Foods Biochemical and Processing Aspects*, Mazza, G. (Ed.), Technomic Publishing Company, Lancaster, PA, 1998, pp. 265–291.

McDonald, R.E. and Mossoba, M.M., *Trans* fatty acids: Labelling, nutrition, and analysis, in *Food Lipids and Health*, McDonald, R.E. and Min, D.B. (Eds.), Marcel Dekker, New York, 1996, pp. 161–197.

McLean, B. and Drake, P., *Review of Methods for the Determination of Fat and Oil in Foodstuffs*, Review No. 37, Camden & Chorleywood Food Research Association Group, Chipping Campden, Gloucestershire, United Kingdom, 2002.

Meier, S., Mjos, S.A., Joensen, H., and Grahl-Nielsen, O., Validation of a one-step extraction/methylation method for determination of fatty acids and cholesterol in marine tissues, *J. Chromatogr. A*, 1104, 291–298, 2006.

Meurens, M., Baeten, V., Yan, S.H., Mignolet, E., and Larondelle, Y., Determination of the conjugated linoleic acids in cow's milk by Fourier transform Raman spectroscopy, *J. Agric. Food Chem.*, 53, 5831–5835, 2005.

Michaud, A.L., Lawrence, P., Adlof, R., and Brenna, J.T., On the formation of conjugated linoleic acid diagnostic ions with acetonitrile chemical ionization tandem mass spectrometry, *Rapid Commn. Mass Spectrom.*, 19, 363–368, 2005.

Michaud, A.L., Yurawecz, M.P., Delmonte, P., Corl, B.A., Bauman, D.E., and Brenna, J.T., Identification and characterization of conjugated fatty acid methyl esters of mixed double bond geometry by acetonitrile chemical ionization tandem mass spectrometry, *Anal. Chem.*, 75, 4925–4930, 2003.

Michelsen, B., Madsen, S.L., and Gotfredsen, P., Polyunsaturated fatty acids, in *Guide to Functional Food Ingredients*, Young, J. (Ed.), Leatherhead Publishing, Leatherhead, Surrey, United Kingdom, 2001, pp. 141–195.

Mickleborough, T.D. and Rundell, K.W., Dietary polyunsaturated fatty acids in asthma- and exercise-induced bronchoconstriction, *Eur J. Clin Nutr.*, 59, 1335–1346, 2005.

Miwa, H., High-performance liquid chromatographic determination of free fatty acids and esterified fatty acids in biological materials as their 2-nitrophenylhydrazides, *Analytica Chim. Acta*, 465, 237–255, 2002.

Mjos, S.A., Identification of fatty acids in gas chromatography by application of different temperature and pressure programs on a single capillary column, *J. Chromatogr. A*, 1015, 151–161, 2003.

Mjos, S.A. and Pettersen, J., A rapid method for the analysis of hydrogenated fats by GC with IR detection, *J. Am. Oil Chem. Soc.*, 80, 839–846, 2003.

Molkentin, J., Bioactive lipids naturally occurring in bovine milk, *Nahrung*, 43, 185–189, 1999.

Morrison, W.R., Tan, S.L., and Hargin, K.D., Methods for the quantitative analysis of lipids in cereal grains and similar tissues, *J. Sci. Food Agric.*, 31, 329–340, 1980.

Mossoba, M.M., Kramer, J.K.G., Yurawecz, M.P., Sehat, N., Roach, J.A.G., Eulitz, K., Fritsche, J., Dugan, M.E.R., and Ku, Y., Impact of novel methodologies on the analysis of conjugated linoleic acid (CLA). Implications of CLA feeding studies, *Fett/Lipid*, 101, 235–243, 1999a.

Mossoba, M.M., McDonald, R.E., Yurawecz, M.P., and Kramer, J.K.G., Application of on-line capillary GC-FTIR spectroscopy to lipid analysis, *Eur. J. Lipid Sci. Technol.*, 103, 826–830, 2001.

Mossoba, M.M., Yurawecz, M.P., Kramer, J.K.G., Eulitz, K.D., Fritsche, J., Sehat, N., Roach, J.A.G., and Ku, Y., Confirmation of conjugated linoleic acid geometric isomers by capillary gas chromatography–Fourier transform infrared spectroscopy, in *Advances in Conjugated Linoleic Acid Research*, Vol. 1, Yurawecz, M.P., Mossoba, M.M., Kramer, J.K.G., Pariza, M.W., and Nelson, G.J. (Eds.), AOCS Press, Champaign, IL, 1999b, pp. 141–151.

Muggli, R., Nutritional aspects of omega-3 long-chain polyunsaturated fatty acids, *Agro Food Ind. Hi-Tech*, January/February, 35–36, 1997.

Muller, A., Dusterioh, K., Ringseis, R., Eder, K., and Steinhart, H., Development of an alternative eluent system for Ag^+-HPLC of conjugated linoleic acid isomers, *J. Sep. Sci.*, 29, 358–365, 2006.

Murrieta, C.M., Hess, B.W., and Rule, D.C., Comparison of acidic and alkaline catalysts for preparation of fatty acid methyl esters from ovine muscle with emphasis on conjugated linoleic acid, *Meat Sci.*, 65, 523–529, 2003.

Nagao, K. and Yanagita, T., Conjugated fatty acids in food and their health benefits, *J. Biosci. Bioeng.*, 100, 152–157, 2005.

Newton, I. and Snyder, D., Nutritional aspects of long-chain omega-3 fatty acids and their use in bread enrichment, *Cereal Foods World*, 42, 126–131, 1997.

Nikolova-Damyanova, B., Momchilova, S., and Christie, W.W., Silver ion high-performance liquid chromatographic separation of conjugated linoleic acid isomers, and other fatty acids, after conversion to *p*-methoxyphenacyl derivatives, *J. High Resol. Chromatogr.*, 23, 348–352, 2000.

Nishimura, K., Suzuki, T., Momchilova, S., Miyashita, K., Katsura, E., and Itabashi, Y., Analysis of conjugated linoleic acids as 9-anthrylmethyl esters by reversed-phase high-performance liquid chromatography with fluorescence detection, *J. Chromatogr. Sci.*, 43, 494–499, 2005.

Ohba, Y., Kuroda, N., and Nakashima, K., Liquid chromatography of fatty acids with chemiluminescence detection, *Analytica Chim. Acta*, 465, 101–109, 2002.

Oomah, B.D., Flaxseed as a functional food source, *J. Sci. Food Agric.*, 81, 889–894, 2001.

Oomah, B.D. and Mazza, G., Flaxseed products for disease prevention, in *Functional Foods Biochemical and Processing Aspects*, Mazza, G. (Ed.), Technomic Publishing Company Inc., Lancaster, PA, 1998, pp. 91–137.

Ostrowska, E., Dunshea, F.R., Muralitharan, M., and Cross, R.F., Comparison of silver-ion high-performance liquid chromatographic quantification of free and methylated conjugated linoleic acids, *Lipids*, 35, 1147–1153, 2000.

Pacetti, D., Hulan, H.W., Schreiner, M., Boselli, E., and Frega, N.G., Positional analysis of egg triacylglycerols and phospholipids from hens fed diets enriched with refined seal blubber oil., *J. Sci. Food Agric.*, 85, 1703–1714, 2005.

Padley, F.B., Gunstone, F.D., and Harwood, J.L., Occurrence and characteristics of oils and fats, in *The Lipid Handbook*, Gunstone, F.D., Harwood, J.L., and Padley, F.B. (Eds.), Chapman and Hall, London, United Kingdom, 1994, pp. 47–223.

Pariza, M.W., The biological activities of conjugated linoleic acid, in *Advances in Conjugated Linoleic Acid Research*, Vol. 1, Yurawecz, M.P., Mossoba, M.M., Kramer, J.K.G., Pariza, M.W., and Nelson, G.J. (Eds.), AOCS Press, Champaign, IL, 1999, pp. 12–20.

Park, S.J., Park, C.W., Kim, S.J., Kim, J.K., Kim, Y.R., Park, K.A., Kim, J.O., and Ha, Y.L., Methylation methods for the quantiative analysis of conjugated analysis of conjugated linoleic acid (CLA) isomers in various lipid samples, *J. Agric. Food Chem.*, 50, 989–996, 2002.

Park, Y. and Pariza, M.W., Evidence that commercial calf and horse sera can contain substantial amounts of *trans*-10, *cis*-12 conjugated linoleic acid, *Lipids*, 33, 817–819, 1998.

Park, Y., Albright, K.J., Cai, Z.Y., and Pariza, M.W., Comparison of methylation procedures for conjugated linoleic acid and artifact formation by commercial (trimethylsilyl)diazomethane, *J. Agric. Food Chem.*, 49, 1158–1164, 2001.

Parodi, P.W., Conjugated linoleic acid in food, in *Advances in Conjugated Linoleic Acid Research*, Vol. 2, Sebedio, J.-L., Christie, W.W., and Adlof, R. (Eds.), AOCS Press, Champaign, IL, 2003, pp. 101–122.

Parrish, F.C., Jr., Wiegand, B.R., Beitz, D.C., Alun, D.U., Du, M., and Trenkle, A.H., Use of dietary CLA to improve composition and quality of animal-derived foods, in *Advances in Conjugated Linoleic Acid Research*, Vol. 2, Sebedio, J.-L., Christie, W.W., and Adlof, R. (Eds.), AOCS Press, Champaign, IL, 2003, pp. 189–217.

Peet, M. and Stokes, C., Omega-3 fatty acids in the treatment of psychiatric disorders, *Drugs*, 65, 1051–1059, 2005.

Peterson, D.G., Kelsey, J.A., and Bauman, D.E., Analysis of variation in *cis*-9, *trans*-11 conjugated linoleic acid (CLA) in milk fat of dairy cows, *J. Dairy Sci.*, 85, 2164–2172, 2002.

Ponnampalam, E.N., Mann, N.J., and Sinclair, A.J., Effect of feeding systems on omega-3 fatty acids, conjugated linoleic acid and *trans* fatty acids in Australian beef cuts: Potential impact on human health, *Asia Pac. J. Clin. Nutr.*, 15, 21–29, 2006.

Priego-Capote, F., Ruiz-Jimenez, J., and Luque de Castro, M.D., Identification and quantification of *trans* fatty acids in bakery products by gas chromatography–mass spectrometry after focused microwave Soxhlet extraction, *Food Chem.*, 100, 859–867, 2007.

Raes, K., De Smet, S., and Demeyer, D., Effect of dietary fatty acids on incorporation of long chain polyunsaturated fatty acids and conjugated linoleic acid in lamb, beef and pork meat: A review, *Anim. Feed Sci. Technol.*, 113, 199–221, 2004.

Ramaa, C.S., Shirode, A.R., Mundada, A.S., and Kadam, V.J., Nutraceuticals—An emerging era in the treatment and prevention of cardiovascular diseases, *Curr. Pharm. Biotechnol.*, 7, 15–23, 2006.

Ratnayake, W.M.N., Analysis of *trans* fatty acids, *in Trans Fatty Acids in Human Nutrition*, Sebedio, J.L. and Christie, W.W. (Eds.), The Oily Press Ltd, Dundee, Scotland, 1998, pp. 115–161.

Ratnayake, W.M.N., Overview of methods for the determination of *trans* fatty acids by gas chromatography, silver-ion thin-layer chromatography, and gas chromatography/mass spectrometry, *J. AOAC Int.*, 87, 523–539, 2004.

Ratnayake, W.M.N., Hansen, S.L., and Kennedy, M.P., Evaluation of the CP-Sil 88 and SP-2560 GC columns used in the recently approved AOCS official method Ce 1h-05: Determination of *cis*-, *trans*-, saturated, monounsaturated and polyunsaturated fatty acids in vegetable or non-ruminant animal oils and fats by capillary GLC method, *J. Am. Oil Chem. Soc.*, 83, 475–488, 2006.

Ratnayake, W.M.N., Plouffe, L.J., Pasquier, E., and Gagnon, C., Temperature-sensitive resolution of *cis*- and *trans*-fatty acid isomers of partially hydrogenated vegetable oils on SP-2560 and CP-Sil 88 capillary columns, *J. Assoc. Off. Anal. Chem. Int.*, 85, 1112–1118, 2002.

Reaney, M.J.T., Liu, Y.-D., and Taylor, W.G., Gas chromatographic analysis of Diels-Alder adducts of geometrical and positional isomers of conjugated linoleic acid, *J. Am. Oil Chem. Soc.*, 78, 1083–1086, 2001.

Richardson, A.J., Long-chain polyunsaturated fatty acids in childhood developmental and psychiatric disorders, *Lipids*, 39, 1215–1222, 2004.

Rickert, R., Steinhart, H., Fritsche, J., Sehat, N., Yurawecz, M.P., Mossoba, M.M., Roach, J.A.G., Eulitz, K., Ku, Y., and Kramer, J.K.G., Enhanced resolution of conjugated linoleic acid isomers by tandem-column silver-ion high performance liquid chromatography, *J. High Resol. Chromatogr.*, 22, 144–148, 1999.

Roach, J.A.G., Identification of CLA isomers in food and biological extracts by mass spectrometry, in *Advances in Conjugated Linoleic Acid Research*, Vol. 1, Yurawecz, M.P., Mossoba, M.M., Kramer, J.K.G., Pariza, M.W., and Nelson, G.J. (Eds.), AOCS Press, Champaign, IL, 1999, pp. 126–140.

Roach, J.A.G., Yurawecz, M.P., Kramer, J.K.G., Mossoba, M.M., Eulitz, K., and Ku, Y., Gas chromatography-high resolution selected-ion mass spectrometric identification of trace 21:0 and 20:2 fatty acids eluting with conjugated linoleic acid isomers, *Lipids*, 35, 797–802, 2000.

Roche, H.M., Noone, E., Nugent, A., and Gibney, M.J., Conjugated linoleic acid: A novel therapeutic nutrient?, *Nutr. Res. Rev.*, 14, 173–187, 2001.

Rose, E.L. and Holub, B.F., Effects of a liquid egg product containing fish oil on selected cardiovascular disease risk factors: A randomized crossover trial, *Food Res. Int.*, 39, 910–916, 2006.

Rosenfeld, J.M., Application of analytical derivatizations to the quantitative and qualitative determination of fatty acids, *Analytica Chim. Acta*, 465, 93–100, 2002.

Ruiz-Jimenez, J., Priego-Capote, F., and Luque de Castro, M.D., Identification and quantification of *trans* fatty acids in bakery products by gas chromatography–mass spectrometry after dynamic ultrasound-assisted extraction, *J. Chromatogr. A*, 1045, 203–210, 2004.

Ruxton, C.H.S., Reed, S.C., Simpson, M.J.A., and Millington, K.J., The health benefits of omega-3 polyunsaturated fatty acids: A review of the evidence, *J. Hum. Nutr. Diet.*, 17, 449–459, 2004.

Salas-Salvado, J., Marquez-Sandoval, F., and Bullo, M., Conjugated linoleic acid intake in humans: A systematic review focusing on its effect on body composition, glucose and lipid metabolism, *Crit. Rev. Food Sci. Nutr.*, 46, 479–488, 2006.

SanGiovanni, J.P. and Chew, E.Y., The role of omega-3 long-chain polyunsaturated fatty acids in health and disease of the retina, *Prog. Retinal Eye Res.*, 24, 87–138, 2005.

Schmid, A., Collomb, M., Sieber, R., and Bee, G., Conjugated linoleic acid in meat and meat products: A review, *Meat Sci.*, 73, 29–41, 2006.

Schmidl, M.K., The role of lipids in medical and designer foods, in *Food Lipids and Health*, McDonald, R.E and Min, D.B. (Eds.), Marcel Dekker Inc., New York, 1996, pp. 417–436.

Schreiner, M., Quantification of long chain polyunsaturated fatty acids by gas chromatography. Evaluation of factors affecting accuracy, *J. Chromatogr. A*, 1095, 126–130, 2005.

Schuhardt, U. and Lopes, O.C., Tetramethylguanidine catalysed transesterification of fats and oils: A new method for rapid determination of their composition, *J. Am. Oil Chem. Soc.*, 65, 1940–1941, 1988.

Sehat, N., Kramer, J.K.G., Mossoba, M.M., Yurawecz, M.P., Roach, J.A.G., Eulitz, K., Morehouse, K.M., and Ku, Y., Identification of conjugated linoleic acid isomers in cheese by gas chromatography, silver ion high performance liquid chromatography and mass spectral reconstructed ion profiles. Comparison of chromatographic elution sequences, *Lipids*, 33, 963–971, 1998a.

Sehat, N., Rickert, R., Mossoba, M.M., Kramer, J.K.G., Yurawecz, M.P., Roach, J.A.G., Adlof, R.O., Morehouse, K.M., Fritsche, J., Eulitz, K.D., Steinhart, H., and Ku, Y., Improved separation of conjugated fatty acid methyl esters by silver-ion high-performance liquid chromatography, *Lipids*, 34, 407–413, 1999.

Sehat, N., Yurawecz., M.P., Roach, J.A.G., Mossoba, M.M., Kramer, J.K.G., and Ku, Y., Silver-ion high performance liquid chromatographic separation and identification of conjugated linoleic acid isomers, *Lipids*, 33, 217–221, 1998b.

Senorans, F.J. and Ibanez, E., Analysis of fatty acids in foods by supercritical fluid chromatography, *Analytica Chim. Acta*, 465, 131–144, 2002.

Shahidi, F., Functional seafood lipids and proteins, in *Functional Foods. Biochemical and Processing Aspects*, Mazza, G. (Ed.), Technomic Publishing Company Inc., Lancaster, PA, 1998, pp. 381–401.

Shahidi, F. and Miraliakbari, H., Omega-3 fatty acids in health and disesase: Part 2—Health effects of omega-3 fatty acids in autoimmune diseases, mental health, and gene expression, *J. Med Food*, 8, 133–148, 2005.

Shahidi, F. and Wanasundara, J.P.D., Extraction and analysis of lipids, in *Food Lipids Chemistry, Nutrition and Biotechnology*, Akoh, C.C. and Min, D.B. (Eds), Marcel Dekker, New York, 1997, pp. 115–136.

Shang, X.G., Wang, F.L., Li, D.F., Yin, J.D., Li, X.J., and Yi G.F., Effect of dietary conjugated linoleic acid on the fatty acid composition of egg yolk, plasma and liver as well as hepatic stearoyl-coenzyme A desaturase activity and gene expression in laying hens, *Poult. Sci.*, 84, 1886–1892, 2005.

Shantha, N.C., Decker, E.A., and Hennig, B., Comparison of methylation methods for the quantitation of conjugated linoleic acid isomers, *J. AOAC Int.*, 76, 644–649, 1993.

Sharma, R., Market trends and opportunities for functional dairy beverages, *Aust. J. Dairy Technol.*, 60, 195–198, 2005.

Shirasawa, S., Sasaki, A., Saida, Y., and Satoh, C., A rapid method for *trans*-fatty acid determination using a single capillary column, *J. Oleo Sci.*, 56, 53–58, 2007.

Sieber, R., Collomb, M., Aeschlimann, A., Jelen, P., and Eyer, H., Impact of microbial cultures on conjugated linoleic acid in dairy products—A review, *Int. Dairy J.*, 14, 1–15, 2004.

Sim, J.S., Designer eggs and their nutritional and functional significance, *World Rev. Nutr. Diet.*, 83, 89–101, 1998.

Simopoulos, A.P., Fatty acids, in *Designer Foods, Pharmafoods, Nutraceuticals*, Goldberg, I. (Ed.), International Thomson Publishing, New York, 1994, pp. 355–392.

Simopoulos, A.P., Human requirement for *n*-3 polyunsaturated fatty acids, *Poult. Sci.*, 79, 961–970, 2000.

Spitzer,V., Marx, F., and Pfeilsticker, K., Electron impact mass spectra of the oxazoline derivatives of some conjugated diene and triene C_{18} fatty acids, *J. Am. Oil Chem. Soc.*, 71, 873–876, 1994.

Stanton, C., Murphy, E., McGrath, E., and Devery, R., Animal feeding strategies for conjugated linoleic acid enrichment of milk, in *Advances in Conjugated Linoleic Acid Research*, Vol. 2, Sebedio, J.-L., Christie, W.W., and Adlof, R. (Eds), AOCS Press, Champaign, IL, 2003, pp. 123–145.

Surai, P.F. and Sparks, N.H.C., Designer eggs: From improvement of egg composition to functional food, *Trends Food Sci. Technol.*, 12, 7–16, 2001.

Szymczyk, B. and Pisulewski, P.M., Effects of dietary conjugated linoleic acid on fatty acid composition and cholesterol content of hen egg yolks, *Br. J. Nutr.*, 90, 93–99, 2003.

Szymczyk, B. and Pisulewski, P.M., Effects of dietary conjugated linoleic acid isomers and vitamin E on fatty acid composition and cholesterol content of hen egg yolks, *J. Anim. Feed Sci.*, 14, 109–123, 2005.

Thurnhofer, S. and Vetter, W., A gas chromatography/electron ionization-mass spectrometry-selected ion monitoring method for determining the fatty acid pattern in food after formation of fattty acid methyl esters, *J. Agric. Food Chem.*, 53, 8896–8903, 2005.

Thurnhofer, S. and Vetter, W., Application of ethyl esters and d_3-methyl esters as internal standards for the gas chromatographic quantification of transesterified fatty acid methyl esters in food, *J. Agric. Food Chem.*, 54, 3209–3214, 2006.

Toschi, T.G., Bendini, A., Ricci, A., and Lercker, G., Pressurized solvent extraction of total lipids in poultry meat, *Food Chem.*, 83, 551–555, 2003.

Troegeler-Meynadier, A. and Enjalbert, F., Conjugated linoleic acids (CLA): Biological interests in nutrition, *Rev. Med. Vet.*, 156, 207–216, 2005.

Uauy-Dagach, R. and Valenzuela, A., Marine oils: The health benefits of *n*-3 fatty acids, *Nutr. Rev.*, 54, S102–S108, 1996.

Ulberth, F. and Henninger, M., One-step extraction/methylation method for determining the fatty acid composition of processed foods, *J. Am. Oil Chem. Soc.*, 69, 174–177, 1992.

Van de Voort, F.R., Sedman, J., and Russin, T., Lipid analysis by vibrational spectroscopy, *Eur. J. Lipid Sci. Technol.*, 103, 815–826, 2001.

Van Elswyk, M., Nutritional and physiological effects of flaxseed in diets for laying fowl, *World's Poult. Sci J.*, 53, 253–264, 1997.

Wahle, K.W.J., Caruso, D., Ochoa, J.J., and Quiles, J.L., Olive oil and modulation of cell signalling in disease prevention, *Lipids*, 39, 1223–1231, 2004.

Wahle, K.W.J., Heys, S.D., and Rotondo, D., Conjugated linoleic acids: Are they beneficial or detrimental?, *Prog. Lipid Res.*, 43, 553–587, 2004.

Wang, Y.W. and Jones, P.J.H., Conjugated linoleic acid and obesity control: Efficacy and mechanism, *Int. J. Obesity*, 28, 941–955, 2004.

Watkins, B.A. and Seifert, M.F., Conjugated linoleic acid and bone biology, *J. Am. Coll. Nutr.*, 19, 478S–486S, 2000.

Watkins, B.A., Feng, S.L., Strom, A.K., DeVitt, A.A., Yu, L.G., and Li, Y., Conjugated linoleic acids alter the fatty acid composition and physical properties of egg yolk and albumen, *J. Agric. Food Chem.*, 51, 6870–6876, 2003.

Watkins, B.A., Li, Y., Lippman, H.E., and Seifert, M.F., Omega-3 polyunsaturated fatty acids and skeletal health, *Exp. Biol. Med.*, 226, 485–497, 2001.

Werner, S.A., Luedecke. L.O., and Schultz, T.D., Determination of conjugated linoleic acid content and isomer distribution in three cheddar-type cheeses: Effects of cheese cultures, processing, and aging, *J. Agric. Food Chem.*, 40, 1817–1821, 1992.

Western, R.J., Lau, S.S.G., Marriott, P.J., and Nichols, P.D., Positional and geometric isomer separation of FAME by comprehensive 2-D GC, *Lipids*, 37, 715–723, 2002.

Willis, W.M., Lencki, R.W., and Marangoni, A.G., Lipid modification strategies in the production of nutritionally functional fats and oils, *Crit. Rev. Food Sci. Nutr.*, 38, 639–674, 1998.

Wilson, R., Lyall, K., Payne, J.A., and Riemersma, R.A., Quantitative analysis of long-chain *trans*-monoenes originating from hydrogenated marine oil, *Lipids*, 35, 681–687, 2000.

Wolff, R.L., Content and distribution of *trans*-18:1 acids in ruminant milk and meat fats. Their importance in Europeans diets and their effect on human milk, *J. Am. Oil Chem. Soc.*, 72, 259–272, 1995.

Wolff, R.L., Bayard, C.C., and Fabien, R.J., Evaluation of sequential methods for the determination of butterfat fatty acid composition with emphasis on *trans*-18:1 acids. Application to the study of seasonal variations in French butters, *J. Am. Oil Chem. Soc.*, 72, 1471–1483, 1995.

Wolff, R.L., Precht, D., and Molkentin, J., Occurrence and distribution of *trans*-18:1 acids in edible fats of natural origin, *in Trans Fatty Acids in Human Nutrition*, Sebedio, J.L. and Christie, W.W. (Eds), The Oily Press Ltd, Dundee, Scotland, 1998, pp. 1–33.

Woods, R.K., Thien, F.C.K., and Abramson, M.J., Dietary marine fatty acids (fish oil) for asthma in adults and children, in *The Cochrane Library*, Issue 2, Update Software, Oxford, United Kingdom, 2003.

Yamasaki, M., Kishihara, K., Ikeda, I., Sugano, M., and Yamada, K., A recommended esterification method for gas chromatographic measurement of conjugated linoleic acid, *J. Am. Oil Chem. Soc.*, 76, 933–938, 1999.

Yang, B., Gunstone, F.D., and Kallio, H., Oils containing oleic, palmitoleic, γ-linolenic and stearidonic acids, in *Lipids for Functional Foods and Nutraceuticals*, Gunstone, F.D. (Ed.), The Oily Press Ltd, Bridgwater, England, 2003, pp. 263–290.

Yang, L., Huang, Y., James, A.E., Lam, L.W., and Chen, Z.Y., Differential incorporation of conjugated linoleic acid isomers into egg yolk lipids, *J. Agric. Food Chem.*, 50, 4941–4946, 2002.

Yurawecz, M.P., Hood, J.K., Roach J.A.G., Mossoba, M.M., Daniels, D.H., Ku, Y., Pariza, W., and Chin S.F., Conversion of allylic hydroxy oleate to conjugated linoleic acid and methoxy oleate by acid-catalysed methylation procedures, *J. Am. Oil Chem. Soc.*, 71, 1149–1155, 1994.

Yurawecz, M.P., Kramer, J.K.G., and Ku, Y., Methylation procedures for conjugated linoleic acid, in *Advances in Conjugated Linoleic Acid Research*, Vol. 1, Yurawecz, M.P., Mossoba, M.M., Kramer, J.K.G., Pariza, M.W., and Nelson, G.J. (Eds), AOCS Press, Champaign, IL, 1999, pp. 64–82.

Yurawecz, M.P., Roach, J.A.G., Sehat, N., Mossoba, M.M., Kramer, J.K.G., Fritsche, J., Steinhart, H., and Ku, Y., A new conjugated linoleic acid isomer, 7 *trans*,9 *cis*-octadecadienoic acid, in cow milk, cheese, beef and human milk and adipose tissue, *Lipids*, 33, 803–809, 1998.

Yurawecz, M.P., Sehat, N., Mossoba, M.M., Roach, J.A.G., Kramer, J.K.G., and Ku, Y., Variations in isomer distribution in commercially available conjugated linoleic acid, *Fett/Lipid*, 101, 277–282, 1999.

Xu, S., Boylston, T.D., and Glatz, B.A., Conjugated linoleic acid content and organoleptic attributes of fermented milk products produced with probiotic bacteria., *J. Agric. Food Chem.*, 53, 9064–9072, 2005.

Zulet, M.A., Marti, A., Parra, M.D., and Martinez, J.A., Inflammation and conjugated linoleic acid: Mechanisms of action and implications for human health, *J. Physiol. Biochem.*, 61, 483–494, 2005.

3 Analysis of Flavonoids in Functional Foods and Nutraceuticals

Laura Bravo and Raquel Mateos

CONTENTS

3.1 INTRODUCTION

Among the functional components naturally occurring in foods, plant flavonoids probably constitute the group of substances that has arisen most interest among the food scientists and nutritionists, as well as from food manufacturers and consumers in the last two decades because of their putative beneficial effects on human health. Most flavonoids are reducing agents that act as antioxidants and free radical scavengers, thus protecting against oxidative reactions, which is largely the main mechanism of action explaining the biological activity of dietary flavonoids, although not the only one. Flavonoids may play a role in the prevention of several chronic diseases such as cancer, cardiovascular disease, inflammation, neurodegenerative disorders, and other pathologies associated with oxidative stress. Epidemiological studies have shown an inverse relationship between the consumption of plant foods rich in flavonoids and the incidence of certain diseases. Thus, the intake of foods like soy, rich in isoflavones, green tea as a source of flavanols, fruits, etc. may protect against different types of cancers,[1–7] while red wide, cocoa, tea, and also high intakes of fruits and vegetables have been related with a reduced incidence of cardiovascular diseases, inflammation, etc.[3,4,8–10] These findings from epidemiological, human and animal intervention studies, as well as from in vitro mechanistic experiments have encouraged the study of these food bioactive components and their use as functional food ingredients and nutraceuticals. A nutraceutical can be defined as any food or part of a food with health beneficial effects, including a reduction on the risk of suffering from certain diseases, as well as with potential use in the treatment against different pathologies.

In this chapter, a brief overview of the chemistry and occurrence of plant flavonoids will be discussed, together with a review of the most common analytical methodologies applied for their identification and quantification in foods.

3.2 CHEMISTRY OF PLANT FLAVONOIDS

3.2.1 General Aspects

With over 5000 compounds currently described, flavonoids constitute the most widely distributed group of phenolic compounds in plants, virtually present in all higher plants.[11] In some cases, flavonoids have even been described in animals (in butterfly wings, in beaver odoriferous glands, or in propolis, a secretion formed by bees), likely derived from plant foods in these animals' diets. Flavonoids are essential to plant physiology, being involved in reproduction (helping to attract animals for pollination and seed dispersion) and plant growth, protecting plants against UV-induced damage, providing them with resistance against pathogens or predators (increasing astringency and thus unpalatability of plants), etc. Flavonoids are natural pigments, being responsible for the wide variety of colors in flowers or fruits; thus, flavonols and flavones are yellow, while anthocyanins are blue, purple, violet, red, or orange.

Flavonoids are polyphenolic compounds formed as secondary plant metabolites, derived from two biosynthetic pathways, namely the shikimate and the acetate pathways[12] (Figure 3.1). Shikimic acid, formed from carbohydrates in the Calvin cycle, gives rise to phenylpropanoids (C_6–C_3 compounds), that combine with simple phenols like resorcinol or phloroglucinol (C_6) originated in the acetate pathway resulting in the formation of chalcones and ultimately flavonoids. These compounds have a common 1,3-diphenylpropane (C_6–C_3–C_6) structure, the flavan nucleus, consisting of two aromatic rings (rings A and B, originated from the acetate and shikimate pathways, respectively) linked by a three-carbon structure that, except in chalcones and dihydrochalcones, forms an oxygenated heterocycle (Figure 3.2). Flavonoids can be divided into 13 different classes, depending on the heterocycle structure and its degree of hydroxylation. Table 3.1 shows the chemical structure of the major classes of flavonoids.

Although existing in free forms (aglycones), flavonoids are most commonly found glycosylated, usually with monosaccharides or disaccharides and also with oligosaccharides, forming *O*-glycosides or *C*-glycosides. Carbon–carbon linkages in *C*-glycosides occur between the anomeric carbon of the sugar moiety and the C6 or C8 carbon of the flavan nucleus, these linkages being resistant to acid hydrolysis in contrast with *O*-glycosides. In addition to glycosylation, acylation with aromatic or aliphatic acids and methylation can also occur.

3.2.2 Major Classes of Flavonoids

3.2.2.1 Flavonols and Flavones

Flavonols and flavones are chemically related compounds, the later lacking the hydroxyl group in the C3 position of the heterocycle characteristic of flavonols. Both types of flavonoids are most commonly found *O*-glycosylated, mainly as 3-glucosides, although 4- and 7-glucosides have also been described, and glycosylation with other sugars (rutinose, neohesperidose) is also frequent. These flavonoids, with a light yellow color, are widespread in the plant kingdom with the exception of algae and fungi, occurring mainly in the aerial parts of plants (leaves and external parts of fruits), but only in trace amounts in roots. Among the natural flavones, apigenin (4′,5,7-trihydroxyflavone) and luteolin (3′,4′,5,7-tetrahydroxyflavone) are the most common compounds (Table 3.1), being found in parsley, celery, thyme, basil, artichoke, and other herbs. Tangeretin is a polymethoxylated flavone (4′,5,6,7,8-pentamethoxyflavone), abundant in the peels of citrus fruits. Chrysin

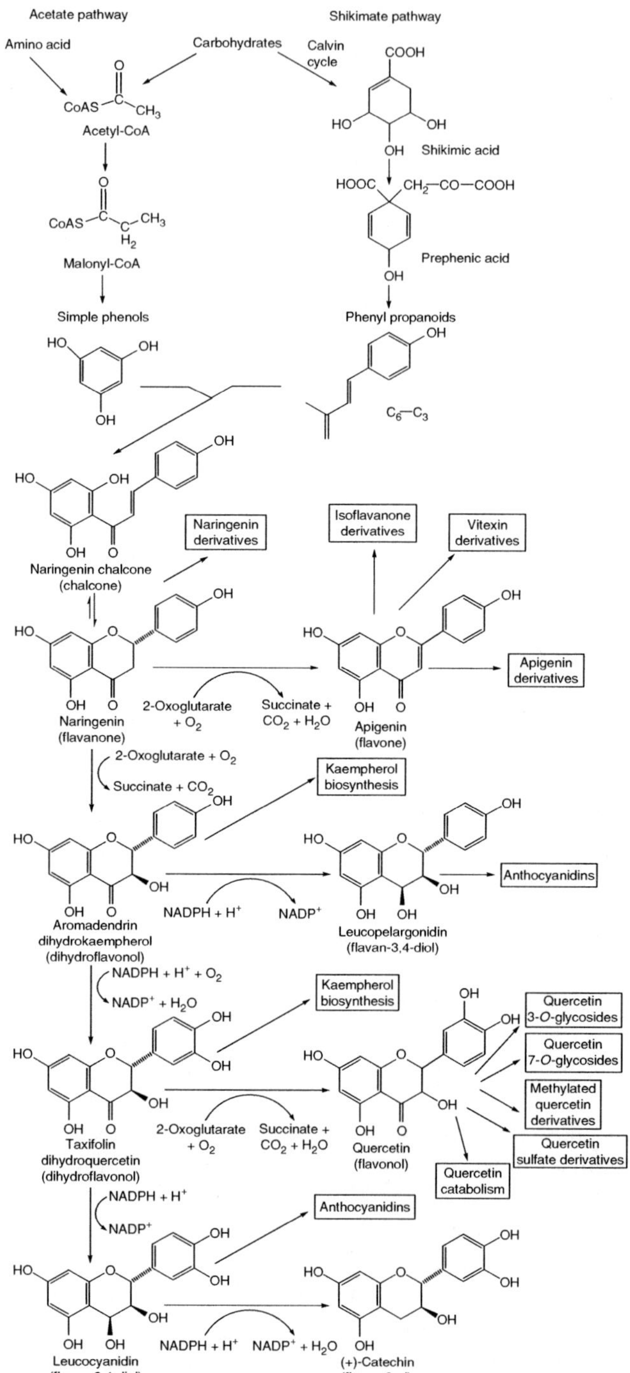

FIGURE 3.1 Pathway of flavonoids biosynthesis.

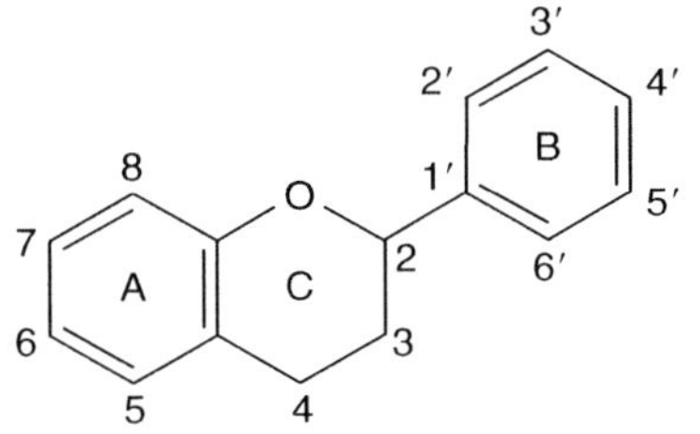

FIGURE 3.2 Basic structure and numbering system of flavonoids.

(5,7-dihydroxyflavone) is found in passion flower, honey, or propolis. Concerning flavonols, quercetin is the most abundant compound, present in many fruits, vegetables, and herbs, especially in onions, apples, berries, tea, or wine. Quercetin may be present either as the aglycone or as glycosides (rutin, quercitrin, isoquercitrin, hyperoside), with other common flavonols being myricetin and kaempferol (Table 3.1). There is a great variability in this group of phenols, with nearly 400 different flavonol glycosides described to date. Table 3.2 shows the presence of the different types of flavonoids in foods and beverages.

3.2.2.2 Isoflavones

Isoflavones differ from flavones in the position of the B-ring, linked to the C3 position of the heterocycle instead of the C2 position as in most flavonoids. Isoflavones are also termed phytoestrogens because of their estrogenic activity in mammals derived from the structural similitude between estrogens and isoflavones. They occur almost exclusively in legumes, mainly in soy, although some isoflavones can also occur in nuts, grain products, etc. The most common isoflavones are daidzein and genistein, together with their 7-*O*-glucosides daidzin and genistin, respectively.

3.2.2.3 Flavanones

Flavanones, also termed citrus bioflavonoids, are bitter compounds found mainly in citrus fruits and prunes, either as aglycones or as *C*- or *O*-glycosides. Hesperetin and naringenin (Table 3.1) are two of the most common flavanones, together with their respective 7-*O*-glycosides hesperidin (hesperetin-7-*O*-rutinoside), neohesperetin (hesperetin-7-*O*-neohesperidoside), naringin (naringenin-7-*O*-neohesperidoside), and narirutin (naringenin-7-*O*-rutinoside). Opening of the heterocycle under basic conditions and further hydrogenation transform the bitter flavanones into sweet chalcones and dihydrochalcones, respectively, although 2′-OH-chalcones are rather unstable and tend to recycle into flavanones at neutral pH. These compounds exist as pairs of diastereomers because of the chirality of the C2 position and the optically active sugar residue, the prevalence of the different isomers varying during maturation or processing.

TABLE 3.1

Chemical Structure of the Main Flavonoid Subclasses and Some of Their More Representative Components Together with Their Spectral Characteristics Recorded Online after Reversed-Phase High-Performance Liquid Chromatography (RP-HPLC) Separation Using Methanol–Water Acidified with Sodium Phosphate (pH 3.3)[a–c]

Flavonoids	Substituents			λ_{max}	Chemical Structure
	—OH	—OCH_3	–X		
Flavonols					
Quercetin	3,5,7,3′,4′			253, 268 sh, 297 sh, 368	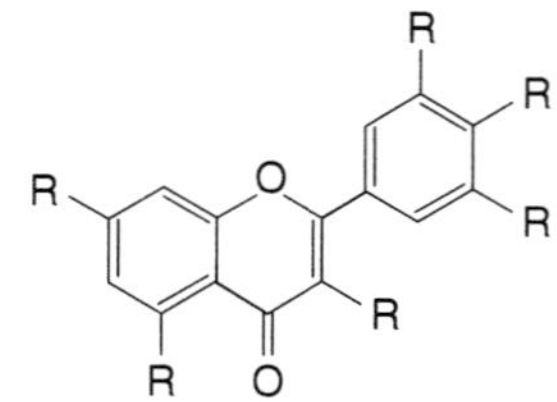
Q-3-*O*-glucoside (isoquercitrin)	5,7,3′,4′		3(Glu)	253, 263 sh, 294 sh, 351	
Q-3-*O*-rutinoside (rutin)	5,7,3′,4′		3(Rut)	255, 265 sh, 294 sh, 352	
Q-3-*O*-rhamnoside (quercitrin)	5,7,3′,4′		3(Rha)	253, 263 sh, 344	
Q-3-*O*-galactoside	5,7,3′,4′		3(Gal)		
Kaempherol	3,5,7,4′			264, 292 sh, 318 sh, 363	
K-3-*O*-glucoside (astragalin)	5,7,4′		3(Glu)	264, 291 sh, 320 sh, 344	
K-3-*O*-rutinoside	5,7,4′		3(Rut)	263, 292 sh, 344	
Isorhamnetin	3,5,7,4′	3′		253, 269 sh, 302 sh, 367	
Myricetin	3,5,7,3′,4′,5′			251, 267 sh, 300 sh, 370	
M-3-*O*-rhamnoside (myricitrin)	5,7,3′,4′,5′		3(Rha)	250 sh, 262, 298 sh, 349	
Flavanones					
Hesperetin	5,7,3′	4′		229 sh, 286, 333 sh	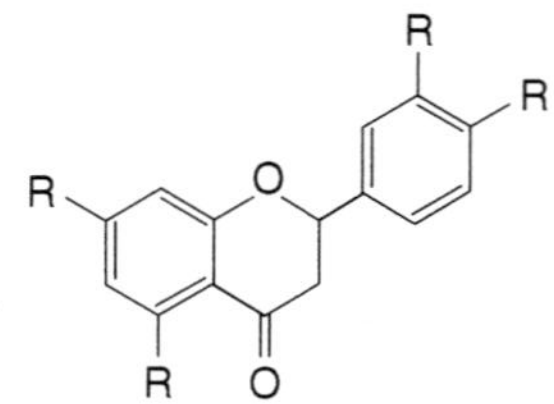
H-7-*O*-rutinoside (hesperidin)	5,3′	4′	7(Rut)	224 sh, 281, 334 sh	
H-7-*O*-neohesperidoside (neohesperetin)	5,3′	4′	7(Neo)		
Narigenin	5,7,4′			226 sh, 288, 331 sh	
Naringenin-7-*O*-neohesperidoside (naringin)	5,4′		7(Neo)	211 sh, 224 sh, 281, 326 sh	
Naringenin-7-*O*-rutinoside (narirutin)	5,4′		7(Rut)		

Flavones					
Luteolin	5,7,3′,4′			252, 265, 290 sh, 347	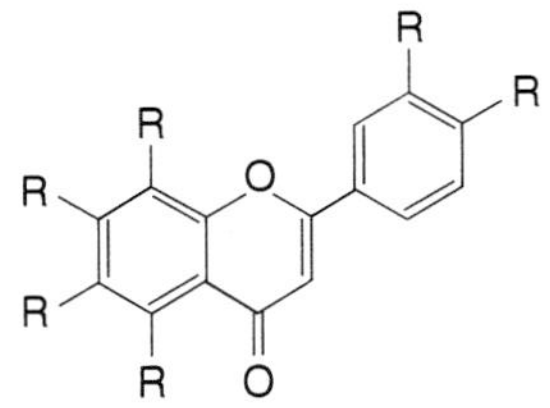
Luteolin-7-*O*-glucoside	5,3′,4′		7(Glu)	253, 265, 346	
Apigenin	5,7,4′			265, 290 sh, 336	
A-7-*O*-glucoside (apigetrin)	5,4′		7(Glu)	265, 333	
Baicalein	5,6,7			251 sh, 274, 321	
Diosmetin	5,7,3′	4′		250, 265, 344	
Tangeretin		5,6,7,8,4′		269, 323	
Chrysin	5,7			243 sh, 267, 311	
Isoflavones (B-ring binds to position 3)					
Daidzein	7,4′			239 sh, 246, 261 sh, 300	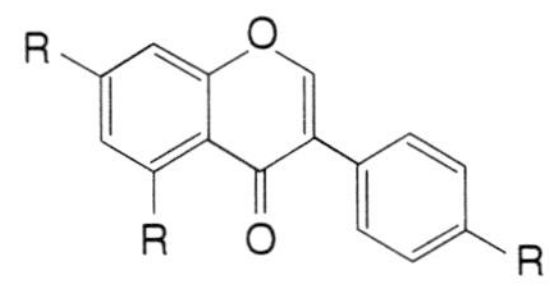
D-7-*O*-glucoside (daidzin)	4′		7(Glu)	248, 258 sh, 299	
Genistein	5,7,4′			259, 329 sh	
G-7-*O*-glucoside (genistin)	5,4′		7(Glu)	259, 325	
Catechins					
(+)-Catechin	3(*trans*),5,7,3′,4′			230 sh, 278	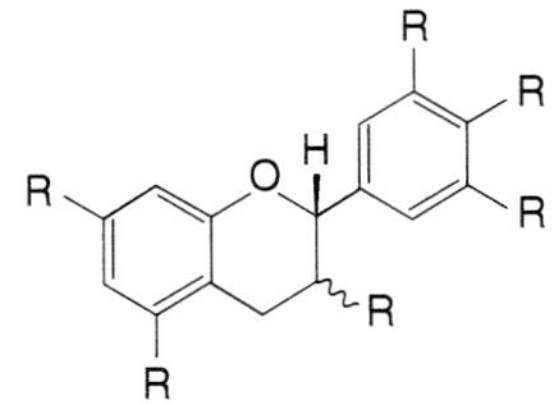
(−)-Gallocatechin	3(*trans*),5,7,3′,4′,5′			269	
(−)-Epicatechin	3(*cis*),5,7,3′,4′			229 sh, 277	
(−)-Epigallocatechin	3(*cis*),5,7,3′,4′,5′			229 sh, 269	
(−)-Gallocatechin gallate	5,7,3′,4′,5′		3(*trans*-OG)	273	
(−)-Epigallocatechin gallate	5,7,3′,4′,5′		3(*cis*-OG)	273	
(−)-Catechin gallate	5,7,3′,4′		3(*trans*-OG)		
(−)-Epicatechin gallate	5,7,3′,4′		3(*cis*-OG)		

(*Continued*)

TABLE 3.1 (Continued)

Chemical Structure of the Main Flavonoid Subclasses and Some of Their More Representative Components Together with Their Spectral Characteristics Recorded Online after Reversed-Phase High-Performance Liquid Chromatography (RP-HPLC) Separation Using Methanol–Water Acidified with Sodium Phosphate (pH 3.3)[a–c]

Flavonoids	Substituents —OH	Substituents —OCH_3	Substituents –X	λ_{max}	Chemical Structure
Anthocyanins					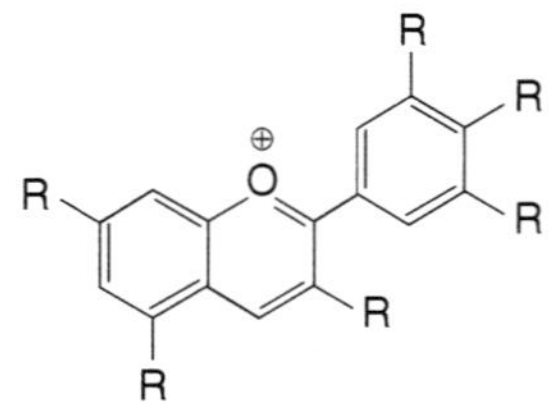
Pelargonidin	3,5,7,4′			267, 325 sh, 409, 503	
Cyanidin	3,5,7,3′,4′			276, 320 sh, 503	
Cyanidin-3-*O*-rutinoside	5,7,3′,4′		3(Rut)	229 sh, 278, 424 sh, 503	
Delphinidin	3,5,7,3′,4′,5′			271, 336 sh, 429, 521	
Malvidin	3,5,7,4′	3′,5′			
Peonidin	3,5,7,4′	3′			
Petunidin	3,5,7,4′,5′	3′			
Proanthocyanidins					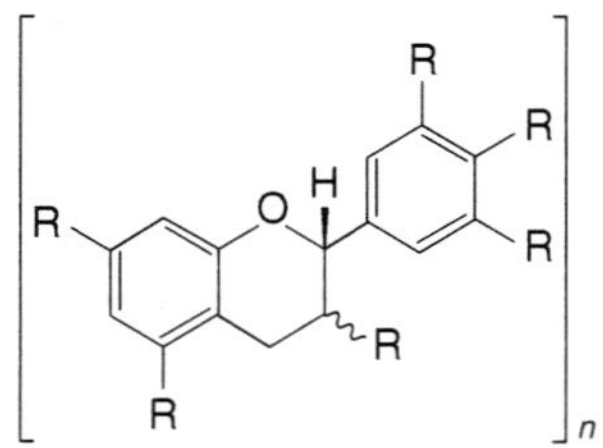
Procyanidins	3,5,7,3′,4′				
Prodelphinidins	3,5,7,3′,4′,5′				
Propelargonidins	3,5,7,4′				

[a] Data from Sakakibara, H. et al., *J. Agric. Food Chem.*, 51, 571, 2003.
[b] Glu, glucose; OG, gallate; Rha, rhamnose; Rut, rhamnoglucoside or rutinoside; Gal, galactoside; Neo, neohesperidoside.
[c] For carbon numbering, see Figure 3.2.

TABLE 3.2
Major Sources of Different Types of Food Flavonoids

Flavonoid	Food Group
Flavonols	Onions, fruits, berries, vegetables, herbs, tea, wine
Flavones	Berries, vegetables, herbs
Isoflavones	Legumes (nuts, cereals)
Flavanones	Citrus fruits, prunes, berries, vegetables
Flavanols	Fruits, vegetables, legumes, cocoa, tea (green), wine, beer
Procyanidins	Fruits, berries, nuts, cereals, legumes, cocoa, wine, cider
Anthocyanins	Berries, cherries, wine (red), colored vegetables, legumes, and cereals (red onions, red cabbage, radish, eggplant, red sorghum, red beans, etc.)
Tannins	Nuts, carrot

3.2.2.4 Flavanols and Flavandiols

Flavanols or flavan-3-ols like catechin, epicatechin, the related compounds trihydroxylated in the B-ring, gallocatechin and epigallocatechin, and their galloylated derivatives catechin gallate, epicatechin gallate, gallocatechin gallate, and epigallocatechin gallate (Table 3.1) are very common polyphenolic compounds occurring in many fruits and vegetables. They are especially relevant in beverages like tea, wine, or in cocoa, where oligomeric procyanidins (from dimers to decamers) occur. Together with flavandiols (flavan-3,4-diols or leucoanthocyanidins), these compounds are found mainly as aglycones. Flavandiols are very unstable, forming a carbo-cation that is easily reduced to the flavan-3-ol structure. In nonreducing conditions, the carbo-cation tends to polymerize in dimeric, oligomeric, and even polymeric structures either by copolymerization or by condensation with flavanols or with anthocyanins, giving rise to condensed tannins or proanthocyanidins.

3.2.2.5 Anthocyanins

Anthocyanins are the glycosylated derivatives of anthocyanidins, which at pH below 2 appear in the form of flavylium cations. Anthocyanins are usually present as *O*-glycosides at the C3 position, although diglycosylation is also common, as the glycosylation with di- or trisaccharides. Anthocyanins can also appear methoxylated, and the sugar moiety in C3 is commonly acylated with other aromatic acids (mainly hydroxycinnamic acids) and mono- and dicarboxylic aliphatic acids. Anthocyanins are responsible for the blue, violet, purple, red, or orange colors of fruits and flowers, thus being the most important group of water-soluble pigments in plants. They are especially abundant in berries, red grapes, and red wine. The six basic anthocyanins (i.e., cyanindin, delphinidin, malvidin, pelargonidin, peonidin, and petunidin) are shown in Table 3.1.

3.2.2.6 Proanthocyanidins

Condensed tannins or proanthocyanidins are oligomeric to polymeric compounds formed from the condensation of flavanols or flavandiols, with degrees of polymerization of over 50. The monomeric flavanols are linked through C4→C8 bonds, although C4→C6 linkages also occur, forming the so-called B-type proanthocyanidins. The less common A-type proanthocyanidins contain double interflavan linkages with an additional bond formed through C2→O7 bonds. The monomeric flavanols can be *O*-glycosylated in the C3 or C5 positions or galloylated in C3. Interflavan linkages are acid-labile, rendering the free anthocyanidins when heated in acid-alcoholic solutions. Depending on the type of the monomeric flavanols, proanthocyanidins are termed procyanidins (consisting of catechin or epicatechin units), prodelphinidins (containing (epi)gallocatechin), or propelargonidins (formed by (epi)afzelechin). Proanthocyanidins are very common flavonoids, actually the second most abundant phenolic in nature after lignin. Fruits (especially berries), nuts, beans, cereals, and spices are rich in proanthocyanidins, together with beverages like red wine, cider, tea, or cocoa, while condensed tannins are less common in vegetables. As mentioned above, the most widely described proanthocyanidins are polymers with degrees of polymerization (DP) up to 10 in beverages, or up to 50 in other foods, although condensed tannins with molecular weights as high as 30,000 have also been described.[13]

3.2.2.7 Other Flavonoids

Other groups of flavonoids are chalcones, intermediates in the biosynthesis of all other flavonoids, as well as dihydrochalcones, aurones, or dihydroflavonols, together with biflavonoids, which are flavonoid dimers different from proanthocyanidins, since they are based on flavones and flavanones linked through a carbon–carbon bond or less frequently by an ether bond.[14] In general, all these flavonoids are less abundant in foods than the compounds mentioned above, although in some cases they may have important applications in the food industry as for the sweetener neohesperidin dihydrochalcone (NHDC).

3.3 FLAVONOIDS IN PLANT FOODS AND NUTRACEUTICALS

3.3.1 Occurrence of Flavonoids in Plant Foods

Flavonoids are present in almost all plant foods (vegetables, cereals, legumes, fruits, nuts, seeds, oils, etc.) as well as in beverages (tea, wine, fruit juices, beer, cider, cocoa, etc.) (Tables 3.2 and 3.3). Their quantitative and qualitative composition, however, varies greatly depending not only on the type of plant, but also within the same crop there are differences depending on the variety (i.e., differently colored varieties of apple, prune, onions, cabbage, etc. present different concentrations of flavonoids), environmental aspects, germination, degree of ripeness upon recollection, processing, storage, etc.[15–19]

The occurrence and distribution of flavonoids in plant organs and tissues is highly variable. Flavonoids usually concentrate in the aerial parts of plants and thus roots and tubers are usually devoid of flavonoids with exceptions like onions or liquorice.[15] The higher contents usually occur in the outer parts of plants (leaves, fruit peels, stems, seed coats, etc.).

TABLE 3.3
Occurrence of the Different Types of Flavonoids in Foods (mg/100 g of Fresh Food or mg/L of Beverage)[a]

Food	Flavonols	Isoflavones	Proanthocyanidins	Flavan-3-ols	Anthocyanins
Fruits					
Apples	2–26 (20), 2.3–7.4 (25)		47.2–141.0 (23), 7.5–35.6 (22)	2.3–7.3 (22)	1.8–17 (28)
Apricots	2.5–5.3 (20), 2.7 (25)		15.6 (23), 0.2 (22)	0.3 (22)	
Avocados			7.4 (23), 0.02 (22)	0.1 (22)	
Bananas			4.0 (23)	0.1 (22)	
Berries	24.9 (24), 0.3–149.0 (20), 1.5–2.4 (25)		8.9–664.0 (23), 1.9–3.7 (22)	3.0–14.8 (22)	1–2147 (28)
Cherries	0.6–8.0 (20), 1.5–4.0 (25)		8.2 (23), 5.7 (22)	6.4 (22)	117 (28)
Grapes	0.2–3.7 (20), 1.4–1.7 (25)		61.0–81.5 (23), 1.3–1.6 (22)	0.4–1.8 (22)	42.7–192.0 (28)
Grape seeds			3532 (23)	0.3 (22)	
Kiwis			3.7–13.9 (23)		
Mangoes			12.8 (23)		
Nectarines			22.8 (23)		9.2 (28)
Peaches	0.3 (25)		67.3 (23), 7.4 (22)	1.2 (22)	4.7 (28)
Pears	0.3–4.5 (20), 0.5–1.2 (25)		31.9–42.3 (23), 1.1–2.2 (22)	0–0.9 (22)	
Plums	0–1.5 (20), 1.1 (25)		216–257 (23), 40.1 (22)	9.6 (22)	12.5–82.2 (28)
Vegetables					
Broccoli	10.2 (25)	13.7 (27)			
Cabbage (red)	0.2–0.6 (20), 0.4–0.6 (25)	24.4 (27)			113 (28)
Cabbage (white)	0–0.1 (20), 0.3 (25)	2.9 (27)			
Carrot		nd (27)			
Cauliflower	0–3.1 (20)	nd (27)			
Celery		nd (27)			
Chives	10.4–30.0 (20)				

(Continued)

TABLE 3.3 (Continued)
Occurrence of the Different Types of Flavonoids in Foods (mg/100 g of Fresh Food or mg/L of Beverage)[a]

Food	Flavonols	Isoflavones	Proanthocyanidins	Flavan-3-ols	Anthocyanins
Eggplant		7.9 (27)			35.1 (28)
Endives	1.8 (24), 0.1–2.6 (20), 1.6–9.6 (25)				
French bean	3.2–4.5 (20), 1.3–5.9 (25)	384 (27)			
Kale	1.2–11 (20); 32.1 (25)				
Leek	33 (24); 1–2.5 (20); 1.2–5.7 (25)	nd (27)			
Lettuce	0.9 (24); 0.2–7.4 (20); 2.2 (23); 0.4–3.2 (25)	nd (27)			
Lollo rosso	91.1 (21)				1.5 (28)
Onion	18–63 (20); 1–135.9 (21); 28.6–48.8 (25)	17.6 (27)			
Red onion	133.7 (21)				38.8 (28)
Radish	0.5–0.9 (25)	2.3 (27)			116 (28)
Spinach	0.3 (25)	nd (27)			
Peas	9.8–14.5 (21); 0.3 (25)	54.1–477 (27)		0.01 (22)	
Tomatoes	0.2–7.4 (20); 0.2–5.5 (21); 0.7–1.3 (25)	32.5 (27)			
Legumes					
Beans		236 (27)		0.1 (22)	
Beans (black)			8.1 (23)		23.1 (28)
Beans (pinto)			796 (23), 2.2 (22)	5.3 (22)	
Beans (red)		223–400 (27)	457–564 (23)		6.2 (28)
Chickpeas		1241 (27)			
Lentils		51–223 (27)	1.1 (22)	0.5 (22)	
Soyabeans		142.1 (27)			
Cereals					
Barley			74.2 (23)		
Sorghum			787–1919 (23)		

Nuts				
Almonds		184 (23)		
Cashews		8.7 (23)		
Hazelnuts		501 (23)		
Peanuts		15.6 (23)		
Pecans		494 (23)		
Pistachios		237 (23)		2.1 (28)
Walnuts		67.3 (23)		
Beverages				
Beer	1 (26)	23 (23), 1.2 (22)	1 (22)	
Chocolate	1.8 (26)	26–246 (23), 0.5–3.9 (22)	1.3–3.4 (22)	
Fruit juice				
Apple	2.5 (20); 3 (26)	9 (23)		
Cranberry		231 (23)		
Grape	4.4 (20); 10.6 (26)	524 (23)		
Grapefruit	4.9 (20); 5.4 (26)			
Orange	3.4–5.7 (20); 3.9–6.2 (26)			
Tomato	11–13 (20); 13.5 (26)			
Tea				
Black	10–25 (20); 21–39 (21); 18–46.2 (26)	8.2 (22)	18.6 (22)	
Green	1.1–2.3 (20); 8–32 (21); 28.3–50 (26)	5.5 (22)	38.4 (22)	
Wine				
Red	2–16 (20); 9.9–58.4 (21); 11.6–24 (26)	313 (20), 3.3 (22)	3.6 (22)	
White	1.7 (21); 1–1.5 (26)	0.03 (22)	0.17 (22)	

[a] Numbers in parenthesis correspond to references; nd, not detected.

As shown above, the structural diversity of flavonoids is enormous, which together with the great variations in the content of flavonoids in plant foods pose a considerable difficulty in the quantification of these compounds in foods. The lack of standardized analytical procedures can also be added to these difficulties. Most data in the literature refer to the total polyphenolic content of foods as determined by the nonspecific Folin–Ciocalteau method, which not only does not discriminates between different classes of plant phenolics, but also measures other nonphenolic reducing compounds like ascorbic acid. Chromatographic estimation of flavonoids in foods is more appropriate, discriminating between different plant polyphenols; however, literature data usually report only on specific types of flavonoids (i.e., only on flavones and flavonols, proanthocyanidins, etc.), sometimes underestimating their content because of different reasons (presence of flavonoids not quantified in trace amounts or compounds not resolved by chromatography like oxidized polyphenols or procyanidin polymers), and thus the total flavonoid content of foods is seldom reported. Table 3.3 shows the amounts of different flavonoids reported in foods and beverages, where the differences in the values given sometimes even for the same food by the same authors are patent.[20–28]

On the other hand, the offer to consumers of functional foods or nutraceuticals rich in flavonoids has increased considerably in recent years, significantly contributing to the daily intake of these phenolic compounds. Although the present monograph aims at the study of functional foods and nutraceuticals, it is not intended to give a comprehensive review on the flavonoids in these kinds of foodstuffs because of the wide range of different types of functional food ingredients, foods rich in flavonoids, and supplements (nutraceuticals) available in the market worldwide.

3.3.2 Effect of Food Processing and Storage

Processing and storage of plant foods can also significantly affect the content and composition of polyphenols. Since most flavonoids are confined to the outer parts of the plant organs and tissues,[15] processes like peeling of fruits and vegetables, refining of cereal flours, and other practices that eliminate external regions of the plant (skins, seed coats, etc.) involve extensive loss of flavonoids.[29,30] On the contrary, pressing of fruits to obtain juices, musts, during jam preparation, olive oil production, or wine and cider elaboration increases the level of soluble flavonoids extracted from peels and pips.[31–34] Minimal processing involving cutting, coring, trimming, etc. not only causes wounding of plant tissues, which triggers the biosynthesis of polyphenols, but also may increase leaching of flavonoids as well as chemical (oxidation) and enzymatic (by polyphenol oxidase and peroxidase) degradation of these compounds, resulting in decreased levels of food phenolics in fresh-cut products.[35–37] Thermal processing inactivates degrading enzymes; however, it could result in losses as reported for citrus peels, air-dried plums, roasted cocoa beans and nuts, and leaching of flavonoids during blanching, boiling, frying, etc.[38–40] Other typical procedures like fermentation give rise to new compounds not present in the original plant material, such as the case of black tea production, cocoa-bean fermentation, wine-making, etc. where oxidative condensation of monomeric and oligomeric flavonoids takes place.[39,41]

Cold storage of fruits and vegetables may induce biosynthesis of flavonoids (e.g., anthocyanins) as an initial defense against abiotic stress, while partial loss of polyphenols may occur as a consequence of chilling injury.[19] Increased biosynthesis of stilbene and anthocyanin has also been reported after UV irradiation of grapes and apples, respectively,[42,43] while gamma irradiation of strawberries caused no significant loss of flavonoid.[44] Storage under modified atmospheres (high CO_2 or O_2 levels) can have marked effects on plant phenolics in general, limiting their biosynthesis and causing precipitation of tannins.[19,45] Other reactions that can take place during storage like depolymerization and further degradation of monomers, release of aglycones, precipitation, polymerization, etc. can alter the phenolic composition of the final product.[19,34,41]

All these modifications in the flavonoid profile and concentration in processed and stored plant foods will alter their organoleptic (color, taste, aroma) and nutritional quality, since these changes may affect the stability, bioactivity, and bioavailability of flavonoids.

3.3.3 Intake of Flavonoids

Estimation of the daily intake of flavonoids is not a straightforward matter because of the mentioned variation in the content and composition of flavonoids in foods. Moreover, the variability and the lack of standardized analytical methods for their estimation further hinder the creation of compositional databases necessary to calculate flavonoids consumption. Intakes will also be affected by seasonal and regional availability of plant foods, as well as by dietary habits dependant of cultural traditions and individual preferences.

Earlier estimates of polyphenols consumption in the United States fixed at 1 g/day the intake of flavonoids, of which 115 mg/day were flavonols, flavones, and flavanones.[46] However, more recent data establish this intake in much lower amounts of these flavonoids ranging from 15 to 23 mg/day, plus 12 mg of anthocyanins and up to 70 mg of procyanidins per day in the United States (Table 3.4).[28,54–56] Similarly, flavonols and flavone intake in other European countries and Japan show figures much lower that those initially estimated by Khünau,[46] as can be seen in Table 3.4.[47–56] To these values, should be added the intake of flavonoids from nutraceuticals and functional foods enriched in these polyphenols, consumption of which has increased considerably in recent years. Therefore, it would be necessary to develop reliable databases of polyphenols/flavonoids in foods and nutraceuticals to obtain more updated estimates of flavonoids intakes.

3.4 SAMPLE PREPARATION

Analysis of flavonoids in foods and beverages requires adequate sample preparation to ensure quantitative extraction of the analytes avoiding degradation or modification of their chemical structure. The initial steps for the isolation and extraction of flavonoids from the complex food matrix usually comprise homogenization of solid samples, either fresh or after freezing, drying, or freeze-drying. In general, it is advisable to avoid high temperatures that might lead to thermal degradation of

TABLE 3.4
Estimated Intakes of Flavonoids in Different Countries (mg/person/day) and Major Sources of Flavonoids in the Diet

Country	mg/person/day	Food Source	References
Denmark	26	Tea, onion, apples	47
Finland	3–10	Fruits (apples), vegetables (onions)	48
	0–41		49
Greece	15	Fruits (apples), vegetables (onions)	48
Italy	23–35	Red wine, fruits, vegetables	48
	35		50
Japan	17	Green tea, soy foods	51
	60–68		48
	45.8		52
Netherlands	23	Tea, onions, apples	53
	33		48
USA	20	Onions, black tea Tea, onions, apples	54
	15–23		55
Procyanidins	49–71	Apples, chocolate, grapes	56
Anthocyanins	12.53	Fruits (berries, grapes, apples), Vegetables (onion, red cabbage), Beverages (grape juice, wine)	28

flavonoids, as well as working with fresh materials since flavonoids can be modified during handling owing to their susceptibility to oxidation or by enzymatic hydrolysis. Crushing of frozen materials, blending, milling, etc. are some strategies used for homogenization of plant materials before solvent extraction of flavonoids. Maceration, stirring, ultrasound-assisted extraction, Soxhlet, and extraction under reflux are some of the most common extraction processes using liquid solvents. Different types of solvents are used, the most common are shown in Table 3.5. In samples with high fat content, prior defatting with hexane or light petroleum in a Soxhlet equipment is often applied;[57–59] alternatively, elimination of liposoluble compounds from the extract by washing with hexane or chloroform can also be used.[58,60–62]

In the case of beverages, these liquid samples can be freeze-dried before extraction and analysis, although they are more frequently used directly after slight prepurification steps (filtration or centrifugation).[24,25,63,64] In both beverages and extracts obtained from solid foods, further purification steps may follow, especially solid-phase extraction (SPE) or liquid–liquid extraction (LLE).

3.4.1 Solvent Extraction and Liquid–Liquid Extraction

The extraction conditions applied for sample preparation can have an important influence on the type of flavonoids isolated as well as on the yield of extraction, and thus factors such as pH and temperature of extraction, and the type of solvent used need to be carefully chosen.

TABLE 3.5
Extraction of Flavonoids from Foods and Beverages[a]

Flavonoids	Food	Sample Preparation	References
Proanthocyanidins	Vegetables, fruits, nuts, cereals, beverages	SE: acetone/water/acetic acid (70:29.5:0.5, v/v/v); SPE: sephadex LH-20 eluted with 30% methanol and 70% acetone; beverages: SPE Sephadex LH-20	23, 56, 57, 84
	Fruits, nuts, beverages	SE: acetone/water/acetic acid (70:29.5:0.5, v/v/v)	85
		SPE: supelcosil Envi-18 eluted with acetone/water/acetic acid (70:29.5:0.5, v/v/v); beverages: SPE Sephadex LH-20	
	Cocoa	SE: 80% aqueous methanol; LLE: hexane; SPE: Sephadex LH-20 eluted with methanol	62
	Wine	Dealcoholized under vacuum; SPE: C18 Sep-Pak eluted with water (pH 7.0), ethyl acetate, and methanol	94
	Wine	Dealcoholized under vacuum; SPE: Sephadex LH-20, eluted with acetone/water/acetic acid (70:29.5:0.5, v/v/v)	59
	Saskatoon berries	SE: 80% aqueous methanol and 80% aqueous acetone; SPE: Sephadex LH-20 eluted with 30% methanol	89
Anthocyanins	Fruits, vegetables, legumes, nuts	SE: methanol/water/acetic acid (85:15:0.5, v/v/v)	28
	Colored tubers	SE: acidified water (pH 0.5); SPE: C18 Sep-Pak eluted with acidified methanol Alkaline hydrolysis	72
	Red blood orange juice	SE: acetone/ethanol/hexane (25:25:50, v/v/v); SPE: C18 Sep-Pak eluted with acidified (0.1% HCl) methanol	80
	Berries	SE: acetonitrile; SPE: polyamide 6S column washed with aqueous acetic acid and eluted with methanol and 0.25% acetic acid	74
	Berries, port wines	SE: 70% aqueous acetone; SPE: Toyopearl HW-40 gel column eluted with aqueous methanol, methanol, and acetone	81, 82
	Berries	LLE: chloroform, diethyl ether, ethyl acetate, and methanol	96
	Berries	SE: methanol; SPE: C18 Sep-Pak eluted with methanol	100

(Continued)

TABLE 3.5 (Continued)
Extraction of Flavonoids from Foods and Beverages[a]

Flavonoids	Food	Sample Preparation	References
Isoflavonoids	Soy food	SE: acetonitrile/water (90:10, v/v)	86
	Soy food	SE: acetonitrile/water (50:50, v/v)	87
	Soy food	SE: reflux 2 h with 1.0–1.1 N HCl in ethanol	67, 68
	Soy food	SE: reflux 2 h with 80% aqueous ethanol; hydrolysis with β-glucosidase	76
	Pueraria lobata roots	SE: ethanol SPE; Polystyrene resin (NKA0) eluted with 20% ethanol	103
	Fruits, vegetables, cereals, legumes	SE: 80% aqueous methanol; hydrolysis: cellulase in 0.1 M acetate buffer pH 5.0; LLE: ethyl acetate	27
Flavanols	Fruits, vegetables, legumes, cereals, beverages	SE: methanol; beverages: filtration and direct injection	22
	Fruits (apples, grapes), legumes (beans)	SE: 90% aqueous methanol (fruits), 70% aqueous methanol (legumes)	65
Flavonols, flavones, and flavanones	Vegetable (artichoke)	SE: 60% aqueous methanol; SPE: C18 (Chromabond cartridge) eluted with 10% methanol and pure methanol	99
Flavonols and flavones	Fruits, vegetables, beverages	SE and hydrolysis: 50% aqueous methanol with 1.2 N HCl containing 1 g/L BHT; beverages: direct treatment	24–26, 69, 70
	Spinach	SE: 70% aqueous methanol; SPE: ODS C18 eluted with 70% aqueous methanol	97
Flavonols, flavones, and anthocyanins	Red onions	SE: methanol estabilized with BHT	66
Flavonols and xanthones	Mango peels	SE: 80% aqueous acetone; SPE: polyamide CC6 column eluted with methanol	83

Flavonols, procyanidins, and dihydrochalcones	Apple cider	Degassing, filtration, and direct injection	63
Flavonoids	Wine	LLE: diethyl ether	95
	Vegetables, fruits, and teas	SE: 90% aqueous methanol containing 0.5% acetic acid	77
	Chinese medicinal herb (*Glyzyrrhiza uralensis*)	SE: 90% aqueous ethanol; LLE: ether and ethyl acetate	90
	Fruits (plums)	SE: 80% methanol; SPE: C18 (Mega Bond Elute cartridge); elution with 50% methanol, 100% methanol, and acetone	98
	Almond skins	SE: 95% aqueous ethanol; LLE: hexane, ethyl acetate, and *n*-butanol; SPE: fractionation in Silica gel column, fractions further purified by chromatography on Sephadex LH-20 or TLC	91
	Chinese medicinal herb (*Patrinia villosa*)	Reflux with 70% aqueous ethanol; SPE: D101 macroporous resin washed with water and eluted with 40% ethanol and 85% ethanol	73
Flavonoid glycosides	Rizoma of *Smilax glabra*	SE: 95% aqueous methanol; LLE: ethyl acetate; SPE: silical gel column, eluted with chloroform/methanol/water (15:4:1, v/v/v)	92
	Chinese medicinal herbs (*Trollius ladebouri*, *Patrinia villosa*)	SE: 60% or 70% aqueous ethanol; SPE: D101 macroporous resin, eluted with water and ethanol	73, 104
	Leaves of *Ampelopsis grossedentata*, Seeds of *Oroxylum indicum*	SE: water; LLE: chloroform and *n*-butanol	61, 93
	Clusia criuva	SE: hexane and methanol; LLE: hexane, chloroform, ethyl acetate, and butanol	60
	Vegetables, strawberries	SE: methanol; SPE: polyamide column eluted with methanol	29, 44

[a] SE, solvent extraction; SPE, solid-phase extraction; LLE, liquid–liquid extraction; BHT, butylated hydroxytoluene; TLC, thin layer chromatography.

The type of solvent to be used for the extraction of flavonoids from solid foods depends mainly on the type of compounds to be extracted, that is whether they are aglycones, glycosides, acylated or methoxylated flavonoids, etc., which might affect their polarity and thus their solubility on the selected solvent. Also, the degree of polymerization of proanthocyanidins or the cationic nature of anthocyanidins should be taken into account when selecting the extraction solvents.

Methanol and methanol–water mixtures (most frequently 50%, 70%, and 80% aqueous methanol) are the most commonly used solvents (Table 3.5), often slightly acidified with hydrochloric acid (1–1.5 N HCl) or acetic acid, or containing antioxidants like BHT, *tert*-butylhydroquinone, sulfites, or ascorbic acid. Alcohols cause instability of cell membranes, facilitating extraction of phenolic compounds; besides, they inactivate enzymes like polyphenol oxidase and thus contribute to an increased stability of the extracted compounds,[65] stability that can be further improved by the presence of antioxidants in the extraction media.[66] On the other hand, hydrolysis of flavonoid glycosides might be required before analysis. This is the case for the quantification of isoflavone aglycones, where reflux in 1–1.1 N HCl in ethanol for 2 h has been applied,[67,68] or for the analysis of flavonols and flavones as their aglycones, where reflux for 2 h at 90°C with 50% aqueous methanol containing BHT and acidified with 1.2 N HCl has been used.[24–26,70,71] Other hydrolyzing strategies like alkaline hydrolysis,[72] reflux in acidified ethanol,[73] or the use of enzymes (e.g., cellulase, alpha-amylase, pectinolytic enzymes, etc.) have also been used during the extraction of phenolic compounds, mostly phenolic and hydroxycinnamic acids, although in some cases also isoflavonoids[27] or anthocyanins.[74] Other hydrolytic enzymes like β-glucosidase or β-glucuronidase are commonly used for the analysis of flavonoid aglycones in biological samples (reviewed by Day and Morgan[75]), although enzymatic hydrolysis of flavonoid glycosides with β-glucosidase has also been used in foods.[76]

Acidification of the solvent is specially important in the case of anthocyanins, since it would determine extraction of the flavylium cation that is stable at acid pH; however, care should be taken since too strong acid conditions may cause the hydrolysis of acylated anthocyanins, and thus reduction of the concentration of hydrochloric acid (between 0.1% and 1% HCl in methanol) or the use of acetic acid,[28,77] or weaker acids like tartaric, oxalic, or citric acids have shown good results.[78,79]

Acetone and acetone–water mixtures have also been used for the extraction of anthocyanins,[31,80–82] flavonols, and xanthones,[83] or flavanols and proanthocyanidins,[57,84,85] as well as other solvents like acetonitrile for the extraction of isoflavones,[86,87] acidified water for anthocyanins,[72] etc.

Sequential extraction with different solvents has also been applied to ensure more complete extraction of the flavonoids, like the use of 1.2 N HCl in 50% methanol followed by 70% acetone,[88] 80% methanol followed by 80% acetone,[89] 80% methanol, 50% methanol, water, and 75% acetone,[79] although the combinations are multiple.

Besides the type of solvent used, other factors to be considered when setting up the extraction procedure are the sample particle size, number of extraction steps, sample to solvent ratio, or the time and temperature of the process. In general, high temperatures should be avoided to prevent degradation of flavonoids, although in many cases temperatures as high as 90°C are used, especially when targeting hydrolysis of flavonoid glycosides.[24–26,70,71] After the extraction, separation from the food sample material can be achieved by filtration or more preferably by centrifugation, followed by concentration under vacuum at low temperature to eliminate excess solvent.

After these initial steps, further LLE might follow, usually with nonpolar solvents like hexane, chloroform, ethyl acetate, or diethyl ether to remove fats, pigments, and other matrix compounds from the crude extract, partition that can be followed by the use of more polar solvents like *n*-butanol or methanol (Table 3.5).[27,61,62,90–93]

In the case of beverages, these liquid samples are often directly treated by LLE after initial treatments like degassing (for gas-containing beverages),[63] concentration under vacuum,[94] or cleaning up procedures like filtration or centrifugation, although in some cases this cleaning up is the only treatment applied to the liquid samples.[22,95,96]

3.4.2 Solid-Phase Extraction

Liquid samples or extracts obtained after solvent extraction with or without further LLE can be submitted to SPE to eliminate interfering components or to enrich the extract in a certain flavonoid fraction.

SPE is the most commonly used purification and preconcentration technique (Table 3.5). Minicartridges filled with C18 reversed-phase (RP) material allow rapid treatments with excellent recoveries, comparable to those of simple filtration.[72,80,85,94,97–100] Diol columns have also been used with excellent results for nonpolar samples like olive oil.[101,102] Also, minicolumns filled with stationary phases, like silica gel,[91,92] Toyopearl HW-40,[81,82] Polyamide gel,[29,44,74,83] Polystyrene resin,[103] or D101 macroporous resin[73,104] among others, have been used, although the stationary phase most widely used is Sephadex LH-20.[23,57,59,62,84,89,91,105] In all cases, conditioning, washing, and elution of the cartridges or minicolumns have to be performed. Sequential elution with different solvents according to their polarity is useful for fractionation of different classes of phenolics and flavonoids. Solvents may be slightly acidified to prevent ionization of flavonoids.

Matrix solid-phase extraction (MSPE) is a recently introduced alternative to SPE in which samples are thoroughly mixed with a solid substrate, then packed into a column and washed and eluted with solvents, so the sample cleanup and analytes extraction are carried out in a single step. The solid support can be C18- or C8-bonded silica[106,107] or Amberlite XAD-2,[105,108] and even the use of sea sand as solid substrate has also been reported recently.[109] As in the traditional chromatographic and SPE procedures, extraction conditions, especially the composition of the eluents, need to be optimized before the extraction.

Solid-phase microextraction (SPME) in open-tubular fused-silica capillary columns or in fused-silica or Carbowax-templated poly(divinylbenzene) resin fibers coated or filled with a stationary phase (polydimethylsiloxane or polyacrylate)[110–112] has recently been used to extract analytes from liquid or gaseous samples. This procedure is more restricted to the extraction of volatile compounds, which are scarce in the case of flavonoids. Also, extraction recoveries are low while extraction times can be long. This extraction method has been used to analyze flavonoids by gas chromatography (GC) or liquid chromatography (LC) coupled with mass spectrometry (MS).

3.4.3 Supercritical and Subcritical Fluid Extraction

In the last years, several alternatives to conventional solvent-based methodologies have been developed involving the use of subcritical or supercritical extraction conditions, such as supercritical fluid extraction, subcritical water extraction, or superheated liquid extraction.

Supercritical fluid extraction (SFE) is the most widespread of these methodologies. SFE is based on the use of a fluid at temperatures and pressures near the critical point. It comprises an extraction phase where the analyte is extracted from the sample matrix, followed by collection or trapping of the analytes, which might be coupled online into an analytical instrument, usually a liquid chromatograph. Off-line collection of the analytes can also be achieved after depressurizing of the supercritical fluid (SF) into a collection device such as an empty vessel, a vessel containing a small volume of solvent, solid-phase or solid–liquid phase traps, or a cryogenically cooled capillary (reviewed by Turner et al.[113]).

SFE is specially recommended to extract analytes of low volatility or labile compounds susceptible to thermal degradation and decomposition since extraction is performed at reduced temperatures and in the absence of light and air, rendering germ-free extracts. Moreover, when the SF is carbon dioxide, this is easily separated after depressurizing because of its high volatility, minimizing sample handling, and yielding extracts devoid of solvents that are considered natural and allowed for food or pharmaceutical applications having the GRAS (generally recognized as safe) status.[114]

The major advantages of SFE are related to the properties of the SFs used. Several solvents can be used in SFE, including ethane, *n*-pentane, propane, nitrous oxide, etc., although the most widely used is carbon dioxide owing to its low toxicity, nonflammability, environmental safety, as well as its availability as a high purity solvent at low cost. The properties of SF include higher diffusion coefficients and lower density, viscosity, and surface tension in comparison with conventional organic solvents. As a consequence, solubility and diffusivity of analytes in SF tend to be much higher than in liquid solvents. Moreover, the properties of SF can be manipulated over a wide range by varying the operational conditions (pressure or temperature), thus improving selectivity of the extraction process, which is especially convenient when handling complex samples. Furthermore, high extraction efficiencies can be achieved when analyte collection has been optimized. Finally, this technique is available fully automated, which together with the sort extraction times and the minimal use of organic solvents are additional advantages.

However, SFE pose several drawbacks. Carbon dioxide, which is the most common SF, behaves as a nonpolar solvent and thus with inadequate solvating power for highly polar analytes. The use of appropriate cosolvents or modifiers can overcome this limitation, by enhancing the solubility of target polar analytes and improving extraction selectivity. Methanol in varying proportions (from 3% to up to 40%) is the modifier of choice in the extraction of flavonoids from foods and medicinal herbs,[115–120] although ethanol (10%–15%) has also been employed.[121,122] The use of these cosolvents allows operation at lower pressures. Modification of the operating conditions permits increased selectivity of the extraction. Thus, sequential SFE of grape seeds with pure CO_2 at high pressure (65 MPa, 80°C), followed by extraction with CO_2/methanol (60:40, v/v) under the same operational conditions, and extraction with 100% methanol at 13.8 MPa allowed the obtention of nonpolar fractions and extracts rich in monomeric (catechin and epicatechin) and oligomeric (procyanidins) flavanols from grape by-products.[115] Pressures typically applied in SFE of polyphenols vary from 9.5 to up to 65 MPa, and temperatures from 40°C to up to 80°C.[115–122] However, recovering efficiencies are not always improved with respect to conventional solvent extractions.[118,121] Moreover, the optimal conditions depend on the type of compounds to be extracted and the nature of the sample matrix, and thus need to be carefully set up for every sample. This is one of the major drawbacks of this technique that has prevented its implementation in analytical routine laboratories. Furthermore, the resulting extracts might carry interfering substances, especially when working with complex matrices and thus subsequent cleanup of the extract might be necessary. Also, when using CO_2 as SF, it can dissolve water to some extent, limiting the application of CO_2-SFE in wet or liquid samples with high water content.[123] Some additional disadvantages of SFE derive from the requirement of expensive equipment and potential hazards associated to the use of high-operating pressures, which has limited the implementation of this technique.

Pressurized-liquid extraction, also known as accelerated solvent extraction (ASE) or more commonly subcritical water extraction (SWE), is a relatively new technology based on the extraction using hot water under pressures high enough to maintain water in the liquid state. This extraction procedure has been recently used to extract flavonoids and other polar compounds from plants like rosemary[124,125] and oregano,[126] as well as from medicinal herbs like American skullcap.[121] SWE has a remarkable capacity to selectively extract different classes of compounds depending on the operating conditions. Thus, more polar compounds are more efficiently extracted at lower temperatures, while less polar analytes are better extracted at higher temperatures. Pressure is usually maintained at 10 MPa, while temperatures may vary between 25°C and 200°C.[121,124–126] Extractions are performed under static conditions, and short operation times (up to 30 min) are required for adequate extractions. However, this technique has seldom been used in foods, only for essential oil[127] and antioxidant[124–130] extraction.

Another recently described methodology for the extraction of polyphenols is the use of superheated ethanol–water extraction (SEWE). This technique is a variation of SWE in which water–ethanol mixtures (20:80 or 40:60, v/v) are used in static extractions at high temperatures (180°C–240°C) and pressure (4 MPa).[131,132] As for SWE, this method allows modification of the extract composition by changing the operation conditions (working pressure, temperature, and water/ethanol ratios). SEWE has been applied for the extraction of phenolic compounds from winery by-products with extraction efficiencies superior to those of conventional solvent-extraction methods.[132] Both SWE and SEWE could have applications for the obtention of extracts with functional properties (e.g., rich in antioxidants) for their use in the food and nutraceutical industries, although, as for SFE, their extended use as alternative to the traditional solvent and SPE methods for routine analysis of flavonoids in foods and food products is not likely in the near future.

3.5 SPECTROPHOTOMETRIC METHODS FOR FLAVONOIDS ANALYSIS

Different spectrophotometric methods for the quantification of phenolic compounds in foods have been developed. Spectrophotometric methods are based on the formation of a compound or colored complex that is measured at a certain wavelength. To avoid interference, an effective extraction of flavonoids is necessary before spectrophotometric measurement.

3.5.1 Acid-Butanol Assay

Several methods have been described in the literature to quantify proanthocyanidins (condensed tannins), however, the most common is the colorimetric method based on acid-butanol assay.

This colorimetric reaction has a long history[133] and uses an acid-catalyzed oxidative depolymerization of condensed tannins to yield red anthocyanidins. Although the method is simple and gives good indication of the presence of condensed tannins, chemical characteristic of the tannins such as position of the interflavan bond and oxygenation pattern affect color yield significantly. Other factors that can affect the color's intensity (violet) are the amount of water in the reaction medium, the presence of transition metal ions in the assay medium, and the ratio of acid-butanol to sample in the reaction mixture. On the other hand, the choice of standards remains an unresolved issue because of the heterogeneity of condensed tannins and the lack of appropriate standards for their quantification. The use of internal standards obtained from the plant materials under study has been considered to minimize the problems derived from the use of inappropriate standards (reviewed in Schofield).[134] Taking into account these considerations, the butanol-HCl reaction should be used with caution as a quantitative assay. The assay's greatest strength lies in its confirmation of the presence of a polymeric interflavan structure. Hydrolyzable tannins do not react in this assay. Despite the shortcomings outlined above, the acid-butanol assay remains the most commonly used method for determination of proanthocyanidins in foods.

3.5.2 Vanillin Assay

The vanillin procedure involves reaction of an aromatic aldehyde, vanillin, with the meta-substituted ring of flavanols to yield a red adduct. Although the vanillin assay is an easy and cheap assay to measure flavanols and proanthocyanidins, several aspects of the procedure need to be carefully considered to obtain reliable results. Catechin is commonly used to standardize the vanillin reaction, although both proanthocyanidins and catechin react with vanillin at different rates. Therefore, the use of internal standards is advisable. On the other hand, the vanillin reaction is very sensitive to temperature as well as to the presence of water, with decreased absorbance when water is present in the reaction mixture, so all standards should be prepared in anhydrous organic solvents (usually methanol). With regard to temperature, it has been reported that an increase of 1.4°C caused an 11% raise in absorbance readings; therefore, strict control of temperature is necessary.[133]

Although the vanillin reaction has been widely used to estimate condensed tannins (proanthocyanidins), the reaction is not specific for this kind of compound. Any appropriately substituted flavanol reacts in the assay as does the monomeric unit of condensed tannins, catechin, that reacts with vanillin to yield a red colored adduct. Moreover, the subunits of the polymeric condensed tannins show variable reactivity with vanillin. In this sense, a modified vanillin method has been proposed to estimate the molecular weight on condensed tannins instead of using it with quantitative purposes.[135,136]

3.5.3 Colorimetric Assays for Total Phenolics

The most commonly used methods for determination of total phenolics in foodstuffs are the Prussian Blue assay[137] and the Folin–Ciocalteau method.[138] Although these procedures have undergone numerous modifications, the version adopted by the American Oil Chemists' Society (AOCS)[139] is most commonly used nowadays.[133] These assays are not specific for phenolic compounds, although they are widely used to get a global concentration of phenolic compounds present in foods.

3.5.3.1 Prussian Blue Assay

This method involves oxidation–reduction reactions in which the phenolate ion is oxidized while $[Fe(CN)_6]^{3-}$ ion is reduced. It is a popular method to quantify total phenolics because it is simple, rapid, and with little interferences by nonphenolic compounds.[140] Critical for the successful use of this assay are temperature, pH, and order in which reagents are added, since this affects the formation of $Fe_4[Fe(CN)_6]_3$ (Prussian Blue). However, this assay presents two important shortcomings that have to be taken into account for the adequate interpretation of results: the formation of a precipitate after short incubation periods and the increase in color density with time.

3.5.3.2 Folin–Ciocalteau Method

This is probably the most popular procedure for quantifying total phenols in foods since it is a rapid and simple procedure. In addition, it is widely used and so it is possible to compare results with values reported in the literature. This method entails oxidation of the phenolate ion coupled to the reduction of the phosphotungstic–phosphomolybdic reagent. The chromophore produced is a blue phosphotungstic–phosphomolybdic complex of undefined structure since the underlying chemistry of this reaction is not well understood yet. Although less prone to interferences by nonphenolics than the original method (Folin–Denis), there are still some compounds like reducing substances such as ascorbic acid that react with the Folin–Ciocalteau reagent, and thus overestimated results are obtained in comparison to chromatographic methods.

3.6 SEPARATION METHODS

The most common techniques used for separation of flavonoids are chromatographic and electrophoretic methods.

3.6.1 CHROMATOGRAPHIC TECHNIQUES

3.6.1.1 Thin Layer Chromatography

Thin layer chromatography (TLC) is a separation technique that lacks quantitative precision, sensitivity, and separation efficiency in comparison with methodologies involving column separation. However, TLC methods have been developed and commonly employed for the separation and determination of flavonoids since TLC is a quick, convenient, and inexpensive technique that does not require complex instrumental equipment, with the ability to assay many samples in parallel on a single TLC plate. It can also be used to control isolation in column chromatography and, to some extent, to identify analytes.

In general, crude extracts of food flavonoids have highly complex profiles consisting on mixtures of many different compounds, and often isocratic separation cannot resolve them satisfactorily. Therefore, multiple mobile phases used in regular or double-development (2D) TLC[141,142] are useful for good separation of flavonoids. Detection is mainly performed using UV light at 350–365 or 250–260 nm or by densitometry at the same wavelengths.

Despite these applications, disadvantages such as requirement of large sample amounts may restrict the use of TLC compared to column methods because such amounts are not always available. On the other hand, recovery of the antioxidant from TLC plates could also be challenging. In this sense, most TLC applications are restricted to the fractionation and preliminary separation of antioxidant flavonoids before they are separated, quantified, and identified by high-performance liquid chromatography (HPLC) or other high-performance separation techniques.

Several studies have been described for the identification of flavonoids in foods where silica gel plates were used,[142–149] although cellulose plates are also common.[143,150] As developing solvent systems, different mixtures have been employed with variable polarity such as butanol/acetic acid/water (4:1:5, v/v/v) for flavonoids present in buckwheat,[150] and proanthocyanidins and flavonol glycosides from the seed coat of a Manteca-type dry bean (*Phaseolus vulgaris* L.);[143] water/methanol/ethyl acetate (3:4:2, v/v/v) to isolate flavonols from onions;[145] benzene/acetone (1:1, v/v) to isolate citrus polymethyloxylated flavones; methanol/acetic acid (85:15, v/v) to isolate flavonoids from peppermint;[151] toluene/ethyl acetate/formic acid/methanol (3:3:0.8:0.2, v/v/v/v) to isolate luteolin from *Cuminum cyminum* fruits,[147] among others.

Preparative layer chromatography (PLC) is carried out on thicker layers with application of larger weights and volumes of sample to separate and recover from 10 to 1000 mg of compounds for further analysis.[144,152]

High-performance TLC (HPTLC) is an instrumentalized, quantitative method that allows fast and more efficient separations, and improved detection limits.

This chromatographic technique is carried out on layers composed of particles with a smaller particle diameter than in conventional TLC.[148,153,154]

Although generally quantification is not the main goal of TLC studies, densitometry is used in several studies with results comparable to those of gas and column liquid chromatographies.[142,147,149,155]

3.6.1.2 Gas Chromatography

Despite the high resolution and sensitivity of GC, its use in the separation of plant flavonoids has not been as popular as HPLC, because of the lack of volatility of most flavonoids. For this reason, flavonoids have to be converted to volatile derivatives, and in this sense silylation is an ideal procedure for the GC analysis of such nonvolatile, thermolabile compounds. Compared with the parent molecules, trimethylsilylether (TMS) derivatives are more volatile, less polar, and more thermotolerant. However, in flavonoids with more than one hydroxyl group, methylation may yield several different derivatives, which hinders quantification. *N,O*-bis-(trimethylsilyl)-trifluoroacetamide (BSTFA)[156–159] and *N*-(*tert*-butyldimethylsilyl)-*N*-methyltrifluoroacetamide (TBDMS)[160–162] are the most commonly used derivatizing agents, although methods not requiring derivatization have been described.[149,163,164]

After TMS derivatization, flavonoids are injected onto a nonpolar capillary column in the split or splitless modes, and separated with a lineal 30–90 min temperature program up to 320°C. MS detector with positive electrospray ionization (ESI) and a temperature source of up to 250°C is the most common detection system used for the identification of flavonoids in the selected-ion monitoring (SIM) mode (Table 3.6). The molecular ion, $[M+H]^+$, and fragment ions formed by the loss of CH_3 or CO, as well as retro-Diels-Alder (RDA) reaction are typically used for identification.

However, to analyze more polar molecules such as flavonoid glycosides, high temperature–high resolution (HT-HR) GC is used, with columns that can withstand temperatures up to 400°C. This technique offers high resolution, faster separation, and is easily coupled with a wide variety of detectors, with the additional advantage that it does not require the use of solvents. Today, HRGC coupled with MS (HRGC-MS) is an important and consolidated method for the systematic analysis of natural products including flavonoids.[158,164,165]

A summary of GC applications on the separation and determination of flavonoids is presented in Table 3.6. In general, although the identification of flavonoids by GC-MS in SIM mode shows low limits of detection, this technique clearly does not outweigh the speed of direct LC procedures, which justifies the scarcity of studies found to analyze flavonoids in foods by GC techniques.

3.6.1.3 High-Performance Liquid Chromatography

HPLC is the most powerful and frequently used technique for the separation of natural compounds such as flavonoids. It has also been used extensively to determine both flavonoid aglycones and glycosides.

HPLC offers significant advantages in terms of simplicity, speed, cost (although this would also depend on the type of detection method), sensitivity, specificity,

TABLE 3.6
Gas Chromatographic Applications for the Separation and Determination of Flavonoids[a]

Flavonoid	Food	Analytical Approach	Derivatization	References
Isoflavones	Fruits and nuts	GC-MS	TBDMS	161
Isoflavones	Soy food	GC-MS	BSTFA	157
Isoflavones	Foods	GC-MS	TBDMS	160
Anthocyanins, flavonones, and flavonols	Sour cherries	GC-MS	HMDS	234
Flavones and flavonols	Propolis	GC-MS	—	163
Flavones and flavonols	Propolis	GC-MS	—	149
Isoprenylated flavonoids	*Vellozia graminifolia*	HT-HRGC-MS	—	164
Flavanone and flavones	*Populus tremuloides*	GC-MS	With/without TBDMS	162
Flavonols	*Ginkgo biloba*	GC-MS	BSTFA	156
Flavonoids	*Lonchocarpus urucu*	HT-HRGC-MS and -FID	BSTFA	158
Flavanols, flavonols, and flavones	Aromatic plants	GC-MS	BSTFA	159

[a] BSTFA, *N,O*-bis-(trimethylsilyl)-trifluoroacetamide; HMDS, hexamethyldisilazane; TBDMS, *N*-(*tert*-butyldimethylsilyl)-*N*-methyl-trifluoroacetamide.

precision, and sample preservation. HPLC is versatile and adaptable for specific requirements through the use of adequate stationary phases, composition of mobile phases, and the possibility to couple with a wide range of selective detectors. In addition, HPLC has the advantage of simultaneous separation and quantification of flavonoids without preliminary derivatization.

However, because of the lack of standard compounds for many flavonoids, especially for flavonoid glycosides, and the enormous number of different flavonoids existing in nature, identification of these compounds is not a straightforward task. In the case of flavonoid glycosides, it has become an accepted practice to hydrolyze them into aglycones before HPLC analysis.

A wide range of stationary and mobile phase combinations has been reported in the literature to obtain adequate resolution, since this is considered to be the main difficulty for the separation of flavonoids in a complex mixture. Normal phases (i.e., silica gel columns), yet seldom used, can be considered appropriate for the separation of nonpolar or weakly polar flavonoid aglycones, such as polymethoxylated flavones, flavanones, or isoflavones. Usually, no gradient can be employed.[166] Thus, most of the different classes of plant flavonoids and their metabolites are separated by RP-HPLC (Table 3.7).

For the separation of flavonoids, common chromatographic conditions include the use of almost exclusively an RP C18 column and a binary solvent system containing acidified water (solvent A) such as aqueous acetic, phosphoric, sulfuric,

TABLE 3.7
Some High-Performance Liquid Chromatography (HPLC) Procedures for the Determination of Multiple Classes of Flavonoids in Foods with Ultraviolet–Visible (UV–Vis) Detector

Flavonoids	Food	Stationary Phase	Mobile Phase	DAD (nm)	References
Isocratic Elution					
Flavonols	Red wine	C18 column	Organic solvent (ACN)/H_2O acidified	265	176
Flavanones	Orange juice		(acetic, formic, phosphoric acids)	280	69
Flavanones enantiomers	—	Chiralpak AD-RH column	ACN/H_2O/H_3PO_4 (42:58:0.01, v/v/v)	298	177
	Orange juice	Cyclobond I 2000	H_2O/MeOH (90:10, v/v)	280	178
Gradient Elution (Two Phases)					
Flavonols	Onion skin, apple peel, wine,	C18 column	Solvent A: acidified water (acetic, sulfuric,	215	170
Isoflavones	apple, pear, green beans,		phosphoric, formic acids)	350	171
Flavones	lentils, fruit juice,		Solvent B: organic solvent (MeOH, ACN)	270	172
Flavanols	sugarcane, vegetables,			280, 370, 525	173
Anthocyanidins	St. John's Wort dietary			284	174
Flavanones	supplements			265	69
				280	167–169
Flavanols	Vegetables	C18 column	Solvent A: mixture of acidified water	205	175
Anthocyanins and flavones	Fruit		(TFA, phosphoric acid) with organic	280, 320, 370,	77
Isoflavones flavanones	Tea		solvent (ACN, MeOH)	510	
			Solv B: acidified organic acid (ACN, MeOH)		
Gradient elution (Three phases)					
Flavonols	Vegetable extracts	C18 column	Solvent A: acidified water ($NaCH_3CO_2$)	356	179
Flavones		Monolithic type			
Flavanones	Tea	C18 column	Solvent B: organic solvent (MeOH)	270, 290, 365	180
Flavonols	Red wine, oranges		Solvent C: organic solvent (ACN)		
Flavanols	Aromatic plants	C18 column	Solvent A: acidified water (Acetic acid)	280	159
Flavonols			Solvent B: acidified water (Acetic Acid)		
Flavones			Solvent C: mixture of water and organic solvent (ACN)		

or formic acids, and a less polar organic solvent (solvent B) such as methanol or acetonitrile, possibly acidified.[69,167–174] Solvent A could contain minor amounts of organic solvents as described by Lee and Ong[175] or Sakakibara et al.[77] Less frequently, runs are isocratic[69,176–178] or tertiary.[159,179,180] The separation normally requires 1 h at a flow rate of 1.0–1.5 mL/min.

Chiral separation of flavonoids has also been carried out by chromatographic systems by using a chemically bonded chiral stationary phase or by the addition of chiral mobile phase additives (reviewed by Yáñez et al.[181]). These chiral polymer phases can be further subdivided into polysaccharide-derived columns,[177] and cyclodextrin and mixed cyclodextrin columns.[178] With regard to chiral mobile phase additives, the addition of an optically active molecule to the mobile phase can facilitate separation of enantiomers on conventional stationary phases. Cyclodextrin as a chiral additive is widely used to separate enantiomers mainly by capillary electrophoresis (CE), as discussed in Section 3.6.2.1. Table 3.7 summarizes the most habitual HPLC procedures employed for the analysis of various classes of food flavonoids.

3.6.1.4 Countercurrent Chromatography

Countercurrent chromatography (CCC) constitutes a modern liquid–liquid chromatography without solid support. The principle of separation involves the partition of a solute between two immiscible phases. In most of the reported modalities of CCC, one of the phases remains stationary while the second phase is passed through the stationary solvent phase.

This technique presents numerous advantages in comparison to other chromatographic methods such as the absence of reversible adsorption of the sample since no solid support is employed, and a high sample capacity because of the large volumes of stationary phase used (up to several hundred milligrams of pure compounds can be separated or recovered in a single CCC run).

A couple of examples have been included in Table 3.8, using as solvent system a mixture of *n*-hexane/ethyl acetate/methanol/water to isolate flavanones,[182] and chloroform/methanol/water to isolate flavones[60] with successful separation of flavonoids from complex matrices.

High-speed countercurrent chromatography (HSCCC) is a relatively new technique. It is a liquid–liquid partitioning chromatographic method in which the stationary phase is immobilized by a centrifugal force. HSCCC is the most advanced form of CCC in terms of partition efficiency and separation time. As solvent system, a mixture in different ratio is usually used, depending on the flavonoids' nature: ethyl acetate/*n*-butanol/water,[103,104] ethyl acetate/*n*-butanol/water/methanol,[61,73] or *n*-hexane/ethyl acetate/methanol/acetonitrile/water[90] have been used between others. In addition, UV detector is necessary to detect flavonoids at specific wavelength (254 or 280 nm, see Table 3.8). Thus, HSCCC has become an attractive method for the preparative separation of analytes, and it has been applied in the measurement of flavonoids, since HSCCC provides many important advantages over the conventional CCC, including an over tenfold increase in sample loading capacity, high concentration fraction, concentration of minor impurities, etc.

TABLE 3.8
Countercurrent Chromatography (CCC) Applications for the Separation and Determination of Flavonoids in Foods[a]

Flavonoid	Food	CCC Mode	Solvent System	Detector	References
Flavanones	Leaves of *Patrinia villosa*	CCC	*n*-Hexane/ethyl acetate/methanol/water (10:13:13:10, v/v/v/v)	UV–Vis detector (280 nm)	182
Flavones	*Clusia criuva*	CCC	Chloroform/methanol/water (7:13:8, v/v/v)	UV–Vis detector (254 nm)	60
Isoflavones	*Pueraria lobata*	HSCCC	Ethyl acetate/*n*-butanol/water (2:1:3, v/v/v)	UV detector (254 nm)	103
Flavones and flavonols	*Trollius ladebouri*	HSCCC	Ethyl acetate/*n*-butanol/water (2:1:3, v/v/v)	UV detector (254 nm)	104
Flavanones	*Patrinia villosa*	HSCCC	*n*-Hexane/ethyl acetate/methanol/water (10:11:11:8, v/v/v/v)	UV detector (280 nm)	73
Flavones	Leaves of *Ampelopsis grossedentata*	HSCCC	*n*-Hexane/ethyl acetate/methanol/water (1:6:1.5:7.5, v/v/v/v)	UV detector (254 nm)	61
Flavanones	*Glycyrrhiza uralensis.*	HSCCC	*n*-Hexane/ethyl acetate/methanol/ACN/water (2:2:1:0.6:2, v/v/v/v/v)	UV detector (280 nm)	90
Anthocyanidin	Koaling seed coat	pH-z-CCC	Diethyl ether/ACN/ water (4:2:5 v/v/v). Mobile phase with ammonium hydroxide (20 mM) and stationary phase with TFA (20 mM).	LC/ESI-MS (negative mode)	183
Anthocyanins	Red wine	pH-z-CCC	MTBE/*n*-butanol/ACN/water (1:3:1:5 v/v/v/v), 15 mM TFA in upper phase and 15 mM NH_3 in lower phase.	UV detector (500 nm)	184

[a] MTBE, *tert*-butyl methyl ether.

In pH-zone refining countercurrent chromatography (pH-z-CCC), a retainer alkali (or acid) in the stationary phase is applied to retain the analytes in the column, and an elute acid (or base) is used to elute the analytes according to their pKa values and hydrophobicity.[183,184]

3.6.2 Electrophoretic Techniques

3.6.2.1 Capillary Electrophoresis

Although HPLC stays as the dominating separation technique for food flavonoids, electrophoretic techniques are gaining popularity, especially capillary electrophoresis. CE is a high efficiency separation technique with low costs and shorter analysis times than other techniques, requiring lower sample and reagent volumes. Separation is based on differences in electromigration between analytes in a given electric field. Several modes of CE are available for separation of different analytes, although capillary zone electrophoresis and micellar electrokinetic chromatography are the most widely used (Table 3.9).

Capillary zone electrophoresis (CZE) is the simplest and most versatile CE mode. Separation in CZE is based on differences in the charge-to-mass ratio that determine the electrophoretic migration time of analytes, and it is only applicable to charged molecules.[96,108,172,175,185–190]

In micellar electrokinetic chromatography (MEKC), a hybrid technique between electrophoresis and chromatography, distinction has to be made between neutral and charged analytes. During MEKC separation, nonpolar portions of neutral solutes are incorporated into micelles and comigrate with them at the same speed, while polar compounds are free and migrate at the electroosmotic flow velocity. The distribution coefficient between the micellar and the nonmicellar phases greatly influences the migration velocity of the analytes. To form micelles, surfactants like sodium dodecyl sulfate (SDS) are added into the electrolyte at amounts above the critical micelle concentration (CMC). The main advantage of MEKC is its much higher separation speed and better resolution in comparison to CZE.[105,191–194]

Flavonoids present at least one phenolic hydroxyl group, ionization of which determines the electrophoretic mobility of the flavonoid. The magnitude of the net negative charge of a particular flavonoid is determined by the position and number of these hydroxyl groups and by the pH of the buffer solution. Also, addition to the running buffer of modifiers such as organic solvents results in the reduction of the sphere of hydration of the analytes, increasing the difference in charge-to-size ratio and providing better resolution of analytes.

Recently, separation of chiral flavonoids has been achieved by CE system with various cyclodextrins as chiral selectors. These chiral additives are added to the mobile phase, allowing the selective interaction with enantiomers and facilitating the formation of transient diastereomers.[187,194]

In spite of the several advantages mentioned above for CE, some aspects of this technique still require to be improved, such as the reproducibility of retention/ migration times and its sensitivity. However, CE constitutes a separation technique complementary to LC.

TABLE 3.9
Capillary Electrophoresis Applications for the Separation and Determination of Flavonoids in Foods[a]

Flavonoids	Food	Mobile Phase and Voltage	Detector	References
CZE				
Flavones	Grape wine	Buffer:	DAD	
Flavanones	Beer	pH = 8.9/Sep volt 15.4 kV	250 nm	172
Flavonols	Soybean seed	pH = 10.25/Sep volt 20 kV	210 nm	185
Flavanols		pH = 10.5/Sep volt 23.5 kV	260 nm	186
Isoflavones				
Anthocyanins (Flavanones enantiomers)	Black currant	Buffer/organic solvent:	DAD	
Flavonols	Lemon juice	pH = 1.5/Sep volt 25kV	520 nm	96
Flavanols	Wine	pH = 9.3/Sep volt 10–30kV	290 nm	187
		pH = 13.5/Sep volt 30kV	230 nm	188
Flavanols	Green and black tea	Buffer + CD + organic solvent:	DAD	
		pH = 7.2/Sep volt 25 kV	205 nm	175
Flavanones and glycosides	Ripe fruit of *Citrus aurantium*	Buffer:	EC	
		pH = 8.45/Sep volt 12 kV	SCD: 0.85 V	189
Flavonols	Rosemary honey	Buffer/organic solvent:	ESI-MS	
Flavanones	Phytomedicine	pH = 10/Sep volt 20 kV	214 nm (negative mode)	108
		pH = 9.5/Sep volt 15–25 kV	254 nm (positive mode)	190
MEKC				
Flavanones, flavonols	Honey, legumes	Buffer + SDS + organic solvent:	UV	
Flavones		pH = 8.0/Sep volt 20kV	280 nm	105
Isoflavones		pH = 9.0/Sep volt 25kV	287 nm	191
Flavanols	Green tea	Buffer + SDS:	UV	
		pH = 7.0/Sep volt 30kV	200 nm	192
Flavonols	Fruits of *Forsythia suspense*	Buffer + SDS + organic solvent:	EC	
		pH = 8.4/Sep volt 12kV	SCD: 1.2 V	193
Flavanones enantiomers	Standards	Buffer + SDS + CD:	DAD	
		pH = 8.3/Sep volt 20kV	220 nm	194

[a] CD, cyclodextrin; SCD, single coulometric detector; SDS, sodium dodecyl sulfate as micellar agent.

3.7 DETECTION SYSTEMS

The detection systems that are described in this section can be coupled to the different instruments where separation of flavonoids is performed.

3.7.1 UV–Vis Detection

The most common method for flavonoids' identification and quantification is UV-visible (UV–Vis) detection. Flavonoids have two characteristic absorption bands in the UV–Vis region, Band I (arising mainly from the B- and C-rings of the flavan nucleus) and Band II (from the A-ring), with absorption maxima in the 300–550 and 240–285 nm range, respectively.[195] Figure 3.3 shows UV–Vis spectra representative

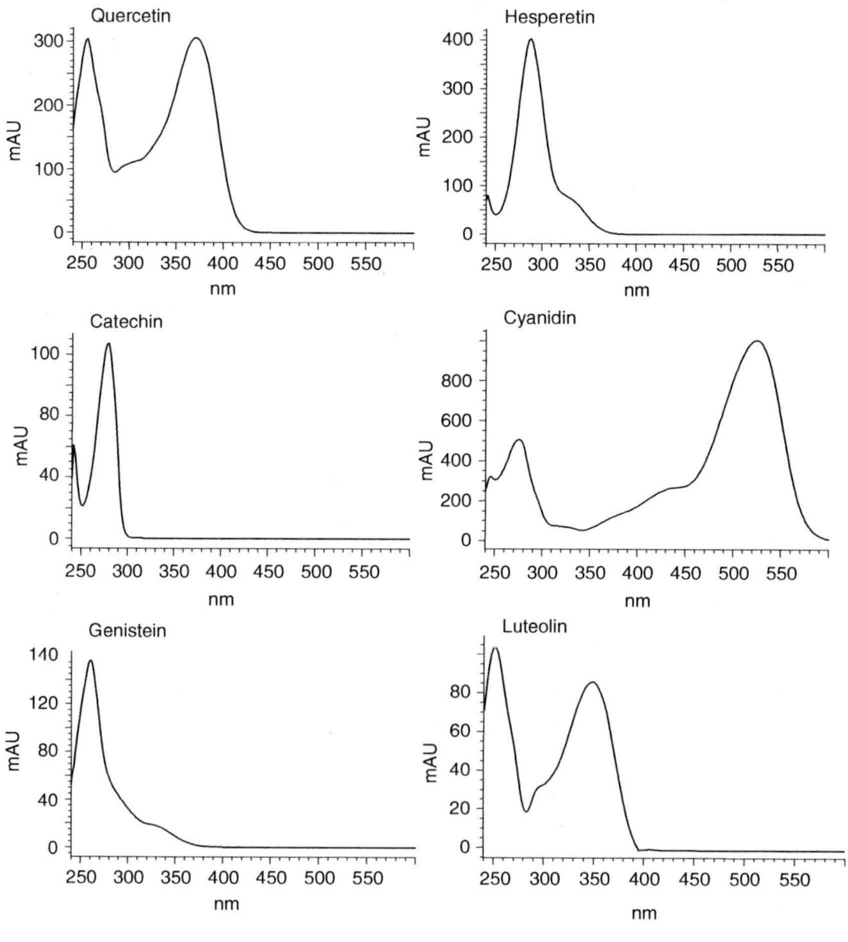

FIGURE 3.3 Ultraviolet (UV)–visible spectra of the main classes of flavonoids recorded online after reversed-phase high-performance liquid chromatography (RP-HPLC) separation using an acetonitrile/4.5% formic acid (v/v) gradient.

of the major classes of flavonoid aglycones. As can been observed, all flavonoids have an absorption maximum in the wavelength range corresponding to Band II, which is less specific and therefore less useful than Band I for the identification of flavonoids. Because of the lack of conjugation between A- and B-rings, flavanols, flavanones, and isoflavones hardly show absorption in Band I, whereas anthocyanidins are characterized for the long wavelength position of this band (around 460–560 nm), and flavones and flavonols have a peak in Band I in the region of 300–380 nm.

The degree of hydroxylation and substitution of the hydroxyls will cause variations in the absorption spectrum. Methylation, methoxylation, and nondissociated hydroxyl groups generally result in minor changes in the position of the absorption maxima. Methylation may have a slight hypsochromic effect, causing band shifts of 1–4 nm to shorter wavelengths when methylation occurs in B-ring, and up to 10 nm when affecting hydroxyls in the A-ring.[196] Similarly, glycosylation usually has a hypsochromic effect irrespectively of the nature of the sugar residue. Some examples of characteristic spectra corresponding to the most representative compounds of different classes of flavonoids are shown in Table 3.1. In Figure 3.4, the bathochromic shift that takes place in delphinidin-3-glucoside is represented with a hydroxyl group in C3′, in comparison to cyanidin-3-glucoside, whose structure lacks such hydroxyl in the B-ring. In this same figure, the effect generated by the presence of sugar residues can be observed by comparing the spectra of cyanidin versus cyanidin-3-glucoside.

In flavones and flavonols, an increase in oxygenation results in a bathochromic shift in the corresponding band maximum. Introduction of methyl ethers on the free hydroxyls has little effect on the UV spectrum with the exception of hydroxyls at positions 3, 5, 6, and 8 in which methylation has a marked hypsochromic effect on

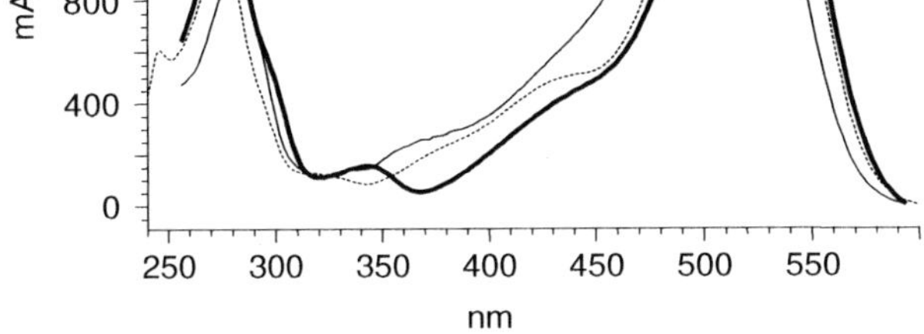

FIGURE 3.4 Ultraviolet (UV)–visible spectra of cyanidin ($\lambda_{max} = 524$ nm), cyanidin-3-glucoside ($\lambda_{max} = 510$ nm), and delphinidin-3-glucoside ($\lambda_{max} = 526$ nm), as recorded online after reversed-phase high-performance liquid chromatography (RP-HPLC) separation using an acetonitrile/4.5% formic acid (v/v) gradient.

the DAD (diode array detection)-recorded spectrum while methylation at position 7, 3′, or 4′ has little effect on the wavelength maxima and spectrum shape.

Flavan-3-ols are usually detected in the UV region at 270–280 nm while gallocatechins and prodelphinidins show absorption maxima at 270 nm, and catechins and procyanidins at 280 nm. Mixed proanthocyanidins show intermediate values and the presence of galloyl moieties produces a broadening of the spectra at about 310 nm.

UV spectra of flavanones do not supply conclusive information on the pattern of B-ring substitution. Removal of the 5-hydroxyl group causes a hypsochromic shift of Band II while increasing oxygenation in the B-ring has no noticeable effect on the UV spectra.

The difference in the chemical structure of isoflavones in comparison to the other flavonoids with a double bond between the C2-C3 atoms and the location of the B-ring at C3, which prevents conjugation of the phenyl group with the pyrone carbonyl group, results in a significantly diminished contribution of the B-ring to the spectrum, relatively unaffected by increased hydroxylation. Nevertheless, Band II is shifted bathochromically by increased oxygenation in the A-ring.

Problems caused by differences in the wavelengths for maximum UV absorption by individual flavonoids can be solved by the use of DAD since it allows simultaneous recording of chromatograms at different wavelengths, thus facilitating the selective detection of different groups of flavonoids at their maximum wavelength to improve sensitivity.

Although UV–Vis detection and DAD provide useful information for the identification of flavonoids, the use of conventional approaches based on spectra is often limited when samples contain very similar compounds. For complete structural identification, other techniques such as MS and nuclear magnetic resonance (NMR) spectroscopy are often necessary.

3.7.2 Electrochemical Detection

Electrochemical detection is a particularly useful method for the determination of electroactive compounds such as flavonoids. Electrochemical detection is based on the induced oxidation/reduction of the analyte elicited by an electrode set at a fixed potential.[197] There are two types of electrochemical detectors, both working under the same principle: the amperometric and the coulometric detectors. In the amperometric detector, only partial oxidation/reduction of the analytes takes place, while in the coulometric detector oxidation/reduction of the analytes is almost complete, which results in an increased sensitivity of this detector in comparison with the amperometric one. Coulometric electrode array detection (CEAD) consisting of a platinum reference electrode that sets the electrochemical zero and working pairs of electrodes placed in series conforming a multichannel detector that enables an enhanced sensitivity and selectivity of the CEAD in comparison with single coulometric detectors based on palladium[198–200] or Ag/AgCl[201] as reference electrodes. The working range of the electrodes in CEAD can vary between −1.0 and +2.0 V, depending on the electrochemical properties of the analyte,[197] which is an additional advantage in comparison with single coulometric detection, where the working potential needs to be selected before analysis based on the maximum response (oxidation/reduction) of the analyte as determined by the study of hydrodynamic voltamograms.[69]

CEAD permits efficiencies in the oxidation of the analyte of nearly 100% oxidation, and lower detection limits are possible.[197] Two analytes could thus be separated if they have different redox properties even though they present similar chromatographic properties. Therefore, CEAD shows high sensitivity in comparison to other detectors, although it is especially dependent on appropriate sample pretreatment.

For flavonoids, there is no clear linear relationship between their antioxidant activity and the electrochemical signal. Thus, glycosylation of the 3-hydroyxl group decreases the antioxidant or antiradical activity in flavonols, but not their electrochemical behavior. This decrease can be attributed to the esteric hindrance posed by the glucose group in C3 or to a reduced solubility of the glycoside in the alkane solvent used in these experiments. Nevertheless, for most analytes the antioxidant efficiency could be related to the value of the first maximum, which corresponds to the lowest energy required to donate an electron.

Examples of electrochemical detection applied to flavonoid analysis are presented in Table 3.10. Acceptable resolution of flavonoids has been achieved with LC working in isocratic conditions[69,200] or with gradient elution, mainly consisting of two phases,[198–201] in RP (C18) columns with electrochemical detector.

3.7.3 Fluorescence Detection

Although the number of flavonoids that are naturally fluorescent is quite limited, fluorescence detection can provide better selectivity and sensitivity than UV detection, permitting selective detection of flavonoids in complex mixtures. Thus, fluorescence detection was reported to be tenfold more sensitive than UV detection.[203] A fluorescence detector, mainly set at 280–270 and 310 nm for excitation and emission wavelengths, respectively, connected in series with a UV detector allows a more complete characterization of fluorescent and nonfluorescent compounds.[95,203–205]

The fluorescence of a particular flavonoid is determined by the nature of the functional groups and their substitution pattern. Several examples have been summarized in Table 3.10. Among isoflavones, only those that lack a hydroxyl group in C5 show strong native fluorescence.[204] Other flavonoids such as catechin[205–208] or flavones with a hydroxyl group in position 3 (3-hydroxyflavone)[209] or methoxylated flavones (3′,4′,5′-trimethoxyflavone)[203] have been described to possess native fluorescence. For nonfluorescent flavonoids, derivatization is used as a strategy to convert these compounds into highly fluorescent derivatives by the formation of complexes with metal cations.[210,211]

In general, considering the complex nature of food flavonoids and their heterogeneous behavior with regard to inherent fluorescence, this kind of detector could constitute an adequate complement for the identification of flavonoids in foods.

TABLE 3.10
Some High-Performance Liquid Chromatography (HPLC) Procedures for the Determination of Multiple Classes of Food Flavonoids with Electrochemical (ECD), Fluorimetric, and Mass Spectrometry (MSD) detectors[a]

Flavonoids	Food	Stationary Phase	Mobile Phase	Detection Conditions	References
Isocratic Elution					
Flavanones	Orange juice	C18 column	Organic solvent (ACN) /Acidified water (formic, phosphoric, acetic acids)	**ECD**	
Flavonols	Fruits			SCD: At 0.8 or 0.7 V	69
Flavones	Phytochemical products			0–300 mV and 600–900 mV	200
Flavanols				**Fluorimetric detector**	
				$\lambda_{ex} = 280$ nm, $\lambda_{em} = 315$ nm	206
				$\lambda_{ex} = 330$ nm, $\lambda_{em} = 440$ nm in series UV and ESI-MS detector	203
				MSD	
				QTRAP: LC/MS/MS (negative mode)	223
Gradient Elution (Two Phases)					
Flavanones	Juice	C18 column	Solvent A: acidified water (TFA, formic, phosphoric acids)	**ECD**	
Flavanols	Beverages			16 arrays, Δ60 mV (Pd ref)	198
Flavones	Fruits		Solvent B: organic solvent acidified (MeOH, ACN, isopropanol/TFA, formic, phosphoric acids)	12 arrays, Δ70 mV (Pd ref)	199
Isoflavones	Vegetables			8 arrays, Δ100 mV	202
Anthocyanidins	Beer				
Flavanols	Chocolate			**MSD**	
Flavones	Citrus juice			LC/APCI-MS (positive mode)	219
Flavanones				LC/ESI (negative mode)	212
Flavanols	Vegetables, legumes, tea, wine, fruit juices, chocolate milk	C18 column	Solvent A: acidified water (TFA, phosphoric acid) with organic solvent (ACN, MeOH)	**Fluorimetric detector**	
Stilbens				$\lambda_{ex} = 280$–270 nm, $\lambda_{em} = 310$ nm	207, 208
				$\lambda_{ex} = 278$–330 nm, $\lambda_{em} = 360$–374 nm in series UV detector	95

Flavonols	St. John's		Solvent B: acidified organic solvent	**MSD**	
Flavones	Wort dietary		(ACN, MeOH)	LC-ESI (positive mode)	174
Isoflavones	supplements			LC/DAD-ESI-MS (SIM) (negative mode)	213
	Tomato				
Isoflavones	Soybean food	C18 column	Solvent A: acidified water	**ECD**	
Flavanols	Sorrel leaf extracts		(phosphoric, formic, acetic acids;	8 arrays, Δ20 mV (Ag/AgCl ref)	201
Isoflavones	Vegetables		ammonium formatted buffer,	**Fluorimetric detector**	
			ammonium acetate)	$\lambda_{ex} = 279$ nm, $\lambda_{em} = 307$nm in series UV detector	205
				$\lambda_{ex} = 250$ nm, $\lambda_{em} = 350$–450 nm in series UV detector	204
Flavanol	Wine		Solvent B: organic solvent (ACN,	**MSD**	
Flavones	Fruits		MeOH, dicloromethane)	LC/API-MS (positive/negative mode)	217
Flavonols	Orange juice			LC-ESI (negative mode)	173
Flavanones	Green tea			LC/ES-MS (positive mode)	214
Anthocyanidins	Propolis			HPLC-ESI/MS (negative mode)	215
Isoflavones	Mate			HPLC-ESI-MS (negative mode)	88
	Vegetables			LC-ESI-MS/MS (negative mode)	216
				HPLC-DAD/API-MS (negative mode)	218

[a] SCD, single coulometric detector.

3.7.4 Mass Spectrometry

MS has proven to be a very powerful technique for flavonoids' analysis because of its high sensitivity and the possibility to combine with different chromatographic techniques, for example, GC/MS, CE/MS, and specially LC/MS, as a stand-alone instrument allowing both qualitative and quantitative determinations.

There are two main types of ionization techniques: the ion-spray techniques such as thermospray ionization, ESI, or atmospheric pressure chemical ionization (APCI); and the ion-desorption techniques that include fast atom bombardment (FAB), plasma desorption (PD), and matrix-assisted laser desorption ionization (MALDI). ESI[88,173,174,212–216] and APCI[217–219] are the two most widely used ionization methods for flavonoids, and coupled to LC techniques (LC-MS) they have been described in numerous studies to identify flavonoids (Table 3.10).

Although there is no clear tendency, ESI is more often used to ionize polar and nonvolatile molecules, like condensed tannins and anthocyanidins, and APCI is used for less polar and nonionic compounds especially flavonols, flavones, flavanones, and chalcones.

APCI and ESI can be operated under both positive (molecular species $[M+H]^+$) and negative ion modes (molecular species $[M-H]^-$). The most frequently used mass analyzers can also be separated into two main groups: analyzers based on ion beam transport such as magnetic field, time-of-flight (TOF), and quadrupole mass filter; and those based on ion trapping technology. These analyzers vary in their capabilities with respect to resolution, accuracy, and mass range. MS detector is critical for the identification of flavonoids because of the complex and diverse structures of these compounds and their low concentrations in foods. Sensitivity and selectivity of detection can be increased using tandem mass spectrometry (MS/MS), that is two (MS/MS) or more (MS^n) mass analyzers coupled in series. MS/MS and MS^n produce more fragmentation of the precursors and daughter ions, therefore providing additional structural information for the identification of flavonoids. Additionally, most of these equipments allow the recording of a desired range of masses in the full scan mode, as well as the recording in SIM mode that give useful results because of its high sensitivity. There are many excellent recent reviews on the applications of LC-MS in quantitative and qualitative analyses of flavonoids.[220,221]

Fragmentation pathways are largely independent of the ionization mode (ESI or APCI) and the type of analyzer used.[222] The RDA reaction is an important fragmentation reaction of flavonoids being RDA fragments especially important for the structural characterization of aglycones and the aglycone part of flavonoid conjugates.

Positive ion collision-induced dissociation (CID) spectra are most frequently used than negative ion CID. However, negative ion mode is more sensitive in flavonoid analysis[88,173,212,213,218,223] since higher collision energies are required to yield adequate fragmentation that could differ in comparison to positive ion mode, providing complementary information.

In addition, losses of small molecules or radicals from the $[M+H]^+$ ion, such as 18 u (H_2O), 56 u (COCO), 42 u (C_2H_2O), 72 u (CO_2CO), and 15 u (CH_3), are commonly observed and are useful to identify the presence of specific functional groups. In the negative ion mode, several molecules such as 46 u (CH_2CHOH), 56 u

TABLE 3.11
Fragment Ions Observed for Selected Flavonoids Ionized in the Negative Ion Mode

Compound	–OH	$-OCH_3$	MW	$[M-H]^-$	Fragment Ions	References
Flavonols						
Quercetin	3,5,7,3′,4′		302	301	107, 179, 121, 151, 229	235–237
Isorhamnetin	3,5,7,4′	3′	316	315	107, 151, 300, 271	235
Morin	3,5,7,2′,4′		302	301	107, 151	235
Rhamnetin	3,5,3′,4′	7	316	315	121, 165, 300, 271	235
Kaempherol	3,5,7,4′		286	285	135, 107, 151, 161	235, 236
Flavanones						
Hesperetin	5,7,3′	4′	302	301	107, 151, 286	235
Naringenin	5,7,4′		272	271	107, 151, 119	235–237
Eriodictyol	5,7,3′,4′		288	287	107, 151, 135, 125	235–237
Isosakurametin	5,7	4′	286	285	151, 125, 270	238
Flavones						
Apigenin	5,7,4′		270	269	107, 151, 117, 149	235–237
Luteolin	5,7,3′,4′		286	285	107, 151, 133, 213	235, 237
Acacetin	5,7	4′	284	283	107, 151, 133, 268	235
Isoflavones						
Daidzein	7,4′		254	253	133, 117, 197, 181	236, 238, 191
Genistein	5,7,4′		270	269	133, 107, 163, 197	236, 238, 191
Flavanols						
Catechin	3(*trans*),5,7,3′,4′		290	289	245	217
Epicatechin	3(*cis*),5,7,3′,4′		290	289	245	217

(COCO), 72 u (CO_2CO), and 15 u (CH_3) are also observed after study of mass spectrum. Tables 3.11 and 3.12 summarize the fragment ions observed for several selected flavonoid classes in the negative and positive ionization modes, respectively.

The identification of prenylated flavonoids determined by the loss of isobutene (C_4H_8) (56 u) is not observed in the negative ion mode,[224] while in the positive ion mode it is detected using ESI or FAB ionization. Also, $[M+H-C_4H_8]^+$ and $[M+H-C_4H_8-C_4H_8]^+$ fragments are characteristic for diprenylated flavonoids.[225]

With respect to flavonoid glycosides, useful information can be obtained from mass spectra such as molecular mass, structure of the aglycone, information about acylation of sugar hydroxyl groups, methylation of the aglycone hydroxyl groups, number of sugars, and in some cases placement of the glycosidic bonds. However, MS does not provide information about stereochemistry of the glycosidic linkage nor distinguish between diastereomeric sugar units.

Differentiation between *O*-glycosides, *C*-glycosides, and *O*,*C*-diglycosides can be made by examining first-order positive ion spectra or low-energy CID spectra.

TABLE 3.12
Fragment Ions Observed for Selected Flavonoids Ionized in the Positive Ion Mode[a]

Compound	–OH	$-OCH_3$	–X	MW	$[M+H]^+$	Fragment Ions	References
Flavonols							
Quercetin	3,5,7,3′,4′			302	303	165, 137, 153, 285, 247, 262, 257, 229	203, 239
Isorhamnetin	3,5,7,4′	3′		316	317	165, 151, 153, 302, 261, 275	203, 239
Myricetin	3,5,7,3′,4′,5′			318	319	165, 153, 153, 301, 263, 277, 273, 245	203, 239
Kaempherol	3,5,7,4′			286	287	165, 121, 153, 269, 231, 245, 241, 213	203, 239
Flavanones							
Naringenin	5,7,4′			272	273	153, 119, 147, 231, 227	236
Flavones							
Apigenin	5,7,4′			270	271	121, 153, 119, 163, 229, 225	239
Luteolin	5,7,3′,4′			286	287	137, 153, 135, 179, 245, 241	239
Acacetin	5,7	4′		284	285	135, 153, 133, 177, 270, 243, 239, 213	203, 239
Isoflavones							
Daidzein	7,4′			254	253	137, 119, 199	222
Flavanols							
Catechin	3(*trans*),5,7,3′,4′			290	291	139	217, 240
Epicatechin	3(*cis*),5,7,3′,4′			290	291	139, 150	217, 240
Epigallocatechin	3(*cis*),5,7,3′,4′,5′			308	307	139	217, 240
Epigallocatechin-3-*O*-gallate	5,7,3′,4′,5′		3(*cis*-OG)	458	459	289, 139	217
Epicatechin-3-*O*-gallate	5,7,3′,4′		3(*cis*-OG)	442	443	273	217
Proanthocyanidins							
Epicatechin-4,8-epicatechin	3,5,7,3′,4′		4(*cis*-EC)	578	579	291, 427	241
Anthocyanidins							
Cyanidin	3,5,7,3′,4′			286	287		242

[a] EC, epicatechin; OG: gallate.

Flavonoids commonly occur as flavonoid-*O*-glycosides, the 3- and 7-hydroxyl groups being the typical glycosylation sites. Table 3.13 shows some examples of fragmentation in the positive ion mode for this kind of glycoside.

MS methods presently available rarely provide complete structural information, so that the search for novel MS approaches to characterize flavonoids and the implementation of new technologies to make the present methods faster, easier, and more selective should continue.

3.7.5 Nuclear Magnetic Resonance

The most powerful technique for the determination of flavonoids structure is NMR spectroscopy, providing unambiguous information on the substitution pattern of the flavan nucleus. Thus, coupling an efficient separation system such as HPLC with NMR detection (LC-NMR) is a highly effective instrument in flavonoids analysis.[226]

Both simple ^1H-NMR spectra and two-dimensional ^1H-^1H correlation experiments (correlation spectroscopy, COSY) can be obtained from the application of LC-NMR, while ^{13}C-NMR spectra are restricted to certain cases in which high concentrations of the analyte of interest are ensured, where ^{13}C-NMR information can be indirectly deduced from inverse detection experiments.[227]

A common drawback to LC-NMR analysis derives from the interference of the resonances of the mobile phase, usually much higher than those of the analyte itself. To overcome this problem, the mobile phase coming from the liquid chromatograph is efficiently suppressed with techniques such as water-suppression enhanced-through T_1 effects (WET).[228] This allows the separation of analytes in typical RP-HPLC conditions using common solvents such as MeOH or acetonitrile, yet usually replacing water by D_2O to achieve better quality spectra; however, in the case of analyte signals coincident with the suppressed solvent resonances, those of the analytes will also be suppressed. On the other hand, in spite of the usefulness of LC-NMR for structural elucidation of flavonoids, this is not a sensitive LC detection technique. Therefore, to improve sensitivity, transients can be accumulated by operation in two different modes: the on-flow and stop-flow modes. In both cases, the analytes from an LC column pass into an NMR microflow probe that typically has an active volume of 60–120 μL, comparable with the conventional NMR sample size of 500–600 μL.

Operation in the on-flow LC-NMR mode implies the continuous acquisition and storage of spectra during separation. These on-flow data are processed as a two-dimensional NMR experiment, one dimension representing the NMR scale (ppm) and the second dimension the time scale. This alternative technique has relatively low sensitivity, which restricts its application mainly to direct measurements of the major constituents of a complex mixture, often under overloaded LC-conditions. Some strategies to enhance sensitivity have been proposed, such as carrying out the analysis at low flow,[229] performing online trace enrichment,[230] or running time-slice experiments over a whole chromatogram,[231] where the flow is stopped at defined time intervals (e.g., each minute) and a large number of acquisitions are performed.

The stop-flow mode is another way to improve the detection limit in LC-NMR. In this operation mode, separation can be halted when an analyte of interest reaches

TABLE 3.13
Fragment Ions Observed for Selected Flavonoid Glycosides in the Positive Ionization Mode[a]

Compound	–OH	$-OCH_3$	–X	MW	$[M+H]^+$	Fragment Ions	References
Flavonols							
Rutin (Quercetin-3-*O*-rutinoside)	5,7,3′,4′		3-*O*-Rut	610	611	465, 303, 449	243, 244
Nicotiflorin (K-3-*O*-rutinoside)	5,7,4′		3-*O*-Rut	594	595	449, 287, 433	243
K-7-*O*-neohesperidoside	5,7,4′		7-*O*-Neo	594	595	449, 287	243
Quercitrin	5,7,3′,4′		3-*O*-Rha	448	449	303, 431	203
Flavonones							
Hesperidin (hesperitin-7-*O*-rutinoside)	5,3′	4′	7-*O*-Rut	610	611	465, 303, 449	243
Naringin (naringenin-7-*O*-neohesperidoside)	5,4′		7-O-Neo	580	581	435, 273, 419	243
Narirutin (naringenin-7-*O*-rutinoside)	5,4′		7-*O*-Rut	580	581	435, 273, 419	243
Flavones							
Rhoifolin (Apig-7-*O*-neohesperidoside)	5,4′		7-*O*-Neo	578	579	433, 271	243
Isorhoifolin (Apig-7-*O*-rutinoside)	5,4′		7-*O*-Rut	578	579	433, 271	243
Anthocyanins							
Cyanidin-3-glucoside	5,7,3′,4′		3-*O*-Glu	448	449	287	242, 245
Malvidin-3-*O*-glucoside	5,7,4′	3′,5′	3-*O*-Glu	492	493	331	245
Peonidin-3-*O*-glucoside	5,7,4′	3′	3-*O*-Glu	462	463	301	245

[a] Glu, glucose; Neo, neohesperidose; Rha, rhamnose; Rut, rhamnoglucoside or rutinoside.

the probe and thus several NMR datasets can be acquired over a longer time. In this alternative operation mode, previous knowledge of the retention times of the analytes of interest is required; alternatively, a sensitive method of detection such as LC-UV or LC-MS can be used before the NMR instrument to trigger the stop-flow analysis. In stop-flow mode, when the concentration of the analyte is high enough (generally more than 100 μg of sample is required), various two-dimensional correlation experiments are also possible, like COSY, HSQC (heteronuclear single quantum coherence), HMBC (heteronuclear multiple bond correlation), or NOESY (nuclear Overhauser effect spectroscopy). Thus, in most publications found for flavonoid analysis, the stop-flow mode (Table 3.14) was used to enable very long scan times to record the NMR spectra, with scan times varying between 1 h and several days per chromatographic peak.

In LC-NMR, separation is usually performed on an RP C18 column and broad acetonitrile or methanol gradients are applied, using D_2O instead of water, and trifluoracetic acid to improve resolution of polyphenols. UV spectra are recorded for all constituents with the aid of a photodiode array detector. However, it should be taken into consideration that chemical shifts in an RP solvent will slightly differ from those recorded in standard deuterated solvents.

Currently, much effort is devoted to the development of micro- or even nano-LC-NMR. This is an attractive development since deuterated solvents are now more easily available. The drawback to this technique posed by its low sensitivity could be resolved with the recent development of new technologies based on a cryoflow NMR probe.[232] This probe, which cools the receiver coils to cryogenic temperature to improve the signal-to-noise ratio of the NMR spectra, has been recently applied for the analysis of oregano extracts.[233] Spraul et al.[232] showed that with cryoflow probes the analyte detectability was about four times better than with conventional probes or, alternatively, the scan time was 16-fold shorter for the same amount of sample. This approach, and the use of online preconcentration using LC-SPE-NMR, is likely to gain popularity in the future.

LC-NMR is still scarcely applied to foodstuffs. However, in spite of practical problems inherent to this technique such as low sensitivity, costly instruments, and long running times, the recent progress in this technique is impressive, providing highly valuable online information on the chemical identification of flavonoids.

3.8 FINAL CONSIDERATIONS

The potential health beneficial effects of food flavonoids has awakened the interest of nutritionists, food scientists, and food manufacturers worldwide, which has evolved in an extended research on these plant constituents. To accurately analyze flavonoids in foods and food products, including functional foods and nutraceuticals, it is critical to develop efficient methods for their extraction and separation from the different and often complex matrices where they are present, to precisely identify and quantify these polyphenolic compounds.

Because of the enormous diversity of structures corresponding to the different classes of flavonoids, it is difficult to establish an ideal method that can efficiently resolve all flavonoid compounds from complex mixtures. Selection of the extraction

TABLE 3.14
Main Applications of Liquid Chromatography–Nuclear Magnetic Resonance (LC-NMR) in the Analysis of Food Flavonoids

Flavonoid	Food	Mode	Experiment	Instrument and Conditions	LC Eluents	References
On-Flow						
Flavanones, isoflavones	*Erythrina vogelii*	On-flow	^{1}H	500 MHz, 60 μL cell	MeCN-D_2O low flow (0.1 mL/min)	229
Flavonoids	*Gnidia involucrata*	Time-slice	^{1}H	500 MHz, 60 μL cell	—	231
Flavonol glycosides	Leaves of *Sorocea bomplandii*	On-flow	^{1}H	600 MHz, 120 μL cell	D_2O, MeOH	246
Flavones	*Humulus lupulus*	Online enrichment	^{1}H TOCSY	500 MHz, 120 μL cell	D_2O, ACN, phosphoric acid	230
Stop-Flow						
Flavone glycosides	*Gentiana ottonis*	Stop-flow	^{1}H	500 MHz, 60 μL cell	D_2O, ACN, TFA	247
Flavones, flavonols	*Hypericum perforatum*	Stop-flow	^{1}H	500 MHz, 120 μL cell	D_2O, ACN, acetic acid	248
Flavonoid	*Maytenus aquifolium*	Stop-flow	^{1}H	600 MHz, 120 μL cell	D_2O-MeOH	249
Isoflavones	Leaves of *Trifolium pratense*	Stop-flow	^{1}H COSY, NOESY	400 MHz, 120 μL cell	D_2O, ACN deuterated, TFA	86
Flavonol glycosides	Apple peel	Stop-flow	^{1}H	500 MHz, 120 μL cell	D_2O, ACN, TFA	250
Flavonols, flavanones glycosides	Tomato	Stop-flow	^{1}H	500 MHz, 120 μL cell	D_2O, ACN, TFA	251
Flavanones	*Monotes engleri*	Stop-flow	^{1}H NOESY GHSQC, GHMBC	500 MHz, 60 μL cell	MeCN-D_2O	227

conditions (types of solvents, further purification steps, etc.) is crucial to ensure complete extraction of the flavonoids from foods without losses or modifications of their chemical structure.

HPLC with UV–Vis detection is the most widely utilized separation and determination technique, although other chromatographic methodologies such as GC, CCC, or even electrophoretic techniques have also been used with success. The hydrophobic or hydrophilic character of the analytes should be considered before selecting the separation procedure. RP columns are widely used in the separation of flavonoids for their excellent performance, avoiding selective losses of analytes onto the stationary phase of the column, which could entail errors in the quantification of flavonoids. Spiking of samples with standards and recovery experiments may be helpful in establishing the extent of error introduced by physical losses not only during separation, but also in the preliminary preparation of the sample. For quantitative analysis of flavonoids, the use of external or internal standards when available is recommended. In the case of internal standards, the election of a suitable standard with similar polarity to the analyte of interest is the most critical step to achieve acceptable results.

Although UV–Vis detector is the most widespread detection system, fluorescence or electrochemical detectors have also proven adequate for flavonoid analysis. However, the lack of active fluorescent or electrochemical groups in many flavonoids may affect the accuracy of these detection methods. MS is a powerful, highly sensitivity technique. Its use is spreading among food scientists because of its effectiveness in the identification of flavonoids and their applicability for quantitative analysis. Common sources of error in flavonoid analysis may derive from nonlinearity of the detector, which should be checked since many detectors are linear over only one or two decades. In addition, the different response factors of different flavonoids should be taken into account since detectors usually are not equally sensitive to all components of a mixture.

REFERENCES

1. Aggarwal, B.A. and Shishodia, S., Molecular targets of dietary agents for prevention and therapy of cancer, *Biochem. Pharmacol.*, 71, 1397, 2006.
2. Block, G., Patterson, B., and Subar, A., Fruit, vegetables, and cancer prevention: A review of the epidemiological evidence, *Nutr. Cancer*, 18, 1, 1992.
3. Duthie, G.G., Duthie, S.J., and Kyle, J.A.M., Plant polyphenols in cancer and heart disease: Implications as nutritional antioxidants, *Nutr. Res. Rev.*, 13, 79, 2000.
4. Kris-Etherton, P.M. et al., Bioactive compounds in foods: Their role in the prevention of cardiovascular disease and cancer, *Am. J. Med.*, 113, 71S, 2002.
5. Lambert, J.D. et al., Inhibition of carcinogenesis by polyphenols: Evidence from laboratory investigations, *Am. J. Clin. Nutr.*, 81S, 284S, 2005.
6. Park, O.J. and Surh, Y.-J., Chemopreventive potential of epigallocatechingallate and genistein: Evidence from epidemiological and laboratory studies, *Toxicol. Lett.*, 150, 43, 2004.
7. Yang, C.S. et al., Inhibition of carcinogenesis by dietary polyphenolic compounds, *Annu. Rev. Nutr.*, 21, 381, 2001.

8. Kris-Etherton, P.M. and Keen, C.L., Evidence that the antioxidant flavonoids in tea and cocoa are beneficial for cardiovascular health, *Curr. Opin. Lipidol.*, 13, 41, 2002.
9. Renaud, S.C. and de Lorgeril, M., Wine, alcohol, platelets, and the French paradox for coronary heart disease, *Lancet*, 339, 1523, 1992.
10. Stocklet, J.-C. et al., Vascular protection by dietary polyphenols, *Eur. J. Pharmacol.*, 500, 299, 2004.
11. Harborne, J.B., *The Flavonoids*, Chapman & Hall, London, United Kingdom, 1994.
12. Harborne, J.B., *Methods in Plant Biochemistry, I: Plant Phenolics*, Academic Press, London, United Kingdom, 1989.
13. Würsch, P. et al., The tannin granules from ripe carob pod, *Lebensm. Wiss. Technol.*, 17, 351, 1984.
14. Markham, K.R., *Techniques of Flavonoid Identification*, Academic Press, London, United Kingdom, 1982.
15. Herrman, H., On the occurrence of flavonol and flavone glycosides in vegetables, *Z. Lebensm. Unters. Forsch.*, 186, 1, 1988.
16. Mazza, G., Anthocyanins in grapes and grape products, *Crit. Rev. Food Sci. Nutr.*, 35, 341, 1995.
17. Naczk, M. and Shahidi, F., Phenolics in cereals, fruits and vegetables: Occurrence, extraction and analysis, *J. Pharm. Biomed. Anal.*, 41, 1523, 2006.
18. Rasmussen, S.E. et al., Dietary proanthocyanidins: Occurrence, dietary intake, bioavailability, and protection against cardiovascular disease, *Mol. Nutr. Food Res.*, 49, 159, 2005.
19. Tomás-Barberán, F.A., Effect of storage and processing on food phenolics, in: *Bioactive Compounds in Plant Foods. Health Effects and Perspectives for the Food Industry*, Ámado, R., Abt, B., Bravo, L., and Saura-Calixto, F. (Eds.), Office for Official Publications of the European Communities, Luxembourg, 2002.
20. Aherne, S.A. and O'Brien, N.M., Dietary flavonols: Chemistry, food content and metabolism, *Nutrition*, 18, 75, 2002.
21. Crozier, A. et al., Antioxidant flavonols from fruits, vegetables and beverages: Measurements and bioavailability, *Biol. Res.*, 33, 79, 2000.
22. De Pascual-Teresa, S., Santos-Buelga, C., and Rivas-Gonzalo, J.C., Quantitative análysis of flavan-3-ols in Spanish foodstuffs and beverages, *J. Agric. Food Chem.*, 48, 5331, 2000.
23. Gu, L. et al., Fractionation of polymeric procyanidins from lowbush blueberry and quantification of procyanidins in selected foods with an optimized normal-phase HPLC-MS fluorescent detection method, *J. Agric. Food Chem.*, 50, 4852, 2002.
24. Hertog, M.G.L., Hollman, P.C.H., and Venema, D.P., Optimization of a quantitative HPLC determination of potentially anticarcinogenic flavonoids in vegetables and fruits, *J. Agric. Food Chem.*, 40, 1591, 1992.
25. Hertog, M.G.L., Hollman, P.C.H., and Katan, M.B., Content of potentially anticarcinogenic flavonoids of 28 vegetables and 9 fruits commonly consumed in the Netherlands, *J. Agric. Food Chem.*, 40, 2379, 1992.
26. Hertog, M.G.L., Hollman, P.C.H., and van de Putte, B., Content of potentially anticarcinogenic flavonoids of tea infusions, wines, and fruit juices, *J. Agric. Food Chem.*, 41, 1242, 1993.
27. Liggins, J. et al., Daidzein and genistein contents of vegetables, *Br. J. Nutr.*, 84, 717, 2000.
28. Wu, X. et al., Concentrations of anthocyanins in common foods in the United States and estimation of normal consumption, *J. Agric. Food Chem.*, 54, 4069, 2006.

29. DuPont, M.S. et al., Effect of variety, processing, and storage on the flavonoid glycoside content and composition of lettuce and endive, *J. Agric. Food Chem.*, 48, 3957, 2000.
30. Gil, M.I., Ferreres, F., and Tomás-Barberán, F.A., Effect of postharvest storage and processing on the antioxidant constituents (flavonoids and vitamin C) of fresh-cut spinach, *J. Agric. Food Chem.*, 47, 2213, 1999.
31. García-Viguera, C. et al., Colour and anthocyanins stability of red raspberry jam, *J. Sci. Food Agric.*, 78, 565, 1998.
32. Gil, M.I. et al., Antioxidant activity of pomegranate juice and its relationship with phenolics, *J. Agric. Food Chem.*, 48, 4581, 2000.
33. Gil-Izquierdo, A., Gil, M.I., and Ferreres, F., Effect of processing techniques at industrial scale on orange juice antioxidant and beneficial health compounds, *J. Agric. Food Chem.*, 50, 5107, 2002.
34. Skrede, G., Wrolstad, R.E., and Durst, R.W., Changes in anthocyanins and polyphenolics during juice processing of highbush blueberries (*Vaccinium corybosum* L.), *J. Food Sci.*, 65, 357, 2000.
35. Ewald, C. et al., Effect of processing on the major flavonoids in processed onions, green beans, and peas, *Food Chem.*, 64, 231, 1999.
36. Makris, D.P. and Rossiter, J.T., Domestic processing of onion bulbs (*Allium cepa*) and asparagus spears (*Asparagus officinalis*): Effect on flavonol content and antioxidant status, *J. Agric. Food Chem.*, 49, 3216, 2001.
37. Tomás-Barberán, F.A. et al., Early wound- and ethylene-induced changes in phenylpropanoid metabolism in harvested lettuce, *J. Am. Soc. Hortic. Sci.*, 122, 399, 1997.
38. Sze-Tao, K.W.C. et al., Effects of processing and storage on walnut (*Juglans regia* L) tannins, *J. Sci. Food Agric.*, 81, 1215, 2001.
39. Wollgast, J. and Anklam, E., Review on polyphenols in *Theobroma cacao*: Changes in composition during the manufacture of chocolate and methodology for identification and quantification, *Food Res. Int.*, 33, 423, 2000.
40. Xu, G. et al., Effect of heat treatment on the phenolic compounds and antioxidant capacity of citrus peel extract, *J. Agric. Food Chem.*, 55, 330, 2007.
41. Santos-Buelga, C., Bravo-Haro, J., and Rivas-Gonzalo, J., Interactions between catechin and malvidin-3-monoglucoside in model solution, *Z. Lebensm. Unters-Forsch.*, 201, 269, 1995.
42. Cantos, E. et al., Effect of postharvest ultraviolet irradiation on resveratrol and other phenolics on cv. Napoleon table grapes, *J. Agric. Food Chem.*, 48, 4606, 2000.
43. Dong, Y.H. et al., Postharvest stimulation of skin color in Royal Gala apple, *J. Am. Soc. Hortic. Sci.*, 120, 95, 1995.
44. Breitfellner, F., Solar, S., and Sontag, G., Effect of gamma irradiation on flavonoids in strawberries, *Eur. Food Res. Technol.*, 215, 28, 2002.
45. Gil, M.I., Holcroft, D.M., and Kader, A.A., Changes in strawberry anthocyanins and other polyphenols in response to carbon dioxide treatments, *J. Agric. Food Chem.*, 45, 1662, 1997.
46. Khünau, J., The flavonoids: A class of semi-essential food components: Their role in human nutrition, *World Rev. Nutr. Diet.*, 24, 117, 1976.
47. Justesen, U., Knuthsen, P., and Leth, T., Determination of plant polyphenols in Danish foodstuffs by HPLC-UV and LC-MS detection, *Cancer Lett.*, 114, 165, 1997.
48. Hertog, M.G.L. et al., Flavonoid intake and long term risk of coronary heart disease and cancer in the Seven Countries Study, *Arch. Intern. Med.*, 155, 81, 1995.
49. Knekt, P. et al., Flavonoid intake and coronary mortality in Finland: A cohort study, *Br. Med. J.*, 312, 478, 1996.

50. Pietta, P. et al., Dietary flavonoids and oxidative stress, in: *Natural Antioxidants and Food Quality in Atherosclerosis and Cancer Prevention*, Kumpaleinen, J.T., and Salonen, J.T. (Eds.), Royal Society of Chemistry, London, United Kingdom, 1996.
51. Arai, Y. et al., Dietary intakes of flavonols, flavones and isoflavones by Japanese women and the inverse correlation between quercetin intake and plasma LDL colesterol concentration, *J. Nutr.*, 130, 2243, 2000.
52. Kimira, M. et al., Japanese intake of flavonoids and isoflavonoids from foods, *J. Epidemiol.*, 8, 168, 1998.
53. Hertog, M.G.L. et al., Intake of potentially anticarcinogenic flavonoids and their determinants in adults in the Netherlands, *Nutr. Cancer*, 20, 21, 1993.
54. Rimm, E.B. et al., Relation between intake of flavonoids and risk for coronary heart disease in male health professionals, *Ann. Intern. Med.*, 125, 384, 1996.
55. Sampson, L. et al., Flavonol and flavone intake in US health professionals, *J. Am. Diet. Assoc.*, 102, 1414, 2002.
56. Gu, L. et al., Concentrations of proanthocyanidins in common foods and estimation of normal consumption, *J. Nutr.*, 134, 613, 2004.
57. Adamson, G.E. et al., HPLC method for the quantification of procyanidins in cocoa and chocolate samples and correlation to total antioxidant capacity, *J. Agric. Food Chem.*, 47, 4184, 1999.
58. Chukwumah, Y.C. et al., Comparison of extraction methods for the quantification of selected phytochemicals in peanuts (*Arachis hypogaea*), *J. Agric. Food Chem.*, 55, 285, 2007.
59. Hammerstone, J.F., Lazarus, S.A., and Schmitz, H.H., Procyanidin content and variation in some commonly consumed foods, *J. Nutr.*, 130, 2086S, 2000.
60. Chedier, L.M. et al., Comparative methodology for isolation of flavonoid glycosides from *Clusia criuva* Cambess, *J. High Resol. Chromatogr.*, 22, 527, 1999.
61. Du, Q. et al., Preparative separation of flavonoid glycosides in leaves extract of *Ampelopsis grossedentata* using high-speed counter-current chromatography, *J. Chromatogr. A*, 1040, 147, 2004.
62. Sánchez-Rabaneda, F. et al., Liquid chromatographic/electrospray ionization tandem mass spectrometric study of the phenolic composition of cocoa (*Theobroma cacao*), *J. Mass Spectrom.*, 38, 35, 2003.
63. Rodriguez-Madrera, R., Picinelli-Lobo, A., and Suarez-Valles, B., Phenolic profile of Asturian (Spain) natural cider, *J. Agric. Food Chem.*, 54, 120, 2006.
64. Revilla, E. and Ryan, J.-M., Analysis of several phenolic compounds with potential antioxidant properties in grape extracts and wines by high performance liquid chromatography–photodiode array detection without sample preparation, *J. Chromatogr. A*, 881, 461, 2000.
65. Arts, I.C.W. and Hollman, P.C.H., Optimization of a quantitative method for the determination of catechins in fruits and legumes, *J. Agric. Food Chem.*, 46, 5156, 1998.
66. Gennaro, L. et al., Flavonoid and carbohydrate contents in tropea red onions: Effects of homelike peeling and storage, *J. Agric. Food Chem.*, 50, 1904, 2002.
67. Mullner, C. and Sontag, G., Determination of some phytoestrogens in soybeans and their processed products with HPLC and coulometric electrode array detection, *Fresenius J. Anal. Chem.*, 364, 261, 1999.
68. Mullner, C. and Sontag, G., Determination of daidzein and genistein in soy based infant food, soy milk and soy based supplements, *Eur. Food Res. Technol.*, 211, 301, 2000.
69. Careri, M. et al., Spectrophotometric and coulometric detection in the high-performance liquid chromatography of flavonoids and optimization of sample treatment for the determination of quercetin in orange juice, *J. Chromatrogr. A*, 881, 449, 2000.

70. Crozier, A. et al., Quantitative analysis of the flavonoid content of commercial tomatoes, onions, lettuce, and celery, *J. Agric. Food Chem.*, 45, 590, 1997.
71. Mattila, P., Astola, J., and Kumpulainen, J., Determination of flavonoids in plant material by HPLC with diode-array and electro-array detections, *J. Agric. Food Chem.*, 48, 5834, 2000.
72. Alcalde-Eon, C. et al., Identification of anthocyanins of pinta boca (*Solanum stenotomum*) tubers, *Food Chem.*, 86, 441, 2004.
73. Peng, J. et al., Preparative isolation and separation of a novel and two known flavonoids from *Patrinia villosa* Juss by high-speed counter-current chromatography, *J. Chromatogr. A*, 1092, 235, 2005.
74. McDougall, G.J. et al., Anthocyanin-flavanol condensation products from black currant (*Ribes nigrum* L.), *J. Agric. Food Chem.*, 53, 7878, 2005.
75. Day, A.J. and Morgan, M.R.A., Methods of polyphenols extraction from biological fluids and tissues, in: *Methods in Polyphenols Analysis*, Santos-Buelga, C. and Williamson, G. (Eds), The Royal Society of Chemistry, Cambridge, United Kingdom, 2003.
76. Setchell, K.D.R., Welsch, M.B., and Lim., C.K., High-performance liquid chromatographic analysis of phytoestrogens in soy protein preparations with ultraviolet, electrochemical and thermospray mass spectrometric detection, *J. Chromatogr.*, 386, 315, 1987.
77. Sakakibara, H. et al., Simultaneous determination of all polyphenols in vegetables, fruits and teas, *J. Agric. Food Chem.*, 51, 571, 2003.
78. Oomah, B.D. and Mazza, G., Flavonoids and antioxidative activities in buckwheat, *J. Agric. Food Chem.*, 44, 1746, 1996.
79. Revilla, E., Ryan, J.-M., and Martín-Ortega, G., Comparison of several procedures used for the extraction of anthocyanins from red grapes, *J. Agric. Food Chem.*, 46, 4592, 1998.
80. Lee, H.S., Characterization of major anthocyanins and the color of red-fleshed budd blood orange (*Citrus sinensis*), *J. Agric. Food Chem.*, 50, 1243, 2002.
81. Mateus, N. et al., Structural diversity of anthocyanin-derived pigments in port wines, *Food Chem.*, 76, 335, 2002.
82. Schmidt, B.M. et al., Effective separation of potent antiproliferation and antiadhesion components from wild blueberry (*Vaccinium angustifolium* Ait.) fruits, *J. Agric. Food Chem.*, 52, 6433, 2004.
83. Schieber, A., Bernardini, N., and Carle, R., Identification of flavonol and xanthone glycosides from mango (*Mangifera indica* L. Cv. "Tommy Atkins") peels by high-performance liquid chromatography–electrospray ionization mass spectrometry, *J. Agric. Food Chem.*, 51, 5006, 2003.
84. Gu, L. et al., Screening of foods containing proanthocyanidins and their structural characterization using LC-MS/MS and thiolytic degradation, *J. Agric. Food Chem.*, 51, 7513, 2003.
85. Lazarus, S.A. et al., High-performance liquid chromatography/mass spectrometry analysis of proanthocyanidins in foods and beverages, *J. Agric. Food Chem.*, 47, 3693, 1999.
86. De Rijke, E. et al., Liquid chromatography coupled to nuclear magnetic resonance spectroscopy for the identification of isoflavone glucoside malonates in *T. pratense* L. leaves, *J. Sep. Sci.*, 27, 1061, 2004.
87. Griffith, A.P. and Collinson, M.W., Improved methods for the extraction and analysis of isoflavones from soy-containing foods and nutritional supplements by reversed-phase high-performance liquid chromatography and liquid chromatography–mass spectrometry, *J. Chromatogr. A*, 913, 397, 2001.

88. Bravo, L., Goya, L., and Lecumberri, E., LC/MS characterization of phenolic constituents of mate (*Ilex paraguariensis*, St. Hil.) and its antioxidant activity compared to commonly consumed beverages, *Food Res. Int.*, 40, 393, 2007.
89. Hellström, J. et al., Isolation and elucidation of procyanidin oligomers from Saskatoon berries (*Amelanchier alnifolia*), *J. Agric. Food Chem.*, 55, 157, 2007.
90. Ma, C.J. et al., One step isolation and purification of liquiritigenin and isoliquiritigenin from *Glycyrrhiza uralensis* Risch. using high-speed counter-current chromatography, *J. Chromatogr. A*, 1078, 188, 2005.
91. Sang, S. et al., Antioxidative phenolic compounds isolated from almond skins (*Prunus amygdalus* Batsch), *J. Agric. Food Chem.*, 50, 2459, 2002.
92. Du, Q., Li, L., and Jerz, G., Purification of astilbin and isoastilbin in the extract of *Smilax glabra* rhizome by high-speed counter-current chromatography, *J. Chromatogr. A*, 1077, 98, 2005.
93. Chen, L.-J., Games, D.E., and Jones, J., Isolation and identification of four flavonoid constituents from the seeds of *Oroxylum indicum* by high-speed counter-current chromatography, *J. Chromatogr. A*, 988, 95, 2003.
94. Sun, B. et al., Separation of grape and wine proanthocyanidins according to their degree of polymerization, *J. Agric. Food Chem.*, 46, 1390, 1998.
95. Rodríguez-Delgado, M.A. et al., Separation of phenolic compounds by high-performance liquid chromatography with absorbance and fluorimetric detection, *J. Chromatogr. A*, 912, 249, 2001.
96. Da Costa, C.T. et al., Separation of blackcurrant anthocyanins by capillary zone electrophoresis, *J. Chromatogr. A*, 799, 321, 1998.
97. Edenharder, R. et al., Isolation and characterization of structurally novel antimutagenic flavonoids from spinach (*Spinacia oleracea*), *J. Agric. Food Chem.*, 49, 2767, 2001.
98. Fang, N., Yu, S., and Prior, R.L., LC/MS/MS characterization of phenolic constituents in dried plums, *J. Agric. Food Chem.*, 50, 3579, 2002.
99. Schütz, K. et al., Quantitative determination of phenolic compounds in artichoke-based supplements and pharmaceuticals by high-performance liquid chromatography, *J. Agric. Food Chem.*, 54, 8812, 2006.
100. Zafrilla, P., Ferreres, F., and Tomás-Barberán, F.A., Effect of processing and storage on the antioxidant ellagic acid derivatives and flavonoids of red raspberry (*Rubus idaeus*) jams, *J. Agric. Food Chem.*, 49, 3651, 2001.
101. Gómez-Caravaca, A.M. et al., Electrophoretic identification and quantification of compounds in the polyphenolic fraction of extra-virgin olive oil, *Electrophoresis*, 26, 3538, 2005.
102. Mateos, R. et al., Determination of phenols, flavones, and lignans in virgen olive oils by solid-phase extraction and high-performance liquid chromatography with diode array ultraviolet detection, *J. Agric. Food Chem.*, 49, 2185, 2001.
103. Cao, X. et al., Separation and purification of isoflavones from *Pueraria lobata* by high-speed counter-current chromatography, *J. Chromatogr. A*, 855, 709, 1999.
104. Zhou, X. et al., Isolation and purification of flavonoid glycosides from *Trollius ladebouri* using high-speed counter-current chromatography by stepwise increasing the flow-rate of the mobile phase, *J. Chromatogr. A*, 1092, 216, 2005.
105. Ferreres, F. et al., Separation of honey flavonoids by micellar electrokinetic capillary chromatography, *J. Chromatogr. A*, 669, 268, 1994.
106. De Rijke, E. et al., Flavonoids in leguminosae: Analysis of extracts of *T. pratense* L., *T. dubium* L., *T. repens* L., and *L. corniculatus* L. leaves using liquid chromatography with UV, mass spectrometric and fluorescence detection, *Anal. Bioanal. Chem.*, 378, 995, 2004.

107. Xiao, H.B. et al., Determination and identification of isoflavonoids in *Radix astragali* by matrix solid-phase dispersion extraction and high-performance liquid chromatography with photodiode array and mass spectrometric detection, *J. Chromatogr. A*, 1032, 117, 2004.
108. Arráez-Román, D. et al., Identification of phenolic compounds in rosemary honey using solid-phase extraction by capillary electrophoresis–electrospray ionization–mass spectrometry, *J. Pharm. Biomed. Anal.*, 41, 1648, 2006.
109. Martins-Teixeira, D. and Teixeira da Costa, C., Novel methods to extract flavanones and xanthones from the root bark of *Maclura pomifera*, *J. Chromatogr. A*, 1062, 175, 2005.
110. Kataoka, H., Lord, H.L., and Pawliszyn, J., Applications of solid-phase microextraction in food análysis, *J. Chromatogr. A*, 880, 35, 2000.
111. Mitani, K., Narimatsu, S., and Kataoka, H., Determination of daidzein and genistein in soybean foods by automated on-line in-tube solid-phase microextraction coupled to high-performance liquid chromatography, *J. Chromatogr. A*, 986, 169, 2004.
112. Satterfield, M. and Brodbelt, J.S., Structural characterization of flavonoid glycosides by collisionally activated dissociation of metal complexes, *J. Am. Soc. Mass Spectrom.*, 12, 537, 2001.
113. Turner, C., Eskilsson, C.S., and Björklund, E., Collection in analytical-scale supercritical fluid extraction, *J. Chromatogr. A*, 947, 1, 2002.
114. Díaz-Reinoso, B. et al., Supercritical CO_2 extraction and purification of compounds with antioxidant activity, *J. Agric. Food Chem.*, 54, 2441, 2006.
115. Ashraf-Khorassani, M. and Taylor, L.T., Sequential fractionation of grape seeds into oils, polyphenols, and procyanidins via a single system employing CO_2-based fluids, *J. Agric. Food Chem.*, 52, 2440, 2004.
116. Peng, J. et al., Efficient new method for extraction and isolation of three flavonoids from *Patrinia villosa* Juss. by supercritical fluid extraction and high-speed counter-current chromatography, *J. Chromatogr. A*, 1102, 44, 2006.
117. Lin, M.C., Tsai, M.J., and Wen, K.C., Supercritical fluid extraction of flavonoids from *Scutellariae radix*, *J. Chromatogr. A*, 830, 387, 1999.
118. Scalia, S., Giuffreda, L., and Pallado, P., Analytical and preparative supercritical fluid extraction of chamomile flowers and its comparison with conventional methods, *J. Pharm. Biomed. Anal.*, 21, 549, 1999.
119. Petsalo, A., Janonen, J., and Tolonen, A., Identification of flavonoids of *Rhodiola rosea* by liquid chromatography–tandem mass spectrometry, *J. Chromatogr. A*, 1112, 224, 2006.
120. Palenzuela, B. et al., Bioguided extraction of polyphenols from grape marc by using an alternative supercritical-fluid extraction method based on a liquid solvent trap, *Anal. Bioanal. Chem.*, 378, 2021, 2004.
121. Bergeron, C. et al., Comparison of the chemical compositon of the extracts from *Scutellaria lateriflora* using accelerated solvent extraction and supercritical fluid extraction versus standard hot water or 70% ethanol extraction, *J. Agric. Food Chem.*, 53, 3076, 2005.
122. Giannozzo, A.N. et al., Supercritical fluid extraction of naringin from the peel of *Citrus paradisi*, *Phytochem. Anal*, 14, 221, 2003.
123. Zougagh, M., Valcárcel, M., and Ríos, A., Supercritical fluid extraction: A critical review of its analytical usefulness, *Trends Anal. Chem.*, 23, 399, 2004.
124. Ibañez, E. et al., Subcritical water extraction of antioxidant compounds from rosemary plants, *J. Agric. Food Chem.*, 51, 375, 2003.
125. Herrero, M. et al., Pressurized liquid extraction–capillary electrophoresis–mass spectrometry for the analysis of polar antioxidants in rosemary extracts, *J. Chromatogr. A*, 1084, 54, 2005.

126. Rodríguez-Meizoso, I. et al., Subcritical water extraction of nutraceuticals with antioxidant activity from oregano. Chemical and functional characterization, *J. Pharm. Biomed. Anal.*, 41, 1560, 2006.
127. Soto-Ayala, R. and Luque de Castro, M.D., Continuous subcritical water extraction as a useful tool for isolation of edible essential oils, *Food Chem.*, 75, 109, 2001.
128. Basile, A., Jiménez-Carmona, M.M., and Clifford, A.A., Extraction of rosemary by superheated water, *J. Agric. Food Chem.*, 46, 5205, 1998.
129. Clifford, A.A., Basili, A., and Al-Saidi, S.H.R., A comparison of the extraction of clove buds with supercritical carbon dioxide and superheated water, *Fresenius J. Anal. Chem.*, 364, 635, 1999.
130. Ammann, A. et al., Superheated water extraction, steam distillation and SFE of peppermint oil, *Fresenius J. Anal. Chem.*, 364, 650, 1999.
131. González-Rodríguez, J., Pérez-Juan, P., and Luque de Castro, M.D., Use of superheated liquids for the extraction of non-volatile compounds from wood: Liquid chromatography studies, *J. Chromatogr. A*, 1038, 3, 2004.
132. Luque-Rodríguez, J.M., Pérez-Juan, P., and Luque de Castro, M.D., Extraction of polyphenols from vine shoots of *Vitis vinifera* by superheated ethanol–water mixtures, *J. Agric. Food Chem.*, 54, 8775, 2006.
133. Waterman, P.G. and Mole, S., *Analysis of Phenolic Plant Metabolites*, Blackwell Scientific Publications, Oxford, United Kingdom, 1994.
134. Schofield, P., Mbugua, D.M., and Pell, A.N., Analysis of condensed tannins: A review, *Animal Feed Sci. Technol.*, 91, 21, 2001.
135. Hagerman, A.E., Rice, M.E., and Ritchard, N.T., Mechanisms of protein precipitation for two tannins, pentagalloyl glucose and epicatechin 16 ($4 > 8$) catechin (procyanidin), *J. Agric. Food Chem.*, 46, 2590, 1998.
136. Schofield, J.A., Hagerman, A.E., and Harold, A., Loss of tannins and other phenolics from willow leaf litter, *J. Chem. Ecol.*, 24, 1409, 1998.
137. Price, M.L. and Butler, L.G., Rapid visual estimation and spectrophotometric determination of tannin content of sorghum grain, *J. Agric. Food Chem.*, 25, 1268, 1277, 1977.
138. Folin, O. and Ciocalteu, V., On tyrosine and tryptophan determinations in proteins, *J. Biol. Chem.*, 27, 627, 1927.
139. AOCS, *Official Methods of Analysis*, 12th ed., Washington DC, WA, 1975.
140. Hagerman, A.E., Zhao, Y., and Johnson, S., Methods for determination of condensed and hydrolyzable tannins, in *Antinutrients and Phytochemicals in Food*, Shahidi, F. (Ed.), American Chemical Society, Washington DC, WA, 1997.
141. Soczewinski, E., Hawryl, M.A., and Hawryl, A., Retention behaviour of some flavonoids in 2D-TLC systems on cyano bonded polar phases, *Chromatographia*, 54, 789, 2001.
142. Soczewinski, E., Wojciak-Kosior, M., and Matysik, G., Analysis of glycosides and aglycones of flavonoid compounds by double-development thin-layer chromatography, *J. Planar Chromatogr.*, 17, 261, 2004.
143. Beninger, C.W. and Hosfield, G.L., Flavonol glycosides from the seed coat of a new manteca-type dry bean (*Phaseolus vulgaris* L.), *J. Agric. Food Chem.*, 46, 2906, 1998.
144. Mouly, P.P. et al., Blood orange juice authentication using cinnamic acid derivatives. Variety differentiations associated with flavanone glycoside content, *J. Agric. Food Chem.*, 45, 373, 1997.
145. Hirota, S., Shimoda, T., and Takahama, U., Tissue and spatial distribution of flavonoid and peroxidase in onion bulbs and stability of flavonol glucosides during boiling of the scales, *J. Agric. Food Chem.*, 46, 3497, 1998.

146. Del Río, J.A. et al., Citrus polymethoxylated flavones can confer resistance against *Phytophthora citrophthora, Penicillium digitatum*, and *Geotrichum* species, *J. Agric. Food Chem.*, 46, 4423, 1998.
147. Srinivasa, H. et al., A rapid densitometric method for the quantification of luteolin in medicinal plants using HPTLC, *Chromatographia*, 60, 131, 2004.
148. Matysik, G. and Wójciak-Kosior, M., Optimized method for the determination of flavonoid glycosides and aglycones, *Chromatographia*, 61, 89, 2005.
149. Medic-Saric, M. et al., Quantitative analysis of flavonoids and phenolic acids in propolis by two-dimensional thin layer chromatography, *J. Planar Chromatogr.*, 17, 459, 2004.
150. Watanabe, M., Ohshita, Y., and Tsushida, T., Antioxidant compounds from buckwheat (*Fagopyrum esculentum* Moench) hulls, *J. Agric. Food Chem.*, 45, 1039, 1997.
151. Fecka, I., Kowalczyk, A., and Cisowski, W., Optimization of the separation of flavonoid glycosides and rosmarinic acid from *Mentha piperita* on HPTLC plates, *J. Planar Chromatogr.*, 17, 22, 2004.
152. Pieroni, A., Heimler, D., and Huang, Y., A TLC method for separation and identification of flavones and flavone glycosides from biflavones in vegetable extract, *J. Planar Chromatogr.—Mod. TLC*, 11, 230, 1998.
153. Chen, J., Montanan, A.M., and Widmer, W.W., Two new polymethoxylated flavones, a class of compounds with potential anticancer activity, isolated from cold pressed Dancy tangerine peel oil solids, *J. Agric. Food Chem.*, 45, 364, 1997.
154. Pereira, C.A.M. et al., A HPTLC densitometric determination of flavonoids from *Passiflora alata, P. edulis, P. incarnata* and *P. caerulea* and comparison with HPLC method, *Phytochem. Anal.*, 15, 241, 2004.
155. Wojciak-Kosior, M., Matysik, G., and Skalska, A., Densitometric determination and kinetics of hydrolysis of flavonoid glycosides, *J. Planar Chromatogr.*, 17, 286, 2004.
156. Deng, F.X. and Zito, S.W., Development and validation of a gas chromatographic–mass spectrometric method for simultaneous identification and quantification of marker compounds including bilobalide, ginkgolides and flavonoids in *Ginkgo biloba* L. extract and pharmaceutical preparations, *J. Chromatogr. A*, 986, 121, 2003.
157. Morton, M. et al., Analysis of phyto-oestrogens by gas chromatography–mass spectrometry, *Environ. Toxicol. Pharmacol.*, 7, 221, 1999.
158. Pereira, A.D.S., Padilha, M.C., and Neto, F.R.D.A., Two decades of high temperature gas chromatography (1983–2003): What's next?, *Microchem. J.*, 77, 141, 2004.
159. Proestos, C., Sereli, D., and Komaitis, M., Determination of phenolic compounds in aromatic plants by RP-HPLC and GC-MS, *Food Chem.*, 95, 44, 2006.
160. Liggins, J. et al., Daidzein and genistein content of fruits and nuts, *J. Nutr. Biochem.*, 11, 326, 2000.
161. Liggins, J. et al., Extraction and quantification of daidzein and genistein in food, *Anal. Biochem.*, 264, 1, 1998.
162. Fernandez, M.P., Watson, P.A., and Breuil, C., Gas chromatography–mass spectrometry method for the simultaneous determination of wood extractive compounds in quaking aspen, *J. Chromatogr. A*, 922, 225, 2001.
163. Maciejewicz, W. et al., GC-MS identification of the flavonoids aglycones isolated from propolis, *Chromatography*, 53, 343, 2001.
164. Branco, A. et al., Further lipophilic flavonols in *Vellozia graminifolia* (Velloziaceae) by high temperature gas chromatography: Quick detection of new compounds, *Phytochem. Anal.*, 12, 266, 2001.

165. Carbonell, S.A. et al., Rapid screening of natural products by high-resolution high-temperature gas chromatography, *J. Chromatogr. Sci.*, 38, 234, 2000.
166. Hostettmann, K. and Hostettmann, M., *The Flavonoids: Advances in Research*, Harbone, J.B. and Mabry, T.J. (Eds.), Champman & Hall, New York, 1982, p. 1.
167. Escarpa, A. and González, M.C., Approach to the content of total extractable phenolic compounds from different food samples by comparison of chromatographic and spectrophotometric methods, *Anal. Chim. Acta*, 427, 119, 2001.
168. Wach, A., Pyrzynska, K., and Biesaga, M., Quercetin content in some food and herbal samples, *Food Chem.*, 100, 699, 2007.
169. Chen, H., Zuo, Y., and Deng, Y., Separation and determination of flavonoids and other phenolic compounds in cranberry juice by high-performance liquid chromatography, *J. Chromatogr. A*, 913, 387, 2001.
170. Shui, G. and Leong, L.P., Separation and determination of organic acids and phenolic compounds in fruit juices and drinks by high-performance liquid chromatography, *J. Chromatogr. A*, 977, 89, 2002.
171. Colombo, R., Lancas, F.M., and Yariwake, J.H., Determination of flavonoids in cultivated sugarcane leaves, bagasse, juice and in transgenic sugarcane by liquid chromatography–UV detection, *J. Chromatogr. A*, 1103, 118, 2006.
172. Wang, S.P. and Huang, K.J., Determination of flavonoids by high-performance liquid chromatography and capillary electrophoresis, *J. Chromatogr. A*, 132, 273, 2004.
173. Franke, A.A. et al., Vitamin C and flavonoids levels of fruits and vegetables consumed in Hawaii, *J. Food Comp. Anal.*, 17, 1, 2004.
174. Liu, F.F. et al., Evaluation of major active components in St. John's Wort dietary supplements by high-performance liquid chromatography with photodiode array detection and electrospray mass spectrometric confirmation, *J. Chromatogr. A*, 888, 85, 2000.
175. Lee, B.L. and Ong, C.N., Comparative analysis of tea catechins and theaflavins by high-performance liquid chromatography and capillary electrophoresis, *J. Chromatogr. A*, 881, 439, 2000.
176. Molinelli, A., Weiss, R., and Mizaikoff, B., Advanced solid phase extraction using molecularly imprinted polymers for the determination of quercetin in red wine, *J. Agric. Food Chem.*, 50, 1804, 2002.
177. Yáñez, J.A. et al., Stereospecific high-performance liquid chromatographic analysis of hesperetin in biological matrices, *J. Pharm. Biomed. Anal.*, 37, 591, 2005.
178. Asztemborska, M., and Żukowskib, J., Determination of diastereomerization barrier of some flavanones by high-performance liquid chromatography methods, *J. Chromatogr. A*, 1134, 95, 2006.
179. Lu, Y. et al., Quality evaluation of *Platycladus orientalis* (L.) Franco through simultaneous determination of four bioactive flavonoids by high-performance liquid chromatography, *J. Pharm. Biomed. Anal.*, 41, 1186, 2006.
180. Repollés, C., Herrero-Martínez, J.M., and Ráfols, C., Analysis of prominent flavonoid aglycones by high-performance liquid chromatography using a monolithic type column, *J. Chromatogr. A*, 1131, 51, 2006.
181. Yáñez, J.A., Andrews, P.K., and Davies, N.M., Methods of analysis and separation of chiral flavonoids, *J. Chromatogr. B*, 848, 159, 2007.
182. Peng, J., Fan, G., and Wu, Y., Preparative isolation of four new and two known flavonoids from the leaf of *Patrinia villosa* Juss. by counter-current chromatography, *J. Chromatogr. A*, 1115, 103, 2006.
183. Wada, K., Koda, T., and Aoki, H., Analytical and preparative separation of Kaoliang and Lac colors by pH-zone-refining CCC, *J. Liq. Chromatogr. Relat. Technol.*, 28, 2097, 2005.

184. Degenhardt, A. and Winterhalter, P., Isolation of natural pigments by high speed CCC. *J. Liq. Chromatogr. Relat. Technol.*, 24, 1745, 2001.
185. Cortacero-Ramírez, S. et al., Simultaneous determination of multiple constituents in real beer samples of different origins by capillary zone electrophoresis, *Anal. Bioanal. Chem.*, 380, 831, 2004.
186. Aussenac, T., Lacombe, S., and Daydé, J., Quantification of isoflavones by capillary zone electrophoresis in soybean seeds: Effects of variety and environment, *Am. J. Clin. Nutr.*, 68, 1480S, 1998.
187. Gel-Moreto, N., Streich, R., and Galensa, R., Chiral separation of six diastereomeric flavanone-7-O-glycosides by capillary electrophoresis and analysis of lemon juice, *J. Chromatogr. A*, 925, 279, 2001.
188. Demianová, Z. et al., Nonaqueous capillary electrophoretic separation of polyphenolic compounds in wine using coated capillaries at high pH in methanol, *Electrophoresis*, 24, 4264, 2003.
189. Peng, Y., Liu, F., and Ye, J., Quantitative and qualitative analysis of flavonoids markers in *Frucus aurantii* of different geographical origin by capillary electrophoresis with electrochemical detection, *J. Chromatogr. B*, 830, 224, 2006.
190. Huck, C.W. et al., Analysis of three flavonoids by CE-UV and CE-ESI-MS. Determination of naringenin from a phytomedicine, *J. Sep. Sci.*, 25, 904, 2002.
191. Baggett, B.R. et al., Profiling isoflavonoids found in legume root extracts using capillary electrophoresis, *Electrophoresis*, 23, 1642, 2002.
192. Bonoli, M. et al., Fast determination of catechins and xanthines in tea beverages by micellar electrokinetic chromatograpy, *J. Agric. Food Chem.*, 51, 1141, 2003.
193. Li, X., Zhang, Y., and Yuan, Z., Determination of rutin and forsythin in fruti of *Forsythia suspensa* (Thunb) Vahl by capillary electrophoresis–electrochemical detection, *Chromatographia*, 56, 171, 2002.
194. Asztemborska, M., Miskiewicz, M., and Sybilska, D., Separation of some chiral flavanones by micellar electrokinetic chromatography, *Electrophoresis*, 24, 2527, 2003.
195. Mabry, T.J., Markham, K.R., and Thomas, M.B., *The Systematic Identification of Flavonoids*, Springer-Verlag, New York, 1970.
196. Santos-Buelga, C., García-Viguera, C., and Tomás-Barberán, F.A., On-line identification of flavonoids by HPLC coupled to diode array detection, in: *Methods in Polyphenols Analysis*, Santos-Buelga, C. and Williamson, G. (Eds.), The Royal Society of Chemistry, Cambridge, United Kingdom, 2003.
197. Peñalvo, J.L. and Nurmi, T., Application of coulometric electrode array detection to the analysis of isoflavonoids and lignans, *J. Pharm. Biomed. Anal.*, 41 1497, 2006.
198. Gamache, P., Ryan, E., and Acworth, I.N., Analysis of phenolic and flavonoid compounds in juice beverages using high-performance liquid chromatography with coulometric array detection, *J. Chromatogr.*, 635, 143, 1993.
199. Guo, C. et al., High-Performance liquid chromatography coupled with coulometric array detection of electroactive components in fruits and vegetables: Relationship to oxygen radical absorbance capacity, *J. Agric. Food Chem.*, 45, 1787, 1997.
200. Peyrat-Maillard, M.N., Bonnely, S., and Berset, C., Determination of the antioxidant activity of phenolic compounds by coulometric detection, *Talanta*, 51, 709, 2000.
201. Klejdus, B. et al., Determination of isoflavones in soybean food and human urine using liquid chromatography with electrochemical detection. *J. Chromatogr. B*, 806, 101, 2004.
202. Rehova, L., Skerikova, V., and Jandera, P., Optimisation of gradient HPLC analysis of phenolic compounds and flavonoids in beer using a CoulArray detector, *J. Sep. Sci.*, 27, 1345, 2004.

203. Huck, C.W. and Bonn, G.K., Evaluation of detection methods for the reversed-phase HPLC determination of 3′,4′,5′-trimethoxyflavone in different phytopharmaceutical products and in human serum, *Phytochem. Anal.*, 12, 104, 2001.
204. de Rijke, E. et al., Natively fluorescent isoflavones exhibiting anomalous Stokes' shifts, *Anal. Chim. Acta*, 468, 3, 2002.
205. Stoggl, W.M., Huck, C.W., and Bonn, G.K., Structural elucidation of catechin and epicatechin in sorrel leaf extracts using liquid-chromatography coupled to diode array-, fluorescence-, and mass spectrometric detection, *J. Sep. Sci.*, 27, 524, 2004.
206. Tsanova-Savova, S., Ribarova, F., and Gerova, M., (+)-Catechin and (−)-epicatechin in Bulgarian fruits, *J. Food Comp. Anal.*, 18, 691, 2005.
207 Arts, I.C.W., Van de Putte, B., and Hollman, P.C.H., Catechin contents of foods commonly consumed in the Netherlands. 1. Fruits, vegetables, staple foods and processed foods, *J. Agric. Food Chem.*, 48, 1746, 2000.
208. Arts, I.C.W., Van de Putte, B., and Hollman, P.C.H., Catechin contents of foods commonly consumed in the Netherlands. 2. Tea, wine, fruit juices and chocolate milk, *J. Agric. Food Chem.*, 48, 1752, 2000.
209. Sengupta, P.K. and Kasha, M., Excited-state proton-transfer spectroscopy of 3-hydroxyflavone and quercetin, *Chem. Phys. Lett.*, 68, 382, 1979.
210. Hollman, P.C.H., van Trijp, J.M.P., and Buysman, M.N.C.P., Fluorescence detection of flavonoids in HPLC by postcolumn chelation with aluminium, *Anal. Chem.*, 68, 3511, 1996.
211. Gutiérrez, A.C. and Gehlen, M.H., Time resolved fluorescence spectroscopy of quercetin and morin complexes with Al^{3+}, *Spectrochim. Acta A*, 58, 83, 2002.
212. Dugo, P. et al., Determination of flavonoids in citrus juices by micro-HPLC-ESI/MS, *J. Sep. Sci.*, 28, 1149, 2005.
213. Mauri, P.L. et al., Liquid chromatography/electrospray ionization mass spectrometric characterization of flavonoids glycosides in tomato extracts and human plasma, *Rapid Commn. Mass Spectrom.*, 13, 924, 1999.
214. Tolonen, A. and Uusitalo, J., Fast screening method for the analysis of total flavonoid content in plant and foodstuffs by high-performance liquid chromatography/electrospray ionization time-of-flight mass spectrometry with polarity switching, *Rapid Commn. Mass Spectrom.*, 18, 3113, 2004.
215. Volpi, N. and Bergonzini, G., Analysis of flavonoids from propolis by on-line HPLC–electrospray mass spectrometry, *J. Pharm. Biomed. Anal.*, 42, 354, 2006.
216. Charrouf, Z. et al., Separation and characterization of phenolic compounds in argan fruit pulp using liquid chromatography–negative electrospray ionization tandem mass spectroscopy, *Food Chem.*, 100, 1398, 2007.
217. Pérez-Magariño, S. et al., Various applications of liquid chromatography–mass spectrometry to the analysis of phenolic compounds, *J. Chromatogr. A*, 847, 75, 1999.
218. Romani, A. et al., HPLC-DAD/MS characterization of flavonoids and hydroxycinnamic derivatives in Turnip Tups (*Brassica rapa* L. Subsp. *Sylvestris* L.), *J. Agric. Food Chem.*, 54, 1342, 2006.
219. Nelson, B.C. and Sharpless, K.E., Quantification of the predominant monomeric catechins in baking chocolate standard reference material by LC/APCI-MS, *J. Agric. Food Chem.*, 51, 531, 2003.
220. Cuyckens, F. and Claeys, M., Mass spectrometry in the structural analysis of flavonoids, *J. Mass Spectrom.*, 39, 1, 2004.
221. de Rijke, E. et al., Analytical separation and detection methods for flavonoids, *J. Chromatogr.*, 1112, 31, 2006.

222. de Rijke, E. et al., Liquid chromatography with atmospheric pressure chemical ionization and electrospray ionization mass spectrometry of flavonoids with triple-quadrupole and ion-trap instruments, *J. Chromatogr. A*, 984, 45, 2003.
223. Ablajan, K. et al., Structural characterization of flavonol 3,7-di-*O*-glycosides and determination of the glycosilation position by using negative ion electrospray ionization tandem mass spectrometry, *J. Mass Spectrom.*, 41, 352, 2006.
224. Stevens, J.F. et al., Prenylflavonoids from *Humulus lupulus*, *Phytochemistry*, 44, 1575, 1997.
225. Takayama, M. et al., Mass spectrometry of prenylated flavonoids, *Heterocycles*, 33, 405, 1992.
226. Albert, K. et al., On-line coupling of separation techniques to NMR, *J. Chromatogr. A*, 856, 199, 1999.
227. Garo, E. et al., Prenylated flavanones from *Nomotes engleri*: On-line structure elucidation by LC/UV/NMR, *Helvetica Chim. Acta*, 81, 754, 1998.
228. Smallcombe, S.H., Patt, S.L., and Keiffer, P.A., WET solvent suppression and its applications to LC-NMR and high-resolution NMR spectroscopy, *J. Magn. Resonance Ser. A*, 117, 295, 1995.
229. Queiroz, E.F. et al., On-line identification of the antifungal constituents of *Erythrina vogelii* by liquid chromatography with tandem mass spectrometry, ultraviolet absorbance detection and nuclear magnetic resonance spectrometry combined with liquid chromatographic micro-fractionation, *J. Chromatogr. A*, 974, 123, 2002.
230. de Koning, J.A. et al., On-line trace enrichment in hyphenated liquid chromatography nuclear magnetic resonance spectroscopy, *J. Chromatogr. A*, 813, 55, 1998.
231. Ferrari, J. et al., Benzophenone glycosides from *Gnidia involucrata*, *Phytochem*istry, 54, 883, 2000.
232. Spraul, M. et al., Advancing NMR sensitivity for LC-NMR-MS using a cryoflow probe: Application to the analysis of acetaminophen metabolites in urine, *Anal. Chem.*, 75, 1536, 2003.
233. Exarchou, V. et al., LC-UV-solid-phase extraction-NMR-MS combined with a cryogenic flow probe and its application to the identification of compounds present in Greek oregano, *Anal. Chem.*, 75, 6288, 2003.
234. Füzfai, Zs., Kovacs, E., and Molnar-Perl, I., Identification and quantification of the main constituents of sour cherries: Simultaneously, as their trimetyl silyl derivatives, by gas chromatography–mass spectrometry, *Chromatography*, 60, S143, 2004.
235. Justensen, U., Negative atmospheric pressure chemical ionisation low-energy collision activation mass spectrometry for the characterisation of flavonoids in extracts of fresh herbs, *J. Chromatogr. A*, 902, 369, 2000.
236. Hughes, R.J. et al., A tandem mass spectrometric study of selected characteristic flavonoids, *Int. J. Mass Spectrom.*, 210, 371, 2001.
237. Fabre, N. et al., Determination of flavone, flavonol, and flavanone aglycones by negative ioin liquid chromatography electrospray ion trap mass spectrometry, *J. Am. Soc. Mass Spectrom.*, 12, 707, 2001.
238. Prasain, J.K. et al., Profiling and quantification of isoflavonoids in kudzu dietary supplements by high-performance liquid chromatography and electrospray ionization tandem mass spectrometry, *J. Agric. Food Chem.*, 51, 4213, 2003.
239. Ma, Y.L. et al., Internal glucose residue loss in protonated *O*-diglycosyl flavonoids upon low-energy collision-induced dissociation, *J. Am. Soc. Mass Spectrom.*, 11, 136, 2000.
240. Holly, A.W. et al., Comparison of proanthocyanidins in commercial antioxidants: Grape seed and pine bark extracts, *J. Agric. Food Chem.*, 55, 148, 2007.

241. De Pascual-Teresa, S., Santos-Buelga, C., and Rivas-Gonzalo, J.C., Prodelphinidins and related flavanols in wine, *Int. J. Food Sci. Technol.*, 35, 33, 2000.
242. Chandra, A., Rana, J., and Li, Y., Separation, identification, quantification, and method validation of anthocyanins in botanical supplement raw materials by HPLC and HPLC-MS, *J. Agric. Food Chem.*, 49, 3515, 2001.
243. Ma, Y.L. et al., Mass spectrometric methods for the characterisation and differentiation of isomeric *O*-diglycosyl flavonoids, *Phytochem. Anal.*, 12, 159, 2001.
244. Lin, Y.Y., Ng, K.J., and Yang, S., Characterization of flavonoids by liquid-chromatography tandem mass-spectrometry, *J. Chromatogr.*, 629, 389, 1993.
244. Pati, S. et al., Simultaneous separation and identification of oligomeric procyanidins and anthocyanin-derived pigments in raw red wine by HPLC-UV-ESI-MS^n, *J. Mass Spectrom.*, 41, 861, 2006.
246. Andrade, F.D.P. et al., Use of on-line liquid chromatography–nuclear magnetic resonance spectroscopy for the rapid investigation of flavonoids from *Sorocea bomplandii*, *J. Chromatogr. A*, 953, 287, 2002.
247. Wolfender, J.L., Rodriguez, S., and Hostettmann, K., Liquid chromatography coupled to mass spectrometry and nuclear magnetic resonance spectroscopy for the screening of plant constituents, *J. Chromatogr. A*, 794, 299, 1998.
248. Hansen, S.H. et al., High-performance liquid chromatography on-line coupled to high-field NMR and mass spectrometry for structure elucidation of constituents of *Hypericum perforatum*, *Anal. Chem.*, 71, 5235, 1999.
249. Vilegas, W. et al., Application of on-line C-30 RP-HPLC-NMR for the analysis of flavonoids from leaf extracts of *Maytenus aquifolium*, *Phytochem. Anal.*, 11, 317, 2000.
250. Lommen, A. et al., Application of directly coupled HPLC-NMR-MS to the identification and confirmation of quercetin glycosides and phloretin glycosides in apple peel, *Anal. Chem.*, 72, 1793, 2000.
251. Le Gall, G. et al., Characterization and content of flavonoid glycosides in genetically modified tomato (*Lycopersicon esculentum*) fruits, *J. Agric. Food Chem.*, 51, 2438, 2003.

4 Exploitation of Residues from Vineyards, Olive Groves, and Wine and Oil Production to Obtain Phenolic Compounds of High-Added Value

María D. Luque de Castro, José M. Luque-Rodríguez, and Rafael Japón-Luján

CONTENTS

4.1 INTRODUCTION

One common cultural trait of the people settled around the Mediterranean Sea is their traditional diet, which has remained virtually unchanged for centuries. Obviously, economic, ethnic, and even religious differences among regions influence the dietary intake. Thus, the Greek diet is dominated by the consumption of olive oil, vegetables, and fruits [1], the Italian version is pasta, and the Spanish version is especially rich in fish. However, a common pattern prevails in all variants that includes the following characteristics [2]:

(a) A high consumption of legumes, cereals, fruits, and vegetables. Legumes contain proteins of a high biological value and complex carbohydrates; cereals provide carbohydrates and contribute to the intake of vitamins in the B group; and fruits and vegetables are the primary sources of water-soluble vitamins (overall, vitamin C), provitamins (α- and β-carotene, β-cryptoxanthine), carotenoids, antioxidants, minerals, and fibers.
(b) A low consumption of meat and meat products. Although animal foods are the main sources of B group vitamins, their intake is supplemented by cereals. Moreover, several Mediterranean countries consume preserved pork liver, which provides vitamin B_{12}. Liver is also a source of folates (also called vitamin B_9).
(c) A moderate consumption of milk and dairy products. The most commonly used dairy products are yoghurt and cottage cheese, which provide calcium without a high saturated fat intake. These are great sources of proteins of a high biological value, minerals, and fat-soluble vitamins (vitamins A and D).
(d) A high monounsaturated/saturated fat ratio. The high consumption of olive oil is one of the most salient features of the Mediterranean diet and provides a large amount of oleic acid. On the other hand, fish and nuts also provide large amounts of $\omega 3$ fatty acids such as linoleic acid. Moreover, olive oil and nuts contain abundant vitamin E.
(e) A moderate consumption of wine, which is usually drunk with meals. This is the other essential feature of the Mediterranean diet. Wine, particularly of the red type, is an important source of polyphenols.

The health benefits of the traditional Mediterranean diet are extensively documented [3,4] and include protective effects against cardiovascular diseases [5–11], diabetes [12,13], rheumatoid arthritis [14], intestinal diseases [15], and several types of cancers [16–18]. Some constituents of wine and olive oil (particularly polyphenols) play a key role in most of such healthy effects.

Studies on the healthy effects of wine began with the discovery of the "French Paradox" in the 1970s [19,20]. This phenomenon is usually defined as a lower-than-expected coronary heart disease mortality rate in France, where classic risk factors are not less prevalent than in other industrialized countries and where, in addition, the diet has always been rich in saturated animal fat [21]. A number of hypotheses have been put forward to explain the phenomenon [21–27], the most widely accepted of which ascribes it to dietary habits (particularly the wine consumption), the crucial role of which has later been confirmed by a number of epidemiological studies [28–32]. Some studies have also associated alcohol or wine consumption with a reduced incidence of other health problems including dementia and Alzheimer's disease [33], age-related macular degeneration [34], kidney stones [35–37], gall-stones [38], and cancer [39].

Although the exact origin of these effects remains unclear, alcohol (ethanol) and polyphenolic compounds are thought to play a key role in them. Thus, ethanol has been reported to increase the levels of good cholesterol (high-density lipoprotein, HDL) and to inhibit platelet aggregation, which protects against coronary heart disease and stroke [40–45]. On the other hand, the health benefits of polyphenolic compounds are ascribed to their antioxidant and free radical scavenging properties [46,47].

Olive oil, which is the principal source of fat in the Mediterranean region, possesses a high content in monounsaturated fatty acids. Oleic acid, its main component, accounts for 55%–83% (mean 75%) of total fatty acids in it.

The unsaponifiable fraction of olive oil contains a variety of minor components including tocopherols (mainly α-tocopherol, in amounts from 12 to 25 mg/100 g), phenols, flavor compounds, hydrocarbons, and sterols. Olive biophenols (OBPs, which are inaccurately referred to as polyphenols in some cases) are used as flavorants, colorants, and antioxidants. The best known olive oil phenols are hydroxytyrosol, tyrosol, oleuropein, apigenin, and luteolin, in both free and derivative forms.

Olive oil thus provides large amounts of stable, not easily oxidized fatty acids, as well as substantial quantities of powerful antioxidant molecules favoring a healthier ageing and increased longevity in its consumers [48].

In recent years, a number of healthy properties of olive oil have been demonstrated. Thus, the regular intake of olive oil reduces the major risk factors for cardiovascular diseases such as the lipoprotein profile, blood pressure, glucose metabolism, and antithrombotic profile [49]. Also, it facilitates the modulation of endothelial function, inflammation, and oxidative stress [50].

Epidemiological studies suggest that olive oil exerts a protective effect against some types of malignant tumors (skin, breast, prostate, endometrium, digestive tract). Prevention of oxidative DNA damage or DNA strand breakage [51], protection against chronic liver disease and the bowel disorder known as Crohn's disease [52], and reduction of incidence of melanoma are but a few of the beneficial effects of olive oil on precancerous lesions [52].

In addition to heart and cancer diseases, olive oil may also help prevent or delay the onset of diabetes [53], protect the immune system [54], reduce the risk of developing rheumatoid arthritis [55], and is extremely helpful in the treatment and prevention of the disorders of the bile ducts, thereby improving hepatobiliary diseases [56]. Also, it is recommended to fight pancreas diseases [57] and facilitate calcium absorption, thereby improving growth and preventing osteoporosis [58].

These properties have boosted olive oil production to a world output exceeding 2.5 million tons in the 2005/06 campaign (see production in Table 4.1).

TABLE 4.1
Olive Oil World Production (×1000 Tm)

Country	2000/01	2001/02	2002/03	2003/04	2004/05	2005/06
Algeria	26.5	25.5	15.0	69.5	33.5	47.5 (1.8%)
Argentina	4.0	10.0	11.0	13.5	10.0	25.0 (1.0%)
Greece	430.0	358.3	414.0	308.0	435.0	424.0 (16.4%)
Italy	509.0	656.7	634.0	685.0	879.0	600.0 (23.2%)
Morocco	35.0	60.0	45.0	100.0	50.0	75.0 (2.9%)
Portugal	24.6	33.7	28.9	31.2	46.4	30.0 (1.2%)
Spain	973.7	1411.4	861.1	1412.0	980.3	880.0 (34.1%)
Syria	165.0	92.0	165.0	110.0	175.0	100.0 (3.9%)
Tunisia	130.0	35.0	72.0	280.0	130.0	200.0 (7.7%)
Turkey	175.0	65.0	140.0	79.0	145.0	112.0 (4.4%)

4.2 POLYPHENOLS FROM VINEYARD AND WINE PRODUCTION RESIDUES

Vineyards produce vast amounts of agricultural residues (particularly vine shoots and leaves). According to Jiménez and Sánchez [59], 1.4–2.0 tons of shoots can be obtained per hectare of vine crop. Since vineyard cultivated area in the world is around 8 million ha (Table 4.2), an estimated total of 11.2–16 million tons of vine shoots are produced each year. On the other hand, the most abundant residue of the wine-making process is the solid waste remaining after grapes are pressed, grape pomace. Since 20%–25% of grape weight processed remains as pomace [60] and about 80% of the world grape output (Table 4.2) is used to make wine, around 10.5–13.1 million tons of grape pomace is produced in the world each year.

Its abundance, and its especial richness in polyphenolic compounds, makes exploitation of grape pomace highly interesting in economic terms as this waste can be used as a raw material for obtaining products of a high-added value, for use in the nutraceutical, cosmetic, pharmacological, oenological, and food additive industries; thus, an increasing number of dietary supplements for disease prevention obtained from grape skins and seeds have been placed on the market in recent years [61].

4.2.1 Polyphenolic Constituents of Vineyard and Wine Production Wastes

Vineyard and wine production residues differ in polyphenolic composition among one another. Thus, even grape pomace (or grape marc) is rather heterogenous as it consists of skins (50%), seeds (25%), and stems (25%). In any case, polyphenols in these residues are of either the flavonoid or nonflavonoid type.

4.2.1.1 Flavonoids

Chemically, all flavonoids possess a $C_6-C_3-C_6$ backbone, with two benzene rings (A and B) connected via a heterocyclic pyrane or a pyrone ring (C) (Figure 4.1).

TABLE 4.2
World Distribution of Vineyard Surface and Grape and Wine Production

Vineyard Surface			Grape Production			Wine Production		
First 10 Countries	**Million ha**	**%**		**Million Tm (Metric Ton)**	**%**		**Million hl**	**%**
Spain	1180	14.9	Italy	8.6	13.1	Italy	50.6	18.3
France	890	11.2	France	6.8	10.3	France	50.5	18.2
Italy	847	10.7	USA	6.3	9.6	Spain	35.3	12.7
Turkey	570	7.2	Spain	5.9	9.0	USA	23.5	8.5
China	487	6.1	China	5.6	8.5	Argentina	15.2	5.5
USA	399	5.0	Turkey	3.7	5.6	Australia	14.0	5.1
Iran	296	3.7	Argentina	2.8	4.3	China	12.0	4.3
Portugal	250	3.1	Iran	2.8	4.3	Germany	9.1	3.3
Rumania	218	2.7	Chile	2.3	3.4	South Africa	8.3	3.0
Argentina	217	2.7	Australia	2.0	3.1	Chile	7.9	2.9
Distribution by Continents								
Europe	4686	59	Europe	30.2	46	Europe	185.6	67
Asia	1747	22	Asia	15.7	24	America	52.6	19
America	953	12	America	13.8	21	Asia	13.9	5
Africa	397	5	Africa	3.9	6	Oceania	13.9	5
Oceania	159	2	Oceania	2.0	3	Africa	11.1	4
World	7943		World	65.6		World	277.0	

Source: Annual report of the International Organization of Vine and Wine (OIV), 2006.

Differences among flavonoid families lie in the type, number, and position of functional groups in this parent structure.

4.2.1.1.1 Anthocyans

Anthocyans account for most polyphenolic compounds remaining in skin from red grape pomace; by contrast, they are virtually absent in skin from white grape pomace. This is practically the only significant difference between red and white grape varieties as regards qualitative polyphenolic composition [62].

Anthocyans are responsible for the color of red grapes, and also for the colors ranging from red to blue in other fruits, vegetables, and flowers. The flavonoid backbone is hydroxylated at positions 3, 5, and 7 (Figure 4.1) and the B-ring exhibits a variable hydroxylation–methoxylation pattern depending on the particular anthocyanidin (Figure 4.2). Also, these molecules can bind to one or several sugar molecules in anthocyanins. The most abundant anthocyans in skin from *Vitis vinifera* grapes are 3-glucosides of malvidin (Mv), peonidin (Pn), delphinidin (Dp), petunidin (Pt), and cyanidin (Cy), in addition to their esters at position 6 in the sugar with acetic, *p*-coumaric, and caffeic acids [63]. Their relative proportions vary widely among varieties [64] and are influenced by particular cultivation conditions [65] and environmental factors (temperature, moisture, isolation conditions, soil type) [66–68].

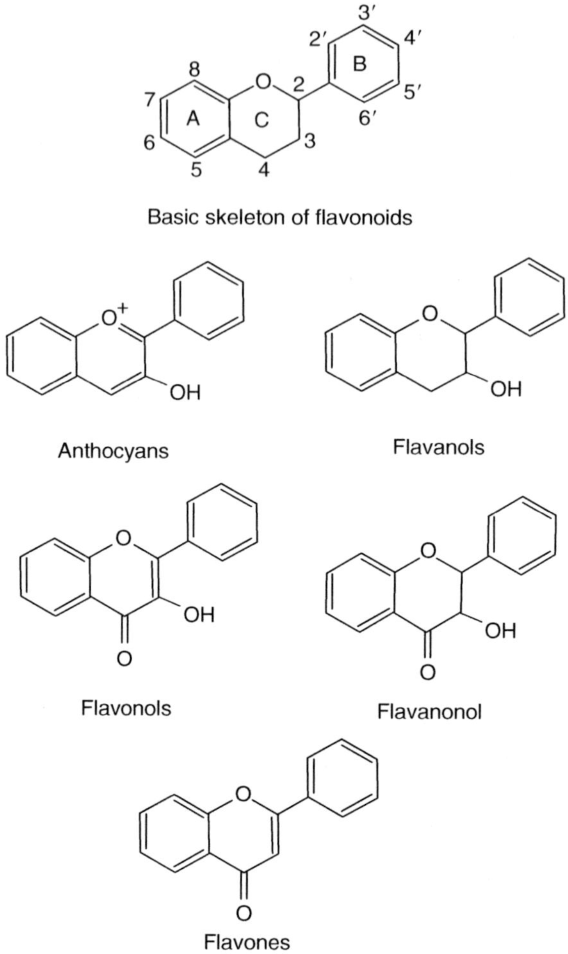

FIGURE 4.1 Basic structure of flavonoid families present in grapevine.

Anthocyanins are also present in other parts of grapevine. Thus, leaves are rich in anthocyans when they turn red in autumn. The 3-glucosides of Cy, Pn, and Mv are the most abundant anthocyanins [69]; however, those of Dp and Pt are also present, as are the 3-(6-*p*-coumaroyl)-glucosides of the five anthocyanidins and the 3-(6-acetyl)-glucosides of Cy and Pn [70]. On the other hand, Mv-3-glucoside and its caffeoyl and *p*-coumaroyl derivatives are the most abundant anthocyanins in stalks, shoots, and stems [69].

Grape pomace also contains anthocyanin-derived pigments, some of which are also found in aged wines. These compounds are produced by two types of reactions during the processing of fresh grapes in the first step of the wine-making process, namely (a) direct and acetaldehyde-mediated anthocyanin–tannin condensation [71,72], which produces polymeric pigments and (b) the C_4/C_5 cycloaddition with

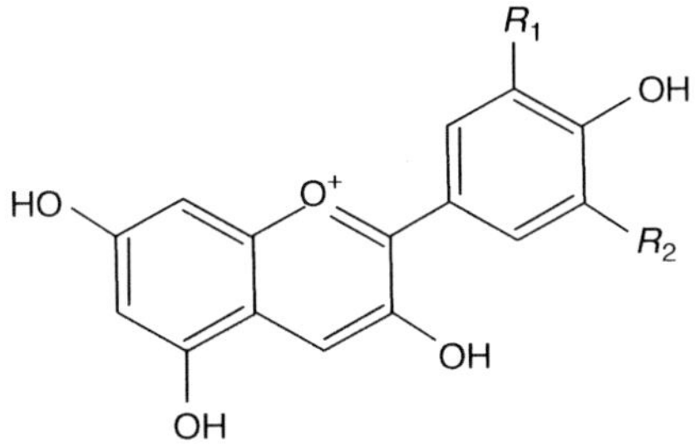

Anthocyanidin	R_1	R_2
Cyanidin	–OH	–H
Delphinidin	–OH	–OH
Peonidin	$-OCH_3$	–H
Petunidin	$-OCH_3$	–OH
Malvidin	$-OCH_3$	$-OCH_3$

FIGURE 4.2 Common anthocyanidins in grapevine.

other molecules bearing a polarizable double bond (e.g., pyruvic acid, 4-vinylphenols, hydroxycinnamic acids, vinylflavanols, acetaldehyde) to give pyranoanthocyanins [73,74]. The pyruvate, vinylcatechol, vinylepicatechin, vinylphenol, and vinylguaiacol derivatives of Mv-3-glucoside and Mv-3-(6-acetyl)-glucoside-vinylepicatechin have been detected in extracts from grape pomace [70], as have Mv-3-(6-*p*-coumaroyl)-glucoside-4-vinylcatechol [75], Cy-3-*O*-glucoside acetaldehyde, Pn-3-*O*-glucoside acetaldehyde, Pt-3-*O*-glucoside acetaldehyde, and Mv-3-*O*-glucoside acetaldehyde [76] (Figure 4.3).

4.2.1.1.2 Flavanols

These compounds occur mainly in the outer coating of seeds [77]; however, they are also present in some skin cells [78] (both free and bound to polysaccharides in cell walls [79]), stems, shoots, and leaves [80].

The most abundant flavanols in grapes are flavan-3-ols. These compounds are flavonoids where the heterocycle (C-ring) is completely saturated (Figure 4.1), which can occur as monomers, oligomers (2–3 units), and polymeric structures (>3 units). The most common monomers in grapes are (+)-catechin, (–)-epicatechin, and epicatechin-3-*O*-gallate. Gallocatechin [81], catechin-3-*O*-gallate, and gallocatechin-3-*O*-gallate [82] have also been detected in some varieties of the genus *Vitis*. The total content, average degree of polymerization, and proportion of constituent units differ depending on the particular variety and cultivation conditions.

The oligomers and polymers of flavan-3-ols in seeds are mainly procyanidins from the B-series consisting of (+)-catechin, (–)-epicatechin, and epicatechin-3-*O*-gallate

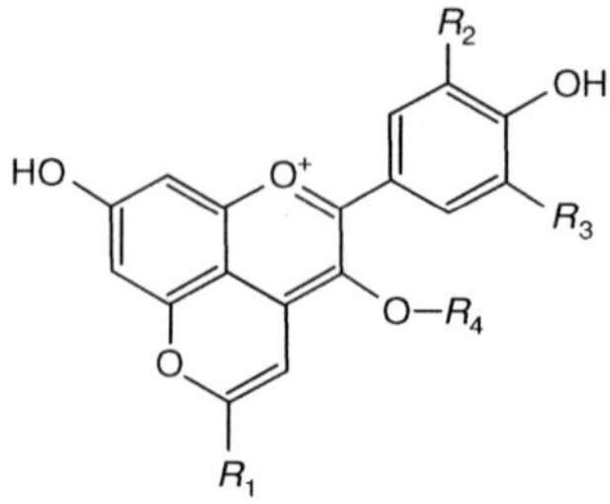

Compound	R_1	R_2	R_3	R_4
Cyanidin-3-glucoside acetaldehyde	–H	–OH	–H	Glucose
Peonidin-3-glucoside acetaldehyde	–H	$–OCH_3$	–H	Glucose
Petunidin-3-glucoside acetaldehyde	–H	$–OCH_3$	–OH	Glucose
Malvidin-3-glucoside acetaldehyde (vitisin B)	–H	$–OCH_3$	$–OCH_3$	Glucose
Malvidin-3-glucoside-pyruvic acid	–COOH	$–OCH_3$	$–OCH_3$	Glucose
Malvidin-3-glucoside-vinylphenol	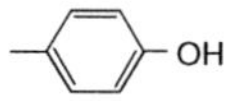	$–OCH_3$	$–OCH_3$	Glucose
Malvidin-3-glucoside-vinylguaiacol	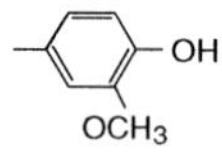	$–OCH_3$	$–OCH_3$	Glucose
Malvidin-3-glucoside-vinylcatechol (Pinotin A)	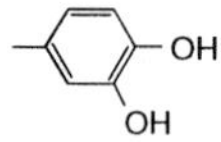	$–OCH_3$	$–OCH_3$	Glucose
Malvidin-3-glucoside-vinylepicatechin	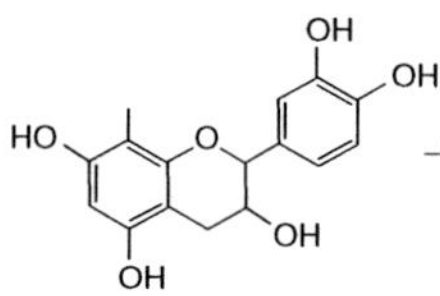	$–OCH_3$	$–OCH_3$	Glucose
Malvidin-3-(6-*p*-coumaroyl)-glucoside-vinylcatechol	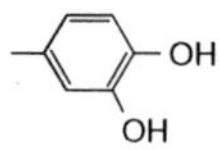	$–OCH_3$	$–OCH_3$	–(6-p-coumaroyl)-glucose
Malvidin-3-(6-acetyl)-glucoside-vinylphenol	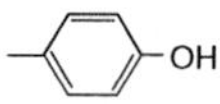	$–OCH_3$	$–OCH_3$	–(6-acetyl)-glucose
Malvidin-3-(6-acetyl)-glucoside-vinylguaiacol	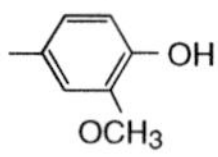	$–OCH_3$	$–OCH_3$	–(6-acetyl)-glucose
Malvidin-3-(6-acetyl)-glucoside-vinylcatechol	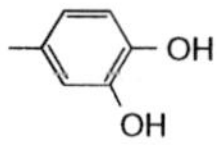	$–OCH_3$	$–OCH_3$	–(6-acetyl)-glucose
Malvidin-3-(6-acetyl)-glucoside-vinylepicatechin	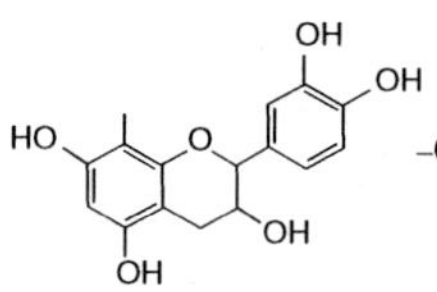	$–OCH_3$	$–OCH_3$	–(6-acetyl)-glucose

FIGURE 4.3 Main pyranoanthocyans detected in grape pomace extracts.

connected by C_4–C_8 or C_4–C_6 bonds [83,84] (Figure 4.4). For polymers, the average degree of polymerization (mDp) is 10 and the major component in the extended chain is (−)-epicatechin. The proportion of galloylated units ranges from 13% to 23% as the

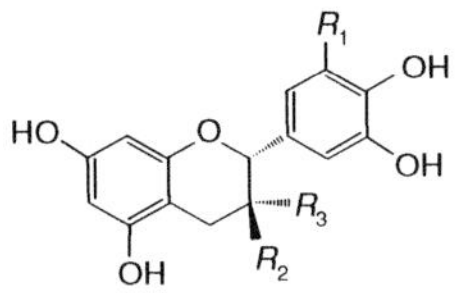

Monomers	R_1	R_2	R_3
(+)-catechin	–H	–OH	–H
(+)-catechin–*O*–gallate	–H	–O-gallate	–H
(+)-gallocatechin	–OH	–OH	–H
(+)-gallocatechin–*O*–gallate	–OH	–O-gallate	–H
(–)-epicatechin	–H	–H	–OH
(–)-epicatechin–*O*–gallate	–H	–H	–O-gallate
(–)-epigallocatechin	–OH	–H	–OH
(–)-epigallocatechin–*O*–gallate	–OH	–H	–O-gallate

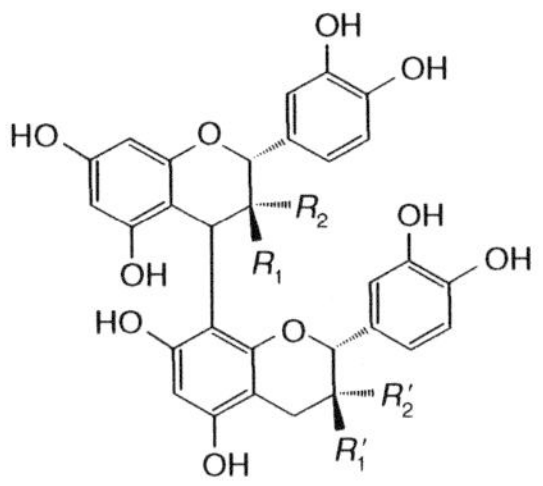

Dimers(4 →8)	R_1	R_2	R_1'	R_2'
Procyanidin B_1 (EC-C)	–H	–OH	–OH	–H
Procyanidin B_2 (EC-EC)	–H	–OH	–H	–OH
Procyanidin B_3 (C-C)	–OH	–H	–OH	–H
Procyanidin B_4 (C-EC)	–OH	–H	–H	–OH

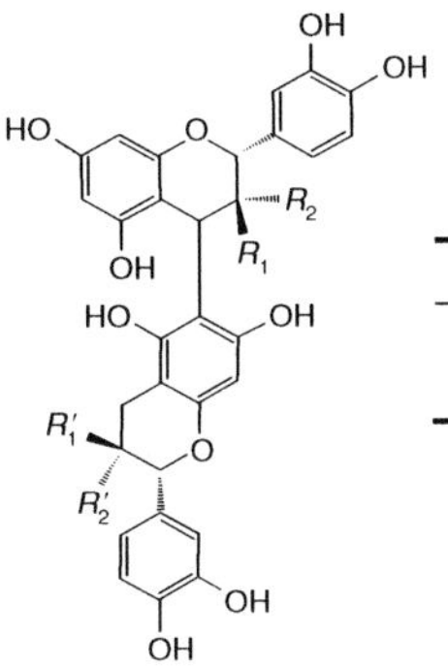

Dimers(4 →6)	R_1	R_2	R_1'	R_2'
Procyanidin B_5 (EC-EC)	–H	–OH	–H	–OH
Procyanidin B_7 (EC-C)	–H	–OH	–OH	–H

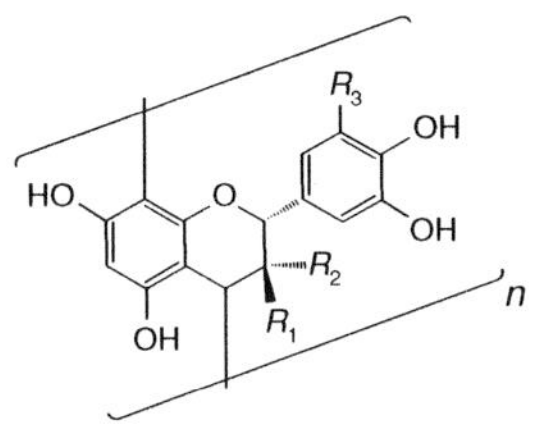

Oligomers and polymers	R_3
Proanthocyanidins	–H
Proanthodelphinidins	–OH

R_1: –H, –OH, or –*O*-gallate
R_2: –H or –OH

FIGURE 4.4 Flavan-3-ols common in grapevine.

average degree of polymerization increases [85]. Thus, gallic acid, monomeric flavan-3-ols [(+)-catechin, (−)-epicatechin, and epicatechin-3-*O*-gallate], procyanidin dimers (B_1, B_2, B_3-3-*O*-gallate, B_4), and trimers (T_2 and C_1) are the main flavanols identified in extracts from grape seeds [70]. Other less common biflavanols have also been identified in grape pomace extracts. Their chemical structures have been identified from long-range *H*,*C*-coupling data as epicatechin-[6′→8]-epicatechin and epicatechin-[6′→8]-catechin [86].

Oligomers and procyanidins in grape skins are similar to those in seeds [87]; the latter, however, have a higher mDp (30 units in some varieties [78,85]). (−)-Epicatechin prevails in the extension units, and catechin in terminal units [78]. Prodelphinidins, also present in skin [67,88], contain epigallocatechin as their main constituent unit, in addition to epigallocatechin-3-*O*-gallate and gallocatechin, the latter as a terminal unit.

Flavan-3-ols from stems are quite similar to those from skins (i.e., they consist of a mixture of procyanidins and prodelphinidins). In general, galloylation levels are low; (−)-epicatechin is virtually absent as a terminal unit; and polymers have a medium mDp value (>5) [89].

Procyanidins in vine shoots consist of partially galloylated units [80]. Thus, two trimeric procyanidins [C_1 and epicatechin-(4β→8)-epicatechin-(4β→8)-catechin], six dimeric procyanidins (B_1, B_2, B_3, B_4, B_7, and 3′-*O*-galloyl-B_2), in addition to 3-*O*-galloyl-(−)-epicatechin, (+)-catechin, and (−)-epicatechin, have been identified in acetone–water and ethyl acetate extracts from vine shoots [90].

On the other hand, proanthocyanidins in leaves contain ungalloylated prodelphinidin and procyanidin units, the procyanidin-to-prodelphinidin ratio increases as leaves turn from green to red in autumn [80].

Finally, leucoanthocyanins (flavan-3,4-diols and flavan-4-ols) have also been found in seeds, stems, and shoots [91], their proportions depending on the particular variety, degree of ripeness of the grapes, and some external factors [92].

4.2.1.1.3 Flavonols

These compounds are flavonoids containing a carbonyl group at position 4, a double bond between C_2 and C_3, and a hydroxyl group at C_3 (Figure 4.1). Flavonols are abundant in the skin of fresh grapes, mainly as 3-*O*-glucosides of kaempferol, quercetin, myricetin, and isorhamnetin [93,94] (Figure 4.5). Glucuronides, galactosides, xilosides, and arabinosides have also been found. Although no significant differences in total content have been found between red and white varieties, the relative composition of flavonols varies considerably. In any case, quercetin derivatives

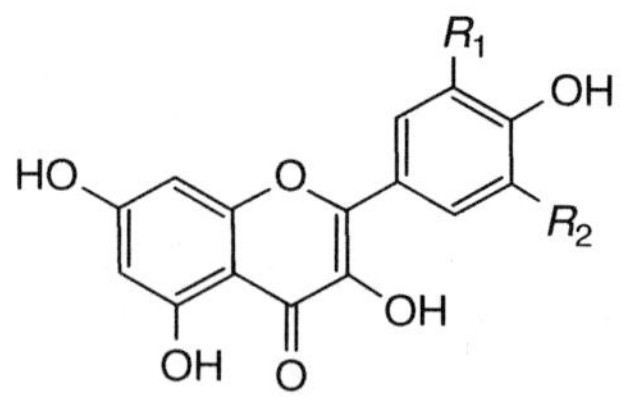

Flavonol	R_1	R_2
Kaempferol	–H	–H
Quercetin	–OH	–H
Myricetin	–OH	–OH
Isorhamnetin	$-OCH_3$	–H

FIGURE 4.5 Aglycon forms of the main flavonols present in grapevine.

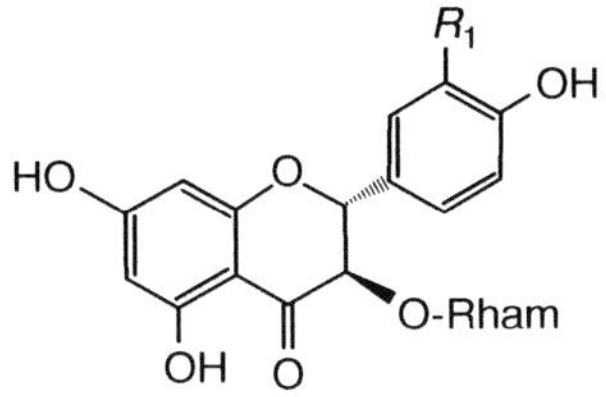

Flavanonol	R_1
Astilbin	–OH
Engeletin	–H

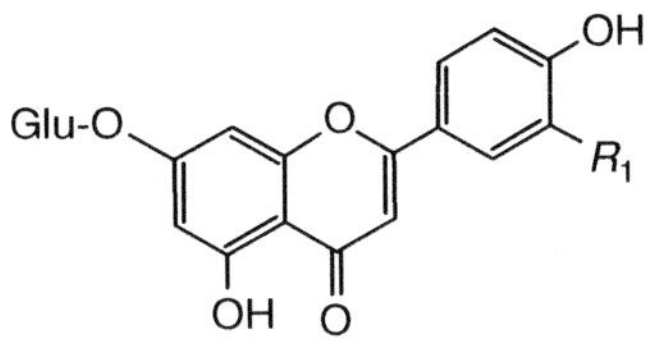

Flavone	R_1
Luteolin-7-*O*-glucoside	–OH
Apigenin-7-*O*-glucoside	–H

FIGURE 4.6 Flavanonols and flavones identified in skins and leaves of grapevine.

predominate, whereas myricetin derivatives and isorhamnetin-3-*O*-glucoside are thought to be specific to red varieties [95].

Concerning other parts of grapevine, quercetin-3-glucuronide and quercetin-3-glucoside are the most important flavonols in both photosynthetic and transport organs (stems) [69,96,97].

Finally, a flavonol-related compound called eucryphin (3,5,7-trihydroxy-chromone-3-α-L-rhamnopyranoside) has also been detected in grape pomace extracts [98].

4.2.1.1.4 Flavanonols (Dihydroflavonols) and Flavones

Flavanonols are 2,3-dihydroflavonols; by contrast, flavones are structurally similar to flavonols but possess no OH group at C_3 (Figure 4.1). Astilbin (dihydroquercetin-3-*O*-rhamnoside) and engeletin (dihydrokaempferol-3-*O*-rhamnoside) have been identified in skin of white grapes (Figure 4.6) [99] and in stems [97], while the flavones apigenin-7-glucoside and luteolin-7-glucoside have been detected in leaves [100].

4.2.1.2 Nonflavonoids

4.2.1.2.1 Phenolic Acids

Phenolic acids in grapevine include mainly gallic, ellagic, and hydroxycinnamic acids. They are usually present as various esters, and are also frequently found in free form in grape pomace and leaf extracts, probably as a result of partial hydrolysis of the esters during extraction.

Hydroxycinnamic acids are found mainly in skin and, to a lesser extent, in leaves [101,102] and stems [97]. The most abundant acids (viz., caffeic, *p*-coumaric, and ferulic) are present as tartaric esters (caftaric, cutaric, and fertaric acids; Figure 4.7) or in acylated anthocyanins (basically, *p*-coumaric and caffeic acids).

Gallic acid is mainly found esterified to flavan-3-ols, and also as a constituent of hydrolysable tannins. These compounds are esters containing one or several molecules of gallic (gallotannins) or ellagic acid (ellagitannins) and a polyol (usually

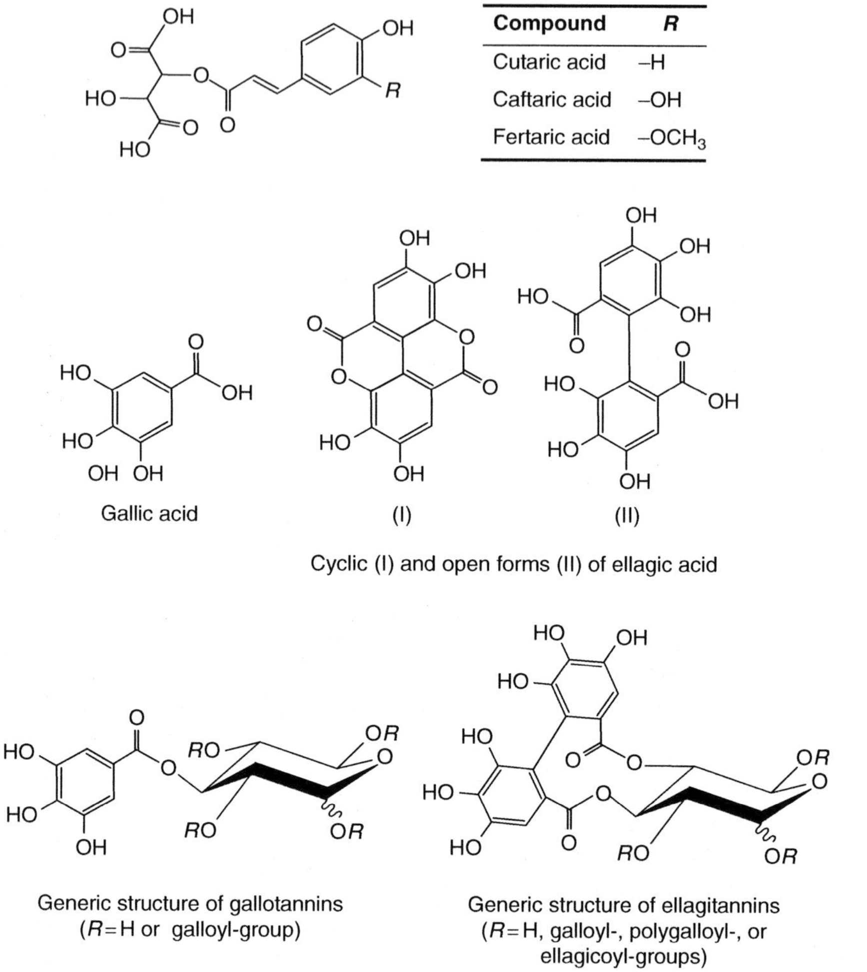

Compound	*R*
Cutaric acid	–H
Caftaric acid	–OH
Fertaric acid	$-OCH_3$

FIGURE 4.7 Chemical structures of the main phenolic acids and derivatives in grapevine.

glucose) (Figure 4.7). Four of them are present in minor amounts in seed extracts [103]. Also, 3- and 4-β-glucopyranosides of gallic acid, as well as ellagic acid and its precursors, have been detected in grape pomace extracts [98,104]. Finally, three ellagitannins have been isolated in leaves, namely brevilagin 1; 1,3-digalloyl-4,6-dehydrohexahydroxydiphenoylglucose (vitilagin); and 3,4-digalloyl-1,6-dehydro-hexahydroxydiphenoylglucose (isovitilagin) [105].

4.2.1.2.2 Stilbenes

Stilbenes in grapevine are present as constituents of the woody organs (roots, vine shoots, and stems), and also as substances of induced production (in leaves and berries) acting as phytoalexins in the mechanisms of grape resistance against certain pathogens such as fungi [106]. Other abiotic elicitors including ultraviolet (UV) light and heavy metals can lead stilbene production [107–109]. Stilbene synthesis in grapes also depends on various viticultural factors such as grape variety, the environment, and cultural practices. Concerning grape variety, red grapes possess higher stilbene levels than white grapes [110].

The many stilbenic compounds found in various parts of grapevine are derived from *trans*-resveratrol (3,5,4′-trihydroxystilbene) and occur as glucosylated derivatives or oligomeric forms of this compound called viniferins (Figure 4.8). The most common among the latter is the dimer ε-viniferin [111]; however, some trimers (α-viniferin) [112] and tetramers have also been isolated and characterized [113].

The greatest amounts of stilbenes in vineyard and wine production residues are found in grape skins, which contains mainly *trans*-resveratrol and piceid (its 3-*O*-β-glucoside) [114]. These compounds, in addition to viniferins and a minor stilbenoid

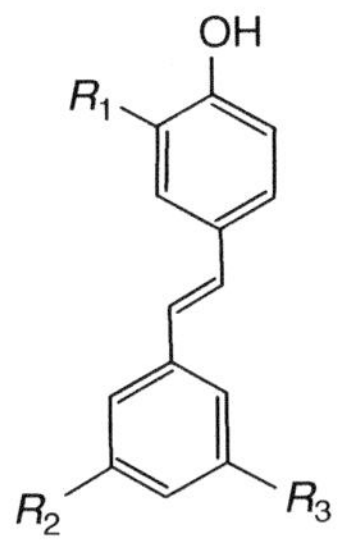

Compound	R_1	R_2	R_3
Trans-resveratrol	–H	–OH	–OH
Trans-pterostilbene	–H	$-OCH_3$	$-OCH_3$
Astringinin	–OH	–OH	–OH
Piceid	–H	–*O*-β-D-glucoside	–OH

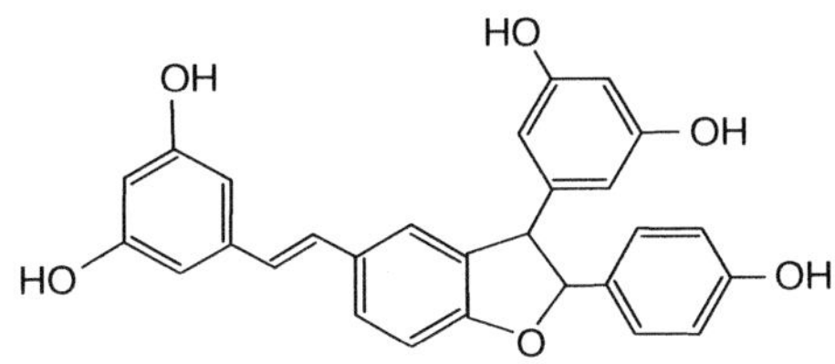

Trans-δ-viniferin

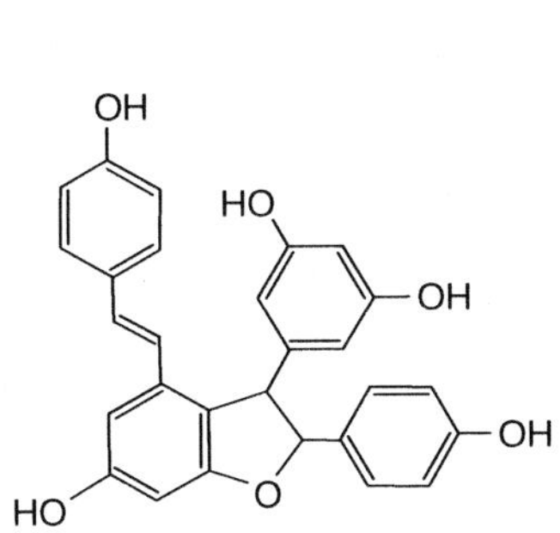

Trans-ε-viniferin

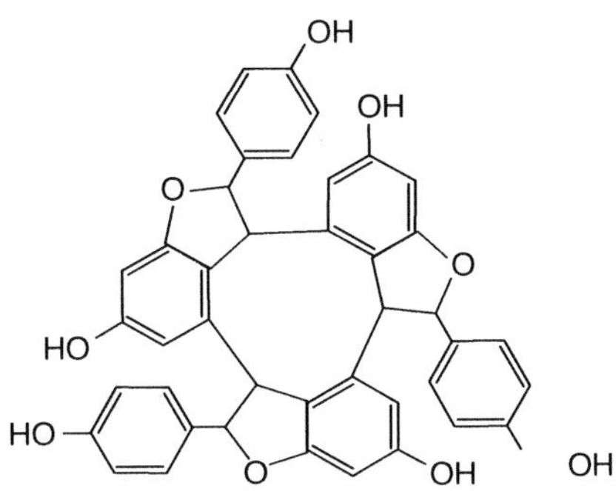

α-Viniferin

FIGURE 4.8 Main stilbenes and viniferins of grapevine.

called piceatannol or astringinin (*trans*-3,3′,4,5′-tetrahydroxystilbene), also occur in stems [110,115] and shoots [116]. In addition to *trans*-resveratrol, its β-D-glucoside and ε-viniferin, several minor components including pterostilbene (whose antifungal activity is high relative to resveratrol and viniferins [117]), *trans*-δ-viniferin and six other resveratrol dimers, two dimethylated resveratrol dimers and a resveratrol trimer have been reported [118].

4.2.1.2.3 Lignin

Lignin serves three key biological functions in plants, namely (a) increasing the mechanical strength of cell walls, (b) facilitating water transport stems, and (c) providing an effective physical barrier against pathogens by virtue of its difficult degradation.

Lignin accounts for about 17%–28% of dry weight of grapevines (20% on average in vine shoots [119]). It possesses a three-dimensional amorphous structure resulting from the polymerization of coniferyl, sinapyl, and *p*-coumaryl alcohol units (Figure 4.9), that is typical of hardwood [120], and its degradation produces low-molecular mass phenolic alcohols, aldehydes, ketones, and acids [121,122], as well

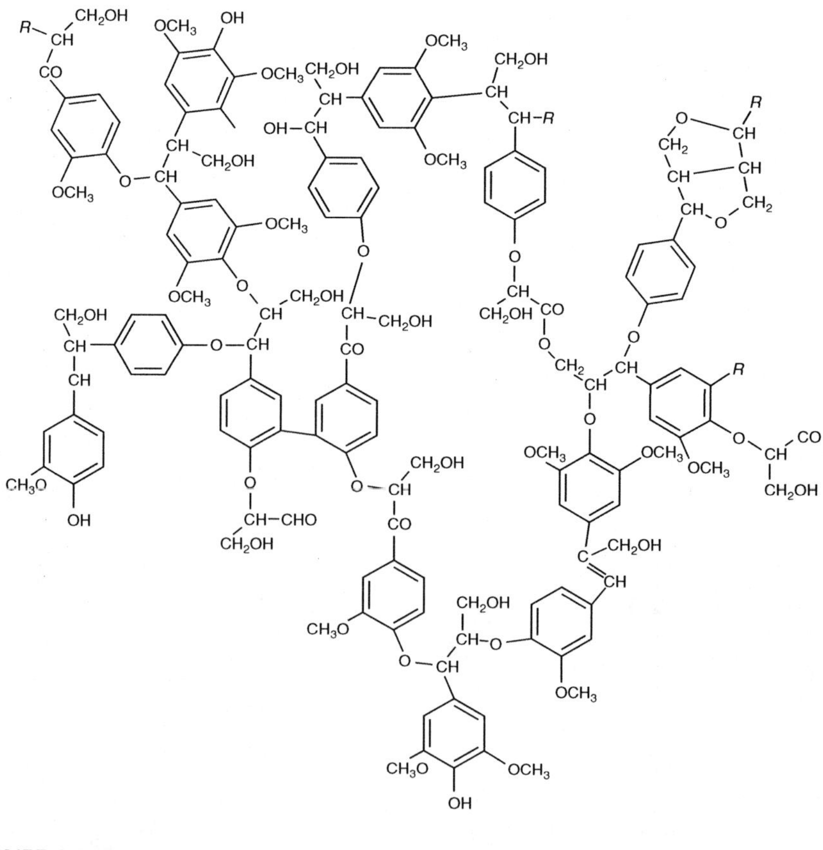

FIGURE 4.9 Example of a proposed structure for lignin.

as a large number of nonphenolic compounds; the yields and proportions of each family involved depending on the particular degradation process.

4.2.2 Present and Potential Uses of Vineyard and Wine Production Residues

4.2.2.1 Grape Pomace

Grape pomace has traditionally been used as a soil conditioner, a cattle feed, and a raw material for producing some spirits. These beverages, obtained by distillation, have a high alcohol content (over 40%–50%) and different characteristics; also they are given different names in different places (e.g., orujo in Spain, grappa in Italy, aguardente in Portugal, and pisco in Peru).

The possibilities of exploiting grape pomace have been investigated since the late nineteenth century [123,124], and several industrial processes have been developed to obtain ethanol, calcium tartrate [125], grape seed oil, and colorants [126]. Investigations have continued with a search for alternative, more efficient [127–130], environmentally friendly, and easily automated processes [75]. In parallel, new potential uses for grape pomace including production of furfural [131], compost [132], citric acid [133–135], and dietary fiber [136–144]; cultivation of yeast for animal feeding [145]; and protein extraction [146] have been studied.

The principal applications regarding polyphenolic compounds, which constitute the focus of this section, are colorants and antioxidant products. Thus, a number of researchers have patented procedures to obtain concentrated aqueous solutions of anthocyanins for use as food additives [126,147] ever since the pigment enocyanin was first isolated in 1879. At present, anthocyanins are common additives in drinks, marmalades, candies, and ice creams, among other products. In addition to their use in the food industry, anthocyanin-based pigments are also of interest to the cosmetic and pharmaceutical industries, in as much as they are completely safe alternatives to synthetic colorants and antioxidants [148]. Also, grape seeds have since 1970s been exploited as industrial sources of oligomeric proanthocyanidins (OPCs). Initially, these polyphenols were used almost exclusively by the pharmaceutical industry to make vascular protectors; through the years, however, new applications have been found. Thus, procyanidin extracts from grape seeds exhibit antioxidant activity both in vitro and in vivo [149–153]. Also, they are currently used as common ingredients in cosmetics (e.g., antiaging lotions, solar protectors, body-milks), and in nutraceutical products (e.g., dietary supplements).

Grape pomace is particularly rich in all these compounds by effect of their extraction during the wine-making process being far from complete (even in red wines, where the crushed grapes remain in contact with the must for several days). Thus, the estimated yield of this raw material is 60% for total phenolics, 30%–40% for anthocyans [154], and 20% for flavan-3-ols [155].

4.2.2.2 Vine Shoots and Leaves

Vine shoots and leaves have never had any significant economic value. The former has traditionally been cast on the ground to rot or used as fuel and, more recently, as a draining material and raw material for manufacturing boards. The tender grapevine leaves are edible and have several culinary uses in some countries of the Mediterranean basin such as Greece, Turkey, Lebanon, and Romania.

The vast amount of vine shoots produced each year has aroused increasing interest in their exploitation in recent years. The most extensively studied uses are probably as a raw material for obtaining paper pulp and ethanol. The former possibility was first examined in the early twentieth century [156], but was not seriously considered until the late 1980s [59,119,157–161]. The interest in the latter use, however, has remained throughout [162–166].

Vine shoots have also been investigated as raw materials for the production of cattle feed [167], methanol and fuels [168–171], lactic acid [172,173], compost [132], active carbon (charcoal) [174–176], and sorbents for pesticide control [177].

The use of vine shoots as source of polyphenols could be a profitable choice. Polyphenols can be obtained from vine shoots and leaves in two compatible ways, namely (a) by extracting the polyphenols that are not constituents of lignin and (b) by degrading lignin to obtain low-molecular mass phenolic compounds. At present, the former choice is being developed by some industries, where extracts rich in resveratrol and viniferins (vineatrol) from young vine shoots and leaves are being obtained. These extracts have proved effective in reducing proliferation of leukemic cells [178] and potentially useful against epilepsy [179]. Also, resveratrol and viniferins have allelopathic properties, so these extracts could be used to control weed growth and to reduce the use of herbicides [180]. In addition, their high antioxidant and free radical scavenging power makes them potentially useful for the cosmetic and food additive industries. Moreover, red leaf extracts are used by the pharmaceutical industry to obtain drugs effective against circulatory disorders [181–183].

Lignin degradation produces solutions containing abundant phenolic compounds with potential uses such as tanning and dyeing of leather [184], synthesis of chemical intermediates of pharmaceutical interest [185], the production of smoke flavorings [186–188], and the obtainment of wood extracts for the wine-making industry as an alternative to oak extracts currently used for wine aging [189]. In addition, lignin degradation can be used as a complementary process for obtaining paper pulp, as lignin must be removed during wood digestion if the end product is to possess an appropriate color and good mechanical properties [122].

4.2.3 Extraction Processes

4.2.3.1 Colorants from Grape Pomace

The industrial methods most widely used to obtain these pigments employ aqueous sulfur dioxide as extraction solvent or adsorptive resins. The most popular among the former is the Sefcal process, in which grape pomace is extracted with a hot 0.2% aqueous SO_2 solution. After decantation of potassium tartrate, the solution is concentrated in vacuo and the remaining sugars are fermented, then the resulting solution is reconcentrated by removing the alcohol and clarified. Finally, the liquid colorant can be atomized to obtain the pigment as a powder. The industrial methods based on adsorptive resins (e.g., the Applexion process) use grape pomace macerated in water. The resulting solution is filtered and passed through a battery of columns packed with adsorptive resins. Subsequently, anthocyanic pigments are eluted with ethanol and the alcoholic solution is concentrated under vacuum to recover the ethanol. The product contains about 250 g dry matter per liter, 95% of the dry matter as polyphenols, and 5%–10% as anthocyanins [190]. The liquid colorant can be dried to obtain a powder.

Other methods use conventional solid–liquid extraction (SLE) with acidified alcohols [191], but have no industrial application. Thus, the most common procedure for extracting anthocyanins at the laboratory scale uses methanol or ethanol solutions acidified with different proportions of 0.1%–1% HCl. With foods, it is preferable to use ethanol because it is less toxic, even though it is less efficient as an extractant and is more difficult to remove. Also, it can generate artifacts by effect of esterification reactions with free carboxylic groups if the extraction process is heat-assisted. On the other hand, the use of HCl as acidifier may cause hydrolysis of acylated anthocyanins (especially those with aliphatic acids) during extraction and subsequent concentration. For this reason, acetic, tartaric, or citric acids are preferred to HCl [192]. Ethyl acetate and acetone have also been used to extract anthocyanins from grape pomace [193] and various plant sources [194,195], respectively. Although extraction with organic solvents is efficient and simple, it is made expensive by the need to use large amounts of solvent and the presence of residual traces in polyphenol extracts, which is undesirable when the product is aimed at human consumption. Moreover, traditional solvent extraction is difficult to automatize and implement at an industrial scale.

Green extraction processes (viz., subcritical water and sub- and supercritical carbon dioxide extraction) are the most promising methods for industrial implementation. In fact, these methods are excellent for isolating anthocyanins and other polyphenols from grape processing wastes [196].

Sub- and supercritical fluid extraction with CO_2 has several advantages. One is that CO_2 is nontoxic, nonflammable, and noncorrosive. Also, the end product is obtained as a powder without the need for drying [197]. However, the relatively high equipment purchase and maintenance pose a major constraint. This technique has been successfully used to extract polyphenols from grape pomace, generally using CO_2 modified with methanol [198–201] or ethanol as cosolvent [202].

Pressurized liquid extraction (PLE) is more affordable than supercritical fluid extraction (SFE). Also, it has two crucial advantages over conventional processes, namely (a) raising the temperature above the boiling point of the solvent increases the diffusion rate, solubility, and mass transfer of the compounds and decreases the viscosity and surface tension of the solvent; these changes result in improved contact of the compounds with the solvent and facilitate extraction, which is thus faster and uses less solvent and (b) the absence of light and air significantly reduces degradation and oxidation of these compounds during extraction [203].

The PLE technique has so far been used almost exclusively as a first step in methods for analyzing several polyphenols from fresh grapes. Thus, Piñeiro et al. [204] determined *trans*-resveratrol in grapes by using a method involving a triple-discrete PLE step with methanol as cosolvent. Also, Ju and Howard [205,206] examined the effect of various solvents and temperature conditions on the PLE of anthocyanins from fresh grape skins obtained from a highly pigmented advanced breeding line of wine grape and concluded that either superheated (110°C) water or sulfured water was similarly effective to plain hot water or a 60:40 methanol–water mixture. PLE methods using grape pomace have rarely been optimized, even though the use of this raw material is of paramount importance as skins from grape pomace are extracted during the wine-making process and they only contain strongly retained polyphenols. González-Rodríguez et al. [207] and Luque-Rodríguez et al. [75] optimized the extraction of anthocyanins and other polyphenols with superheated ethanol–water mixtures, using a batch mode with grape pomace and a continuous mode with skins from grape pomace. The yields obtained with PLE exceeded those of conventional solvent extraction; obviously, however, total amounts obtained were lower than those provided by fresh grapes.

One other extraction method which may be a useful industrial alternative to conventional processes is the enzyme-assisted extraction of polyphenols [208], which has been tested at a pilot-plant scale. With the results, pre-extracting the pomace with hot water, followed by treatment with cell-wall degrading enzymes, provided better extraction rates comparable to those obtained by extracting grape pomace with sulfite.

4.2.3.2 Antioxidants from Grape Seeds (Oligomeric Procyanidins)

Grape seeds, even those obtained as an end by-product after the color extraction and alcohol distillation of wine pomace, retain substantial flavanol concentrations and significant antioxidant activity even upon heating at high temperatures [209]. In the usual industrial extraction process, dry seeds are extracted with hot water, then high-molecular weight compounds (tannins) are precipitated by addition of NaCl and the aqueous phase is extracted with organic solvents (frequently ethyl acetate). Finally, the organic phases are freeze-dried or concentrated at a reduced pressure to dryness.

Some variants of this method have been patented. Thus, Shrikhande et al. [210] developed a process in which seeds are extracted with hot water and the extracts treated with a pectolytic enzyme at pH 2.5, the extract being kept at 4°C for 3 weeks, after which the pH is raised to 4.5, the extract decanted and filtered and polyphenols recovered by using a column of XAD-7HP adsorbent resin.

Some methods based on aqueous acetone or ethanol extraction have also been patented. In one, the extract is repeatedly fractionated with ethyl acetate to extract monomers and oligomers, leaving polymers behind, then the ethyl acetate is evaporated and the oligomers are precipitated using methylene chloride, which is subsequently evaporated while the phenols are dissolved in water. Finally, the aqueous solution of phenols is spray-dried to a phenol powder [211]. In other process intended to obtain grape seed extracts with low contents in monomeric phenols, seeds are extracted with 30:70 ethanol–water, the extract is purified in a chromatographic step with a polystyrene resin such as XAD-16, XAD-4, or Diaion HP-20, and spray-dried [212]. These processes are more complex and expensive than the classical process; also, they use toxic solvents such as methylene chloride.

As in colorant production, SFE and SLE may be effective industrial alternatives to conventional solvent extraction of oligomeric procyanidins (OPCs). However, the potential use of these techniques for OPC production from grape seeds has scarcely been explored. In fact, only one analytical SFE method for this purpose has been developed [213,214], which involves sequential extraction of flavanols with supercritical CO_2 modified with methanol as cosolvent. On the other hand, SLE studies in this direction have grown substantially over the last 5–6 years. Thus, the stability of flavanols during extraction with superheated methanol has been demonstrated [215] and an analytical method for determining catechins that includes SLE with methanol has been developed [216]. García-Marino et al. [217] conducted a comprehensive study of the extraction of catechins and procyanidins from winery by-products by SLE and found superheated water to provide better efficiency than conventional extraction with methanol–water mixtures at atmospheric pressure. Also, selective extraction of compounds in variable degrees of polymerization can be achieved by one-step extraction at different temperatures.

4.2.3.3 Phenolics from Ligneous Organs of Vine

Although the content of resveratrol in *Polygonum cuspidatum*, a herbaceous plant native to eastern Asia (viz., Japan, China, and Korea), exceeds that in grapevine stems, young vine shoots, and leaves, these wastes constitute an effective alternative source, particularly in those regions where the vine cultivation is very extensive. In fact, some European and American companies market products manufactured with extracts from these parts of grapevine, which are usually obtained by solvent extraction with ethanol–water mixtures. Several studies have been carried out in recent years with a view to improving the efficiency or accelerating the extraction of resveratrol. Thus, ultrasound-assisted conventional solvent extraction has been found to provide increased yields (24%–30%) relative to conventional solvent extraction with 80:20 ethanol–water at 60°C for 30 min [218].

The isolation of chemicals from Kraft lignin, the polymer by-product of the Kraft pulping process for paper production, is an established industrial process [219]. There have been several studies on alternative degradation processes such as hydrogenolysis [220], biodegradation with white-rot fungi [221], and chemical oxidation [121,122,222]. These processes can in principle be used with residual lignin obtained from vine shoots employed for paper pulp production.

However, little research has been conducted toward the exploitation of lignin from vine shoots for polyphenol production. Volf et al. [223] have proposed a method for polyphenols, which are separated from wood and stems of *Vitis sp.* by global alkaline extraction, and developed a mathematical model for predicting the amount of polyphenols extracted under specific operational conditions with a view to transfer their method to a pilot or even industrial scale.

Finally, Luque-Rodríguez et al. [189] have studied the nonvolatile fraction of extracts from vine shoots obtained with superheated ethanol–water mixtures, a method which had previously proved effective for obtaining oak wood extracts [224,225]. The authors confirmed the superiority of PLE over conventional solvent extraction for this purpose showed that the extracts obtained with 80% (v/v) ethanol at pH 3.0 and 240°C for 60 min contained most of the compounds usually present in commercial wood extracts.

4.3 PHENOLS FROM OLIVE TREES AND OLIVE OIL PRODUCTION

4.3.1 INTRODUCTION

The olive oil industry produces vast amounts of direct waste from the production line. Depending on the particular technology used for oil production, mill waste such as alpechín (an aqueous liquor coming from the plant water and soft tissues of the olive fruit plus water added during processing), orujo (a solid waste consisting of olive pulp and stones), or alperujo (an emulsion of alpechín and orujo) is produced with a three- or two-phase process. One indirect waste of olive oil production consists of leaves accompanying the olive fruits. Olive leaves also contain substantial amounts of biophenols, which makes them a raw material with a high added value.

4.3.2 CHARACTERISTICS OF MAJOR BIOPHENOLS FROM OLIVE BY-PRODUCTS

A number of olive biophenols (OBPs) including oleuropein, hydroxytyrosol, tyrosol, verbacoside, apigenin-7-glucoside, and luteolin-7-glucoside (Figure 4.10) are present in variable amounts in by-products; however, only those possessing the most interesting properties and potential applications are considered in this section.

Oleuropein is the most abundant BP in olive leaves; however, it has also been found throughout the tree, including all constituent parts of the fruit (peel, pulp, and bone), and also in alperujo as a result. This secoiridoid, which, interestingly, must be removed from table olives owing to its bitter taste, prevents cardiac diseases by avoiding membrane lipid oxidation [226], acting on coronary dilation; also, it has antiarrhythmic action [227], improves lipid metabolism and helps reduce obesity problems as a result [228], protects enzymes and hypertensive cell death in cancer patients [226], and possesses antiviral properties [229,230].

Hydroxytyrosol is an *o*-diphenol present mainly in alperujo, but also in olive leaves [231], in free and esterified forms. It is a better scavenger or antioxidant than typical antioxidants such as vitamins C and E or 2,6-di-*tert*-butyl-4-methylphenol (BHT) [232]; as a result, this OBP prevents a number of cardiac (through anti-aggregating platelet action [228] or antimycoplasmal activity [233], among others) and tumoral diseases (e.g., by avoiding the protein damage induced by long-wave ultraviolet radiation in melanoma cells [234]). This phenolic compound can also be used in interesting applications such as the prevention of macrophage activation [235], inhibition of bacterial plaque accumulation on teeth, and prevention and treatment of diseases of the dental cavity and periodontium [236].

Although tyrosol does not surpass the antioxidant capacity of hydroxytyrosol, it exhibits cytostatic activity against McCoy cells and potential activity as an antitumoral compound [237]. Also, it exerts antiarrhythmic [238] and cardioprotective actions by preventing stressor-induced change at the level of cyclic nucleotides in the myocardium [239].

Verbacoside, a hydroxycinnamic derivative present both in olive leaves and in alperujo, exhibits moderate cytotoxic and cytostatic activity against skin tumors by inhibiting protein kinase C activity [240,241] and decreases the tumorigenicity of human gastric adenocarcinoma by 75% [242]. This compound has been used to repair brain oxidative damage caused by heroin consumption [243].

Phenolic compound	Chemical formula
Oleuropein	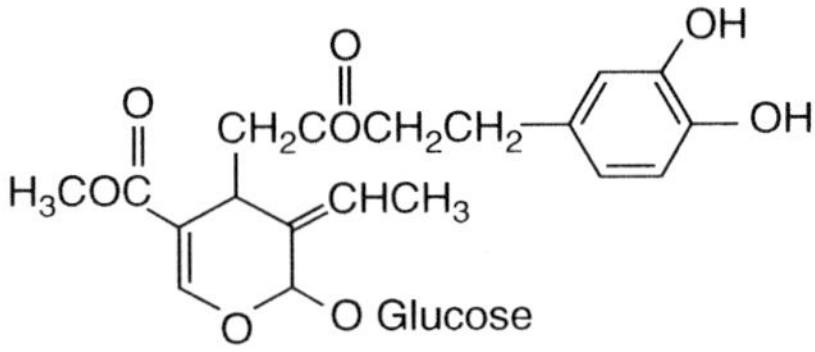
Tyrosol	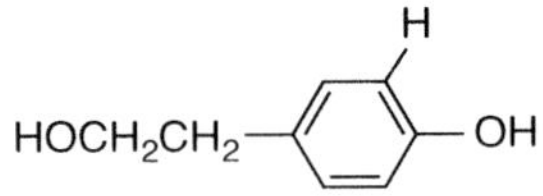
Hydroxytyrosol	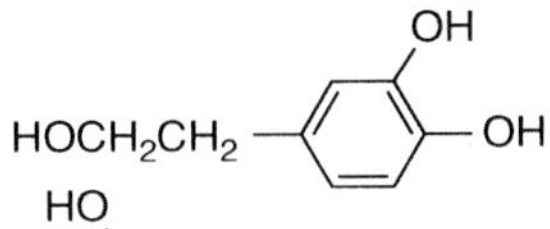
Verbascoside	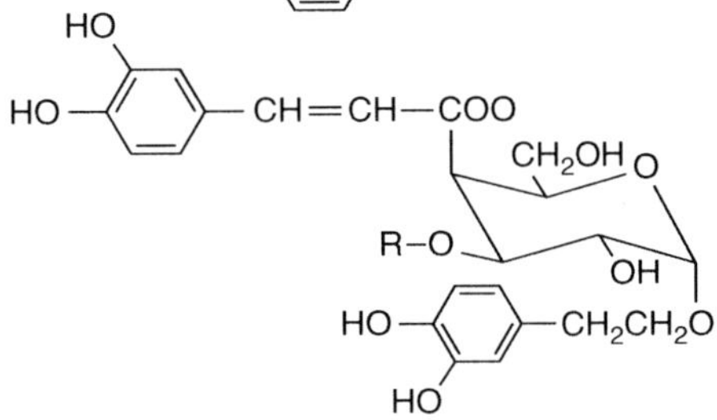
Apigenin-7-glucoside	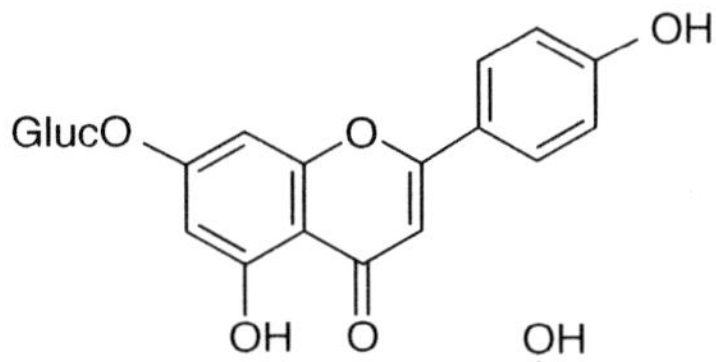
Lutcolin-7-glucoside	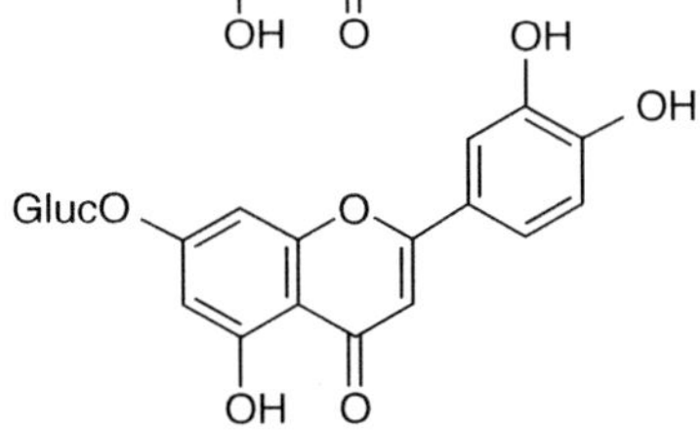

FIGURE 4.10 Main olive biophenols.

Apigenin-7-glucoside and luteolin-7-glucoside are two flavones present in relatively large amounts in olive leaves that possess healthy effects. Thus, the former is used to fight Alzheimer's [244] and liver diseases [56], and the latter to avoid the abnormal proliferation of aortic vascular smooth muscle cells, which is a frequent cause of pathogenesis such as atherosclerosis and restenosis [245].

The growing importance of OBPs warrants the development of new methods based on updated techniques for transfer to the industry with a view to their large-scale production in foods, drugs, and cosmetics.

4.3.3 Main Raw Materials and Methods for Obtaining Olive Phenols

4.3.3.1 Olive Leaves

Olive leaves are an agricultural residue resulting from the pruning of olive trees; however, they may also be considered an industrial by-product as they find their way to the olive mill in substantial amounts (around 10% of the total weight of olives arriving to mills [246]). The scientifically proven healthy properties of olive leaf extracts are leading to the filing of new patents for functional foods or cosmetics based on them on a nearly daily basis [247–251].

4.3.3.1.1 Olive Leaf Varieties

Hundreds of olive tree varieties have been selected over centuries for their adaptation to different microclimates and soil types. Some cultivars are typical of a specific zone, whereas others can be found in several countries. Regarding names, the same name is in some cases given to similar but clearly different varieties; also, different names are occasionally used for identical varieties.

Properly classifying varieties and cultivation zones of olives, which also includes the tree and its different parts, has emerged as a new problem in the process of controlling the quality and designation of the origin of olive oils as each variety and cultivation zone results in a different biophenolic composition [252–254]. With the aim of solving these problems, chemometric methods such as principal component analysis, hierarchical cluster analysis, K nearest neighbors, and soft independent modelling of class analogy have been used to differentiate and classify both varieties of olive trees cultivated in the same geographical zone and samples of the same variety of olive trees (arbequina) cultivated in zones with a different climate. Peak areas obtained by high-performance liquid chromatography–diode array detector (HPLC–DAD) analysis of the major BPs in olive leaves, extracted with microwave assistance for expeditiousness, have been used as chemotaxonomic markers to construct appropriate chemometric models [255]. The relative concentration of BPs from olive leaves in each olive variety and cultivation zone can be useful with a view to ensuring traceability in olive oil.

4.3.3.1.2 Conventional Methods for Extracting Biophenols from Olive Leaves

The characteristics of laboratory-scale extraction methods (e.g., time, extraction efficiency, extractant nature, working temperature) are poorly documented as researchers in this field have so far focussed on the determination of oleuropein and its derivatives [256,257], and the antioxidant activity of the phenolics extracted [231], but not on optimizing the extraction process itself. The extraction methods reported so far are complex and take hours or even days to complete. Sequential extraction of powdered leaves with 1:1 methanol–water, chloroform, and ethyl acetate was attempted by Savournin et al. [256]. In any case, such toxic extractants should be avoided if the products are ultimately aimed at human use.

Briante et al. [258] used a simple maceration method to extract BPs from leaves (15 min extraction with a 1:1 ethanol–water mixture) and obtained large amounts of hydroxytyrosol by biotransforming the extract with a partially purified hyperthermophilic β-glycosidase enzyme for 15 h.

Other nonphenolic compounds have also been extracted from olive leaves; such is the case with terpenic acids including oleanolic, ursolic, and maslinic acids, which were extracted by Albi et al. [259]. Both maceration and Soxhlet extraction using hexane as extractant were used to remove terpenic dialcohols, triglycerides, α-tocopherol, ester waxes, squalene, and β-carotene, among other substances [260]. The extraction efficiency was similar, the differences between the two methods never exceeding 5%.

4.3.3.1.3 Present Methods for Extracting Biophenols from Olive Leaves

The use of auxiliary energies and less toxic extractants for this purpose has been explored in continuous, batch, and batch–continuous operation. A continuous approach for ultrasound-assisted extraction using ethanol–water as extractant, which is less toxic than those used in conventional extractions, was optimized by using multivariate methodology with the experimental setup shown in Figure 4.11 (100% extraction efficiency was achieved in 25 min) [261]. The use of microwaves in batch operation to accelerate the process ensured complete extraction of the target analytes (namely, oleuropein, verbacoside, apigenin-7-glucoside, and luteolin-7-glucoside) within 8 min [262]. No degradation of the target compounds was caused by any of the energy types used to improve extraction.

A static–dynamic superheated extraction approach (shown in Figure 4.12) has also been used to extract BPs from olive leaves. A time of 13 min sufficed to extract up to 23 g/kg oleuropein with no degradation of the target analytes under the working conditions used (140°C and 6 bar).

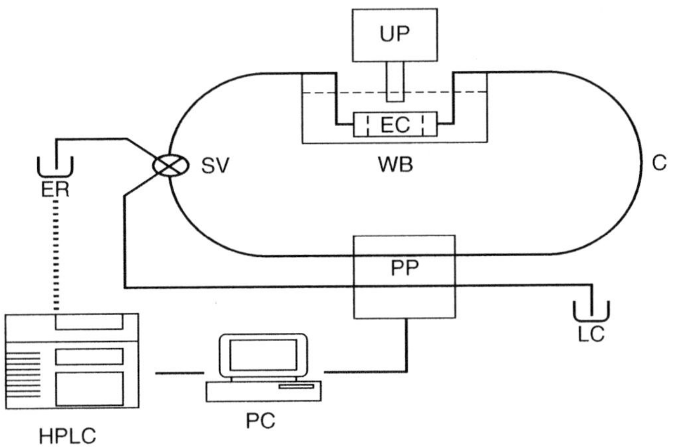

FIGURE 4.11 Experimental setup for the dynamic ultrasound-assisted extraction of phenolic compounds from olive leaves and alperujo (C, extraction coil; EC, extraction chamber; ER, extract reservoir; HPLC, high-performance liquid chromatography; LC, leaching carrier; PC, personal computer; PP, peristaltic pump; SV, selection valve; UP, ultrasonic probe; WB, thermostatic water-bath).

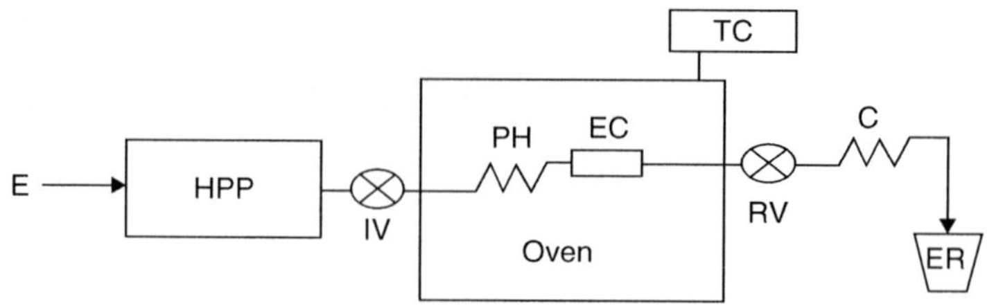

FIGURE 4.12 Experimental setup for the static–dynamic superheated liquid extraction of phenolic compounds from olive leaves and alperujo (C, cooler; E, extractant; EC, extraction chamber; ER, extract reservoir; HPP, high-pressure pump; IV, inlet valve; PH, preheater; RV, restrictor valve; TC, temperature controller).

4.3.3.1.4 Cleanup of Extracts, and Identification and Determination of Olive Leaf Biophenols

Cleanup is not an essential operation in the identification and determination of BPs from olive leaves. Solid-phase extraction (SPE) with four SPE cartridges (e.g., C18, C18 ec, C18 Hydra, and HR-P) and liquid–liquid extraction (LLE) with ethyl acetate as extractant were used to clean up and preconcentrate phenolic compounds in extracts. Both treatments proved ineffective owing to incomplete retention in SPE and poor partitioning in LLE [261]. Of the two polar sorbents HR-P and C18 Hydra, the latter was found to produce better results, which, however, were still worse than those obtained by direct injection of the extract into the chromatograph as shown in Figure 4.13.

Low-pressure liquid and thin layer chromatographies have been used with success to collect and examine, respectively, BPs fractions from the extracts [256].

One unsual contribution as regards current technologies for isolating BPs in extracts from olive leaves is that of a method based on countercurrent SFE with CO_2 containing ethanol as modifier (or a high pressure instead), by fractionating the raw extract obtained by macerating olive leaves with hexane. Although the system was optimized by using multivariate methodology, separation was poor, and waxes and α-tocopherol remained in the raffinate fraction, whereas hydrocarbons remained in the separator [263].

The individual separation of OBPs has most often been addressed using HPLC [231,257,261,262], and almost always C18 as solid phase [231,256,261]; by exception, one method used C8 and provided good results in all instances [257]. Gradient methods have frequently been used [231,257,261,262], but in some cases isocratic regimes have also been employed [256]. The mobile phases have been diluted with solutions of acetic acid [231,261] or ammonium acetate [257], methanol [256], or acetonitrile [231,257,261,262]. C18 guard columns have frequently been used to protect the chromatographic columns [231,256,257,261,262].

Phenolic compounds in olive leaves have usually been identified by comparing their retention times with those of appropriate standards or from their UV spectra as obtained with a diode array spectrometer.

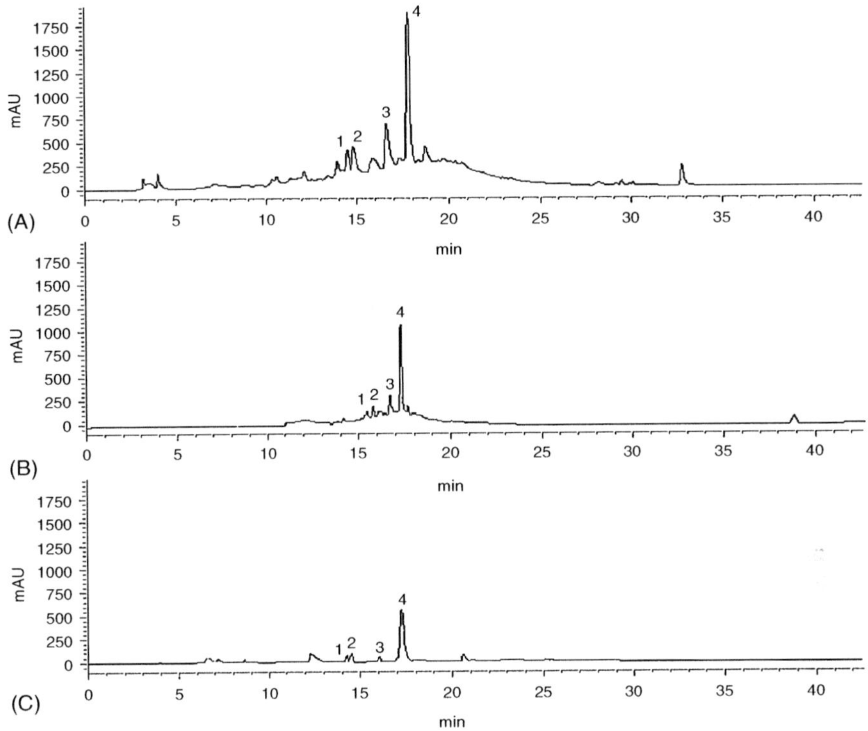

FIGURE 4.13 Chromatogram at 280 nm of an extract from olive leaves obtained under the optimal working conditions. (A) Injected directly in the liquid chromatograph. (B) After SPE with C_{18} Hydra cartridges. (C) Aqueous phase from LLE. Peak identification: 1, verbacoside; 2, luteolin-7-glucoside; 3, apigenin-7-glucoside; 4, oleuropein.

Other less frequently used techniques for the individual separation and identification of simple biophenols include gas chromatography (GC) with [258] and without prior derivatization [231,257,261], ^{13}C- and ^{1}H-NMR (nuclear magnetic resonance) spectroscopies [256], and liquid chromatography–mass spectrometry (LC-MS) [257].

Savournin et al. [256] used coumarin as internal standard for quantitation in their chromatographic method; however, if precision in the injections was high enough, no internal standard was necessary and BPs were quantified in area units.

4.3.3.2 Olive Oil Industry Wastes

The manufacturing process of olive oil has undergone revolutionary changes. Different olive oil production methods produce also different types of wastes depending essentially on the particular extraction technology favored in each country. The classic production method for olive oil generates three phases and two wastes, namely olive oil (20%), orujo (30%), and alpechín (50%). The use of a two-phase decanter to which no water is added produces oil and alperujo, the former being of higher quality than the oil obtained with the three-phase process as the overall aqueous volume separated from the oil is smaller, so the partition equilibrium enables a higher proportion of OBPs to remain in the oil fraction. In Spain, over 90% of olive oil mills operate with this system and produce 2.5–6 million tons of alperujo, depending on the crop [264].

4.3.3.2.1 Conventional Methods for Extracting Biophenols from OIW

LLE has traditionally been the technique of choice for extracting biophenols from olive oil industry wastes (OIW), using organic solvents as extractants.

Obied et al. [265] used aqueous mixtures of methanol, ethanol, ethyl acetate, *n*-propanol, acetonitrile, and acetone to extract BPs from alpechín by conventional LLe, blending with Ultra Turrax, and stirring with a magnetic bar. Although extraction was maximal with aqueous propanol and acetonitrile, they chose a 60:40 methanol–water mixture based on technical grounds.

Soxhlet extraction was compared with conventional stirring extraction (using ethanol, ethanol–water, methanol, and ethyl acetate as extractants) to remove BPs from alpechín, the first extractant being selected as it provided a higher extraction efficiency of verbacoside and its derivatives and also because ethanol (a food-safe solvent) is a better choice with a view to upscaling the process [266].

Tyrosol and other BPs have been extracted from spiked samples of alpechín by using liquid–liquid microextraction with ethyl acetate, which was preferred to ethyl ether, *n*-hexane, dichloromethane, and trichloromethane by virtue of its increased recoveries and low boiling point, which would facilitate its removal [267].

4.3.3.2.2 Present Methods for Extracting Biophenols from OIW

The use of auxiliary energies to accelerate or increase the extraction efficiency of OBPs from OIW has scarcely been explored even though ultrasound and superheated extractants have been found to result in improved extraction. An attempt to extract BPs from alperujo was made by using the continuous ultrasound-assisted method based on the approach illustrated in Figure 4.11, where the sample, held in the extraction chamber, was subjected to the action of an ultrasound probe; meanwhile, the extractant (1:3 methanol–water) was recirculated in a closed circuit to ensure adequate mass transfer to the liquid phase under optimal working conditions. The direction of the extractant was changed at 40 s intervals to minimize dilution of the extract and compaction in the extraction cell, which could result in overpressure in the dynamic system [268].

A static–dynamic superheated extraction approach has also been used to address the previous objective. With the approach depicted in Figure 4.12, only 27 min was needed for the quantitative extraction of the target BPs from alperujo (2.8 and 1.5 g/kg of hydroxytyrosol and tyrosol, respectively, were obtained) [269]. The higher extraction efficiency of this proposed method as compared with conventional methods can be ascribed to its ability to disrupt matrix–analyte interactions by effect of the high temperature and pressure used. Moreover, the surface tension and the viscosity of the extractant are reduced by the increased temperatures, an elevated pressure is used to maintain the solvent in the liquid state, leading to improved sample wetting and matrix penetration, and increasing mass transfer and enhancing extraction as a result [270]. No degradation of the target compounds under the optimal working conditions for extraction with both approaches was observed.

4.3.3.2.3 Cleanup of Extracts, and Identification and Determination of OBPs

Cleanup is usually required before individual separation and detection of the target analytes in the analysis of alpechín and alperujo to ensure adequate precision and sensitivity. However, these pretreatments are redundant with olive leaves [261,262].

The most usual treatment of the extracts is concentration to dryness under reduced pressure and subsequent reconstitution in methanol–water [255] or ethanol [266]. Priego-Capote et al. [268] subjected alperujo extracts to SPE treatments similar to those described in Section 4.2.3.4; however, they found direct analysis to provide better results.

Mulinacci et al. [266] rinsed the extract with hexane to remove the lipid fraction with a view to avoiding technical problems in the identification and determination steps.

Derivatization of the analytes is inevitable when GC-MS is used for identification and quantification of OBPs. A 4:1 mixture of *N-O*-bis(trimethylsilyl)-trifluoroacetamide (BSTFA) and pyridine was found to provide better results than other mixtures such as BSTFA diluted in ethyl acetate or BSTFA containing 1% trimethylchlorosilane (TMCS) [267].

LC has been used to separate and identify OBPs in these residues for their subsequent photometric diode array [261,262,265] or MS detection [266]. C18 columns ensured optimal separation in all cases.

Capillary electrophoresis (CE) has also been tested for the individual separation of target OBPs in the extracts. Tyrosol, oleuropein, and other 18 OBPs were thus identified and quantified by using a CE method optimized with multivariate methodology. The time required for individual separation by CE was only 11 min as compared with 1 h in HPLC.

REFERENCES

1. Trichopoulou, A. and Vasilopoulou, E., Mediterranean diet and longevity, *Br. J. Nutr.*, 84, 205, 2000.
2. Trichopoulou, A. and Lagiou, P., Healthy traditional Mediterranean diet—an expression of culture, history and lifestyle, *Nutr. Rev.*, 55, 383, 1997.
3. Kushi, L.H., Lenart, E.B., and Willett, W.C., Health implications of Mediterranean diets in light of contemporary knowledge. 1. Plant foods and dairy products. 2. Meats, wine, fats, and oils, *Am. J. Clin. Nutr.*, 61, S1407, 1995.
4. Gerber, M., Health benefits of the Mediterranean diet model, in: *Mediterranean Diet and Health. Current News and Prospects*, Agropolis, John Libbey, Paris, France, 3, 2001.
5. de Lorgeril, M. et al., Mediterranean alpha-linolenic acid-rich diet in secondary prevention of coronary heart disease, *Lancet*, 343, 1454, 1994.
6. de Lorgeril, M., Mediterranean diet in the prevention of coronary heart disease, *Nutrition*, 14, 55, 1998.
7. de Lorgeril, M. et al., Mediterranean diet, traditional risk factors and the rate of cardiovascular complications after myocardial infraction, Final Report of the Lyon Diet Heart Study, *Circulation*, 99, 779, 1999.

8. Marchioli, R. et al., Mediterranean dietary habits and risk of death after myocarcial infarction, *Circulation*, 102, 379, 2000.
9. de Lorgeril, M. and Salen, P., Diet as preventive medicine in cardiology, *Curr. Opin. Cardiol.*, 15, 364, 2000.
10. Reddy, K.S. and Katan, M.B., Diet, nutrition and the prevention of hypertension and cardiovascular diseases, *Public Health Nutr.*, 7, 167, 2004.
11. Kok, F.J. and Kromhout, D., Atherosclerosis. Epidemiological studies on the health effects of a Mediterranean diet, *Eur. J. Nutr.*, 43, I/2, 2004.
12. Feskens, E.J.M. et al., Dietary factors determining diabetes and impaired glucose tolerance, *Diabetes Care*, 18, 1104, 1995.
13. Biesalski, H.K., Diabetes preventive components in the Mediterranean diet, *Eur. J. Nutr.*, 43, I/26, 2004.
14. Choi, H.K., Dietary risk factors for rheumatic diseases, *Curr. Opin. Rheumatol.*, 17, 141, 2005.
15. Dryden, G.W., Song, M., and McClain, C., Polyphenols and gastrointestinal diseases, *Curr. Opin. Gastroenterol.*, 22, 165, 2006.
16. Tzonou, A. et al., Diet and cancer of the prostate: A case control study in Greece, *Int. J. Cancer*, 80, 704, 1999.
17. Trichopoulou, A. et al., Cancer and Mediterranean dietary traditions, *Cancer Epidemiol. Biomar.*, 9, 869, 2000.
18. La Vecchia, C., Mediterranean diet and cancer, *Public Health Nutr.*, 7, 965, 2004.
19. Keys, A., Coronary heart disease in seven countries, *Circulation*, 41, 1, 1970.
20. Keys, A. et al., *Seven Countries. A Multivariate Analysis of Death and Coronary Heart Disease*, Keys, A. (Ed.), Harvard University Press, Cambridge, MA, 1980.
21. Renaud, S. and de Lorgeril, M., Wine, alcohol, platelets and the French paradox for coronary heart disease, *Lancet*, 100, 1253, 1992.
22. Tunstall-Pedoe, H., Autres pays, autres moeurs. Theories on why the French have less heart disease than the British, *Br. Med. J.*, 297, 1559, 1998.
23. Criqui, M.H. and Ringel, B.L., Does diet or alcohol explain the French paradox? *Lancet*, 344, 1719, 1994.
24. Parodi, P.W., The French paradox unmasked: The role of folate, *Med. Hypotheses*, 49, 313, 1997.
25. Law, M. and Wald, N., Why heart disease mortality is low in France: The time lag explanation, *Br. Med. J.*, 318, 1471, 1999.
26. Yarnell, J.W.G. and Evans, A.E., The Mediterranean diet revisited—towards resolving the (French) paradox, *Q.J. Med.*, 93, 783, 2000.
27. de Lorgeril, M. et al., Mediterranean diet and the French paradox: Two distinct biogeographic concepts for one consolidated scientific theory on the role of nutrition in coronary heart disease, *Cardiovasc. Res.*, 54, 503, 2002.
28. Fuhrman, B., Lavy, A., and Aviram, M., Consumption of red wine with meals reduces the susceptibility to human plasma and low density lipoprotein to lipid peroxydation, *Am. J. Clin. Nutr.*, 61, 549, 1995.
29. Pearson, T.A., Alcohol and heart disease, *Circulation*, 94, 3023, 1996.
30. Rimm, E.B. et al., Review of moderate alcohol consumption and reduced risk of coronary heart disease: Is the effect due to beer, wine or spirits? *Br. Med. J.*, 312, 731, 1996.
31. Rotondo, S., Di Castelnuovo, A., and de Gaetano, G., The relationship between wine consumption and cardiovascular risk: From epidemiological evidence to biological plausibility, *Ital. Heart. J.*, 2, 1, 2001.
32. Wollin, S.D. and Jones, P.J.H., Alcohol, red wine and cardiovascular disease, *J. Nutr.*, 131, 1401, 2001.

33. Orgogozo, J.M. et al., Wine consumption and the elderly: A prospective community study in the Bordeaux area, *Rev. Neurol—Fr.*, 153, 185, 1997.
34. Obisesan, T. and Hirsch, R., Moderate wine consumption is associated with decreased odds of developing age-related macular degeneration in HNAES-1, *J. Am. Geriatr. Soc.*, 46, 1, 1998.
35. Curhan, G.C. et al., Prospective study of beverage use and the risk of kidney stones, *Am. J. Epidemiol.*, 143, 240, 1996.
36. Curhan, G.C. et al., Beverage use and risk for kidney stones in women, *Ann. Intern. Med.*, 128, 534, 1998.
37. Hirvonen, T. et al., Nutrient intake and use of beverages and the risk of kidney stones among male drinkers, *Am. J. Epidemiol.*, 150, 187, 1999.
38. Leitzman, M.F. et al., Prospective study of alcohol consumption patterns in relation to symptomatic gallstone disease in men. Alcoholism, *Clin. Exp. Res.*, 23, 835, 1999.
39. Tavani, A. et al., Alcohol intake and risk of cancers of the colon and rectum, *Nutr. Cancer*, 30, 213, 1998.
40. Renaud, S. et al., Alcohol and platelet aggregation: The Caerphilly prospective heart disease study, *Am. J. Clin. Nutr.*, 55, 1012, 1992.
41. Gaziano, J.M. et al., Moderate alcohol intake, increased levels of high-density lipoprotein and its subfractions, and decreased risk of myocardial infarction, *New Engl. J. Med.*, 329, 1829, 1993.
42. Goldberg, D.M., Hahn, S.E., and Parkes, J.G., Beyond alcohol: Beverage consumption and cardiovascular mortality, *Clin. Chim. Acta*, 237, 155, 1995.
43. Rimm, E.B. et al., Moderate alcohol intake and lower risk of coronary heart disease: Meta-analysis of effects on lipids and haemostatic factors, *Br. Med. J.*, 319, 1523, 1999.
44. Klatsky, A.L., Wine, alcohol and cardiovascular diseases, in: *Wine: A Scientific Exploration*, Sandler, M. and Pinder, R. (Eds.), Taylor & Francis, London, United Kingdom, 108, 2003.
45. Vogel, R.A., Vintners and vasodilators. Are French red wines more cardioprotective? *J. Am. Coll. Cardiol.*, 41, 479, 2003.
46. Hung, L.M. et al., Cardioprotective effect of resveratrol, a natural antioxidant derived from grapes, *Cardiovasc. Res.*, 47, 549, 2000.
47. van de Wiel, A. et al., Blessings of the grape, *Eur. J. Int. Med.*, 12, 484, 2001.
48. Battino1, M. and Ferreiro, M.S., Ageing and the Mediterranean diet: A review of the role of dietary fats, *Public Health Nutr.*, 7, 953, 2004.
49. Servili, M. and Montedoro, G., Contribution of phenolic compounds to virgen olive oil quality, *Eur. J. Lipid Sci. Technol.*, 104, 602, 2002.
50. Perona, J.S., Cabello-Moruno, R., and Ruiz-Gutiérrez, V., The role of virgen olive oil components in the modulation of endothelial function, *J. Nutr. Biochem.*, 17, 429, 2006.
51. Panza, F. et al., Mediterranean diet and cognitive decline, *Public Health Nutr.*, 7, 959, 2004.
52. Owen, R.W. et al., The antioxidant/anticancer potential of phenolic compounds isolated from olive oil, *Eur. J. Cancer*, 36, 1235, 2000.
53. Leger, C.L. et al., A thromboxane effect of a hydroxytyrosol-rich olive oil wastewater extracts in patients with uncomplicated type I diabetes, *Eur. J. Clin. Nutr.*, 59, 727, 2005.
54. Puertollano, M.A. and De Pablo, M.A., Clinical application of fatty acids as modulating agents in immune funtions, *Rev. Clínica Esp.*, 202, 215, 2002.
55. Berbert, A.A. et al., Supplementation of fish oil and olive oil in patients with rheumatoid arthritis, *Nutrition*, 21, 131, 2005.

56. Zheng, Q.S. et al., Mechanisms of apigenin-7-glucoside as a hepatoprotective agent, *Biomed. Environ. Sci.*, 18, 65, 2005.
57. Martínez, M.A. et al., Dietary virgin olive oil enhances secretagogue-evoked calcium signaling in rat pancreatic acinar cells, *Nutrition*, 20, 536, 2004.
58. Pérez-Granados, A.M., Vaquero, M.P., and Navarro, M.P., Calcium absorption in rats consuming olive oil or sunflower oil unused or used in frying, *J. Food Sci.*, 65, 892, 2000.
59. Jiménez, L. and Sánchez, I., Obtención de pastas celulósicas a partir de residuos agrícolas, *Inv. Téc. Papel*, 101, 514, 1989.
60. Kammerer, D. et al., Recovery of anthocyanins from grape pomace extracts (*Vitis vinifera* L. cv. Cabernet Mitos) using a polymeric adsorber resin, *Eur. Food Res. Technol.*, 220, 431, 2005.
61. Shrikhande, A.J., Wine by-products with health benefits, *Food Res. Int.*, 33, 469, 2000.
62. Kammerer, D. et al., Polyphenol screening of pomace from red and white grape varieties (*Vitis vinifera* L.) by HPLC–DAD–MS/MS, *J. Agric. Food Chem.*, 52, 4360, 2004.
63. Wulf, L.W. and Nagel, C.W., High-pressure liquid chromatographic separation of anthocyanins of *Vitis vinifera*, *Am. J. Enol. Vitic.*, 29, 42, 1978.
64. Mazza, G. and Miniati, E., Grapes, in: *Anthocyanins in Fruits, Vegetables and Grains*, CRC Press, London, United Kingdom, 149, 1993.
65. Vanden Heuvel, J.E., Timing of partial defoliation affects carbohydrate concentration of vegetative tissues and concentration of phenolics in berries of potted De Chaunac (*Vitis sp.*) grapevines, *Can. J. Plant Sci.*, 85, 703, 2005.
66. Larice J.L. et al., Composition anthocyanique des cépages. II-Essai de classification sur trois ans par analyse en composantes principales et étude des variations annuelles de cépages de meme provenance, *Rev. Fr. Oenol.*, 121, 7, 1989.
67. Rigaud J. et al., Caractérisation des flavonoides de la baie de raisin. Application à une étude terroir, in: *Oenologie 95*, Paris, France, 137, 1996.
68. Ojeda, H. et al., Influence of pre- and post-veraison water deficit on synthesis and concentration of skin phenolic compounds during berry growth of *Vitis vinifera* var. Shiraz, *Am. J. Enol. Vitic.*, 53, 261, 2002.
69. Di Stefano, R. and Maggiorotto, G., Anthocyanins, hydroxycinnamic acids and flavonols in berries, leaves, stems and shoots of grape vines, *Riv. Vitic. Enol.*, 48, 51, 1995.
70. Monagas, M. et al., Chemical characterization of commercial dietary ingredients from *Vitis vinifera* L., *Anal. Chim. Acta*, 563, 401, 2006.
71. Somers, T.C., The polymeric nature of wine pigments, *Phytochemistry*, 10, 2175, 1971.
72. Timberlake, C.F. and Bridle, P., Interactions between anthocyanins, phenolic compounds, and acetaldehyde and their significance in red wines, *Am. J. Enol. Vitic.*, 27, 97, 1976.
73. Bakker, J. and Timberlake, C.F., Isolation, identification, and characterization of new color-stable anthocyanins occurring in some red wines, *J. Agric. Food Chem.*, 45, 35, 1997.
74. Fulcrand, H. et al., A new class of wine pigments generated by reaction between pyruvic acid and grape anthocyanins, *Phytochemistry*, 47, 1401, 1998.
75. Luque-Rodríguez, J.M., Luque de Castro, M.D., and Pérez-Juan, P., Dynamic superheated liquid extraction of anthocyanins and other phenolics from red grape skins of winemaking residues, *Bioresour. Technol.*, 98, 2705, 2007. 10.1016/j.biortech.2006.09.019, 2006.
76. Amico, V. et al., Constituents of grape pomace from the Sicilian cultivar "Nerello Mascalese", *Food Chem.*, 88, 599, 2004.

77. Thorngate, J.H. and Singleton, V.L., Localization of procyanidins in grape seeds, *Am. J. Enol. Vitic.*, 45, 259, 1994.
78. Souquet, J.M. et al., Polymeric proanthocyanidins from grape skins, *Phytochemistry*, 43, 509, 1996.
79. Amrani-Joutei, K., Glories, Y., and Mercier, M., Localisation des tannins dans la pellicule de baie de raisin, *Vitis*, 33, 133, 1994.
80. Boukharta, M., Girardin, M., and Metche, M., Comparative study of polymeric proanthocyanidins from shoots and leaves of grape (*Vitis vinifera*), *Bull. Liaison-Groupe Polyphenols*, 15, 41, 1990.
81. Piretti, M.V., Ghendini, M., and Serrazanetti, G., Isolation and identification of the polyphenolic and terpenoid constituents of *Vitis vinifera*, *Ann. Chim.—Rome*, 66, 429, 1976.
82. Lee, C.Y. and Jaworski, A.W., Identification of some phenolics in white grapes, *Am. J. Enol. Vitic.*, 41, 87, 1990.
83. Boukharta, M., Girardin, M., and Metche, M., Procyanidines galloylées du sarment de vigne (*Vitis vinifera*). Séparation et identification par chromatographie en phase liquide à haute performance et chromatographie en phase gazeuse, *J. Chromatogr.*, 455, 406, 1988.
84. Da Silva, J.M.R. et al., Procyanidin dimmers and trimers from grape seeds, *Phytochemistry*, 30, 1259, 1991.
85. Prieur, C. et al., Oligomeric and polymeric procyanidins from grape seeds, *Phytochemistry*, 36, 781, 1994.
86. Foo, L.Y., Lu, Y., and Wong, H., Biphenyl-linked biflavanoids from grape pomace, *Phytochemistry*, 47, 1137, 1998.
87. Escribano-Bailón, M.T. et al., Procyanidin composition in skin and seed from grapes, *JIEP*, 94, 225, 1994.
88. Czochanska, Z., Foo, L.Y., and Porter, L., Compositional changes in lower molecular weight flavans during grape maturation, *Phytochemistry*, 18, 1819, 1979.
89. Vivas, N., Nonier, M.F., and Vivas De Gaulejac, N., Structural characterization and analytical differentiation of grape seeds, skins, stems and Quebracho tannins, *Bull. OIV*, 77, 643, 2004.
90. Boukharta, M., Girardin, M., and Metche, M., Oligomeric procyanidins of grapevine shoots of cultivar Carignan, *Vitis*, 33, 5, 1994.
91. Cantarelli, C. and Peril, C., The leucoanthocyanins in white grapes: Their distribution, amount, and fate during fermentation, *Am. J. Enol. Vitic.*, 15, 146, 1964.
92. Darne, G., Characterization and interpretation of different phases in the evolution of total phenols and leucoanthocyanins in the main shoots of *Vitis vinifera* L., *C.R. Seances Acad. Sci., Ser. D: Sci. Naturelles*, 284, 441, 1977.
93. Ribéreau-Gayon, P., Les composes phénoliques du raisin et du vin II. Les flavonosides et les anthocyanosides, *Ann. Physiol. Vég.*, 6, 211, 1964.
94. Cheynier, V. and Rigaud, J., HPLC separation and characterization of flavonols in the skins of *Vitis vinifera* var. Cinsault, *Am. J. Enol. Vitic.*, 37, 248, 1986.
95. Cheynier, V. and Rigaud, J., Identification et dosage de flavonols du raisin, *JIEP*, 442, 1986.
96. Revilla, E., Carpena, O., and Mataix, J., Leaf polyphenolic composition of two grapevine cultivars (Airen and Cencibel) during ripening, *An. Edafología Agrobiología*, 44, 787, 1985.
97. Souquet, J.M. et al., Phenolic composition of grape stems, *J. Agric. Food Chem.*, 48, 1076, 2000.
98. Lu, Y. and Foo, L.Y., The polyphenol constituents of grape pomace, *Food Chem.*, 65, 1, 1999.

99. Trousdale, E.K. and Singleton, V.L., Astilbin and engeletin in grapes and wine, *Phytochemistry*, 22, 619, 1983.
100. Hmamouchi, M. et al., Flavones and flavonols in leaves of some Moroccan *Vitis vinifera* cultivars, *J. Agric. Food Chem.*, 47, 186, 1996.
101. Ribéreau-Gayon, P., Identification of cinnamic acid esters of tartaric acid in the limbs and berries of *Vitis vinifera*, *Fr. Compt. Rend.*, 260, 341, 1965.
102. Vogel, F., Evolution of phenolic compounds in the leaves of grape *Vitis vinifera* L. cv. Chasselas during vegetation, *Schweiz. Landwirtsch. Forsch.*, 17, 57, 1978.
103. Santos-Buelga, C., Francia-Aricha, E.M., and Escribano-Bailón, M.T., Comparative flavan-3-ol composition of seeds from different grape varieties, *Food Chem.*, 53, 197, 1995.
104. Lin, T.Y., Ellagic acid precursors in muscadine grapes, *Zhonghua Nongxue Huibao*, 164, 70, 1993.
105. Karl, C., Mueller, G., and Pedersen, P.A., Ellagitannins from the leaves of *Vitis vinifera*, *J. Biosci.*, 38C, 13, 1983.
106. Bavaresco, L. et al., Stilbene compounds: From the grapevine to wine, *Drugs Exp. Clin. Res.*, 25, 57, 1999.
107. Langcake, P. and Pryce, R.J., The production of resveratrol and the viniferins by grapevines in response to ultraviolet irradiation, *Phytochemistry*, 16, 1193, 1977.
108. Douillet-Breuil, A.C. et al., Changes in the phytoalexin content of various *Vitis spp.* in response to ultraviolet C elicitation, *J. Agric. Food Chem.*, 47, 4456, 1999.
109. Keller, M., Steel, C.C., and Creasy, G.L., Stilbene accumulation in grapevine tissues: Developmental and environmental effects, *Acta Hortic.*, 514, 275, 2000.
110. Bavaresco, L., Role of viticultural factors on stilbene concentrations of grapes and wine, *Drugs Exp. Clin. Res.*, 29, 181, 2003.
111. Langcake, P. and Pryce, R.J., A new class of phytoalexins from grapevines, *Experientia*, 33, 151, 1977.
112. Pryce, R.J. and Langcake, P., α-Viniferin: An antifungal resveratrol trimer from grapevines, *Phytochemistry*, 16, 1452, 1977.
113. Bezhuashvili, M.G. et al., Stilbene tetramers from one-year grapevine shoots, *Bioorganicheskaya Khimiya*, 23, 979, 1997.
114. Waterhouse, A.L. and Lamuela-Raventos, R.M., The occurrence of piceid, a stilbene glucoside in grape berries, *Phytochemistry*, 37, 571, 1994.
115. Puessa, T. et al., Survey of grapevine *Vitis vinifera* stem polyphenols by liquid chromatography–diode array detection–tandem mass spectrometry, *J. Agric. Food Chem.*, 54, 7488, 2006.
116. Amalfitano, C. et al., Phenols and stilbene polyphenols in the wood of esca diseased grapevines, *Phytopathol. Mediterranea*, 39, 178, 2000.
117. Langcake, P., Cornford, C.A., and Pryce, R.J., Identification of pterostilbene as a phytoalexin from *Vitis vinifera* leaves, *Phytochemistry*, 18, 1025, 1979.
118. Jean-Denis, J.B., Pezet, R., and Tabacchi, R., Rapid analysis of stilbenes and derivatives from downy mildew-infected grapevine leaves by liquid chromatography–atmospheric pressure photoionisation mass spectrometry, *J. Chromatogr. A*, 1112, 263, 2006.
119. Angulo, V. and García, E., Characterization of wood from the vine shoots for paper use, *Vitic./Enol. Prof.* 2005, 96, 34–42.
120. Chakar, F.S. and Ragauskas, A.J., Review of current and future softwood Kraft lignin process chemistry, *Ind. Crop. Prod.*, 20, 131, 2004.
121. Maman, O. et al., Separation of phenolic aldehydes, ketones and acids from lignin degradation by capillary zone electrophoresis, *J. Chromatogr. A*, 755, 89, 1996.

122. Javor, T., Buchberger, W., and Tanzcos, I., Determination of low molecular mass phenolic and non-phenolic lignin degradation compounds in wood digestion solution by capillary electrophoresis, *Microchim. Acta*, 135, 45, 2000.
123. Ventre, J., The exploitation and utilization of grape marc, *Exp. Stn. Rec.*, U.S. Department of Agriculture, 46, 509, 1919.
124. Rabak, F. and Shrader, J.H., Commercial utilization of grape pomace and stems from the grape-juice industry, *Bur. Plant Ind.*, U.S. Department of Agriculture Bullentin, 952, 1, 1921.
125. Galoppini, C. and Catastini, C., Grape marc, a precious by-product of the enological industry, *Vignevini*, 4, 35, 1977.
126. Mazza, G., Anthocyanins in grapes and grape products, *Crit. Rev. Food Sci.*, 35, 341, 1995.
127. Hang, Y.D., Lee, C.Y., and Woodams, E.E., Solid-state fermentation of grape pomace for ethanol production, *Biotechnol. Lett.*, 8, 53, 1986.
128. Nurgel, C. and Canbas, A., Production of tartaric acid from pomace of some Anatolian grape cultivars, *Am. J. Enol. Vitic.*, 49, 95, 1998.
129. Da Porto, C., Cortella, G., and Freschet, G., Preliminary study on a cooling practice of grape pomace during storage on an industrial scale, *Ital. J. Food Sci.*, 16, 87, 2004.
130. Da Porto, C. and Freschet, G., Grape pomace storage on an industrial scale in containers: Effects of refrigeration on the volatile composition of grappa brandy phlegm, *Industrie delle Bevande*, 33, 331, 2004.
131. Flanzy, C., *Enología: Fundamentos científicos y tecnológicos*, 1st edition, AMV & Mundi Prensa, Madrid, Spain, 708, 2000.
132. Nogales, R., Cifuentes, C., and Benitez, E., Vermicomposting of winery wastes: A laboratory study, *J. Environ. Sci. Health, Part B: Pest. Food Contaminants Agric. Wastes*, B40, 659, 2005.
133. Hang, Y.D. and Woodams, E.E., Grape pomace: A novel substrate for microbial production of citric acid, *Biotechnol. Lett.*, 7, 253, 1985.
134. Hang, Y.D. and Woodams, E.E., Utilization of grape pomace for citric acid production by solid-state fermentation, *Am. J. Enol. Vitic.*, 37, 141, 1986.
135. Yazdanparast, R. and Bahrani, A., Citric acid production by Aspergillus niger isolated from the microflora of Iran, *J. Sch. Pharm.*, Teheran University of Medical Sciences and Health Services, 5, 34, 1996.
136. Saura-Calixto, F., Goni, I., Manas, E. and Abia, R., Condensed tannins and resistant protein as dietary fiber constituents, *Spec. Publ.—R. Soc. Chem.*, 83, 109, 1990.
137. Saura-Calixto, F. et al., Klason lignin, condensed tannins and resistant protein as dietary fiber constituents: Determination in grape pomaces, *Food Chem.*, 39, 299, 1991.
138. Valiente, C. et al., Grape pomace as a potential food fiber, *J. Food Sci.*, 60, 818, 1995.
139. Martín-Carrón, N. and Goni, I. Prior exposure of cecal microflora to grape pomaces does not inhibit in vitro fermentation of pectin, *J. Agric. Food Chem.*, 46, 1064, 1998.
140. Saura-Calixto, F., Antioxidant dietary fiber product: A new concept and a potential food ingredient, *J. Agric. Food Chem.*, 46, 4303, 1998.
141. Bravo, L. and Saura-Calixto, F., Characterization of dietary fiber and the in vitro indigestible fraction of grape pomace, *Am. J. Enol. Vitic.*, 49, 135, 1998.
142. Stredansky, M. and Conti, E., Xanthan production by solid state fermentation, *Process Biochem.*, 34, 581, 1999.
143. Saura-Calixto, F.D., Antioxidant dietary fiber, *Electron. J. Environ., Agric. Food Chem.*, 2, 2003.
144. Llobera, A. and Canellas, J., Dietary fibre content and antioxidant activity of Manto Negro red grape (*Vitis vinifera*): Pomace and stem, *Food Chem.*, 101, 659, 2007.

145. Lo Curto, R.B. and Tripodo, M.M., Yeast production from virgin grape marc, *Bioresour. Technol.*, 78, 5, 2001.
146. Igatuburce, J.M. et al., Study of agricultural by products. Extractability and amino acid composition of grape seed (*Vitis vinifera*) proteins, *J. Food Agric.*, 54, 489, 1991.
147. Marakakis, P., *Anthocyanins Food Colors*, Academic Press, New York, 1982.
148. Giusti, M.M. and Wrolstad, R.E., Acylated anthocyanins from edible sources and their applications in food systems, *Biochem. Eng. J.*, 14, 217, 2003.
149. Da Silva, J.M.R., Oxygen free radical scavenger capacity in aqueous models of different procyanidins from grape seeds, *J. Agric. Food Chem.*, 39, 1549, 1991.
150. Tebib, K., Besancon, P., and Rouanet, J.M., Dietary grape seed tannins affect lipoproteins, lipoprotein lipases and tissue lipids in rats fed hypercholesterolemic diets, *J. Nutr.*, 124, 2451, 1994.
151. Plumb, G.W. et al., Antioxidant properties of catechins and proanthocyanidins: Effect of polymerisation, galloylation and glycosylation, *Free Radic. Res.*, 29, 351, 1998.
152. Bomser, J.A. et al., Inhibition of TPA-induced tumor promotion in CD-1 mouse epidermis by a polyphenolic fraction from grape seeds, *Cancer Lett.*, 135, 151, 1999.
153. Sato, M. et al., Grape seed proanthocyanidin reduces cardiomyocyte apoptosis by inhibiting ischemia/reperfusion-induced activation of JNK-1 and C-JUN, *Free Radic. Biol. Med.*, 31, 729, 2001.
154. Van Balen, J., Recovery of anthocyanins and other phenols from converting grapes into wine, *M.S. Thesis*, University of California, Davis, CA, 1984.
155. Augustin, M.M., Étude de l'influence de certains facteurs sur les composes phénoliques du raisin et du vin, Ph.D. Thesis, University of Bordeaux II, 1986.
156. Groud, Ch., Suitability of vine shoots for the preparation of paper pulp, *Papier*, 24, 114, 1921.
157. Jiménez, S., Cartagena, M.C., and Arce, A., Tracking variables in crude and treated vine shoots digestion process, *Environ. Technol.*, 12, 349, 1991.
158. Jiménez, L. et al., Vine shoots as a raw material for the pulp and paper industry, *Revue ATIP*, 46, 85, 1992.
159. Jiménez, L., López, F., and Ferrer, J.L., Evaluation of agricultural residues for paper manufacture, *Tappi J.*, 76, 169, 1993.
160. Jiménez, L. et al., Cellulosic pulp from vine tendrils, *Afinidad*, 61, 194, 2004.
161. Jiménez, L. et al., Comparison of various pulping processes for producing pulp from vine shoots, *Ind. Crop Prod.*, 23, 122, 2006.
162. Carriere, A., Production of ethyl alcohol from vegetable matter, particularly from vine shoots, *Bull. Soc. Encouragement Ind. Natl.*, 126, 612, 1927.
163. Faucounau, M., Vine shoots for the production of ethyl alcohol, *Chimie Ind.*, 23, 1509, 1930.
164. de Fabregues y Soler, J.M., Estalella, T., and García, J., The utilization of grape-vine shoots and marc, *Bol. Inst. Nac. Invest. Agron.*, 15, 531, 1955.
165. Puppo, S. and Settineri, D., Use of microbial fermentation to increase the nutritive value of agricultural cellulosic wastes, *Ann. Ist. Sper. Zootec.*, 18, 81, 1985.
166. Jiménez, L., Martin, A., and Bonilla, J.L., Kinetics of the acid hydrolysis of vine shoots, *An. Quím.*, 88, 175, 1992.
167. Giuliani, R., Vine shoots as a cattle feed, *Intern. Rev. Sci. Pract. Agric.*, 3, 836, 1925.
168. Jiménez, S., Cartagena, M.C., and Arce, A., Influence of lignin on the methanization of lignocellulosic wastes, *Biomass*, 21, 43, 1990.
169. Jiménez, L. and González, F., Study of the physical and chemical properties of lignocellulosic residues with a view to the production of fuels, *Fuel*, 70, 947, 1991.

170. Jiménez, L., Bonilla, J.L., and Ferrer, J.L., Exploitation of agricultural residues as a possible fuel source, *Fuel*, 70, 223, 1991.
171. Ganan, J. et al., Energetic exploitation of vine shoot by gasification processes, *Fuel Process. Technol.*, 87, 891, 2006.
172. Bustos, G. et al., Production of fermentable media from vine-trimming wastes and bioconversion into lactic acid by *Lactobacillus pentosus*, *J. Sci. Food Agric.*, 84, 2105, 2004.
173. Bustos, G. et al., Influence of the metabolism pathway on lactic acid production from hemicellulosic trimming vine shoots hydrolyzates using Lactobacillus pentosus, *Biotechnol. Prog.*, 21, 793, 2005.
174. Corcho-Corral, B. et al., Preparation and textural characterisation of activated carbon from vine shoots (*Vitis vinifera*) by H_3PO_4-chemical activation, *Appl. Surf. Sci.*, 252, 5961, 2006.
175. Corcho-Corral, B. et al., Development of activated carbon using vine shoots (*Vitis vinifera*) and its use for wine treatment, *J. Agric. Food Chem.*, 53, 644, 2005.
176. Valdés-Sánchez, E., Treatment of red Spanish wines with activated charcoal prepared from grapevine shoots: Evaluation of the effect on ochratoxin A content and polyphenol composition, *Enólogos*, 7, 46, 2005.
177. Romero, E. et al., Effect of vermicomposting process on pesticide sorption capability using agro-industrial wastes, *Int. J. Environ. Anal. Chem.*, 86, 289, 2006.
178. Billard, C. et al., Comparative antiproliferative and apoptotic effects of resveratrol, ε-viniferin and vine shoots derived polyphenols (vineatrols) on chronic B lymphocytic leukemia cells and normal human lymphocytes, *Leukemia Lymphoma*, 43, 1991, 2002.
179. Gupta, Y.K. and Briyal, S., Protective effect of vineatrol against kainic acid-induced seizures, oxidative stress and on the expression of heat-shock proteins in rats, *Eur. Neuropsychopharmacol.*, 16, 85, 2006.
180. Strack D., Phenolic metabolism, in: *Plant Biochemistry*, Dey, P.M. and Harborne, J.B. (Eds.), Academic Press, London, United Kingdom, 387, 1997.
181. Laparra, J. and Darne, G., Red grapevine leaves in phytotherapy: Study of the anthocyanin compounds in some pharmaceutical products, *Plantes Medicinales Phytother.*, 23, 129, 1989.
182. Kiesewetter, H. et al., Efficacy of orally administered extract of red vine leaf AS 195 (folia *Vitis viniferae*) in chronic venous insufficiency (stages I-II). A randomized, double-blind, placebo–controlled trial, *Arzneimittel-Forsch.*, 50, 109, 2000.
183. Schaefer, E. et al., Edema protective properties of the red vine leaf extract AS 195 (foli *Vitis viniferae*) in the treatment of chronic venous insufficiency: A 6-week observational clinical trial, *Arzneimittel-Forsch.*, 53, 243, 2003.
184. Suparno, O., Covington, A.D., and Evans, C.S., Kraft lignin degradation products for tanning and dyeing of leather, *J. Chemical Technol. Biotechnol.*, 80, 44, 2005.
185. Sales, F.G., Abreu, C.A.M., and Pereira, J.A.F.R., Catalytic wet-air oxidation of lignin in a three-phase reactor with aromatic aldehyde production, *Braz. J. Chem. Eng.*, 21, 211, 2004.
186. Guillén, M.D. and Ibargoitia, M.L., Volatile components of aqueous liquid smokes from *Vitis vinifera* L. shoots and *Fagus sylvatica* L. wood, *J. Sci. Food Agr.*, 72, 104, 1996.
187. Guillén, M.D. and Ibargoitia, M.L., Relationships between the maximum temperature reached in the smoke generation processes from *Vitis vinifera* L. shoot sawdust and composition of the aqueous smoke flavoring preparations obtained, *J. Agric. Food Chem.*, 44, 1302, 1996.
188. Guillén, M.D., Manzanos, M.J., and Ibargoitia, M.L., Carbohydrate and nitrogenated compounds in liquid smoke flavorings, *J. Agric. Food Chem.*, 49, 2395, 2001.

189. Luque-Rodríguez, J.M., Perez-Juan, P., and Luque de Castro, M.D., Extraction of polyphenols from vine shoots of *Vitis vinifera* by superheated ethanol–water mixtures, *J. Agric. Food Chem.*, 54, 8775, 2006.
190. Lancrenon, X., The problem of natural red dyes: Advances brought about in this field by the APPLEXION procedure, *Ind. Aliment. Agric.*, 95, 965, 1978.
191. Salgues, M., La matière colorante du raisin. Son extraction, sa purification en vue de son utilisation dans diverses industries, *Bull. OIV*, 53, 286, 1980.
192. Strack, D. and Wray, V., Anthocyanin glycosides, in: *Methods in Plant Biochemistry, Volume I. Plant Phenolics*, Harborne, J.B. (Ed.), Academic Press, New York, 325, 1989.
193. Bonilla, F. et al., Extraction of phenolic compounds from red grape marc for use as food lipid antioxidants, *Food Chem.*, 66, 209, 1999.
194. Giusti, M.M. et al., Anthocyanin pigment composition of red radish cultivars as potential food colorants, *J. Food Sci.*, 63, 219, 1998.
195. García-Viguera, C., Zafrilla, P., and Tomás-Barberán, F.A., Use of acetone as an extraction solvent for anthocyanins from strawberry fruit, *Phytochem. Anal.*, 9, 274, 1998.
196. Howard, L.R., Green processes to extract phytochemicals from grape processing waste, Abstracts of Papers, 231st ACS National Meeting, Atlanta, GA, March, 2006.
197. Shi, J., Extraction of polyphenolics from plant material for functional foods—engineering and technology, *Food Rev. Int.*, 21, 139, 2005.
198. Mantell, C., Martínez de la Ossa, E., and Rodríguez, M., Supercritical fluid extraction of anthocyanins from grape pomace, in: *Solvent Extraction for the 21st Century, Proceedings of ISEC '99*, Cox, M., Hidalgo, M., and Valiente, M. (Eds.), Barcelona, Spain, 1999.
199. Mantell, C., Rodríguez, M., and Martínez de la Ossa, E., Kinetics and mathematical modeling of anthocyanin extraction with carbon dioxide and methanol at high pressure, *Sep. Sci. Technol.*, 38, 3689, 2003.
200. Mantell, C., Rodríguez, M., and Martínez de la Ossa, E., A screening analysis of the high-pressure extraction of anthocyanins from red grape pomace with carbon dioxide and cosolvent, *Eng. Life Sci.*, 3, 38, 2003.
201. Palenzuela, B. et al., Bioguided extraction of polyphenols from grape marc by using an alternative supercritical-fluid extraction method based on a liquid solvent trap, *Anal. Bioanal. Chem.*, 378, 2021, 2004.
202. Chafer, A. et al., Supercritical fluid extraction and HPLC determination of relevant polyphenolic compounds in grape skin, *J. Sep. Sci.*, 28, 2050, 2005.
203. Escribano-Bailón, M.T. and Santos-Buelga, C., Polyphenol extraction from foods, in: *Methods in Polyphenols Analysis*, Santos-Buelga, C. and Williamson, G. (Eds.), RSC, Cambridge, United Kingdom, 2, 2003.
204. Piñeiro, Z., Palma, M. and Barroso, C.G., Determination of *trans*-resveratrol in grapes by pressurized liquid extraction and fast high-performance liquid chromatography, *J. Chromatogr. A*, 1110, 61, 2006.
205. Ju, Z. and Howard, L., Effect of solvent and temperature on pressurized liquid extraction of anthocyanins and total phenolics from dried red grape skin, *J. Agric. Food Chem.*, 51, 5207, 2003.
206. Ju, Z.Y. and Howard, L.R., Subcritical water and sulfured water extraction of anthocyanins and other phenolics from dried red grape skin, *J. Food Sci.*, 70, 270, 2005.
207. González-Rodríguez, J., Pérez-Juan, P., and Luque de Castro, M.D., Superheated liquids for extraction of solid residues from winemaking processes, *Anal. Bioanal. Chem.*, 377, 1190, 2003.
208. Kammerer, D. et al., A novel process for the recovery of polyphenols from grape (*Vitis vinifera* L.) pomace, *J. Food Sci.*, 70, C157, 2005.

209. González-Paramas, A., Flavanol content and antioxidant activity in winery byproducts, *J. Agric. Food Chem.*, 52, 234, 2004.
210. Shrikhande, A.J. et al., Process for extraction, purification and enrichment of polyphenolic substances from whole grapes, grape seeds and grape pomace, Patent No. U.S. 6544581-B1, 2003.
211. Frangi, E. et al., Process for preparing grape seed extracts enriched in procyanidol oligomers. U.S. Patent No. 5484594, 1996.
212. Angeletti, M. and Sparapani, L., Process for producing a grape seed extract having a low content of monomeric polyphenols, *PCT Int. Appl.*, 2005.
213. Palma, M. and Taylor, L.T., Extraction of polyphenolic compounds from grape seeds with near critical carbon dioxide, *J. Chromatogr. A*, 849, 117, 1999.
214. Palma, M. et al., Fractional extraction of compounds from grape seeds by supercritical fluid extraction and analysis for antimicrobial and agrochemical activities, *J. Agric. Food Chem.*, 47, 5044, 1999.
215. Palma, M., Piñeiro, Z., and Barroso, C.G., Stability of phenolic compounds during extraction with superheated solvents, *J. Chromatogr. A*, 921, 169, 2001.
216. Piñeiro, Z., Palma, M., and Barroso, C.G., Determination of catechins by means of extraction with pressurized liquids, *J. Chromatogr. A*, 1026, 19, 2004.
217. García-Marino, M. et al., Recovery of catechins and proanthocyanidins from winery by-products using subcritical water extraction, *Anal. Chim. Acta*, 563, 44, 2006.
218. Cho, Y.J. et al., Ultrasonication-assisted extraction of resveratrol from grapes, *J. Food Eng.*, 77, 725, 2006.
219. Sarkanen, K.V., Renewable resources for the production of fuels and chemicals, *Science*, 191, 773, 1976.
220. Goheen, D.W., Hydrogenation of lignin by the Noguchi process, in: *Lignin Structure and Reaction*, Marton, J. (Ed.), American Chemical Society, Washington DC, WA, 205, 1966.
221. Kirk, T.K. and Farell, R.L., Enzymatic "combustion": The microbial degradation of lignin, *Annu. Rev. Microbiol.*, 41, 465, 1987.
222. Suparno, O., Covington, A.D., and Evans, C.S., Biomimetic degradation of Kraft lignin, in: *Proceedings of the 8th Indonesian Students' Scientific Meeting*, Gautama, H. and Roseno, S. (Eds), ISTECS-Europe, 35, 2003.
223. Volf, I., Mamaliga, I., and Popa, V.I., Active principia with antioxidant character from vegetal by-products. II. Equilibrium study of the polyphenolic compounds extraction from wood of *Vitis* sp., *Cellulose Chem. Technol.*, 40, 211, 2006.
224. González-Rodríguez, J., Pérez-Juan, P., and Luque de Castro, M.D., Extraction of wood compounds by use of subcritical fluids, *Chromatographia*, 57, 363, 2003.
225. González-Rodríguez, J., Pérez-Juan, P., and Luque de Castro, M.D., Use of superheated liquids for the extraction of non-volatile compounds from wood: Liquid chromatography studies, *J. Chromatogr. A*, 1038, 3, 2004.
226. Bonoli, M. et al., Qualitative and semiquantitative analysis of phenolic compounds in extra virgin olive oils as a function of the ripening degree of olive fruits by different analytical techniques, *J. Agric. Food Chem.*, 52, 7026, 2004.
227. Petrov, V. and Manolov, P., Pharmacological studies on substances of plant origin with coronary dilatating and antiarrhythmic action, *Comp. Med. East-West*, 6, 123, 1978.
228. Iwai, K. et al., Appl. JP 2004-342612 20041126, *Jpn. Kokai Tokkyo Koho*, 2005.
229. Ucella, N., Olive biophenols: Biomolecular characterization, distribution and phytoalexin histochemical localization in the drupes, *Trends Food Sci. Technol.*, 11, 315, 2001.

230. Kubo, A., Lunde, C.S., and Kubo, I., Antimicrobial activity of the olive oil flavor compounds, *J. Agric. Food Chem.*, 43, 1629, 1995.
231. Benavente-García, O. et al., Antioxidant activity of phenolics extracted from *Olea europaea* L. leaves, *Food Chem.*, 68, 457, 2000.
232. Fabiani, R. et al., Cancer chemoprevention by hydroxytyrosol isolated from virgin olive oil through G1 cell cycle arrest and apoptosis, *Eur. J. Cancer Prev.*, 11, 351, 2002.
233. Furneri, P.M. et al., Antimycoplasmal activity of hydroxytyrosol, *Antimicrobial agents and chemotherapy*, 48, 4892, 2004.
234. D'Angelo, S. et al., Hydroxytyrosol, a natural antioxidant from olive oil, prevents protein damage induced by long-wave ultraviolet radiation in melanoma cells, *Free Radic. Biol. Med.*, 38, 908, 2005.
235. Maiuri, M.C. et al., Hydroxytyrosol, a phenolic compound from virgin olive oil, prevents macrophage activation, *Narunyn-Schmiedeberg's Arch. Pharmacol.*, 371, 457, 2005.
236. Kim, M.M. and Kim, S.N., App: KR 2000-27738 200000523, *Repub. Korean Kongkae Taeho Kongbo*, 2002.
237. Saenz, M.T. et al., Cytostatic activity of some compounds from the unsaponifiable fraction obtained from virgin olive oil, *Farmaco*, 53, 448, 1998.
238. Maimeskulova, L.A. and Maslov, L.N., Study of the antiarrhythmic effect of *Rhodiolae fluidum* extract and *n*-tyrosol on experimental arrhythmia models, *Eksp. Klin. Farmakol.*, 61, 37, 1998.
239. Maslova, L.V., Lishmanov, I.U.B., and Maslov, L.N., Cardioprotective effects of adaptogens of plant origin, *Bull. Eksp. Biol. Med.*, 115, 269, 1993.
240. Lee, H.S. et al., Isolation and biological activity of verbacoside, a potent inhibitor of protein kinase C from the calyx of *Campsis grandiflora, Yankhak Hoeji*, 37, 598, 1993.
241. Saracoglu, I. et al., Studies on constituents with citotoxic and cytostatic activity of two Turkish medicinal plants *Phlomis armeniaca* and *Scutellaria salviifolia*, *Biol. Pharm. Bull.*, 18, 1396, 1995.
242. Ji, L. et al., Differentiation of human gastric adenocarcinoma cell line MGc80-3 induced by verbacoside, *Plant Med.*, 63, 499, 1997.
243. Zheng, Q. et al., Effects of verbascoside and luteolin on oxidative damage in brain of heroin treated mice, *Pharmazie*, 60, 539, 2005.
244. Patil, S.C. et al., Protective effect of flavonoids against aging- and lipopolysaccharide-induced cognitive impairment in mice, *Pharmacology*, 69, 59, 2003.
245. Kim, T.J. et al., The inhibitory effect and mechanism of luteolin 7-glucoside on rat aortic vascular smooth muscle cell proliferation, *Arch. Pharm. Res.*, 29, 67, 2005.
246. Espínola, F., Cambios tecnológicos en la extracción del aceite de oliva virgen, *Alimentación Equipos Tecnología*, 21, 314, 1997.
247. Scheneder, L., Patent No. FR 2588159-A1, 1985.
248. Coll, D., Mathonnet, J.P., and Zannini, G., Patent No. FR 2792831-A1, 1999.
249. Shtukatur, I., No. WO 2004080203-A2, 2003.
250. Stueckler, F., No. WO 9948386-A1, 1998.
251. Amari, G., No. EP 0937455-A1, 1998.
252. Ranalli, A. et al., Incidence of olive cultivation area on the analytical characteristics of the oil, Note 1, *Riv. Ital. delle Sostanze Grasse*, 74, 501, 1997.
253. Sharma, R., Kaushal, B.B.L., Sharma, P.C., and Singh, R.P., Evaluation of olive (*Olea europaea* L.) cultivars grown in different regions of Himachal Pradesh for oil extraction, *Acta Hortic.*, 696, 559, 2005.
254. Cerretani, L. et al., Preliminary characterisation of virgin olive oils obtained from different cultivars in Sardinia, *Eur. Food Res. Technol.*, 222, 354, 2006.

255. Japón-Luján, R., Ruiz-Jiménez, J., and Luque de Castro, M.D., Discrimination and classification of olive tree varieties and cultivation zones by biophenol contents, *J. Agric. Food Chem.*, 54, 9706, 2006.
256. Savournin, C. et al., Rapid high-performance liquid chromatography analysis for the quantitative determination of oleuropein in olea europaea leaves, *J. Agric. Food Chem.*, 49, 618, 2006.
257. Agalias, A. et al., Quantitation of oleuropein and related metabolites in decoctions of Olea Europaea leaves from ten greek cultivated varieties by HPLC with diode array detection, *J. Liq. Chromatogr. Relat. Technol.*, 28, 1557, 2006.
258. Briante, R. et al., Production of highly purified hydroxytyrosol from olea europaea leaf extract biotransformed by hyperthermophilic β-glycosidase, *J. Biotechnol.*, 111, 67, 2004.
259. Albi, T., Guinda, A., and Lanzón, A., Procedure for isolation and determination of terpenic acids from the olive leaf (*Olea europaea*), *Grasas Aceites*, 52, 275, 2001.
260. Guinda, A. et al., Aislamiento y cuantificación de los componentes de la hoja del olivo: Extracto de hexano, *Grasas Aceites*, 53, 419, 2002.
261. Japón-Luján, R., Luque-Rodríguez, J.M., and Luque de Castro, M.D., Dynamic ultrasound-assisted extraction of oleuropein and related biophenols from olive leaves, *J. Chromatogr. A*, 1108, 76, 2006.
262. Japón-Luján, R., Luque-Rodríguez, J.M., and Luque de Castro, M.D., Multivariate optimisation of the microwave-assisted extraction of oleuropein and related biophenols from olive leaves, *Anal. Bioanal. Chem.*, 385, 753, 2006.
263. Tabera, J. et al., Countercurrent supercritical fluid extraction and fractionation of high-added-value compounds from a hexane extract of olive leaves, *J. Agric. Food Chem.*, 52, 4774, 2004.
264. Aragón, J.M. and Palancar, M.C., *Present and Future of Alpeorujo*, Editorial Complutense, S.A., Madrid, Spain, 2001.
265. Obied, H.K. et al., Investigation of Australian olive mill waste for recovery of biophenols, *J. Agric. Food Chem.*, 53, 9911, 2005.
266. Mulinacci, N. et al., Solid olive residues, *J. Agric. Food Chem.*, 53, 8963, 2005.
267. Zafra, A. et al., Determination of polyphenolic compounds in wastewater olive oil by gas chromatography–mass spectrometry, *Talanta*, 70, 213, 2006.
268. Priego-Capote, F., Ruiz-Jiménez, J., and Luque de Castro, M.D., Fast separation and determination of phenolic compounds by capillary electrophoresis–diode array detection. Application to the characterisation of alperujo after ultrasound-assisted extraction, *J. Chromatogr. A*, 1045, 239, 2004.
269. Japón-Luján, R. and Luque de Castro, M.D., Static–dynamic superheated liquid extraction of hydroxytyrosol and other biophenols from alperujo, unpublished data, 2006.
270. Morales-Muñoz, S., Luque-García, J.L., and Luque de Castro, M.D., Approaches for accelerating sample preparation in environmental analysis, *Crit. Rev. Environ. Sci. Technol.*, 33, 391, 2003.

5 Anthocyanins

Gary Takeoka and Lan Dao

CONTENTS

5.1 STRUCTURE AND CHEMISTRY

Anthocyanins, the largest group of water-soluble pigments in the plant kingdom, are responsible for most of the red, blue, and purple colors of fruits, vegetables, flowers, and other plant tissues (Harborne, 1967). The anthocyanins are all based chemically on a single aromatic structure, cyanidin, and all are derived from this compound by the addition or subtraction of hydroxyl groups, by the degree of methylation of these hydroxyl groups, and by the nature and number of sugars and their position on the aglycon (Harborne, 1998). Additionally, sugars can be esterified with aliphatic or aromatic acids leading to mono- and polyacylated anthocyanins. There are six common anthocyanins whose structures are shown in

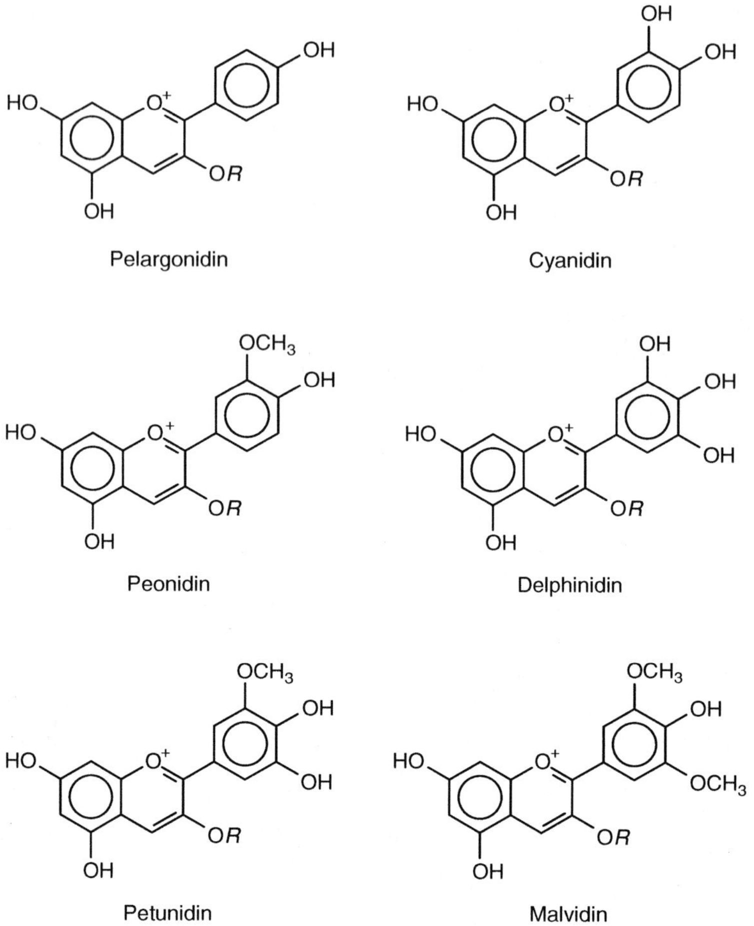

FIGURE 5.1 Structures of six common anthocyanins. R = sugar moiety. This is often glucose but may also be galactose, rhamnose, xylose, or arabinose.

Figure 5.1. With a few exceptions, anthocyanins are always glycosylated at C3. The most common sugar is glucose but arabinose, galactose, rhamnose, or xylose may also be present. Many di- and trisaccharides are known to occur with combinations of the previously mentioned sugars. Besides the C3 position, other sugars can also be attached at any one of the hydroxyls at C5, C7, C3′, C5′, and even C4′ (Brouillard, 1998). Sugars may be acylated with acetic, malonic, malic, oxalic, succinic, *p*-hydroxybenzoic, or hydroxycinnamic (*p*-coumaric, caffeic, ferulic, or sinapic) acids with acylation most commonly occurring with the sugar moiety at the C3 position. Acylation with aliphatic dicarboxylic acids such as malic, malonic, oxalic, and succinic acids renders the cationic anthocyanin a zwitterion that allows the distinction between these pigments and other anthocyanins by paper electrophoresis in a weakly acidic buffer (Harborne and Grayer, 1988). Another important feature of anthocyanins acylated with aliphatic acids is the labile nature of the ester bond. When malonated anthocyanins are subjected to standard extraction procedures using methanol acidified with 0.1%–1.0% HCl, there is an intermediate methyl ester formation at the free carboxyl group, while the main reaction is the loss of the malonyl group in a short time (Harborne, 1988).

An important distinction between anthocyanins and other flavonoids is not only that they are red, orange, violet, purple, or blue while most flavones and flavonols are yellow, but also that in aqueous solution they give rise to reactions such as proton transfer, isomerization, and tautomerization (Brouillard, 1988). Under the same conditions, the colorless flavonoids do not react and thus appear in only one chemical state. The structural transformations of anthocyanins in aqueous solution have been summarized by Brouillard (1988) as follows (Figure 5.2): At pHs below 2, anthocyanins exist primarily in the form of the red (glycosylated at C3) flavylium cations (AH^+) (1). Solvation of a flavylium salt in a slightly acidic or neutral aqueous solution results in the immediate formation of the neutral (2–4) or ionized quinonoidal bases (5–7). However, the common 3-glycosides and 3,5-diglycosides change more or less rapidly to the more stable carbinol (8) and chalcone pseudobases (9). Rapid and almost complete hydration of the flavylium cation (AH^+) occurs almost exclusively at position 2 to give the colorless carbinol pseudobase at pH values ranging from 3 to 6. This in turn can equilibrate, at a slower rate, to an open form, the chalcone pseudobase, which is also colorless.

Anthocyanins help to attract animals which lead to seed dispersal and pollination (Strack and Wray, 1994). There is evidence that anthocyanins may play a role in protecting plants against ultraviolet (UV)-induced damage (Bohm, 1998). They may also be beneficial to human health because of their possible role as dietary antioxidants and anti-inflammatory agents (Tsuda et al., 1994, 1998; Wang et al., 1997, 1999).

An impressive compilation of the qualitative and quantitative composition of anthocyanins in a wide variety of fruits, cereal grains, legumes, and vegetables was produced by Mazza and Miniati (1993).

Analytical methods for the isolation, separation, and characterization of anthocyanins have been described (Markham, 1982; Jackman et al., 1987; Harborne, 1998; Rivas-Gonzalo, 2003; Andersen and Francis, 2004). A comprehensive and highly recommended source for anyone involved in anthocyanin analysis is that of Strack and Wray (1989). Further details of flavonoid chemistry can be found in volumes of *The Flavonoids* series (Harborne et al., 1975; Harborne and Mabry, 1982; Harborne, 1988, 1994). Recent advances in flavonoid research are thoroughly described in the book *Flavonoids: Chemistry, Biochemistry and Applications*, edited by Andersen and Markham (2006). Extensive information on the occurrence of anthocyanins in various natural products reported after 1992 was presented by Andersen and Jordheim (2006). One of the most useful sources of current protocols of anthocyanin analysis is the *Handbook of Food Analytical Chemistry, Pigments, Colorants, Flavors, Texture, and Bioactive Food Components*, edited by Wrolstad et al. (2004). This book is a practical how to manual that contains detailed in-structions on the following topics: (1) extraction, isolation, and purification of anthocyanins, (2) characterization and measurement of anthocyanins by UV–Vis

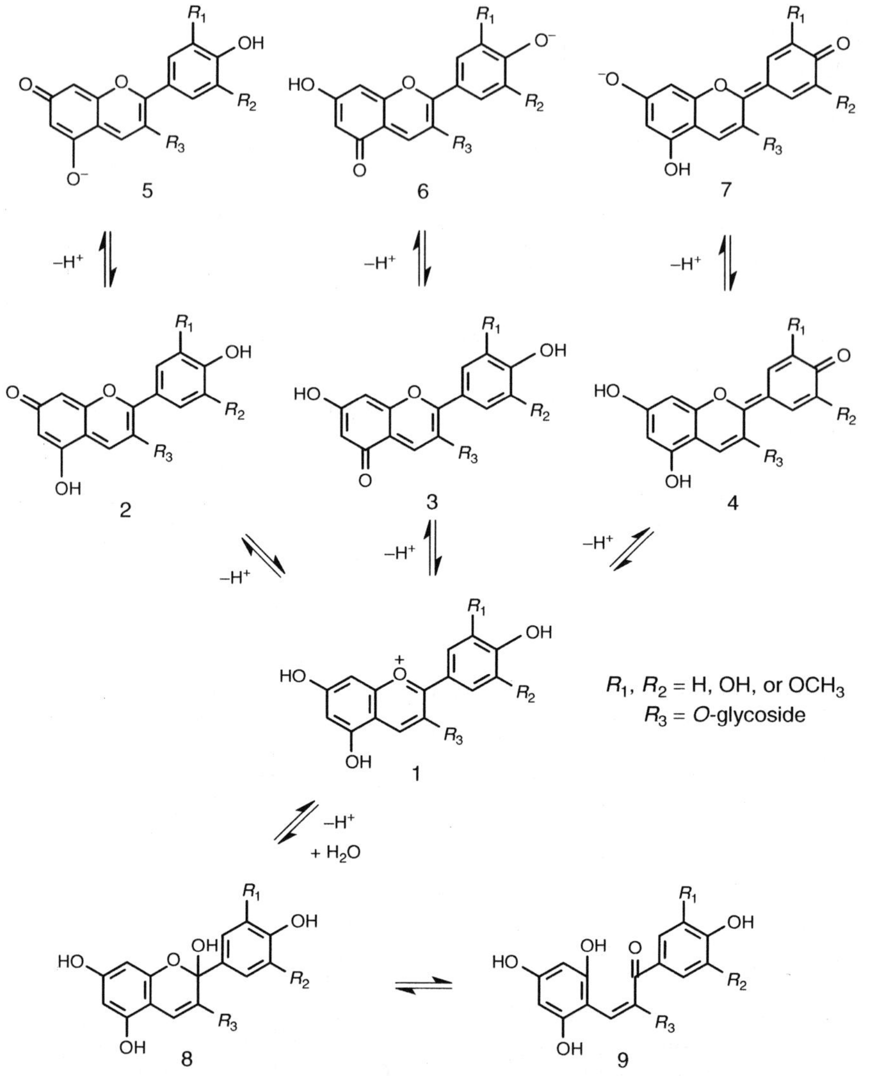

FIGURE 5.2 Structural transformations of anthocyanins in aqueous solution at varying pH. 1 = flavylium cation (red to orange color), 2–4 = neutral quinonoidal bases (purple to violet color), 5–7 = ionized quinonoidal bases (blue color), 8 = carbinol pseudobase (colorless), 9 = chalcone pseudobase (colorless). (Adapted from Brouillard, R., *The Flavonoids: Advances in Research Since 1980*, Chapman and Hall, London, United Kingdom, 1988, 525–538.)

spectroscopy, (3) separation and characterization of anthocyanins by high-performance liquid chromatography (HPLC), (4) characterization of anthocyanins by nuclear magnetic resonance (NMR).

Andersen and coworkers at the University of Bergen in Norway have examined the anthocyanin composition of many products and their generalized analytical

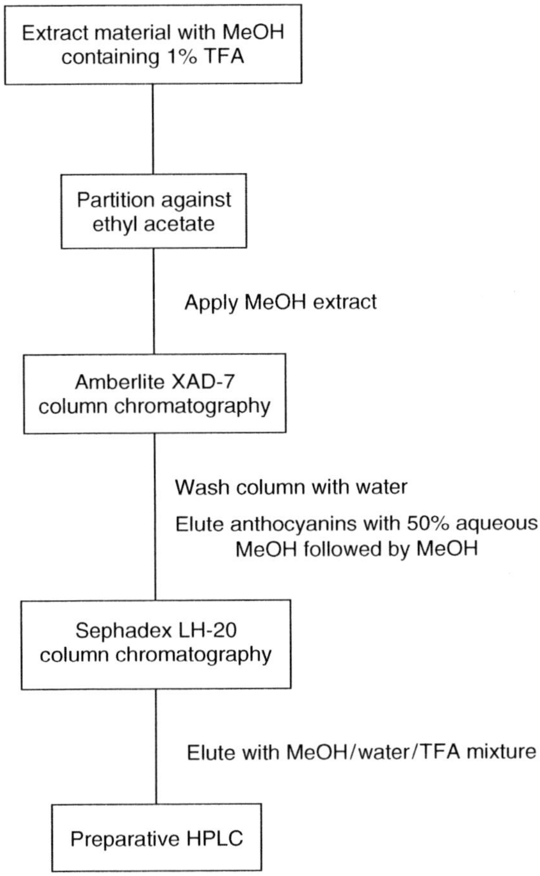

FIGURE 5.3 Generalized scheme for the isolation and purification of anthocyanins. (From Fossen, T., Rayyan, S., Holmberg, M.H., Nateland, H.S., and Andersen, Ø.M., Acylated anthocyanins from leaves of *Oxalis triangularis, Phytochemistry*, 2005, 66, 1133–1140.)

approach is shown in Figure 5.3. The material of interest is extracted with methanol containing 1% trifluoroacetic acid (TFA). The extract is partitioned against ethyl acetate to remove interfering phenolic acids and flavonoids. The methanolic extract is applied to an Amberlite XAD-7 column. The column is washed with water and the anthocyanins are eluted with 50% aqueous methanol and 100% methanol (both containing 0.1%–0.5% TFA). The sample is then fractionated on a Sephadex LH-20 gel filtration column eluting with a MeOH–water–TFA mixture. The homogeneity of the separated anthocyanins is checked by thin-layer chromatography (TLC) and analytical HPLC. If further purification is required, preparative HPLC on a reversed-phase C-18 column is used. Characterization of purified anthocyanins is achieved by ^{1}H- and ^{13}C-NMR spectroscopy. Further details of each step are provided in the following sections.

5.1.1 Authentic Standards

Delphinidin 3-glucoside and cyanidin 3-glucoside can be isolated from black currant (*Ribes nigrum*) while whortleberry (*Vaccinium myrtillus*) and strawberry (*Fragaria ananassa*) are sources of petunidin 3-glucoside and pelargonidin 3-glucoside, respectively (Kidøy et al., 1997). A limited variety of anthocyanins is available from Indofine Chemical Company (Somerville, NJ), Funakoshi Co. (Tokyo, Japan), and Extrasynthese S.A. (Genay Cedex, France), while a more extensive selection is available from Polyphenols AS, Hanabryggene Technology Center, Hanaveien 4–6, N-4327 Sandnes, Norway (www.polyphenols.com).

5.2 ISOLATION

5.2.1 Extraction

Since anthocyanins are reactive compounds, it is recommended that plant material be frozen with liquid nitrogen and powdered using a Waring blender or mill suitable for use under extremely low temperatures (Wrolstad et al., 2002). Anthocyanin degradation is minimized by the low temperature and nitrogen environment. The fine powder helps to maximize pigment recovery because of its high surface area. Because of their instability in neutral or basic solutions, anthocyanins have been most commonly extracted with methanol or ethanol containing HCl (approximately, 0.01%–1%). Labile acylated anthocyanins, particularly those containing aliphatic acids, that is, malonic acid, should be extracted with weak acids such as acetic (Harborne and Boardley, 1985; Gläßgen et al., 1992a, b; Markham, 1996), tartaric (Philip, 1974), or citric acids (Main et al., 1978; Strack et al., 1986; Fossen and Andersen, 1997a, b). Researchers performing chemotaxonomic studies have observed that the extraction of some acylated anthocyanins under acidic conditions may result in their partial or total hydrolysis (Van Sumere et al., 1985; Van Wyk and Winter, 1994). Similarly, Revilla et al. (1998) have shown that the use of solvent containing up to 1% of 12 N HCl caused the partial hydrolysis of some acylated anthocyanins during extraction. Artifact formation has also been observed using methanol containing 2% formic acid as the extracting solvent (Bakker and Timberlake, 1985). It is recommended that extraction protocols using organic and mineral acids be tested against extraction procedures that use neutral solvents to verify that anthocyanins are not hydrolyzed during the extraction step. TFA (boiling point 72.4°C) has also been used and has the advantage that it is easily removed by evaporation under reduced pressure. Kondo et al. (1985) employed 3% aqueous TFA to isolate malonated pigments from *Monarda didyma*.

Wrolstad and coworkers (Giusti and Wrolstad, 1996; Giusti et al., 1999a) have developed and used a novel method to extract anthocyanins from various foods. The method involves powdering the sample with liquid nitrogen followed by blending with one volume of acetone. The sample is then filtered on a Buchner funnel using Whatman No. 1 paper. The filter cake residue is re-extracted with aqueous acetone (30:70, v/v) until a clear solution is obtained. The combined filtrates are shaken in a separatory funnel with chloroform (acetone/chloroform, 1:2, v/v) and stored overnight at 1°C. The upper aqueous layer is collected and residual acetone is removed by rotary evaporation. The aqueous extract is then made up to a known volume with distilled water. The aqueous extract is free of lipids, carotenoids, and chlorophyll pigments. This method prevents hydrolysis of acyl substituents that may occur when using methanolic HCl. Wrolstad et al. (2002) compared the efficiency of the acetone/chloroform method with the acidified methanol extraction procedure for 20 samples of juices, extracts, and dried powders derived from cranberries, elderberries, chokeberries, and bilberries. Anthocyanin recoveries were higher for all samples, except one, with the use of acetone. Recoveries were typically twice as high with the acetone extraction. The authors speculated that the higher recoveries were due to the lower temperature during the concentration step and the extraction efficiency of aqueous acetone. The use of acetone as an extraction solvent for anthocyanins (in strawberries) has also been investigated by Garcia-Viguera et al. (1998) who compared its efficiency to that of classical acidified aqueous or methanolic solvents (i.e., 3% formic acid in methanol, 3% formic acid in water, and methanol/acetic acid/water, 25:1:24). Both acetone and 3% formic acid in methanol gave clear extracts while the other solvents (3% formic acid in water and methanol/acetic acid/water, 25:1:24) produced turbid extracts that required multiple filtrations. This is probably due to the fact that water extracts extraneous materials including pectins (Timberlake and Bridle, 1980). Acetone gave more reproducible extractions (lower standard deviations) than the other solvents tested. The less efficient extractions with the other solvents may be due to more difficult sample preparation and filtration, rather than due to differences in extraction efficiencies.

5.3 METHODS OF SEPARATION AND PURIFICATION

5.3.1 Paper Chromatography

Paper chromatography (PC) offers the advantages of being inexpensive and being able to work with crude extracts. Additionally, the R_f values on paper are more reproducible than by TLC and identification (at least tentative) can often be achieved by comparison with published values of known standards (Harborne, 1967). However, since R_f values can vary depending on a number of factors such as temperature, composition of the developing solvent, and amount of material spotted, it is preferable to run standards along with the sample. It is quite convenient to carry out the separations on filter paper, typically Whatman No. 1. In preparative work, Whatman No. 3 filter paper has been widely used for the separation of individual anthocyanins. Descending PC is a useful method since the developing solvent can be easily overrun if desired (Harborne, 1998). The most popular solvent system consists of a mixture of butanol/acetic acid/water (4:1:5; BAW). This mixture exists as a two-phase system that is prepared in a separatory funnel. The upper phase is used to develop the paper while the lower phase is used to saturate the atmosphere in the chromatography chamber. The Forestal solvent consisting of acetic acid/concentrated HCl/water (30:3:10) has also been widely used. It was originally developed for the separation of anthocyanidins (Bate-Smith, 1954) for which it gives efficient separations. It possesses the additional advantage of providing increased anthocyanidin stability compared to the previously mentioned system. Two other systems that have

TABLE 5.1
R_f Values of Common Anthocyanins and Anthocyanidins by Paper Chromatography

Constituent	R_f Values (×100) in			
	BAW	BuHCl	1% HCl	Forestal
Pelargonidin 3-glucoside	44	38	14	
Cyanidin 3-glucoside	38	25	17	
Peonidin 3-glucoside	41	30	09	
Delphinidin 3-glucoside	26	11	03	
Petunidin 3-glucoside	35	14	04	
Malvidin 3-glucoside	38	15	06	
Pelargonidin	80			68
Cyanidin	68			49
Peonidin	71			63
Delphinidin	42			32
Petunidin	52			46
Malvidin	58			60

Source: Data from Harborne, J.B., *Comparative Biochemistry of the Flavonoids*, Academic Press, London, United Kingdom, 1967.

Note: Solvents are BAW, *n*-butanol/acetic acid/water (4:1:5, upper layer); BuHCl, *n*-butanol/HCl (1:1, upper layer); 1% HCl, water/concentrated HCl (97:3); and Forestal, acetic acid/concentrated HCl/water (30:3:10).

been used in anthocyanin analysis include butanol/2 M HCl (1:1, top layer; BuHCl) and 1% HCl (concentrated HCl/H_2O, 3:97). In general, as the number of sugar units present increases, anthocyanins have decreasing R_f values in BAW and BuHCl and increasing values in 1% HCl. In contrast, addition of acyl groups results in a decrease in R_f values in 1% HCl and increasing values in BAW and BuHCl (Harborne, 1967). Spot color provides information regarding the identity of anthocyanin glycosides since pelargonidin glycosides are orange, cyanidin and peonidin glycosides magenta, and delphinidin glycosides mauve (Harborne, 1998). Examination under UV light is useful since certain anthocyanins, that is, the 3,5-diglycosides of pelargonidin, peonidin, and malvidin, are distinguished from the 3-glycosides by their distinct fluorescence (Strack and Wray, 1989). However, Ribéreau-Gayon (1972) has noted that diglycosides of other anthocyanidins also show weak fluorescence on strongly acid chromatograms. Table 5.1 lists the R_f values of some common anthocyanins and their aglycons.

5.3.2 Thin-Layer Chromatography

Thin-layer chromatography (TLC) has several advantages over PC. It is more rapid, gives better resolution, has greater sensitivity, and is more versatile that a variety of adsorbents may be used. A useful mixture for the separation of anthocyanidins on silica gel is ethyl acetate/formic acid/2 M HCl (85:6:9) (Harborne, 1967). The use of

TABLE 5.2
R_f Values of Common Anthocyanidins by Thin-Layer Chromatography (TLC)

Constituent	R_f Values (×100) in BAW	Forestal
Pelargonidin	82	61
Cyanidin	68	45
Peonidin	71	63
Delphinidin	42	30
Petunidin	52	43
Malvidin	56	58

Source: Data on microcrystalline cellulose from Strack, D., Wray, V., *Methods in Plant Biochemistry*, Academic Press, New York, 1989, 325–356.

Note: Solvents are BAW, *n*-butanol/acetic acid/water (4:1:5, upper layer) and Forestal, acetic acid/concentrated HCl/water (30:3:10).

standard-grade silica gel is recommended since the trace metals present aid the separation of peonidin and malvidin from cyanidin and delphinidin (Harborne, 1998). Preparative TLC of anthocyanins using 20 × 20 cm chromatoplates with 1-mm layers of a mixture of 2/3 silica gel (adsorbosil-2) and 1/3 cellulose powder (MN-300, gypsum free) was described by Asen (1965). He employed the following solvent systems to purify milligram quantities of anthocyanins from plant tissues: *n*-butanol/water (1:1), water/HCl/formic acid (8:4:1), 1% HCl, and acetone/0.5 N HCl (1:3). Table 5.2 lists R_f values of some common anthocyanidins on microcrystalline cellulose (Strack and Wray, 1989). The simultaneous analysis of anthocyanidins and anthocyanins on cellulose layers using a solvent system of concentrated HCl/formic acid/water (24.9:23.7:51.4) has been demonstrated by Andersen and Francis (1985).

5.3.3 Column Chromatography

Column chromatography has long been an important tool in isolating natural products. Crude extracts are typically subjected to column chromatography before applying HPLC. Two widely used packings used to fractionate and isolate anthocyanins are Amberlite XAD-7 and Sephadex LH-20. The typical procedure is straightforward. A crude extract is loaded onto the top of a column of adsorbent and compounds are eluted by varying the polarity of the solvent.

5.3.3.1 Amberlite Polymeric Adsorbents

Amberlite XAD-7 is a nonionic moderately polar acrylic resin that has been used to purify anthocyanins. Lacking ionic groups, it presumably separates compounds through hydrophobic and polar interactions. Andersen (1988) employed Amberlite XAD-7 column chromatography (18 × 2.6 cm) as a fractionation step (to eliminate polar nonphenolic compounds) in his elucidation of anthocyanins from *Dacrycarpus dacrydioides*. After extracts had been loaded onto the column, it was washed with 2 L H_2O followed by 300 mL each of 50% aqueous MeOH and anhydrous MeOH (both containing 0.5% CF_3COOH) to elute the anthocyanins. In their investigation of anthocyanins from the flowers of Crocus, Nørbæk and Kondo (1998) applied extracts onto an Amberlite XAD-7 column first washing with 0.5% TFA and then eluting anthocyanins stepwise with 4%–20% aqueous CH_3CN containing 0.5% TFA. Catalano et al. (1998) applied a concentrated methanolic extract of the flowers of *Vicia villosa* (first purified by partition [three times] against 0.3 L of EtOAc) to an Amberlite XAD-7 column. The anthocyanins were partially purified by washing with about 3 L of water before being eluted from the resin with about 1 L of methanol/acetic acid (19:1, v/v).

5.3.3.2 Sephadex Gel Column Chromatography

Sephadex is a crosslinked dextran gel with a narrow molecular weight range. A particularly useful material in anthocyanin purification is Sephadex LH-20 (Saito et al., 1985). Sephadex LH-20 is produced by treating Sephadex G-25 with propylene oxide which replaces each hydroxyl hydrogen in the polymer with a 2-propanol group. This treatment results in a gel with increased hydrophobicity. Because of its dual lipophilic and hydrophilic nature, Sephadex LH-20 swells in a range of solvents from weak and medium polarity to strongly polar (Henke, 1995). This material, designed to be used with organic solvents and water/solvent mixtures, possesses an exclusion limit (at maximum swelling) of 4000 (Henke, 1995). Column chromatography on this packing has been used to fractionate crude extracts and has additionally been useful for the purification of individual anthocyanins (Andersen, 1988). Andersen and coworkers (Fossen and Andersen, 1997a, b; Cabrita and Andersen, 1999; Torskangerpoll et al., 1999) have used both isocratic ($MeOH/H_2O/TFA$, 49.5:50:0.5 or $H_2O/MeOH/TFA$, 80:19.6:0.4, v/v) and step elution (20%–60% $MeOH/H_2O$ [0.1% TFA]) on Sephadex LH-20 columns (100 × 5.0 cm or 100 × 1.0 cm) to purify anthocyanins. Bakker and Timberlake (1997) used Sephadex LH-20 column chromatography as their first fractionation step in their isolation and purification of anthocyanins from red wines. Wine (concentrated by rotary evaporation) was applied to a 60 × 5 cm column packed with Sephadex LH-20. The column was eluted with up to 2500 mL of aqueous 3% formic acid. Though this packing material is moderately expensive it can be used repeatedly.

5.3.4 Solid-Phase Extraction

Solid-phase extraction (SPE) is one of the most commonly used sample preparation techniques in liquid chromatography (LC). Small disposable cartridges packed with silica gel, alumina, ion-exchange resins, C8 and C18 silica-based material, etc. (sorbent mass typically ranging from 10 mg to 20 g) are used to isolate the compounds of interest. As a rule, the sample components to be determined or isolated are retained quantitatively on the SPE cartridge while interfering species are washed from the cartridge. Finally, the retained sample components are eluted from the cartridge with an appropriate solvent.

SPE cartridges provide a simple and convenient means of isolating anthocyanins. Kraemer-Schafhalter et al. (1998) compared the efficiency of 16 solid-phase materials in purifying anthocyanins from *Aronia melanocarpa* var Nero. They found that reversed-phase silica gels and macroreticular nonionic acrylic polymer adsorbents such as Serolit PAD IV and Amberlite XAD-7 were the most effective in separating the anthocyanins from the sugars. Anthocyanins are typically loaded onto C18 cartridges which have been previously activated by washing with methanol followed by 0.01% aqueous HCl (Hong and Wrolstad, 1990b). The cartridges are then treated with two volumes of 0.01% aqueous HCl to remove sugars, acids, and other water-soluble compounds. Less polar phenolics such as phenolic acids and flavonoids are removed by washing with two volumes of ethyl acetate (Oszmianski and Lee, 1990). This step has been shown to be an important cleanup procedure that produces mass spectra with very little interferences from other compounds (Giusti et al., 1999). Anthocyanins are finally eluted with methanol containing 0.01% HCl.

SPE cartridges can also be used for the stabilization of anthocyanidins. Anthocyanidins (the aglycons of anthocyanins) have limited solubility in water, are rapidly destroyed by alkali, and are very unstable compared to their corresponding glycosides (anthocyanins). The half-life of a typical anthocyanin, that is, cyanidin 3-glucoside, is about 65 days at room temperature in 0.01 M citric acid, pH 2.8. In contrast, the corresponding aglycon, cyanidin, has a half-life of only 12 h in the same solution (Iacobucci and Sweeney, 1983). Ohta et al. (1980) have also shown that peonidin and malvidin are much less stable than their corresponding 3-glucosides at pH values of 2.5 and 4.5, respectively. Because of the limited commercial availability of anthocyanidin standards, naturally occurring anthocyanins have been used as the main source in most studies to identify anthocyanidins in vitro. In practice, authentic anthocyanidins are needed for all anthocyanin-related studies. A novel approach to the stabilization of anthocyanidins involves their deposition in C18 SPE cartridges. Dao et al. (1998) have shown that the amount of anthocyanidins, delphinidin, petunidin, and malvidin was unchanged for the first 7 days of storage in SPE cartridges under nitrogen atmosphere at 2°C. After 45 days of storage, the levels of delphinidin, petunidin, and malvidin were reduced to 82%, 63%, and 49% of the original content, respectively. In contrast, the level of delphinidin, petunidin, and malvidin decreased to 17%, 10%, and 3%, respectively, of the original amount when these anthocyanidins were stored for 3 days in acidic methanol (0.01% HCl) at 2°C.

5.3.5 Countercurrent Chromatography

Countercurrent chromatography (CCC) is a support-free all-liquid chromatographic technique that relies on the partition of a sample between two immiscible phases. This technique has the following advantages over other chromatographic techniques: (1) no irreversible adsorption of the sample (since no solid support is employed), (2) quantitative recovery of the sample, and (3) sample capacity is high due to the large volumes of stationary phase used (up to several hundred milligrams of pure compounds can be achieved in a single CCC run).

Droplet countercurrent chromatography (DCCC) with acid-containing solvents has been employed for the separation of anthocyanins. Francis and Andersen (1984) tested the suitability of three solvent systems: CMH, chloroform/methanol/0.1 N HCl (5:6:4); PAH, *n*-propanol/acetic acid/water (3:4:3); and BAW (4:1:5) for the separation of anthocyanins by DCCC. The latter system (BAW) was found to be the most effective in separating anthocyanins from black currants (*Ribes nigrum* L.) and raspberries (*Rubus idaeus* L.). They suggested that the upper phase of this solvent system (BAW) be used for the separation of less polar anthocyanins, whereas the lower phase be employed for the separation of more polar anthocyanins. Andersen and coworkers (1991) used the following procedure to isolate cyanidin 3-sambubioside from the fruits of the European elderberry *Sambucus nigra* (Caprifoliaceae), fruits were extracted with acidified methanol and then partitioned against ethyl acetate. The polar fraction was chromatographed on an Amberlite XAD-7 column and then injected onto an Eyela Tokyo Rikakikai Model A DCCC instrument (containing 300 glass columns, 40 cm × 2 mm). Elution was achieved with the lower phase of *n*-butanol/acetic acid/water (4:1:5). Final purification of the anthocyanin was performed by another Amberlite XAD-7 step followed by Sephadex LH-20 column chromatography.

Centrifugal partition chromatography (CPC), a type of CCC that uses discrete partition cells inside a rotor, was used by Renault et al. (1997) to separate anthocyanins from black currant and grape skins. They used two instruments from Sanki Engineering Ltd (Kyoto, Japan): a laboratory model LLB-M and a pilot-scale model LLI-7, both fitted with a stacked disks type rotor.

The separation of anthocyanins using high-speed countercurrent chromatography (HSCCC) was impressively demonstrated by Degenhardt et al. (2000). A high-speed model CCC-1000 (Pharma-Tech Research Corp., Baltimore, MD), equipped with three preparative coils, connected in series (total volume = 850 mL, tubing diameter = 2.6 mm) was employed. These researchers used a biphasic mixture of *tert*-butyl methyl ether/*n*-butanol/acetonitrile/water (2:2:1:5) acidified with TFA as the solvent system and a revolution speed of 1000 rpm to effectively fractionate anthocyanin mixtures from red cabbage, black currant, black chokeberry, and roselle. Figure 5.4 shows the HSCCC separation of black currant anthocyanins. The elution mode was head to tail with the upper layer being the stationary phase. They were able to separate anthocyanins from extracts without any sample pretreatment. However, minor impurities were observed in the fractionated anthocyanins (by NMR) so the authors recommended cleanup of the extracts by XAD-7 column chromatography before separation by HSCCC. Vidal et al. (2004) used multilayer coil countercurrent chromatography (MLCCC) with step gradient elution (*tert*-butyl methyl ether/*n*-butanol/acetonitrile/water acidified with TFA, 2:2:*x*:5) to fractionate grape anthocyanins as a series of glucosides, and the corresponding acetylated, coumaroylated, and caffeoylated derivatives.

5.3.6 High-Performance Liquid Chromatography

High-performance liquid chromatography (HPLC) has revolutionized anthocyanin studies because of its high efficiency, sensitivity, speed, and accurate quantitation

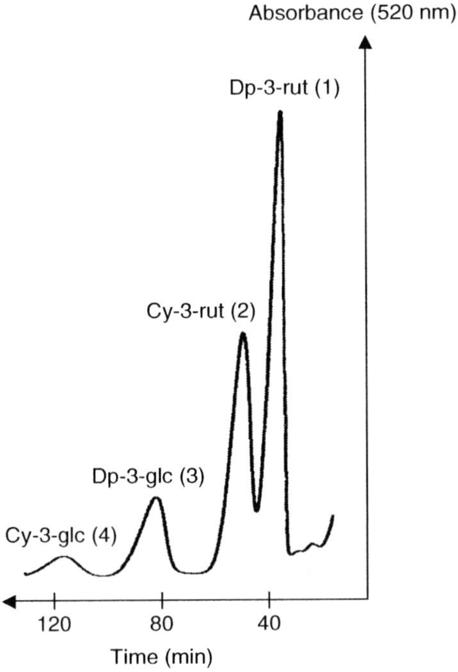

FIGURE 5.4 High-speed countercurrent chromatography (HSCCC) separation of black currant (*Ribes nigrum* L.) anthocyanins on a Pharma-Tech Research Corp. model CCC-1000 (three coils, total volume = 850 mL). Solvent system, *tert*-butyl methyl ether/*n*-butanol/acetonitrile/water (2:2:1:5) acidified with trifluoroacetic acid; speed, 1000 rpm; elution mode, head to tail; stationary phase, upper layer; peak identities, (1) delphinidin 3-rutinoside, (2) cyanidin 3-rutinoside, (3) delphinidin 3-glucoside, (4) cyanidin 3-glucoside. (Reprinted from Degenhardt, A., Knapp, H., Winterhalter, P., *J. Agric. Food Chem.*, 48, 338–343, 2000. With permission.)

(Wilkinson et al., 1977; Wulf and Nagel, 1978; Baj et al., 1983; Casteele et al., 1983; Andersen, 1985, 1987a, b; Hong and Wrolstad, 1986, 1990a, b; Strack and Wray, 1989). Preparative HPLC has been shown to be a rapid and gentle means of purifying anthocyanins from complex mixtures, while analytical HPLC is useful in determining the purity of isolated fractions and the quantitative distribution of anthocyanins in natural extracts. Applications are typically performed on C18 (silica-based) columns using acidic solvents such as formic, acetic, TFA, or phosphoric acids in water–methanol or water–acetonitrile mixtures as the mobile phase. The addition of small amounts of acid is desirable since above pH 2 peak broadening occurs due to slow interconversion between the red flavylium cation and the blue quinoidal species (Hale et al., 1986), leading to poor resolution and lower detection limits. Polymeric reversed-phase columns (polystyrene-divinylbenzene; PLRP-S column, Polymer Laboratories Inc., Amherst, MA) used by Wrolstad and coworkers (Hong and Wrolstad, 1990b; Giusti et al., 1998b; Rodriguez-Sanona et al., 1998) have the advantage of low pH stability (chemically stable through the pH range from 1 to 13). These columns have a high surface area (~500 m^2/g) and a homogeneous surface which eliminates the need for a hydrophobic ligand such as C_{18} to be bound to the base matrix. These columns provide better resolution of complex anthocyanin mixtures, typically those containing acylated anthocyanins (Wrolstad et al., 2002). Red grapes, red cabbage, and blueberries contain complex anthocyanin mixtures and hence would be appropriate samples for these columns (Durst and Wrolstad, 2004). Polymeric reversed-phase columns can be used with phosphoric acid/water and acetonitrile mobile phases which have a lower UV cutoff than acetic acid/water/methanol mixtures. This may be less of a problem for reversed-phase columns available today which feature improved pH stability because of three innovations: (1) introduction of bulky hydrophobic groups on the silane (to hinder access of the mobile phase to the siloxane bond), (2) use of bonded phases with multiple attachment points to the silica surface (making hydrolysis at a given anchoring site less likely to cause phase loss), and (3) use of hybrid organic–inorganic particles (alkyl groups are incorporated into the silica matrix improving both low and high pH stability) (Wehr, 2000). Various HPLC and sample preparation methods for the measurement of food flavonoids including anthocyanins and anthocyanidins have been tabulated in a recent review of Merken and Beecher (2000). Structural effects on HPLC retention (C18 silica-based columns) can be summarized as follows: hydroxylation and glycosylation result in shorter retention times while methylation and acylation cause longer retention times. Under typical conditions, the order of elution is delphinidin derivatives, followed by cyanidin, petunidin, pelargonidin, peonidin, and malvidin derivatives. The presence of sugars decreases the retention of the anthocyanins. Triglycosides typically elute before diglycosides which usually elute before monoglycosides. The 3-rutinosides are an exception to this generalization as they have longer retention times than the corresponding 3-glucosides because of the nonpolarity imparted by the C6 methyl group of rhamnose (Durst and Wrolstad, 2004). Acylated anthocyanins have longer retention times compared to their nonacylated counterparts. The effect of anthocyanin structure on HPLC retention time is shown in Figure 5.5.

5.3.7 Capillary Electrophoresis

In 1996, the capillary zone electrophoresis (CZE) separation of a model mixture of six anthocyanins was reported using a 150-mM borate buffer at pH 8 (Bridle et al., 1996). It was proposed that the separation was influenced by three factors: electroosmotic flow (EOF), charge/mass ratio, and complex formation with borate. Bridle and Garcia-Viguera (1997) compared HPLC and CZE for the separation of strawberry and elderberry anthocyanins. They used fused-silica capillaries (57 cm × 75 μm i.d.), run at 25°C with 150 mM sodium borate buffer (pH 8.0). A more concentrated sample (87 times) was needed for equivalent CZE response at pH 8 compared to HPLC at pH 1.8 because of the much smaller sample introduction volume, the shorter detector cell path length, and the smaller proportion of colored anthocyanin species at pH 8.0. The separation of strawberry anthocyanins by CZE was satisfactory but the separation of elderberry anthocyanins by CZE was marred by the presence of a large baseline hump. The baseline rise may have been caused by interfering compounds in the extract (i.e., colored polymers) which did not affect the

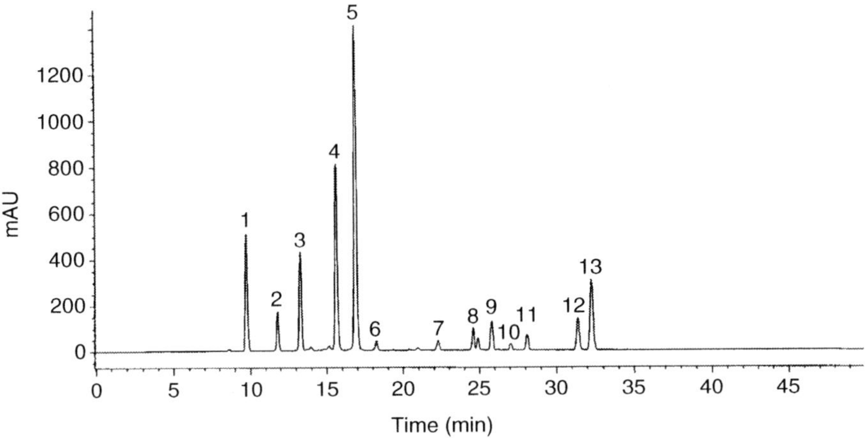

FIGURE 5.5 High-performance liquid chromatography (HPLC) of anthocyanins from a Cabernet Mitos extract monitored at 520 nm. Peak identities: 1, delphinidin 3-*O*-glucoside; 2, cyanidin 3-*O*-glucoside; 3, petunidin 3-*O*-glucoside; 4, peonidin 3-*O*-glucoside; 5, malvidin 3-*O*-glucoside; 6, delphinidin 3-*O*-acetylglucoside; 7, petunidin 3-*O*-acetylglucoside; 8, peonidin 3-*O*-acetylglucoside; 9, malvidin 3-*O*-acetylglucoside; 10, cyanidin 3-*O*-*p*-coumaroylglucoside; 11, petunidin 3-*O*-*p*-coumaroylglucoside; 12, peonidin 3-*O*-*p*-coumaroylglucoside; 13, malvidin 3-*O*-*p*-coumaroylglucoside. Column, Phenomenex Aqua C18 250 × 4.6 mm i.d., 5 μm particle size; solvent A, water/formic acid/acetonitrile (87:10:3, v/v/v); solvent B water/formic acid/acetonitrile (40:10:50, v/v/v); from 10% to 25% B (10 min), from 25% to 31% B (5 min), from 31% to 40% B (5 min), from 40% to 50% B (10 min), from 50% to 100% B (10 min), from 100% to 10% B (5 min); flow rate of 0.8 mL/min. (Reprinted from Kammerer, D., Claus, A., Carle, R., Schieber, A., *J. Agric. Food Chem.*, 52, 4360–4367, 2004. With permission.)

HPLC separation. Ichiyanagi and coworkers (2000, 2004) separated blueberry anthocyanins in 10 min using a 30-mM sodium borate buffer (pH 8.78) containing 7.5 mM *trans*-1,2-diaminocyclohexane-*N,N,N,N*-tetraacetic acid monohydrate (CyDTA). CyDTa served three purposes: (1) it permitted fine adjustment of buffer pH, (2) it stabilized the current to increase analytical reproducibility, and (3) it served as a metal chelator to minimize changes in anthocyanin electrophoretic mobility. Glucosides moved faster than galactosides, followed by arabinosides, when the mobility of the anthocyanins carrying the same aglycon was compared under the same pH conditions. As expected, the mobility of the anthocyanins with the same conjugated sugar was primarily determined by the aglycon structure. The mobility was in the following order, malvidin > peonidin > petunidin > cyanidin > delphindin, in both the glucoside and galactoside series as long as the buffer pH was constant. Anthocyanin separation was markedly different in the narrow pH range between 8.6 and 9.2. It was observed that at a low buffer pH (pH 8.65), cyanidin 3-glucoside and delphinidin 3-glucoside were not separated and cyanidin 3-galactoside and delphinidin 3-galactoside were also not resolved. When the pH was decreased further, migration time, peak resolution, and sensitivity were significantly decreased since the migration of undissociated anthocyanins is mainly determined by the EOF at an acidic pH. A limitation of methods employing basic pH buffers is the instability of anthocyanins under these alkaline conditions. Four major black currant anthocyanins were separated by CZE using a fused-silica capillary column with a phosphate buffer containing 30% (v/v) acetonitrile at an apparent pH of 1.5 (da Costa et al., 1998). An attempt to perform the CZE analysis at a pH value of less than 1.5 was not successful since the current was excessive (over 100 μA) at the voltage used (25 kV). Acetonitrile gave a better separation than methanol in terms of peak efficiency and resolution, baseline stability, and analysis time. The improvement in resolution observed with the addition of organic solvents has been attributed to decreasing the EOF and changing the electrophoretic mobility of the analytes (da Costa et al., 1998). The authors evaluated the interaction of anthocyanins with the capillary wall by comparing the separation of anthocyanins on an uncoated fused-silica capillary column to that on a linear polyacrylamide (LPA) coated capillary (using a phosphate buffer containing 30% acetonitrile at an apparent pH of 1.8). The peak shape and resolution on the two columns were similar though the migration times of the analytes were 1–3 min longer on the LPA column. This indicated that there was almost no interaction between the silanols on the uncoated capillary and the anthocyanins. Bicard and coworkers (1999) also used acidic media (pH 2.1) to separate four anthocyanins by CZE. The separation was performed with a 72 cm × 50 μm i.d. fused-silica capillary column with a running buffer of 0.25 mM of the cationic surfactant cetyltrimethylammonium bromide (CTAB) in 160 mM NaH_2PO_4/H_3PO_4. CTAB, used below its critical micelle concentration, served to reverse the direction of EOF (the system was configured to run from cathode to anode). Surfactants are widely used buffer additives in CE and can act as solubilizing agents for hydrophobic solutes, as ion-pairing reagents, or as wall modifiers (Heiger, 2000). Watanabe and coworkers (1998) used a different mode of CE called micellar electrokinetic chromatography (MEKC) to separate elderberry anthocyanins in candy, juice, and jelly. The four anthocyanins were separated (in less than 10 min) on a 36 cm (31.4 cm to the detector) × 50 μm i.d. uncoated fused-silica capillary column using 30 mM sodium dodecyl sulfate (SDS) in 30 mM phosphate/60 mM borate buffer at pH 7.0. An advantage of MEKC over other modes of CE is that it can be used to separate neutral solutes as well as charged ones. Bednář et al. (2005) used capillary electrophoresis–mass spectrometry (CE-MS^n) to monitor anthocyanins in wine and wine must. CE-MS was performed with both acidic (200 mM chloroacetate-ammonium, pH 2) and basic (200 mM borate-ammonium, pH 9) electrolytes with a sheath liquid of methanol/water (80:20 + 0.25% (v/v) acetic acid). Sensitivity was higher with the acidic electrolyte than the basic electrolyte (i.e., limit of detection (LOD) for malvidin 3-glucoside of 0.9 and 4 mg/L, respectively) though the basic electrolyte is valuable when separation of diastereomers is required (i.e., resolution of malvidin 3-glucoside from malvidin 3-galactoside).

5.4 METHODS OF IDENTIFICATION

5.4.1 Hydrolysis

The first step in anthocyanin characterization is the identification of the anthocyanidin moiety (aglycon) and the sugar(s). This can be accomplished by acid hydrolysis followed by chromatography of the products with reference standards. Hydrolysis of anthocyanins can be accomplished by heating the sample with 2 N HCl in methanol (1:1) under reflux for 60 min and then cooling it in an ice bath. The anthocyanidins can then be extracted with ethyl acetate, evaporated to dryness under a stream of nitrogen, redissolved in 10% formic acid, filtered through a 0.5-μm disposable membrane filter, and promptly analyzed by HPLC (Takeoka et al., 1997). Ichiyanagi and coworkers (2001) used a CZE method to study acid-catalyzed hydrolysis of anthocyanins. They found that anthocyanin hydrolysis in TFA solution was determined mainly by the type of conjugated sugar and not by the aglycon structure. The rate constant of anthocyanin hydrolysis was in the following order, arabinoside > galactoside > glucoside, regardless of the aglycon structure. Anthocyanins acylated with cinnamic acid are resistant to acid hydrolysis and it is advisable to treat acylated anthocyanins with 10% aqueous KOH for 8–10 min at room temperature in the dark to remove the acylating acids (Wrolstad et al., 2002). The anthocyanin can then be subjected to acid hydrolysis to generate the aglycon. Alternatively, the anthocyanin can be neutralized with HCl, isolated by SPE, and analyzed by HPLC. The acylated acids liberated by saponification can be isolated, analyzed by HPLC, and characterized by their retention times and spectra (Wrolstad et al., 2002).

5.4.2 Chromatography

TLC, HPLC, and capillary gas chromatography (GC) are effective methods of identifying carbohydrates liberated by the acid hydrolysis of anthocyanins. Carbohydrate samples must be derivatized before analysis by capillary GC and gas chromatography–mass spectrometry (GC-MS). We have employed methyloxime–trimethylsilyl (MO-TMS) derivatives of carbohydrates for these analyses (Takeoka et al., 1997) using the following procedure: A mixture of dry pyridine (125 μL) and 1 mg of *O*-methylhydroxyamine-HCl was added to 1 mg of each carbohydrate sample and standard. The mixtures were mixed well, heated for 2 h at 40°C, and allowed to stand overnight at room temperature. Pyridine was removed in a stream of nitrogen while the sample was heated to 40°C. When the solution approached dryness, two drops of benzene were added to azeotropically remove the last traces of water formed in the oximation reaction. A 100 μL of a 99:1 mixture of *N,O*-bis(trimethylsilyl)trifluoroacetamide (BSTFA) plus trimethylchlorosilane (TMCS) (Sylon BFT, Supelco, Inc., Bellefonte, PA) was added to each sample. The mixtures were shaken well, heated at 40°C for 2 h, and allowed to stand overnight at room temperature. This procedure is a modification of the method of Laine and Sweeley (1971, 1973) for the preparation of MO-TMS derivatives of aldoses, partially methylated aldoses, deoxyaldoses, and ketoses. Samples can be subsequently analyzed by capillary GC using a nonpolar stationary phase such as DB-1 (J&W Scientific, Folsom, CA). Polar stationary phases such as DB-Wax are not recommended due to possible reaction of the silylating agent with the stationary phase.

Hong and Wrolstad (1990a, b) have effectively used a combination of retention properties on reversed-phase HPLC and spectral properties to characterize anthocyanins in various colorants and fruit juices.

5.4.3 Spectroscopy

5.4.3.1 UV–Vis Adsorption Spectroscopy

The spectral characteristics of anthocyanins give important information with regard to the nature of the aglycon and the type of glycosylation. The spectra of anthocyanins are typically measured in methanol containing 0.01% HCl. The type of aglycon present can be inferred from the visible λ_{max}. Pelargonidin 3-glucosides have a visible λ_{max} at about 505 nm, cyanidin and peonidin 3-glucosides exhibit their λ_{max} at 520–526 nm, while all delphindin derivatives (delphinidin, malvidin, and petunidin) display their λ_{max} at 532–537 nm (Harborne, 1967). Anthocyanidins typically display a 6–12 nm larger visible λ_{max} in 0.01% methanolic HCl than the corresponding anthocyanins (Harborne, 1958). Pelargonidin derivatives have a visible λ_{max} at 512 nm, cyanidin and peonidin derivatives display their λ_{max} at 523–525 nm, while delphinidin derivatives display their λ_{max} at 529–533 nm (Hong and Wrolstad, 1990). When measuring the spectrum online (i.e., diode array detector [DAD]) as the anthocyanin elutes from the HPLC column, the solvent composition is limited by the mobile phase composition. Since the spectral characteristics of the anthocyanins are dependent on both the pH and nature of the solvent, direct comparison of spectral characteristics in different solvent systems may not be appropriate. In general, there is a 15 nm shift toward shorter wavelengths when water is substituted for methanol (Harborne, 1958). Using HPLC combined with DAD, Hong and Wrolstad (1990a) have shown that acylated (with hydroxy aromatic acids) cyanidin glycosides tended to have a much higher wavelength maximum than their nonacylated counterparts. They attributed this difference (as much as 20 nm in some cases) to the predominantly aqueous HPLC solvent system since earlier studies with anthocyanins in acidified methanol showed only small differences in the visible wavelength maximum between acylated and nonacylated anthocyanins with the same aglycon (Hrazdina et al., 1977; Wulf and Nagel, 1978). The glycosidic substitution pattern of anthocyanins can be elucidated by absorption in the 400–460 nm region since 3,5- and 5-glycosides exhibit ratios of $A_{440}/A_{\lambda max}$ about 50% of those of 3-glycosides and the free anthocyanidins (Harborne, 1967). Increasing the solvent polarity changes the spectral characteristics of anthocyanins by increasing the $A_{440}/A_{\lambda max}$ ratio and decreasing the visible λ_{max} (Andersen, 1985, 1987a; Hong and Wrolstad, 1990b). It has been reported that compounds possessing a disaccharide as a substituent group in position 3 of the chromophore exhibit a large drop in absorptivity (Elhabiri et al., 1995; Figueiredo et al., 1996a, b). These workers found that cyanidin 3-sambubioside-5-glucoside, delphinidin 3-gentiobioside, and cyanidin 3-rutinoside showed a marked hypsochromic effect when compared to the corresponding 3-glucosides and 3,5-diglucosides with a similar substitution pattern. In their study of pelargonidin derivatives, Giusti et al. (1999b) reported a small decrease in the molar absorptivity when a glucose unit was attached to position 3 as compared to the aglycon. Addition of other glucose units produced a hyperchromic effect as pelargonidin 3-sophoroside-5-glucoside had a substantially higher molar absorptivity than pelargonidin 3-glucoside (Table 5.3). The presence of acylation with hydroxylated aromatic acids can be deduced by the appearance of a peak in the 310–335 nm

TABLE 5.3
Absorptivity Coefficients (L cm^{-1} mol^{-1}) of Different Pelargonidin Derivatives

Anthocyanin	λ_{max}		Molar Absorptivity	
	MeOH[b]	Buffer[a]	MeOH[b]	Buffer[a]
Pg aglycon	524	505	19,780	18,420
Pg 3-glu	508	496	17,330	15,600
Pg 3-soph-5-glu	506	497	30,700	25,370
Pg 3-soph-5-glu + *p*-coumaric	508	506	34,890	28,720
Pg 3-soph-5-glu + ferulic	507	506	29,640	24,140
Pg 3-soph-5-glu + *p*-coumaric and malonic	508	508	39,780	33,020
Pg 3-soph-5-glu + ferulic and malonic	508	508	39,380	31,090
Pg 3-rut-5-glu + *p*-coumaric	511	504	39,590	32,080

Source: Adapted from Giusti, M.M., Rodríguez-Saona, L.E., Wrolstad, R.E., *J. Agric. Food Chem.*, 1999b, 47, 4631–4637.

Pg, pelargonidin; glu, glucose; soph, sophoroside.

[a] 0.2 N KCl, pH 1.0.
[b] 0.1% HCl in methanol.

range. The position of the acyl λ_{max} gives an indication of the nature of the acylating acid, while the $A_{\lambda max}(\text{acyl})/A_{\lambda max}(\text{visible})$ ratio is a measure of the number of aromatic acids present in the anthocyanin (Harborne, 1967). Harborne (1958) noted that in acidified methanolic solutions, a $A_{\lambda max}(\text{acyl})/A_{\lambda max}(\text{visible})$ ratio of 48%–71% is characteristic of a 1:1 molar ratio of cinnamic acid to anthocyanin, while a $A_{\lambda max}(\text{acyl})/A_{\lambda max}(\text{visible})$ ratio of 83%–107% is indicative of a 2:1 molar ratio of cinnamic acid to anthocyanin.

5.4.3.2 Mass Spectrometry

Fast atom bombardment (FAB)-MS became a useful technique for anthocyanin structural studies in the 1980s (Harborne and Grayer, 1988; Strack and Wray, 1989). The sample is dissolved in a nonvolatile matrix such as glycerol or a mixture of thioglycerol/diglycerol (1:1) and placed on a direct insertion probe. The matrix facilitates desorption as well as ionization. FAB desorption is accomplished by bombarding the sample solution with a high-energy beam of xenon atoms or cesium ions causing sample desorption (often as an ion) by momentum transfer. Saito et al. (1983) was the first group to perform FAB-MS on anthocyanins. This technique has been widely used for molecular weight determination of anthocyanins (Takeda et al., 1986; Kim et al., 1989; Baublis et al., 1994; Saito et al., 1995; Fossen et al., 1996; Bakker and Timberlake, 1997; Bakker et al., 1997). Tamura et al. (1994) reported the use of LC-MS with continuous-flow FAB (CF-FAB) to separate and characterize anthocyanins in Muscat Bailey A grapes. The anthocyanins (1 mg/mL) were separated by gradient elution on a C18 column using acetonitrile, water, and 0.1% TFA at a flow rate of 1.0 mL/min. A matrix solvent of 3% glycerol and 1000 ppm dimethyl sulfoxide in methanol was added to the column effluent at a flow rate of 0.3 mL/min. The column effluent and matrix solvent were allowed to mix before reaching a splitter that introduced about 5 μL/min of solution into the mass ion source. It was possible to obtain the molecular weights of the grape anthocyanins from one LC-MS run. Plasma desorption mass spectrometry (PDMS) has been shown to be effective in the analysis of anthocyanidins that are hydroxylated at position 3 (Wood et al., 1993). In PDMS, the sample is deposited on a thin metal foil typically nickel. Spontaneous fission of a radioactive ^{252}Cf source produces fission fragments such as $^{142}Ba^{18+}$ and $^{106}Tc^{22+}$ which have kinetic energies of roughly 79 and 104 MeV, respectively (Williams and Fleming, 1995). Passage of a high-energy fission fragment through the sample foil results in extremely rapid localized heating that desorbs the sample molecules and generates ions. These ions are then accelerated out of the source and into the mass analyzer. PDMS has also been successfully used to identify and characterize both anthocyanins and 3-deoxyanthocyanidins though it has been found that acetic acid interferes with the analysis (Wood et al., 1994). With regard to sensitivity, it has been shown that 1 ng levels of pelargonidin chloride could be detected with an observed ion, m/z 271 (corresponding to the flavylium cation) using PDMS (Wood et al., 1993). It was necessary in most cases to partially purify the samples by HPLC or by partitioning with organic solvents before PDMS analysis. Phippen and Simon (1998) used PDMS to determine the molecular weight of various purified anthocyanins in basil. The high concentrations (0.5–1.0 mg/mL) of anthocyanins needed for optimal detection necessitated the use of preparative HPLC.

Matrix-assisted laser desorption/ionization (MALDI) coupled with time-of-flight (TOF) mass analysis has emerged as a particularly useful tool for the analysis of biomolecules (Cotter, 1992). In this method, the sample is first co-crystallized with a large molar excess of a matrix compound that is selected to strongly absorb laser light. Typical matrix compounds are 2,5-dihydroxybenzoic acid and 3,5-dimethoxy-4-hydroxycinnamic acid (sinapinic acid). Pulse UV laser radiation of the sample–matrix mixture results in vaporization of the matrix which carries the sample with it. Sample degradation is avoided since energy transfer is indirect and in a controlled manner. MALDI-TOF has had its greatest impact in the area of protein research since it can give approximate molecular weight determinations up to 100,000–200,000 Da (Williams and Fleming, 1995). MALDI-TOF has been employed in the analysis of mixtures of 3-deoxyanthocyanins and anthocyanins (Sugui et al., 1998). For sample preparation, these workers solubilized the matrix, α-cyano-4-hydroxycinnamic acid (10 mg) in 1 mL of a mixture of 1% TFA/H_2O/CH_3CN (1:4:5). The samples were mixed with the matrix for 1 min then the resulting mixture was applied to the sample plate. The solvents were allowed to evaporate at room temperature. Sensitivities of 15 pmol/μL were achieved for 3-deoxyanthocyanidins in extracts. MALDI-MS with a Proflex III linear mode, Bruker Analytical Systems, Inc., was used to analyze anthocyanins in red wine and fruit juice (Wang and Sporns, 1999). These researchers used 2,4,6-trihydroxyacetophenone as the matrix. It was shown that monoglucoside anthocyanins had a similar response, while a diglucoside anthocyanin, e.g., malvidin 3,5-diglucoside, or an anthocyanin with a disaccharide attached to position 3, e.g., cyanidin 3-rutinoside, both had relative molar responses only one-fourth that of the monoglucosylated anthocyanins

(Wang and Sporns, 1999). Anthocyanins were extracted from wine and fruit juice samples using Sep-pak C18 cartridges before MALDI-MS analysis.

Atmospheric pressure chemical ionization (APCI) was investigated as a possible technique for MS analysis of anthocyanins but, unfortunately, showed poor sensitivity for anthocyanin standards (Revilla et al., 1999). Electrospray ionization mass spectrometry (ESI-MS) may be the most promising ionization technique for the characterization of anthocyanins (Giusti et al., 1999a). ESI is the most useful and versatile ionization technique in MS (Covey, 1995). The only absolute requirement for ionization is that the constituents of interest must be soluble in some solvent. ESI generates ions directly from solution by creating a spray or fine mist of highly charged droplets in the presence of a strong electrical field (3–6 kV). If the spray contains sample molecules, a molecular ion of the sample molecules can be generated by evaporation of the solvent. This is achieved by passing a drying gas across the spray. In the process of evaporation, droplets are reduced in size while the electric charge density on their surface increases. Desolvation may then be aided by repulsive Coulombic forces overcoming the cohesive forces of the droplet. Sample molecules eventually leave the liquid phase and become gas-phase ions in the atmospheric region of the ion source. The ions are then electrostatically directed into the mass analyzer. The positive charge of anthocyanins at low pH values allows for their easy detection at low voltages since other interfering compounds are usually not ionized (Giusti et al., 1999a). Alternatively, samples can be introduced with minimal cleanup (C18 SPE) directly into the mass spectrometer allowing the effective molecular weight determination of anthocyanins present in extracts. However, an intermediate step of washing the C18 SPE cartridge with ethyl acetate (to remove interfering phenolic acids and flavonoids) before elution with acidified methanol was found to produce much cleaner MS profiles and was recommended (Giusti et al., 1999a). LC-MS may be the ideal technique for the rapid characterization of anthocyanins (Miller et al., 1998). Standard formic acid/water/methanol gradients are readily adapted to LC-MS with formic acid facilitating ionization. The natural molecular cations M^+ are readily observed in the positive ion mode. In source collision-induced dissociation (CID) can be used to fragment the anthocyanin ion to elucidate the aglycon. Alternatively, ESI with an ion trap (Piovan et al., 1998; Degenhardt et al., 2000) or a triple-stage quadrupole (Giusti et al., 1999a) may be employed to elucidate characteristic fragments of the molecular ion. Giusti et al. (1999a) performed CID experiments of anthocyanins on a triple quadrupole using argon as the target gas. MS/MS of purified anthocyanins using voltages ranging from 15 to 25 eV resulted in clear, characteristic, and reproducible fragmentation patterns. The collision energy did influence the relative abundances of the fragment ions. The use of energies higher than 25 eV (30 eV) resulted in complete fragmentation of the parent ion, with observation of only the aglycon ion. In the case of anthocyanins substituted with rutinose, MS/MS revealed cleavage of the 1→6 linkage between rhamnose and glucose, while other anthocyanins containing the disaccharides, sambubiose (xylose 1→2 glucose), or sophorose (glucose 1→2 glucose) with more stable sugar bonds did not undergo cleavage at that bond. With acylated anthocyanins, ester bond cleavage was not observed in tandem mass spectroscopy (MS/MS). Figure 5.6 shows the positive ion ESI-MS/MS product

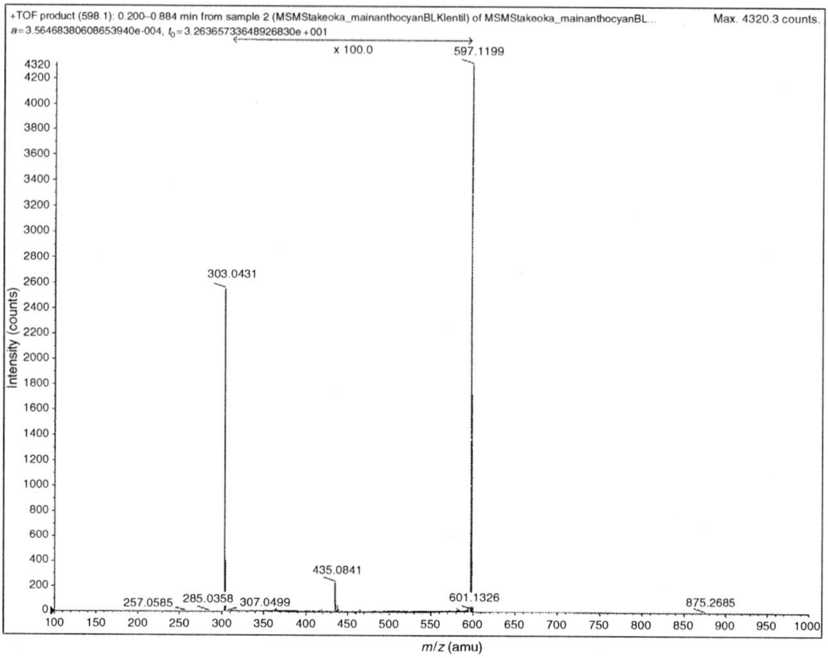

FIGURE 5.6 Positive ion electrospray ionization mass spectrometry (ESI-MS)/MS product ion spectrum of delphinidin 3-*O*-(2-*O*-β-D-glucopyranosyl-α-L-arabinopyranoside), the major anthocyanin of Beluga black lentils. The MS/MS spectrum was obtained by isolating and fragmenting the molecular ion $[M]^+$ at *m*/*z* 597.1199. The ion at 303.0431 corresponds to the aglycon delphindin. (Reprinted from Takeoka, G.R., Dao, L.T., Tamura, H., Harden, L.A., *J. Agric. Food Chem.*, 53, 4932–4937, 2005. With permission.)

ion spectrum of delphinidin 3-*O*-(2-*O*-β-D-glucopyranosyl-α-L-arabinopyranoside) (Figure 5.7) obtained on a hybrid quadrupole time-of-flight mass spectrometer (Q-TOF; API QSTAR Pulsar equipped with a nanoelectrospray ion source). The purified anthocyanin was diluted in 50:50 (v/v) methanol/water with 0.1% TFA. The electrospray mass spectrum of the anthocyanin exhibited a molecular ion $[M]^+$ as the flavylium cation at *m*/*z* 597.1199 (the exact mass calculated for $C_{26}H_{29}O_{16}$ was *m*/*z* 597.1456). MS/MS of the molecular ion gave an ion at *m*/*z* 303.0431 corresponding to the aglycon delphinidin (the exact mass calculated for $[C_{15}H_{11}O_7]^+$ was *m*/*z* 303.0505), which was formed by the loss of arabinose and glucose. One 100-fold magnification of the region between 303 and 597 revealed a small ion at *m*/*z* 435.0841 (the exact mass calculated for $C_{20}H_{19}O_{11}$ was *m*/*z* 435.0927), indicating loss of glucose (162, glucose–H_2O). This showed that arabinose was connected directly to delphinidin. Gläßgen et al. (1992a) applied NMR and pneumatically assisted electrospray (ionspray) to characterize six anthocyanins from cell cultures of an Afghan cultivar of *Daucus carota*. They used a Sciex API III triple quadrupole mass spectrometer equipped with an ionspray ion source (PE Sciex, Foster City, CA). Extracts and purified anthocyanins were dissolved in

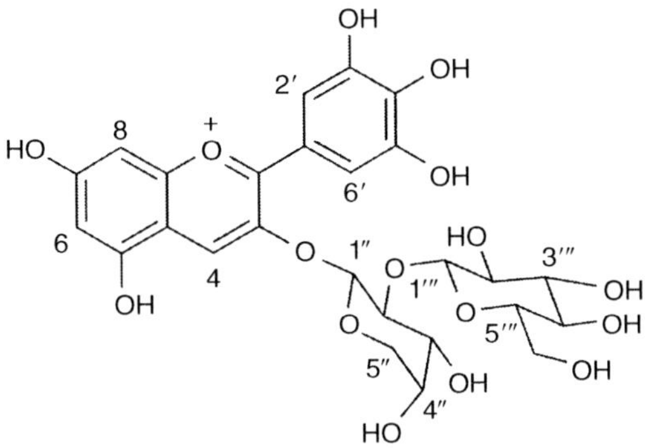

FIGURE 5.7 Structure of delphinidin 3-*O*-(2-*O*-β-D-glucopyranosyl-α-L-arabinopyranoside). (Reprinted from Takeoka, G.R., Dao, L.T., Tamura, H., Harden, L.A., *J. Agric. Food Chem.*, 53, 4932–4937, 2005. With permission.)

MeOH/HOAc/H_2O (50:8:42) and introduced directly into the ionspray source at a constant flow rate of 5 μL/min with a medical infusion syringe pump (Harvard Apparatus, Southnatick, MA). The system could detect 4.49 ng (10 pmol) of cyanidin 3-galactoside (molecular ion) with a signal-to-noise ratio of 3:1 under full scan conditions (m/z 100–1200 at a scan rate of 5 s/scan) (Gläßgen et al., 1992b).

Cooper and Marshall (2001) used positive- and negative-ion ESI Fourier transform ion cyclotron resonance (FT-ICR) MS to characterize five wines, without any separation or purification steps. The high mass resolving power, $m/\Delta m_{50\%} > 80{,}000$ (in which m is ion mass and $\Delta m_{50\%}$ is the mass spectral peak full width at half-maximum peak height) and mass accuracy (≤ 1 ppm) of FT-ICR-MS make it ideal for analyzing complex mixtures since the constituents are simultaneously resolved and their elemental compositions are determined. Sample preparation was very simple. Wine (10 μL) was diluted in 90 μL of 1:1 methanol/water (for positive ESI) or in 90 μL of 1:1 methanol/5 mM ammonium acetate (for negative ESI). Peonidin 3-glucoside, delphinidin 3-glucoside, petunidin 3-glucoside, and malvidin 3-glucoside were identified in the flavylium form in the four red wines (Inglenook California Chianti, Chorus Red Corbiere (1998), Rancho Zabaco California Red Zinfandel (unfiltered, 1997), and Cuvee Plateau de Bel-Air, Brouilly Red Beaujolais (1999)) studied. The flavylium ion *p*-coumarates of peonidin 3-glucoside, petunidin 3-glucoside, and malvidin 3-glucoside were identified in all four red wines. The protonated carbinol forms of delphinidin-3-glucoside and mavidin-3-glucoside were identified in all wines except the Zinfandel.

MS does not differentiate between different diastereomeric forms of sugars, nor does it reveal the positions of attachment of sugar and acyl substituents. However, Favretto and Flamini (2000) had a clever solution for differentiating between two isobaric compounds, malvidin 3-*O*-(6-*O*-caffeoyl)glucoside and malvidin 3,5-*O*-diglucoside, which have identical molecular weights and identical aglycon moieties. Samples were dissolved in D_2O and different mass shifts were observed by MS which agreed with the number of exchangeable, acidic protons present in the molecules.

5.4.3.3 Nuclear Magnetic Resonance Spectroscopy

Modern one- and two-dimensional nuclear magnetic resonance (NMR) spectroscopy represents the most powerful tool in the structural elucidation of complex anthocyanins. Andersen and coworkers (cited in Andersen and Jordheim, 2006) have published extensive studies on the use of NMR to deduce anthocyanin structures. They stated that this technique is particularly valuable for the determination of the sugar moieties and the sites of acyl and sugar substitutions (Andersen and Fossen, 2004). Andersen and coworkers (1991) were the first to apply a heteronuclear shift correlation experiment to the structural characterization of an anthocyanin. Giusti et al. (1998a) recently used one- and two-dimensional NMR techniques for structural conformation of two diacylated anthocyanins, pelargonidin 3-*O*-[2-*O*-(β-glucopyranosyl)-6-*O*-(*trans-p*-coumaroyl)-β-glucopyranoside] 5-*O*-(6-*O*-malonyl-β-glucopyranoside) and pelargonidin 3-*O*-[2-*O*-(β-glucopyranosyl)-6-*O*-(*trans*-feruloyl)-β-glucopyranoside] 5-*O*-(6-*O*-malonyl-β-glucopyranoside) isolated from red radish (*Raphanus sativus*). Nuclear Overhauser effect spectroscopy (NOESY) revealed spatial proximity between hydrogens from the cinnamic acid acylating group and the C4 of the pelargonidin. These results suggest folding of the acylated anthocyanins which is an important factor in their increased color stability.

Detailed ^{1}H-NMR spectroscopy of flavonoids and their glycosides (including anthocyanins) was reviewed in a chapter by Markham and Geiger (1994). Other useful information on NMR of anthocyanins can be found in Strack and Wray (1989, 1994) and Andersen and Fossen (2004).

REFERENCES

Andersen, Ø.M., Chromatographic separation of anthocyanins in Crowberry (Lignonberry) *Vaccinium vites-Idaea* L., *J. Food Sci.*, 1985, 50, 1230–1232.

Andersen, Ø.M., Anthocyanins in fruits of *Vaccinium uliginosum* L. (bog whortleberry), *J. Food Sci.*, 1987a, 52, 665–667.

Andersen, Ø.M., Anthocyanins in fruits of *Vaccinium japonicum*, *Phytochemistry*, 1987b, 26, 1220–1221.

Andersen, Ø.M., Semipreparative isolation and structure determination of pelargonidin 3-*O*-α-L-rhamnopyranosyl-(1→2)-β-D-glucopyranoside and other anthocyanins from the tree *Dacrycarpus dacrydiodes*, *Acta Chem. Scand.*, 1988, B42, 462–468.

Andersen, Ø.M., Aksnes, D.W., Nerdal, W., Johansen, O.P., Structural elucidation of cyanidin-3-sambubioside and assignments of the ^{1}H and ^{13}C NMR resonances through two-dimensional shift-correlated NMR techniques, *Phytochem. Anal.*, 1991, 2, 175–183.

Andersen, Ø.M., Fossen, T., Characterization of anthocyanins by NMR, in *Handbook of Food Analytical Chemistry: Pigments, Colorants, Flavors, Texture, and Bioactive Food Components*, Wrolstad, R.E., Acree, T.E., Decker, E.A., Penner, M.H., Reid, D.S., Schwartz, S.J., Shoemaker, C.F., Smith, D.M., Sporns, P. (Eds.), John Wiley & Sons, Inc., Hoboken, NJ, 2004, F.1.4.1–F.1.4.23.

Andersen, Ø.M., Francis, G.W., Simultaneous analysis of anthocyanins and anthocyanidins on cellulose thin layers, *J. Chromatogr.*, 1985, 318, 450–454.

Andersen, Ø.M., Francis, G.W., Techniques of pigment identification, in *Plant Pigments and Their Manipulation*, Davies, K. (Ed.), Blackwell Publishing, London, United Kingdom, 2004, 293–326.

Andersen, Ø.M., Jordheim, M., The anthocyanins, in *Flavonoids: Chemistry, Biochemistry and Applications*, Andersen, Ø.M., Markham, K.R. (Eds.), CRC Press, Taylor & Francis Group, Boca Raton, FL, 2006, 471–551.

Andersen, Ø.M., Markham, K.R. (Eds.), *Flavonoids: Chemistry, Biochemistry and Applications*, CRC Press, Taylor & Francis Group, Boca Raton, FL, 2006.

Asen, S., Preparative thin-layer chromatography of anthocyanins, *J. Chromatogr.*, 1965, 18, 602–603.

Baj, A., Bombardelli, E., Gabetta, B., Martinelli, E.M., Qualitative and quantitative evaluation of *Vaccinium myrtillus* anthocyanins by high-resolution gas chromatography, *J. Chromatogr.*, 1983, 279, 365–372.

Bakker, J., Bridle, P., Honda, T., Kuwano, H., Saito, N., Terahara, N., Timberlake, C.F., Identification of an anthocyanin occurring in some red wines, *Phytochemistry*, 1997, 44, 1375–1382.

Bakker, J., Timberlake, C.F., The distribution of anthocyanins in grape skin extracts of Port wine cultivars as determined by high performance liquid chromatography, *J. Sci. Food Agric.*, 1985, 36, 1315–1324.

Bakker, J., Timberlake, C.F., Isolation, identification, and characterization of new color-stable anthocyanins occurring in some red wines, *J. Agric. Food Chem.*, 1997, 45, 35–43.

Bate-Smith, E.C., Leuco-anthocyanins. 1. Detection and identification of anthocyanins formed from leuco-anthocyanins in plant tissues, *Biochem. J.*, 1954, 58, 122–125.

Baublis, A., Spomer, A., Berber-Jiménez, M.D., Anthocyanin pigments: Comparison of extract stability, *J. Food Sci.*, 1994, 59, 1219–1221.

Bednář, P., Papoušková, B., Müller, L., Barták, P., Stávek, J., Pavioušek, P., Lemr, K., Utilization of capillary electrophoresis/mass spectrometry (CE/MS^n) for the study of anthocyanin dyes, *J. Sep. Sci.*, 2005, 28, 1291–1299.

Bicard, V., Fougerousse, A., Brouillard, R., Analysis of natural anthocyanins by capillary zone electrophoresis in acidic media, *J. Liq. Chromatogr. Relat. Technol.*, 1999, 22, 541–550.

Bohm, B.A., *Introduction to Flavonoids*, Harwood Academic Publishers, Amsterdam, The Netherlands, 1998.

Bridle, P., García-Viguera, C., Analysis of anthocyanins in strawberries and elderberries. A comparison of capillary zone electrophoresis and HPLC. *Food Chem.*, 1997, 59, 299–304.

Bridle, P., García-Viguera, C., Tomás-Barberán, F.A., Analysis of anthocyanins by capillary zone electrophoresis, *J. Liq. Chromatogr. Relat. Technol.*, 1996, 19, 537–545.

Brouillard, R., Flavonoids and flower color, in *The Flavonoids: Advances in Research Since 1980*, Harborne, J.B. (Ed.), Chapman and Hall, London, United Kingdom, 1988, 525–538.

Cabrita, L., Andersen, Ø.M., Anthocyanins in blue berries of *Vaccinium padifolium*, *Phytochemistry*, 1999, 52, 1693–1696.

Casteele, K.V., Gieger, H., De Loose, R., Van Sumere, C.F., Separation of some anthocyanidins, anthocyanins, proanthocyanidins and related substances by reversed-phase high-performance liquid chromatography, *J. Chromatogr.*, 1983, 259, 291–300.

Catalano, G., Fossen, T., Andersen, Ø.M., Petunidin 3-*O*-α-rhamnopyranoside-5-*O*-β-glucopyranoside and other anthocyanins from flowers of *Vicia villosa*, *J. Agric. Food Chem.*, 1998, 46, 4568–4570.

Cooper, H.J., Marshall, A.G., Electrospray ionization Fourier transform mass spectrometry analysis of wine, *J. Agric. Food Chem.*, 2001, 49, 5710–5718.

Cotter, R.J., Time-of-flight mass spectrometry for the structural analysis of biological molecules, *Anal. Chem.*, 1992, 64, 1027A–1039A.

Covey, T., Analytical characteristics of the electrospray ionization process, in *Biochemical and Biotechnological Applications of Electrospray Ionization Mass Spectrometry*, Snyder, A.P. (Ed.), ACS Symposium Series 619, American Chemical Society, Washington DC, WA, 1995, 21–59.

da Costa, C.T., Nelson, B.C., Margolis, S.A., Horton, D., Separation of black currant anthocyanins by capillary zone electrophoresis, *J. Chromatogr. A*, 1998, 799, 321–327.

Dao, L.T., Takeoka, G.R., Edwards, R.H., Berrios, J.D.J., An improved method for the stabilization of anthocyanidins, *J. Agric. Food Chem.*, 1998, 46, 3564–3569.

Degenhardt, A., Knapp, H., Winterhalter, P., Separation and purification of anthocyanins by high-speed countercurrent chromatography and screening for antioxidant activity, *J. Agric. Food Chem.*, 2000, 48, 338–343.

Durst, R.W., Wrolstad, R.E., Separation and characterization of anthocyanins by HPLC, in *Handbook of Food Analytical Chemistry: Pigments, Colorants, Flavors, Texture, and Bioactive Food Components*, Wrolstad, R.E., Acree, T.E., Decker, E.A., Penner, M.H., Reid, D.S., Schwartz, S.J., Shoemaker, C.F., Smith, D.M., Sporns, P. (Eds.), John Wiley & Sons, Inc., Hoboken, NJ, 2004, F.1.3.1–F.1.3.13.

Elhabiri, M., Figueiredo, P., Fougerousse, A., Brouillard, R., A convenient method for conversion of flavonoids into anthocyanins, *Tetrahedron Lett.*, 1995, 36, 4611–4614.

Favretto, D., Flamini, R., Application of electrospray ionization mass spectrometry to the study of grape anthocyanins, *Am. J. Enol. Vitic.*, 2000, 51, 55–64.

Figueiredo, P., Elbahiri, M., Saito, N., Brouillard, R., Anthocyanin intramolecular interactions. A new mathematical approach to account for the remarkable colorant properties of the pigments extracted from *Matthiola incana*, *J. Am. Chem. Soc.*, 1996a, 118, 4788–4793.

Figueiredo, P., Elbahiri, M., Toki, K., Saito, N., Dangles, O., Brouillard, R., New aspects of anthocyanin complexation. Intramolecular copigmentation as a means for colour loss? *Phytochemistry*, 1996b, 41, 301–308.

Fossen, T., Rayyan, S., Holmberg, M.H., Nateland, H.S., and Anderson, Ø.M., Acylated anthocyanins from leaves of *Oxalis triangularis, Phytochemistry*, 2005, 66, 1133–1140.

Fossen, T., Andersen, Ø.M., Acylated anthocyanins from leaves of the water lily, *Nymphaé* X *marliacea*, *Phytochemistry*, 1997a, 46, 353–357.

Fossen, T., Andersen, Ø.M., Malonated anthocyanins of garlic *Allium sativum* L., *Food Chem.*, 1997b, 58, 215–217.

Fossen, T., Andersen, Ø.M., Øvstedal, D.O., Pedersen, A.T., Raknes, Å., Characteristic anthocyanin pattern from onions and other *Allium spp.*, *J. Food Sci.*, 1996, 61, 703–706.

Francis, G.W., Andersen, Ø.M., Droplet counter-current chromatography of anthocyanins. *J. Chromatogr.*, 1984, 283, 445–448.

Garcia-Viguera, C., Zafrilla, P., Tomás-Barberán, F.A., The use of acetone as an extraction solvent for anthocyanins from strawberry fruit, *Phytochem. Anal.*, 1998, 9, 274–277.

Giusti, M.M., Ghanadan, H., Wrolstad, R.E., Elucidation of the structure and conformation of red radish (*Raphanus sativus*) anthocyanins using one-and two-dimensional nuclear magnetic resonance techniques, *J. Agric. Food Chem.*, 1998a, 46, 4858–4863.

Giusti, M.M., Rodríquez-Saona, L.E., Baggett, J.R., Reed, G.L., Durst, R.W., Wrolstad, R.E., Anthocyanin pigment composition of red radish cultivars as potential food colorants, *J. Food Sci.*, 1998b, 63, 219–224.

Giusti, M.M., Rodríquez-Saona, L.E., Griffin, D., Wrolstad, R.E., Electrospray and tandem mass-spectroscopy as tools for anthocyanin characterization, *J. Agric. Food Chem.*, 1999a, 47, 4657–4664.

Giusti, M.M., Rodríquez-Saona, L.E., Wrolstad, R.E., Molar absorptivity and color characteristics of acylated and non-acylated pelargonidin-based anthocyanins, *J. Agric. Food Chem.*, 1999b, 47, 4631–4637.

Giusti, M.M., Wrolstad, R.E., Characterization of red radish anthocyanins, *J. Food Sci.*, 1996, 61, 322–326.

Gläßgen, W.E., Wray, V., Strack, D., Metzger, J.W., Seitz, H.U., Anthocyanins from cell suspensions of *Daucus carota*, *Phytochemistry*, 1992a, 31, 1593–1601.

Gläßgen, W.E., Seitz, H.U., Metzger, J.W., High-performance liquid chromatography/electrospray mass spectrometry and tandem mass spectrometry of anthocyanins from plant tissues and cell cultures of *Daucus carota* L., *Biol. Mass Spectrom.*, 1992b, 21, 271–277.

Hale, M.L., Francis, F.J., Fagerson, I.S., Detection of enocyanin in cranberry juice cocktail by HPLC anthocyanin profile, *J. Food Sci.*, 1986, 51, 1511–1513.

Harborne, J.B., Spectral methods of characterizing anthocyanins, *Biochem. J.*, 1958, 70, 22–28.

Harborne, J.B., *Comparative Biochemistry of the Flavonoids*, Academic Press, London, United Kingdom, 1967.

Harborne, J.B. (Ed.), *The Flavonoids: Advances in Research Since 1980*, Chapman and Hall, London, United Kingdom, 1988.

Harborne, J.B. (Ed.), *The Flavonoids: Advances in Research Since 1986*, Chapman & Hall, London, United Kingdom, 1994.

Harborne, J.B., *Phytochemical Methods: A Guide to Modern Techniques of Plant Analysis*, 3rd edition, Chapman & Hall, London, United Kingdom, 1998.

Harborne, J.B., Boardley, M., The widespread occurrence in nature of anthocyanins as zwitterions, *Z. Naturforsch.*, 1985, 40c, 305.

Harborne, J.B., Mabry, T.J. (Eds.), *The Flavonoids: Advances in Research*, Chapman and Hall, London, United Kingdom, 1982.

Harborne, J.B., Grayer, R., The anthocyanins, in *The Flavonoids: Advances in Research Since 1980*, Harborne, J.B. (Ed.), Chapman & Hall, New York, 1988, 1–20.

Harborne, J.B., Mabry, T.J., Mabry, H. (Eds), *The Flavonoids*, Part 1, Academic Press, New York, 1975.

Heiger, D., *High Performance Capillary Electrophoresis*, Agilent Technologies, Santa Clara, California, 2000, 51.

Henke, H., *Preparative Gel Chromatography on Sephadex LH-20*, Hüthig, Heidelberg, Germany, 1995.

Hong, V., Wrolstad, R.E., Cranberry juice composition, *J. Assoc. Off. Anal. Chem.*, 1986, 69, 199–207.

Hong, V., Wrolstad, R.E., Characterization of anthocyanin-containing colorants and fruit juices by HPLC/photodiode array detection, *J. Agric. Food Chem.*, 1990a, 38, 698–708.

Hong, V., Wrolstad, R.E., Use of HPLC/photodiode array detection for the characterization of anthocyanins, *J. Agric. Food Chem.*, 1990b, 38, 708–715.

Hrazdina, G., Iredale, H., Mattick, L.R., Anthocyanin composition of *Brassica oleracea* cv. Red Danish, *Phytochemistry*, 1977, 16, 297–299.

Iacobucci, G.A., Sweeney, J.G., The chemistry of anthocyanins, anthocyanidins and related flavylium salts, *Tetrahedron*, 1983, 39, 3005–3038.

Ichiyanagi, T., Kashiwada, Y., Ikeshiro, Y., Hatano, Y., Shida, Y., Horie, M., Matsugo, S., Konishi, T., Complete assignment of bilberry (*Vaccinium myrtiillus* L.) anthocyanins separated by capillary zone electrophoresis, *Chem. Pharm. Bull.*, 2004, 52, 226–229.

Ichiyanagi, T., Oikawa, K., Tateyama, C., Konishi, K., Acid mediated hydrolysis of blueberry anthocyanins, *Chem. Pharm. Bull.*, 2001, 49, 114–117.

Ichiyanagi, T., Tateyama, C., Oikawa, K., Konishi, T., Comparison of anthocyanin distribution in different blueberry sources by capillary zone electrophoresis, *Biol. Pharm. Bull.*, 2000, 23, 492–497.

Jackman, R., Yada, R.Y., Tung, M.A., A review: Separation and chemical properties of anthocyanins used for their qualitative and quantitative analysis, *J. Food Biochem.*, 1987, 11, 279.

Kammerer, D., Claus, A., Carle, R., Schieber, A., Polyphenol screening of pomace from red and white grape varieties (*Vitis vinifera* L.) by HPLC-DAD-MS/MS, *J. Agric. Food Chem.*, 2004, 52, 4360–4367.

Kidøy, L., Nygård, A.M., Andersen, Ø.M., Pedersen, A.T., Aksnes, D.W., Kiremire, B.T., Anthocyanins in fruits of *Passiflora edulis* and *P. suberosa*, *J. Food Comp. Anal.*, 1997, 10, 49–54.

Kim, J.H., Nonaka, G.-I., Fujieda, K., Uemoto, S., Anthocyanin malonylglucosides in flowers of *Hibiscus syriacus*, *Phytochemistry*, 1989, 28, 1503–1506.

Kondo, T., Nakane, Y., Tamura, H., Goto, T., Eugster, C.H., Structure of monardaein, a bis-malonylated anthocyanin isolated from golden balm, *Monarda didyma*, *Tetrahedron Lett.*, 1985, 26, 5879–5882.

Kraemer-Schafhalter, A., Fuchs, H., Pfannhauser, W., Solid-phase extraction (SPE)—a comparison of 16 materials for the purification of anthocyanins from *Aronia melanocarpa* var Nero, *J. Sci. Food Agric.*, 1998, 78, 435–440.

Laine, R.A., Sweeley, C.C., Analysis of trimethylsilyl *O*-methyloximes of carbohydrates by combined gas-liquid chromatography-mass spectrometry, *Anal. Biochem.*, 1971, 43, 533–538.

Laine, R.A., Sweeley, C.C., *O*-Methyl oximes of sugars. Analysis as *O*-trimethylsilyl derivates by gas-liquid chromatography and mass spectrometry, *Carbohydr. Res.*, 1973, 27, 199–213.

Main, J.H., Clydesdale, F.M., Francis, F.J., Spray drying anthocyanin concentrates for use as food colorants, *J. Food Sci.*, 1978, 43, 1693–1694, 1697.

Markham, K.R., *Techniques of Flavonoid Identification*, Academic Press, New York, 1982.

Markham, K.R., Novel anthocyanins produced in petals of genetically transformed lisianthus, *Phytochemistry*, 1996, 42, 1035–1038.

Markham, K.R., Geiger, H., ^{1}H nuclear magnetic resonance spectroscopy of flavonoids and their glycosides in hexadeuterodimethylsulfoxide, in *The Flavonoids: Advances in Research Since 1986*, Harborne, J.B. (Ed.), Chapman & Hall, London, United Kingdom, 1994, 441–497.

Mazza, G., Miniati, E., *Anthocyanins in Fruits, Vegetables, and Grains*, CRC Press, Boca Raton, FL, 1993.

Merken, H.M., Beecher, G.R., Measurement of food flavonoids by high-performance liquid chromatography: A review, *J. Agric. Food Chem.*, 2000, 48, 577–599.

Miller, C.A., Fischer, S.M., Kalt, W., McDonald, J.E., Ricker, R.D., LC/MS analysis of anthocyanins, Presented at ASMS '98, 1998, Orlando, FL.

Nørbæk, R., Kondo, T., Anthocyanins from flowers of Crocus (Iridaceae), *Phytochemistry*, 1998, 47, 861–864.

Ohta, H., Akuta, S., Osajima, Y., Stability of anthocyanin pigments and related compounds in acidic solutions, *Nippon Shokuhin Kogyo Gakkaishi*, 1980, 27, 81–85.

Oszmianski, J., Lee, C.Y., Isolation and HPLC determination of phenolic compounds in red grapes, *Am. J. Enol. Vitic.*, 1990, 41(3), 204–206.

Philip, T., An anthocyanin recovery system from grape wastes, *J. Food Sci.*, 1974, 39, 859.

Phippen, W.E., Simon, J.E., Anthocyanins in basil (*Ocimum basilicum* L.), *J. Agric. Food Chem.*, 1998, 46, 1734–1738.

Piovan, A., Filippini, R., Favretto, D., Characterization of the anthocyanins of *Catharanthus roseus* (L.) G. Don in vivo and in vitro by electrospray ionization trap mass spectrometry, *Rapid Commn. Mass Spectrom.*, 1998, 12, 361–367.

Renault, J.-H., Thépenier, P., Zèches-Hanrot, M., Le Mon Olivier, L., Durand, A., Foucault, A., Margraff, R., Preparative separation of anthocyanins by gradient elution centrifugal partition chromatography, *J. Chromatogr. A*, 1997, 763, 345–352.

Revilla, I., Pérez-Margariño, S., González-SanJosé, M.L., Beltrán, S., Identification of anthocyanin derivatives in grape skin extracts and red wines by liquid chromatography with diode array and mass spectrometric detection, *J. Chromatogr. A*, 1999, 847, 83–90.

Revilla, E., Ryan, J.-M., Martin-Ortega, G., Comparison of several procedures used for the extraction of anthocyanins from red grapes, *J. Agric. Food Chem.*, 1998, 46, 4592–4597.

Ribéreau-Gayon, P., University reviews in botany, in *Plant Phenolics*, Heywood, V.H. (Ed.), Vol. 3, Oliver and Boyd, Edinburgh, Scotland, 1972.

Rivas-Gonzalo, J.C., Analysis of anthocyanins, in *Methods of Polyphenol Analysis*, Santos-Buelga, C., Williamson, G. (Eds.), Royal Society of Chemistry, Cambridge, United Kingdom, 2003, 338–358.

Rodríguez-Sanona L.E., Giusti, M.M., Wrolstad, R.E., Anthocyanin pigment composition of red-fleshed potatoes, *J. Food. Sci.*, 1998, 63, 458–465.

Saito, N., Abe, K., Honda, T., Timberlake, C.F., Bridle, P., Acylated delphinidin glucosides and flavonols from *Clitoria ternatea*, *Phytochemistry*, 1985, 24, 1583–1586.

Saito, N., Tatsuzawa, F., Yoda, K., Yokoi, M., Kasahara, K., Iida, S., Shighara, A., Honda, T., Acylated cyanidin glycosides in the violet-blue flowers of *Ipomoea purpurea*, *Phytochemistry*, 1995, 40, 1283–1289.

Saito, N., Timberlake, C.F., Tucknott, O.G., Lewis, I.A.S., Fast atom bombardment mass spectrometry of the anthocyanins violanin and platyconin, *Phytochemistry*, 1983, 22, 1007–1009.

Strack, D., Busch, E., Wray, V., Grotjahn, L., Klein, E., Cyanidin 3-oxalylglucoside in orchids, *Z. Naturforsch.*, 1986, 41c, 707.

Strack, D., Wray, V., Anthocyanins, in *Methods in Plant Biochemistry*, Dey, P.M., Harborne, J.B. (Eds.), Academic Press, New York, 1989, 325–356.

Strack, D., Wray, V., The anthocyanins, in *The Flavonoids: Advances in Research Since 1986*, Harborne, J.B. (Ed.), Chapman & Hall, London, United Kingdom, 1994, 1–22.

Sugui, J.A., Bonham, C., Lo, S.-C., Wood, K.V., Nicholson, R.L., MALDI-TOF analysis of mixtures of 3-deoxyanthocyanidins and anthocyanins, *Phytochemistry*, 1998, 48, 1063–1066.

Takeda, K., Harborne, J.B., Self, R., Identification and distribution of malonated anthocyanins in plants of the Compositae, *Phytochemistry*, 1986, 25, 1337–1342.

Takeoka, G.R., Dao, L.T., Full, G.H., Wong, R.Y., Harden, L.A., Edwards, R.H., Berrios, J.D.J., Characterization of black bean (*Phaseolus vulgaris* L.) anthocyanins, *J. Agric. Food Chem.*, 1997, 45, 3395–3400.

Takeoka, G.R., Dao, L.T., Tamura, H., Harden, L.A., Delphinidin 3-*O*-(2-*O*-β-D-glucopyranosyl-α-L-arabinopyranoside): A novel anthocyanin identified in Beluga black lentils, *J. Agric. Food Chem.*, 2005, 53, 4932–4937.

Tamura, H., Hayashi, Y., Sugisawa, H., Kondo, T., Structure determination of acylated anthocyanins in Muscat Bailey A grapes by homonuclear Hartmann-Hahn (HOHAHA) spectroscopy and liquid chromatography–mass spectrometry, *Phytochem. Anal.*, 1994, 5, 190–196.

Timberlake, C.F., Bridle, P., Anthocyanins, in *Developments in Food Colours*, Walford, J. (Ed.), Vol. 1, Applied Science Publishing Ltd, London, United Kingdom, 1980, 115–149.

Torskangerpoll, K., Fossen, T., Andersen, Ø.M., Anthocyanin pigments of tulips, *Phytochemistry*, 1999, 52, 1687–1692.

Tsuda, T., Horio, F., Osawa, T., Dietary cyanidin 3-*O*-β-D-glucoside increases ex vivo oxidation resistance of serum in rats, *Lipids*, 1998, 33, 583–588.

Tsuda, T., Watanabe, M., Ohshima, K., Norinobu, S., Choi, S.-W., Kawakishi, S., Osawa, T., Antioxidative activity of the anthocyanin pigments cyanidin 3-O-β-D-glucoside and cyanidin, *J. Agric. Food Chem.*, 1994, 42, 2407–2410.

Van Sumere, C.F., Vande Casteele, K., De Loose, R., Heursel, J., Reversed phase HPLC analysis of flavonoids and the biochemical identification of cultivars of evergreen Azelea, in *Annual Proceedings of the Phytochemical Society of Europe*, Van Sumere, C.F., Lea, P.J. (Eds.), Vol. 25, Clarendon Press, Oxford, United Kingdom, 1985.

Van Wyk, B.E., Winter, P.J.D., Chemotaxonomic value of anthocyanins in Podalyria and Virgilia (Tribe Podalyrieae: Fabaceae), *Biochem. Syst. Ecol.*, 1994, 22, 813–818.

Vidal, S., Hayasaka, Y., Meudec, E., Cheynier, V., Skouroumounis, G., Fractionation of grape anthocyanin classes using multilayer coil countercurrent chromatography with step gradient elution, *J. Agric. Food Chem.*, 2004, 52, 713–719.

Wang, H., Cao, G., Prior, R.L., Oxygen radical absorbing capacity of anthocyanins, *J. Agric. Food Chem.*, 1997, 45, 304–309.

Wang, H., Nair, M.G., Strasburg, G.M., Chang, Y.-C., Booren, A.M., Gray, J.I., DeWitt, D.L., Antioxidant and antiinflammatory activities of anthocyanins and their algycon, cyanidin, from tart cherries, *J. Nat. Prod.*, 1999, 62, 294–296.

Wang, J., Sporns, P., Analysis of anthocyanins in red wine and fruit juice using MALDI-MS, *J. Agric. Food Chem.*, 1999, 47, 2009–2015.

Watanabe, T., Yamamoto, A., Nagai, S., Terabe, S., Analysis of elderberry pigments in commercial food samples by micellar electrokinetic chromatography, *Anal. Sci.*, 1998, 14, 839–844.

Wehr, T., Configuring HPLC systems for LC-MS, *LCGC*, 2000, 18, 406–416.

Wilkinson, M., Swenny, J.G., Iacobucci, G.A., High-pressure liquid chromatography of anthocyanins, *J. Chromatogr.*, 1977, 132, 349–351.

Williams, D.H., Fleming, I., *Spectroscopic Methods in Organic Chemistry*, 5th edition, McGraw-Hill Publishing Company, Berkshire, England, 1995.

Wood, K.V., Bonham, C., Hipskind, J., Nicholson, R.L., Analysis of anthocyanins and 3-deoxyanthocyanidins by plasma desorption mass spectrometry, *Phytochemistry*, 1994, 37, 557–560.

Wood, K.V., Bonham, C.C., Ng, J., Hipskind, J., Nicholson, R.L., Plasma desorption mass spectrometry of anthocyanidins, *Rapid Commn. Mass Spectrom.*, 1993, 7, 400.

Wrolstad, R.E., Acree, T.E., Decker, E.A., Penner, M.H., Reid, D.S., Schwartz, S.J., Shoemaker, C.F., Smith, D.M., Sporns, P. (Eds.), *Handbook of Food Analytical Chemistry: Pigments, Colorants, Flavors, Texture, and Bioactive Food Components*, John Wiley & Sons, Inc., Hoboken, NJ, 2004.

Wrolstad, R.E., Durst, R.W., Giusti, M.M., Rodríquez-Saona, L.E., Analysis of anthocyanins in nutraceuticals, in *Quality Management of Nutraceuticals*, Ho, C.-T., Zheng, Q.Y. (Eds.), ACS Symposium Series 803, American Chemical Society, Washington DC, WA, 2002, 42–62.

Wulf, L., Nagel, C.W., HPLC separation of anthocyanins in *Vitis vinifera*, *Am. J. Enol. Vitic.*, 1978, 29, 42–49.

6 Carotenoids and Provitamin A in Functional Foods

María Isabel Mínguez-Mosquera, Dámaso Hornero-Méndez, and Antonio Pérez-Gálvez

CONTENTS

The definition of functional food can be very wide and variable, but we should not forget that all foods have at least one main function, alimentation. A functional food should be considered a dietary ingredient that can have above-normal nutritional value. According to Gibson and Roberfroid, a useful definition of functional food is a dietary ingredient that affects its host in a targeted manner so as to exert positive effects that may, in due course, justify certain health claims.[1] This definition leads directly to the question: how can foods be made more functional? We might consider four possible mechanisms: (i) elimination of a component having a negative physiological effect, (ii) increasing the concentration of components that contribute to the beneficial aspects, (iii) addition of a new ingredient observed to have general advantages, and (iv) partial substitution of a negative component by another, positive one, without adversely affecting the nutritional value of the food.[2]

Current dietetic recommendations tending to increase the consumption of fruits and vegetables are based on a series of epidemiological evidence relating such diets with a longer and better-quality life. Fruits and vegetables are considered rich sources of antioxidants such as vitamin C, vitamin E, carotenoids, flavonoids, etc. However, they are not the only source of dietary antioxidants, and, for example, we can find high contents of vitamin E in nuts, seeds, and cereals in particular, eggs, margarines, vegetable oils, and dairy products. Similarly, the carotenoids are not restricted to fruits and vegetables, so that dairy products, vegetable oils, and some animal products can be considered an important source of these compounds.[3]

Research carried out in the last 70 years has shown that the carotenoid pigments present in fruits and vegetables are the main dietary source of vitamin A for most people, especially in poorer countries, where vitamin A deficiency is a serious problem. β-Carotene is the main compound with provitamin A activity.[4] When incorporated in the diet, it is broken down into two molecules of retinol (vitamin A) by action of the enzyme β-carotene-15,15′-dioxygenase in the intestine. However, β-carotene is not the only carotenoid with provitamin A activity, and, as we will discuss in detail, any carotenoid with at least one unsubstituted β-ring can undergo similar cleavages and give rise to a vitamin A molecule. Carotenoids such as α-carotene and β-cryptoxanthin can thus contribute substantially to the nutritional value of fruits and vegetables.

In recent years, however, growing evidence has indicated that this may not be the only contribution of carotenoids to health. Numerous epidemiological studies indicate that carotenoid-rich diets are correlated with lower risk of contracting certain types of cancers, heart disease, and other important human diseases.[5–8] While such facts by themselves do not show that the carotenoids are the active factors in this beneficial effect, experiments and tests with humans and other animals have shown that carotenoids do have a direct effect. Current attention is centered on the action of β-carotene as antioxidant, as it may interfere in free radical oxidation (such as the peroxidation of lipids), typical of many degenerative diseases. Although it has been clearly demonstrated that β-carotene has a significant antioxidant effect in vitro, there is still no real proof that this is its in vivo function at the low concentrations in which it is found and under physiological conditions.[9,10]

Recent research has demonstrated that the carotenoids can modify membrane structure and properties, with substantial effects on the human immune response system and on the process of gap–junction communication between neighboring cells.[11,12] This opens up a range of exciting new possibilities of research and applications.

To be able to have any beneficial action, the carotenoids must be absorbed, transported, and deposited in certain tissues. All carotenes and many xanthophylls can be used, although it seems that the epoxyxanthophylls such as violaxanthin and neoxanthin, which are abundant in plants (particularly in green tissues), cannot be incorporated into and used by the human body. This brings up the important aspect of the bioavailability of dietary carotenoids, which depends on other dietetic factors, in particular the contribution of dietary fats. Carotenoids are absorbed more efficiently from cooked foods than from fresh ones, probably because cooking and processing liberates the carotenoids from their usual molecular environment, making them more accessible for solubilization. The carotenoids with *Z*-conformation are generally more bioavailable than the all-*E* forms, apparently because the *Z*-forms have a lower tendency to produce microcrystalline aggregates, and thus a higher solubility.

Furthermore, the carotenoid pigments, either in isolation or jointly with other natural pigments such as chlorophylls and anthocyanins, are responsible for food color. Color is the first characteristic the consumer perceives of a food, and confers expectations of quality and flavor. Food quality is judged firstly on color, and the consumer will reject foods with an external color other than that established as correct. The food industry, knowing well this natural relation of color–quality (and vice versa), tries to adjust the industrial processes of transformation and preparation of foods to preserve the integrity of the compounds responsible for an acceptable color. This is not always possible, and it is normal practice to add coloring matter to enhance, homogenize, or even modify color to make the food more attractive to the consumer.[13]

Carotenoids are thus compounds of great dietetic importance, not only as precursors of vitamin A, but also as molecules that take part in cell protection and consumer attraction. In consequence, carotenoid content and composition are important factors in the nutritional evaluation of fruits, vegetables, and foods in general.[14] The latest advances show that all the carotenoids, not only β-carotene and others having provitamin A value, are of considerable benefit and should be included in the composition of any functional food. Current trends in legislation and use of pigments or colorants in the food industry are toward the progressive exclusion of synthetic pigments in favor of natural ones, used not only for their coloring power but also for the added value of contributing nutritional and physiological properties beneficial for health, such as in the case of carotenoids and anthocyanins. These aspects are discussed at the International Symposia on Pigments in Food. For instance, the conference entitled "More than just colours," is a good reflect of the importance of the pigments in the food field.[15]

6.1 INTRODUCTION TO THE CAROTENOIDS

Among all the pigments present in living organisms, there is no doubt that the carotenoids are the most widely distributed in nature. They are found throughout the plant kingdom—in both photosynthetic and nonphotosynthetic tissues—in bacteria, in fungi, and in animals, although the latter are unable to synthesize them, so incorporate them from dietary plants. It is estimated that the annual production of carotenoids in nature is around 10^8 tons.[16]

In 1831, Wackenroder isolated an orange pigment from the carrot root (*Daucus carota*), and coined the term "carotene" from the Latin word carota.[17] Six years later, Berzelius assigned the name "xanthophylls" to the yellow pigment of autumn leaves. Nevertheless, it was not until 1906 that Tswett was able to separate the pigments from an extract of leaves, inventing chromatography.[18] Around 1929, the works of von Euler, Karrer, and Moore demonstrated the relationship between the carotenoids and vitamin A, revealing their nutritional value. From then on, interest in these compounds has grown; thanks to the continuous development of analytical techniques, the number of known naturally occurring carotenoids has risen from 11 in 1934 to 32 in 1948, 230 in 1971, 450 in 1987, and more than 650 today.[19]

Most of the carotenoids found naturally in fruits and vegetables present a skeleton of 40 carbon atoms (C_{40}), and are biosynthesized from two molecules of an intermediary C_{20} (geranylgeranyl diphosphate), giving rise to a phytoene as generic precursor of the whole wide range of carotenoids present in the plant kingdom. The phytoene molecule undergoes a series of successive desaturations (up to four), introducing new double bonds into the carbon chain, resulting in spreading of the double bond conjugation and thus of the chromophore that is typical of these natural pigments and responsible for their chromatic properties. Successive structural changes, such as cyclization of one or both ends, hydroxylation or introduction of other oxygenated functions, give rise to the great range of carotenoid structures found in nature.

In general, the carotenoids can be classified into two great groups: carotenes, which are strictly hydrocarbons, and xanthophylls, which are derived from the former and contain oxygenated functions. Structurally, the carotenoids may be acyclic (e.g., lycopene) or contain a ring of five or six carbons at one or both ends of the molecule (e.g., β-carotene). Figure 6.1 shows the structure and the system of numbering using lycopene and β-carotene as models of acyclic and bicyclic carotenoids, respectively. Figure 6.2 includes the structures of some representative carotenes and xanthophylls, and Appendix 6.1 shows the structures of the carotenes and xanthophylls commonly found in foods.

Traditionally, trivial names have been given to new carotenoids after discovering, which in most cases referred to the natural source from which it had first

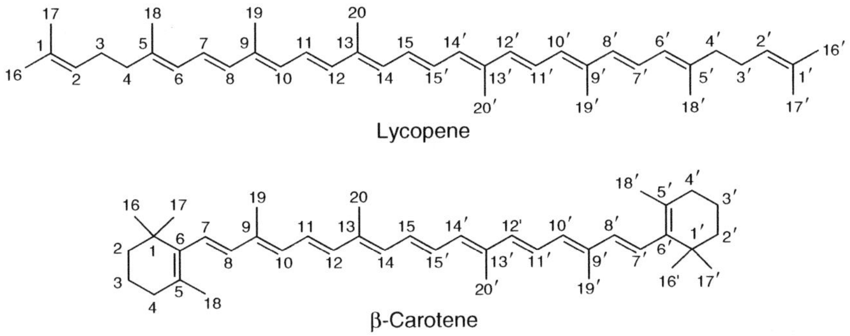

FIGURE 6.1 C_{40} skeleton and numbering scheme of carotenoids.

been isolated, for example, zeaxanthin from maize (*Zea mays*) and lycopene from tomato (*Lycopersicon esculentum*). Because this system does not include any structural information, a semisystematic nomenclature was defined that does give

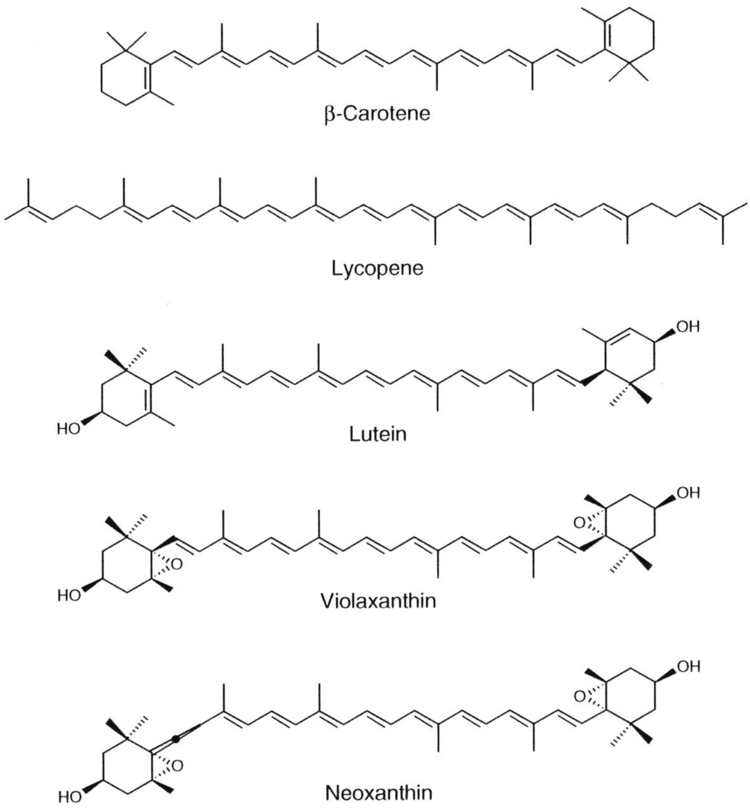

FIGURE 6.2 Some common carotenes and xanthophylls.

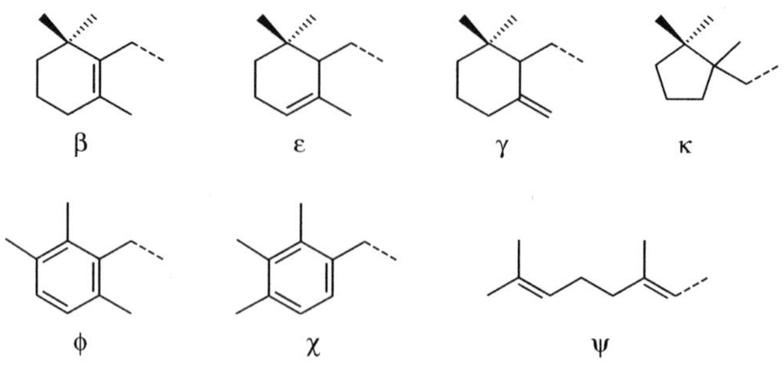

FIGURE 6.3 Chemical structures for the seven end groups found in natural carotenoids.

structural information and additionally refers to the parent carotene. This system was approved by the International Union of Pure and Applied Chemistry (IUPAC), which publish in 1975 a compendium of rules for the nomenclature of carotenoids.[20,21] Greek letters are used to designate the end groups that may be present in the carotenoid molecule. Those carotenoids lacking an end group are denominated apocarotenoids. Figure 6.3 shows the seven end groups that have been found in natural carotenoids, namely β, ε, γ, κ, ϕ, χ, and ψ. In addition, Table 6.1 shows trivial and semisystematic names for the carotenes and xanthophylls most commonly found in foods.

The presence of a large number of conjugated double bonds in the carotenoid molecule makes possible numerous geometric isomers (*Z–E* isomers, also called cis-trans). In practice, however, most of the carotenoids are naturally present as all-*E*. A few carotenoids are present in the *Z*-form in nature, such as bixin present in the annatto seeds (*Bixa orellana*), or prolycopene (a carotenoid with several double bonds and *Z* configuration, poly-*Z*) present in certain varieties of tomato.

6.2 PRESENCE, DISTRIBUTION, LOCALIZATION, AND FUNCTIONS

Carotenoid pigments are widespread among living organisms, both plants and animals, but are found in greater concentration and variety in the former, which are the only organisms (together with certain bacteria) able to biosynthesize them.[16] The distribution of carotenoids among the different groups of higher plants does not follow a single pattern. In green plant tissues, the class and content of carotenoid pigments follows the general model associated with the presence of chloroplasts, with β-carotene being the predominant carotene, followed by the xanthophylls, lutein, violaxanthin, and neoxanthin. Zeaxanthin, γ-carotene, β-cryptoxanthin, and antheraxanthin are found in small amounts. In the case of fruits, the xanthophylls are normally found in greater amounts. Exceptions are found in maize, the predominant pigments being lutein and zeaxanthin, while in mango (*Mangifera indica*) and persimmon (*Diospyros kaki*), they are β-cryptoxanthin and zeaxanthin. In contrast, in tomato, the major carotenoid is lycopene, a carotene. In certain fruits, a carotenoid, besides being the major one, is limited totally or almost totally to a single plant species. Capsanthin and capsorubin are found almost exclusively in ripe fruits of the

TABLE 6.1
Trivial and Semisystematic Names for Some Carotenes and Xanthophylls

Trivial Name	Semisystematic Name
Carotenes	
α-Carotene	β,ε-Carotene
β-Carotene	β,β-Carotene
δ-Carotene	ε,ψ-Carotene
γ-Carotene	β,ψ-Carotene
ε-Carotene	ε,ε-Carotene
ζ-Carotene	7,8,7′,8′-Tetrahydro-ψ,ψ-carotene
Lycopene	ψ,ψ-Carotene
Neurosporene	7,8-Dihydro-ψ,ψ-carotene
Phytoene	7,8,11,12,7′,8′,11′,12′-Octahydro-ψ,ψ-carotene
Phytofluene	7,8,11,12,7′,8′-Hexahydro-ψ,ψ-carotene
Xanthophylls	
Antheraxanthin	5,6-Epoxy-5,6-dihydro-β,β-carotene-3,3′-diol
Astaxanthin	3,3′-Dihydroxy-β,β-carotene-4,4′-dione
Auroxanthin	5,8,5′,8′-Diepoxy-5,8,5′,8′-tetrahydro-β,β-carotene-3,3′-Diol
Bixin	Methyl hydrogen 9′-Z-6,6′-diapocarotene-6,6′-dioate
Canthaxanthin	β,β-Carotene-4,4′-dione
Capsanthin	3,3′-Dihydroxy-β,κ-carotene-6′-one
Capsorubin	3,3′-Dihydroxy-κ,κ-carotene-6,6′-dione
Crocetin	8,8′-Diapocarotene-8,8′-dioic acid
α-Cryptoxanthin	β,ε-Carotene-3-ol
β-Cryptoxanthin	β,β-Carotene-3-ol
Cryptoxanthin-5,6-epoxide	5,6-Epoxy-5,6-dihydro-β,β-carotene-3-ol
Cucurbitaxanthin A	3′,6′-Epoxy-5′,6′-dihydro-β,β-carotene-3,5′-diol
Lactucaxanthin	ε,ε-Carotene-3,3′-diol
Lutein	β,ε-Carotene-3,3′-diol
Luteoxanthin	5,6,5′,8′-Diepoxy-5,6,5′,8′-tetrahydro-β,β-carotene-3,3′-Diol
Mutatoxanthin	5,8-Epoxy-5,8-dihydro-β,β-carotene-3,3′-diol
Neochrome	5′,8′-Epoxy-6,7-didehydro-5,6,5′,8′-tetrahydro-β,β-Carotene-3,5,3′-triol
Neoxanthin	5′,6′-Epoxy-6,7-didehydro-5,6,5′,6′-tetrahydro-β,β-Carotene-3,5,3′-triol
Norbixin	6,6′-Diapocarotene-6,6′-dioic acid
Violaxanthin	5,6,5′,6′-Diepoxy-5,6,5′,6′-tetrahydro-β,β-carotene-3,3′-Diol
Zeaxanthin	β,β-Carotene-3,3′-diol

genus Capsicum, and are responsible for their attracting red color.[22–24] The orange (*Citrus sinensis*) contains varying amounts of β-citraurin and β-citranaxanthin (both apocarotenoids), together with β-cryptoxanthin, lutein, antheraxanthin, violaxanthin, and traces of their carotene precursors.[25] The presence and distribution of the most common carotenoid pigments found in fruit and vegetables, and in general in foods, are shown in Table 6.2.

In green plant tissues, the xanthophylls are in the free form. However, as a consequence of leaf senescence and the ripening of many fruits, coinciding with the transformation of the chloroplasts into chromoplasts, the carotenoid pigments undergo esterification with different fatty acids. The esterification is related to the capacity of the plant (in particular fruits and flowers) to overproduce and accumulate carotenoid pigments. The change in esterification profile of the xanthophylls in fruits of red pepper has been proposed recently as a ripening index.[26] The function of esterification, which does not affect the chromophore properties of the pigment, seems to be related, in the case of fruits and flowers, with attracting animals that act as vehicle for the dissemination of seeds and pollen for increasing reproductive success.[27]

TABLE 6.2
Natural Occurrence of Some Common Carotenes and Xanthophylls

Carotenoid	Natural Occurrence
Carotenes	
α-Carotene, β-carotene, δ-carotene, γ-carotene, ε-carotene, and ζ-carotene	Fruits and vegetables, especially in carrots, sweet potato, and palm tree fruit. Delta tomato mutant has δ-carotene as major carotene. Rose hips are good source for γ-carotene
Lycopene and neurosporene	Tomato (*Licopersicon esculentum*), water melon, and rose hips (*Rosa spp.*)
Phytofluene and phytoene	Carotenoid-rich fruits, flowers, and roots (carrot)
Xanthophylls	
Antheraxanthin	Anthers and petals of many yellow flowers. Also in fruits and vegetables
Astaxanthin	Bird feathers, fish (salmon), and invertebrate animals (Lobster, *Homarus gammarus*)
Bixin and norbixin	Annatto (*Bixa orellana*) seeds
Canthaxanthin	Mainly synthetic, naturally found in some cyanobacteria and green algae
Capsanthin, capsanthin-5,6-epoxide, and capsorubin	Capsicum annuum ripe fruits
Crocetin	Saffron (*Crocus sativus*) flowers
Cucurbitaxanthin A	Pumpkin (*Cucurbita maxima*) flesh
Lactucaxanthin	Lettuce (*Lactuca sativa*) leaves
Lutein, violaxanthin, neoxanthin, and mutatoxanthin (as minor carotenoid)	Green fruits, vegetables, and flowers
Luteoxanthin, neochrome, and auroxanthin	Vegetables and fruits processed under acid conditions and fermentation
Rubixanthin	Rose hips (*Rosa spp.*)
Zeaxanthin, β-cryptoxanthin, α-cryptoxanthin, and cryptoxanthin-5,6-epoxide	Seeds (corn), flowers, and fruits: mango, papaya, persimmon

In the animal kingdom, carotenoids are incorporated via the diet, and stored in different tissues. The egg yolk owes its yellow color to xanthophylls such as lutein and zeaxanthin, and to traces of β-carotene. While the presence and distribution of carotenoids in mammals is very limited, other vertebrates, such as birds, fish, reptiles, and amphibians, and, above all, the invertebrates show a great diversity of carotenoid pigments, and even have the capacity to modify structurally some of the carotenoids ingested in the diet.[28]

In the case of the invertebrates, the carotenoids can be intimately associated to proteins, giving rise to a very important group of compounds known as carotenoproteins. Such association results in changes of the chromatic characteristics of the carotenoid, which presents coloration including green, blue, purple, and gray.[29] The presence of carotenoproteins in crustaceans has been known for some time, the first mention probably being that of the French biologist Pouchet in 1872.[30] They are normally found in the exoskeleton, and in the eggs and ovaries, suggesting their participation in the animal's development and in the nutrient reserve. An important function associated to the presence of carotenoproteins is protective coloring, used as a means of camouflage in the environment (cryptic function), or as a form of protection from the harmful effects of external agents such as radiation.[31]

In plants, the carotenoids are located and accumulated in specialized subcellular organelles called plastids, namely chloroplasts and chromoplasts.[32] The chloroplasts are present in all photosynthetic tissues (mainly leaves), where practically all the carotenoids are present in the form of chlorophyll–carotenoid–protein complexes (photosystems) at the level of the thylakoid membranes. In this environment, the carotenoids have their prime natural function as assistant collectors of light energy (antenna pigments) in the photosynthetic process because, owing to their absorption spectrum, they are able to capture photons that escape the reach of the chlorophylls. Nevertheless, the chromoplasts present in flowers, ripe fruits, and certain roots and tubercles are the organelles specializing in the massive accumulation of carotenoids, and have the greatest variety of structural forms. In the case of the chromoplasts, the carotenoids are usually accumulated in lipid-rich structures, the plastoglobules, as for example in the fruits of the genus Capsicum and many flowers. In certain cases, such as tomato, carrot (*Daucus carota*), and pumpkin (*Cucurbita maxima*), the presence of carotenoid crystals has also been reported, mainly carotenes, immersed in the stromatic space.[33] The change from chloroplast to chromoplast, which is associated to the fruit ripening process, is especially important in the case of the fruit-denominated carotenogenic (e.g., pepper and tomato), characterized by a massive synthesis of carotenoids during ripening, which is usually accompanied by a change in the carotenoid profile of the fruit. It is noteworthy that the chromoplast xanthophylls are usually esterified with different fatty acids, increasing their lipophilic character and facilitating their accumulation in the plastoglobules. In the case of fruits and flowers, the main function of carotenoids is undoubtedly to attract animals (insects, birds, and mammals) so that they cooperate in seed dispersion and pollen transport.[27]

The carotenoids also play a very important role as protectors of the chlorophylls and the photosynthetic apparatus in general by blocking (the quenching effect) very reactive forms of triplet chlorophylls (^{3}Chl) and singlet oxygen ($^1O_2$) formed during the capture of light energy. The carotenoids also take an active part in the plant's photoprotective and antioxidant action.[34]

6.3 GENERAL PROPERTIES

The carotenoids are lipophilic substances, and thus insoluble in aqueous medium, except in certain cases where highly polar functional groups are present, as in norbixin, a carotenoid with dicarboxyl acid structure. The presence of the long, extensive system of conjugated double bonds (or polyene chain) is responsible for one of the most distinctive characteristics of the carotenoids, light absorption. A chromophore with seven or more double bonds gives the capacity of absorbing light in the visible range, and consequently, the observation of colors spanning from yellow to red, via a great variety of orange tones. Moreover, the polyene chain makes the carotenoid molecule extremely susceptible to isomerizing and oxidizing conditions such as light, heat, or acids. These properties regarding its structural lability largely determine the mode of operation and the precautions to be taken when working on the isolation and identification of carotenoids in the laboratory.

The properties of carotenoid pigments in vitro, that is, once extracted and dissolved, may be different to those in vivo because of the interaction with the physicochemical environment (mainly, lipids and proteins) surrounding the pigments. This can be particularly critical regarding the functionality and action of the carotenoids in vivo.

6.4 SPECTROSCOPIC PROPERTIES

The characteristic visible light absorption spectrum of the carotenoid pigments is due to the system of conjugated double bonds of their hydrocarbon chain (polyene). For a given carotenoid, the position of the bands of maximum light absorption (λ_{max}) is a function of the number of conjugated double bonds in the molecule (Figure 6.4).

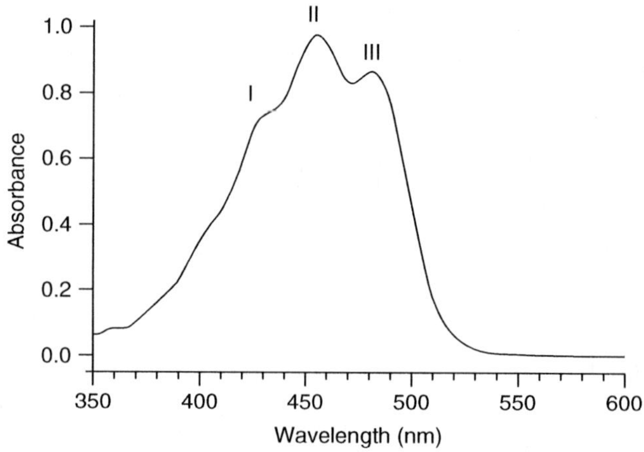

FIGURE 6.4 General ultraviolet (UV)–visible light absorption spectrum of carotenoids.

The absorption maxima are usually referred to using roman numbers (I, II, III). The introduction of a new conjugated double bond into the chromophore causes a bathochromic displacement (to higher wavelength) of 20–22 nm in the absorption maxima (λ_{max}), although this effect may be modified by the presence of different end groups. If the end group is β-ring, the contribution to the chromophore is only 9–11 nm. In the case of a ε-ring, the conjugation of its double bond is lost and it does not participate in the chromophore. If the double bond of the β-ring is replaced by a 5,6-epoxide group, there is a hypsochromic displacement of 6–9 nm in λ_{max}.

The conversion in acid medium of the 5,6-epoxide group to 5,8-epoxide causes a new hypsochromic displacement of 20–22 nm because of the loss of a conjugated double bond at positions 7 and 8. The introduction of a hydroxyl group into the cyclic end group (normally positions 3 and 3′) has almost no effect on the position of the absorption maxima, and the same happens with ketone groups not conjugated with the polyene chain, whereas the conjugated ones cause a bathochromic displacement of 5–10 nm in the maxima. Figure 6.5 illustrates some of these changes in the electron absorption spectrum depending on the nature of the chromophore. The *Z*/*E* isomerism has a profound effect on the absorption spectrum, with a new maximum appearing in the near ultraviolet zone, around 320–340 nm.

The shape of the absorption spectrum of the carotenoid, and the positions of the absorption maxima, can vary depending on the interactions of the molecule with the solvent or lipid environment in which it is dissolved.[35] In general, solvents with low polarity have little effect on the position of the absorption maxima, so that for a given carotenoid, the values of λ_{max} are almost identical in hexane, light petroleum, diethyl ether, methanol, and ethanol. Acetone, commonly used in carotenoid extraction, causes a bathochromic displacement of around 2–6 nm in the maxima compared with the aforementioned. In contrast, very polar solvents such as chloroform, benzene, and pyridine cause very significant bathochromic displacements (10–25 nm), which are extreme in the case of carbon disulfide (30–40 nm).

The acyclic carotenoids usually present greater persistence or a finer structure than the cyclic (monocyclic and bicyclic) ones. This is related with the noncoplanarity of the end rings with the central polyene chain. The first of the absorption maxima in carotenoids with two β end groups, for example β-carotene, appears as an inflexion. Most of the ketocarotenoids, such as capsanthin, where the carbonyl group is conjugated with the polyene chain, lose the fine structure, and only a wide main band is observed with very weak inflexions on each side.[36] Figure 6.6 compares the fine structure of the ultraviolet (UV)–visible spectra for various carotenoids. When comparing the absorption spectrum of a given carotenoid, it is important to compare not only the positions of the absorption maxima (λ_{max}), but also the shape and fine structure (defined by percent of III/II).

Carotenoid pigments also present other spectroscopic properties such as fluorescence and absorption of energy in the infrared (IR) region. Fluorescence is a property rarely present in the carotenoids, and in fact, only a few carotenoids fluoresce when they are excited at appropriated wavelengths (e.g., phytofluene). Therefore, fluorescence spectroscopy is not frequently used in carotenoid studies. Similarly, the use of

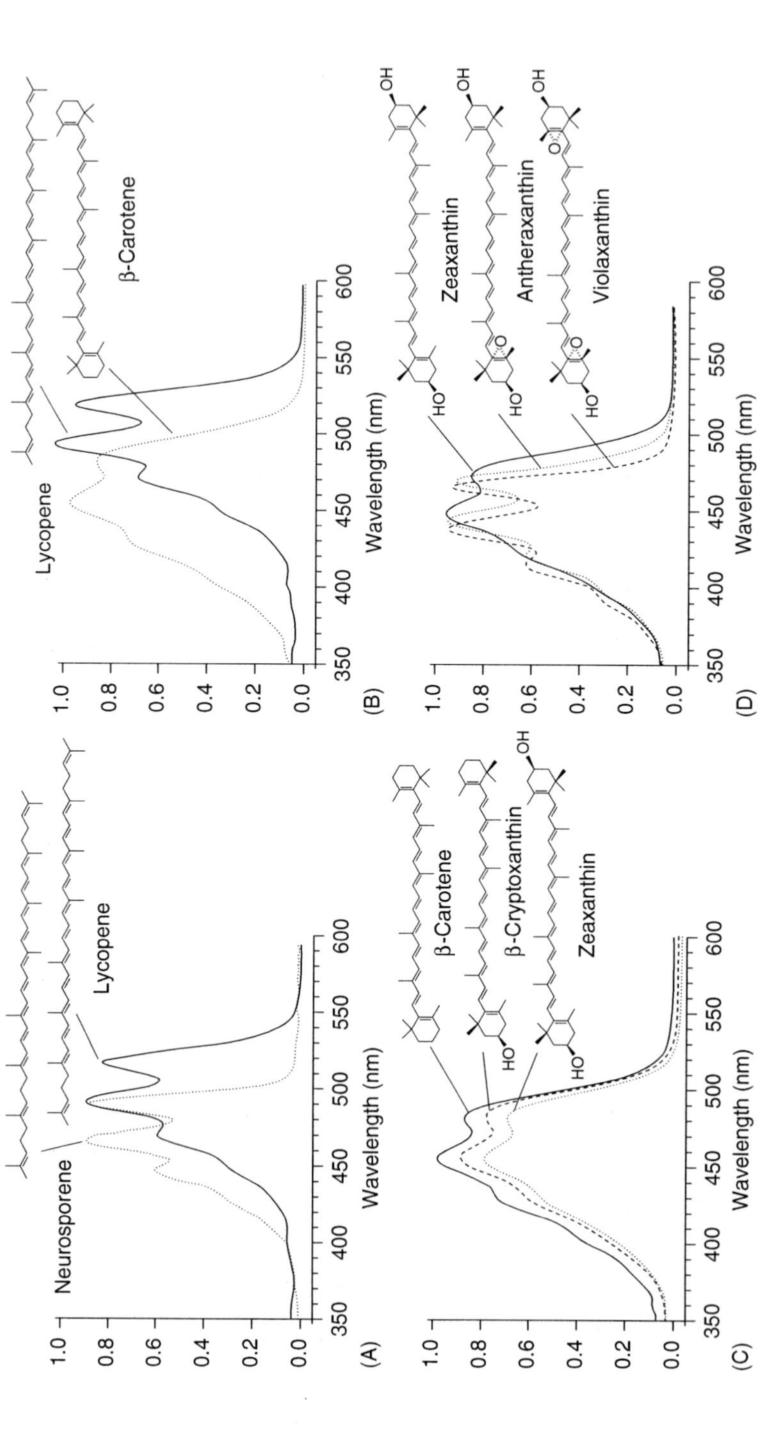

FIGURE 6.5 Effect of some structural changes of the chromophore on the ultraviolet (UV)–visible light absorption spectrum. (A) Increase of the chromophore length; (B) Cyclization; (C) Hydroxylation; (D) Epoxydation.

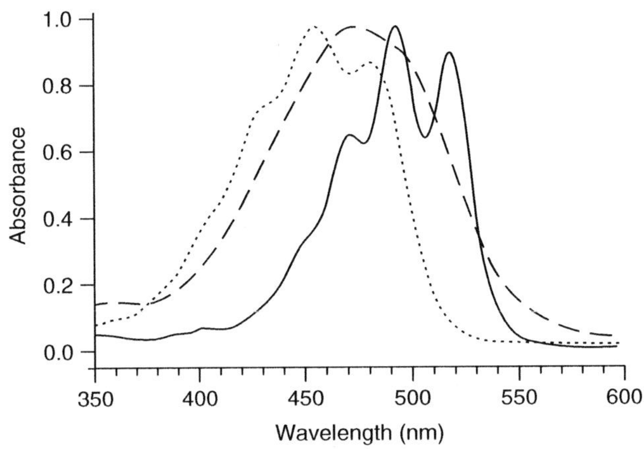

FIGURE 6.6 Effect of structure and end groups on the spectral fine structure. Lycopene (—), β-carotene (·····), and capsanthin (– –).

IR spectroscopy is restricted essentially to the identification functional groups, mainly hydroxyle, carbonyle, and allene.

6.5 NUTRITIONAL AND BENEFICIAL PROPERTIES OF THE CAROTENOIDS

Nutritionally, the main physiological function of the carotenoids is their capacity as precursors of vitamin A, so that they are said to have provitamin A value. This important quality has led some authors to propose their classification according to nutritional activity (depending on their provitamin A character) and their biological activity (antiulcer, anticancer, immunological regulators, antenna photosynthetic pigments, etc.).[9,37]

The condition for a carotenoid to have such activity is that it possesses at least one unsubstituted end group with a β-ring. β-Carotene presents the greatest potential activity, since the central enzymatic cleavage of its molecule originates two molecules of vitamin A (Figure 6.7). Other carotenes such as α-carotene, γ-carotene, β-apo-8′-carotenal, and β-cryptoxanthin give rise to only one molecule of vitamin A, as they possess only one β-ring in their structure. Table 6.3 shows the relative vitamin A activity of some carotenoids.

The conversion of carotenoid to retinol takes place in the intestinal mucosa by action of the enzyme β-carotene-15,15′-dioxygenase on β-carotene, giving rise to two molecules of retinal, subsequently reduced to retinol (vitamin A), which is esterified with long-chain fatty acids, transported, and stored in the liver. Although one molecule of β-carotene can be metabolized into two of retinol, the in vitro assays carried out in 1967 by the FAO/WHO (Food and Agriculture Organization/World Health Organization) established that only one-half of the β-carotene is converted to retinol, and only one-third of the carotenoid is absorbed in the intestine, therefore 1:6 of the β-carotene ingested is metabolically available as vitamin A.[38]

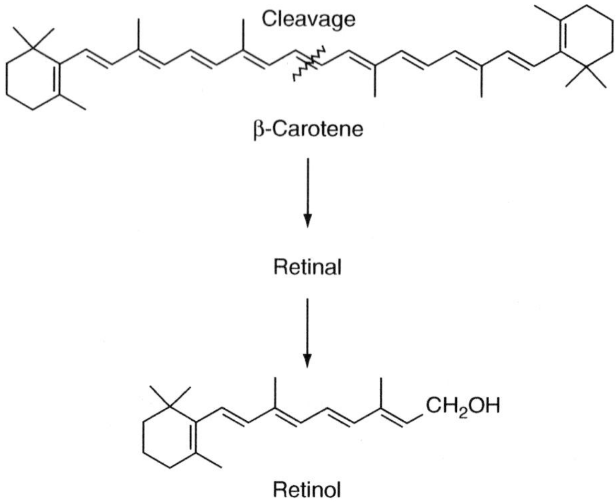

FIGURE 6.7 Central enzymatic cleavage of β-carotene molecule to give two molecules of vitamin A (retinol).

The term "retinol equivalent" (RE) was introduced to express the content in vitamin A (1 RE = 6 μg β-carotene). Although a joint FAO/WHO expert consultation, held in 1988, confirmed those conversion factors, some recent evidences indicate that

TABLE 6.3
Relative Provitamin A Activity of Some Representative Carotenes and Xanthophylls

Carotenoid	Activity (%)
all-*trans*-β-Carotene	100
9-*cis*-β-Carotene	38
13-*cis*-β-Carotene	53
all-*trans*-α-Carotene	53
9-*cis*-α-Carotene	13
13-*cis*-α-Carotene	16
all-*trans*-β-Cryptoxanthin	57
9-*cis*-β-Cryptoxanthin	27
15-*cis*-β-Cryptoxanthin	42
β-Carotene-5,6-epoxide	21
Mutatochrome	50
γ-Carotene	42–50
β-Zeacarotene	20–40

Sources: From Bauernfeind, J.C., *J. Agric. Food Chem.*, 20, 456, 1972; Zechmeister, L., *Vitam. Horm.*, 7, 57, 1949.

provitamin A activity of carotenoids could have been overestimated, which forced to review the conventional conversion factors. The report of a joint FAO/WHO expert consultation on human vitamin and mineral requeriments[39] also corrected the conventional conversion factors following the suggestion of Van het Hof et al.[40] considering that conversion factor from usual mixed vegetable diets should be 1:14 for β-carotene and 1:28 for other provitamin A carotenoids. But this is not the only standard of conversion factors. The U.S. Institute of Medicine (IOM) coined in 2001 the term "retinol activity equivalent" (RAE).[41] The conversion factor was changed on the basis of a lower vitamin A activity of β-carotene (one-sixth instead of one-third), the absorption by enterocytes being the limiting step. The bioconversion ratio was maintained to the previous value. Therefore, the RAE for β-carotene is 1:12 (1 μg RAE = 12 μg β-carotene) and 1:24 for the rest of provitamin A carotenoids.

In any case, taking into account the recommendations of the FAO/WHO or the standards of the IOM, consumption of larger amounts of fruits and vegetables containing provitamin A carotenoids is needed to meet the vitamin A requirements.

In humans, and in mammals in general, it has been shown that the carotenoids ingested in the diet are partly absorbed as such, and deposited in various tissues, such as adipose and plasma, and in the macula, where lutein and zeaxanthin have been found. The highest concentration of carotenoids is found in the plasma, always associated to lipoproteins, and mainly to the low-density fraction (low-density lipoprotein (LDL) and very low-density lipoprotein (VLDL)). The carotenoids most commonly found in the plasma are α-carotene, β-carotene, lycopene, zeaxanthin, lutein, canthaxanthin, and β-cryptoxanthin.[42,43]

Besides being provitamin A precursors, the carotenoid pigments present a functional property: they are liposoluble antioxidants.[44,45] This property is not linked with the provitamin A nature, and all the carotenoid pigments can potentially develop it. This is because most have a carbonated central polyene structure responsible for the antioxidant nature of these compounds.

There are various factors in the effectiveness of the antioxidant action. Among these are the presence of oxygenated functional groups in the structure of the pigment,[46] the conditions of the medium where the pigment acts,[47,48] and the nature of the prooxidant substance.[49] Any of these factors may cause a self-oxidizing effect in place of the expected antioxidant beneficial one. Nevertheless, different in vitro and in vivo studies have concluded that the antioxidant action of pigments such as β-carotene and lycopene is effective.

Other health-benefiting effects of the carotenoid pigments derive from their antioxidant action, which can protect against certain cancers and tumors related with the appearance of free radicals (pro-oxidant substances). By intercepting these harmful substances, the carotenoid pigments become chemiprotectors or anticancerigenic substances. Numerous recent epidemiological studies have shown the positive relationship between the level of lycopene ingested in the diet and the lower probability of the appearance of prostate cancer.[50–52] Any other process involving prooxidant substances (cell aging, appearance of ulcers, etc.) can be attenuated by this carotenoid function.[53]

An important aspect is bioavailability, defined as the carotenoid fraction ingested that is available for use under normal physiological conditions or for storage. As mentioned before, the assimilation of carotenoid pigments involves their absorption, transport, and metabolization. During this complex process, part of the ingested carotenoids is lost from, and another fraction is incorporated into, the cell structure where the function takes place. A primordial factor in bioavailability is the liposoluble nature of these substances, obviously important in the absorption and transport stages.[54] Carotenoid absorption and transport are reduced when fat consumption is low, so that a minimum fat consumption is required to increase the absorption and subsequent transport of carotenoids. The consumption of fiber leads to a decreased absorption of fats and liposoluble substances, and a decreased bioavailability of carotenoids.[55,56]

The presence of oxygenated functional groups also modifies the bioavailability of these compounds. It has been demonstrated recently that some ketocarotenoids are more rapidly absorbed and metabolized than other carotenes such as, for instance, lycopene.[57] These xanthophylls do not present provitamin A activity, but their antioxidant action is more effective than that of β-carotene. The incorporation of the carotenoid pigments into cell structures is affected by the pigment structure and the presence of functional groups that may modify the interaction with other molecules. Such structure, as mentioned above, determines the effectiveness of the pigment's action.

6.6 USES OF CAROTENOID PIGMENTS

6.6.1 Carotenoids as Food Colorants

The attractive colors of carotenoid pigments, ranging from yellow to deep red, make the fruits that contain large amounts of them to be a good raw material for the natural colorant processing industry. Pigment extracts obtained from natural sources are used to enhance, correct, or contribute color of foods. This is of special importance as consumer criteria of an ideal product are based on its organoleptic properties, among which color is a prime indicator of quality.[13]

The use of food colorants, including carotenoids, is the subject of legislation, with compulsory rules that are more or less restrictive from country to country. In the European Union, current legislation lays special emphasis on the protection of consumer interests and health, promoting the use of natural pigments. In the case of the carotenoids, their use is permitted in all those foods in which the addition of colorants is permitted. Table 6.4 shows the carotenoids most commonly used, and the code assigned that must appear on the product label. In the United States, legislation does not require certification for certain natural colorants, including the carotenoids in isolated form and various preparations or extracts rich in them. Table 6.5 shows carotenoid-rich colorants that are currently permitted by the FDA (Food and Drug Administration).[58]

The advantage of using carotenoids, rather than other substances, as colorants is their natural origin, which neutralizes any rejection (especially on the part of the consumer) when they are used to color foods. The external addition of carotenoids to

TABLE 6.4
Natural Carotenoids (or Nature-Identical Forms) Listed by the European Union and Approved for Use in Foods

Colorant Code	Pigment Composition
E160a	β-Carotene and mixtures of carotenes
E160b	Annatto, bixin, and norbixin
E160c	Paprika extract (oleoresin), capsanthin, and capsorubin
E160d	Lycopene
E160e	β-Apo-8′-carotenal
E161b	Lutein
E161g	Canthaxanthin

Source: From Henry, B.S., *Natural Food Colorants*, Blackie Academic and Professional, London, United Kingdom, 1996, 40–79.

foods achieves a triple effect. The first is the color given. Second, the nutritional value of the food increases if pigments with provitamin A activity are added. Third, most carotenoids increase the liposoluble antioxidant content. Of the three effects, the most important for the industry is the first, but the benefits of the other two should always be borne in mind.

TABLE 6.5
Natural Carotenoids (or Nature-Identical Forms) and Rich-Carotenoid Sources Listed by the Food and Drugs Administration (FDA) for Food and Beverage Use

Annatto extract
β-Apo-8′-carotenal
β-Carotene
Canthaxanthin
Carrot oil
Paprika and paprika oleoresin
Saffron

Source: From Henry, B.S., *Natural Food Colorants*, Blackie Academic and Professional, London, United Kingdom, 1996, 40–79.

The liposoluble nature of these compounds determines, a priori, the type of food in which they can be incorporated to dissolve efficiently. β-Carotene is added to fatty foods, such as butter, cheeses, and oils, although other pigments such as bixin and apocarotenals are also employed. Paprika and oleoresins industrially obtained from red pepper are used directly or as ingredient in the manufacture of sauces and meat products. Saffron, which contains crocin as a major pigment (a diester of crocetin with the disaccharide gentiobiose), is used as a hydrosoluble condiment for soups and to color foods and drinks. Other hydrosoluble preparations of β-carotene and other pigments such as canthaxanthin and apocarotenoids are used to color drinks. Norbixin, a product derived from the saponification of bixin, is hydrosoluble, and used to color ice cream, cereals, and cheese.

The addition of carotenoids to the diet of different animals is commonly used as a method to incorporate certain carotenoids into products that will be obtained from such animals. For instance, β-carotene is added to cattle foodstuff to increase the concentration of provitamin A in milk. In poultry, alfalfa and maize respectively are used as lutein- and zeaxanthin-rich sources, pigments that are incorporated for the pigmentation of the skin and, in particular, egg yolk. The red and yellow coloring of the feathers of certain birds is due to the presence of dietary carotenoids. In the case of salmon, the red color of the flesh is due to pigmentation with dietary astaxanthin and canthaxanthin, which can be introduced artificially in animals bred on fish farms.

6.6.2 Other Uses: Pharmaceuticals, Clinical, and Cosmetic

The main use of carotenoids in drugs is to correct the levels of vitamin A in patients with hypovitaminosis or requiring an extra supply of vitamins. However, other applications have been developed not related with the provitamin A character but rather with photochemical and antioxidant properties. For many years, these pigments have been used in therapy to reduce the effects of erythropoietic protoporphyria, a skin disease related with metabolism of the porphyrins. Carotenoids have also been considered as protectors in certain cancer treatments, above all those requiring radiotherapy. Carotenoid pigments are used in cosmetic products in the form of suspensions, emulsions, or lotions, lipsticks, and makeup foundations.

6.7 INDUSTRIAL PRODUCTION OF CAROTENOIDS

The first synthetic carotenoid was β-carotene, produced on industrial scale from 1954, following the patent of Hoffmann–La Roche. Since then, the synthesis of carotenoids for their sale and use as food colorants has reached an annual production of 500 tons, with a value of 300 million US dollars.

Six main carotenoids are produced industrially by chemical synthesis: β-carotene, canthaxanthin, astaxanthin, β-apo-8′-carotenal, β-apo-8′-carotenoic ethyl ester, and citranaxanthin. The pigment is usually presented in microcrystalline form for its use in fatty foods, or microencapsulated as a hydrophilic colloid for use in aqueous media.

Current trends in industrial production are toward the introduction of environmentally safer biotechnological processes, using microorganisms such as bacteria

(*Flavobacterium multivorum* and *Brevibacterium linens*), yeasts (*Phaffia rhodozyma*), fungi (Phycomyces), and microalgae (*Haematococus pluvialis* and *Dunaliella salina*) for carotenoid production at industrial scale. Recent developments in the molecular biology of carotenoid biosynthesis have provided a variety of genes that can be employed, once introduced in a proper host organism, for the biotechnological production of selected carotenoids.[59–61]

Extraction of carotenoids from natural rich sources is another industrial alternative for the production of these pigments. The raw material could be fruits and vegetables with a high content of carotenoids with interest from the industrial or nutritional point of view, considering the coloring capacity or the provitamin A activity or functional properties, respectively. The extraction could be also performed from food- and agro-industry by-products (peel, remains, etc.) obtained from the processing of fruits and vegetables for other purposes. The use of organic solvent as extraction fluid presents several drawbacks. The trend of the directives and standards of the governments is to increase the restrictions in the solvent-residue levels in the final product. This technique presents a low selectivity, all the lipophilic compounds are co-extracted (flavor, aroma, and glicerides) with the carotenoids what supposes a serious disadvantage if we consider that the functional food market demands pure extract of the functional compounds. Consequently, the use of an alternative to conventional solvent extraction is required and the supercritical fluid extraction represents a promising technique.[62] This procedure has been applied to the obtention of β-carotene from carrots,[63] lycopene from tomato,[64] and different carotenoid mixtures from paprika.[65]

6.8 ANALYSIS OF CAROTENOIDS IN FOODS

6.8.1 INTRODUCTION

The high number of carotenoid pigments in nature (more than 650) and their structural variability make it practically impossible to describe a general methodology for their analysis. Fortunately for the analyst, the number of carotenoid pigments in the food field is relatively small, even though their complexity remains.

For many years, the analysis of carotenoids in foods has focussed on the determination of provitamin A value.[66] Many of the existing data are based on a poor identification and structural assignation of the pigments, causing great errors in the calculation of provitamin A. Calculation from the total carotenoid content of a food overestimates the provitamin A value. On the other hand, underestimation is frequent when only β-carotene is considered as provitamin A source. Currently, it is clear that the nutritional or physiological functionality of the carotenoids goes beyond provitamin A activity, so that those carotenoids qualified as and long considered inactive (from the provitamin point of view) should not be qualified as afunctional.

The continuous advances in instrumental techniques for organic compound analysis enable us to be rigorous in the analysis of carotenoid pigments. The following sections describe the main stages in the procedures of extraction, isolation, identification, and quantification of carotenoid pigments in foods of plant and animal origin.

6.8.2 General Precautions

Work with carotenoid pigments is based on a series of general principles and strategies, which in practice can be modified depending on the work to be carried out, on the techniques to be applied, and, above all, on the raw material. In any case, the existence of an extensive polyene chromophore makes the carotenoids highly sensitive to heat, oxygen, light, and, in some cases, acids and alkalis. This means that the precautions taken with other natural products have to be stretched to the maximum when working with carotenoid pigments. Whenever possible, a quick and careful manipulation will minimize possible losses from destruction and the appearance of artifacts.

The oxidative degradation of carotenoids in the presence of molecular oxygen, and in general of any oxidant species, and the potentiation of their effects when combined with light or heat will indicate the main precautionary measures to be taken. Whenever possible, samples from the extraction of carotenoids should be stored in vacuo or under inert atmosphere (Ar or N_2). To minimize structural isomerizations during isolation and chromatographic analysis of the pigments, an essential precaution is their protection from heat and light, above all the latter when the sample includes photosensitizing substances (e.g., chlorophylls). Similarly, the sample should not be subjected to excessive heat, so that the use of solvents with high boiling point is generally unadvisable when evaporation is envisaged.

Practically, all the carotenoids undergo dehydration, isomerization, and finally decomposition when subjected to acid action. Isomerization by acid action is shown particularly in the case of carotenoids with 5,6-epoxide groups, which are quantitatively transformed to 5,8-epoxide. Pigments containing 5,6-epoxide groups, such as antheraxanthin, neoxanthin, and violaxanthin, are widely distributed in plants, which also contain organic acids in amounts that vary, though are always enough to cause isomerization to 5,8-epoxide. The appearance of such artifacts is usually prevented by adding neutralizing agents, such as sodium bicarbonate ($NaHCO_3$), during the extraction and disruption of the plant material. Whenever possible, the use of organic solvents which may contain traces of acids is avoided, such is the case of chloroform which usually contains a small amount of HCl.

In contrast, most carotenoids are considerably stable to the action of alkalis, enabling saponification as a routine procedure to eliminate fatty matter and chlorophylls, or for hydrolysis of the esters of xanthophylls with fatty acids. Certain carotenoids are, however, alkali-sensitive, notably astaxanthin and in general the carotenoids containing 3-hydroxy-4-oxo-β end groups. When analyzing samples containing such carotenoids, contact with alkalis should be avoided.

6.8.3 Extraction

The natural ubiquity of carotenoids in both the animal and plant kingdoms means that there is an enormous variety of sources and materials containing them, with very different characteristics. Consequently, no single, universally applicable extraction method can be established. In each case, the extraction system must be adapted to the characteristics of the tissue or source from which the pigments will be extracted, always carefully observing the general precautions required in the analysis of that type of substance, especially with regard to its photo- and thermolability and, if applicable, to the presence of acids or alkalis.

6.8.3.1 Preparation of the Sample

The sample to be analyzed should have been taken as recently as possible and should be damage free, to ensure that the pigment fraction has not been modified. If the analysis is not going to be performed immediately, the sample should be stored in the refrigerator, or even frozen (at −30°C) for prolonged storage periods. In samples with a high content in water, lyophilization may be useful to remove the water, rehydrating with a small amount of the same in the later extraction step. If the sample is already lyophilized or dehydrated, it will also have to be rehydrated for the extraction.

The sample should be as representative as possible, with the removal of any damaged material and those tissues either not containing pigmentation or whose presence might interfere in the analysis. The weight of the sample for analysis will depend on the carotenoid content. For samples having high carotenoid concentration, 2–3 g are usually taken, increasing this to 10 g when the water content is high.

6.8.3.2 Choice of Solvent and Extraction

Because of the lipophilic nature of the carotenoids, and because most foods contain a certain amount of water, the organic solvent used for the extraction must be miscible with water, as are for example methanol and ethanol. After one or two extractions with the solvent, the material to be extracted can be treated with another organic solvent (acetone, THF, hexane, diethyl ether), not necessarily miscible with water.

When the material to be extracted does not contain a large amount of fat, the number of possible extraction solvents is higher. The most frequent are acetone, methanol, ethanol, mixtures of these, and even mixtures with water (acetone/water, 80:20).

A high fat content can interfere in the later stages of analysis, and must be removed. This is done directly by saponification (using KOH/methanol 20% (w/v)) or using phase distribution techniques with two solvents: one lipophilic and the other selectively retaining the pigments. The phase distribution is carried out after the homogenization and filtration stage described next.

To facilitate contact between solvent and material to be extracted, homogenization of the sample using appliances is advisable in all cases. Traditionally, this was done in a mortar with sand, disgregating and homogenizing the sample with the extraction solvent. More sophisticated and better are the Ultra-Turrax or Polytron homogenizers, which give a greater disgregation of the material. The addition of $NaHCO_3$ during this stage helps to neutralize acids liberated during disgregation, and their harmful effect on the carotenoids. The extract obtained can be filtered with vacuo or centrifuged, and the residue collected to continue its extraction until colorless.

6.8.3.3 Removal of Fatty Matter and Final Preparation of the Extract

Any high lipid content present after the extraction stage must be removed. Two techniques are available: phase distribution and saponification. Phase distribution is particularly useful when analyzing chlorophylls, which are destroyed by alkali action, so that saponification cannot be used to remove the fatty matter. The same technique is necessary when the sample contains alkali-sensitive carotenoid pigments (astaxanthin and bixin), and when studying xanthophyll esters present in ripe fruits having a high lipid content. An example of phase distribution is that used for the analysis of chlorophylls and carotenoids in extracts obtained from olives (*Olea europaea*) or their oils, although it can also be used in many other samples having high fat content.[67,68] The solvents of extraction and distribution used are *N,N*-dimethylformamide and hexane, both immiscible. The hexane phase recovers the carotenes and fats, and the *N,N*-dimethylformamide phase recovers the xanthophylls and chlorophylls.

Saponification is the technique most used for the removal of fatty matter and other components such as chlorophylls (when their analysis is not required). In addition, saponification hydrolyzes the fatty acid esters of xanthophylls present in many ripe fruits, facilitating subsequent stages of analysis (such as isolation, identification, and quantification). The general procedure of pigment extract saponification is usually preceded by a step of transfer to diethyl ether, which is immiscible with water and has a low boiling point (below 35°C), simplifying water removal and its own removal by evaporation. This transfer not only helps saponification, but also prevents the formation of saponification artifacts, above all by reaction between ketones (usually because of the presence of acetone in the extract) and apocarotenal aldehyde groups.

Transference is carried out by adding a sufficient amount of diethyl ether to the extract and shaking in a decanting funnel, followed by the addition of aqueous 10% (w/v) NaCl solution. Two clearly different phases are obtained: the ether phase (containing the pigments) and the aqueous one. The ether phase is washed with an aqueous solution of 2% (w/v) Na_2SO_4 to remove traces of water. Saponification is carried out by adding KOH/methanol, usually at a concentration of 20% (w/v), in a volume similar to that of the ether extract, the reaction being complete in 1 or 2 h, preferably in darkness and under inert atmosphere of N_2. The reaction is stopped by adding distilled water and the phases are left to separate. The ether phase is washed repeatedly with distilled water to neutrality, and finally washed with an aqueous solution of 2% (w/v) Na_2SO_4. The extract is filtered through a solid bed of anhydrous Na_2SO_4 to completely remove water, and taken to dryness in rotary evaporator using temperatures below 30°C. The final result is a residue containing the carotenoid fraction, ready for subsequent analytical operations.

6.8.4 Separation and Isolation of Pigments

The method used to isolate carotenoid pigments depends mainly on their properties and their relative amounts in the food. After the operations of extraction and saponification (if the latter is necessary), the extract obtained will contain the carotenoid mixture to be separated. Normally, the extract will comprise of pigments of very different polarity (carotenes and xanthophylls), requiring an initial separation by phase distribution. Solvents of polarity similar to each pigment group will thus fractionate the extract before other separation techniques. Traditionally, the distribution and determination of the partition coefficients of carotenoids in systems of petroleum ether (85%–95% aqueous)/methanol has been used to distinguish the fractions of carotenes and mono-, di-, and polyhydroxylated carotenoids, and the bibliography contains extensive and detailed information.[69,70] The use of these techniques has been pushed into the background by current techniques of chromatographic separation.

Chromatographic techniques in general, and that of column chromatography (CC) in particular, were born out of the experiments of Tswett in 1906 to separate chlorophylls and carotenoids from leaf extracts.[18] The continuous development of different materials of adsorption, together with the appropriate use of solvents or mixtures of them, made this technique one of the most satisfactory isolation methods. Today, CC and thin-layer chromatography (TLC) have become rather left behind by the enormous improvement in separations by high-performance liquid chromatography (HPLC), although their use is still necessary on many occasions. In all cases, the aforementioned general precautions required in the analysis of these compounds must be followed.

6.8.4.1 Column Chromatography (CC)

CC is used mainly in the separation of a mixture of carotenoids at semipreparative or preparative scale, although subsequently TLC is required. The choice of a stationary phase in which the carotenoid mixture is adsorbed and separated depends on its selectivity and nonreactivity with the pigments or mobile phase. Table 6.6 shows some of the stationary phases most commonly used in CC of carotenoids. Pigment

TABLE 6.6
List of Adsorbents Used for Column Chromatography (CC) of Carotenoids

Adsorbent
Cellulose
Sucrose
Starch
$CaCO_3$
$Ca_3(PO)_4$
$ZnCO_3$
Al_2O_3
MgO
$Ca(OH)_2$
CaO
Silica gel

TABLE 6.7
Solvents Most Commonly Used for Column Chromatography (CC) and Thin-Layer Chromatography (TLC) of Carotenoids

Solvent
Light petroleum
n-Hexane
Cyclohexane
Carbon tetrachloride
Benzene
Toluene
Diethyl ether
Acetone
Ethyl acetate
Dichloromethane
tert-Butyl alcohol
n-Propanol
Ethanol
Methanol
Pyridine
Acetic acid in ethanol (1%–10%, v/v)

Note: Listed from low to high polarity.

polarity partly determines the stationary phase to be used. Thus, the carotenes are separated better on columns of calcium hydroxide or alumina (deactivated to grade III). If the carotenoid fraction is of medium polarity, the use of calcium carbonate and magnesium oxide is recommended. Highly polar xanthophylls require a stationary phase of weaker adsorption, such as cellulose.

The mobile phase chosen also depends on its polarity and that of the carotenoid mixture to be separated. Table 6.7 shows some of the solvents used as mobile phase in CC of carotenoids. Some appropriate mixtures are diethyl ether, benzene, or acetone in light petroleum ether, ethanol in diethyl ether, or ethyl acetate in benzene. The recovery of each pigment isolated is helped by using a solvent with a very low boiling point.

One parameter to consider after the choice of each phase is the length/diameter ratio of the column—the higher the ratio, the higher the efficiency of the chromatographic separation. Values of 8:1–20:1 give the best results. Generally, one-third of the column is packed with adsorbent material and the amount of sample is calculated with reference to the weight of packing (normally at a ratio of 1:100). A minimum amount of sample to be chromatographed is dissolved in a low-polarity solvent and added to the column. Once the mixture has been adsorbed by the packing, the chosen solvent is added for the separation at an outlet flow rate of 1–2 drops/s.

A precaution to bear in mind is that this technique does not allow uncolored compounds to be distinguished. Therefore, there is a considerable possibility that any carotenoid fraction isolated may be contaminated with other compounds of similar polarity but without color.

As a concrete example of CC separation of carotenoid pigments present in plants, this can be carried out by a prior separation on silica column using methanol as eluent.[71,72] In this first separation, three fractions of different polarity (carotenes, mono- and, polyhydroxylated xanthophylls) are obtained. Each fraction can be rechromatographed to give a second separation. Thus, for instance, the carotene fraction of carrots, tomatoes, and maize is separated on a column of MgO–Hyflo Super Cel. Other adsorbents ($CaCO_3$, $ZnCO_3$, polyethylene, cellulose, etc.) have been used to separate the carotenoids of pepper, tomato, carrot, etc.

6.8.4.2 Thin-Layer Chromatography (TLC)

The popularity of this technique lies in the versatility and efficiency of the separation achieved, enabling the subsequent quantification of each isolated pigment.[73,74] Such characteristics, together with the ease of use, make this a technique still widely used, even in laboratories with more-advanced analytical systems such as HPLC. In the particular case of the carotenoids, it can be considered a fundamental tool in identification. The use of TLC has been described in numerous publications, and it is common as a preliminary method of separation of carotenoid mixtures, for the purification of carotenoids previously separated by CC, and for the tentative identification of carotenoids depending on their chromatographic properties (especially the R_f value). The literature widely describes the properties of chromatographic separation and the R_f value for many pigments.[69,70,75]

Table 6.8 shows some adsorbents used to prepare the stationary phase in the chromatographic separation of carotenoids by TLC. The choice between them depends on the solvent or mixture of solvents to be used as eluent phase. The adsorbent layer is placed on the glass plate (normally 20×20 cm) as a slurry, with a thickness that is variable but small (0.2–0.7 mm). The adsorbent is allowed to air-dry and is activated in the oven at 110°C. The pigment extract is applied to the base of the plate, and the plate is put into a tank containing the eluent. Development is usually carried out upwards, and when complete, the band or bands of interest are selected, scraped off, and eluted from the silica with either diethyl ether (in the case of polar carotenoids) or acetone or ethanol (if the polarity is medium), and filtered to remove the silica.

Among the general methods to separate carotenoids by TLC, that of Gross should be mentioned.[76] This uses a development on silica gel with acetone/light petroleum ether (30:70), giving a first separation into fractions of different polarity (carotenes, mono-, di-, and, polyhydroxylated xanthophylls). Each group is then separated into the individual components by rechromatography on MgO/diatomite (1:1 w/w) with the same eluent mixture but with higher proportion of acetone (4%–30% volume) depending on the polarity of the pigment group to be separated.

Table 6.9 summarizes some TLC methods used for the analysis of carotenes and xanthophylls in various vegetables, fruits, and foods in general. Some commonly used conditions are the following. For the separation of chlorophylls and carotenoids

TABLE 6.8
List of Adsorbents Used for Thin-Layer Chromatography (TLC) of Carotenoids

Cellulose
Sucrose
Mannitol
Diatomite
$CaCO_3/MgO/Ca(OH)_2$ (30:6:5)
$Mg_3(PO_4)_2$
$ZnCO_3/Al_2O_3$ (deactivated)
$Mg_2(OH)_2CO_3$
MgO/diatomite (1:1)
MgO
MgO/silica gel (1:1)
$Ca(OH)_2$
$Ca(OH)_2$/silica gel (6:1)
Silicic acid
Silica gel
Al_2O_3

in an olive extract, a mixture of light petroleum ether/acetone/diethylamine (10:4:1) can be used, with silica gel 60 GF_{254} as stationary phase.[77] This method can also be used for the chromatographic separation of samples from green plants in general, and for the separation of the pigments present in saponified extracts of various fruits. For the separation of the carotenoid mixture in a direct extract from fruits, the presence of esters of xanthophylls with fatty acids requires the use of mobile phases of lower polarity, such as the mixture hexane/ethyl acetate/ethanol/acetone (95:3:2:2) used for the analysis of esterified xanthophylls from the direct extract from fruits of red pepper.[78] Carotenes from tomato are separated using the mixture hexane/isopropyl alcohol/methanol (100:2:0.2) on plates covered with MgO/Hyflo Super Cel/cellulose (10:9:1).[79]

Considerable improvements have been made in this separation technique. Adsorbents of high quality and more-uniform particle size, including chemically bonded (C_{18}, C_8, C_2) reversed phases, have been introduced, which allow an increased capacity of separation, resolution, and effectiveness, so that high-performance TLC (HPTLC) can be carried out.

6.8.4.3 High-Performance Liquid Chromatography (HPLC)

HPLC methods have been used for the separation of carotenoids since the beginnings of this technique in the 1970s. One of the first separations of carotenoids by HPLC was carried out by Stewart and Wheaton in 1971, from extracts of citrus fruits.[80] Since then, important advances have been made in both the technique of HPLC and

TABLE 6.9
Thin-Layer Chromatography (TLC) Methods for Quantitative Determination of Carotenoids in Foods

Sample Type	Analyzed Carotenoids	Stationary Phase	Mobile Phase	References
Green and red pepper fruits	Antheraxanthin, capsanthin, capsanthin-5, 6-epoxide, capsorubin, β-carotene, β-carotene-5, 6-epoxide, hydroxy-α-carotene, cryptocapsin, β-cryptoxanthin, lutein, neoxanthin, violaxanthin, and zeaxanthin	Silica gel G	Benzene/ethyl acetate/methanol (75:20:5)	23
Green and processed vegetables, and ripe fruits (saponified extract)	β-Carotene, β-cryptoxanthin, lutein, neoxanthin, violaxanthin, neocrome, auroxanthin, zeaxanthin, capsanthin, capsorubin, and cucurbitaxanthin A	Silica gel G	Light petroleum ether/acetone/ diethylamine (10:4:1)	78
Ripe fruits (direct extract)	β-Carotene, β-cryptoxanthin, zeaxanthin, capsanthin, and capsorubin	Silica gel G	Hexane/ethyl acetate/ethanol/ acetone (95:3:2:2)	79
Tomato fruits	Lycopene, prolycopene, violaxanthin, neoxanthin, *cis*-mutatoxanthin, and lutein	Silica gel 60	Hexane/benzene/ acetone/acetic acid (80:10:5:5)	115
Tomato fruits	Lycopene, prolycopene, violaxanthin, neoxanthin, *cis*-mutatoxanthin, and lutein	MgO/Hyflo Super Cel/cellulose (10:9:1)	Hexane/isopropyl alcohol/methanol (100:2:0.2)	79

the development of chromatographic methods for the separation and detection of carotenoids.

In any chromatographic separation by HPLC, a series of factors must be added to the general precautions of management and analysis of carotenoid pigments. The solvents must be of high quality, and before their use must be degassed to prevent the entrance of air into the chromatographic system and minimize noise on the base line. This is achieved using a prior sonication, filtration in vacuo, or a purging with helium. The mobile phase should always be filtered at 0.45 μm to remove any suspended particle. The use of a precolumn of the same packing as the main column is advisable, to delay deterioration and the consequent decrease in separation efficiency of the analytical column. For the same reasons, the samples should be filtered at 0.45 μm, or centrifuged at 12,000 rpm before being analyzed.

6.8.4.3.1 *Instrumentation and Chromatographic Conditions*

If a mixture of solvents is used as eluent, and whenever a gradient is employed, a binary, ternary, or quaternary pump is necessary. For isocratic separations, a simple pump is sufficient. Detection is usually carried out with a UV–visible spectrophotometric detector that can be of fixed or variable wavelength, or of diode array. Detection is generally performed at around 450 nm, which is the region of maximum absorption for most carotenoids. The increasingly widespread use of diode array detectors allows recording the complete absorption spectrum (range 350–550 nm), which, together with the chromatographic properties of the pigment, can provide very useful information for the subsequent identification. Other detection systems have improved the detection limits achieved with the classical UV–visible detectors. The coulometric electrochemical detector (CED) has been successfully used for the analysis of carotenoids in human plasma.[81] The use of CED significantly increases sensitivity and enables routine use of microsamples representing a 10–100-fold increase over the UV–visible detection levels. Thermal lens spectrometry is another ultra-sensitive method for the detection of carotenoids applied in their analysis in fish oil.[82] The detection limits with this technique supposed a 100-fold increase over the UV–visible detection method.

Carotenoid separation can be carried out using both normal-phase HPLC (NP-HPLC) and reversed-phase HPLC (RP-HPLC). In the former, the stationary phase has a polar nature, with silica gel as the most commonly used packing, although it is also possible to use nitrile- or amino-type linked phases. The normal phase tolerates the presence of glycerides and any nonpolar matter, as it does not adsorb them strongly and can be removed later in a washing step. This type of stationary phase requires a mobile phase that is nonpolar or of low polarity. Thus, the most-used solvent is hexane together with small amounts of other solvents of higher polarity (methanol, propanol, etc.), while water is not recommended. RP-HPLC uses nonpolar stationary phases such as octyl silane (C_8), octadecyl silane (C_{18}), and polymers (polystyrene, divinyl benzene, and polymethacrylate). The recent introduction of the C_{30} stationary phase into RP-HPLC, enabling a significant increase in the interaction between analyte and solid phase compared with C_{18}, has meant a considerable advance in chromatographic resolving power in the separation of carotenoids.[83,84] The mobile phase comprises a mixture of polar solvents, generally methanol, acetonitrile, dichloromethane, and water, although nonaqueous systems are preferred for the chromatographic separation of carotenes. This type of chromatography requires the removal of nonpolar compounds (glycerides and fats in general) that occasionally accompany the sample, so that a step of washing the column is required to ensure equal chromatographic conditions from one analysis to another.

In both types of chromatography (normal and reversed), the effectiveness of separation depends largely on particle size, which ranges between 3 and 10 μm, and is commonly 5 μm. The quality of separation increases if the size is uniform throughout the packing. The column is usually of steel, typically 25 cm in length

TABLE 6.10
General Chromatographic Conditions for Carotenoid Pigments Separation in Foods

System	Column	Mobile Phase	Elution Gradient Linear Gradient		Flow Rate (mL/min)
			0–10 min	40 min	
A	Octadecyl (C_{18})-silylated silica, 5 μm, 25 × 0.46 cm	Acetonitrile	85%	45%	0.7
		Dichloromethane/hexane (1:1)	5%	45%	
		Methanol	10%	10%	
B	Octadecyl (C_{18})-silylated silica, 5 μm, 25 × 0.46 cm		Isocratic		1
		Acetonitrile	55%		
		Dichloromethane	23%		
		Methanol	22%		
C	Octadecyl (C_{18})-silylated silica, 5 μm, 25 × 0.46 cm		Isocratic		0.7
		Acetonitrile	85%		
		Methanol	10%		
		Dichloromethane/hexane (1:1)	5%		

Source: Proposed by Khachik, F. et al., *Methods Enzymol.*, 213, 347, 1992.

and 0.4 cm internal diameter, although other dimensions are also used, above all for preparative or semipreparative HPLC applications.

Both NP-HPLC and RP-HPLC can either use an isocratic elution system that is of constant composition during the whole separation, or use an elution gradient that changes the composition of the eluent during analysis. Although there are many chromatographic conditions to achieve the separation of any mixture of carotenoid pigments, Khachick et al. described three systems of general use for the separation of carotenoids present in foods (Table 6.10) using RP-HPLC, with C_{18} stationary phase and employing elution gradients or isocratic systems depending on the pigment fraction to be separated.[85] Table 6.11 summarizes some methods used for the HPLC analysis of carotenes and xanthophylls in foods.

Some of the methodologies used are outlined. For the separation of carotenes, interesting methods have been described.[86–89] O'Neil et al. evaluate the resolving capacity and quantification of *Z/E*–β-carotene isomers in four chromatographic columns and with various solvent systems, finding the $Ca(OH)_2$ column the best.[90] Biacs et al. employed the isocratic mixture acetonitrile/2-propanol/water (39:57:4) to separate esterified carotenoids of paprika.[91] Khachik et al. quantify the major carotenoids and their esters in apricot, peach, melon, and grape.[92] Heinonen et al. use this last method to analyze the carotenoid content of 69 types of vegetables and fruits, both fresh and processed.[93]

Currently, the coupled combination of HPLC and mass spectrophotometric detectors (MS) enables the identification of carotenoid pigments without the need for the prior stages of isolation and purification, with the advantage of lower detection levels.[84] The atmospheric pressure ionization and electrospray ionization

TABLE 6.11
High-Performance Liquid Chromatographic (HPLC) Methods for Quantitative Determination of Carotenoids in Foods

Sample Type	Analyzed Carotenoids	Stationary Phase	Mobile Phase	Detection	References
Green vegetables: potato leaves, cassava leaves, salad pea leaves, and pumpkin leaves	β-Carotene	ODS-Hypersil, 3 μm	Methanol/acetonitrile/chloroform/water (200:250:90:11)	Absorbance at 450 nm	116
Sweet potato, carrot, squash, collard, cucumber, tomato, peach, apricot, nectarine, and plum	9-*cis*-, 13-*cis*-, and all-*trans*-β-Carotene	$Ca(OH)_2$	Acetone/hexane (3:997)	Absorbance at 436 nm, and 340 nm for cis isomers	117
Carrot, blueberry, and potato	α-Carotene and β-carotene	Partisil 5 ODS	Acetonitrile/tetrahydrofurane/water (85:12.5:2.5)	Absorbance at 470 nm	86
Peas, carrot, sweet potato, kale, spinach, squash, apricot, and peach	9-*cis*-, 13-*cis*-, and all-*trans*-β-Carotene	Vydac 201 TP, 5 μm	Methanol/chloroform (94:6)	Absorbance at 475 nm Uses Sudan 1 as internal standard	118
Margarine	Carotenes and vitamin A	LiChrosorb Si-60, 5 μm	Hexane/diethyl ether (92:8)	Absorbance at 453 nm	119
Red pepper	Capsorubin, violaxanthin, capsanthin-5,6-epoxide, capsanthin, antheraxanthin, luteoxanthin mutatoxanthin, capsolutein, zeaxanthin, lutein, cryptoflavin, cryptocapsin, β-cryptoxanthin, and β-carotene	Spherisorb Silica, 5 μm	A = Light petroleum B = Acetone 95% A to 75% A in 30 min and kept for 5 min Flow rate: 1.0 mL/min	Absorbance at 460 nm	120

Red pepper, paprika, and oleoresins	Capsorubin, violaxanthin, capsanthin-5,6-epoxide, capsanthin, 9-*cis*-capsanthin, 13-*cis*-capsanthin, antheraxanthin, mutatoxanthin, cucurbitaxanthin A (capsolutein), zeaxanthin, 9-*cis*-zeaxanthin, 13-*cis*-zeaxanthin, β-apo-8′-carotenal, cryptocapsin, β-cryptoxanthin, β-carotene, and *cis*-β-carotene	Spherisorb ODS2, 5 μm	A = Acetone B = Water 75% A for 5 min, to 95% A in 5 min, 95% A by 7 min, to 100% A in 3 min Flow rate: 1.5 mL/min	Absorbance at 450 nm Uses β-Apo-8′-carotenal as internal standard	108
Olive and olive oil, and green vegetables (fresh and processed) in general	Neoxanthin, violaxanthin, lutein, antheraxanthin, mutatoxanthin, β-carotene, neochrome, auroxanthin, luteoxanthin, and chlorophylls	Spherisorb ODS2, 5 μm	A = Water/ion-pair reagent/methanol (1:1:8) B = Acetone/methanol (50:50) Ion-pair reagent: Tetrabutylammonium acetate 0.05 M/ammonium acetate 1 M 75% A to 75% B in 8 min, kept isocratic for 2 min, then to 90% B in 8 min (convex profile), to 100% B in 5 min (concave profile) Flow rate: 2 mL/min	Absorbance at 410, 430, and 450 nm	110

(Continued)

TABLE 6.11 (Continued)
High-Performance Liquid Chromatographic (HPLC) Methods for Quantitative Determination of Carotenoids in Foods

Sample Type	Analyzed Carotenoids	Stationary Phase	Mobile Phase	Detection	References
Orange juice	α-Cryptoxanthin, β-cryptoxanthin, zeaxanthin, α-carotene, β-carotene, and cis isomers	Vydac 201 TP, 5 μm	Methanol/chloroform (90:10) Flow rate: 1.0 mL/min	Absorbance at 475 nm	121
Vegetables	Canthaxanthin, β-cryptoxanthin, α-carotene, γ-carotene, β-carotene, cis isomers, and lycopene	Vydac 218 TP, 5 μm	Methanol/acetonitrile/tetrahydrofurane (52:40:8) Flow rate: 1.0 mL/min	Absorbance at 470 nm	122
Paprika	Capsorubin, violaxanthin, capsanthin, 9-*cis*-capsanthin, 13-*cis*-capsanthin, antheraxanthin, zeaxanthin, cryptocapsin, cryptoxanthin, β-carotene, and more than 30 esters	Zorbax ODS, 5 μm	A = Acetone/water (75:25) B = Acetone/methanol (75:25) 100% A to 65% B in 10 min, to 80% B by 30 min, to 100% B by 60 min Flow rate: 1 mL/min	Absorbance at 460 nm	123

Vegetables	Neoxanthin, *cis*-neoxanthin, violaxanthin, neochrome, lutein 5,6-epoxide, lutein, *cis*-lutein, β-apo-8′-carotenal, β-carotene, 15-*cis*-β-carotene, luteoxanthin, auroxanthin, and some esters	Microsorb C_{18}, 5 μm	A = Methanol/acetonitrile/dichloromethane/hexane (15:75:5:5) B = Methanol/acetonitrile/dichloromethane/hexane (15:40:22.5:22.5) 100% A for 12 min, to 100% B in 15 min (linear) Flow rate: 0.5 mL/min	Absorbance at 450 nm	98
Fruits and vegetables	β-Carotene, α-carotene, and β-cryptoxanthin	Partisil ODS, 5 μm	Acetonitrile/tetrahydrofurane/water (85:12.5:2.5)	Absorbance at 470 nm	124
Fruits and vegetables	Lutein, zeaxanthin, and rest of chloroplastic pigments	Shandon Hypersil, 5 μm	Tetrahydrofurane/water (51:49) Flow rate: 1 mL/min	Absorbance at 450 nm	125

are the common interfaces used in the HPLC-MS methods recently developed. The combination of HPLC and MS has facilitated the direct analysis of carotenoid esters. Esterification of hydroxycarotenoids by fatty acids has been an issue that has limited direct identification and quantification of xanthophyll esters when they were present in extracts. This circumstance forced to apply a saponification process to the extract previous to analysis. However, with the development of coupled HPLC-MS methods, it has been possible to analyze directly the carotenoid extracts including xanthophyll esters and nowadays there is a wide series of analytical methods for carotenoid esters.[94–97]

6.8.4.4 Preparation of Standards

Excepting the carotenoids obtained at industrial scale, and thus commercially available (β-carotene, canthaxanthin, astaxanthin, β-apo-8′-carotenal, β-apo-8′-carotenoic ethyl ester and, citranaxanthin), carotenoid standards must be obtained either by total or partial synthesis, or from natural sources in which their presence is confirmed using the extraction and separation techniques described above. Table 6.2 (presence and distribution of the most common carotenoids in foods) can be used to choose the natural source from which carotenoid pigment standards can be obtained.

Carotenoids such as lutein, antheraxanthin, violaxanthin, and neoxanthin can be readily obtained from a saponified extract of pigments from spinach (*Spinacia oleracea*), alfalfa (*Medicago sativa*), or any other green plant. Lycopene, phytofluene, and ζ-carotene are isolated from a pigment extract of tomato. Zeaxanthin and β-cryptoxanthin are obtained from fruit extracts of peach (*Prunus armeniaca*) and papaya (*Carica papaya*), and other fruits. If the carotenoid pigments required are exclusively synthesized by and present in a plant genus, it is necessary to use their natural source. Such is the case of capsanthin and capsorubin present in the red pepper (*Capsicum annuum*), bixin and norbixin in the annatto seeds, cucurbitaxanthin A in pumpkin, lactucaxanthin in lettuce (*Lactuca sativa*), or crocin in saffron anthers (*Crocus sativus*).

Some carotenoids are obtained in the laboratory from another related carotenoid, by means of partial synthesis, which include the reactions described in the identification of functional group section. For example, auroxanthin and luteoxanthin are obtained from violaxanthin by acidification. Neochrome and mutatoxanthin are obtained from neoxanthin and antheraxanthin, respectively, using the same procedure.[98] The *Z*–β-carotene isomers are prepared by reflux heating at 200°C an acetone solution of all-*E* β-carotene obtained from a pigment extract of carrot.[99]

6.8.5 Identification

When the carotenoids have been isolated and purified, they are identified by their physicochemical and spectroscopic properties. Generally, carotenoid pigment identification has been carried out essentially from chromatographic and spectroscopic characteristics. According to Schiedt and Liaaen-Jensen,[100] for an identification to be considered acceptable today, it must include a series of minimum identification criteria: (i) the UV–visible absorption spectrum (λ_{max} and fine structure) must be in concordance with the proposed chromophore; (ii) the chromatographic properties must be identical in two different systems, preferably TLC (R_f) and HPLC (t_R), and where possible, co-chromatography with standards should be performed; (iii) the mass spectrum of pigments must be obtained to know at least the exact molecular weight. Modern assignation and complete elucidation of the structure of a carotenoid pigment includes the use of sophisticated spectroscopic techniques such as ^{1}H-NMR (nuclear magnetic resonance), ^{13}C-NMR, CD (circular dichroism), ORD (optical rotatory dispersion), and Raman spectroscopies, requiring the qualified use of complex instruments. Their application to carotenoid pigments has been the subject of specific monographs of essential reading for both newcomers and specialists.[101–103]

Routinely, the identification of a pigment begins with the study of the adsorption properties on chromatographic supports, with TLC, and occasionally CC, being the most-used techniques. TLC gives preliminary information on color and, above all, mobility (R_f) of the pigment under the assay conditions (support and eluents). As the affinity of adsorption depends on many external factors, the R_f value is of little use except when a co-chromatography is performed with pure standards. If these are not available, they will have to be isolated in the laboratory from natural sources that contain them, as stated above in Section 6.8.4.4. The literature lists R_f values for many carotenoids under defined TLC conditions, and these can be useful in a first inspection when little is known about the nature of the pigment.[69,70,75]

The chromatographic study must be complemented spectroscopically, with UV–visible spectroscopy being the most commonly used because of its availability and simplicity. The UV–visible absorption spectrum of a pure pigment is usually recorded in various solvents, and the values obtained for the absorption maxima and fine structure are compared with those listed in the bibliography, and if possible, with the spectrum of a pure standard recorded under the same conditions.[36,69,70] In practice, it is normal to find differences of a few nanometers with respect to the λ_{max} values in the literature, because basically of instrument-related factors. Table 6.12 details the absorption maxima in different solvents for the carotenoid pigments commonly found in foods of plant and animal origin.

Currently, the use of HPLC coupled with detection systems such as spectrophotometric diode array detectors has meant a considerable advance in the identification of carotenoid pigments, enabling chromatographic and spectroscopic information to be obtained at the same time. Furthermore, the greater resolving power of HPLC and the diversity of chromatographic supports have enabled the tackling of problems difficult to solve by classic techniques, such as the separation of optic or chiral isomers. The use of coupled HPLC-MS and HPLC-NMR allows identification and determination of carotenoids using extracts with a very low concentration of the compounds to be identified. These coupling techniques have been used satisfactorily in the identification of stereoisomers and optimal separation and identification of different cis/trans isomers.[104,105]

In addition, the identification of carotenoid pigments includes the characterization of the functional groups, which can occasionally be inferred from the UV–visible spectrum. IR spectroscopy, although not very applicable to the identification of carotenoid pigments, is especially selective in determining the presence of particular

TABLE 6.12
Absorption Maxima for the Visible Light Spectrum of Carotenoids in Different Solvents

Carotenoid	Absorption Maxima λ_{max} (nm)			Solvent
Antheraxanthin	430	456	484	Chloroform
	422	445	472	Light petroleum
	422	444	472	Ethanol
Astaxanthin		468		Light petroleum
		478		Ethanol
		480		Acetone
Auroxanthin	378	400	424	Light petroleum
	380	400	422	Ethanol
Bixin	432	456	490	Light petroleum
	433	470	502	Chloroform
Canthaxanthin		466		Light petroleum
		474		Ethanol
		482		Chloroform
Capsanthin	450	475	505	Light petroleum
	460	483	518	Benzene
Capsorubin	445	479	510	Light petroleum
	460	489	524	Benzene
α-Carotene	422	444	473	Light petroleum
	422	444	472	Ethanol
	424	448	476	Acetone
	433	457	484	Chloroform
β-Carotene	425	449	476	Light petroleum
		450	476	Ethanol
	429	452	478	Acetone
	435	461	485	Chloroform
δ-Carotene	431	456	489	Light petroleum
	440	470	503	Chloroform
γ-Carotene	437	462	494	Light petroleum
	439	461	491	Acetone
ε-Carotene	416	440	470	Light petroleum
	417	440	470	Ethanol
ζ-Carotene	378	400	425	Light petroleum
	377	399	425	Ethanol
Crocetin	400	422	450	Light petroleum
	413	435	462	Chloroform
α-Cryptoxanthin	421	445	475	Light petroleum
	434	456	485	Chloroform
β-Cryptoxanthin	425	449	476	Light petroleum
	428	450	478	Ethanol
Cryptoxanthin-5,6-epoxide	418	443	470	Light petroleum
	424	447	476	Ethanol

TABLE 6.12 (Continued)
Absorption Maxima for the Visible Light Spectrum of Carotenoids in Different Solvents

Carotenoid	Absorption Maxima λ_{max} (nm)			Solvent
Cucurbitaxanthin A	423	445	473	Ethanol
	434	458	486	Benzene
Lactucaxanthin		438	468	Light petroleum
	419	440	470	Ethanol
Lutein	421	445	474	Light petroleum
	422	445	474	Ethanol
	435	458	485	Chloroform
Luteoxanthin	402	426	448	Light petroleum
	400	420	446	Ethanol
Lycopene	444	470	502	Light petroleum
	446	472	503	Ethanol
	448	474	505	Acetone
	458	484	518	Chloroform
Mutatoxanthin	402	424	448	Light petroleum
	410	434	460	Chloroform
Neochrome	399	418	446	Light petroleum
	402	426	454	Chloroform
Neoxanthin	416	438	467	Light petroleum
	415	439	467	Ethanol
	423	448	476	Chloroform
Neurosporene	414	439	467	Light petroleum
	416	440	470	Hexane
	416	440	469	Ethanol
	424	451	480	Chloroform
Norbixin	442	474	509	Chloroform
Phytoene	276	286	297	Light petroleum
Phytofluene	331	348	367	Light petroleum
Violaxanthin	416	440	465	Light petroleum
	419	440	470	Ethanol
	426	449	478	Chloroform
Zeaxanthin	424	449	476	Light petroleum
	428	450	478	Ethanol
	430	452	479	Acetone
	433	462	493	Chloroform

functional groups such as acetylenic, allenic, hydroxyl, and carbonile. In contrast, mass spectrometry of carotenoid pigments has become an essential technique because of the valuable information it gives about functional groups and substituents in general. Mass spectrometry of carotenoid pigments is normally performed using electron impact (EI) as method of ionization. This is applied directly to the sample (direct probe) subjected to a temperature of around 200°C. The presence of functional groups in a carotenoid will be shown by a characteristic fragmentation: for instance, hydroxyl groups give rise to the loss of water (18 mu), especially intense in the case of hydroxyl groups at allyl positions. The presence of 5,6- and 5,8-epoxide groups gives rise to the abundant appearance of fragments of 80, 165, and 205 mu. The presence of β and ε end rings can also be determined from the appearance of characteristic fragments. In the case of the central polyene chain, characteristic fragments of it are usually present in the mass spectra of most carotenoids, very frequently fragments of 92 mu (loss of toluene) and 106 mu (loss of *m*-xylene). Most carotenoids subjected to ionization by EI give rise to an abundant molecular ion that enables knowledge of the molecular mass. The application of mass spectrometry has been discussed in detail in various articles and reviews.[106,107]

The presence of the most characteristic and common functional groups can also be established using specific chemical reactions such as those described below.

6.8.5.1 Test for 5,6-Epoxide Groups

The test is based on chromophore modification resulting from the transformation of a 5,6-epoxide group to 5,8-epoxide in acid medium. This structural transformation means the loss of the conjugated double bond at positions 7 and 8 of the central polyene, causing a hypsochromic displacement of the absorption spectrum of 15–20 nm. Chromatographically, the 5,8-epoxide derivatives are more polar, so that chromatographic development by TLC using silica gel as support reveals bands with lower R_f values.

In practice, the test can be performed in situ on the chromatographic plate, subjecting the pigments to HCl fumes after the chromatographic separation. The appearance of a characteristic blue color identifies the presence of 5,6-epoxide groups: diepoxides give a deep blue color, and the monoepoxides a greenish-blue one. The test is more useful when carried out in the spectrophotometric cuvette, recording the electron absorption spectrum before and after adding a few drops of dilute HCl to the pigment dissolved in ethanol. A hypsochromic displacement of 15–20 nm indicates the presence of a single 5,6-epoxide group, while a displacement of 35–40 nm indicates the presence of two 5,6-epoxide groups (Figure 6.8).

6.8.5.2 Test for Reduction of Carbonyl Groups

$LiAlH_4$ or $NaBH_4$ is usually used as reducing agents of carbonyl groups, forming the corresponding alcohols. The higher reducing power of $LiAlH_4$ allows the reduction not only of aldehydes and ketones, but also of acids, esters, and other polar functional groups.

The test is made starting from a solution of the pigment in ethanol. Some crystals of $NaBH_4$ are added and the reaction is kept in the refrigerator and darkness for 3 h. The pigment is transferred to diethyl ether and the electron absorption spectrum is recorded. In the case of carbonyl groups and those conjugated with the central polyene, reduction is shown by a hypsochromic displacement of the absorption maxima in the electron adsorption spectrum and a considerable increase in the fine structure (Figure 6.9). Larger hypsochromic displacements are characteristic of

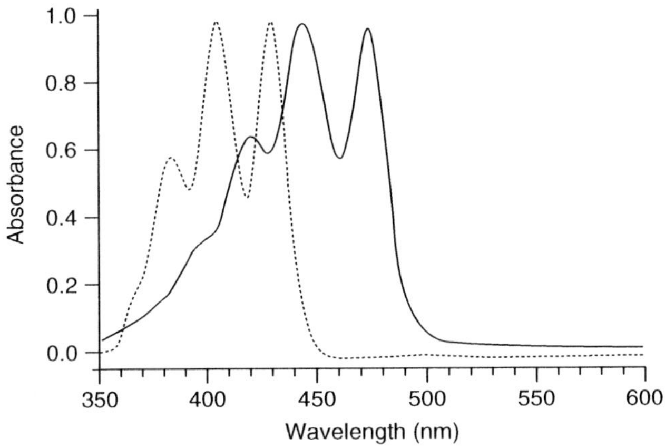

FIGURE 6.8 Spectroscopic test for 5,6-epoxy groups. Light absorption spectra of violaxanthin (—) and auroxanthin (·····) in ethanol.

carbonyl groups situated on the polyene chain and whose reduction largely cleaves the conjugation. At chromatographic level, the reduction of carbonyl groups results in greater polarity of the reaction product.

6.8.5.3 Test for Acetylation of Hydroxyl Groups

The test is based on the conversion of alcohols to esters by reaction with carboxylic acids or acid chlorides.

The test is made starting from a solution of pigment in pyridine (2 mL), to which is added acetic anhydride (0.2 mL). The reaction is kept in darkness for 12 h and

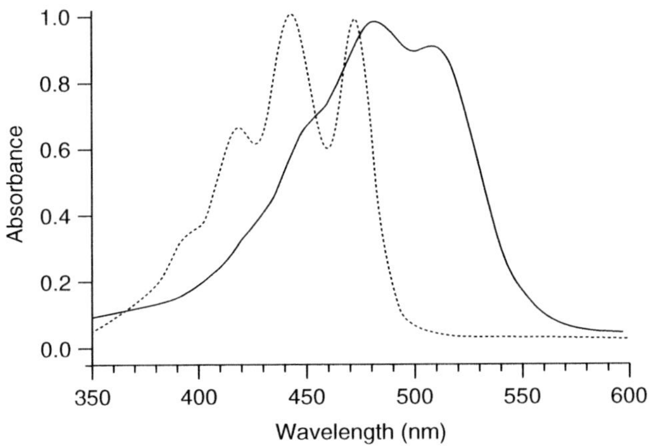

FIGURE 6.9 Spectroscopic test for carbonyl groups. Light absorption spectra of capsorubin (—) and the reducted product capsorubol (·····) in ethanol.

stopped by adding water. The pigments are transferred to petroleum ether. Chromatographic analysis by TLC or HPLC of the reaction mixture shows the number of hydroxyl groups in the carotenoid. The presence of a single hydroxyl group gives rise to a single acetylated derivative, and the presence of two results in one diacetylated derivative and one or two monoacetylated derivatives, depending on whether the position of hydroxyl groups are symmetric or not.

6.8.5.4 Test for Allyl Hydroxyl Groups

This functional group is identified by causing a dehydration, which introduces an additional double bond into the chromophore and originates a bathochromic displacement of 10–16 nm. The test is carried out subjecting the pigment, dissolved in chloroform, to the action of dilute HCl.

6.8.6 Quantification

6.8.6.1 Quantitative Determination by UV–Visible Spectrophotometry

As discussed above, the existence of an extensive system of conjugated double bonds making up the structural chromophore of the carotenoids is responsible for the characteristic absorption of light in the near UV range and, above all, in the visible range. This property has long been used for the routine quantification of carotenoids in solution, according to the Beer–Lambert law:

$$A = A^{1\%}_{1\,\mathrm{cm}} \times L \times C$$

In practice, the main problem when applying this law is to know the exact value of the specific absorption coefficient ($A^{1\%}_{1\,\mathrm{cm}}$) or molar extinction coefficient ($\varepsilon_{\mathrm{mol}}$) for a given carotenoid. The estimation of these values is not simple, and requires isolating and completely purifying the pigment, exactly weighing 1–2 mg, and making the spectrophotometric recording in a 1% (w/v) solution. By definition, the specific absorption coefficient ($A^{1\%}_{1\,\mathrm{cm}}$) represents the theoretical absorbance of the solution of 1 g of pigment in 100 mL of solvent (*C*, concentration), measured in a cuvette of 1 cm light path (*L*).

Values of absorption coefficients for most carotenoids commonly found in foods of plant and animal origin have been published in the various monographs on carotenoids.[36,69,70] Nevertheless, caution is advisable when using this type of bibliographic information: the values of $A^{1\%}_{1\,\mathrm{cm}}$ or $\varepsilon_{\mathrm{mol}}$ should be contrasted with other sources, as it has been demonstrated that there are discrepancies because mainly of the fact that in the beginnings of working with carotenoids, isolation and purification of pigments was not totally satisfactory, and the presence of impurities (mainly uncolored ones) caused errors in the estimation of the coefficients. Currently, the need to revise the data published in the bibliography is under discussion.

When the absorption coefficient is not known, an average value for $A^{1\%}_{1\,\mathrm{cm}}$ of 2500 is commonly taken. This is also applied to calculate the carotenoid content in an extract with a mixture of pigments. Another approach that can be used is to remember that carotenoids with the same chromophore must in theory have the same ε_{mol} value. Thus, if we know this value for a carotenoid, we can apply it to another having the same chromophore.

As a rule, the carotenoid (or carotenoid mixture in the case of an extract) to be quantified is dissolved in a known volume of an appropriate solvent (normally hexane, light petroleum ether, acetone, or ethanol). As in all spectrophotometric measurements, the spectrophotometric reading should be between 0.3 and 0.7 units to ensure linearity of the measurement and minimize instrument errors. One of the most common sources of error is the insufficient or poor solution of the pigments, mainly when they are in crystalline form, and may require the use of a small amount of a strong solvent such as dichloromethane, chloroform, or tetrahydrofuran. When the carotenoid has been dissolved, a portion is placed in the quartz spectrophotometric cuvette (1 cm light path) and the value of absorbance (A) at the appropriate wavelength is recorded. The measurement is usually performed at the wavelength of the greatest absorption maximum (normally the centra, II). Today, with the spreading use of diode array spectrophotometers, it is possible to obtain instantly the absorption spectrum over the whole visible range of wavelength with a precision of 1–2 nm, enormously facilitating the localization of the absorption maxima.

The amount X (mg) of a carotenoid present in a volume V (mL) of solution can be calculated from the following equation:

$$X = (A \times V \times 1000)\Big/\left(A_{1\ \text{cm}}^{1\%} \times 100\right)$$

If the value of ε_{mol} is known it is useful to apply the mathematical relationship with $A_{1\ \text{cm}}^{1\%}$:

$$\varepsilon_{mol} = \left(A_{1\ \text{cm}}^{1\%} \times \text{molecular weight}\right)\Big/10$$

Table 6.13 shows the values of $A_{1\ \text{cm}}^{1\%}$ for some of the most common carotenes and xanthophylls.

6.8.6.2 Quantitative Determination by Separation by TLC and UV–Visible Spectrophotometry

After chromatographic separation of the carotenoids present in a sample, the bands corresponding to individual pigments can be recovered from the chromatographic support by scraping each one off, followed by elution of the pigment with diethyl ether or acetone. The chromatographic adsorbent is removed by filtration, and the pigment is taken to known volume, normally 5–25 mL. Next, the electronic absorption spectrum is obtained and quantified spectrophotometrically as described above. To know the carotenoid content in a food sample, the analyst need to know exactly the weight of the sample extracted [W (g)], the final volume of the extract [V_e (mL)], the

TABLE 6.13
Specific Absorption Coefficients ($A_{1\ cm}^{1\%}$) Used for Quantitative Spectrophotometric Determination of Carotenoids

Carotenoid	$A_{1\ cm}^{1\%}$	λ (nm) for Measurement	Solvent
Antheraxanthin	2350	446	Ethanol
Astaxanthin	2100	470	Hexane
Bixin	4200	456	Light petroleum
Canthaxanthin	2200	466	Light petroleum
Capsanthin	2072	483	Benzene
Capsorubin	2200	489	Benzene
α-Carotene	2800	444	Light petroleum
β-Carotene	2592	449	Light petroleum
δ-Carotene	3290	456	Light petroleum
γ-Carotene	3100	462	Light petroleum
ε-Carotene	3120	440	Light petroleum
ζ-Carotene	2555	400	Hexane
Crocetin	4320	450	Light petroleum
α-Cryptoxanthin	2636	445	Light petroleum
β-Cryptoxanthin	2386	449	Light petroleum
Lutein	2550	445	Ethanol
Lycopene	3450	470	Light petroleum
Neoxanthin	2243	439	Ethanol
Neurosporene	2918	440	Hexane
Phytoene	1250	286	Light petroleum
Phytofluene	1350	348	Light petroleum
Violaxanthin	2250	440	Ethanol
Zeaxanthin	2348	449	Light petroleum

volume of extract chromatographed [V_{cr} (mL)], and the final volume of elution of the pigment chromatographed [V_f (mL)]. The following equation gives the content (mg/kg) for a determined pigment:

$$C = \frac{A \times V_e \times V_f \times 10000}{A_{11\ cm}^{1\%} \times W \times V_{cr}}$$

where

C = Concentration (mg/kg)
V_e = Initial volume of pigment extract (mL)
V_f = Final volume of pigment eluent (mL)
W = Weight of sample (g)
V_{cr} = Volume chromatographed (mL)

6.8.6.3 Quantitative Determination by HPLC and UV–Visible Spectrophotometric Detection

The now-routine chromatographic technique of HPLC has been in use since the 1970s for the quantification of carotenoid pigments. It is particularly sensitive when coupled with detection methods based on UV–visible spectrophotometry. When the pigments have been separated, they are quantified by detection at wavelengths as close as possible to λ_{max} of each pigment, and the recording of the corresponding chromatogram. The ratio between absorbance and amount (concentration) of pigment defined by the Beer–Lambert law is calculated to relate chromatographic peak area and amount of pigment. A detection wavelength of 450 nm is used for the routine detection of carotenoid pigments when using a fixed-wavelength detector. In the case of multiple-wavelength detectors, several chromatograms can be recorded at different detection wavelengths, choosing those that coincide with λ_{max} of the major pigments. The use of variable-wavelength detectors, and the modern diode array ones, helps to acquire chromatograms at λ_{max} of each pigment, above all when using automated systems that include computerized data treatment and storage.

The determination of each pigment concentration in a given sample requires a calibration to be carried out. Often, many workers have opted for a single calibration using β-carotene or another commercially available carotenoid as standard. This simplification is bad practice, and in many cases has resulted in serious quantitative errors. For instance, quantitative determination of capsanthin using the calibration plot obtained for β-carotene, using 450 nm as detection wavelength, causes an underestimation (more than 30%) of capsanthin, whose λ_{max} is around 475 nm (Figure 6.10). Thus, although more complex and tedious, the best option is to perform an individual calibration for each pigment present in the sample, requiring

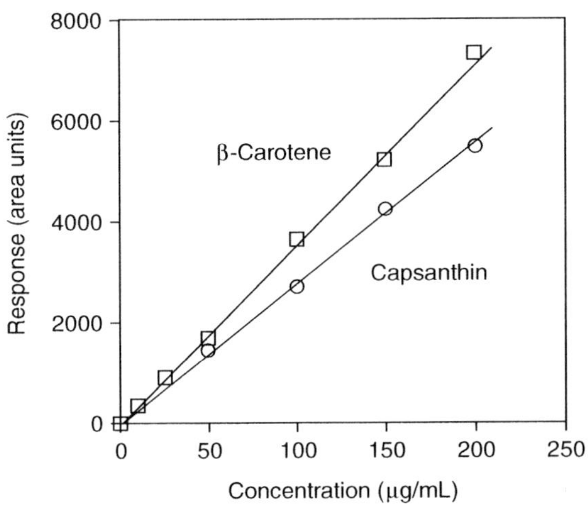

FIGURE 6.10 Calibration plots for ultraviolet (UV)–visible spectrophotometric determination of capsanthin and β-carotene, after high-performance liquid chromatography (HPLC) separation, using 450 nm as detection wavelength.

a further extraction, separation, isolation, and purification. The stock solution of each pigment is quantified spectrophotometrically as described above, so that it will be essential to know $A^{1\%}_{1cm}$ exactly. Calibration plot is obtained from the representation of concentration or amount of pigment versus the peak area measured after injecting aliquots (normally 10–20 μL) of solutions of increasing concentration.

To avoid quantification errors caused by the multiple manipulations of the sample during the various steps of extraction and preparation, the use of an internal standard (IS) in combination with the external calibration is advisable. The IS must be chosen carefully, as it has to meet a series of minimum requirements. It must be a carotenoid pigment not present in the sample to be analyzed, it must be chromatographically separable from the others under the analytical conditions used, it must have a λ_{max} of absorption as close as possible to the λ of detection employed, and it must be as stable as possible. Various ISs have been proposed: β-apo-8′-carotenal and canthaxanthin are commonly used in the analysis of vegetable foods.[108] The use of artificial colorants, such as Congo red and Sudan 1, and of synthetic carotenoids not present in natural samples, such as C45-β-carotene, has also been proposed.[80,109]

As a practical example and reference, we now describe in detail the HPLC method for separation and quantification of carotenoid pigments in foods of different origin: green vegetables, ripe fruits, processed vegetables, and animal products.

6.8.6.3.1 Analysis of Carotenoid Pigments in Green Vegetables

Not every chromatographic method can be used to analyze carotenoid pigments in green plants, as not all of them separate the chlorophylls present in the resulting extract. In such a case, saponification is normally used to remove them. The joint analysis of carotenoids and chlorophylls is extremely complex: when the latter are the object of study, their chromatographic analysis will have to be from a direct extract (see Chapter 7). Regarding the carotenoids, their analysis from a direct extract in green plants has certain advantages, among which stands out the lack of need for saponification which, if not carried out carefully, can give rise to quantitative loss and the appearance of artifacts. The method used routinely in the authors' laboratory for the analysis of chlorophylls and carotenoids in green plant tissues is described below.[110] This method was initially proposed for the study of chloroplast pigments in olive, but has subsequently been used with success for the analysis of these pigments in many green plants.

Separation is performed on a reversed-phase column (Spherisorb ODS2, 5 μm, 25 × 0.46 cm). The eluents used are (A) water/ion-pair reagent/methanol (1:1:8) and (B) acetone/methanol (1:1). The ion-pair reagent, a solution of tetrabutylammonium acetate (0.05 M) and ammonium acetate (1 M) in water, is added to improve the separation of chlorophyll derivatives.[111] The pigment extract is prepared from 10 g of fresh sample, and taken to a final volume with acetone (2–5 mL). Injection (20 μL) into the chromatograph is done after centrifuging the sample at 12,000 rpm. Elution is carried out at a flow rate of 2 mL/min, with the following gradient system: from an initial 75% A linearly to 25% A in 8 min, the composition is kept constant for 2 min, and then to 10% A in 8 min following a convex profile (curve 4 in the gradient profile of the Waters 600 E quaternary pump), then to 100% B in 5 min via a concave profile (curve 10). The column is conditioned between successive injections, allowing

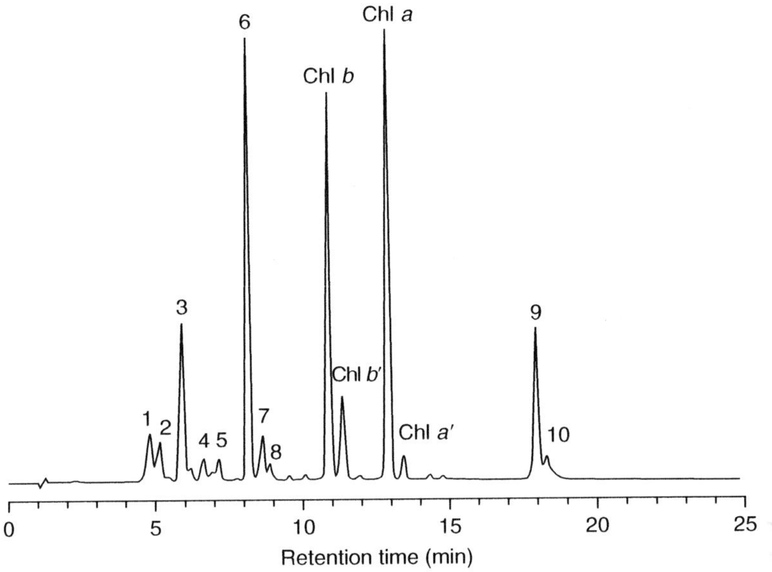

FIGURE 6.11 Carotenoid high-performance liquid chromatography (HPLC) profile of a pigment extract from a fresh green vegetable (spinach). Peaks: 1, neoxanthin; 2, neoxanthin isomer; 3, violaxanthin; 4, luteoxanthin; 5, antheraxanthin; 6, lutein; 7, 9-*cis*-lutein; 8, 13-*cis*-lutein; 9, β-carotene; 10, *cis*-β-carotene isomers. Sample also included chlorophyll pigments: chlorophyll *a* (Chl *a* and Chl *a*′) and chlorophyll *b* (Chl *b* and Chl *b*′). Mobile phase A: water/ion-pair reagent (tetrabutylammonium acetate 0.05M/ammonium acetate 1M)/methanol (1:1:8) and B: acetone/methanol (50:50); flow rate, 2 mL/min; detection at 450 nm. (From Mínguez-Mosquera, M.I. et al., *J. Chromatogr.*, 585, 259, 1991.)

7–10 min to reestablish the initial conditions. Detection is performed with a diode array spectrophotometric detector (Waters 996) at two wavelengths, 430 and 450 nm, to enable joint monitoring of chlorophylls and carotenoids. Figure 6.11 shows a typical chromatogram for the separation of chloroplast pigments included in an extract of a green plant tissue (e.g., fresh spinach).

6.8.6.3.2 *Analysis of Carotenoid Pigments in Ripe Fruits*

The analysis of carotenoid pigments in fruits is more complex than in green plants. The number of possible carotenoids is higher, and there is no typical composition profile, because of the wide structural diversity of carotenes and (especially) xanthophylls present. Moreover, fruit ripening generally involves three simultaneous processes: the disappearance of chlorophylls, the massive biosynthesis of carotenoid pigments, and esterification of the xanthophylls with fatty acids. This makes chromatographic separation and, in many cases, pigment identification extremely complicated, with saponification being necessary in the preparation of the extract for its analysis by HPLC. We now describe the method used routinely by the authors for the analysis and quantification of carotenoids in ripe fruits of red pepper and its industrial derivatives, paprika and oleoresins.[108] This method has been used successfully for the analysis of carotenoids in other carotenogenic fruits such as tomato, persimmon, carrot, melon, and pumpkin, and for quality control and authentication of concentrates and juices.

The chromatographic separation is carried out on a reversed-phase column (Spherisorb ODS2, 5 μm, 25 × 0.46 cm). The eluents used are (A) acetone and (B) water. The pigment extract is prepared from some 10 g of fresh sample taken to volume with acetone (10–25 mL). Injection (10 μL) into the chromatograph is done after centrifuging the sample at 12,000 rpm. Elution is performed at a flow rate of 1.5 mL/min, using the following gradient system: from an initial 75% A, kept constant for 5 min, linearly to 95% A in 5 min, where the composition is kept constant for 7 min, and finally to 100% A in 3 min. The column is conditioned between successive injections, allowing 5 min to reestablish the initial conditions. Detection is performed with a spectrophotometric detector at 450 nm. Quantification is carried out using β-apo-8′-carotenal as IS, which is added in known amount at the beginning of extraction. Figure 6.12 shows the chromatograms for a direct extract and a saponified one of carotenoid pigments of ripe fruits of red pepper.

6.8.6.3.3 *Analysis of Carotenoid Pigments in Processed Vegetables*

During the different stages in the processing of vegetables, the structure and stability of the chloroplast pigments may be affected, resulting in changes in the qualitative and quantitative composition of the final product. The transformations will depend on the type of process (fermentation, freezing, blanching, dehydration, etc.) and on the severity of the conditions applied. In the case of the carotenoid pigments, the most usual transformations are those involving fermentation, treatment with acids, or heating. In the case of fermentation and acid treatment, the most frequent change is the appearance of 5,8-epoxide derivatives from carotenoids having 5,6-epoxide groups, which are generally present in all green plants. In the case of heat processes, the main consequence is the formation of cis isomers and quantitative losses, directly affecting the provitamin A value.

For the analysis of chlorophylls and carotenoids, and their derivatives, in processed green plants, the previously described method can also be used.[110] Figure 6.13 shows the chromatogram for the separation of chloroplast pigments of a processed green plant, green table olives, whose processing includes a fermentation step.

6.8.6.3.4 *Analysis of Carotenoid Pigments in Foods of Animal Origin*

Foods of animal origin contain very variable amounts of carotenoids, always incorporated via the diet because of the incapacity to synthesize them. Nevertheless, some animals are able to transform certain carotenoids, so that together with the carotenoids coming from ingested foods, we find their metabolites. Although the carotenoid profile found in a food of animal origin is determined by the type of diet, it is the bioavailability of each carotenoid that determines its presence and levels. Foods of animal origin that contain carotenoids include fish (salmon, trout, etc.), crustaceans (lobster, prawn, etc.), butter and other animal fats, milk, and birds' eggs. The carotenoids most commonly found in animal products are lutein, zeaxanthin, β-cryptoxanthin, β-carotene, α-carotene, canthaxanthin, lycopene, astaxanthin, and different apocarotenoids coming from the metabolization of these.

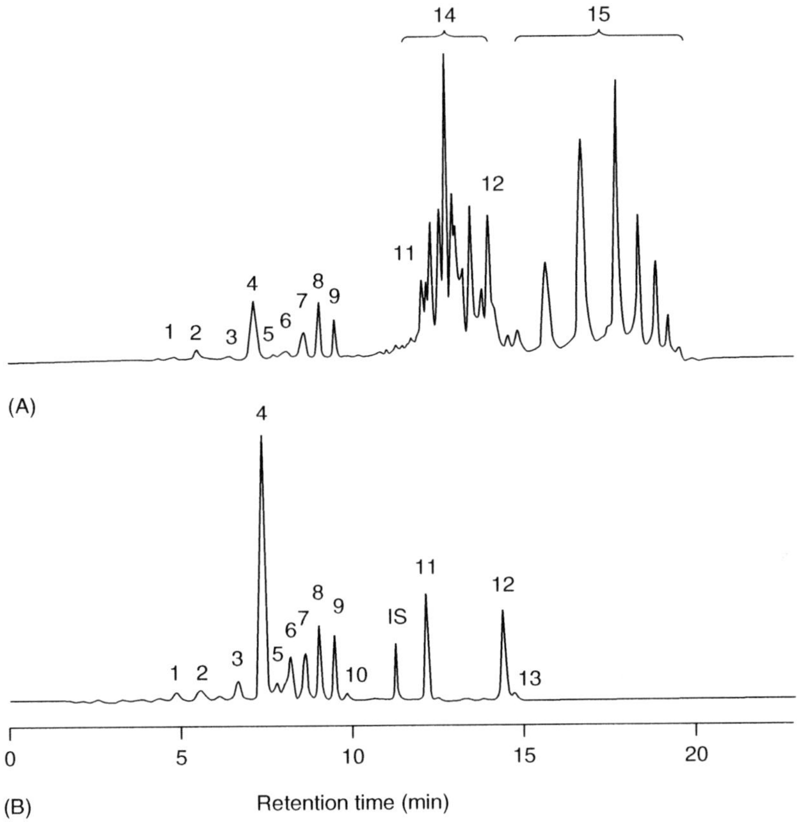

FIGURE 6.12 Carotenoid high-performance liquid chromatography (HPLC) profile of a pigment extract from a ripe fruit (red pepper, *Capsicum annuum*). (A) Direct extract (non-saponified) and (B) Saponified extract. Peaks: 1, capsorubin; 2, violaxanthin; 3, capsanthin-5, 6-epoxide; 4, capsanthin; 5, 13-*cis*-capsanthin; 6, 15-*cis*-capsanthin; 7, antheraxanthin; 8, cucurbitaxanthin a; 9, zeaxanthin; 10, 9-*cis*- and 13-*cis*-zeaxanthin; 11, β-cryptoxanthin; 12, β-carotene; 13, 9-*cis*- and 13-*cis*-β-carotene; 14, partially esterified xanthophylls; 15, totally esterified xanthophylls; IS, internal standard. Mobile phase (A) acetone and (B) water, gradient elution; flow rate, 1.5 mL/min; detection at 450 nm. (From Mínguez-Mosquera, M.I. and Hornero-Méndez, D., *J. Agric. Food Chem.*, 41, 1616, 1993.)

For the analysis of carotenoids in animal products, saponification is normally advisable to remove the fats, although this should be avoided in the case of astaxanthin, because of its lability to alkalis. The HPLC methods for this type of analysis are varied, and although those described above can be used in many cases, there are specific methods listed in the References section at the end of the chapter. Once the pigment extract has been prepared in each particular case, any HPLC method for the analysis of carotenoids in blood plasma can be used. The method employed by Khachik's group is ideal, with the separation of a wide range of carotenoids and their metabolites.[112] The chromatographic separation is performed on a reversed-phase column (Rainin Microsorb 5 μm C18, 25 × 0.46 cm).

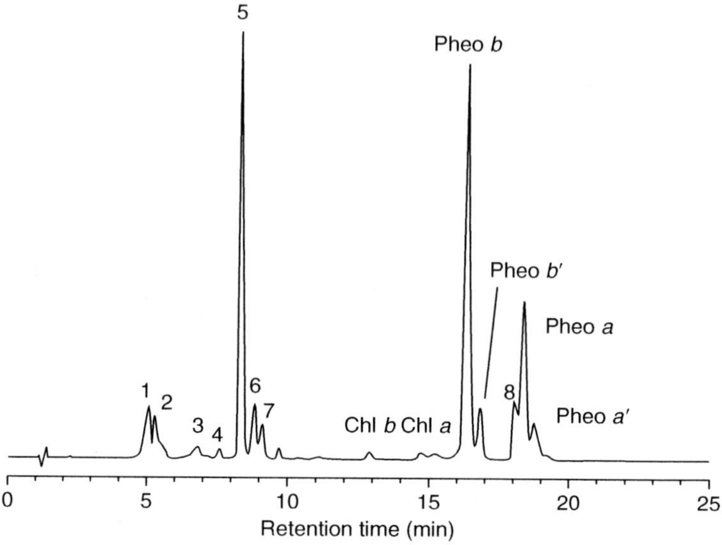

FIGURE 6.13 Carotenoid high-performance liquid chromatography (HPLC) profile of a pigment extract from a processed green vegetable (olive). Peaks: 1, neochrome; 2, neochrome isomer; 3, luteoxanthin; 4, auroxanthin; 5, lutein; 6, 9-*cis*-lutein; 7, 13-*cis*-lutein; 8, β-carotene. Sample also included chlorophyll derived pigments: chlorophyll *a* (Chl *a*), chlorophyll *b* (Chl *b*), pheophythin a (Pheo *a* and Pheo *a'*), and pheophythin *b* (Pheo *b* and Pheo *b'*). Mobile phase A: water/ion-pair reagent (tetrabutylammonium acetate 0.05M/ammonium acetate 1M)/methanol (1:1:8) and B: acetone/methanol (50:50); flow rate, 2 mL/min; detection at 450 nm. (From Mínguez-Mosquera, M.I. et al., *J. Chromatogr.*, 585, 259, 1991.)

The eluents used are (A) acetonitrile, (B) dichloromethane/hexane (1:1), and (C) methanol. Elution is carried out at a flow rate of 0.7 mL/min, with the following gradient system: from an initial 85% A–5% B, kept constant for 10 min, then linearly to 45% A–45% B in 30 min. Detection is performed simultaneously at 470, 445, 400, 350, and 290 nm, using a spectrophotometric detector. Quantification is done using β-apo-8′-carotene-3,8′-diol as IS. Figure 6.14 shows a chromatogram for an extract of carotenoid pigments from egg yolk, using the methodology described.

6.8.6.4 Determination of Provitamin A Value

The provitamin A equivalency of a food could be calculated using the conversion factors proposed by the FAO/WHO on the basis of the REs or following the recommendation of the Food and Nutrition Board of the IOM with the RAE. On the first case, 1 RE is equal to 1 μg of retinol and corresponds to 14 μg of β-carotene or to 28 μg of other provitamin A carotenoids. The second standard suggests factors of 1:12 for β-carotene and 1:24 for other provitamin A carotenoids.

To calculate the vitamin A equivalency of a food, the sum of the weight of the retinol portion plus the weight of the β-carotene divided by its conversion factor plus

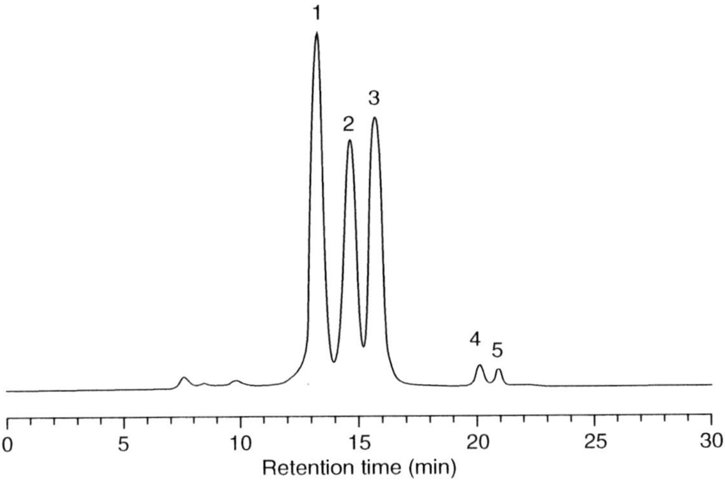

FIGURE 6.14 Carotenoid high-performance liquid chromatography (HPLC) profile of a pigment extract from animal product (egg-yolk). Peaks: 1, canthaxanthin; 2, lutein; 3, zeaxanthin; 4, α-carotene; 5, β-carotene. Mobile phase A: acetonitrile, B: dichloromethane/hexane (1:1), and C: methanol; flow rate, 0.7 mL/min; detection at 470 nm. (From Khachik, F. et al., *Methods Enzymol.*, 213, 205, 1992.)

the weight of other provitamin A carotenoids divided by their conversion factor should be made. Apart from β-carotene, with maximum provitamin A activity, the other pigments meeting the above-discussed requirements for possession of provitamin A activity have to be considered. Of these, the most commonly found in foods are α-carotene, β-cryptoxanthin, α-cryptoxanthin, γ-carotene, mutatochrome, cis isomers of β-carotene, β-zeacarotene, and β-apo-8′-carotenal, which is added as colorant to many foods, although it can be found naturally in oranges and other citrus fruits. As these carotenoids present, at most, 50% of the activity of β-carotene (see Table 6.3), the vitamin A equivalency of a food can be calculated from the following expression:

$$\text{RE} = [\mu\text{g of } \beta\text{-carotene}/14 + \mu\text{g of the rest of the active carotenoids}/28]$$

Old food composition tables report the provitamin A carotenoid content of food as international units (IU). The following conversion factors can be used to obtain values in micrograms:

$$1 \text{ IU} = 0.36\ \mu\text{g of retinol}$$

$$1 \text{ IU of } \beta\text{-carotene} = 0.6\ \mu\text{g of } \beta\text{-carotene}$$

$$1 \text{ IU of retinol} = 3 \text{ IU of } \beta\text{-carotene}$$

APPENDIX 6.1 CHEMICAL STRUCTURES FOR CAROTENES AND XANTHOPHYLLS COMMONLY FOUND IN FOODS

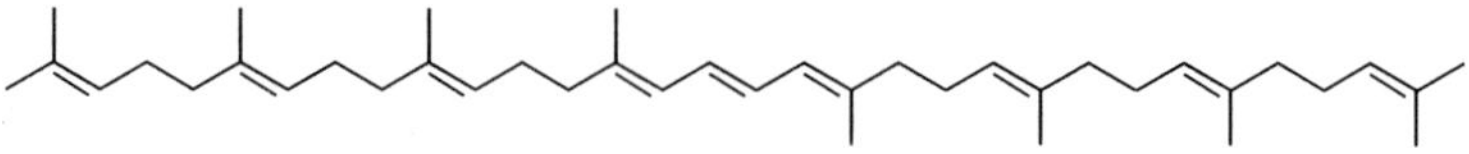

Phytoene

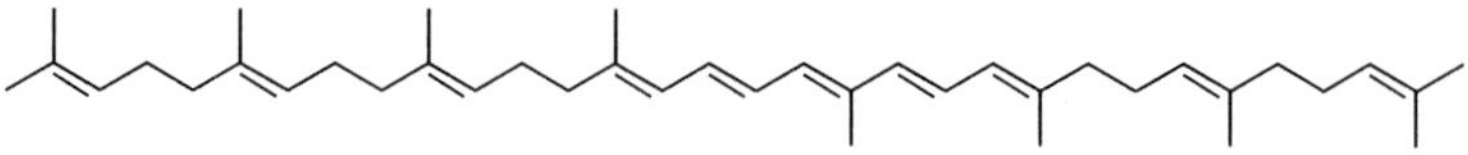

Phytofluene

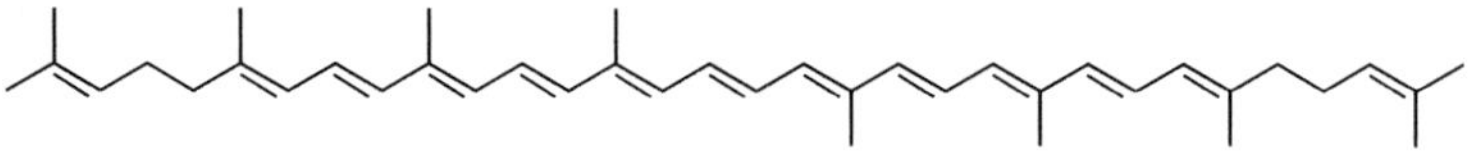

Lycopene

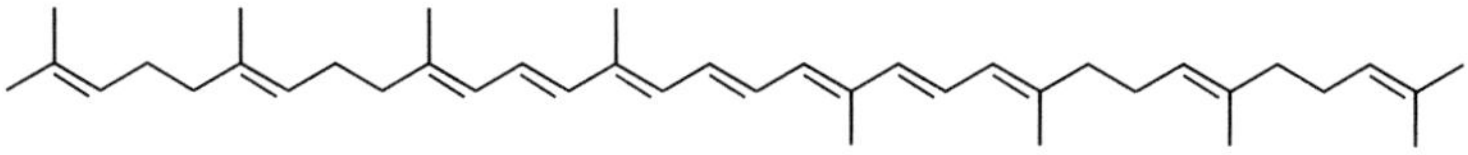

ζ-Carotene

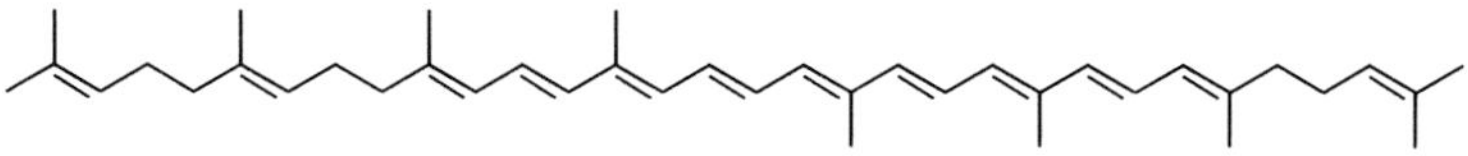

Neurosporene

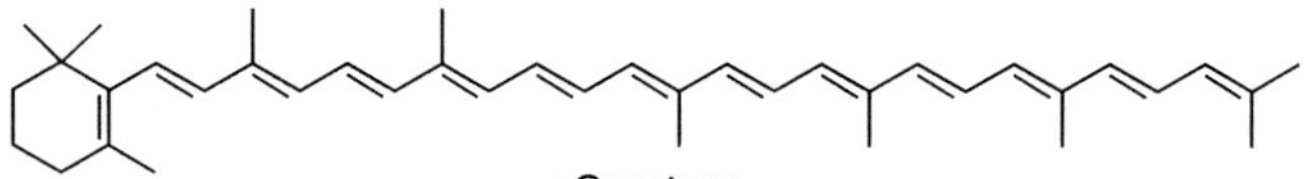

γ-Carotene

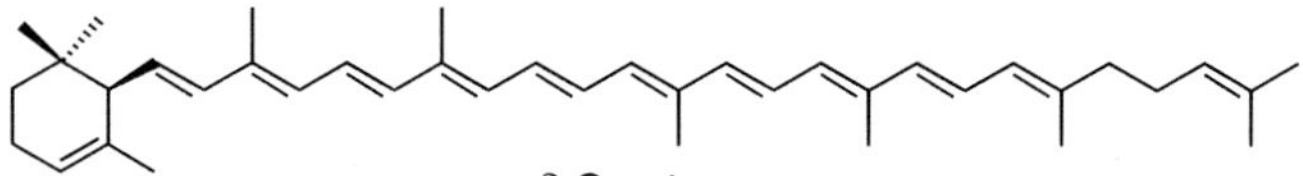

δ-Carotene

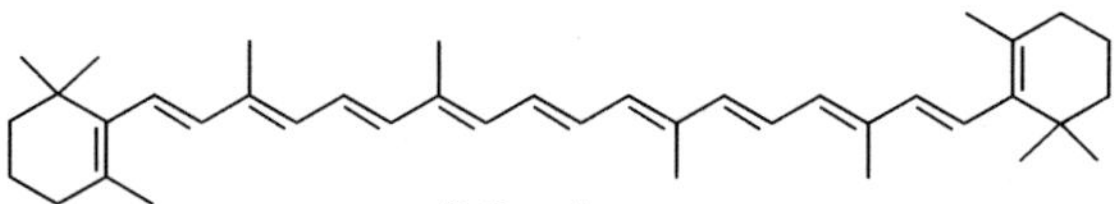

β-Carotene

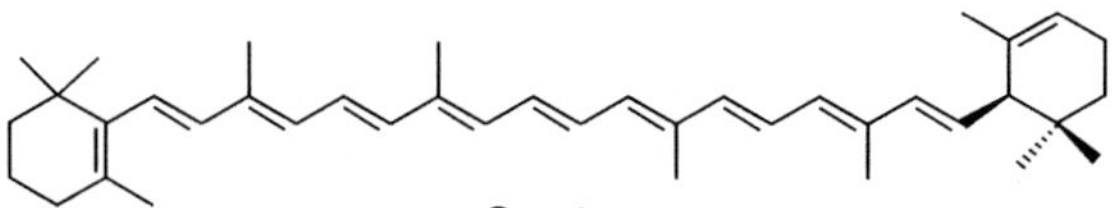

α-Carotene

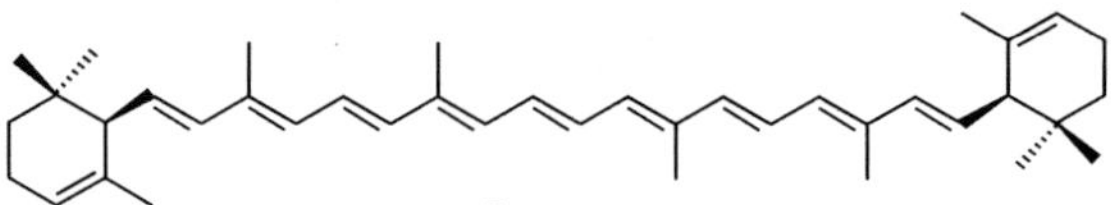

ε-Carotene

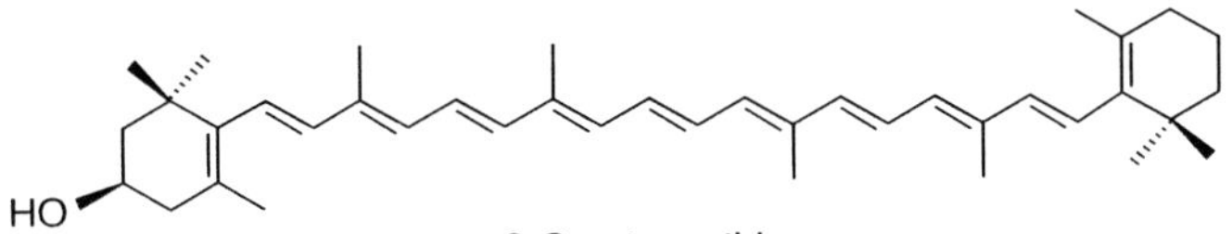

β-Cryptoxanthin

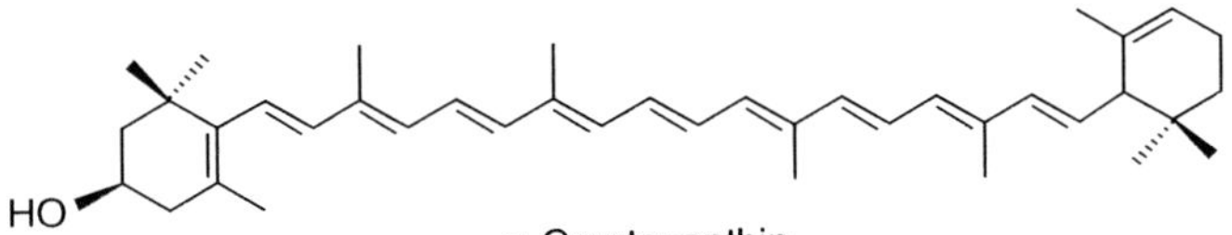

α-Cryptoxanthin

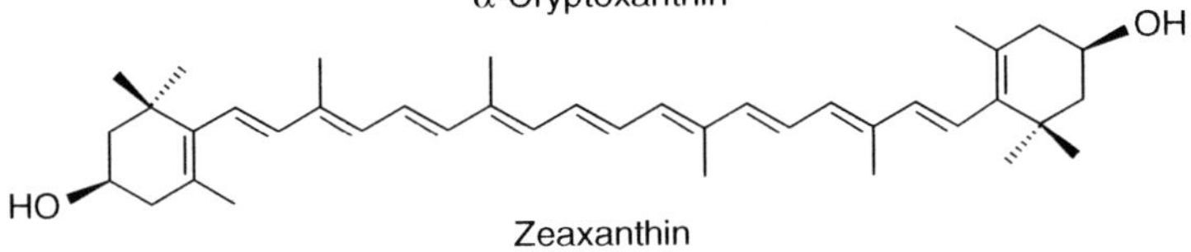

Zeaxanthin

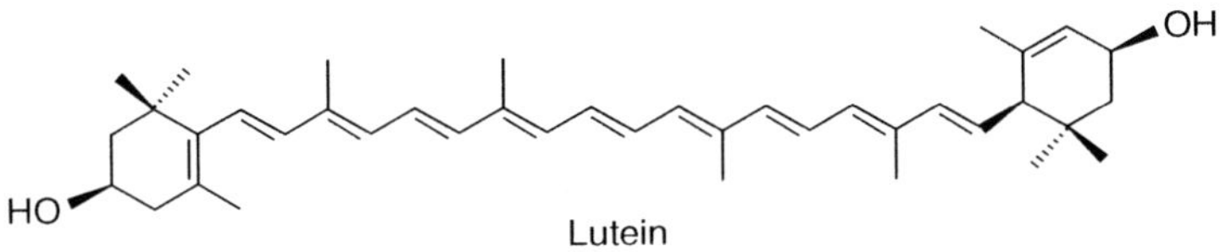

Lutein

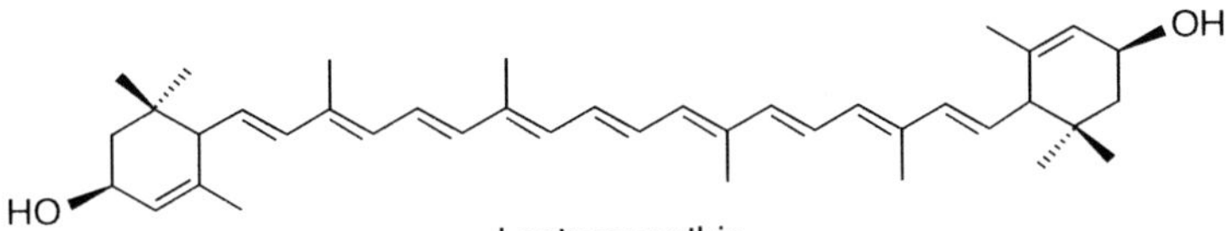

Lactucaxanthin

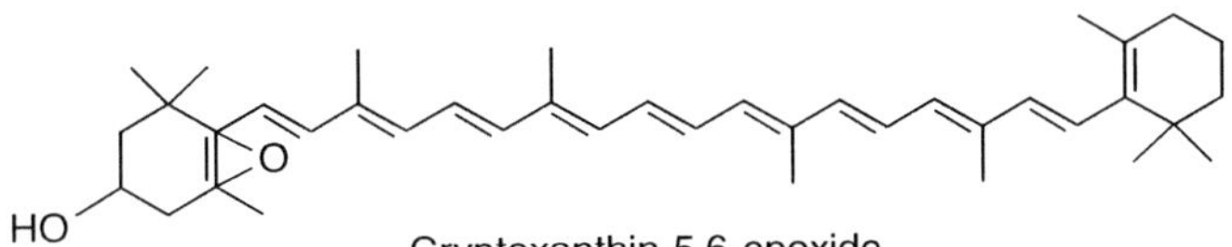

Cryptoxanthin-5,6-epoxide

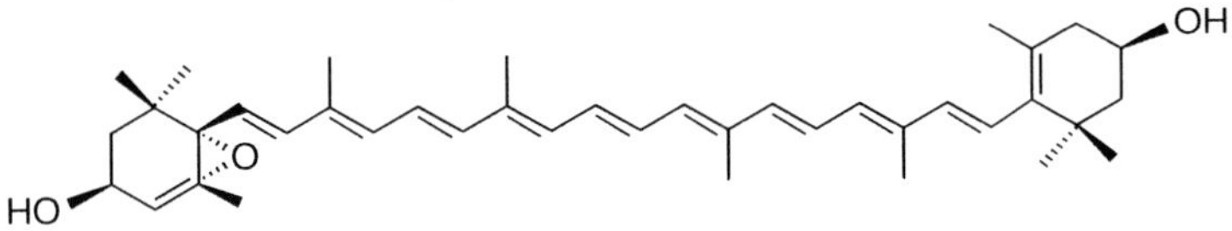

Antheraxanthin

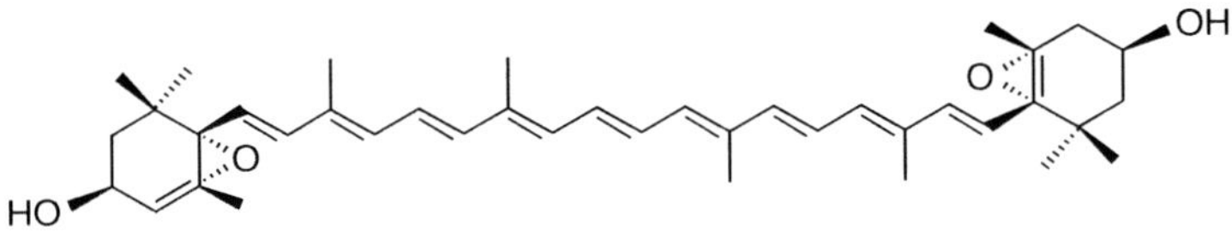

Violaxanthin

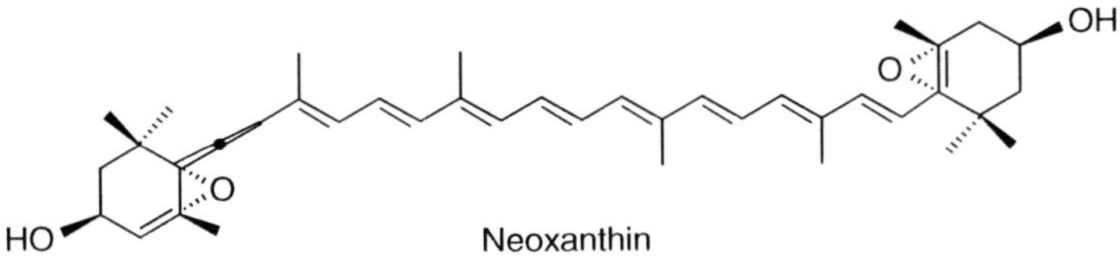

Neoxanthin

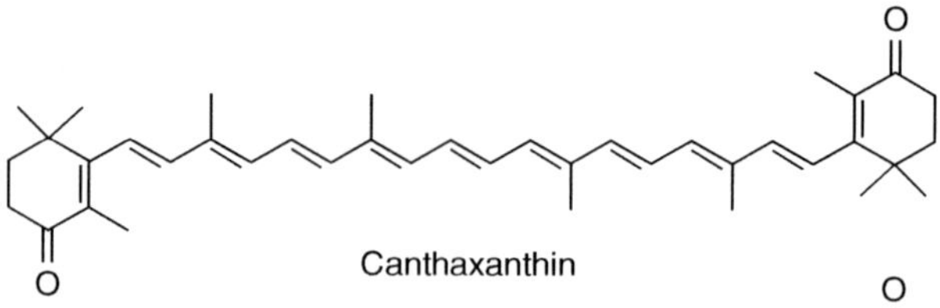

Canthaxanthin

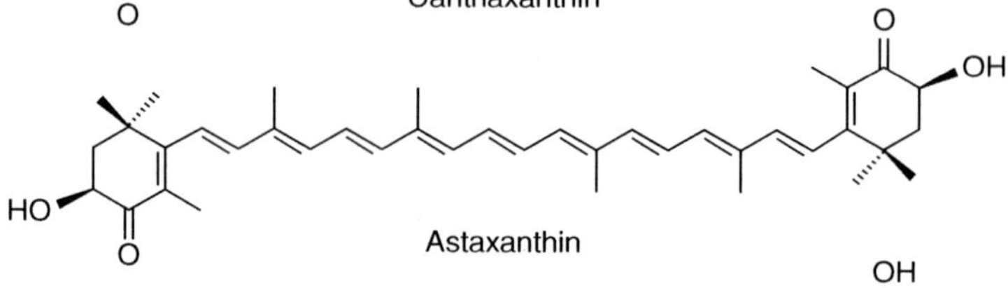

Astaxanthin

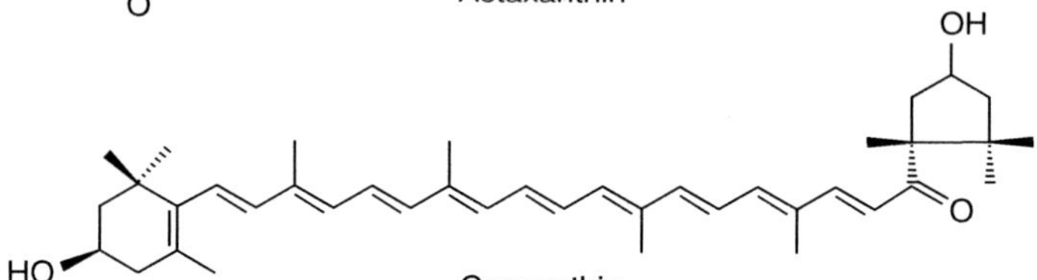

Capsanthin

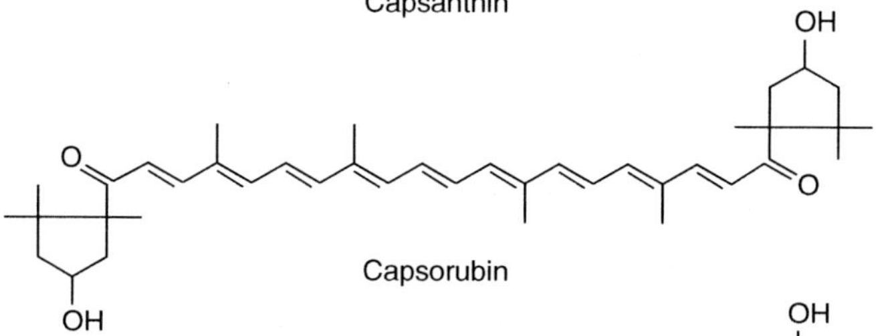

Capsorubin

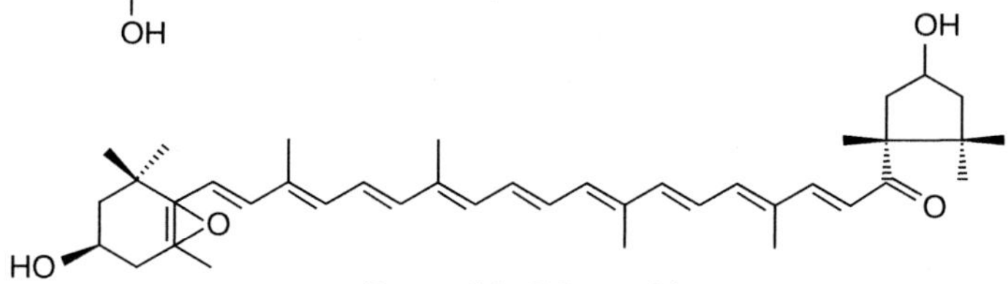

Capsanthin-5,6-epoxide

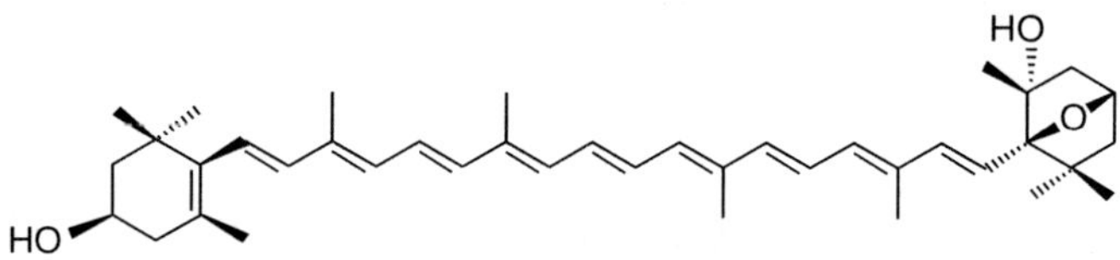

Cucurbitaxanthin A

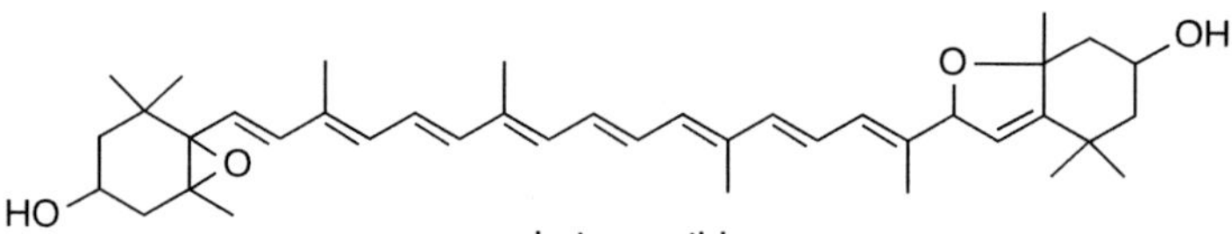

Luteoxanthin

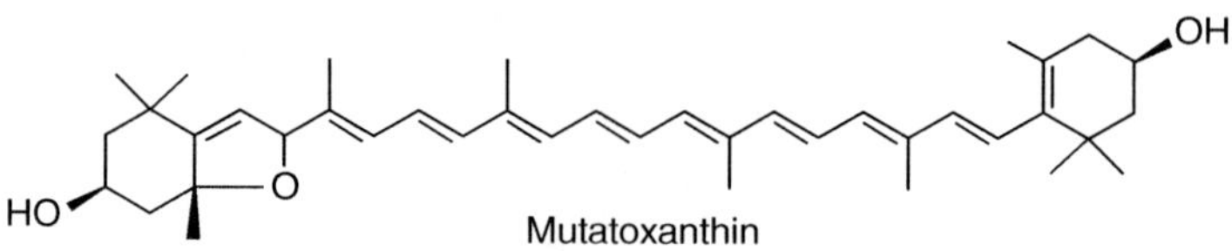

Mutatoxanthin

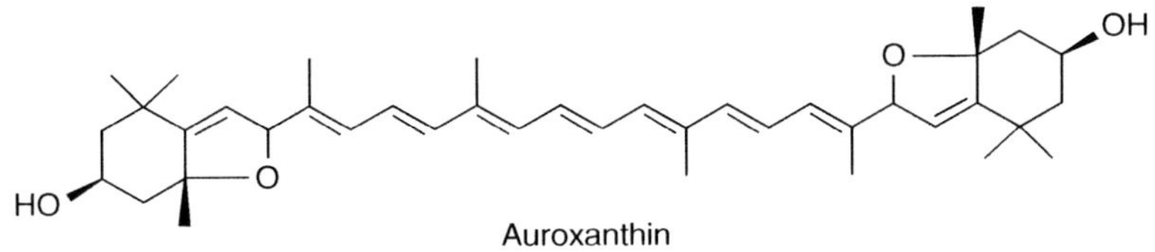

Auroxanthin

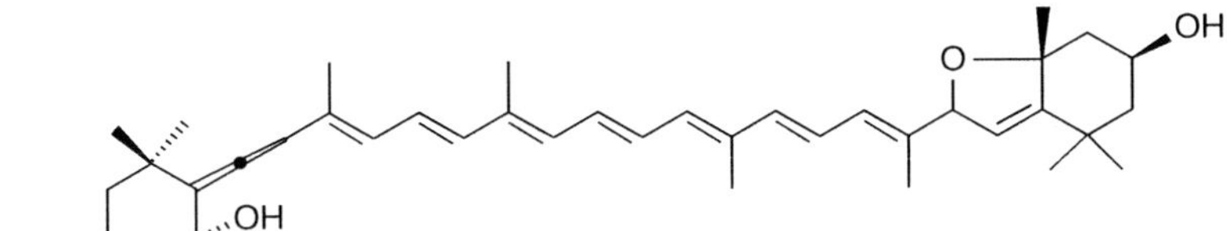

Neochrome

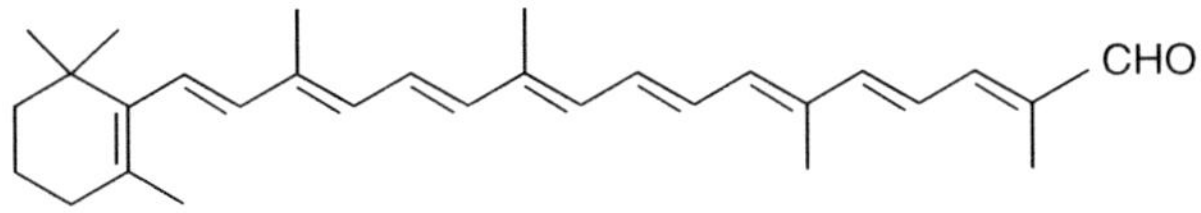

β-Apo-8′-carotenal

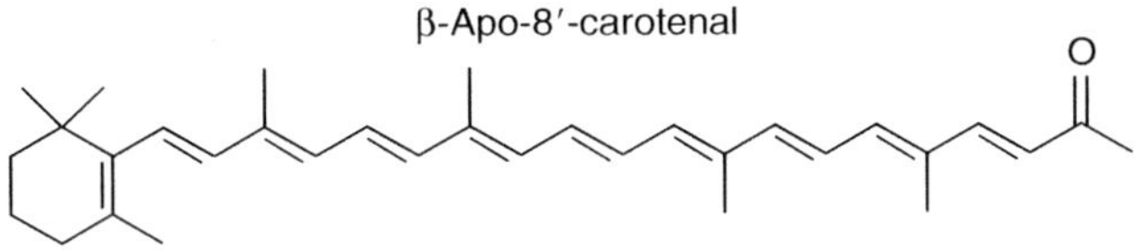

Citranaxanthin

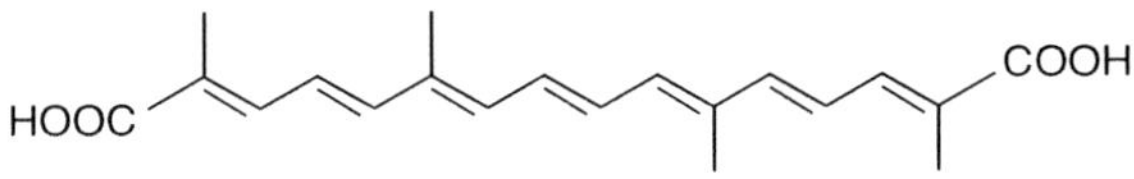

Crocetin

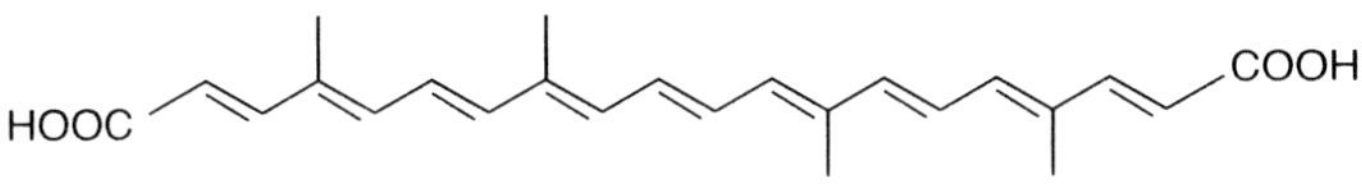

Norbixin

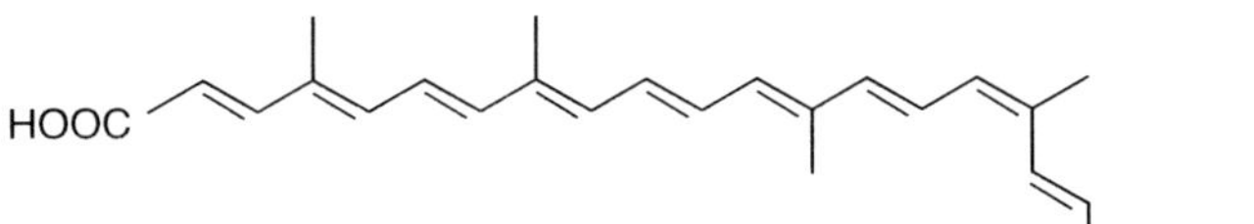

Bixin

REFERENCES

1. Gibson, G.R. and Roberfroid, M.B., Dietary modulation of the human colonic microbiota—introducing the concept of prebiotics, *J. Nutr.*, 125, 1401, 1995.
2. Gibson, G.R. and Fuller, R., The role of probiotics and prebiotics in the functional food concept, in *Functional Foods. The Consumer, the Products and the Evidence*, Sadler, M.J. and Saltmarsh, M. (Eds.), The Royal Society of Chemistry, Cambridge, United Kingdom, 1998, 3–4.

3. Strain, J.J. and Benzie, I.F.F., Antioxidant nutrients, in *Functional Foods. The Consumer, the Products and the Evidence*, Sadler, M.J. and Saltmarsh, M. (Eds.), The Royal Society of Chemistry, Cambridge, United Kingdom, 1998, 74–79.
4. Olson, J.A., Vitamin A and carotenoids as antioxidants in a physiological context, *J. Nutr. Sci. Vitaminol.*, 39, S57, 1993.
5. Moon, R.C., Comparative aspects of carotenoids and retinoids as chemopreventive agents for cancer, *J. Nutr.*, 119, 127, 1989.
6. Ziegler, R.G., A review of epidemiologic evidence that carotenoids reduce the risk of cancer, *J. Nutr.*, 119, 116, 1989.
7. Colditz, G.A., Branch, L.G., and Lipnick, R.J., Increased green and yellow vegetable intake and lowered cancer deaths in an elderly population, *Am. J. Clin. Nutr.*, 41, 32, 1985.
8. Mathews-Roth, M.M., Antitumor activity of beta-carotene, canthaxanthin and phytoene, *Oncology*, 39, 33, 1982.
9. Bendich, A. and Olson, J.A., Biological actions of carotenoids, *FASEB J.*, 3, 1927, 1989.
10. Burton, G.W. and Ingold, K.U., Beta-carotene—An unusual type of lipid antioxidant, *Science*, 224, 569, 1984.
11. Zhang, L.X., Conney, R.V., and Bertram, J.S., Carotenoids enhance gap junctional communication and inhibit lipid peroxidation in C3H/10T1/2 cells: Relationship to their cancer chemopreventive action, *Carcinogenesis*, 12, 2109, 1991.
12. Bendich, A., Carotenoids and the immune-response, *J. Nutr.*, 119, 112, 1989.
13. Henry, B.S., Natural food colours, in *Natural Food Colorants*, Hendry, G.A.F. and Houghton, J.D. (Eds.), Blackie Academic and Professional, London, United Kingdom, 1996, 40–79.
14. Britton, G. and Hormero-Méndez, D., Carotenoids and colour in fruit and vegetables, in *Phytochemistry of Fruit and Vegetables*, Tomás-Barberán, F.A. and Robins, R.J. (Eds.), Clarendon Press, Oxford, United Kingdom, 1997, 11–27.
15. Britton, G., Pigments in food: More than just colours, in *Proc. First Int. Congr. Pigments Food Technol.*, Mínguez-Mosquera, M.I., Jarén-Galán, M., and Hornero-Méndez, D. (Eds.), PFT, Sevilla, Spain, 1999, 3–9.
16. Weedon, B.C.L., Occurrence, in *Carotenoids*, Isler, O. (Ed.), Birkhäuser Verlag, Basel, Switzerland, 1971, 29–59.
17. Wackenroder, H.W.F., Über das oleum rad. *Danci aethereum*, das carotin, den carotlenzucher und den officinellum succus Danci, *Mag. Pharm.*, 33, 144, 1831.
18. Tswett, M., Adsorptionsanalyse and chromatographische method: Anwendung auf die chemie des chlorophylls, *Berl. Dtsch. Bot. Ges.*, 24, 384, 1906.
19. Eugster, C.H., History: 175 years of carotenoid chemistry, in *Carotenoids, Vol. 1A: Isolation and Analysis*, Britton, G., Liaaen-Jensen, S., and Pfander, H. (Eds.), Birkhäuser Verlag, Basel, Switzerland, 1995, 1–12.
20. Weedon, B.C.L. and Moss, G.P., Structure and nomenclature, in *Carotenoids. Vol. 1A: Isolation and Analysis*, Britton, G., Liaaen-Jensen, S. and Pfander, H. (Eds.), Birkhäuser Verlag, Basel, Switzerland, 1995, 27–70.
21. Britton, G., Structure and nomenclature, in *Carotenoids in Photosynthesis*, Young, A. and Britton, G. (Eds.), Chapman and Hall, London, United Kingdom, 1993, 1–15.
22. Goodwin, T.W., Distribution of carotenoids, in *Chemistry and Biochemistry of Plant Pigments, Vol. 1*, Goodwin, T.W. (Ed.), Academic Press, London, New York, and San Francisco, 1976, 225–261.
23. Davies, B.H., Matthews, S., and Kirk, J.T.O., The nature and biosynthesis of the carotenoids of different colour varieties of *Capsicum annuum*, *Phytochemistry*, 9, 797, 1970.

24. Goodwin, T.W. and Goad, L.J., Carotenoids and triterpenoids, in *The Biochemistry of Fruits and their Products, Vol. 1*, Hulme, A.C. (Ed.), Academic Press, London, New York and San Francisco, 1970, 305–368.
25. Farin, D., Ikan, R., and Gross, J., The carotenoid pigments in the juice and flavedo of mandarin hybrid (*Citrus reticulata*) cv michal during ripening, *Phytochemistry*, 22, 403, 1983.
26. Hornero-Méndez, D. and Mínguez-Mosquera, M.I., Xanthophyll esterification accompanying carotenoid overaccumulation in chromatoplast of *Capsicum annuum* ripening fruits is a constitutive process and useful for ripeness index, *J. Agric. Food Chem.*, 48, 1617, 2000.
27. Bartley, G.E. and Scolnik, P.A., Plant carotenoids: Pigments for photoprotection, visual attraction, and human health, *Plant Cell*, 7, 1027, 1995.
28. Schiedt, K., Isolation and analysis, in *Carotenoids. Vol. 3: Biosynthesis*, Britton, G., Liaaen-Jensen, S., and Pfander, H. (Eds.), Birkhäuser Verlag, Basel, Switzerland, 1998, 285–355.
29. Zagalsky, P.F., Invertebrate carotenoproteins, *Methods Enzymol.*, 111, 216, 1985.
30. Pouchet, G., Des changements de coloration sous l4influence des nerfs, *J. Anat. Physiol. Paris*, 12, 199, 1876.
31. Lee, W.L. (Ed.), *Carotenoproteins in Animal Coloration*, Dowdens, Hutchinson & Ross, Inc., Stoudsburg, PA, 1977.
32. Goodwin, T.W. and Britton, G., Distribution and analysis of carotenoids, in *Plant Pigments*, Goodwin, T.W. (Ed.), Academic Press, London, United Kingdom, 1988, 61–132.
33. Sitte, P., Falk, H., and Liedvogel, B., Chromoplasts, in *Pigments in Plants*, Czygan, F. Ch. (Ed.), Gustav Fischer, Stuttgart, Germany, 1980, 117–148.
34. Frank, H.A. and Cogdell, R.J., The photochemistry and function of carotenoids in photosynthesis, in *Carotenoids in Photosynthesis*, Young, A. and Britton, G. (Eds.), Chapman and Hall, London, United Kingdom, 1993, 252–326.
35. Britton, G., UV/Visible spectroscopy, in *Carotenoids. Vol. 1B: Spectroscopy*, Britton, G., Liaaen-Jensen, S., and Pfander, H. (Eds.), Birkhäuser Verlag, Basel, Switzerland, 1995, 13–62.
36. Goodwin, T.W. (Ed.), *The Biochemistry of Plant Pigments, Vol. 1*, Chapman and Hall, London, United Kingdom, 1980.
37. Swanson, J.E. and Parker, R.S., Biological effects of carotenoids in humans, in *Handbook of Antioxidants*, Cadenas, E. and Packer, L. (Eds.), Marcel Dekker Inc., New York, Basel, Hong Kong, 1996, 337–367.
38. FAO/WHO Joint Expert Committee on Food Additives, WHO Tech. Rep. Ser. 362, 1967.
39. FAO/WHO expert consultation on human vitamin and mineral requirements vitamin and mineral requirements in human nutrition: Report of a joint FAO/WHO expert consultation, Bangkok, Thailand, 21–30 September, 1998.
40. Van het Hof, K.H. et al., Bioavailability of lutein from vegetables is 5 times higher than that of β-carotene, *Am. J. Clin. Nutr.*, 70, 261, 1999.
41. U.S. Institute of Medicine, *Dietary Reference Intakes for Vitamin A, Vitamin K, Arsenic, Boron, Chromium, Copper, Iodine, Iron, Manganese, Molybdenum, Nickel, Silicon, Vanadium, and Zinc*, National Academy Press, Washington DC, WA, 2001.
42. Clevidence, B.A. and Bieri, J.G., Association of carotenoids with human plasma lipoproteins, *Methods Enzymol.*, 214, 33, 1992.
43. Krinsky, N.I., Cornwell, D.G., and Oncley, J.L., The transport of vitamin A and carotenoids in human plasma, *Arch. Biochem. Biophys.*, 73, 233, 1958.

44. Britton, G., Structure and properties of carotenoids in relation to function, *FASEB J.*, 9, 1551, 1995.
45. Krinsky, N.I., Membrane antioxidants, *Ann. N.Y. Acad. Sci.*, 551, 17, 1988.
46. Terao, J., Antioxidant activity of β-carotene-related carotenoids in solution, *Lipids*, 24, 659, 1989.
47. Martin, H.D. et al., Anti- and prooxidant properties of carotenoids, *J. Prakt. Chem.*, 341, 302, 1999.
48. Liebler, D.C., Antioxidant reactions of carotenoids, *Ann. N.Y. Acad. Sci.*, 669, 20, 1993.
49. Everett, S.A. et al., Scavenging of nitrogen dioxide, thiyl, and sulfonyl free radicals by the nutritional antioxidant β-carotene, *J. Biol. Chem.*, 271, 3988, 1996.
50. Johnson, E.J., The role of lycopene in health and disease, *Nutr. Clin. Care*, 3, 35, 2000.
51. Gann, P.H. et al., Lower prostate cancer risk in men with elevated plasma lycopene levels: results of a prospective analysis, *Cancer Res.*, 59, 1225, 1999.
52. Stahl, W. and Sies, H., Perspectives in biochemistry and biophysics. Lycopene: A biological important carotenoid for humans?, *Arch. Biochem. Biophys.*, 336, 1, 1996.
53. Handelman, G.J., Carotenoids as scavengers of active oxygen species, in *Handbook of Antioxidants*, Cadenas, E. and Packer, L. (Eds.), Marcel Dekker Inc., New York, Basel, Hong Kong, 1996, 259–313.
54. Van het Hof, K.H. et al., Dietary factors that affect the bioavailability of carotenoids, *J. Nutr.*, 130, 503, 2000.
55. Castenmiller, J.J.M. and West, C.E., Bioavailability and bioconversion of carotenoids, *Annu. Rev. Nutr.*, 18, 19, 1998.
56. Faulks, R.M. and Southon, S., Challenges to understanding and measuring carotenoid bioavailability, *Biochim. Biophys. Acta*, 1740, 95, 2005.
57. Oshima, S. et al., Accumulation and clearance of capsanthin in blood plasma after the ingestion of paprika juice in men, *J. Nutr.*, 127, 1475, 1997.
58. Downham, A. and Collins, P., Colouring our foods in the last and next millennium, *Int. J. Food Sci. Tech.*, 35, 5, 2000.
59. Sandmann, G. et al., The biotechnological potential and design of novel carotenoids by gene combination in *Escherichia coli*, *Trends Biotechnol.*, 17, 233, 1999.
60. Scolnik, P.A. and Bartley, G.E., A table of some cloned plant genes involved in isoprenoid biosynthesis, *Plant Mol. Biol. Rep.*, 14, 305, 1996.
61. Armstrong, G.A., Eubacteria show their true colors—Genetics of carotenoid pigment biosynthesis from microbes to plants, *J. Bacteriol.*, 176, 4795, 1994.
62. Lang, Q. and Wai, C.M., Supercritical fluid extraction in herbal and natural product studies—a practical review, *Talanta*, 53, 771, 2001.
63. Vega, P.J. et al., Supercritical carbon dioxide extraction efficiency for carotenes from carrots by RSM, *J. Food Sci.*, 61, 757, 1996.
64. Vasapollo, G. et al., Innovative supercritical CO_2 extraction of lycopene from tomato in the presence of vegetable oil as co-solvent, *J. Supercrit. Fluids*, 29, 87, 2004.
65. Jarén-Galán, M., Nienaber, U., and Schwartz, S.J., Paprika (*Capsicum annuum*) oleoresin extraction with supercritical carbon dioxide, *J. Agric. Food Chem.*, 47, 3558, 1999.
66. Rodríguez-Amaya, D.B., Critical review of provitamin A determination in plant foods, *J. Micronutr. Anal.*, 5, 191, 1989.
67. Mínguez-Mosquera, M.I. and Garrido-Fernández, J., Lipids removal during the extraction phase of chloroplast pigments in olive fruit (*Olea europaea*), *Grasas Aceites*, 36, 376, 1985.
68. Mínguez-Mosquera, M.I. et al., Decoloration of vegetable oils and oleoresins with recovery of unaltered pigments, *J. Am. Oil Chem. Soc.*, 68, 809, 1991.

69. Davies, B.H., Carotenoids, in *Chemistry and Biochemistry of Plant Pigments, Vol. 2*, Goodwin, T.W. (Ed.), Academic Press, London, New York, San Francisco, 1976, 38–165.
70. Foppen, F.H., Tables for identification of carotenoid pigments, *Chromatogr. Rev.*, 14, 133, 1971.
71. Lee, C.Y., Changes in carotenoid content of carrots during growth and post-harvest storage, *Food Chem.*, 20, 285, 1986.
72. Lee, C.Y., McCoon, P.E., and LeBowitz, J.M., Vitamin A value of sweet corn, *J. Agric. Food Chem.*, 29, 1294, 1981.
73. Sherma, J., Thin-layer chromatography in food and agricultural analysis, *J. Chromatograph. A*, 880, 129, 2000.
74. Stahl, E., *Thin Layer Chromatography. A Laboratory Handbook*, Springer, Berlin, 1967.
75. Davies, B.H. and Köst, H.P., Carotenoids, in *Handbook of Chromatography, Vol.1*, Köst, H.P. (Ed.), CRC Press, Boca Raton, FL, 1988.
76. Gross, J., A rapid separation of citrus carotenoids by thin-layer chromatography, *Chromatographia*, 13, 572, 1980.
77. Mínguez-Mosquera, M.I. and Garrido-Fernández, J., Chlorophyll and carotenoid presence in olive fruit (*Olea europaea*), *J. Agric. Food Chem.*, 37, 1, 1989.
78. Mínguez-Mosquera, M.I., Garrido-Fernández, J., and Pereda-Marín, J., Paprika (*Capsicum annuum*). Ratio between the red and yellow carotenoid pigments, *Grasas Aceites*, 35, 4, 1984.
79. Johjima, T. and Ogura, H., Analysis of tomato carotenoids by thin-layer chromatography and a cis-form γ-carotene newly identified in tangerine tomato, *J. Jpn. Soc. Hortic. Sci.*, 52, 202, 1983.
80. Stewart, I. and Wheaton, T.A., Continuous separation of carotenoids by liquid chromatography, *J. Chromatogr.*, 55, 325, 1971.
81. Ferruzzi, M.G. et al., Carotenoid determination in biological microsamples using liquid chromatography with a coulometric electrochemical array detector, *Anal. Biochem.*, 256, 74, 1998.
82. Luterotti, S., Frankob, M., and Bicanicc, D., Ultrasensitive determination of β-carotene in fish oil-based supplementary drugs by HPLC-TLS, *J. Pharm. Biomed. Anal.*, 21, 901, 1999.
83. Sander, L.C., Sharpless, K.E., and Pursch, M., C_{30} stationary phases for the analysis of food by liquid chromatography, *J. Chromatograph. A*, 880, 189, 2000.
84. Van Breemen, R.B., Innovations in carotenoid analysis using LC/MS, *Anal. Chem.*, 68, 299A, 1996.
85. Khachik, F. et al., Separation and quantification of carotenoids in foods, *Methods Enzymol.*, 213, 347, 1992.
86. Bushway, R.J. and Wilson, A.M., Determination of α- and β-carotene in fruits and vegetables by HPLC, *Can. Inst. Food Sci. Technol. J.*, 15, 165, 1982.
87. Bushway, R.J., Determination of α- and β-carotene in some raw fruits and vegetables by high-performance liquid chromatography, *J. Agric. Food Chem.*, 34, 409, 1986.
88. Lesellier, E. et al., Optimization of the isocratic non-aqueous reverse phase (NARP) HPLC separation of trans/cis α- and β-carotenes, *J. High Res. Chromatogr.*, 12, 447, 1989.
89. Saleh, M.H. and Tan, B., Separation and identification of cis/trans carotenoids isomers, *J. Agric. Food Chem.*, 39, 1438, 1991.
90. O'Neil, C.A., Schwartz, S.J., and Catignani, G.L., Comparison of liquid chromatographic methods for determination of cis-trans isomers of β-carotene, *J. Assoc. Off. Anal. Chem.*, 74, 36, 1991.

91. Biacs, P.A. et al., Studies on the carotenoid pigments of paprika (*Capsicum annuum* L. var Sz-20), *J. Agric. Food Chem.*, 37, 350, 1989.
92. Khachik, F., Beecher, G.R., and Lusby, W.R., Separation, identification and quantification of the major carotenoids in extracts of apricots, peaches, cantaloupe and pink grapefruit by liquid chromatography, *J. Agric. Food Chem.*, 37, 1465, 1989.
93. Heinonen, M.I. et al., Carotenoid in Finnish foods: Vegetables, fruits and berries, *J. Agric. Food Chem.*, 37, 655, 1989.
94. Pott, I., Breithaupt, D.E., and Carle, R., Detection of unusual carotenoid esters in fresh mango (*Mangifera indica* L. cv. 'Kent'), *Phytochemistry*, 64, 825, 2003.
95. Breithaupt, D.E. and Schwack, W., Determination of free and bound carotenoids in paprika (*Capsicum annuum* L.) by LC/MS, *Eur. Food Res. Technol.*, 211, 52, 2000.
96. Kaufmann, R. et al., Analysis of carotenoids and carotenol fatty acid esters by matrix-assisted laser desorption ionization (MALDI) and MALDI-post-source-decay mass spectrometry, *Anal. Biochem.*, 238, 117, 1996.
97. Biacs, P.A. and Daood, H.G., High-performance liquid chromatography with photodiode-array detection of carotenoids and carotenoid esters in fruits and vegetables, *J. Plant. Physiol.*, 143, 520, 1994.
98. Khachik, F., Beecher, G.R., and Whittaker, N.F., Separation, identification, and quantification of the major carotenoid and chlorophyll constituents in extracts of several green vegetables by liquid chromatography, *J. Agric. Food Chem.*, 334, 603, 1986.
99. Tsukida, K. et al., Separation and determination of cis/trans β-carotenes by HPLC, *J. Chromatogr.*, 245, 359, 1982.
100. Schiedt, K. and Liaaen-Jensen, S., Isolation and analysis, in *Carotenoids. Vol. 1A: Isolation and Analysis*, Britton, G., Liaaen-Jensen, S., and Pfander, H. (Eds.), Birkhäuser Verlag, Basel, Switzerland, 1995, 81–108.
101. Buchecker, R. and Noack, K., Circular dichroism, in *Carotenoids. Vol. 1B: Spectroscopy*, Britton, G., Liaaen-Jensen, S., and Pfander, H. (Eds.), Birkhäuser Verlag, Basel, Switzerland, 1995, 63–116.
102. Englert, G., NMR spectroscopy, in *Carotenoids. Vol. 1B: Spectroscopy*, Britton, G., Liaaen-Jensen, S. and Pfander, H. (Eds.), Birkhäuser Verlag, Basel, Switzerland, 1995, 147–260.
103. Koyama, Y., Resonance Raman spectroscopy, in *Carotenoids. Vol. 1B: Spectroscopy*, Britton, G., Liaaen-Jensen, S., and Pfander, H. (Eds.), Birkhäuser Verlag, Basel, Switzerland, 1995, 135–146.
104. Strohschein, S., Pursch, M., and Albert, K., Hyphenation of high performance liquid chromatography with nuclear magnetic resonance spectroscopy for the characterization of β-carotene isomers employing a C_{30} stationary phase, *J. Pharm. Biomed. Anal.*, 21, 669, 1999.
105. Dachtler, M. et al., Combined HPLC-MS and HPLC-NMR on-line coupling for the separation and determination of lutein and zeaxanthin stereoisomers in spinach and in retina, *Anal. Chem.*, 73, 667, 2001.
106. Enzell, C.R. and Back, S., Mass spectrometry, in *Carotenoids. Vol. 1B: Spectroscopy*, Britton, G., Liaaen-Jensen, S., and Pfander, H. (Eds.), Birkhäuser Verlag, Basel, Switzerland, 1995, 261–320.
107. Vetter, W., Englert, G., Rigassi, N., and Schwieter, U., Spectroscopic methods, in *Carotenoids*, Isler, O. (Ed.), Birkhäuser Verlag, Basel, Switzerland, 1971, 189–266.
108. Mínguez-Mosquera, M.I. and Hornero-Méndez, D., Separation and quantification of the carotenoid pigments in red peppers (*Capsicum annuum*, L.), paprika and oleoresin by reversed-phase HPLC, *J. Agric. Food Chem.*, 41, 1616, 1993.

109. Philip, T. and Chen, T.S., Development of a method for the quantitative estimation of provitamin A carotenoids in some fruits, *J. Food Sci.*, 53, 1703, 1988.
110. Mínguez-Mosquera, M.I. et al., Determination of chlorophylls and carotenoids by HPLC during olive lactic fermentation, *J. Chromatogr.*, 585, 259, 1991.
111. Mantoura, R.F.C. and Llewellyn, C.A., The rapid determination of algal chlorophyll and carotenoid pigments and their breakdown products in natural waters by reverse-phase high-performance liquid chromatography, *Anal. Chim. Acta*, 151, 297, 1983.
112. Khachik, F. et al., Separation and quantification of carotenoids in human plasma, *Methods Enzymol.*, 213, 205, 1992.
113. Bauernfeind, J.C., Carotenoid vitamin A precursors and analogs in foods and feeds, *J. Agric. Food Chem.*, 20, 456, 1972.
114. Zechmeister, L., Stereoisomeric provitamins A, *Vitam. Horm.*, 7, 57, 1949.
115. Daood, H.G. et al., Separation and identification of tomato fruit pigments by TLC and HPLC, *Acta Alimentaria*, 16, 339, 1987.
116. Speek, A.J., Temalilwa, C.R., and Schrijver, J., Determination of beta-carotene content and vitamin-A activity of vegetables by high-performance liquid chromatography and spectrophotometry, *Food Chem.*, 19, 65, 1986.
117. Chandler, A.L. and Schwartz, S.J., HPLC separation of cis-trans carotene isomers in fresh and processed fruits and vegetables, *J. Food Sci.*, 52, 669, 1987.
118. Quackenbush, F.W., Reverse phase HPLC separation of cis-carotenoids and trans-carotenoids and its application to beta-carotenes in food materials, *J. Liq.Chromatograph.*, 10, 643, 1987.
119. Thompson, J.N., Hatina, G., and Maxwell, W.B., High performance liquid chromatographic determination of vitamin A in margarine, milk, partially skimmed milk and skimmed milk, *J. Assoc. Off. Anal. Chem.*, 63, 894, 1980.
120. Almela, L. et al., Separation and determination of individual carotenoids in a capsicum cultivar by normal-phase high-performance liquid chromatography, *J. Chromatograph.*, 502, 95, 1990.
121. Quackenbush, F.W. and Smallidge, R.L., Nonaqueous reverse phase liquid-chromatographic system for separation and quantitation of provitamins-A, *J. Assoc. Off. Anal. Chem.*, 69, 767, 1986.
122. Bushway, R.J., Separation of carotenoids in fruits and vegetables by high-performance liquid chromatography, *J. Liq. Chromatogr.*, 8, 1527, 1985.
123. Fisher, C. and Kocis, J.A., Separation of paprika pigments by HPLC, *J. Agric. Food Chem.*, 35, 55, 1987.
124. Bureau, J.L. and Bushway, R.J., HPLC determination of carotenoids in fruits and vegetables in the United States, *J. Food Sci.*, 51, 128, 1986.
125. Juhler, R.K. and Cox, R.P., High-performance liquid chromatographic determination of chloroplast pigments with optimized separation of lutein and zeaxanthin, *J. Chromatograph.*, 508, 232, 1990.

7 Chlorophylls

María Isabel Mínguez-Mosquera, Beatriz Gandul-Rojas, Lourdes Gallardo-Guerrero, María Roca, and Manuel Jarén-Galán

CONTENTS

7.1 INTRODUCTION

Numerous components present in our normal diet cannot be considered nutrients, or drugs, but are on the dividing line between the two. Their absence from the diet does not cause a nutritional deficit, but their presence, above all those of plant origin, helps to prevent certain diseases in the long term.

Components of our diet are normally grouped in five great families: carbohydrates, lipids, proteins, oligoelements, and vitamins, each having a particular biological function in nutrition. Besides these basic nutrients, there are many other minor constituents that are not members of these great families, such as chlorophylls, flavonoids, coumarins, etc. They are not essential for correct nutrition, and are therefore not nutrients. Nor can they be considered substances having direct therapeutic activity, although some do at concentrations much higher than those found in our normal diet. However, when foods containing these compounds are a regular part of the diet, the incidence and prevalence of certain diseases are changed.

The popularization of advances in knowledge about the properties of dietary components has given rise to a more rational consumer culture. Technology guarantees that any consumer product meets hygiene-health standards in nutritional function, and consumers are now more interested in quality and the preservation of the whole potential of the foodstuff, conceiving it as rather more than a simple source of proteins or carbohydrates. From the technological and consumer viewpoint, this is a culture in which food serves not only for replacement and to repair wear and tear of the organism, but can also help to prevent future disorders. It is the beginning of consumption based on the interrelationship between product quality and food quality.

Knowing the functional activity of certain compounds means that their presence in foods is valued positively, and gives the consumer an additional parameter for product choice. Compounds whose presence are appreciated possibly for their esthetic function or as indicators of the state of ripeness and technological treatment of the foodstuff, as is the specific case of the chlorophylls, are given new scope.

We use our visual sense unconsciously to judge the quality of a product. We expect a fresh vegetable to be bright green, the color of chlorophylls, while one subjected to fermentation will be olive green, typical of pheophytins and pheophorbides. Colors other than those expected make us suspect the quality of the product, and analysis will corroborate the presence of chlorophyll transformation compounds improper of such product. The initial judgment of product quality, perhaps without our knowing it, depends on the pigments responsible for the color. Chlorophyll pigments have a double role in alimentation: esthetic and technological. Being pigments, and possessing chromatic properties, they make a product attractive and appetizing, and give important visual and analytical information about the history of the product. However, chlorophyll compounds are not inert components of our diet, they fulfill certain biological functions that are effective as long as the chlorophyll derivative conserves the basic porphyrin ring structure. Thus, the presence of chlorophylls and their transformation products contributes added value to the plant product.

Although there has been no deep research into the role of chlorophyll compounds in the diet, there are numerous epidemiological studies on the incidence of certain types of cancers between populations with similar dietary habits but differ essentially in the ingestion of chlorophyll-rich plant food. Statistically, there seems to be a relationship between high intake of chlorophyll in the diet and a decreased risk of certain types of cancers. However, the possibility cannot be excluded that the factor responsible for the anticancer activity is some other compound not yet detected.

Although the epidemiological results indicate a relationship between the ingestion of chlorophyll-rich plants and a decreased risk of certain types of cancers, error in assuming cause–effect cannot be ruled out. In experiments in vitro, the relationship is clear and overwhelming. The chlorophyll compounds are able to inhibit the genotoxic action of certain substances, particularly those with aromatic structure, and act as protectors against ionizing radiation, because of a high affinity in the link between DNA (deoxyribonucleic acid) and the porphyrin ring of the chlorophyll molecule.

This seems to indicate that, in fact, the role of dietary chlorophyll compounds is perhaps not as nutrients in the strict sense, since they are not necessary for our proper physiological functioning, but to prevent disorders and as protectors against aggression. Currently, the study of these compounds in our diet is moving away from purely technological aspects, toward an area midway between nutrition and pharmacology. From the alimentary point of view, the presence of chlorophyll compounds in green plants is important in every possible sense. We require the food to be visually attractive and its color appropriate for the treatment received, and that we know beforehand its activity in our organism, so that we can make the best judgment of the product.

Probably without our being conscious of the fact, the chlorophyll compounds that are a regular and widespread part of the human diet have an important protective role. A much deeper study is required of these compounds, which are not essential in our nutrition and whose lack does not mean alterations in the physiological functions, but which do contribute to protection against toxic compounds.

7.2 GENERAL ASPECTS

7.2.1 Structure, Location, Function, and Distribution

Numerous works carried out in the last 100 years have elucidated the chlorophyll structure. Willstatter, in 1906, was the first to achieve separation of chlorophyll *a* in pure state [1]. Later, its structure was established by degradation studies and proved by total synthesis.

Chemically, chlorophyll is classified within the porphyrin group. The common structure of this series of compounds is porphyn, comprising four units of pyrrole, whose alpha positions are linked by methine bridges, forming a new planar aromatic system that is very stable (Figure 7.1). The most noteworthy characteristic of the porphyrins is their readiness to form chelates (intramolecular bonds) with metal ions, leaving the metal firmly bonded in the space bounded by the four nitrogen atoms in the planar system. This basic structure comprises 11 conjugated double bonds

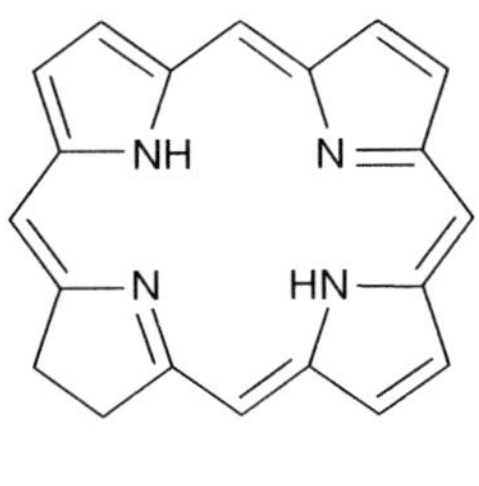

FIGURE 7.1 Structure of porphyn.

making up the chromophore group, able to absorb radiation in the visible spectrum. The porphyrins are generally brightly colored; the basic porphyrin system is orangey, with the color of the porphyrin compound depending on the nature of both the ring substituents and the central atom.

In the case of chlorophyll, the metal ion forming the complex is Mg (Figure 7.2). Chlorophyll also contains a modified propionic acid chain in the form of cyclic β-ketoester (isocyclic ring) and, on C-17, a chain of propionic acid esterified with a diterpene alcohol of C_{20} structure, phytol, making the molecule liposoluble. The difference between chlorophyll *a* and *b* lies in that *a* has a methyl group on carbon C_7, whereas *b* has a formyl group. Chlorophyll *c* has a residue of propionic acid on C-17, and chlorophyll *d* lacks the unsaturation between C-17 and C-18 of ring IV (Figure 7.3). In the bacteriochlorophylls, the alcohols that esterify on C-17^3 are, apart from phytol, farnesil (C_{12}) and geranyl geraniol (C_{16}).

The chlorophyll pigments of higher plants are located, together with a series of carotenoids, in the plastids, which are differentiations of the endoplasmic reticulum, isolated from the rest of the cytoplasm by their own membranes, and with their own genetic information. Plastids that have chlorophyll content, and thus photosynthetic capacity, are denominated chloroplasts. This organelle has a system of double membranes in the form of sacs, the thylakoids, which contain all the chlorophylls. Some are large and isolated (stroma thylakoids), and others are smaller and piled (grana thylakoids). Toward the interior of the chloroplast is a gelatinous and highly viscous phase, the stroma, which contains a group of enzymes, occasionally large granules of assimilation starch and small dark bodies, denominated plastoglobules, considered reservoirs of lipids [2].

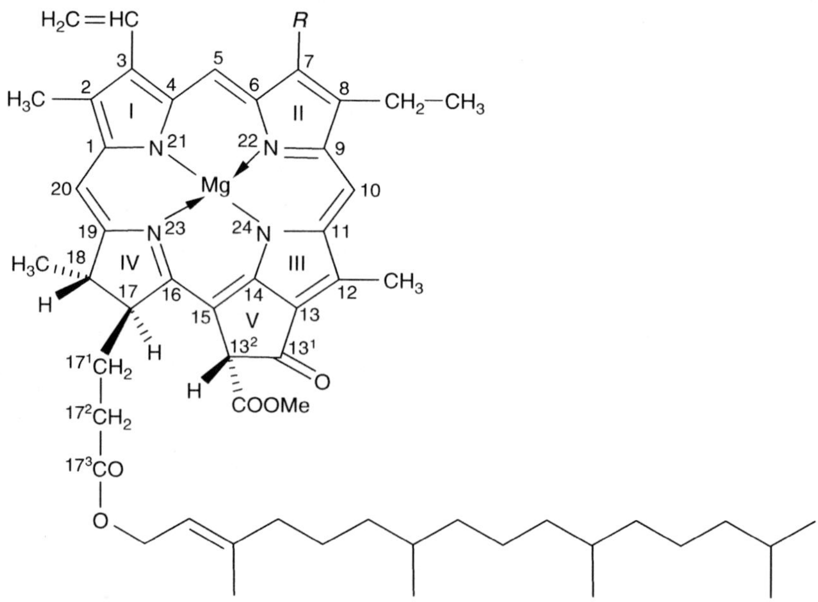

FIGURE 7.2 Structures of chlorophylls *a* ($R = CH_3$) and *b* ($R = CHO$).

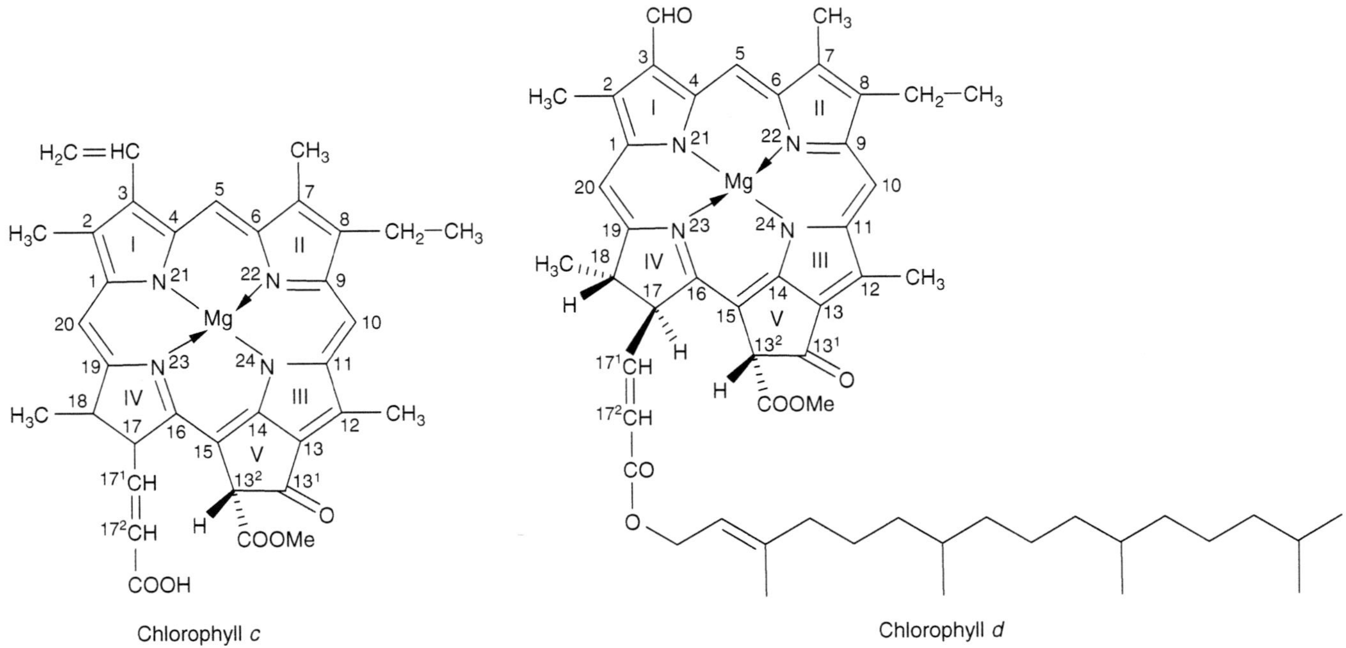

FIGURE 7.3 Structures of chlorophylls *c* and *d*.

The thylakoid membrane is a lipid bilayer composed of galactolipids and small amounts of phospholipids. The amphoteric nature of the chlorophylls allows the phytol chain to be embedded in the membrane, with the porphyrin ring linked by noncovalent bonds to proteins [3,4], constituting supermolecular structures known as photosystems (PS). In higher plants, there are always two photosystems, PSI and PSII, with PSII being responsible for photolysis of water, and PSI for the reduction of $NADP^+$. It is noteworthy that chlorophyll *b* is found mainly in PSII, exclusive to organisms with oxygenic photosynthesis, and evolutively more developed [5].

Chlorophyll–protein interaction enables the capture of radiation energy within the whole visible range, because the protein bonding modifies the excitation energy. Consequently, up to six types of chlorophyll *a*–protein groupings have been found in vivo, with characteristic electron absorption maxima (Cla661, Cla669, Cla677, Cla684, Cla691, and Cla700–720). In the case of chlorophyll *b*, only two groupings have been found (Clb640 and Clb650) [5].

It is difficult to conceive the origin or existence of life without the presence of photosynthetic pigments able to absorb radiation energy and transform it into chemical energy. The whole process is based on the existence of photoreceptive molecules, the chlorophylls, which translate light energy into chemical energy. The set of reactions occurring from the photo-excitation of the chlorophylls until the storage of the light energy in the form of sugars is denominated photosynthesis [6].

All photosynthetic cells contain one or more types of pigments, but not all are green. The photosynthetic algas and the bacteria can be brown, red, or purple. This variety of colors is because, besides chlorophyll, many photosynthetic cells contain other pigments, called accessories, that act as supplementary receptors of light for portions of the visible spectrum that are not covered completely by chlorophyll, thus cooperating in the transfer of light energy. These are the yellow carotenoids, the blue phycocyanins, and the red phycoerythrins. The carotenoids also protect chlorophyll from the degradative attack of molecular oxygen, produced during the photosynthetic process.

As chlorophyll *a* is common to all photosynthetic organisms, it has been postulated that it is the only pigment supplying energy directly to the photosynthetic reaction, and that all the others transfer the absorbed energy to chlorophyll *a* [7,8].

It is not easy to define how plant aging affects the pigments. Certainly, there is a sequence of transformations during fruit ripening, accompanied by changes in flavor, texture, color, etc. However, many fruits are chlorophyll-free before ripeness, with the chloroplasts replaced by chromoplasts. The degradation of chlorophylls during the biosynthesis of carotenoids or anthocyanins and betalains is complex [2,9]. As the fruits approach ripeness, there is a stage in which biochemical changes are initiated by the autocatalytic production of ethylene. This increase in respiration marks the point between development and ripening, with a variation in fruit skin coloration [10]. The yellow of autumn leaves in many plants is the result of the elimination (by the conductor routes) of chlorophylls that have decomposed, while the carotenoids remain. The same does not happen with the red coloration of autumn leaves. In this case, the cell fluid is tinted by anthocyanins. In general, the yellow or orange color of most ripe fruits, such as lemons and oranges (genus *Citrus*), is due to an enrichment in carotenoid pigments accompanied by disappearance of chlorophyll.

Chlorophylls *a* and *b* are widely distributed in nature, and are the best-known types. Chlorophyll *a* is found in all photosynthetic organisms except some groups of bacteria. Chlorophyll *b* is present in all higher plants, and in algas of the divisions Chlorophyta and Euglenophyta, always accompanying chlorophyll *a* as accessory pigment in the photosynthetic process. Other chlorophyll types (*c*, *d*, and *e*) are found only in algas, always in combination with chlorophyll *a*. The bacteriochlorophylls are the chlorophylls of photosynthesizing bacteria, and structurally are slightly different to those typical of higher plants.

In higher plants, only chlorophylls *a* and b are present. Their ratio normally varies between 3 and 1, depending on a multitude of factors, both genetic (species, variety, etc.) and environmental (luminosity, water stress, mineral nutrition, etc.) [11]. Thus, for instance, plants exposed to the sun tend to have higher *a*/*b* ratio than plants in the shade. The chlorophylls generally contribute 0.6%–1.2% of dry weight [12].

7.2.2 Chemical Properties

7.2.2.1 Structural Modification of Chlorophyll

In their natural location within the chloroplast, the chlorophylls are quite stable compounds, but when this organelle is damaged and loses its physiological elements, they become extremely labile, susceptible to a wide range of structural modifications by factors such as temperature, pH, enzyme action, molecular oxygen, and light.

There are four points at which the main structural modifications of the chlorophyll molecule start: (1) the chelate, (2) the ester bond of the phytol alcohol (C-17^3), (3) the isocyclic ring (C-13^2), and (4) the basic porphyrin structure. Figure 7.4 shows the possible transformations of the chlorophyll molecule.

7.2.2.1.1 Chelate Modification

The reaction substituting the coordination ion, Mg, by two H-atoms is known as pheophytinization. It modifies the properties of the chromophore, and the resulting compound has a different color. In the case of chlorophyll *a*, which is bluish-green, grey pheophytin *a* is formed, and in that of chlorophyll *b*, the yellow-green color turns to the brown of pheophytin *b*. There is evidence that in vivo the enzyme magnesium dechelatase catalyzes this reaction, playing an important role in chlorophyll catabolism [4,13], although its existence has yet to be confirmed. Chemically, this reaction takes place under acid conditions. Thus, there will be pheophytins whenever there is loss of cell integrity and acids are liberated. All the kinetic studies of the reaction agree that the rate of pheophytinization of chlorophyll *a* is always higher than that of chlorophyll *b*: between 2.5- and 10-fold [14–16].

The substitution of H by another divalent metal originates a new coordination complex, which may be even more stable than the original Mg chelate, modifying the characteristics of the chromophore. If the substitution is by Cu or Zn, the complex recovers the greenish color and is more stable [17–20]. As with the former reaction, kinetic studies show a higher rate of formation of the metal complex for pheophytin *a* than for pheophytin *b*. The formyl group present on C-7 of the *b*-series derivatives supplies a salient electron to the substituent, resulting in a decrease in electron density on the pyrrole nitrogens, and thus a lower reactivity with the

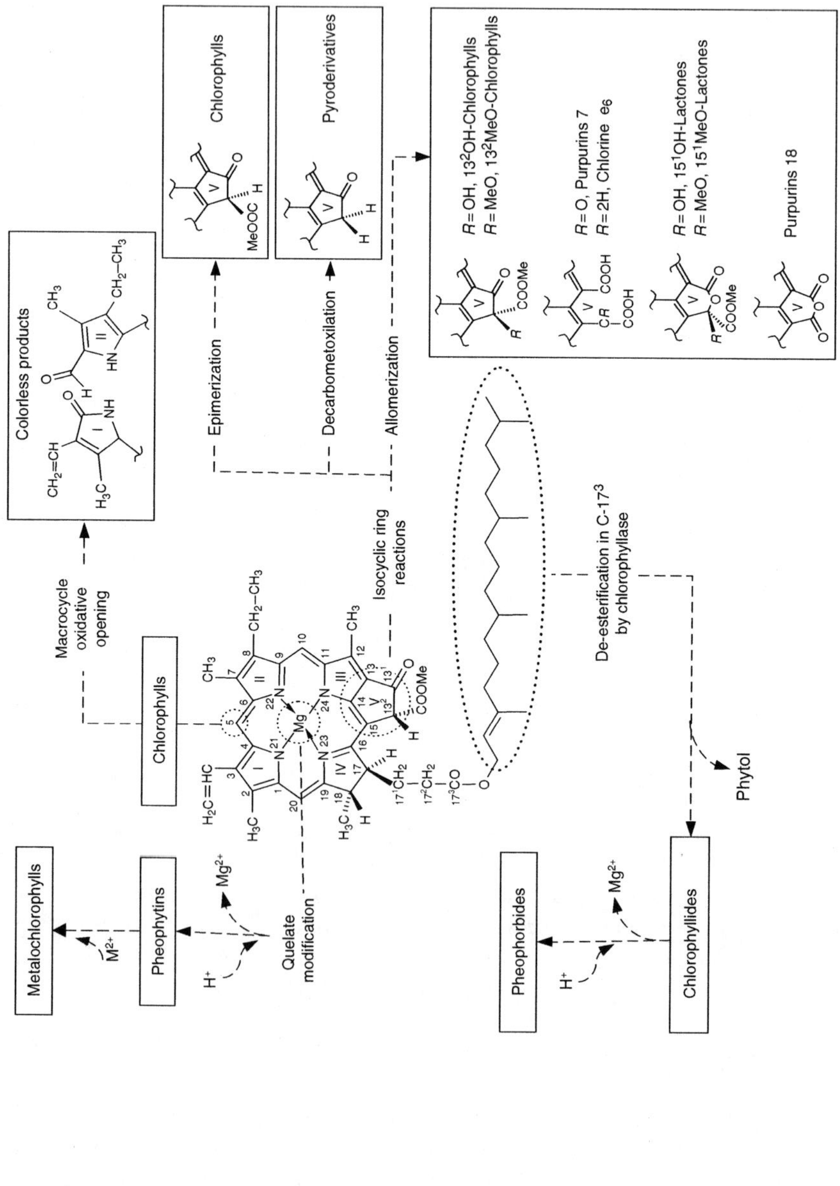

FIGURE 7.4 Transformations of the chlorophyll molecule.

metal cations. At the same time, it seems that the phytol chain interferes sterically in the complexation reaction, with the highest rate of insertion of the metal ion in chlorophyll derivatives de-esterified at C-17^3 [21].

The formation has also been described of other metallochlorophylls—Ni(II), Co(II), Fe(II) and Fe(III), Mn(II) and Mn(III), and La(II)—which are of use in the pharmaceutical industry and in the chemotherapy of cancer [22–26]. Resitting the Mg atom in the porphyrin ring is difficult, except with the use of the Grignard reagent.

7.2.2.1.2 De-esterification at C-17^3

This is a specific reaction catalyzed by the endogenous enzyme chlorophyllase (chlorophyll-chlorophyllide-hydrolase) (EC 3.1.1.14) giving rise to chlorophyllides. The basic macrocyclic structure is not modified, so that the chromophore properties are maintained and the color is unaltered. However, phytol (which makes the chlorophyll molecule liposoluble) is lost, resulting in a considerably increased polarity of the product. Pheophytin can also be the substrate of the enzyme chlorophyllase, with the reaction product of de-esterification being pheophorbide. Chemical de-esterification of phytol under acid or alkaline conditions is not a specific reaction, and is usually accompanied by oxidative side reactions [27].

7.2.2.1.3 Reactions of the Isocyclic Ring

Epimerization on C-13^2—This is the mildest alteration possible. It is simply a spatial isomerization, and does not involve changes in the color or the electron absorption spectrum, only a slight increase in polarity of the molecule. These compounds are not found naturally in fresh plant tissue, although they are often detected as a result of solvent extraction processes. Polar solvents containing exchangeable hydrogens assist the formation of such derivatives, as do Lewis bases, which act as receptors of hydrogen ions from the porphyrin molecule. In general, and as in all reactions, this increases with temperature.

Decarbomethoxylation on C-13^2—This is an oxidative reaction in which the carbomethoxy group of C-13^2 (COOCH3) is substituted by H, with no effect on the basic porphyrin structure. The compounds formed are denominated pyroderivatives, and have the same coloration and spectroscopic properties as their precursors (chlorophylls, pheophytins, chlorophyllides, and pheophorbides), being differentiated only in a slight decrease in polarity. The progress of this reaction is governed by a combination of the factors, time and temperature.

Allomerization—These are the reactions in which the isocyclic ring is oxidized by triplet molecular oxygen ($^3O_2$). They include a complex series of oxidation on C-13^2, in which the H is substituted by an oxygen or oxygenated chemical species. If the oxidation takes place in aqueous medium, the substituent is an OH-group, and 13^2OH-chlorophyll is formed. If the reaction takes place in alcoholic solution, for example methanol, the substituent is a methoxyl group (MeO-), and originates 13^2-MeO-chlorophyll. Different reaction mechanisms have been proposed: the most widely accepted is via free radicals, with the enolate anion as intermediary, although details of the reaction remain unknown. These reactions can also be catalyzed by oxidative enzymes, such as chlorophyll-oxidase or peroxidases. No changes are shown in color or electron absorption spectrum, since none of them modifies the porphyrin ring, but there is a slight increase in polarity.

The opening of the isocyclic ring and the oxidation of C-13^2 originate the purpurins 7, which, according to their structure, are also denominated 15^1-glyoxylic acid-chlorophyll *a*. Although this reaction does not affect the basic macrocyclic structure, the cleavage of the isocyclic ring considerably modifies the electron absorption spectrum and the adsorption properties, as the compounds are much more polar.

In these oxidation reactions, the formation of a cyclic ester originates a lactone ring. Under mild oxidation conditions, OH-lactones of chlorophylls are formed, and under more severe conditions, the purpurins 18. Although the basic chromophore structure is conserved, there are changes in the electron absorption spectrum and the polarity.

If the allomerization reaction takes place in a slightly alkaline medium and under inert atmosphere, there is only solvolysis of the isocyclic ring, and the resulting components are denominated esterified chlorin e_6 and rhodin g_7, depending on whether they come from the *a*- or *b*-series, respectively. Chlorophyll *a* yields Mg-phytol-chlorin e_6 while pheophytin *a* or pheophorbide *a* yields phytol-chlorin e_6 or free chlorin e_6, respectively. The same series of transformations can be described for chlorophyll *b*. The structural difference between these compounds and the purpurins 7 is solely in the substituent of C-15^1, which, in place of an O, is an H, so that their coloration, polarity, and electron absorption properties are very similar. Other techniques, such as mass spectroscopy (MS), must be used for their differentiation [28].

7.2.2.1.4 Cleavage of the Macrocycle

This takes place in a series of oxidative reactions beginning with cleavage of the porphyrin ring, which loses its original chromophore properties, giving rise to colorless products. Cleavage begins at C-5, and the first derivatives maintain chromophore groups that emit fluorescence but do not absorb light in the visible spectrum (Fluorescent chlorophyll catabolite (FC)). There is some evidence that a dioxygenase is the enzyme responsible for this reaction [29–31]. Finally, these derivatives are converted into nonfluorescent products (red chlorophyll catabolite (RCC)) on losing the delocalized double bond system in the pyrrole rings [32].

7.2.2.2 Functional Properties

Chlorophylls and their derivatives are a group of compounds with recognized biological activity that can take place at the levels of concentration normally found in the diet. Studies until now have focussed essentially on the antimutagenic and antigenotoxic activity exhibited by all chlorophyll derivatives, including water-soluble salts [33].

7.2.2.2.1 Antimutagenic Activity

Such activity of chlorophyllin has been demonstrated with respect to heterocyclic amines [34], benzo(a)pyrene [35], aflatoxin [36], heavy metals [37], and ionizing radiation [38]. Such diversity of agents points out that different protective mechanisms may be involved. One way is the formation of complexes of chlorophyllin with the aromatic mutagens and neutralizes them preventing their interaction with DNA. It was proposed to define the agents with such properties as the interceptor molecules [39].

Specifically, it has been shown that chlorophyllin forms complexes with three intercalators (acridine orange, quinacrine mustard, and doxorubicin) [40,41].

In vivo studies using different administration routes for both carcinogen and protective chlorophyll show a marked disparity in protective activity of the chlorophyll compounds depending on administration route. The highest effectiveness is achieved when the two substances are ingested together. A recent work carried out in Drosophila demonstrated that oral, joint administration of a mutagenic agent and a chlorophyll concentrate obtained from spinach and algas suppressed the genotoxic activity of the mutagen [42]. Topical treatments also gave high protective activity. Tests in vivo and in vitro have demonstrated that the use of pheophytins *a* and *b* obtained from the nonphenolic fraction of green tea markedly suppressed the development of skin tumors in mice [43]. With oral or topical administration, the time of contact between the mutagen and the chlorophyll compound before they are absorbed is high, enabling the two to interact chemically, inactivating the carcinogenic capacity, and consequently reducing the rate of tumor appearance. These results directly support the hypothesis of molecular complexation between the two species.

When the carcinogen is injected and the chlorophyll compounds are administered orally, the chemoprotective activity is low, probably because the carcinogen reaches its target cell before interacting with the chlorophyll compounds. In this case, there are also differences in the protective activity of the different chlorophyll compounds, possibly because of the different rate of intestinal absorption of each specific chlorophyll compound, depending on its structure and polarity. The results obtained are evidently insufficient, and in some cases contradictory, because during the passage of the chlorophyll compounds through the intestine, numerous coexisting transformation products are originated, making it difficult to assign a specific activity to a particular compound.

The chemoprotective activity against carcinogens has been used as a preventative in chemotherapy treatments of tumor. Most drugs used in tumor treatment can have side effects and induce genotoxic effects in healthy cells. Tests both in vitro [44] and in vivo with mice [45] have demonstrated the efficacy of chlorophyll derivatives in decreasing the collateral mutagenic activity of the drug in all cases without modifying the antitumoral effect.

When using drugs with collateral mutagenic capacity to treat cancer in human patients, food supplements are commonly administered to diminish the side effects of the therapy. Such foods are often rich in antimutagenic compounds, and for many years, chlorophylls, chlorins, and other porphyrins have been used in clinical treatments [38].

7.2.2.2.2 Anticarcinogenic Activity and Chemoprevention of Cancer

The first studies relating chlorophylls with cancerous processes date from the 1930s to the 1940s, when accumulations of porphyrin derivatives were detected in sarcomas and carcinomas of the breast. Current lines of research focus on the preventive activity exhibited by chlorophyll derivatives against tumor development. An exhaustive review of the anticarcinogenic activity of chlorophyll compounds was published in 1997 by Dr. Dashwood [46].

In relation to skin cancer, and although the initial hypothesis was the putative cancer chemoprevention action of chlorophyllin in the diet, the results have demonstrated augmentation of the ultraviolet (UV)-induced cutaneous carcinogenic process by dietary chlorophyllin [47] in mouse.

Colon cancer was the second leading cause of cancer death in Western countries in 2000, diets being high in red and processed meat, in contrast to white meat, associated with increased colon cancer risk. It has been proposed [48] that heme, the iron porphyrin pigment of red meat, might be an important dietary risk factor as it increases the cytotoxicity of the fecal waters, consequently an increased exposure of the colonic epithelial cells to luminal irritants, colonic epithelial hyperproliferation, and finally increase the risk of endogenous mutations in cell turnover genes. Although the mechanisms explaining the dietary modulation of the risk of colon cancer by intake of red meat and vegetables are still under debate, recently papers associate such an effect to dietary chlorophyll. Balder et al. [49] have developed a cohort study in the Netherlands, with 120.852 subjects aged 55–69 years to verify the existence of such association, and the results obtained suggested the existence of an elevated risk of colon cancer in men with increasing intake of heme iron and decreasing intake of chlorophyll. In vitro, de Vogel et al. [50] have demonstrated that chlorophyll prevents the detrimental, cytotoxic, and hyperproliferative colonic effects of dietary heme in rat colon. The hypothesis is that as chlorophyll could be considered a structural analog of heme, it might compete with heme for solubilization in the gastrointestinal tract and thus prevent the formation of cytotoxic heme metabolites. A forward step has been the demonstration that such preventive action of chlorophylls is specific of natural chlorophylls and cannot be mimicked by water-soluble chlorophyllins [51].

But in addition to being an effective blocking agent during initiation phase, it is under investigation the role of chlorophyllin as a suppressing agent and a possible novel therapeutic strategy directed toward aberrant cell proliferation in the colon. Recent results [52,53] concluded the ability of chlorophyllin to alter tissue homeostasis in the colon, and so suggested an investigation of apoptosis mechanisms in human colon cancer cells. But as a difference of other chemopreventive agents, it has been shown [54] that chlorophyllin induces apoptosis in HCT116 human colon cancer cells without the release of cytochrome *c* from mitochondria or activation of the Apaf-1/caspase-9 pathway. As an alternative mechanism, chlorophyllin activated a pathway involving caspase-6 and the release of AIF (apoptosis-inducing factor) from mitochondria, resulting in the cleavage of nuclear lamins. A further step it has been the finding that chlorophyllin also induces *E*-cadherin expression (an indicator of cell differentiation). The induction of differentiation markers may be important in limiting cancer-cell invasion and metastasis, and there is much interest in understanding the underlying mechanisms [55].

At other level, it has implicated chlorophyllins in conjunction with chitosan (chl-chitosan) in the reduction of the body burden of environmental dioxins, by excretion in the feces [56]. Dioxins are the most toxic artificial compounds that mainly enter in the human body through the food. They have usually long half-lives and accumulate in the adipose tissue. The adverse effects of dioxins in the human body are the following: teratogenicity, reproductive toxicity, immunosuppression, hepatotoxicity, and carcinogenicity. The results of the experiment with volunteers eating for 40 days different amounts of chitosan and chl-chitosan, and examining in the feces and sebum the excretion of 29 congeners of dioxins, concluded that chl-chitosan can prevent accumulation of dioxin, at least by the intake of normal foods.

7.2.2.2.3 Other Properties

One of the beneficial properties of the copper-chlorophyllin is supposedly derived from its capacity to bind strongly amine-containing compounds in the gut [57]. The trimethylaminuria (TMAU) is a metabolic disorder characterized by the inability to oxide and convert dietary-derived trimethylamine (TMA) to the odorless metabolite trimethylamine *N*-oxide (TMAO). Patients suffering from TMAU may emanate a body odor, which in the extreme form, may be unpleasant and fish-like. It has been suggested that the daily intake of charcoal or copper-chlorophyllin may be of significant use in improving the quality of life of individuals suffering from TMAU [58]. The dietary copper-chlorophyllin (180 mg/day for 3 weeks) reduced the free urinary TMA concentration and increased TMAO to those of concentrations present in normal individuals in the TMAU patients studied. The hypothesized mechanism is the sequestration of TMA in the gut by the copper-chlorophyllin.

7.2.2.2.4 Antioxidant Properties

Despite being the most abundant pigment in the nature, chlorophylls have not been included in the majority of the research about the antioxidant activity of phytochemicals. In fact, there have been papers suggesting a prooxidant activity of chlorophylls as well as an antioxidant activity.

The most recent investigations have shown an antioxidant activity by the ability of each compound to scavenge the long-lived free radicals 1,1-diphenyl-2-picrylhydrazyl (DPPH) and 2,2′-azinobis-(3-ethylbenzothiazoline-6-sulfonate) ($ABTS^{+}$) by Ferruzzi et al. 2002. [60] Also the β-carotene bleaching method has been tested for chlorophylls [59], showing that all chlorophyll derivatives presented a dose-dependent response.

Derivatives of chlorophyll *a* were found to be more effective radical quenchers than those of chlorophyll *b*. Furthermore, metal-free derivatives such as chlorins, pheophytins, and pyropheopytins exhibited significantly lower antiradical capacity than metallo-derivatives such as Mg-chlorophylls, Zn-pheophytins, Zn-pyropheophytins, Cu-pheophytin *a*, and Cu-chlorophyllins. It seems that dietary chlorophyll derivatives prevalent in both fresh and processed foods and dietary supplements have antioxidant and antimutagenic activities [60].

7.2.2.2.5 Digestion, Absorption, and Metabolism

The first result demonstrating the uptake of chlorophyll derivatives by human intestinal cells was in 2001 by Schwartz's group [61]. Up to then, only a few studies focussed on the isolation of chlorophyll derivatives in feces of humans and animals assuming negligible absorption.

The investigation was carried out with an in vitro digestion method which simulates both the gastric and small intestinal phases of the process. During the digestion, the native chlorophylls were converted to Mg-free pheophytin derivatives and the micellarization of chlorophyll *a* derivatives was significantly more efficient than chlorophyll *b* derivatives in general. The intestinal cell uptake of the micellarized pigments was also investigated by Caco-2 human cell model, quantifying between 5% and 10% of chlorophyll accumulation [61].

Later, the identification of Cu-chlorin e_4 and Cu-chlorin ethyl ester in serum of human subjects ingesting sodium-copper-chlorophyllin provided evidence that such compounds can be absorbed by humans [62]. In fact, Ferruzzi et al. [63] demonstrated the stability of Cu(II)-chlorine e_4 during in vitro digestion and the effective uptake by Caco-2 enterocyte-like cells, supporting the likelihood that a portion of sodium-copper-chlorophyllin component or its metabolites are absorbed from the human intestine.

7.2.3 Spectroscopic Properties

7.2.3.1 Visible/Ultraviolet Electron Adsorption Spectrum

The electron absorption spectrum of the chlorophylls and their derivatives is very characteristic and is attributed to the system of conjugated double bonds making up the basic porphyrin structure. Electron transitions in the chlorophyll molecules, detected by photosensitive optical equipment, produce absorption bands [8,64,65]. The intensity of a particular absorption band is normally expressed by its molar specific coefficient of extinction (ε), which, according to the Beer–Lambert law, depends on the optic density (A), the molar concentration (C), and the optical pathway of the light beam through the solution, expressed in centimeters (L).

The porphyrins and related compounds have long been of interest because of the sharpness of the absorption bands in the visible and ultraviolet regions. This feature has been very useful in supplying information about both the overall structure of the macrocycle and the nature of the groups included in it.

The absorption spectra of all the porphyrin pigments are characterized by a number of relatively pronounced bands in the yellow, red, and near-infrared regions, and a band of strong absorption in the violet or near-violet region denominated the "Soret band." The presence of the latter indicates that there has been no cleavage of the basic porphyrin structure [66].

The dielectric properties of the solvent also affect the spectrum. The spectral bands of chlorophylls in ether solution are very sharp because of the weak interaction between this solvent and the pigment molecules. In solvents that are more polar, such as acetone, methanol, ethanol, or mixtures of these with water, the interaction with the pigment causes displacement of the absorption maxima toward higher wavelengths, while the intensity (peak height) and specificity (peak width) of the absorption band decrease.

Both chlorophylls *a* and *b* present absorption maxima in the blue-violet region, with characteristic peaks at 428 and 454 nm, and smaller ones at 410 and 430 nm, respectively. In addition, there are four pronounced bands between 500 and 700 nm, in the red region. This variance is the result of the small structural difference between the two chlorophylls. The displacement toward the green region of the spectrum changes the greenish-blue color of the chlorophyll *a* to yellowish-green in chlorophyll *b*.

The substitution of Mg by hydrogens in chlorophyll *a* produces a marked hypsochromic displacement in the Soret band, from 428 to 408 nm, whereas the displacement in the secondary band (Q band) is bathochromic and less marked, from 662 to 666 nm, with an increase in the peak ratio (Figure 7.5). In the case of chlorophyll *b*, there are similar changes in the adsorption spectrum with the pheophytinization reaction. The Soret band is displaced from 454 to 430 nm and the secondary band from 646 to 656 nm (Figure 7.6). Insertion of other divalent cations such as Cu or Zn modifies the adsorption spectrum in the opposite sense. The shape of the spectrum, the location of the absorption maxima, and the peak relation are very similar to those of the spectrum of the original complex with Mg, although as they are not identical, they can be readily distinguished. As an example, Figure 7.7 shows the electron adsorption spectra of Cu-pheophytin *a* and pheophytin *a*. The most-conspicuous changes in the adsorption spectrum are caused by structural modifications affecting the chelate, which alter the chromophore properties.

In contrast, de-esterification of phytol in the chlorophyll molecule does not affect the chromophore structure, so the electron adsorption spectrum is unaltered. Thus, chlorophyllides and pheophorbides have the same spectroscopic properties as their respective precursors, chlorophylls and pheophytins. The same is true of some of the reactions that affect the isocyclic ring, such as epimerization and decarbomethoxylation on C-13^2.

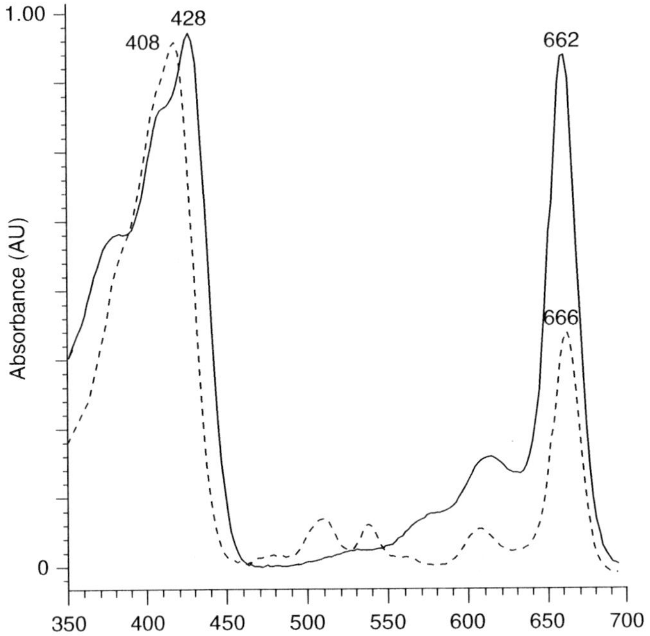

FIGURE 7.5 Electronic absorption spectra of chlorophyll *a* (—) and pheophytin *a* (----).

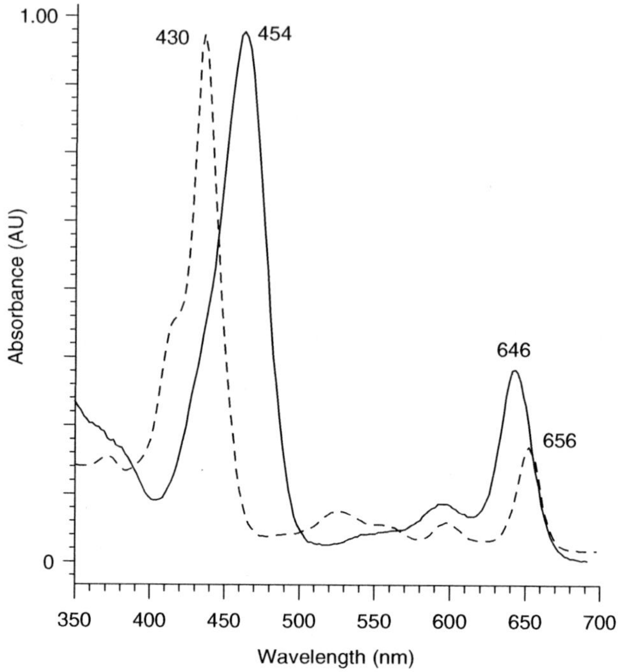

FIGURE 7.6 Electronic absorption spectra of chlorophyll *b* (—) and pheophytin *b* (----).

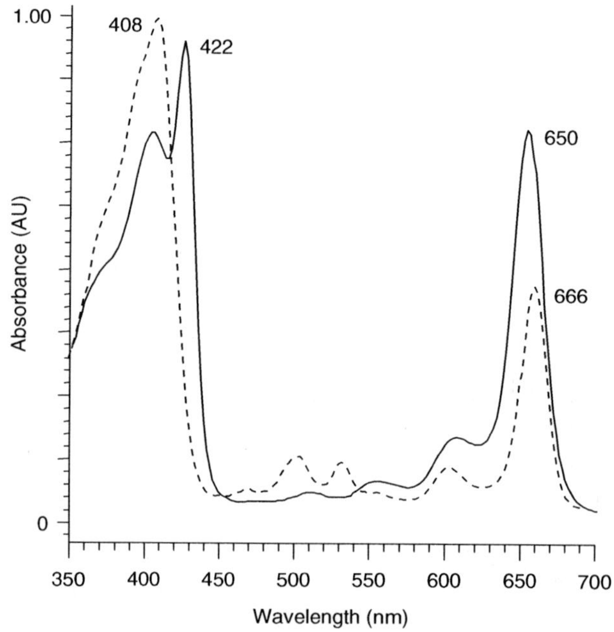

FIGURE 7.7 Electronic absorption spectra of pheophytin *a* (----) and Cu-pheophytin *a* (—).

Primary allomerization reactions that give rise to hydroxy- or methoxy-chlorophylls also do not affect the spectroscopic properties. In contrast, other oxidative reactions involving opening of the isocyclic ring or the formation of a lactone ring do cause important changes in the electron absorption spectrum.

Figure 7.8 shows the specific differences between the spectrum of chlorophyll *a* and that of Mg-phytol-chlorin e_6, the product of solvolysis of the isocyclic ring of chlorophyll *a*. There is a hypsochromic displacement of 14 nm in the Soret band, and of 16 nm in the maximum of the red region. A similar change takes place in the corresponding derivative of *b*-series (Figure 7.9). Figure 7.10 shows the change in the adsorption spectrum of pheophytin *a* caused by solvolysis of the isocyclic ring. There is a hypsochromic displacement of some 8 nm in the Soret band, and another, much less marked, in the red band (4 nm), with considerable changes in the shape of the spectrum.

The spectral changes caused by the inclusion of a lactone ring in the structure of pheophytin *a* can be observed in Figure 7.11. The spectrum of 15^1-OH-lactone-pheophytin *a* shows a 10-nm hypsochromic displacement of the Soret band. In the case of phytol-purpurin 18 *a*, the most-striking displacement is that of the maximum of the red region, a bathochromic change of around 30 nm.

Finally, chlorophyll derivatives in which there has been a cleavage of the porphyrin macrocycle show no electron absorption band in the visible region, and are therefore colorless compounds.

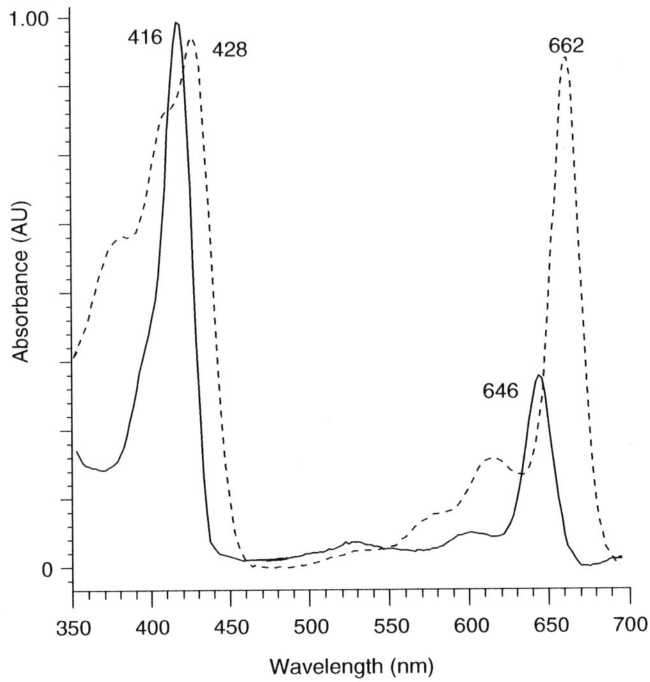

FIGURE 7.8 Electronic absorption spectra of chlorophyll *a* (----) and Mg-phytol-chlorin e_6 (—).

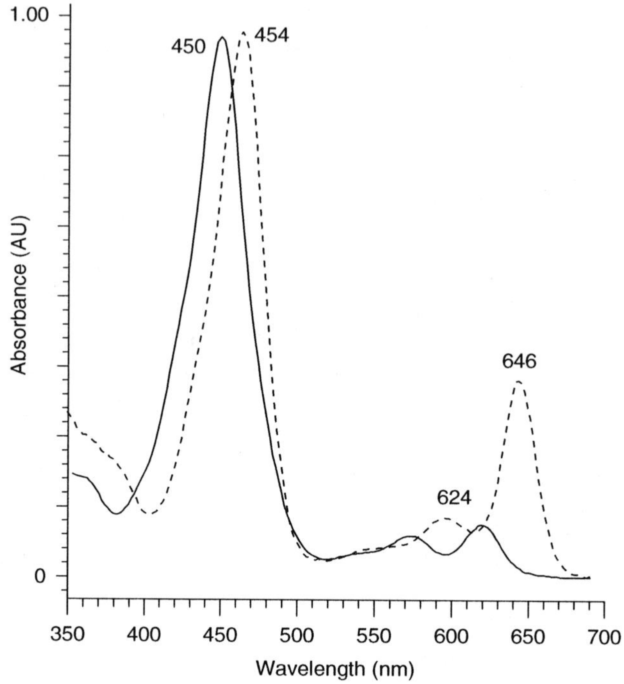

FIGURE 7.9 Electronic absorption spectra of chlorophyll *b* (----) and Mg-phytol-rhodin g_7 (—).

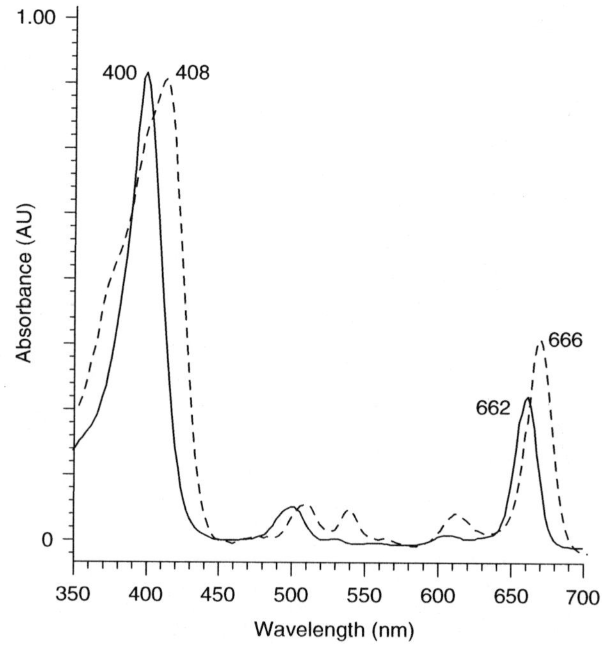

FIGURE 7.10 Electronic absorption spectra of pheophytin *a* (----) and phytol-chlorin e_6 (—).

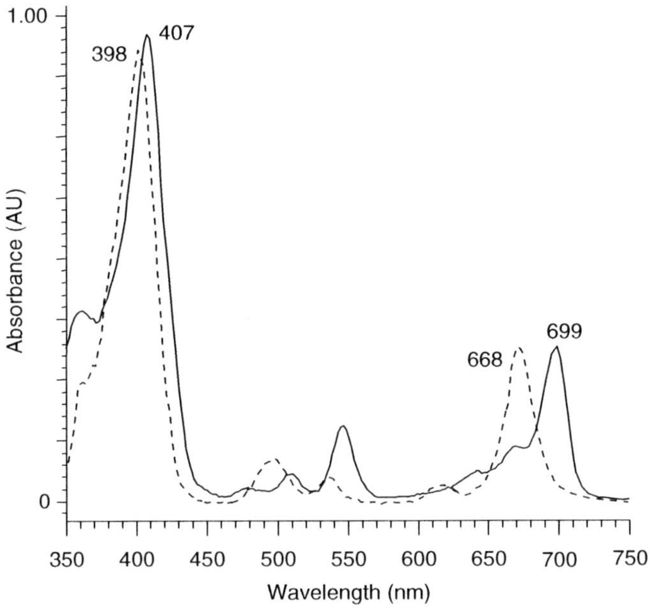

FIGURE 7.11 Electronic absorption spectra of 15^1-OH-lactone-pheophytin *a* (—) and phytolpurpurin 18 *a* (----).

7.2.3.2 Fluorescence Spectrum

The fluorescence spectra of chlorophylls are similar to those of visible absorption. Chlorophyll *a* shows a higher sensitivity at maximum and minimum wavelengths than that of chlorophyll *b*. The spectra of pheophytins *a* and *b* are similar to those of the corresponding chlorophylls [67]. The fluorescence spectrum of the chlorophylls is affected by solvents, with fluorescence being lost in pure and dry hydrocarbons [68]. It is also affected by temperature, molar concentration, and the degree of solvation [69].

Fluorimetric methods have not been used as widely as ultraviolet–visible (UV–Vis) spectrophotometric ones, although they are much more sensitive: detection requires only 1% of the amount of pigment necessary in the latter method [70,71]. With this technique, chlorophylls or their derivatives have been detected quantitatively in picomole amounts [72–77]. Very low amounts of chlorophylls *a* and *b* can also be determined from their mixtures, as the fluorescence of one is independent of the other [78].

Although UV–Vis spectroscopy has been more widely used than fluorescence, the latter is becoming the preferred technique for studies involving chlorophylls in photosynthesis. Different systems of equations have been developed for the fluorimetric determination of the concentration of chlorophylls *a* and *b*, or more-complex mixtures that also include pheophytins, chlorophyllides, and pheophorbides [5,79,80]. Fluorescence is measured in the Soret band, where the differences for the various components are greater.

These compounds show a characteristic strawberry fluorescence under $UV_{254\ nm}$ light, a property used to detect chlorophylls and pheophytins on thin layer chromatography (TLC) plates [75]. Although this property is very generalized in most chlorophyll derivatives, there are some in which it is absent, making it a very useful test of identification. This is the case of the metallochlorophylls of Cu [28] and the purpurins 18 [81]. Fluorimetric detectors are also used as selective monitors coupled with high-performance liquid chromatography (HPLC) [74,82,83].

7.2.3.3 Other Spectroscopic Properties

Nuclear magnetic resonance (NMR) spectroscopy, MS, and infrared (IR) spectroscopy have been used successfully in the structural elucidation of chlorophylls and derivatives [8,84–86]. Studies of hydroporphyrins and chlorophylls also have been carried out with Raman resonance spectroscopy [87].

Specific functional groups have been identified by IR spectroscopy [88]. The IR spectrum of chlorophylls and their derivatives has been determined in emulsions in mineral oil, in potassium bromide disks, and in solution. The ceric properties of chlorophylls hinder the preparation of emulsions and disks; moreover, the spectrum is resolved better in solid state than in solution [66]. The presence of functional groups on the side chain of chlorophylls increases the IR absorption bands in the expected regions of the spectrum. The spectra of chlorophylls *a* and *b* in nonpolar solvents, such as carbon tetrachloride, are surprisingly similar, considering that the two molecules are different in that on C-3, chlorophyll *b* has a formyl group instead of methyl. In contrast, in more-polar solvents, such as tetrahydrofuran, the two chlorophylls have very different spectra. This phenomenon is particularly marked in the carbonyl region. It has been known for some time that the spectrum of these compounds in nonpolar solvents depends on their concentration.

The first authors to obtain complete structural information using NMR were Closs et al. in 1963 [89]. ^{1}H-NMR has been the most-convincing method, thanks to the resonances of the protons in the porphyrin ring structure. A factor in the success of the technique is that it requires only a small amount of the pure pigment (100 μg) [72,90], compared with ^{13}C-NMR [91], which requires a high concentration of pigment.

It was not possible to use MS for the characterization and identification of chlorophylls and its derivatives until the development of desorption methods (desorption ionization) appropriate for nonvolatile and thermolabile compounds. The mass spectrum of chlorophylls has been obtained using laser desorption [92–94], field desorption [95], plasma desorption [96,97], fast atom bombardment (FAB) [28, 98–101], in-beam electron ionization [102], and electrospray ionization (ESI) [103]. A combination of the techniques of desorption and tandem mass spectroscopy (MS/MS) has also been used for the characterization of chlorophylls and derivatives [104–107]. The latest research in this field coupled HPLC with MS, using as ionization source FAB [108–110] or atmospheric pressure chemical ionization (APCI) [111–115].

7.2.4 Chlorophyll Stability during Food Processing

Chlorophylls form part of the natural human diet as integral components of edible plants. However, once the plants have been harvested, the chlorophylls begin to degrade, and can become transformed into other colored derivatives, or be degraded totally to colorless compounds, if the material is stored for a long period. Vegetable processing, particularly by heating, also leads to transformation of the chlorophylls into their derivatives.

To a greater or lesser extent, all processing systems cause changes in the chloroplast pigments present, and thereby in the apparent color of the food. Numerous studies have been carried out in an attempt to prevent these changes and preserve the bright green color that makes green vegetables look appetizingly fresh. Nevertheless, there is no evidence demonstrating that the structural changes occurring in chlorophylls during food processing cause deterioration in their functional properties. On the contrary, the derivatives formed, mainly pheophytins, can preserve or even strengthen such properties [43,116,117]. In contrast, chlorophyll transformation reactions that lead to colorless products play a very negative role in color and in functional activity of the food, since the cleavage of the porphyrin macrocycle presumably suppresses the capacity of these compounds to bond with DNA and mutagenic agents [46].

The whole range of chlorophyllic pigments can be found in processed vegetables, to a greater or lesser extent depending on the characteristics of the plant material and the physicochemical conditions under which the process takes place. Factors such as temperature, pH, enzyme action, oxygen, light, and storage time determine the transformation of these compounds [118].

Although the natural concentration of chlorophyll *a* is higher than that of chlorophyll *b* in all green tissues, their transformation products do not follow the same pattern, as the reaction rate is not the same in the two chlorophylls. The many kinetic studies of chlorophyll transformation reactions all agree that those involving the chelate, such as pheophytinization or the insertion of divalent cations, are more rapid in the *a*-series compounds [23,119–121]. In contrast, the enzymatic de-esterification of phytol or the formation of pyroderivatives is more rapid in the *b*-series compounds [15,122,123].

The purpose of conservation treatments is to conserve the properties of the material as unaltered as possible, destroying microorganisms that might alter the food or cause disease under certain storage conditions. Necessarily, the different technological alternatives for the conservation of foods can have different effects on the type and proportion of chlorophyll derivatives formed. Table 7.1 summarizes the pigments that may be present in plant foods, depending on the type of processing to which they have been subjected.

Postharvest conservation of vegetables normally results in a loss of quality, caused by water loss, biological oxidation or other chemical changes, and physiological ruptures. The generalized liberation of acids following breakdown of cell integrity is very common in plants, in which the pH of the internal medium can reach 5.5 units. The most common reaction under such conditions is pheophytinization, since the central atom of Mg of the chlorophyll molecule is readily liberated under slightly acid conditions and replaced by hydrogen. The rate of chlorophyll conversion to pheophytin will differ with the degree of acidity of the food. The breakdown of cell integrity allows contact of endogenous enzymatic systems with their substrates, giving rise to hydrolytic or oxidative reactions in the chlorophylls. In the case

TABLE 7.1
Profile of Chlorophyll Pigments in Vegetable Products as a Function of the Technological Process Applied

Pigment	Green Vegetable	Green Vegetable after Treatment[a]			
		Freezing	Blanching	Sterilization	Fermentation
Chlorophylls					
Chlorophyll *a*	++++	++++	+++	–	–
Chlorophyll *b*	++	++	++		
Mg-free derivatives					
Pheophityn *a*	–	tr	+	++	+++
Pheophityn *b*	–	tr	tr	+	++
Phytol-free derivatives					
Chlorophyllide *a*	–	–	*	–	–
Chlorophyllide *b*	–	–	*	–	–
Mg-phytol-free derivatives					
Pheophorbide *a*	–	–	*	**	**
Pheophorbide *b*	–	–	*	*	*
Oxidized derivatives					
Pyroderivatives[b]	–	–	–	+++	tr
Isomers					
Epimers C_{13}[c]	–	tr	+	+	+

[a] Codes: Number of symbols denotes relative abundance; +, pigments generally present in fresh or processed green vegetables; *, pigments present in vegetables with chlorophyllase activity; –, pigment absent; tr, pigment in amount of traces.
[b] Include the pyroderivatives of all the chlorophyll pigments.
[c] Include the epimers of all the chlorophyll pigments.

of materials with chlorophyllase activity, the chlorophylls will be converted partially to chlorophyllides, in which the Mg can also be substituted by hydrogens, with transformation to pheophorbides. On the other hand, oxidative reactions catalyzed by lipoxygenase, peroxidase, or chlorophyll-oxidase lead to chlorophyll degradation to colorless products. Storage in controlled atmosphere, combined with refrigeration, slows such reactions [118,124–127].

Blanching is one of the most common initial steps in the preparation of frozen, canned, or dehydrated vegetables to preserve the natural green color and thereby the quality. It is a mild heat treatment, not exceeding a temperature of 100°C, normally carried out with steam or boiling water. Modern methods also use micro-ovens or convection ovens. However, it is not possible to establish a norm for the blanching method that achieves the greatest retention of chlorophylls. The process inactivates the enzymes present, shrinks the material, expels gases from the tissues, reduces any initial infection, and prevents the formation of off-flavors. The chlorophyll derivatives commonly found in blanched vegetables are pheophytins, chlorophyllides, pheophorbides, and the respective C-13^2 epimers [100,128–131].

The extent of the reaction converting chlorophylls into pheophytins is directly related with the temperature and the time that the blanching treatment lasts [132]. Blanching at low temperature (65°C–80°C) favors activation of the enzyme chlorophyllase, and thus the formation of dephytylated chlorophyll derivatives (chlorophyllides and pheophorbides). At higher temperatures (around 100°C), the enzyme is partially inactivated, but is favored the parallel reaction of pheophytinization. The extent of chlorophyllide and pheophorbide formation will depend not only on the specific conditions of the technological treatment, but also on the level of chlorophyllase activity in the plant material [133].

Another type of food processing that involves mild heat treatment is dehydration. This method of conservation reduces water activity by removing the water from the product by evaporation or lyophilization. The conversion of chlorophyll *a* into pheophytin *a* takes place even at very low concentrations of water [134], but can be retarded by reducing the moisture content of the vegetable or decreasing its activity. As in blanching, the temperature used for food dehydration determines the percentage of pheophytin formation: greater amounts of this pigment are found when samples are heated to higher temperatures [118,135].

During the process of sterilization by heating, which involves a stronger heat treatment (to temperatures above 100°C), the vegetables undergo a marked color change from bluish-green to olive green, by the generalized conversion of chlorophylls to Mg-free derivatives. Sterilization is a more drastic process for the chlorophylls, in which, besides the chelate reaction there is also oxidation on the isocyclic ring, with complete transformation to pheophytins and pyropheophytins. The degradation mechanisms suggested include two reactions in series: pheophytinization and pyropheophytinization [136].

The extent of pyropheophytin formation is considered to be related with the severity of the heat treatment. In processes setting up convection currents, the heating rate is higher and the treatment time is lower than when heat is transmitted by conduction. The percentage of pyroderivatives formed is thus higher in the latter. In the system known as HTST (high temperature and short time), which achieves sterilization using high temperatures (150°C) and short times, the percentage of these derivatives formed is lower [121]. This system improves product quality, but may leave a certain residual enzyme activity that causes deterioration in color and flavor during storage [137].

In vegetable foods subjected to lactic acid fermentation, activity of the enzyme chlorophyllase is very often favored during the process, and dephytylated chlorophyll derivatives are found [133,138–141]. The degradation mechanism proposed is as follows:

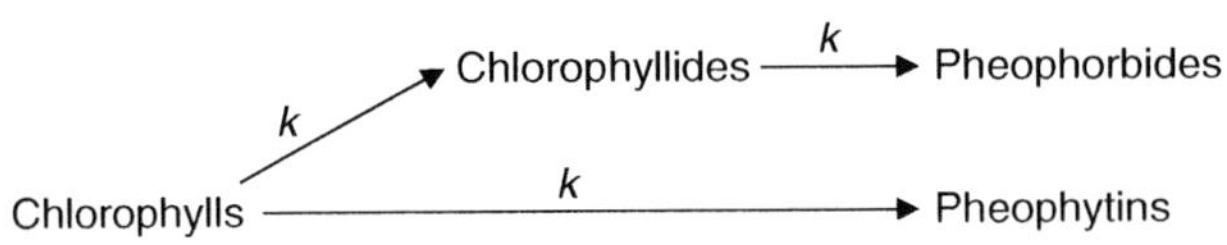

While the enzyme chlorophyllase is active, some of the chlorophylls give rise to chlorophyllides. The remaining chlorophylls have not at this stage been affected by the action of the enzyme, and are a substrate available for the subsequent pheophytinization reaction. This takes place in parallel with the formation of pheophorbides [123]. Pyropheophytins are also detected in fermented vegetables, their presence being attributed to the heat generated in the process [142–144].

During food conservation, chlorophyll degradation may reach the stage of oxidation to colorless products. One of the oxidative enzymes apparently involved in the decoloration of vegetables is lipoxygenase, which catalyzes the coupled oxidation of polyunsaturated lipids and pigments [132,145,146]. The free radicals formed as intermediate products in the catalysis of peroxidation are directly responsible for the co-oxidation of chlorophylls. The heat treatment of blanching inhibits lipoxygenase activity and prevents development of these oxidative reactions during the conservation of frozen vegetables [147]. Heat and light also have a destructive effect on chlorophyll [132].

In vegetable oils, the most common reaction is again pheophytinization as a result of the generalized liberation of acids produced by mechanical breakdown of the plant tissue. Pheophytin *a* is the major pigment in this product, although slight differences are found, depending on whether the oil has been extracted by physical procedures, as in the case of olive oil [146,148,149], or chemical ones using solvents [150,151]. The conditions during physic extraction processes favor activation of the enzyme chlorophyllase, and dephytylated chlorophyll derivatives, mainly pheophorbide *a*, can be found in the oil [152,153]. During oil storage, oxidative reactions also take place, and pyropheophytins and allomerized chlorophyll derivatives are found [150,154]. During the refining of edible oils from oil seeds, the chlorophyll compounds are removed almost entirely by processes of adsorption on decoloring clays (bleaching) or treatment at high temperature (deodorization) [151].

7.2.5 Chlorophylls as Additive

Of all the possible transformations of the chlorophyll molecule during food processing and conservation, those affecting the chelate cause the most-striking changes in the color of the product. The substitution of the Mg by hydrogens in the porphyrin ring means a considerable reduction in quality, as it causes a dramatic color change from bright green to olive green, and the disappearance of the attractive aspect of a fresh vegetable. Chlorophylls, chlorophyllides, or any other product of chlorophyll oxidation are susceptible to this transformation, whenever the essential condition of acidity is present.

Numerous studies have been carried out to prevent this degradation, although none has achieved an optimum result. They include the control of pH with alkalinizing substances [155,156], the favoring of chlorophyllase action to form chlorophyllides, which are more thermostable than chlorophylls [155], or the heat treatment at HTST [157]. Most of the methods tested were able to retain more chlorophylls, but this effect did not last during prolonged storage.

The only alternative for obtaining processed foods with a stable green color is the addition of a colorant. This is normally used to strengthen the green color of a vegetable, or to supply green coloration to a food that does not possess color or has lost it during processing. Examples of such foods are confectionery, desserts, snacks, and drinks, in which colorants are also used to help the consumer to identify the product, or simply to give it a more attractive appearance.

Chlorophyll itself is not particularly appropriate as an additive—it is not only unstable, but also insoluble in water. The stability of the molecule can be enhanced by substitution of the Mg ion in the chlorophyll molecule by other divalent cations, such as copper or zinc, which makes the complex less liable to the modification of the chelate under acid conditions, and a stable green color is conserved [158]. The first observations in this sense referred to the regreening of canned vegetables during storage, which was attributed to the limited formation of metallochlorophyll complexes of copper and zinc [159,160]. As a result of these observations, the Veri-Green process was patented in 1984 [161], whereby Zn ions are incorporated into the lining of the metal containers, stabilizing the color during blanching by the formation of zinc complexes of pheophytin and pyropheophytin [162]. Veri-Green products have not been successful in the United States because the amount of Zn required to achieve an optimum color exceeds 75 ppm, the upper limit admitted in the Unites States for this metal. In Europe, there are no set limits. Subsequent patents add solutions of Cu or Zn salts during food processing [163–165], and very recently, improvement of the color in green vegetables has been achieved using zinc salts coupled with continuous flow aseptic processing technology [165].

As a result of the capacity of these chlorophyll metallocomplexes to preserve the green color, the copper complex of a water-soluble chlorophyll derivative obtained by alkaline treatment was marketed for the first time in 1984. This chlorophyll derivative, known as chlorophyllin, is the result of chemical hydrolysis of the alcohol phytol, of the methyl esters, and of the cyclic β-ketoester (ring V) in the chlorophyll structure. The food colorant, known as sodium-copper-chlorophyllin (E-141ii) is the sodium salt of the chlorophyll saponification product in which the Mg^{2+} has been replaced by Cu^{2+}. The commercial product is a mixture of different Cu(II)-chlorophyll derivative complexes that can include Cu-pheophorbide *a*, Cu-chlorin e_6, Cu-chlorin e_4, Cu-rhodin g_7, and other degradation products [166–168]. This product is bluish-green in color, is moderately soluble in water, and resists the heating conditions used during canning. Such mixture has gained importance as a food colorant and dietary supplement with apparent chemopreventive activities, but the implication of a rapid loss of Cu(II)-chlorin e_4, a reported bioactive component of sodium-copper-chlorophyllin, upon heating may result in alteration of potential dietary benefits such as antimutagenic and antioxidant [169].

For liposoluble foods, there is colorant E-141i, known as copper complexes of chlorophylls. It is obtained by addition of a Cu salt to the solvent-extracted product from an edible plant source. Its main coloring material is Cu-pheophytin, although other compounds are also present, such as carotenoids and lipids from the raw material.

Currently, sodium-copper-chlorophyllin is the most widely used natural green colorant. This chlorophyll derivative has been the basis of many of the studies carried out on the functional value of chlorophyll compounds, and important antimutagenic and anticarcinogenic properties have been found. Research has been conducted into the capacity of chlorophyllin to protect DNA against ionizing radiation, and its considerable potential for prevention of mutation-based health impairment, including cancer and other degenerative diseases [38]. Other studies seem to demonstrate that chlorophyllin is effective as antimutagen against the genotoxic side effects that antitumoral agents can have on healthy cells [44,45].

7.3 ANALYSIS OF CHLOROPHYLLS

7.3.1 EXTRACTION TECHNIQUES

For the extraction of chlorophylls from plant tissues, the AOAC (Association of Official Analytical Chemists) [170] recommends an initial step of milling or blending (homogenization) of the tissue, which can be improved by the addition of quartz sand to the mortar. Pigment extraction is normally carried out by adding small amounts of solvent to the triturate. Solvent and plant material are usually left in contact for a time to assist the extraction by maceration, but precautions are necessary, as the chlorophylls can be easily degraded to other products. The solid material is removed by filtration through paper, glass wool, inert diatomaceous clay, or by centrifugation, although filtration is preferred for its better guarantee of completely recovering the extract without solid residues. The extraction procedure is repeated until there are no traces of pigment in the sample. Finally, the filtrates are combined and taken to volume. All these operations are carried out rapidly, in darkness or under weak green light, to prevent photodegradation or allomerization of the pigments.

Although the extraction process is simple, each plant product requires an appropriate method, and the need often arises to vary the methodology, adapting it to the particular case. Many methods are listed in the References section at the end of the chapter—some quantitative, others qualitative, while others are semiquantitative—to achieve an adequate degree of exhaustion in the extraction using a simple, rapid method. The techniques are varied: percolation, maceration, sonication, diacolation, etc., and are generally useful with almost any inert organic solvent, although most use methanol or acetone because they are extraction solvents of medium–high polarity, little selectivity, and wide range of solubilization [65,136,147,149,158,170–185].

Although the extraction stage is largely mechanical, the operation must be correctly performed to guarantee the consistence of the subsequent analysis. It is essential that the pigment extraction system chosen is complete and reproducible, and that there is no formation of degradation compounds during the process. Chlorophyll pigments are by nature extraordinarily labile. The wrong choice of extraction system, or inappropriate analytical conditions (excessive light, high temperature, long extraction time, etc.), may mean epimerization on C-13^2 of the isocyclic ring, or even the formation of pheophytins by loss of the central atom of Mg. In solvents with exchangeable protons, for instance methanol or ethanol, epimerization is favored, so that if they are used as extractants, the operation should be carried out as quickly as possible. Furthermore, if the buffer capacity of the solvent is low, the liberation of acids from the tissue could increase the rate of pheophytinization. To avoid such possibility, the buffer capacity of the extractant is normally increased by the addition of basic salts [170]. In tissues with high chlorophyllase activity, enzymatic de-esterification of chlorophylls and pheophytins can take place during the extraction phase.

All these products are formed under relatively mild conditions, so that manipulation of samples containing chlorophyll compounds must be carried out at temperatures below 40°C, under weak or green light, preferably with inert atmosphere, and using solvents that do not allow alterations of the natural compounds. All the circumstances mentioned previously mean that work with chloroplast pigments must be performed with the greatest care, as they can rapidly be altered, forming artifacts.

For the extraction of pigments in foods with scant lipid content, any organic solvent can generally be used, since all the pigments will be soluble to a greater or lesser extent, and the exhaustion of color with extraction is only a question of repeating the process. When the aim is not very specific, acetone achieves extraction in a reasonable time and with little expenditure of solvent.

The end of the process is normally signaled by the obtaining of colorless extracts. Theoretically, this point should in every case yield extracts with identical pigment content, both qualitatively and quantitatively. In practice, it does not normally happen. For solvents with a high penetration capacity, extraction is usually easy, with no need to triturate the sample exhaustively. For solvents that are desiccant (acetone or ethanol) or hydrophobic (hexane), access to the internal structures containing the pigments is hindered by dehydration of the sample and subsequent compacting, or by repulsion of the solvent. This drawback is avoided using an adequate and repeated trituration.

A particular case is the study of samples with high content of water. Its presence in the extract may be due to the tissues having either high moisture or little pigmentation, so that a substantial weight of starting sample will be necessary. In either case, the contribution of water will be considerable and will impede the evaporation of solvent. One simple, useful solution is to transfer the acetone extract to ethyl ether.

Another type of sample requiring specific treatment is that with a high-lipid content. The presence of such compounds interferes in the whole process of isolation and subsequent purification. They can be removed by means of selective phase distribution, using two solvents: one with high affinity for fat and the other with selective pigment-retention capacity.

7.3.1.1 Methodology

7.3.1.1.1 Simple Material

The sample (0.5–10 g, depending on the pigment content) is triturated for 1 min with 50 mL of acetone saturated with $MgCO_3$ to minimize pheophytinization during maceration. The mixture is filtered, and the solid residue collected and treated repeatedly with acetone until the extract is colorless. The acetone extracts are combined and the volume is reduced to dryness in a rotavapor at temperatures below 30°C.

In the case of samples with high-moisture content, the final volume of acetone extract is mixed with ether to form a single phase. The solution is separated by adding aqueous solution of NaCl (10%) and the ether phase is washed repeatedly. Re-extraction with ether is continued until the aqueous phase is completely colorless. If there are no highly polar pigments present, the first transfer normally contains all the pigmentation. The ether phase is finally filtered through a bed of anhydrous Na_2SO_4 to remove the water completely, and taken to dryness in a rotavapor.

7.3.1.1.2 Complex Material: High-Lipid Content

Between 5 and 10 g of sample are weighed and transferred quickly to a precipitating flask containing some 50 mL of *N,N*-dimethylformamide (*N,N*-DMF), saturated with $MgCO_3$. The mixture is triturated for 1 min and the homogenate is filtered in vacuo through coarse filter paper. The solid residue is collected and the procedure described above is performed until the filtrate is colorless. Normally, four extractions are sufficient.

The filtrates are combined in a 500-mL decanting funnel and treated with 70 mL of hexane to extract the fatty matter. The mixture is shaken for 1 min and allowed to stand until complete separation of phases. The upper layer, slightly yellow, retains lipids, carotenes, and di-esterified xanthophylls. The *N,N*-DMF hypophase, containing the rest of the pigments in solution, is treated twice more with hexane to remove completely any lipid remains. The three phases of hexane are combined and washed with 50 mL of *N,N*-DMF.

The *N,N*-DMF solution is then transferred to a 1000-mL decanting funnel containing 400 mL of 10% NaCl solution at a temperature close to 0°C. Hexane (70 mL) is added, together with 70 mL of ethyl ether, the mixture is shaken, and then allowed to stand until complete separation of phases (30 min). The aqueous phase is again treated with ethyl ether (50 mL). The remaining water-soluble compounds are retained in the aqueous phase, which is discarded. The organic phases are combined and washed several times with an aqueous solution of Na_2SO_4 (2%), then filtered through a bed of anhydrous Na_2SO_4. The water-free solution is concentrated in a rotavapor to dryness, at a temperature below 30°C [186].

7.3.1.1.3 Solid-Phase Distribution

Besides LPD (liquid-phase distribution) techniques, SPD (solid-phase distribution) methods by means of cartridges can also be used, which reduces the time of extraction. Comparing both techniques, SPD seems to be slightly less effective than LPD, although the recoverable capabilities are similar [76]. Recently, also diol-phase cartridges (diol-SPE) have been described with optimistic published results [187].

Specifically, a routine and sensitive method has been published for quantitative determination of pheophytin and pyropheophytin *a* in olive oils [188], using RP-SPD (reversed-phase SPD) extraction. Under the conditions assayed, more than 96% of oil is efficiently removed, and retention of pheophytin plus pyropheophytin is over 97%.

7.3.1.2 Chromatographic Separation Techniques

The analysis of chlorophylls must be completed as soon as possible after extraction, as artifacts can be formed during storage of the samples [189]. It is essential that storage is under N_2 atmosphere and at −30°C. The precautions already mentioned for the manipulation of samples during extraction must be continued during separation. It is also important that the storage conditions are appropriate for the chromatographic supports, as indicated by the manufacturer [190].

7.3.1.2.1 *Column Chromatography*

The classic chromatography experiments carried out by Tswett [191] on a column of calcium carbonate allowed the separation of chlorophylls *a* and *b* from the rest of the carotenoids, and continue to be widely used today for their large-scale separation (1–10 mg). The adsorbents are chosen for their selectivity and nonreactivity with the pigments or solvents [192,193]. Some inorganic adsorbents alter the chlorophyll molecule, causing isomerization, and those having a slightly acid nature induce degradation to pheophytins [65,175,192]. Adsorbents used include starch, cellulose, certain simple sugars, polyethylene, silica, calcium hydroxide, calcium phosphate, and alumina [157,194–202]. The ion-exchange and molecular sieve methods are commonly employed analytically, especially for the pretreatment of samples. DEAE-Sepharose CL-6B chromatography is particularly useful for the separation (and removal) of carotenoids from plant extracts and for the separation of esterified and nonesterified chlorophylls and their derivatives. The chromatography should be carried out at low temperature and in a short time to prevent isomerization reactions [203,204].

7.3.1.2.2 *Chromatography on Paper*

For many years, this has been a simple technique widely used for the separation and analysis of chlorophylls. An outstanding advantage is its great versatility in the mode of use (unidimensional, bidimensional, radial, radial with acceleration centrifugation, etc.). At the same time, it is proven that pigment recovery following this technique often exceeds that obtained with TLC. Nevertheless, the procedure is no longer used in analytical separations, as its resolution power has been surpassed by TLC or HPLC techniques. For more detail, the References section at the end of the chapter includes reviews of paper chromatography methods used for the recovery of chloroplast pigments, with information on the quality of the paper, solvent systems, and separation conditions [205,206].

7.3.1.2.3 *Thin Layer Chromatography (TLC)*

The low price and simplicity of this technique allow it to be used widely for the separation and monitoring of pigments. The choice of support is vital, as many authors have reported the formation of artifacts or decomposition of products during TLC, caused by light, acidity of the silica, drying methods, oxidation, or solvents. Given the lability of chlorophylls, recommended adsorbents are few: silica gel or other silicates, sugar (powdered sucrose), cellulose, polyethylene, and, more recently, chemically bonded C_8 and C_{18} silica gel, specifically for RP-TLC. However, for qualitative identification, and with due precautions, it is an unsubstitutable technique [65,192,207–211]. It has been reported that saccharose plates produce fewer degradation products [201,212,213]. Recently, the efficiency of this technique has been improved with the development of high-performance (HP) plates in both normal and reversed phases, which have been used for the general separation of chloroplast pigments [27,81,214–216].

7.3.1.2.4 High-Performance Liquid Chromatography (HPLC)

The main advantage of this technique is a high resolving power of complex mixtures in a short time, together with the small sample size required. The time of analysis, including sample preparation, is normally less than with other chromatographic methods. In some cases, separation is completed between 15 and 30 min. Until now, no reference has reported sample degradation during the chromatographic analysis, an important criterion when choosing a particular analytical method. The interaction of chlorophyll pigments with light makes HPLC especially appropriate for their analysis, as spectrophotometry is normally used as detection technique. These advantages, together with the high sensitivity of UV–Vis and fluorescence detectors, have enabled the quantitative analysis of individual chlorophylls in highly complex mixtures to be tackled successfully. This chromatographic technique, in the preparative version, is also particularly useful for collecting amounts of pure individual components for their identification [21,28,217,218].

HPLC was first used to analyze chlorophyll pigments in 1975 [219]. Since then, numerous works have applied it to the separation of carotenoids, chlorophylls, and chlorophyll derivatives in plant tissues and marine phytoplankton. Some methods have used normal-phase chromatography, mainly on silica gel columns, and others reversed-phase, generally on C_{18} functionality columns. This type of column appears to have certain advantages over the former, avoiding problems of pigment degradation and long conditioning times [82,220]. And also there have been described methods using reverse phase C-30 columns [221,222] with very good results in the pigment separation.

Numerous works have been cited that use normal-phase silica columns and isocratic mixtures of solvents such as acetone/ligroin (20:80, v/v) for the quantification of chlorophylls and derivatives in phytoplankton [223], or iso-octane/98% ethanol (9:1, v/v) in spinach [224]. Watanabe et al. [225] separated chlorophylls and pheophytins using the isocratic mixture 2-propanol/*n*-hexane (3:97, v/v). This method yields chlorophylls with levels of purity above 99%. Abaychi and Riley [226], using the mixture petroleum ether/acetone/dimethylsulfoxide/diethylamine (75:23.25:1.5:0.25, v/v) as mobile phase, detected and quantified 16 pigments of chlorophylls, derivatives, and carotenoids.

Gradient solvent systems generally improve the resolution of this chromatographic technique. Outstanding mixtures are isopropyl alcohol in hexane from 1% to 10% [227], *n*-heptane/ethyl ether/acetone [228], and acetone in hexane at 8%, 10%, and 12% [229].

Evans et al. [219] were the first to use RP-HPLC to separate these pigments, with a mixture of ethyl acetate in petroleum ether as mobile phase. They achieved a good resolution of porphyrins obtained from different natural sources, and of the derivative pheophytins *a* and *b*. After collection of the pigment fractions, the mass desorption spectrum was obtained. The data of chromatographic retention and of the mass spectrum provided a complete characterization of the pigments. Subsequently, Eskins et al. [217] developed a preparative method for the separation of chlorophylls and carotenoids in the diatom *Nitzschia closterium*, using a gradient system from 80% aqueous methanol to a methanol/ether mixture. A good resolution was obtained for chlorophylls *a*, *b* and *c*, pheophytins, and carotenoids such as neofucoxanthin, diadinoxanthin, diatoxanthin, and β-carotene, and others typical of plants.

Reversed-phase methods have generally used different mixtures of methanol/acetone [230], methanol/water [189,231], or methanol/acetone/water [232] in isocratic or gradient form. Other workers add to the methanol/water mixture a phase modifier in the form of a gradient that can be acetone [82], ethyl acetate [17,136,233,234], or tetrahydrofuran [235]. Polyethylene powder has been used as reversed phase in the separation of bacteriochlorophylls, and can also be used for chlorophylls in plants [236].

The difficulty in separating acid chlorophylls has prompted considerable effort. To improve resolution of the dephytylated chlorophyll derivatives, ionic suppressing reagents have been added to the mobile phase. These include ammonium and tetrabutylammonium salts [82,237], acetic acid [21,190], and 0.005 M NaCl [215]. Mantoura and Llewellyn [82] include the ionic reagent in both the sample and in the mobile phase to obtain a good resolution of chlorophyllides and pheophorbides. This method enables qualitative and quantitative monitoring of numerous chlorophyll derivatives and carotenoids in acetone extracts of algas from both cultures and natural waters.

The modifications introduced by Mínguez-Mosquera et al. [185] to the eluent system described above enable resolution of the chlorophylls and carotenoids present in olive fruits, and of the degradation products originated during the fruits' lactic fermentation to prepare table olives. The same system is applied to the characterization of pigments of the algal flora growing under immobilization conditions in the treatment of wastewaters [238,239]. Subsequently, the resolution of mixtures of oxidized derivatives of pheophorbides, which include pyropheophorbides *a* and *b*, chlorin, rhodin, and purpurins *a* and *b*, has been achieved introducing small modifications in the elution gradient [81].

At the same time, nonaqueous reversed-phase methods have been developed. Wright and Shearer [220] used a linear gradient from 90% of acetonitrile to 100% of ethyl acetate to separate 44 pigments that included carotenes, xanthophylls, chlorophylls, and derivatives, in marine phytoplankton. Khachik et al. [240] combined an isocratic elution and gradient of methanol, acetonitrile, methylene chloride, and *n*-hexane to separate the major constituents (xanthophylls, chlorophylls, and carotenes) in different vegetables.

All these methods, in normal or reversed phase, are currently used for the HPLC separation of chloroplast pigments, with only slight modifications in the solvent systems and in the form of elution, depending on the raw material to be analyzed [100,101,241].

Detection is normally by absorption or fluorescence spectrophotometry. The high coefficients of extinction of the chlorophyll Soret band enable sensitive detection between 380 and 445 nm. This region of the spectrum also includes the carotenoids, which accompany the chlorophylls in plant pigment extracts and whose analysis and quantification are also usually of interest (see Chapter 6). When these compounds are not of interest, and their possible interference must be excluded, a selective detection of chlorophylls and derivatives can be carried out at 654 [136] or 667 nm [230], where there is no absorption of these pigments. Detection of chlorophylls and derivatives by fluorescence emission has been proposed. This is much more sensitive than absorption, and also avoids interference from the carotenoids. Normally, excitation is done in the Soret band absorption region, and the emission is measured between 650 and 670 nm [242,243].

Although much research effort has been put into the development of modern techniques of HPLC, more is needed to refine the technique for analysis of complex samples of pigments. Advances are expected with the development of coupled HPLC-MS systems.

7.3.1.3 Isolation and Purification

The pigments to be identified can be isolated and purified by preparative TLC. This technique isolates the highest number of pigments with an initial general chromatography, each one in sufficient amount to carry out the subsequent steps of purification before identification by specific tests.

Chromatographic development separates the pigments in inverse relation to their polarity. The hydrocarbons, which are the least-polar pigments, advance with the solvent, leaving those of higher polarity retained close to the base. The separate bands in the resulting general chromatogram show the following characteristics: order and value of retention factor (R_f) indicating pigment polarity and color under white and ultraviolet ($UV_{254,360}$) light, and enabling an initial assignation as chlorophyll or carotenoid pigments. R_f values are defined as distance from origin to spot center/distance from origin to solvent front.

Each band is scraped from the plate and eluted with an appropriate solvent. In principle, acetone is recommended—it solubilizes all the pigments acceptably, and its low viscosity means that the adsorbent decants much quicker, so that well-clarified solutions are easily obtained. The pigment solution is filtered through a small amount of anhydrous Na_2SO_4. The volume of solvent is reduced in a rotavapor and the concentrated solution of pigment is purified by subsequent chromatographic developments.

Each band from the TLC plate is rechromatographed on the same support and with the same developer. After this initial purification, HPLC will show the number of pigments comprising each problem band. Normally, at this point, one will have some idea of the structure of such pigments, to guide their separation by TLC, using different supports and developers depending on the case.

If there are no indications of the nature of the problem substances, the initial polarity shown by the compound in the general development can be used in choosing developers and supports to expand the region. The R_f value of the pigment in a TLC development is directly related to the polarity of the pigment, which in turn depends on the functional groups making up its structure.

A general separation of the acetone extract of pigments can be done on glass plates of 20 × 20 cm, covered in the laboratory with silica gel $60GF_{254}$ to a thickness of 1 mm. These are air-dried and activated for 1 h in a stove at 120°C. If preferred, the plates can be acquired commercially from a wide variety of sources. The usual developer used is the mixture petroleum ether (65°C–95°C)/acetone/diethylamine (10:4:1) (system I) [186]. The development chamber must be saturated, and the solvent must rise to three-quarters the height of the plate, which usually takes between 15 and 30 min.

This system does not allow separation of the individual dephytylated chlorophyll derivatives, which, because of their high polarity, are all retained at the base of the chromatogram ($R_f = 0$). This drawback is circumvented either by modifying the polarity of developer I with a substitution of diethylamine by pyridine at a higher proportion (10:4:2.5) (system II) [211], or by using reversed-phase chromatography. In this case, kieselguhr plates impregnated with maize oil (14%) or commercial plates of silica gel C_{18} are used, with the eluent mixture methanol/acetone/water (20:4:6) (system III) [81,244]. The recovery of these compounds from the adsorbent is very difficult, because the usual solvents do not completely solubilize them. Their elution is achieved with the mixture acetone/pyridine (1:1) [211]. If some pigments coincide in R_f value under these conditions, specific eluent systems are used for their individual separation (Table 7.3)

HPLC can also be used in the isolation of the different standard chlorophyll derivatives in their pure form, employing a semipreparative column of 25 cm in length and 1 cm internal diameter [28]. Each fraction collected in semipreparative HPLC is transferred to ethyl ether, and aqueous NaCl solution (10%) is added. The solvent is removed in a rotavapor, and each pure pigment fraction is subjected to different identification tests. This chromatographic development in reversed phase separates the pigments in direct relation to their polarity. The retention of a compound in HPLC is expressed by the capacity factor k, defined as

$$k = \frac{(t_x - t_m)}{t_m}$$

where

t_x is the retention time of each component

t_m is the time taken to elute the nonretained solvent

However, the most practical way to express the retention of a compound in HPLC is referring to that of a standard substance [245]: relative retention is given as

$$r_{i,st} = \frac{k_i}{k_{st}}$$

where the subscripts i and st refer to the respective values for the peak of the component under study and that of the substance chosen as standard.

7.3.1.4 Identification

Having proven the purity of each pigment, the next stage is to study its physicochemical characteristics. The typical coloration shown by these compounds in TLC under white and UV_{254nm} light, and the corresponding R_f values in TLC—in both normal and reversed phase—are the first guiding data in their characterization. Subsequently, absorption spectra are obtained in different solvents, whose dielectric properties affect the spectrum. The shape, location of the absorption maxima, and the ratio between the Soret band and the maximum in the red region contribute very valuable information about the molecular structure of the compound. Finally, co-chromatography with the corresponding standards, acquired commercially or prepared in the laboratory as detailed above, adds to the identification. Tables 7.2 and 7.3 summarize these characteristics for different pigments.

Table 7.4 shows the chromatographic ($r_{i,st}$) and spectroscopic (electron absorption maxima in the eluent) characteristics obtained from the separation of chlorophylls and derivatives by HPLC and UV–Vis detection. Pigment identification is completed with assignation of IR spectrum bands and mass spectrum.

7.3.1.5 Obtaining of Standards

Chlorophylls a *and* b—A widely used source of chlorophylls *a* and *b* is spinach leaves, because of the abundant presence of these compounds and their easy isolation. This is carried out by extraction of the pigments with acetone, followed by the separation and purification from other pigments by TLC on silica gel with the eluent mixture petroleum ether (65°C–95°C)/acetone/diethylamine (10:4:1) [65,183]. Standard chlorophylls *a* and *b* can be obtained commercially, for instance from Sigma Chemical Co. (St. Louis, MO).

Chlorophyll c—This is obtained from an acetone extract of pigments from algas of the genus *Phaeodactilum*. They are separated and purified by TLC under the same conditions as the other two chlorophylls. The strong adsorption of this pigment on the support requires elution from the silica with acetone/pyridine (1:1) [238].

Pheophytins a *and* b—These are prepared from the respective pure solutions of chlorophylls in ethyl ether, by acidification with 2–3 drops of 13% HCl (v/v). The mixture is shaken for 5 min, and the acid is removed by washing with 2% Na_2SO_4 solution. Finally, the ether phase is dried by filtration through a bed of anhydrous Na_2SO_4 [65,246].

Chlorophyllides a *and* b—Incubation of pure solutions of the corresponding chlorophylls with active enzymatic extract (chlorophyllase) de-esterifies the chlorophylls to form chlorophyllides. Fresh leaves of *Ailanthus altissima*, whose tissue is considered one of the richest sources of chlorophyllase [247], are used as biological material. The protein precipitate is obtained starting from 5 g of fresh, chopped leaves. Acetone (100 mL at −20°C) is added and the mixture is triturated, using Ultraturrax in a crushed-ice bath to keep the temperature down and minimize solubilization of the enzyme. The homogenization is carried out for 1 min at the lowest possible rate to prevent the formation of foam, which is a symptom of enzyme denaturalization. The residue is collected by filtration in vacuo and treated with acetone until the extracts are colorless. Normally, four extractions are sufficient. The protein precipitate obtained is left to dry at room temperature, and stored at −20°C until use. The conversion of chlorophylls to chlorophyllides is carried out under the optimum conditions of enzyme activation. The reaction is performed in glass flasks with PVL screw caps and Teflon washer. Pure solutions of chlorophyll *a* or *b* in acetone are incubated for 24 h at 30°C with 0.5 g of acetone powder and

TABLE 7.2
Chromatographic Characteristics in Thin Layer Chromatography (TLC) of Chlorophylls and Derivatives

Pigment[b]	R_f Values[a]				
	I	II	III	IV	V
Frequent					
Chlorophyll *a*	0.51	—	0.00	—	—
Chlorophyll *b*	0.44	—	0.00	—	—
Chlorophyll *c*	0.00	0.00	—	—	—
Pheophytin *a*	0.57	0.91	0.00	0.48	—
Pheophytin *b*	0.53	0.87	0.00	0.33	—
Chlorophyllide *a*	0.00	0.36	0.75	—	—
Chlorophyllide *b*	0.00	0.24	0.96	—	–11
Pheophorbide *a*	0.00	0.45	0.30	—	0.69
Pheophorbide *b*	0.00	0.34	0.48	—	0.75
Pyropheophytin *a*	0.63	—	—	0.67	—
Pyropheophytin *b*	0.51	—	—	0.45	—
Pyropheophorbide *a*	—	—	—	—	0.58
Pyropheophorbide *b*	—	—	—	—	0.62
Not frequent					
Chlorin e_6	0.00	—	—	—	0.84
Rhodin g_7	0.00	—	—	—	0.88
15^1-OH-lactone-pheophytin *a*	0.00	0.63	—	—	—
15^1-OH-lactone-pheophytin *b*	0.00	0.59	—	—	—
15-Glyoxylic acid-pheophytin *a*	0.00	0.56	—	—	—
15-Glyoxylic acid-pheophytin *b*	0.00	0.41	—	—	—
Zn-pheophytin *a*	0.56	—	—	0.49	—
Zn-pheophytin *b*	0.55	—	—	0.34	—
Zn-pyropheophytin *a*	0.61	—	—	0.68	—
Zn-pyropheophytin *b*	0.51	—	—	0.46	—
Cu-pheophytin *a*	0.57	0.91	—	0.43	—
Cu-pheophytin *b*	0.55	0.87	—	0.30	—
Cu-pyropheophytin *a*	0.64	—	—	0.65	—
Cu-pyropheophytin *b*	0.52	—	—	0.40	—
Cu-15-glyoxylic acid-pheophytin *a*	0.00	0.48	—	—	—
Cu-15-glyoxylic acid-pheophytin *b*	0.00	0.46	—	—	—

Adsorbents: Systems I, II, and IV, silica gel 60 GF_{254}; System III, Kieselgur (with fluorescent indicator UV_{254}) impregnated with 14%; (v/v) of maize oil in light petroleum (40°C–60°C); System V, silica gel C_{18} (HPTLC plates).

[a] Frequent: Pigments frequently found in foods of plant origin.
Not frequent: Pigments occasionally found in foods of plant origin.

[b] Eluents: I, petroleum ether (65°C–95°C)/acetone/diethylamine (10:4:1); II, petroleum ether (65°C–95°C)/acetone/pyridine (10:4:2.5); III, methanol/acetone/water (20:4:6); IV, hexane/acetone/diethylamine (7:4:0.5); V, methanol/acetone/water (20:4:3).

TABLE 7.3
Color on Plate in Thin Layer Chromatography (TLC) and Spectroscopic Characteristics in Different Solvents of Chlorophylls and Derivatives

Pigment[b]	Color on Plate: Daylight	Color on Plate: UV_{366} Radiation[c]	Spectral Data: λ_{max}[a] Acetone	Spectral Data: λ_{max}[a] Diethyl Ether	Spectral Data: λ_{max}[a] Acetone/Pyridine (1:1)
Frequent					
Chlorophyll *a*	Blue-green	Red F	(384) (408) 428 536 580 616 662	(382) (410) 430 533 578 615 662	
Chlorophyll *b*	Yellow-green	Red F	(432) 454 595 646	(430) 456 549 595 644	
Chlorophyll *c*	Yellow-green	Red F	444 580 630	438 579 630	
Pheophytin *a*	Grey	Red F	408 466 504 534 554 608 666	408 471 505 534 609 668	
Pheophytin *b*	Brown	Red F	(414) 430 523 602 656	(412) 434 525 555 599 655	
Chlorophyllide *a*	Blue-green	Red F	(384) (408) 428 536 580 616 662		(390) 420 442 618 640 670
Chlorophyllide *b*	Yellow-green	Red F	(432) 454 595 646		(446) 470 556 502 654
Pheophorbide *a*	Grey	Red F	408 466 504 534 554 608 666		412 470 506 536 610 668
Pheophorbide *b*	Brown	Red F	(414) 430 523 602 656		(438) 526 556 600 656
Pyropheophytin *a*	Grey	Red F	408 466 504 534 554 608 666	408 471 505 534 609 668	
Pyropheophytin *b*	Brown	Red F	(414) 430 523 602 656	(412) 434 525 555 599 655	
Pyropheophorbide *a*	Grey	Red F	408 466 504 534 554 608 666		412 470 506 536 610 668
Pyropheophorbide *b*	Brown	Red F	(414) 430 523 602 656		(438) 526 556 600 656

Not frequent															
Chlorin e_6	Grey	Red F	400	500	(532)	602	660		400	500	(530)	560	610	666	
Rhodin g_7	Brown	Red F	(408)	426	(530)	(570)	(596)	648	(408)	427	(524)	(560)	(597)	652	
15^1-OH-lactone-pheophytin *a*	Grey	Red F	398	498	530	612	668								
15^1-OH-lactone-pheophytin *b*	Brown	Red F	(414)	424	518	(556)	(596)	648							
15-Glyoxylic acid-pheophytin *a*	Grey	Red F	400	500	(532)	602	660								
15-Glyoxylic acid-pheophytin *b*	Brown	Red F	(408)	426	(530)	(570)	(596)	648							
Zn-pheophytin *a*	Green	Red F							(376)	(406)	425	522	564	606	656
Zn-pheophytin *b*	Green	Red F							445	637					
Zn-pyropheophytin *a*	Green	Red F							(376)	(406)	425	522	564	606	656
Zn-pyropheophytin *b*	Green	Red F							445	637					
Cu-pheophytin *a*	Green	No F							398	422	504	548	606	650	
Cu-pheophytin *b*	Green	No F							438	520	584	628			
Cu-pyropheophytin *a*	Green	No F							398	422	504	548	606	650	
Cu-pyropheophytin *b*	Green	No F							438	520	584	628			
Cu-15-glyoxylic acid-pheophytin *a*	Blue	No F	(398)	406	502	(598)	630								
Cu-15-glyoxylic acid-pheophytin *b*	Yellow-green	No F	428	566		614									

[a] The values in parentheses indicate inflection points in the absorption spectrum.

[b] Same as in Table 7.2.

[c] F, fluorescence.

TABLE 7.4
Chromatographic and Spectroscopic Characteristics in High-Performance Liquid Chromatography (HPLC) of Chlorophyll and Derivatives

Pigment[b]	k^c	$r_{i,st}$[d]	Spectral Data in the HPLC Eluent: λ_{max}[a]							
			Soret	I	II	III	IV	V	VI	*R*
Frequent										
Chlorophyllide *b*	0.20	0.02	466					600	650	3.3
Chlorophyllide *a*	1.11	0.12	432	(384)	(412)		(580)	616	664	1.3
Chlorophyll *c*	2.02	0.22	444					582	632	7.0
Pheophorbide *b*	2.10	0.23	436		(412)	524	(558)	598	654	4.3
Pheophorbide *a*	3.09	0.34	409	(400)	(478)	506	534	608	666	1.8
Pyropheophorbide *b*	4.04	0.44	436		(412)	524	(558)	598	654	4.3
Pyropheophorbide *a*	4.50	0.49	409	(400)	(478)	506	534	608	666	1.8
Chlorophyll *b*	7.96	0.87	466					600	650	3.3
Chlorophyll *b′*	8.24	0.90	466					600	650	3.3
Chlorophyll *a*	9.18	1.00	432	(384)	(412)		(580)	616	664	1.3
Chlorophyll *a′*	9.47	1.04	432	(384)	(412)		(580)	616	664	1.3
Pheophytin *b*	11.21	1.22	436		(412)	524	(558)	598	654	4.3
Pheophytin *b′*	11.91	1.30	436		(412)	524	(558)	598	654	4.3
Pheophytin *a*	13.22	1.44	409	(400)	(478)	506	534	608	666	1.8
Pheophytin *a′*	13.75	1.50	409	(400)	(478)	506	534	608	666	1.8
Pyropheophytin *b*	13.90	1.51	436		(412)	524	(558)	598	654	4.3
Pyropheophytin *a*	16.05	1.75	409	(400)	(478)	506	534	608	666	1.8

Not frequent										
Rhodin g_7	0.17	0.02	426	(408)		(530)	(570)	(596)	648	6.5
Chlorin e_6	0.72	0.08	400			500	(532)	602	660	3.8
15-Glyoxylic acid-pheophytin *b*	6.47	0.70	426	(408)		(530)	(570)	(596)	651	6.5
15-Glyoxylic acid-pheophytin *a*	7.21	0.79	400			500	(532)	607	662	2.9
15^1-OH-lactone-pheophytin *b*	8.46	0.92	427		(416)	522	(560)	(600)	651	4.5
Zn-pheophytin *b*	8.89	0.97	456					590	640	3.1
Zn-pyropheophytin *b*	9.84	1.07	456					590	640	3.1
Zn-pheophytin *a*	9.93	1.08	428	(376)	(410)		(560)	610	656	1.2
15^1-OH-lactone-pheophytin *a*	10.33	1.13	400			499	531	614	670	2.6
Zn-pyropheophytin *a*	10.58	1.15	428	(376)	(410)		(560)	610	656	1.2
Cu-pheophytin *b*	11.10	1.21	428				(564)		612	5.6
Cu-pheophytin *a*	14.17	1.54	424	(384)	400	506	548	606	654	1.0
Cu-pyropheophytin *b*	14.41	1.57	428				(564)		612	5.6
Cu-pyropheophytin *a*	17.75	1.93	424	(384)	400	506	548	606	654	1.0

[a] *R*, quotient of absorbance at Soret band divided by absorbance at wavelength VI. The values in parentheses indicate inflection points in the absorption spectrum.

[b] Same as in Table 7.2.

[c] Retention factor, $k = (t_R - t_M)/t_M$, where t_R is the retention time of the pigment peak and t_M is the retention time of an unretained component.

[d] Relative retention, $r_{i,st} = k_i/k_{st}$; *st* = chlorophyll *a*.

chlorophyllase buffer such that the final concentration of acetone is 50%. The chlorophyllase buffer is tris-20 mM HCl at pH 8 which also contains 0.5 M NaCl and 10 mM $MgCl_2$ [248]. At the end of incubation, the pigments are transferred to ethyl ether and completion of the reaction is verified by TLC on silica gel [211].

Pheophorbides a *and* b—These can be prepared similarly to pheophytins, by acidification of the corresponding ether solution of chlorophyllide, or to chlorophyllides, by enzymatic de-esterification of the respective pheophytins.

C_{10} epimers—These are formed by dissolving the corresponding pigment in chloroform and keeping the solution in the refrigerator for 2 h [225].

Pyropheophytins and pyropheophorbides a *and* b—These are prepared from the corresponding pheophytins and pheophorbides by heat treatment of these compounds in pyridine at 100°C for 24 h with reflux [136].

Cu-pheophytins and pheophorbides a *and* b—The copper–chlorophyll complexes are obtained by adding 20 mL of 1 M $CuCl_2$ to a solution of the corresponding pheophytin in acetone (80 mL). Ascorbic acid crystals are added to the reaction mixture to prevent oxidative changes, and the chelation is kept under N_2 atmosphere for 2 h, as described by Jones et al. [249].

Zn pheophytins and pheophorbides a *and* b—These are prepared similarly to the copper complexes, except that crystals of $ZnCl_2$ (5 g) are added directly to the respective solution of pheophytin in acetone (80 mL), and the mixture is stirred magnetically for 2 h [249].

Methyl esters of chlorin e_6 and rhodin g_7—These are formed by saponification of pheophorbides *a* and *b*, respectively, with a solution of 0.5% KOH in methanol [250].

Free chlorin e_6 and rhodin g_7—These are obtained from their respective methyl esters by saponification with methanolic solution of KOH (30%) at room temperature and under N_2 atmosphere [250].

Purpurins18 a *and* b—These are obtained by oxidation of pheophorbide *a* and *b*, respectively, by alkaline treatment with 30% KOH solution in methanol in the presence of atmospheric oxygen [250].

Allomerized chlorophylls—These are prepared by dissolving 1–2 μmol of chlorophyll *a* or *b* in 5 mL of methanol, and keeping the solution at room temperature and in darkness with magnetic stirring for 3 days. The progress of the reaction is followed by HPLC [218].

7.3.1.6 Quantification

7.3.1.6.1 Quantitative Determination of Pigments by UV–Vis Spectrophotometry

The simplest case is an overall evaluation of the chlorophyll pigments contained in a sample. For this, the starting point is a total extract of pigments, which is analyzed spectrophotometrically at the wavelength of maximum absorption in the red region. Such quantification is feasible because in this region of the spectrum, specifically at 662–666 nm, there is no absorption of the carotenoid pigments that normally accompany the chlorophyll pigments in extracts. This method is the one most used in control laboratories. Its foremost advantage is that measurement is almost immediate, and requires minimum sample treatment and no analytical operation (with the exception of absorbance measurement). Its basic disadvantage is that quantification assumes a theoretical major pigment, which may not always be the case.

The relationship between the molar concentration of a chlorophyll solution and the optic density or absorbance at a particular wavelength (λ) fits a linear response defined by the Beer–Lambert law:

$$E = \varepsilon \times C \times L$$

where

E is the optic density, absorbance, or extinction of the sample at λ
C is the molar concentration of pigment
L is the optic pathway through the solution, expressed in centimeter
ε is the molar coefficient of extinction

The intensity of an absorption band can also be expressed as specific coefficient of extinction, $\varepsilon^{1\%}$, defined as the absorbance of 1 g of pigment in 100 mL of a specific solvent at a specific λ.

The relationship between the two coefficients of extinction is the following:

$$\varepsilon\left(M^{-1} \times cm^{-1}\right) = 0.1 \times \varepsilon^{1\%} \times \mathrm{MW}$$

where MW is the molecular weight. The values of both coefficients for each pigment are available in [90].

The evaluation method according to this technique is as follows: the pigment extract is taken to a known volume with a solvent appropriate for spectrophotometric measurement (ethanol, acetone, etc.). The absorbance at the wavelength of measurement must be between 0.2 and 0.8, and quantification is performed adapting the formula of the Beer–Lambert law:

$$C = \frac{E \times V_{\mathrm{f}}}{\varepsilon^{1\%} \times W_{\mathrm{s}}} \times 10^4$$

where

C is the concentration (mg/kg)
V_{f} is the final volume of pigment extract (mL)
W_{s} is the sample weight (g)
$\varepsilon^{1\%}$ is the specific extinction of a 1% solution (1 g/100 mL) measured with an optical pathway of 1 cm
E is the extinction at λ of measurement

Normally, in a complex solution of pigments, $\varepsilon^{1\%}$ is arbitrarily taken as that of the major pigment. The results are expressed as milligram per kilogram of sample.

7.3.1.6.2 *Quantitative Determination of Pigments by TLC and UV–Vis Spectrophotometry*

At the end of the TLC development of an extract containing pigments previously identified, each component identified is scraped from the plate individually and eluted with acetone to a determinate volume whose absorbance lies between 0.2 and 0.8. The value of extinction ε, at the wavelength of maximum absorption, is substituted in the Beer–Lambert equation. After the appropriate operations to express the result in milligram per kilogram of sample, the resulting equation is

$$C = \frac{E \times V_i \times V_f}{\varepsilon^{1\%} \times W_s \times V_{cr}} \times 10^4$$

where

C is the concentration (mg/kg)
V_i is the initial volume of pigment extract (mL)
V_f is the final volume of elution (mL)
W_s is the weight of sample (g)
V_{cr} is the volume chromatographed (mL)
$\varepsilon^{1\%}$ is the specific extinction of a 1% solution (1 g/100 mL) measured with an optical pathway of 1 cm
E is the extinction at λ of measurement

The coefficients of extinction in acetone for each pigment are calculated from those given in the bibliography for chlorophylls and pheophytins in ethyl ether [194] by using a pigment solution of known concentration. It is assumed that the coefficients of extinction of compounds with different chemical structure but identical electron absorption spectrum (e.g., chlorophyll/chlorophyllide) do not differ significantly [251]. Thus, the specific coefficient of extinction ($\varepsilon^{1\%}$) can be calculated from a known molar coefficient (ε_i) of another compound (i) with an identical spectrum:

$$\varepsilon^{1\%} = \frac{\varepsilon_i \times \mathrm{MW}}{0.1}$$

Table 7.5 lists those coefficients.

7.3.1.6.3 *Quantitative Determination of Pigments by HPLC and UV–Vis Spectrophotometry*

Separation of pigments—The proposed chromatographic system considers jointly both the pigments normally found in green tissues and the possible degradation products resulting from the various technological treatments. Pigment separation is achieved using a reversed-phase C_{18} column (Spherisorb ODS-2) of 25 cm in length, 4.6 mm internal diameter, and 5 μm particle size, protected with a precolumn of 3 cm × 4 mm packed with the same material. The eluents used are (A) water/ion-pair reagent/methanol (1:1:8) and (B) acetone/methanol (1:1). The ion-pair reagent consists of a solution of tetrabutylammonium acetate (0.05 M) and ammonium acetate (1 M) in water. The addition of these salts to the eluent improves separation

TABLE 7.5
Specific Absorption Coefficients of Chlorophylls and Derivatives

Pigment[a]	λ_{max}	$\varepsilon^{1\%}_{\lambda\ max}$	$\varepsilon^{1\%}_{430\,nm}$ [b]
Frequent		*Acetone*	
Chlorophyll *a*	428	840	840
Chlorophyll *b*	454	1450	356
Chlorophyll *c*	446	1400	—
Pheophytin *a*	406	1290	268
Pheophytin *b*	434	2060	1545
Pyropheophytin *a*[c]	406	1382	287
Pyropheophytin *b*[c]	434	2205	1653
		Acetone/pyridine	
Chlorophyllide *a*	438	1181	1220
Chlorophyllide *b*	468	2122	514
Pheophorbide *a*	410	1818	394
Pheophorbide *b*	436	3027	2254
Pyropheophorbide *a*[c]	410	2015	437
Pyropheophorbide *b*[c]	436	3347	2492
Not frequent			
Chlorin e_6[d]	—	—	192
Rhodin g_7[d]	—	—	2775
15^1-OH-lactone-pheophytin *a*[d]	—	—	127
15^1-OH-lactone-pheophytin *b*[d]	—	—	1848
15-Glyoxylic acid-pheophytin *a*[d]	—	—	129
15-Glyoxylic acid-pheophytin *b*[d]	—	—	1877
		Ethyl ether[e]	
Zn-pheophytin *a*	423	1335	803
Zn-pheophytin *b*	446	1869	568
Zn-pyropheophytin *a*[c]	423	1424	856
Zn-pyropheophytin *b*[c]	446	1991	606
Cu-pheophytin *a*	421	991	646
Cu-pheophytin *b*	438	1333	1333
Cu-pyropheophytin *a*[c]	421	1057	689
Cu-pyropheophytin *b*[c]	438	1420	1420
Cu-15-glyoxylic acid-pheophytin *a*[d]	—	—	60
Cu-15-glyoxylic acid-pheophytin *b*[d]	—	—	1308

[a] Same as in Table 7.2.
[b] Values of $\varepsilon^{1\%}_{430nm}$ calculated from the absorption spectrum obtained with the photodiode array detector and from the values of $\varepsilon^{1\%}_{\lambda\ max}$.
[c] Values of $\varepsilon^{1\%}$ calculated from the molar absorbance coefficients of the respective pheophytins.
[d] Values based on molar absorbance coefficients of the respective pheophytins.
[e] Values according to Jones, I.D. et al., *J. Agric. Food Chem.*, 16, 80, 1968.

of the de-esterified chlorophyll derivatives [82]. After use, the column is stored in methanol/water (1:1, v/v) to prevent deterioration of the silica particles by the ionic reagent.

Before separating the acetone extract of pigments by HPLC, the sample is filtered through a nylon membrane of 0.45 μm pore diameter, or centrifuged at 12,000 g. Then it is injected (20 μL loop), and eluted at a flow rate of 2 mL/min (initial pressure 3800 psi) with the following gradient scheme:

Time (min)	A (%)	B (%)	Curve
Initial	75	25	
8	25	75	Linear
10	25	75	Isocratic
18	10	90	Convex
23	0	100	Concave
30	75	25	Concave

Depending on the pump model, convex or concave curve profile may be modified, and therefore those steps would need optimization. Detection is carried out at 430 nm, using a programmable diode array detector [185]. Figure 7.12 shows the typical chromatogram for the separation of pigments from a green tissue, and Figure 7.13 that for a mixture of standards of the possible pigments present in an extract of algal origin. The following figures show the chlorophyllic and carotenoid pigments separation obtained in different processed vegetable products: frozen peas (Figure 7.14), canned peas (Figure 7.15), cucumbers in brine (Figure 7.16), and virgin olive oil (Figure 7.17). In addition, Figure 7.18 shows a mixture of standards of oxidized pheophytins [28]. If the pigment mixture contains carotenoids, their possible interference can be resolved by detecting the chlorophyll pigments at 656 nm, a wavelength at which the other pigments do not absorb.

The chromatogram obtained on separating pheophytins *a* and *b* from their respective Cu and Zn complexes (Figure 7.19) [27] enables detection of a possible alteration or adulteration of food color. As can be deduced from the retention times of these complexes, their separation from the rest of the pigments is perfectly practicable.

Besides reverse phase, the References section at the end of the chapter describes methods in relation with the use of normal phase, normally associated with sample dilution and direct injection, without previous pigment extraction. In this sense, Psomiadou and Tsimidou [252] proposed a method, specific for olive oils, for simultaneous determination of chlorophylls and carotenoids. The pigment separation is carried out using *n*-hexane/2-propanol gradient within 20 min. Variations in the mobile phases (in which the sample is diluted and for the gradients) are varied [253,254].

Quantification of pigments—The best method for performing chromatographic quantification is the use of an internal standard, which must be a pigment absent from the material under study and separable from the other pigments. The useful standard

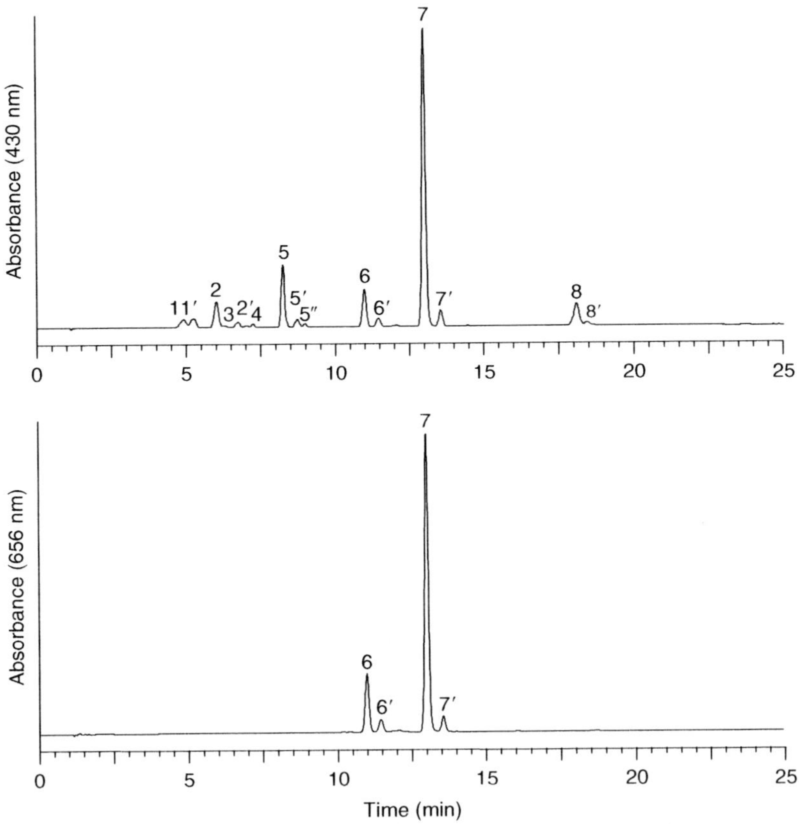

FIGURE 7.12 High-performance liquid chromatography (HPLC) separation of pigments present in a green tissue. Peaks: 1, neoxanthin; 1′, neoxanthin isomer; 2, violaxanthin; 2′, violaxanthin isomer; 3, luteoxanthin; 4, anteraxanthin; 5, lutein; 5′ and 5″, lutein isomers; 6, chlorophyll *b*; 6′, chlorophyll *b* C-132 epimer; 7, chlorophyll *a*; 7′, chlorophyll *a* C-13^2 epimer; 8, β-carotene; 8′, *cis*-β-carotene isomer.

pigments have been described in the chapter devoted to carotenoids. Calibration and response factor calculations are carried out in the same way.

If an assay with internal standard such as β-apo-8'-carotenal or canthaxanthin does not achieve a satisfactorily sharp separation between it and the chlorophyll or carotenoid pigments, other strategies will have to be adopted for the calibration. An external standard can be used, involving the preparation of a standard mixture in which the concentration of each component is known. This mixture is injected periodically to check possible instrumental variations. The calibration curve of each component is obtained by injecting different amounts of each pure pigment and making a linear fit from the plot of each peak area against the amount injected. At the same time, calibration curves in function of peak height are obtained, to calculate the approximate limit of detection for each pigment, taking as lower limit of detection the peak height equal to twice the noise signal.

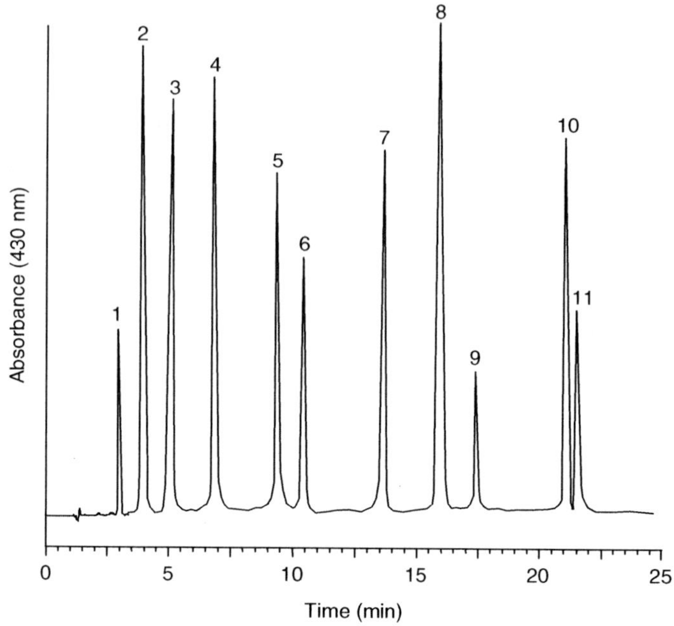

FIGURE 7.13 High-performance liquid chromatography (HPLC) separation of a mixture of standards of possible pigments present in an extract of algal origin. Peaks: 1, chlorophyllide *a*; 2, chlorophyll *c*; 3, pheophorbide *a*; 4, fucoxanthin; 5, diadinoxanthin; 6, lutein; 7, chlorophyll *b*; 8, chlorophyll *a*; 9, asteroidenone; 10, β-carotene; 11, pheophytin *a*.

The slow and laborious obtaining of standards for chlorophyll derivatives in the laboratory, together with the unreliable conservation of the standard mixtures because of the sensitivity of these pigments to light, temperature, etc., limits the use of this method of quantification. The chlorophylls are degraded very readily to pheophytins, totally changing their spectrophotometric characteristics. Consequently, the certainty that the standard mixture is at its initial concentration is quickly lost. This means the continual preparation of pure pigments, excessively lengthening the analysis time. The best way to avoid this complication is to use another method that estimates the concentrations of the components from an extension of the Beer–Lambert law, adapted to nonuniform systems [255], and governed by the following equation:

$$N_o = \frac{A_i \times F}{\varepsilon_\lambda \times L \times 10^3}$$

where

N_o is the number of moles per cm^2
A_i is the peak area (equal to absorbance × time [min])
F is the flow rate (mL/min)
ε_λ is the molar extinction at the λ of detection (cm^2 × mol^{-1})
L is the length of cell (cm), specific to each detector

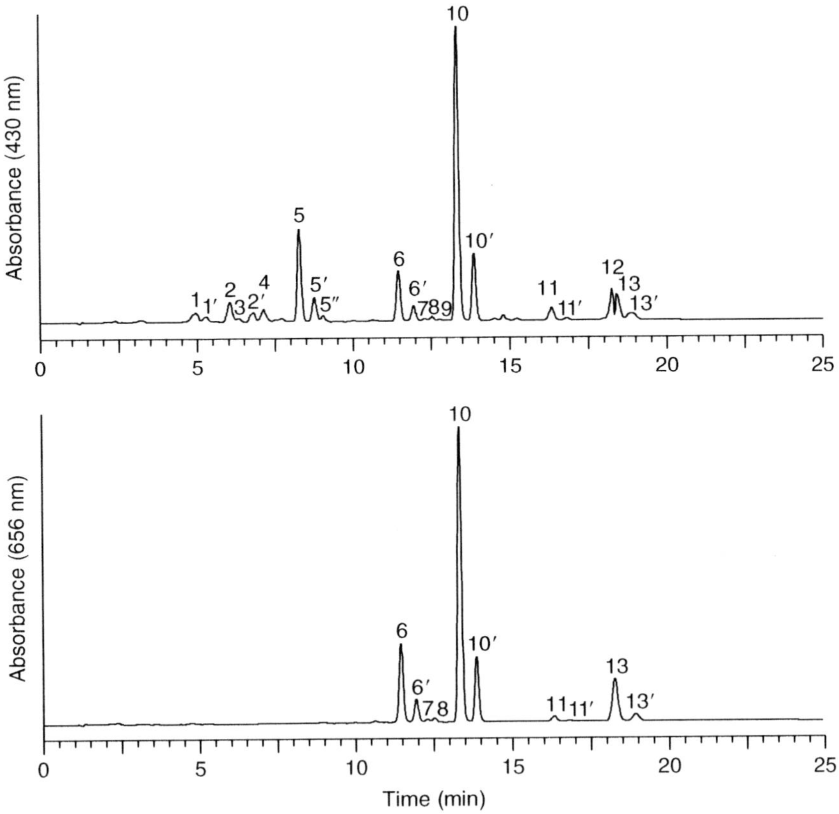

FIGURE 7.14 High-performance liquid chromatography (HPLC) separation of pigments present in frozen peas. Peaks: 1, neoxanthin; 1′, neoxanthin isomer; 2, violaxanthin; 2′, violaxanthin isomer; 3, luteoxanthin; 4, anteraxanthin; 5, lutein; 5′ and 5″, lutein isomers; 6, chlorophyll *b*; 6′, chlorophyll *b* C-132 epimer; 7, OH-lactone-chlorophyll *a*; 8, OH-chlorophyll *a*; 9, β-cryptoxanthin; 10, chlorophyll *a*; 10′, chlorophyll *a* C-13^2 epimer; 11, pheophytin *b*; 11′, pheophytin *b* C-13^2 epimer; 12, β-carotene; 13, pheophytin *a*; 13′, pheophytin *a* C-13^2 epimer.

Making the appropriate conversions of units to express the result in milligram per kilogram, we have

$$C = \frac{A \times F \times V_f \times 10}{\varepsilon_\lambda^{1\%} \times V_i \times \mathrm{MW} \times L} \qquad (7.1)$$

where
- C is the concentration (mg/kg)
- A is the peak area (equal to absorbance × time [min])
- F is the flow rate (mL/min)
- V_f is the volume in which the extract is collected (mL)

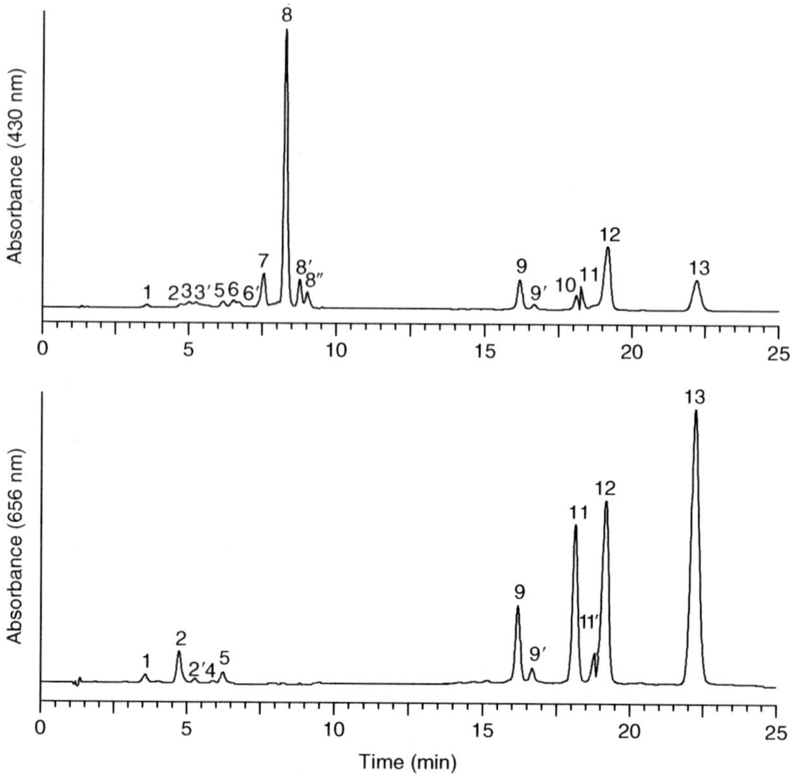

FIGURE 7.15 High-performance liquid chromatography (HPLC) separation of pigments present in canned peas. Peaks: 1, pheophorbide *b*; 1′, pheophorbide *b* C-13^2 epimer; 2, pheophorbide *a*; 2′, pheophorbide *a* C-13^2 epimer; 3, neochrome; 3′, neochrome isomer; 4, piropheophorbide *b*; 5, piropheophorbide *a*; 6, auroxanthin; 6′, auroxanthin isomer; 7, mutatoxanthin; 8, lutein; 8′ and 8″, lutein isomers; 9, pheophytin *b*; 9′, pheophytin *b* C-13^2 epimer; 10, β-carotene; 11, pheophytin *a*; 11′, pheophytin *a* C-13^2 epimer; 12, pyropheophytin *b*; 13, pyropheophytin *a*.

$\varepsilon_{\lambda}^{1\%}$ is the specific extinction at the detection wavelength
V_i is the volume of sample injected into the chromatograph (μL)
MW is the weight of sample (g)

The values of $\varepsilon_{\lambda}^{1\%}$ are calculated for each pigment from the adsorption spectrum obtained with the detector and from the values of $\varepsilon_{\lambda\ \mathrm{max}}^{1\%}$ given in the bibliography [220,185]. For a given concentration,

$$C = \frac{\varepsilon_{\lambda}}{\varepsilon_{\lambda}^{1\%}} = \frac{\varepsilon_{\lambda\ \mathrm{max}}}{\varepsilon_{\lambda\ \mathrm{max}}^{1\%}}$$

In practice, however, there may be some variations in instrumental sensitivity (deterioration of the lamp, changes in the index of refraction with elution gradient, etc.)

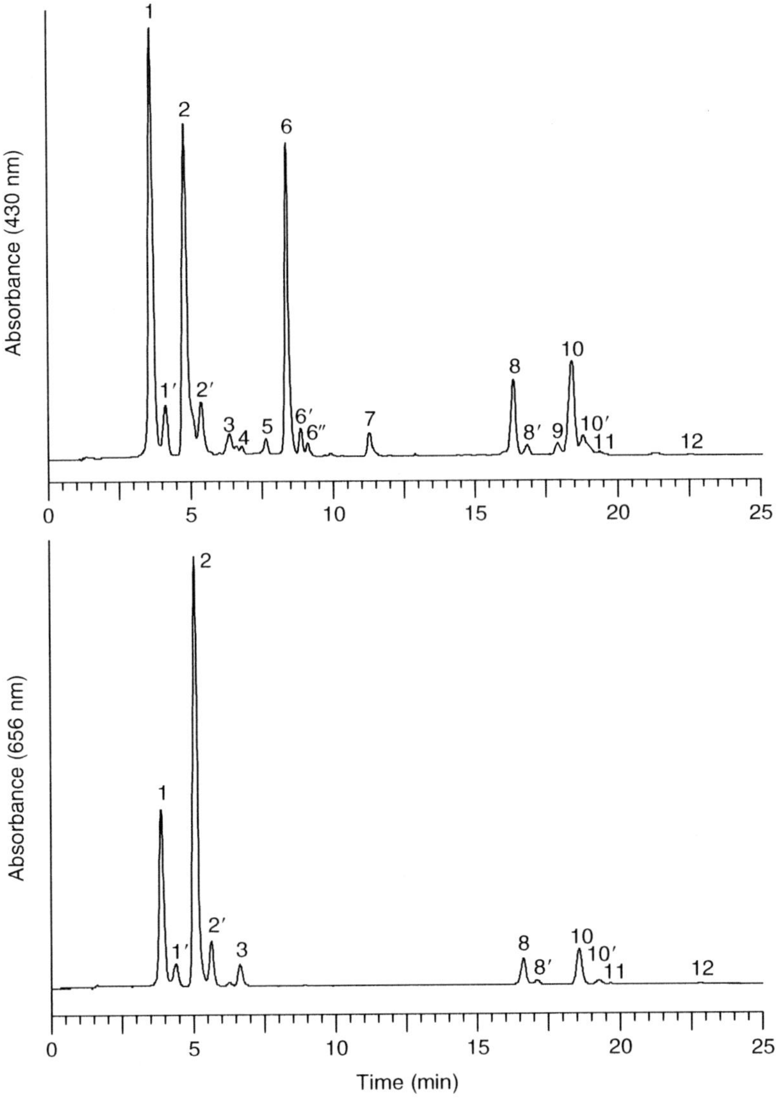

FIGURE 7.16 High-performance liquid chromatography (HPLC) separation of pigments present in cucumbers in brine. Peaks: 1, pheophorbide *b*; 1′, pheophorbide *b* C-13^2 epimer; 2, pheophorbide *a*; 2′, pheophorbide *a* C-13^2 epimer; 3, pirophcophorbide *a*; 4, auroxanthin; 5, mutatoxanthin; 6, lutein; 6′ and 6″, lutein isomers; 7, carotenoid; 8, pheophytin *b*; 8′, pheophytin *b* C-13^2 epimer; 9, β-carotene; 10, pheophytin *a*; 10′, pheophytin *a* C-13^2 epimer; 11, pyropheophytin *b*; 12, pyropheophytin *a*.

that affect absorption of the light and thereby affect quantification. To control such changes, Equation 7.1 is multiplied by a correction factor (F_c) obtained from the calibration curves of chlorophyll *a*.

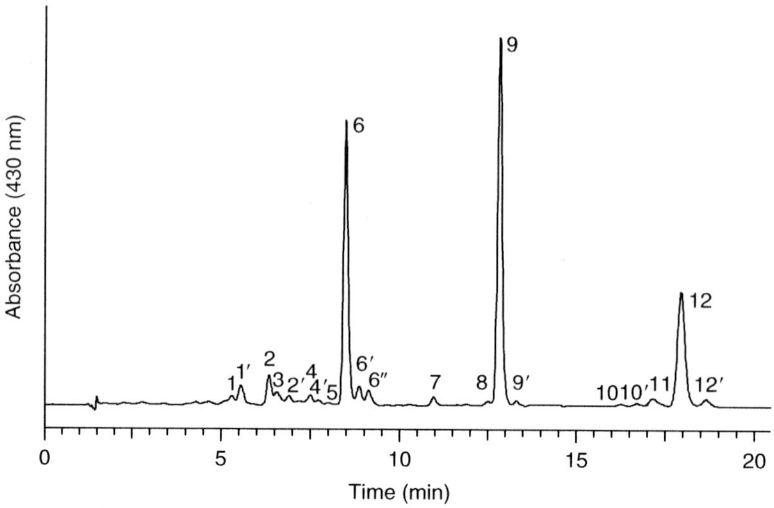

FIGURE 7.17 High-performance liquid chromatography (HPLC) separation of pigments present in virgin olive oil. Peaks: 1, neoxanthin; 1′, neoxanthin isomer; 2, violaxanthin; 2′, violaxanthin isomer; 3, luteoxanthin; 4, anteraxanthin; 4′, anteraxanthin isomer; 5, mutatox-anthin; 6, lutein; 6′ and 6″, lutein isomers; 7, chlorophyll *b*; 8, β-criptoxanthin; 9, chlorophyll *a*; 9′, chlorophyll *a*′; 10, pheophytin *b*; 10′, pheophytin *b*′; 11, β-carotene; 12, pheophytin *a*; 12′, pheophytin *a*′.

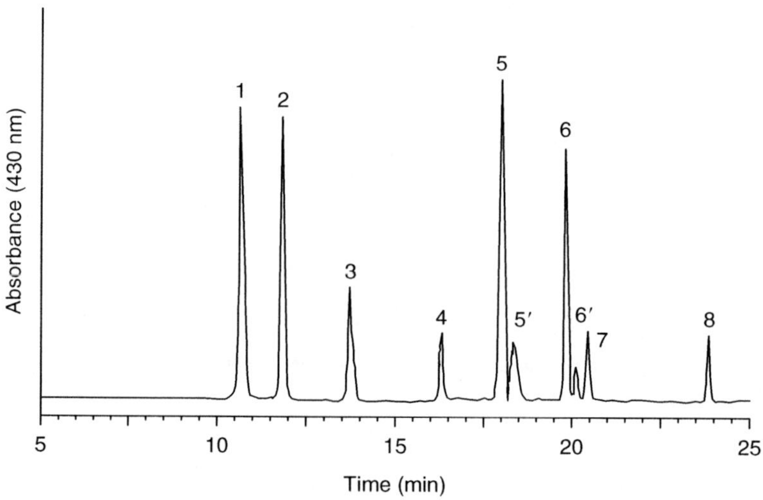

FIGURE 7.18 High-performance liquid chromatography (HPLC) separation of a mixture of standards of pheophytins and their oxidation products. Peaks: 1, 15-glyoxilic acid-pheophytin *b*; 2, 15-glyoxilic acid-pheophytin *a*; 3, 15-OH-lactone-pheophytin *b*; 4, 15-OH-lactone-pheophytin *a*; 5, pheophytin *b*; 5′, pheophytin *b* C-13^2 epimer; 6, pheophytin *a*, 6′, pheophytin *a* C-13^2 epimer; 7, phytol-purpurin 18 *b* ester; 8, phytol-purpurin 18 *a* ester.

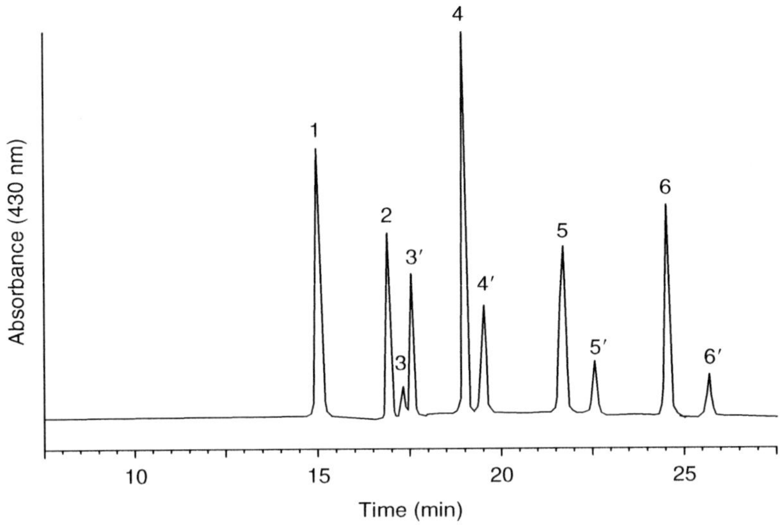

FIGURE 7.19 High-performance liquid chromatography (HPLC) separation of pheophytins *a* and *b* from their respective Cu and Zn complexes. Peaks: 1, Zn-pheophytin *b*; 2, Zn-pheophytin *a*; 3, Cu-pheophytin *b*; 3′, Cu-pheophytin *b* isomer; 4, pheophytin *b*; 4′, pheophytin *b* C-13^2 epimer; 5, pheophytin *a*; 5′, pheophytin *a* C-13^2 epimer; 6, Cu-pheophytin *a*; 6′, Cu-pheophytin *a* isomer.

The proposed method enables a correct quantification of pigments, needing only as periodic control a solution of chlorophyll *a* as standard for the chlorophyll derivatives [185].

7.3.1.6.4 Quantitative Determination of Pigments by HPLC and Coulometric Electrochemical Array Detection

This is a not very common method for chlorophyll quantifications. Puspitasari-Nienaber et al. [222] proposed a method to simultaneous determinations of carotenoids, tocopherols, and chlorophylls, using a C_{30} RP-HPLC. The detection limits (injecting 25 μL) were 1 fmol, 0.15 pmol, and 0.5 pmol, respectively, representing 1000-, 25-, and 5-fold enhancement over the UV–Vis methodology.

REFERENCES

1. Vernon, L.P. and Seely, G.R. (Eds.), *The Chlorophylls*, Academic Press, New York, 1966.
2. Strasburger, E. et al., Plastidios mitocondrios *Tratado de Botánica*, 6th ed., Marín, M. (Eds.), Cia, Barcelona, Spain, 1974.
3. Salisbury, F.B. and Ross, C.W., Photosynthesis: Cloroplasts and light, in *Plant Physiology*, 3rd ed., Salisbury, F.B. and Ross, C.W. (Eds.), Wadsworth Publishing Co., Belmont, CA, 1985, 179.
4. Heaton, J.W. and Marangoni, A.G., Chlorophyll degradation in processed foods and senescent plant tissues, *Trends Food Sci. Tech.*, 7, 8, 1996.

5. Gross, J., Chlorophylls, in *Pigments in Fruits*, Schweigert, B.S. (Ed.), Academic Press Inc., London, United Kingdom, 1987, 1.
6. Devlin, R.M., Fotosíntesis, *Fisiología Vegetal*, Devlin, R.M. (Ed.), Omega, Barcelona, Spain, 1975.
7. Stainer, R.Y., Duodoroff, M., and Alderberg, E.A. (Eds.), *El Mundo de los Microbios*, Aguilar, S.A., Madrid, Spain, 1965.
8. Jackson, A.H., Structure, properties and distribution of chlorophylls, in *Chemistry and Biochemistry of Plant Pigments*, Vol. 2, 2nd ed., Goodwin T.W. (Eds.), Academic Press, New York, 1976, 1.
9. Simpson, K.L. et al., Metabolism in senescent and store tissues, in *Chemistry and Biochemistry of Plant Pigments*, Vol. 1, Goodwin, T.W. (Ed.), Academic Press, London, United Kingdom, 1976.
10. Bauernfeind, J.C. et al., Use of carotenoids, in *Carotenoids*, Isler, O. (Ed.), Birkhäuser Verlag Basel, Stuttgart, Germany, 1971, 743.
11. Lichtenthaler, H.K., Verbreitung und reletive Konzentration der lipophilen Plastidenchinone in grunen Pflanzen, *Planta*, 81, 140, 1968.
12. Willstatter, R. and Stoll, A., *Investigations on Chlorophyll*, The Science Press Printing, Lancaster, PA, 1928, 385.
13. Langmeier, M., Ginsburg, S., and Matile, P., Chlorophyll breakdown in senescent leaves: Demonstration of Mg-dechelatase activity, *Physiol. Plant.*, 189, 347, 1993.
14. Schwartz, S.J. and Lorenzo, T., Chlorophylls in food, *Crit. Rev. Food Sci. Nutr.*, 29, 1, 1990.
15. Mínguez-Mosquera, M.I., Gandul Rojas, B., and Mínguez-Mosquera, J., Mechanism and kinetic of the degradation of chlorophylls during the processing of green table olives, *J. Agric. Food Chem.*, 42, 1089, 1994.
16. Weemaes, C. et al., Pressure–temperature degradation of color in broccoli juice, *J. Food Sci.*, 64, 504, 1999.
17. Schwartz, S.J., High performance liquid chromatography of zinc and copper pheophytins, *J. Liq. Chromatogr.*, 7, 1673, 1984.
18. Schaber, P.M. et al., Preparation and determination of zinc (II) chlorophylls by reversed-phase high-performance liquid chromatography, *J. Chromatogr.*, 316, 25, 1984.
19. Inoue, H. et al., Preparation and determination of Cu (II) chlorophylls by reversed-phase high-performance liquid chromatography, *Anal. Sci.*, 4, 599, 1988.
20. Inoue, H. et al., Preparation and determination of zinc (II) chlorophylls by reversed-phase high-performance liquid chromatography, *J. Chromatogr.*, 645, 259, 1993.
21. Tonnuci, L.H. and von Elbe, J.H., Kinetics of the formation of zinc complexes of chlorophyll derivatives, *J. Agric. Food Chem.*, 40, 2341, 1992.
22. Furuya, K., Inoue, H., and Shirai, T., Preparation and determination of Fe (II) and Fe (III) chlorophylls by reversed-phase high-performance liquid chromatography, *Anal. Sci.*, 3, 353, 1987.
23. Furuya, K. et al., Preparation and determination of Ni(II) chlorophylls by reversed-phase high-performance liquid chromatography, *Chromatographia*, 25, 319, 1988.
24. Nonomura, Y. et al., Preparation and determination of cobalt (II) chlorophylls by high-performance liquid chromatography, *Anal. Sci.*, 10, 117, 1994.
25. Li, S. and Inoue, H., Preparation and determination of Mn(II) and Mn(III) chlorophylls by reversed-phase high-performance liquid chromatography, *Chromatographia*, 33, 567, 1992.
26. Li, S.J. et al., Spectroscopic characteristic of La (II) chlorophyll, *Prog. Nat. Sci.*, 7, 442, 1997.

27. Mínguez-Mosquera, M.I. and Gandul-Rojas, B., High-performance liquid chromatographic study of chlorophyll alkaline treatment, *J. Chromatogr.*, 690, 161, 1995.
28. Mínguez-Mosquera, M.I., Gandul-Rojas, B., and Garrido-Fernandez, J., Preparation of Cu (II) complexes of oxidized chlorophylls and their determination by thin-layer and high-performance liquid chromatography, *J. Chromatogr.*, 731, 261, 1996.
29. Schellenberg, M., Matile, Ph., and Thomas, H., Production of a presumptive chlorophyll catabolite in vitro: Requirement of reduced ferredoxin, *Planta*, 191, 417, 1993.
30. Ginsburg, S., Schellenberg, M., and Matile P., Cleavage of chlorophyll–porphyrin. Requirement for reduced ferredoxin and oxygen, *Plant Physiol.*, 105, 545, 1994.
31. Hörtensteiner, S., Vicentini, F., and Matile, Ph., Chlorophyll breakdown in senescent leaves: Enzymic cleavage of pheophorbide a in vitro, *New Phytol.*, 129, 237, 1995.
32. Ginsburg, S. and Matile, P., Identification of catabolites of chlorophyll-porphyrin in senescent rape cotyledons, *Plant Physiol.*, 102, 521, 1993.
33. Lin, R.S.I., Phytochemicals and antioxidants, in *Functional Food*, Goldberg, I. (Ed.), Chapman & Hall, New York, 1994, 393.
34. Hernaez, J., Xu, M., and Dashwood, R.H., Effects of tea and chlorophyllin on the mutagenicity of *N*-hydroxy-IQ: Studies of enzyme inhibition, molecular complex formation, and degradation/scavenging of the active metabolites, *Environ. Mol. Mutagen.*, 30, 468, 1997.
35. Reddy, A.P. et al., Inhibition of dibenzo[a,l]pyrene-induced multiorgan carcinogenesis by dietary chlorophyllin in rainbow trout, *Carcinogenesis*, 20, 1919, 1999.
36. Egner, P.A., Muñoz, A., and Kensler, T.W., Chemoprevention with chlorophyllin in individuals exposed to dietary aflatoxin, *Mutat. Res.*, 523, 209, 2003.
37. García-Rodríguez, M.C., López-Satniago, V., and Altamirano-Lozano, M., Effect of chlorophyllin on chromium trioxide-induced micronuclei in polychromatic erythrocytes in mouse peripheral blood, *Mutat. Res.*, 496, 145, 2001.
38. Kumar, S.S. et al., Inhibition of radiation-induced DNA damage in plasmid pBR322 by chlorophyllin and possible mechanism(s) of action, *Mutat. Res.*, 425, 71, 1999.
39. Hartman, P.E. and Shankel, D., Antimutagens and anticarcinogens: A survey of putative interceptor molecules, *Environ. Mol. Mutagen.*, 5, 145, 1990.
40. Pietrzak, M. et al., Interaction of chlorophyllin with acridine orange, quinacrine mustard and doxorubicin analyzed by light absorption and fluorescence spectroscopy, *Biophys. Chem.*, 104, 305, 2003.
41. Pietrzak, M. et al., The "interceptor" properties of chlorophyllin measured within the three-component system: Intercalator-DNA-chlorophyllin, *Biophys. Chem.*, 123, 11, 2006.
42. Negishi, T., Rai, H., and Hayatsu, H., Antigenotoxic activity of natural chlorophylls, *Mutat. Res.*, 376, 97, 1997.
43. Higashi-Okai, K., Otani, S., and Okai Y., Potent suppressive activity of pheophytin *a* and *b* from the non-polyphenolic fraction of green tea (*Camellia sinensis*) against tumor promotion in mouse skin, *Cancer Lett.*, 129, 223, 1998.
44. Gentile, J.M. et al., Effect of selected antimutagens on the genotoxicity of antitumor agents, *Mutat. Res.*, 402, 289, 1998.
45. Te, C. et al., In vivo effects of chlorophyllin on the antitumor agent cyclophosphamide, *Int. J. Cancer*, 70, 84, 1997.
46. Dashwood, R.H., Chlorophylls as anticarcinogens, *Int. J. Oncol.*, 10, 721, 1997.
47. Cope, R.B. et al., Ultraviolet radiation-induced non-melanoma skin cancer in the Crl: SKH1:hr-BR hairless mouse: Augmentation of tumor multiplicity by chlorophyllin and protection by indole-3-cabinol, *Photochem. Photobiol. Sci.*, 5, 499, 2006.

48. Sesink, A.L.A. et al., Red meat and colon cancer: The cytotoxic and hyperproliferative effects of dietary heme, *Cancer Res.*, 59, 5704, 1999.
49. Balder, H.F. et al., Heme and chlorophyll intake and risk of colorectal cancer in the Netherlands cohort study, *Cancer Epidemiol. Biomarkers Prev.*, 15, 717, 2006.
50. de Vogel, J. et al., Green vegetables, red meat and colon cancer: Chlorophyll prevents the cytotoxic and hyperproliferative effects of haem in rat colon, *Carcinogenesis*, 26, 387, 2005.
51. de Vogel, J. et al., Natural chlorophyll but no chlorophyllin prevents heme-induced cytotoxic and hyperproliferative effects in rat colon, *J. Nutr.*, 135, 1995, 2005.
52. Blum, C.A. et al., β-Catenin mutation in rat colon tumors initiated by 1,2-dimethyl-hydrazine and 2-amino-3-methylimidazo[4,5-f]quinoline, and the effect of post-initiation treatment with chlorophyllin and indole-3-carbinol, *Carcinogenesis (London)*, 22, 315, 2001.
53. Dashwood, R.H. et al., Colonic cell proliferation, apoptosis and aberrant crypt foci development in rats given 2-amino-3-methylimidazo[4,5-f]quinoline, *Eur. J. Cancer Prev.*, 10, 139, 2001.
54. Díaz, G.D., Li, Q.J., and Dashwood, R.H., Caspase-8 and apoptosis-inducing factor mediate a cytochrome c-independent pathway of apoptosis in human colon cancer cells induced by the dietary phytochemical chlorophyllin, *Cancer Res.*, 63, 1254, 2003.
55. Carter, O., Bailey, G.S., and Dashwood, R.H., The dietary phytochemical chlorophyllin alters E-cadherin and β-catenin expression in human colon cancer cells, *J. Nutr.*, 134, 3441S, 2004.
56. Kitamura, K. et al., Effect of chlorophyllin-chitosan on excretion of dioxins in healthy man, *Environ. Sci. Technol.*, 39, 1084, 2005.
57. Dashwood, R.H., Yamane, S., and Larsen, R., A study of the forces stabilizing complexes between chlorophylls and heterocyclic amine mutagens, *Environ. Mol. Mutagen.*, 27, 211, 1996.
58. Yamazaki, H. et al., Effects of the dietary supplements, activated charcoal and copper chlorophyllin, on urinary excretion of trimethylamine in Japanese trimethylaminuria patients, *Life Sci.*, 74, 2739, 2004.
59. Lanfer-Marquez, U.M., Barros, R.M.C., and Sinnecker, P., Antioxidant activity of chlorophylls and their derivatives, *Food Res. Int.*, 38, 885, 2005.
60. Ferruzzi, M.G. et al., Antioxidant and antimutagenic activity of dietary chlorophyll derivatives determined by radical scavenging and bacterial reverse mutagenesis assays, *J. Food Sci.*, 67, 2589, 2002.
61. Ferruzzi, M.G., Failla, M.L., and Schwartz, S.J., Assessment of degradation and intestinal cell uptake of carotenoids and chlorophyll derivatives from spinach puree using an in vitro digestion and caco-2 human cell model, *J Agric. Food Chem.*, 49, 2082, 2001.
62. Egner, P.A. et al., Characterization of a chlorophyllin metabolite in serum during a chemoprevention trial in Qidong, People's Republic of China, *Proc. Am. Assoc. Cancer Res.*, 40, 2856, 1999.
63. Ferruzzi, M.G., Failla, M.L., and Schwartz, S.J., Sodium copper chlorophyllin: In vitro digestive stability and accumulation by Caco-2 human intestinal cells, *J Agric. Food Chem.*, 50, 2173, 2002.
64. Clydesdale, F.M. and Francis, F.J., Chlorophylls, in *Principles of Food Science*, Vol. 1, Fennema, O.R. (Ed.), Marcel Dekker, New York, 1976, 386.
65. Holden, M., Analytical methods—chlorophylls, in *Chemistry and Biochemistry of Plant Pigments*, Vol. 2, 2nd ed., Goodwin, T.W. (Ed.), Academy Press, New York, 1976.

66. Schwartz, S. et al., Determination of porphyrin in biological material, in *Methods of Biochemical Analysis*, Vol. 3, Glick, D. (Ed.), Interscience Publishers, Inc., New York, London, 1957, 221.
67. Goedheer, J.C., Visible absorption and fluorescence of chlorophyll and its aggregates in solution, in *The Chlorophylls*, Vernon, L.P. and Seeley, G.R. (Eds.), Academic Press, New York, 1966, 21.
68. Livingston, R., Watson, W.F., and McArdle, J., Activation of the fluorescence of chlorophyll solutions, *J. Am. Chem. Soc.*, 71, 1542, 1949.
69. Kirk, J.T.O. and Tilney-Basset, R.A.E., *The Plastids*, 2nd ed., Elsevier, Amsterdam, the Netherlands, 1978.
70. Virgin, H.I., Protochlorophyll formation and greening in etiolated barley leaves, *Physiol. Plant.*, 8, 630, 1955.
71. Virgin, H.I., Action spectrum for the elimination of the lag phase in chlorophyll formation in previously dark grown leaves of wheat, *Physiol. Plant.*, 14, 439, 1961.
72. Bazzaz, M.B. and Rebeiz, C.A., Chloroplast structure. V. Spectrofluorometric determination of chlorophyll(ide) *a* and *b* and pheophytin (or pheophorbide) *a* and *b* in unsegregated pigment mixtures, *Photochem. Photobiol.*, 30, 209, 1979.
73. Falkowski, P.G. and Sucher, J., Rapid quantitative separation of chlorophylls and their degradation products by high-performance liquid chromatography, *J. Chromatogr.*, 213, 349, 1981.
74. Shioi, Y., Fukae, R., and Sasa, T., Chlorophyll analysis by high-performance liquid chromatography, *Biochim. Biophys. Acta*, 722, 72, 1983.
75. Zonneveld, H., Klop, W., and Gorin, N., Analysis of unboiled and boiled beans for chlorophylls and pheophytins by thin-layer chromatography and fluorimetry comparison with spectrometry, *Z. Lebensm. Unters. Forsch.*, 178, 20, 1984.
76. Cichelli, A. and Pertesana, G.P., High-performance liquid chromatographic analysis of chlorophylls, pheophytins and carotenoids in virgin olive oil: Chemometric approach to variety classification, *J. Chromatogr. A*, 1046, 141, 2004.
77. Zandomeneghi, M., Carbonaro, L., and Caffarata, C., Fluorescence of vegetable oils: Olive oils, *J. Agric. Food Chem.*, 53, 759, 2005.
78. Goodwin, R.H., Fluorometric method for estimating small amounts of chlorophyll *a*, *Anal. Chem.*, 19, 789, 1947.
79. Boardman, N.K. and Thorne, S.W., Sensitive fluorescence method for the determination of chlorophyll *a*/chlorophyll *b* ratios, *Biochim. Biophys. Acta*, 253, 222, 1971.
80. White, R.C. et al., Fluorometric estimation of chlorophylls, chlorophyllides, pheophytins and pheophorbides in mixtures, *J. Agric. Food Chem.*, 20, 773, 1972.
81. Mínguez-Mosquera, M.I., Gallardo-Guerrero, L., and Gandul-Rojas, B., Characterization and separation of oxidized derivatives of pheophorbides *a* and *b* by TLC and HPLC, *J. Chromatogr.*, 633, 295, 1993.
82. Mantoura, R.F.C. and Llewellyn, C.A., The rapid determination of algal chlorophyll and carotenoid pigments and their breakdown products in natural waters by reverse-phase high-performance liquid chromatography, *Anal. Chim. Acta*, 151, 297, 1983.
83. Endo, Y., Thorsteinson, C.T., and Daun, J.K., Characterization of chlorophyll pigments presents in cannola seeds, meals and oils, *J. Am. Oil Chem. Soc.*, 69, 564, 1993.
84. Katz, J.J., Dougherty, R.C., and Boucher, L.J., Infrared and nuclear magnetic resonance spectroscopy of chlorophyll, in *The Chlorophylls*, Vernon, L.P. and Seely, G.R. (Eds), Academic Press, New York, 1966, 186.
85. Scheer, H. and Katz, J.J., Nuclear magnetic resonance spectroscopy of porphyrins and metalloporphyrins, in *Porphyrins and Metalloporphyrins*, 2nd ed., Smith, K.M. (Ed.), Elsevier, New York, 1975.

86. Janson, T.R. and Katz, J.J., NMR spectra of diamagnetic porphyrins, in *The Porphyrins*, Vol. 4, Dolphin, D. (Ed.), Academic Press, New York, 1978, 1.
87. Schick, G.A. and Bocian, D.F., Resonance Raman studies of hydroporphyrins and chlorophylls, *Biochim. Biophys. Acta*, 895, 127, 1987.
88. Katz, J.J. et al., Infrared spectra, molecular weights, and molecular association of chlorophylls *a* and *b*, methyl chlorophyllides, and pheophytins in various solvents, *J. Am. Chem. Soc.*, 85, 3801, 1963.
89. Closs, G.L. et al., Nuclear magnetic resonance spectra and molecular association of chlorophylls *a* and *b*, methyl chlorophyllides, pheophytins and methylpheophorbides, *J. Am. Chem. Soc.*, 85, 3809, 1963.
90. Scheer, H., Chlorophylls, in *CRC Handbook of Chromatography*, Vol. I, Kost, H.P. (Ed.), CRC Press, Boca Raton, FL, 1988.
91. Katz, J.J., Chlorophyll function in photosynthesis, in *The Chemistry of Plant Pigments*, Chichester, C.O. (Ed.), Academic Press, New York, 1972, 103.
92. Posthumus, M.A. et al., Laser desorption–mass-spectrometry of polar nonvolatile bio-organic molecules, *Anal. Chem.*, 50, 985, 1978.
93. Grotemeyer, J. et al., Multiphotonionization mass spectroscopy of native chlorophylls, *J. Am. Chem. Soc.*, 108, 4233, 1986.
94. Tabet, J.C. et al., Time resolved laser desorption. III. The metastable decomposition of chlorophyll *a* and some derivatives, *Int. J. Mass Spectrom. Ion Phys.*, 65, 105, 1985.
95. Dougherty, R.C. et al., Hydration behavior of chlorophyll *a*: A field desorption mass spectra study, *J. Am. Chem. Soc.*, 102, 416, 1980.
96. Hunt, J.E. et al., High-energy fragmentation of chlorophyll *a* and its fully deuterated analogue by 252Cf plasma desorption mass spectrometry, *J. Am. Chem. Soc.*, 103, 6775, 1981.
97. Chait, B.Y. and Field, F.H., 252Cf fission fragment ionization mass spectrometry of chlorophyll *a*, *J. Am. Chem. Soc.*, 106, 1931, 1984.
98. Baber, M. et al., Fast atom bombardment mass spectrometry, *Anal. Chem.*, 54, 645, 1982.
99. Brereton, R.G. et al., Positive and negative ion fast atom bombardment mass-spectrometric studies on chlorophylls: Structure of 4-vinyl-4-desethyl chlorophyll *b* tetrahedron, *Letters*, 24, 5775, 1983.
100. Teng, S.S. and Chen, B.H., Formation of pyrochlorophylls and their derivatives in spinach leaves during heating, *Food Chem.*, 65, 367, 1999.
101. Hyvarinen, K. and Hynninen, P.H., Liquid chromatographic separation and mass spectrometric identification of chlorophyll *b* allomers, *J. Chromatogr. A*, 837, 107, 1999.
102. Constantin, E. et al., Electron-impact and chemical ionization mass-spectrometry of chlorophylls, pheophytins and phaeophorbides by fast desorption on a gold support, *Bull. Soc. Chim. Fr.*, 7–8, 303, 1981.
103. Gandul-Rojas, B., Gallardo-Guerrero, M.L., and Mínguez-Mosquera, M.I., Identification of oxidized chlorophylls and metallochlorophyllic complexes of copper in table olives (cv. Gordal) with green staining alteration, *J. Food Prot.*, 62, 1172, 1999.
104. Jackson A.H., Applications of mass spectrometry in studies of porphyrins and related tetrapyrroles, *Philos. Trans. R. Soc. London A*, 293, 21, 1979.
105. Bricker, D.L. and Russell, D.H., Study of target gas excitation accompanying collisional activation of large polyatomic ions, *J. Am. Chem. Soc.*, 108, 6174, 1986.
106. Van Breemen, R.B., Canjura, F.L., and Schwartz, S.J., Identification of chlorophyll derivatives by mass spectrometry, *J. Agric. Food Chem.*, 39, 1452, 1991.
107. Grese, R.P. et al., Determination of structure and properties of modified chlorophylls by using fast atom bombardment combined with tandem mass spectrometry, *J. Am. Soc. Mass Spectrom.*, 1, 72, 1990.

108. Van Breemen, R.B, Canjura, F.L., and Schwartz, S.J., High-performance liquid chromatography–continuous-flow fast atom bombardment mass spectrometry of chlorophyll derivatives, *J. Chromatogr.*, 542, 373, 1991.
109. Grese, R.P. and Gross, M.L., The role of the matrix in fast atom bombardment desorption of chlorophyll, *Int. J. Mass Spectrom. Ion Proc.*, 122, 51, 1992.
110. Kostiainen, R., Hyvärinen, K., and Hynninen, P.H., Fast-atom bombardment mass spectra of the 132-epimers of 132-chlorophyll *a*, *Rapid Commn. Mass Spectrom.*, 9, 555, 1995.
111. Eckardt, C.B., Keely, B.J., and Maxwell, J.R., Identification of chlorophyll transformation products in a lake sediment by combined liquid chromatography–mass spectrometry, *J. Chromatogr.*, 557, 271, 1991.
112. Harris, P.G. et al., Identification of chlorophyll transformation products in zooplankton faecal pellets and marine sediment extracts by liquid chromatography/mass spectrometry atmospheric pressure chemical ionization, *Rapid. Commn. Mass Spectrom.*, 9, 1177, 1995.
113. Lee, S.M. et al., Identification of antimutagenic compounds from kale by high performance liquid chromatography and mass spectrometry, *J. Food Sci. Nutr.*, 3, 334, 1998.
114. Verzegnassi, L. et al., Analysis of tetrapyrrolic pigments and derivatives in sediments by high performance liquid chromatography/atmospheric pressure chemical ionization mass spectrometry, *J. Anal. Chem.*, 364, 249, 1999.
115. Zissis, K.D., Dunkerley, S., and Brereton, R.G. Chemometric techniques for exploring complex chromatograms: Application of diode array detection high performance liquid chromatography electrospray ionization mass spectrometry to chlorophyll *a* allomers, *Analyst*, 124, 971, 1999.
116. Chernomorsky, S., Segelmen, A., and Poretz, R.D., Effect of dietary chlorophyll derivatives on mutagenesis and tumor cell growth, *Teratog. Carcinog. Mutagen.*, 19, 313, 1999.
117. Yang, H. et al., Characteristics of photosensitization of pheophorbide *a* in liposomal media, *Sci. China, Ser. C: Life Sci.*, 42, 471, 1999.
118. Gross, J., Chlorophylls and carotenoids, in *Pigments in Vegetables*, Gross, J. (Ed.), Van Nostrand Reinhold, New York, 1991.
119. Mackinney, G. and Joslyn, M.A., The conversion of chlorophyll to pheophytin, *J. Am. Chem. Soc.*, 62, 231, 1940.
120. Lajollo, F.M., Tannenbaum, S.R., and Labuza, T.P., Reaction at limited water concentration–chlorophyll degradation, *J. Food Sci.*, 36, 850, 1971.
121. Schwartz, S.J. and Lorenzo, T.V., Chlorophyll stability during aseptic processing and storage, in *Annual Institute of Food Technology Meeting*, Chicago, IL, 1989.
122. Schwartz, S.J. and von Elbe, J.H., Kinetics of chlorophyll degradation to pyropheophytin in vegetables, *J. Food Sci.*, 48, 1303, 1983.
123. Mínguez-Mosquera, M.I., Gandul-Rojas, B., and Gallardo-Guerrero, M.L., Measurement of chlorophyllase activity in olive fruit (*Olea europaea*), *J. Biochem.*, 116, 263, 1994.
124. Perrin, P.W. and Gaye, M.M., Effects of simulated retail display and overnight storage treatments on quality maintenance in fresh broccoli, *J. Food Sci.*, 51, 146, 1986.
125. Shewfelt, R.L., Batal, K.M., and Heaton, E.K., Broccoli storage: Effect of N6-benzyladenine, packaging and icing on color of fresh broccoli, *J. Food Sci.*, 48, 1594, 1983.
126. Wang, S.S., Haard, N.F., and Dimarco, G.R., Chlorophyll degradation during controlled atmosphere storage of asparagus, *J. Food Sci.*, 36, 657, 1971.

127. Stewart, J.K. and Uota, M., Carbon dioxide injury and market quality of lettuce held in controlled atmosphere, *J. Am. Soc. Hortic. Sci.*, 96, 27, 1971.
128. Muftugil, N., Effect of different types of blanching on the color and the ascorbic acid and chlorophylls content of green beans, *J. Food Process. Preservation*, 10, 69, 1986.
129. Drake, S.R., Spaid, S.E., and Thompson, J.B., The influence of blanch and freezing method on the quality of selected vegetables, *J. Food Qual.*, 4, 271, 1981.
130. Sistrunk, W.A., Mahon, M.K., and Freeman, D.W., Relationship of processing methodology to quality attributes and nutritional value of canned spinach, *Hortic. Sci.*, 12, 59, 1977.
131. Dietrich, W.C. and Neumann, H.J., Blanching Brussels sprouts, *J. Food Technol.*, 19, 150, 1965.
132. Walker, G.C., Color deterioration in frozen French beans (*Phaseolus vulgaris*), *J. Food Sci.*, 29, 383, 1964.
133. Jones, I.D., White, R.C., and Gibbs, E., Influence of blanching or brining treatments on the formation of chlorophyllides, pheophytins, and pheophorbides in green plant tissue, *J. Food Sci.*, 28, 437, 1963.
134. Lajollo, F.M. and Lanfer Marquez, U.M., Chlorophyll degradation in a spinach system at low and intermediate water activities, *J. Food Sci.*, 47, 1995, 1982.
135. Berset, C. and Caniaux, P., Relationship between color evaluation and chlorophyllian pigment content in dried parsley leaves, *J. Food Sci.*, 48, 1854, 1983.
136. Schwartz, S.J., Woo, S.L., and von Elbe, J.H., High performance liquid chromatography of chlorophylls and their derivatives in fresh and processed spinach, *J. Agric. Food Chem.*, 29, 533, 1981.
137. Clydesdale, F.M., Lin, Y.D., and Francis, F.J., Formation of 2-pyrrolidone-5-carboxylic acid from glutamine during processing and storage of spinach puree, *J. Food Sci.*, 37, 45, 1972.
138. White, R.C., Jones, I.D., and Gibbs, E., Determination of chlorophylls, chlorophyllides, pheophytins and pheophorbides in plant material, *J. Food Sci.*, 28, 431, 1963.
139. Ruskov, P. and Malchev, E., Changes in chlorophyll of green pepper during lactic acid fermentation, *Food Sci. Technol. Abs.*, 8J, 1048, 1971.
140. Blanc, M., Chlorophyllic pigments of tea; distribution and behaviour during black tea manufacture, *Food Sci. Technol. Abs.*, 1h, 89, 1973.
141. Mínguez-Mosquera, M.I., Garrido-Fernández, J., and Gandul-Rojas, B., Pigment changes in olives during fermentation and brine storage, *J. Agric. Food Chem.*, 37, 8, 1989.
142. Schwartz, S.J., von Elbe, J.H., and Lindsay, R.C., Influence of processing on the pigmentation of wild rice grain, *J. Agric. Food Chem.*, 31, 349, 1983.
143. Takeda, Y., Microbial formation of pheophorbide *a* and pyropheophorbide *a* in Takana during brining and fermentation, *J. Food Hyg. Soc. Jpn.*, 30, 228, 1990.
144. Mínguez-Mosquera, M.I. and Gallardo-Guerrero, M.L., Anomalous transformation of chloroplastic pigments in gordal variety olives during processing for table olives, *J. Food Prot.*, 58, 1241, 1995.
145. Holden, M., Chlorophyll bleaching by legume seeds, *J. Sci. Food Agric.*, 16, 312, 1965.
146. Mínguez-Mosquera, M.I. et al., Pigments present in virgin olive oil, *J. Am. Oil Chem. Soc.*, 67, 192, 1990.
147. Buckle, K.A. and Edwards, R.A., Chlorophyll degradation and lipid oxidation in frozen unblanched peas, *J. Sci. Food Agric.*, 21, 307, 1970.
148. Rahmani, M. and Csallany, A.S., Chlorophyll and β-carotene pigments in maroccan virgin olive oils measured by high-performance liquid chromatography, *J. Am. Oil Chem. Soc.*, 68, 672, 1991.

149. Mínguez-Mosquera, M.I., Gandul-Rojas, B., and Gallardo-Guerrero, L., Rapid method of quantification of chlorophylls and carotenoids in virgin olive oil by HPLC, *J. Agric. Food Chem.*, 40, 60, 1992.
150. Ward, K. et al., Effects of processing and storage on chlorophyll derivatives in commercially extracted canola oil, *J. Am. Oil Chem. Soc.*, 71, 811, 1994.
151. Usuki, R. et al., Residual amounts of chlorophylls and pheophytins in refined edible oils, *J. Am. Oil Chem. Soc.*, 61, 785, 1984.
152. Gandul-Rojas, B. and Mínguez-Mosquera, M.I., Chlorophyll and carotenoids composition in virgin olive oils from various Spanish olive varieties, *J. Sci. Food Agric.*, 72, 31, 1996.
153. Gandul-Rojas, B. and Mínguez-Mosquera, M.I., Chlorophyllase activity in olive fruits and its relationship with the loss of chlrorophyll pigments in the fruits and oils, *J. Sci. Food Agric.*, 72, 291, 1996.
154. Gandul-Rojas, B. et al., Stability of chlorophyll pigments during storage of virgin olive oil, in *Proceeding of 1st International Congress on Pigments in Food Technology*, Mínguez-Mosquera, M.I., Jarén-Galán, M., and Hornero-Méndez, D. (Eds.), Sevilla, Spain, 1999, 271.
155. Clydesdale, F.M. and Francis, F.J., Chlorophyll changes in thermally processed spinach as influenced by enzyme conversion and pH adjustment, *Food Technol.*, 22, 793, 1968.
156. Eheart, M.S. and Odland, D., Use of ammonium compounds for chlorophyll retention in frozen green vegetables, *J. Food Sci.*, 38, 202, 1973.
157. Tan, C.T. and Francis, F.J., Effect of processing temperature on pigments and color of spinach, *J. Food Sci.*, 27, 232, 1962.
158. Humphrey, A.M., Chlorophyll, *Food Chem.*, 5, 57, 1980.
159. Fishbach, H., Microdeterminations for organically combined metals in pigment of okra, *J. Assoc. Off. Agric. Chem.*, 26, 139, 1943.
160. Schanderl, S.H., Marsh, G.L., and Chichester, C.O., Color reversion in processed vegetables. I. Studies on regreened pea puree, *J. Food Sci.*, 30, 312, 1965.
161. Segner, W.P. et al., Process for the preservation of green color in canned vegetables, U.S. Patent No. 4,473,591, 1984.
162. von Elbe, J.H. et al., Pigment composition and color of conventional and Veri-Green® canned beans, *J. Agric. Food Chem.*, 34, 52, 1986.
163. Leake, L.H. and Kirk, L.K., Method for color preservation in canned green vegetables, U.S. Patent No. 5,114,24, 1992.
164. Laborde, L.F. and von Elbe J.H., Method for improving the color of containerized green vegetables, U.S. Patent No. 5,482,727, 1996.
165. Canjura, F.L., Watkins, R.H., and Schwartz, S.J., Color improvement and metallo-chlorophyll complexes in continuous flow aseptically processed peas, *J. Food Sci.*, 64, 987, 1999.
166. Chernomorsky, S. et al., Evaluation of commercial chlorophyllin copper complex preparations by liquid chromatography with photodiode array detection, *J. AOAC Int.*, 80, 433, 1997.
167. Yasuda, K. et al., Investigation to find and indicator substance for the analysis of sodium copper chlorophyllin in foods, *J. Food Hyg. Soc. Jpn.*, 36, 710, 1995.
168. Inoue, H. et al., Determination of copper (II) chlorophyllin by reversed-phase high-performance liquid chromatography, *J. Chromatogr.*, 679, 99, 1994.
169. Ferruzzi, M.G. and Schwartz, S.J., Thermal degradation of commercial grade sodium copper chlorophyllin, *J. Agric. Food Chem.*, 53, 7098, 2005.
170. AOAC Official Methods of Analysis, Chlorophyll in plants, *Method 3*, 135, 59, 1984.

171. Strain, H.H., Oxidation and isomerization reactions of the chlorophylls in killed leaves, *J. Agric. Food Chem.*, 2, 1122, 1954.
172. Sweeney, J.P. and Martin, M.E., Determination of chlorophyll and pheophytin in broccoli heated by various procedures, *Food Res.*, 23, 635, 1958.
173. Brunisma, J., The quantitative analysis of chlorophylls *a* and *b* in plant extracts, *Photochem. Photobiol.*, 2, 241, 1963.
174. Strain, H.H. and Svec, W.A., Extraction, separation, estimation and isolation of the chlorophylls, in *The Chlorophylls*, Vernon, L.P. and Seely, G.R. (Eds.), Academic Press, New York, 1966.
175. Bacon, M.F. and Holden, M., Changes in chlorophylls resulting from various chemical and physical treatments of leaves and leaf extracts, *Phytochemistry*, 6, 193, 1967.
176. Clydesdale, F.M., Fleischman, D.L., and Francis, F.J., Maintenance of color in processed green vegetables, *J. Food Prod. Dev.*, 4, 127, 1970.
177. Strain, H.H., Cope, B.T., and Svec, W.A., Analytical procedures for the isolation, identification, estimation and investigation of the chlorophylls, *Methods Enzymol.*, 23, 452, 1971.
178. Svec, W.A., The isolation, preparation, characterization, and estimation of the chlorophylls and the bacteriochlorophylls, in *The Porphyrins*, Vol. 5, Physical Chemistry, Part C, Dolphin, D. (Ed.), Academic Press, New York, 1978.
179. O'Neill, E.J. and Crener, M.S., Use of vibratory ball-mill for the extraction of plant pigments, *Anal. London*, 105, 625, 1980.
180. Oquist, G. and Samuelsson, G., Sequential extraction of chlorophyll from chlorophyll-protein complexes in lyophilized pea thylakoids with solvents of different polarity, *Physiol. Plant.*, 50, 57, 1980.
181. Moran, R. and Porath, D., Chlorophyll determination in intact tissues using *N,N*-dimethylformamide, *Plant Physiol.*, 65, 478, 1980.
182. Van Buren, J.P., Extraction of chlorophylls *a* and *b* from different binding sites on thylakoid chlorophyll–proteins, *J. Agric. Food Chem.*, 33, 204, 1985.
183. Mínguez-Mosquera, M.I. and Garrido-Fernández, J., Eliminación de compuestos lipídicos durante la fase de extracción de pigmentos cloroplásticos en aceitunas (*Olea europaea*), *Grasas Aceites*, 36, 376, 1985.
184. Mínguez-Mosquera, M.I. et al., Decoloration of vegetable oils and oleoresins with recovery of unaltered pigments, *J. Am. Oil Chem. Soc.*, 68, 809, 1991.
185. Mínguez-Mosquera, M.I. et al., Determination of chlorophylls and carotenoides by HPLC during olive lactic fermentation., *J. Chromatogr.*, 585, 259, 1991.
186. Mínguez-Mosquera, M.I. and Garrido-Fernández, J., Chlorophyll and carotenoid presence in olive fruit (*Olea europaea*, L.), *J. Agric. Food Chem.*, 37, 1, 1989.
187. Mateos, R. and García-Mesa, J.A., Rapid and quantitative extraction method for the determination of chlorophylls and carotenoids in olive oil by high-performance liquid chromatography, *Anal. Bioanal. Chem.*, 385, 1247, 2006.
188. Hornero-Méndez, D., Gandul-Rojas, B., and Mínguez-Mosquera, M.I., Routine and sensitive SPE-HPLC method for quantitative determination of pheophytin *a* and pyropheophytin *a* in olive oils, *Food Res. Int.*, 38, 1067, 2005.
189. Braumann, T. and Grimme, L.H., Single-step separation and identification of photosynthetic pigments by high-performance liquid chromatography, *J. Chromatogr.*, 170, 264, 1979.
190. Shioi, Y., Analytical chromatography of chlorophylls, in *Chlorophylls*, Scheer, H. (Ed.), CRC Press, Boca Raton, Ann Arbor Boston, London, 1991.
191. Tswett, M., Adsorptionsanalyse and chromatographishe methods, anwendung auf die chemie des chlorophylls, *Ber. Dtsch. Bot. Ges.*, 24, 384, 1906.

192. Strain, H.H. and Svec, W.A., Some procedures for the chromatography of the fat-soluble chloroplast pigments, *Adv. Chromatogr.*, 8, 118, 1969.
193. Strain, H.H. and Sherma, J., Investigations of the chloroplast pigments of higher plants, green algae and brown algae and their influence upon the invention, modifications, and applications of Tsweet's chromatographic method, *J. Chromatogr.*, 73, 371, 1972.
194. Smith, J.H.C. and Benítez, A., Chlorophylls: Analysis in plant materials, in *Modern Methods of Plant Analysis*, Vol. 4, Paech, K. and Tracey, M.V. (Eds.), Springer-Verlag, New York, 1955, 142.
195. Sweeney, J.P. and Martin, M.E., Stability of chlorophyll in vegetables as affected by pH, *Food Technol.*, 15, 263, 1961.
196. Anderson, A.F.H. and Calvin, M., An improved method for the separation and purification of chlorophyll *a*, *Nature*, 194, 285, 1962.
197. Perkins, H.J. and Roberts, D.W.A., Purification of chlorophylls, pheophytins and pheophorbides for specific activity determinations, *Biochim. Biophys. Acta*, 58, 486, 1962.
198. Jones, O.T.G., The inhibition of bacteriochlorophyll biosynthesis in *Rhodopseudomonas sphaeroides* by 8-hydroxyquinoline, *Biochim. J.*, 85, 335, 1963.
199. Bacon, M.F., Separation of chlorophylls *a* and *b* related compounds by thin-layer chromatography on cellulose, *J. Chromatogr.*, 17, 322, 1965.
200. Stobart, A.K., McLaren, I., and Thomas, D.R., Chlorophylls and carotenoids of colourless callus, green callus and leaves of *Kalanchoe crenata*, *Phytochemistry*, 6, 1467, 1967.
201. Jeffrey, S.W., Quantitative thin-layer chromatography of chlorophylls and carotenoids from marine algae, *Biochim. Biophys. Acta*, 162, 271, 1968.
202. De la Mar, R.R. and Francis, F.J., Carotenoids degradation in bleached paprika, *J. Food Sci.*, 34, 287, 1969.
203. Omata, T. and Murata, N., A rapid and efficient method to prepare chlorophyll *a* and *b* from leaves, *Photochem. Photobiol.*, 31, 183, 1980.
204. Omata, T. and Murata, N., Preparation of chlorophyll *a*, chlorophyll *b*, and bacteriochlorophyll *a* by column chromatography with DEAE-Sepharose CL-6B and Sepharose CL-6B, *Plant Cell Physiol.*, 24, 1093, 1983.
205. Šesták, Z., Paper chromatography of chloroplast pigments, *J. Chromatogr.*, 1, 293, 1958.
206. Šesták, Z., Paper chromatography of chloroplast pigments (chlorophylls and carotenoids), Part 3, *Photosynthetica*, 14, 239, 1980.
207. Bacon, M.F., Artifacts from chromatography of chlorophylls, *Biochim. J.*, 101, 34, 1966.
208. Strain, H.H., Sherma, J., and Grandolfo, M., Alteration of chloroplast pigments by chromatography with siliceous adsorbents, *Anal. Chem.*, 39, 926, 1967.
209. Lord, C.E.C. and Tirimanna, A.S.L., A qualitative study of the carotenoid pigments of Sri Lanka chillies (*Capsicum annuum*), *Mikrochim. Acta*, 1, 469, 1976.
210. Buckle, K.A. and Rahman, F.M.M., Separation of chlorophyll and carotenoid pigments of *Capsicum cultivars*, *J. Chromatogr.*, 171, 385, 1979.
211. Mínguez-Mosquera, M.I., Gandul-Rojas, B., and Garrido-Fernández, J., Determinación cuantitativa de clorofilidas y feoforbidas, *Grasas Aceites*, 40, 114, 1989.
212. Chan, A.S.K. et al., Purification of dihydroporphyrins for specific activity determination by thin-layer chromatography, *J. Chromatogr.*, 47, 395, 1970.
213. Sahlberg, I. and Hynninen, P.H., Thin-layer chromatography of chlorophylls and their derivatives on sucrose layers, *J. Chromatogr.*, 291, 331, 1984.

214. Ziegler, R. et al., Enzymatic formation of pheophorbide and piropheophorbide during chlorophyll degradation in a mutant of *Chlorella fusca* Shihira et Kraus, *J. Plant Physiol.*, 132, 327, 1988.
215. Suzuki, N., Saitoh, K., and Adachi, K., Reversed-phase high-performance thin-layer chromatography and column liquid chromatography of chlorophylls and their derivatives, *J. Chromatogr.*, 408, 181, 1987.
216. Sherma, J. and Latta, M., Reversed phase thin-layer chromatography of chloroplast pigments on chemically bonded C18 plates, *J. Chromatogr.*, 154, 73, 1978.
217. Eskins, K., Scholfield, C.R., and Dutton, H.J., High-performance liquid chromatography of plant pigments, *J. Chromatogr.*, 135, 217, 1977.
218. Kuronen, P. et al., High-performance liquid chromatographic separation and isolation of the methanolic allomerization products of chlorophyll *a*, *J. Chromatogr.*, 654, 93, 1993.
219. Evans, N. et al., Applications of high-pressure liquid chromatography and field desorption mass spectrometry in studies of natural porphyrins and chlorophyll derivatives, *J. Chromatogr.*, 115, 325, 1975.
220. Wright, S.W. and Shearer, J.D., Rapid extraction and high-performance liquid chromatography of chlorophylls and carotenoids from marine phytoplankton, *J. Chromatogr.*, 294, 281, 1984.
221. De Sio, F. et al., A chromatographic procedure for the determination of carotenoids and chlorophylls in vegetable products, *Acta Aliment.*, 30, 395, 2001.
222. Puspitasari-Nienaber, N.L., Ferruzzi, M.G., and Schwartz, S.J., Simultaneous detection of tocopherols, carotenoids, and chlorophylls in vegetable oils by direct injection C-30 RP-HPLC with coulometric electrochemical array detection, *J. Am. Oil Chem. Soc.*, 79, 633, 2002.
223. Jacobsen, T.R., A quantitative method for the separation of chlorophylls *a* and *b* from phytoplankton pigments by high pressure liquid chromatography, *Mar. Sci. Commn.*, 4, 33, 1978.
224. Stransky, H., Die quantitative bestimmung von chloroplastenpigmenten im picomolbereich mit hilfe einer isochratischen HPLC-methode, *Z. Naturforsch.*, 33, 836, 1978.
225. Watanabe, T. et al., Preparation of chlorophylls and pheophytins by isocratic liquid chromatography, *Anal. Chem.*, 56, 251, 1984.
226. Abaychi, J.K. and Riley, J.P., The determination of phytoplankton pigments by high-performance liquid chromatography, *Anal. Chim. Acta*, 107, 1, 1979.
227. Iriyama, K., Yoshiura, M., and Shiraki, M., Micro-method for the qualitative and quantitative analysis of photosynthetic pigments using high-performance liquid chromatography, *J. Chromatogr.*, 154, 302, 1978.
228. De Jong, D.W. and Woodlief, W.C., High speed, low pressure liquid chromatography of chloroplast pigments from tobacco mutants, *J. Agric. Food Chem.*, 26, 1281, 1978.
229. Yoshiura, M., Iriyama, K., and Shiraki, M., High-performance liquid chromatography of chlorophylls and some of their derivatives, *Chem. Lett.*, 281, 1978.
230. Schoch, S. et al., High-performance liquid chromatography of tetrapyrrole pigments, *J. Chromatogr.*, 157, 3567, 1978.
231. Shoaf, W.T., Rapid method for the separation of chlorophylls *a* and *b* by high-pressure liquid chromatography, *J. Chromatogr.*, 152, 247, 1978.
232. Rebeiz, C.A., Bazzaz, M.B., and Belanger, F., The separation of chlorophyll and pheophytins by reversed phase HPLC, *Chromatogr. Rev.*, 4, 2, 1978.
233. Eskins, K. and Dutton, H.J., Sample preparation for high-performance liquid chromatography of higher plant pigments, *Anal. Chem.*, 51, 1885, 1979.

234. Eskins, K. and Harris, L., High-performance liquid chromatography of etioplast pigments in red kidney bean leaves, *Photochem. Photobiol.*, 33, 131, 1981.
235. Craft, N.E., Wise, S.A., and Soares, J.H., Optimization of an isocratic high-performance liquid chromatographic separation of carotenoides, *J. Chromatogr.*, 589, 171, 1992.
236. Chow, H.C., Caple, M.B., and Strouse, C.E., Polyethylene powder as a stationary phase for preparative-scale reversed-phase high-performance liquid chromatography, *J. Chromatogr.*, 151, 357, 1978.
237. Davies, D. and Holdsworth, E.S., The use of high pressure liquid chromatography for the identification and preparation of pigments concerned in photosynthesis, *J. Liq. Chromatogr.*, 3, 123, 1980.
238. Mínguez-Mosquera, M.I. et al., Chlorophylls and carotenoids composition of the algal flora developed under immobilization conditions. Influence of support, *Biomass Bioenergy*, 1, 347, 1991.
239. Mínguez-Mosquera, M.I., Jarén-Galán, M., and Gandul-Rojas, B., Pigments in the algal biomass from aerobic waste water depuration ponds, *Water, Air and Soil Pollut.*, 65, 195, 1992.
240. Khachik, F., Beecher, G.R., and Whittaker, N.F., Separation, identification, and quantification of the major carotenoid and chlorophyll constituents in extracts of several green vegetables by liquid chromatography, *J. Agric. Food Chem.*, 334, 603, 1986.
241. Almela, L., Fernández-López, J.A., and Roca, M.J., High-performance liquid chromatographic screening of chlorophyll derivatives produced during fruit storage, *J. Chromatogr.*, 870, 483, 2000.
242. Brown, L.M., Hargrave, B.T., and MacKinnon, M.D., Analysis of chlorophyll *a* in sediments by high-pressure liquid chromatography, *Can. J. Fish. Aquatic Sci.*, 38, 205, 1981.
243. Shioi, Y., Michio, D., and Sasa, T., Separation of non-esterified chlorophylls by ion-suppression high-performance liquid chromatography, *J. Chromatogr.*, 298, 141, 1984.
244. Jones, I.D. et al., An evaluation of reversed phase partition for thin-layer chromatographic identification of chlorophylls and derivatives, *J. Chromatogr.*, 70, 87, 1972.
245. Yost, R.W., Ettre, L.S., and Conlon, R.D., *Introducción a la Cromatografía Líquida Práctica*, Perkin-Elmer, Norwalk, Connecticut, 1980.
246. Sievers, G. and Hynninen, P.H., Thin-layer chromatography of chlorophylls and their derivatives on cellulose layers, *J. Chromatogr.*, 134, 359, 1977.
247. McFeeters, R.F., Chichester, C.O., and Whitaker, J.R., Purification and properties of chlorophyllase from *Ailanthus altissima*, *Plant Physiol.*, 47, 609, 1971.
248. Terpstra, W. and Lambers, J.W.J., Interactions between chlorophyllase, chlorophyll *a*, plants lipids and Mg 2+, *Biochim. Biophys. Acta*, 746, 23, 1983.
249. Jones, I.D. et al., Absorption spectra of copper and zinc complexes of pheophytins and pheophorbides, *J. Agric. Food Chem.*, 16, 80, 1968.
250. Hynninen, P.H., Chlorophylls. IV. Preparation and purification of some derivatives of chlorophylls *a* and *b*, *Acta Chem. Scand.*, 27, 1771, 1973.
251. Jones, I.D., White, R.C., and Gibbs, E., Some pigment changes in cucumbers during brining and brine storage, *Food Technol.*, March, 96, 1962.
252. Psomiadou, E. and Tsimidou, M. Simultaneous HPLC determination of tocopherols, carotenoids, and chlorophylls for monitoring their effect on virgin olive oil oxidation, *J. Agric. Food Chem.*, 46, 5132, 1998.
253. Nakamura, A., Tanaka, S., and Watanabe, T., Normal-phase HPLC separation of possible biosynthetic intermediates of pheophytin *a* and chlorophyll *a′*, *Anal. Sci.*, 17, 509, 2001.

254. Seppanen, C.M., Rahmani, M., and Csallany, A.S., Simultaneous determination of chlorophylls, pheophytins, beta-carotene, tocopherols, and tocotienols in olive and soybean oils by high-performance liquid chromatography. *J. Food Sci.*, 68, 1644, 2003.
255. Torsi, G., Chiavari, G., and Lippolis, M.T., Quantitation without calibration curves using HPLC with nondestructive detectors, *LC-GC Intl.*, 5, 37, 1992.

8 Water-Soluble Vitamins

Miguel Ángel Fernández-Muiño, María Teresa Sancho-Ortiz, and Felicidad Valls-García

CONTENTS

8.1 INTRODUCTION

Vitamins are organic substances with very high biological activity that are required in small amounts for the growth and maintenance of human health. As they cannot be synthesized in the human body in the quantities required, they need to be supplied in the diet.

Although new trends in nutrition classify vitamins according to their metabolic functions, traditional classification based on their solubility is very useful in food science since they have similar natural sources and similar extraction procedures, mainly the fat-soluble vitamins.

In the last 20 years much attention has been paid to the determination of vitamins in foodstuffs due to their nutritional implications and the need for an adequate quality control of food products. Vitamin analysis in these matrices is difficult due to the high variability in their composition and the great number of components, which implies the risk of interference and the low vitamin content. Additional problems are the existence of different chemical compounds with vitamin effects and their stability influenced by light, heat, oxygen, or pH conditions. The development of new analytical techniques has led in recent years to analytical methods that are more rapid, accurate, and sensitive.

This chapter reviews the analytical techniques for water-soluble vitamins analysis, focusing on the main problems associated with each analytical methodology.

8.2 GENERAL CONSIDERATIONS ON VITAMIN ANALYSIS METHODOLOGY

As in all analyses, special attention needs to be paid to the sampling procedure in order to obtain a representative and homogeneous sample. Another critical point is to maintain the optimal stability conditions for each vitamin that guarantees lower vitamin losses over the analytical procedure; otherwise all the analyses are valueless.

Vitamin analytical methods can be classified into three categories:

1. Physicochemical methods are the most commonly employed and include the phases of extraction, purification, and final analysis. All these steps are critical in the vitamin evaluation. Extraction steps can include several treatments, such as heat, acid or alkali conditions, enzymes, and solvents, and these treatments have several purposes, such as vitamin stabilization and their release from other food components. Cleanup steps remove interfering compounds and are not necessary in many methods. Final determination and quantification can be mainly carried out by chromatographic [Table 8.1 summarizes some common high-performance liquid chromatography (HPLC) methods applied to water-soluble vitamin analysis], spectrometric, enzymatic, inmunological, or radiometric techniques.
2. Microbiological methods use test microorganisms such as bacteria, protozoa, or yeast where growth is proportional to the presence of a specific vitamin. These microorganisms are cultivated in the presence of known quantities of the studied vitamin and in the presence of a food extract were the vitamin needs to be evaluated. The turbidity measurement and other parameters such as gravimetry, acid production, or gas production generally monitor the growth. An interesting related technique is the radiometric-microbiological assay based on the measurement of a ^{14}C-labeled metabolite, normally $^{14}CO_2$ formed by the test microorganism from a ^{14}C-labeled substrate.

   Microbiological assays have the advantage of specificity and sensitivity, but they are highly time consuming if compared with the physicochemical methods and the analytical protocol must be strictly followed due to the variability in microbiological techniques.
3. Biological methods use laboratory animals. A control group is fed with a diet lacking in the studied vitamin and a second group is fed with the food in which the vitamin is being evaluated, or an extract of the food. Vitamin contents are measured by monitoring a physiological effect related to the vitamin intake. Ethical problems related to animal experimentation and the variability of these studies limit the use of biological methods to cases where no satisfactory methods are available.

The generalities of each vitamin with regard to structure, chemistry, stability, sources, bioavailability, metabolic functions, deficiency, and requirements are very well detailed in nutrition articles and treatises. For this chapter, we have summarized the structure, importance, stability, and sources according to the Merck Index (1989), Basu and Dickerson (1996), Thurnham et al. (2000), and Cuellar-Rodríguez (2000).

8.3 THIAMIN

8.3.1 GENERALITIES

Thiamin, also called aneurin and vitamin B_1, is present in most plant and animal tissues. A pyrimidine ring and a thiazole moiety linked by a methylene bridge are

TABLE 8.1
Representative HPLC Methods for Water-Soluble Vitamin Analysis

Vitamin	Sample Preparation	Stationary Phase	Mobile Phase	Detection	Matrix	Reference
Vitamin B_1	Acid hydrolysis with 0.2 *N* H_2SO_4, autoclavation, enzymatic hydrolysis, filtration, oxidation with potassium ferricyanide, extraction with isobutyl alcohol	Lichrosorb RP-8	Methanol/acetonitrile/ isobutyl alcohol (80:10:10)	FL 370:425 nm	Meat and vegetables	Bognar (1981)
Vitamin B_1	Acid hydrolysis with 0.1 *M* H_2SO_4, enzymatic hydrolysis, centrifugation, oxidation	Nucleosil NH_2, 5 μm	25% K_2HPO_4 buffer pH 4.4 in acetonitrile	FL 370:425 nm	Dietetic foods	Bötticher & Bötticher (1986)
Vitamin B_1	Acid hydrolysis with 0.1 and 6 *M* HCl, autoclavation, enzymatic hydrolysis, Amberlite GC-50 and Sep-pack C_{18} clean-up	μ-Bondapak C_{18}, 10 μm	Several mixtures of metanol, acetic acid, hexane sulfonate and heptane sulfonate	UV 254 nm	Foodstuffs	Vidal-Valverde & Reche (1990a)
Vitamin B_2	Acid hydrolysis with 0.1 *M* HCl, autoclavation, enzymatic hydrolysis, protein precipitation, Sep-pack C_{18} clean-up	Nucleosil C_{18}, 10 μm	0.01 *M* K_3PO_4 buffer pH 7.0/acetonitrile (89.5:10.5)	UV 268 nm	Baby foods	Barná (1991)
Vitamin B_2	Acid hydrolysis with 0.1 *M* HCl, autoclavation, enzymatic hydrolysis protein precipitation, filtration	Spherisorb ODS-2 5 μm	5 m*M* heptane sulfonic acid, pH 2.7/acetonitrile (3:1)	FL 227:520 nm	Cooked sausages	Valls et al. (1998)
Vitamins B_1 and B_2	Acid hydrolysis with 0.1 *N* H_2SO_4 10 min in boiling water, enzymatic hydrolysis, centrifugation, oxidation	μ-Bondapak C_{18}	Methanol/water (60:40)	FL 360:415 nm	Cereal products	Mauro & Wetzel (1984)

Vitamin B_3-nicotinic acid	Acid hydrolysis with 0.1 and 6 *N* HCl, autoclavation, enzymatic hydrolysis, dilution, filtration and Douvex 1-X8 column purification	μ-Bondapak C_{18}, 10 μm	Methanol/0.01 *M* sodium acetate buffer pH 4.66 with 5 m*M* tetrabutylammonium	UV 254	Legumes and meats	Vidal-Valverde and Reche (1991)
Vitamin B_3-nicotinic acid and nicotinamide	Water extraction at 100°C, cooling and dilution	Partisil SCX, 10 μm	0.05 *M* K_3PO_4 buffer, pH 3.0	UV 260 nm	Meat	Hamano et al. (1988)
Vitamin B_3-nicotinic acid and nicotinamide	Water extraction, centrifugation, deproteinization with zinc hydroxide, filtration	Spherisorb ODS-2	5 m*M* heptane sulfonic acid, pH 3.3/acetonitrile (75:25)	UV 261 nm	Cooked sausages	Valls et al. (1999)
Vitamin B_6	Extraction with HPO_3, centrifugation, filtration	BioSil ODS-5S C_{18}	0.066 *M* H_3PO_4 buffer, pH 3.0/acetonitrile (90:10)	FL 290:395 nm	Meat	Ang et al. (1988)
Vitamin B_6	Extraction with 5% TCA, dilution, centrifugation, enzymatic hydrolysis, protein precipitation, centrifugation, filtration	Hypersil ODS	Methanol/0.1 *M* KH_2PO_4 buffer, pH 2.15 containing octanesulfonic acid (3:97)	FL 333:375 nm	Foods	Schoonhoven et al. (1994)
Biotin	37°C incubation of samples with 1% reduced glutatione solution, 1% EDTA solution, citrate buffer pH 5.7 and papain solution (in samples with high starch content, Takadiastase), filtration, postcolumn derivatization with avidin FITC	Lichrospher 100 RP-18	0.1 *M* H_3PO_4 buffer, pH 6.0/methanol (81:19)	FL 490:520 nm	Foods	Alelí et al. (1999)

(*Continued*)

TABLE 8.1 (Continued)
Representative HPLC Methods for Water-Soluble Vitamin Analysis

Vitamin	Sample Preparation	Stationary Phase	Mobile Phase	Detection	Matrix	Reference
Folates	Extraction with 0.1 *M* acetate buffer containing 1% of ascorbic acid, pH 4.5, nitrogen flush, strong anion-exchange column cleanup	Merck RP-18	8% Acetonitrile in acetic acid pH 2.3	FL 310:352 nm	Milk and blood samples	Wigertz and Jägerstad (1995)
Folates	Extraction with 0.15 *M* phosphate buffer pH 6.0 containing 2% sodium ascorbate and 0.1% mercaptoethanol, deconjugation with hog kidney deconjugase at pH 4.9, strong anion-exchange column cleanup	Hypersil ODS and Spherisorb ODS	Acetonitrile/H_3PO_4 buffer, pH 2.2. gradient 9% acetonitrile to 24% acetonitrile within 8 min lag phase 4 min	FL 290:356 nm FL 360:460 nm UV 290 nm	Fish, meat, eggs, and dairy products	Vahteristo et al. (1997)
Vitamin B_{12}	Extraction with HCl; preconcentration on an octadecylsilica cartridge	Octylsilica	Acetonitrile/aqueous ammonium phosphate solution, pH 3.0 (5:95 to 30:70)	Radioassay	Dairy products	Fie et al. (1994)
Vitamin B_{12}	HLC-disk 25, polyvinyldifluoride filtration for removing oily particles	Inertsil ODS	50 m*M* KH_2PO_4, pH 2.1/acetonitrile (90:10)	VIS 550 nm	Foods and biological samples	Iwase and Ono (1997)

Vitamin C	Ascorbic acid oxidation, 2,4-dinitrophenylhydrazine (DNFH) derivatization	Nucleosil Si, 3 μm	*n*-Hexane/ethyl acetate/acetic acid/ 2-propanol (40:30:10:20)	UV 495 nm	Foods	Kodaka et al. (1985)
Vitamin C	Extraction with 250 m*M* $HClO_4$ and 5% trichloroacetic acid, centrifugation	Partisil 10 SAX	60 m*M* sodium acetate buffer, pH 4.6	Electrochemical detection 0.75 V vs. Ag/AgCl	Animal tissues	Carr et al. (1983)
Vitamin C	Water dilution	μ-Bondapak-NH_2	Methanol/0.25% KH_2PO_4 buffer pH, 3.5 (50:50)	UV 244 nm	Beverages	Dennison et al. (1981)
Vitamin C	Extraction with 0.1 *M* citric acid, EDTA 5 m*M* and *n*-hexane, centrifugation, filtration	Ultrasphere ODS C_{18}	0.1 *M* NaH_2PO_4 buffer pH 5.0 containing EDTA 5 m*M* and 5 m*M* tetrabutylamonnium (50:50)	FL 350:430 nm	Foods	Vanderslice and Higgs (1988)
Vitamin C	Extraction with 80 g/l HPO_3 and ethanol (35:65), centrifugation, 12 g/l NaSH solution addition and filtration	Cosmosil 5 C_{18}	2 g/l PO_3H acid	UV 243 nm	Juices	Sawamura et al. (1990)

the constituents of the thiamin molecule. Thiamin diphosphate (TDP) and triphosphate (TTP) appear to be present in neural membranes. Thiamin is involved in the energy metabolism of proteins, lipids, and carbohydrates. This vitamin is also believed to have a role in nerve conduction and in some functions of the brain. Mostly in the liver and brain, thiamin is converted to its pyrophosphate form (TPP) to become biologically active as a coenzyme.

Thiamin is stable in slightly acidic water up to the boiling point, but because of its water solubility, it can be leached out of food by boiling. This vitamin is unstable in alkaline solution. Thiamin is destroyed by ultraviolet (UV) irradiation. The loss of thiamin can be also enhanced by using rapidly boiling water in cooking vegetables, as the amount of oxygen in contact with the food increases. Thiaminases (thiamin-degrading enzymes) present in uncooked freshwater fish, shellfish, and some bacteria destroy thiamin by displacing the pyrimidine methylene group with a nitrogenous base or SH-compound. Furthermore, there also exist heat-stable anti-thiamin factors that generally bind to thiamin, rendering the vitamin biologically unavailable. These substances oxidize the thiazole ring to yield thiamin disulfide (which cannot be absorbed); among them, *o*- and ρ-hydroxy polyphenols of tea, betel nuts, many vegetables, and some animal tissue have been found. The primary dietary sources of thiamin are unrefined cereal grains or starchy roots and tubers; meat and meat products, especially pork, are also good thiamin sources. However, much of the dietary thiamin in developed countries appears in fortified cereals and breads, as well as in many functional foods.

8.3.2 Analytical Methods

8.3.2.1 Physicochemical Determination

In these methods, special care should be taken to protect samples and standards from light and heat at alkaline pH. The proposed methods are based on fluorimetry and HPLC.

- **Fluorimetric determination**: After extraction and enzymatic hydrolysis of the phosphate esters of thiamin, the vitamin is cleaned up by ion-exchange chromatography and oxidized to thiochrome by treating with potassium ferricyanide [$K_3Fe(CN)_6$] or other oxidant reagents such as $KMnO_4$, MnO_2, CNBr $HgCl_2$, and H_2O_2. The thiochrome is finally extracted into an organic solvent such as isobutyl alcohol and analyzed by fluorimetry employing 360 to 365 nm as the excitation wavelength and 460 to 480 nm as the emission wavelength.
- **HPLC determination**: The most common techniques nowadays. In most food sample matrices, acid and/or enzyme treatment is needed to release thiamin from phosphate and proteins. Different acid conditions and enzymes have been proposed, including hydrochloric, trichloracetic, and sulfuric acids; single enzymes such as papain and pepsin; or enzyme mixtures such as takadiastase or claradiastase. The last enzyme hydrolysis coupled to hydrochloric acid has proved to be the most efficient treatment.

Although some methods determine the B_1 vitamin directly by HPLC with UV detection, the low vitamin contents and the high quantity of interfering compounds in foodstuffs make its determination better by conversion of thiamin to fluorescent thiochrome and carrying out the final analysis by HPLC with fluorescence detection. This technique also increases the sensitivity and reproducibility. The thiochrome determination has two important problems: (1) the instability of the thiochrome itself, and (2) the chromatographic problems related to the excess of oxidizing reagent. The former problem can be solved by the addition of orthophosphoric acid which minimizes the formation of thiamin disulfide and stabilizes the thiochrome. The last problem can be solved by solid-phase chromatography cleanup using C_{18} cartridges or by extracting the thiochrome into isobutyl alcohol as soon as it is formed. This can be achieved by adding the isobutyl alcohol before the ferricyanide reagent.

For the final analysis, reversed-phase HPLC offers the best results. Columns most commonly employed are C_{18} and C_8, but others such as amino have also been employed. Mobile phases, including organic solvents, ion pair, and organic-aqueous buffer mixtures, have been used.

8.3.2.2 Microbiological Studies

Microbiological studies take from 4 to 72 h and detect thiamin amounts between 5 and 50 ng. The species most commonly employed for thiamin determination are *Phycomyces blakeslekanus*, *Kloeckera brevis* (ATCC 9774), *Ochromonas danica*, *Neurospora crassa*, and some Lactobacilli such as *Lactobacillus fermentum* and *L. viridescens* (ATCC 1270 C).

8.3.2.3 Biological Methods

Biological methods are hardly used, as they are time consuming (6 to 8 weeks) and poorly reproducible. These studies are based on the effect of thiamin on the growth and evolution of diseases related to vitamin B_1-lacking effects. Experimental animals include rats, pigeons, and chickens.

8.4 RIBOFLAVIN

8.4.1 Generalities

Riboflavin, also called vitamin B_2, is structurally composed of an isoalloxazine ring with a ribityl side chain at the nitrogen at position 10. This vitamin functions metabolically as the essential component of two flavin coenzymes, flavin adenine dinucleotide (FAD) and flavin mononucleotide (FMN), complexed with proteins, which act as intermediaries in transfers of electrons in biological oxidation-reduction reactions. Both FAD and FMN function as coenzymes for flavoproteins of flavoenzymes. Flavoproteins are essential for the metabolism of carbohydrates, amino acids, and lipids and for pyridoxine and folate conversion to their respective coenzyme forms.

Even though riboflavin is described as water soluble, its solubility in water is actually very low but can be increased by the addition of urea and by formation of a complex with boron. It is insoluble in lipid solvents. Riboflavin is stable at increases of temperature and in acid solutions. UV light causes irreversible decomposition of this vitamin. One of the most important problems of riboflavin analysis is the lability of flavins to light and alkaline conditions, so all analytical procedures have to be performed avoiding both factors. Acidification and enzyme hydrolytic treatment guarantee the complete hydrolysis of FAD and FMN to riboflavin. Riboflavin occurs in good amounts in dairy products. Animal protein sources such as meats, liver, kidney, and eggs are good sources of riboflavin, as well as some green leafy vegetables. Very good sources include yeast and meat extract.

8.4.2 Analytical Methods

8.4.2.1 Physicochemical Determination

The extraction and cleanup steps for riboflavin are similar to those proposed for thiamin. Special care should be taken to protect samples and standards from UV light and alkaline conditions. Strict control in the oxidation process is also necessary to avoid riboflavin decomposition. Although riboflavin is classified as a water-soluble vitamin, some problems can arise when dissolving the vitamin in water, and this must be taken into consideration when preparing the standard solutions.

As in thiamin determination, proposed methods for riboflavin are based on fluorimetry and HPLC.

- **Fluorimetric determination**: After extraction and cleanup riboflavin fluorescence is measured employing 400 to 420 nm as the excitation wavelength and 550 to 570 nm as the emission wavelength. Despite its good sensitivity, this method has the disadvantage of interference from other fluorescent compounds.
- **HPLC determination**: Acid and/or enzyme treatments similar to those applied in thiamin analysis are normally required for the transformation of FMN and FAD to riboflavin. A few methods avoid this hydrolytic treatment, proposing direct dilution of the sample in water or acetate buffer followed by cleanup on a C_{18} cartridge.

Final analysis is carried out by reversed-phase liquid chromatography on C_{18}, C_8, or C_{22} columns, as well as by ion pair reversed-phase chromatography. Even though riboflavin, because of its structure, does not occur in ion form, the latter technique has proved to be very effective in separating riboflavin from disturbing components. Detectors used in both HPLC systems are mainly fluorescence or UV ones. Fluorescence excitation wavelengths range from 425 to 470 nm and emission wavelengths range from 510 to 570 nm. For UV detection 254, 268, and 270 nm are the selected wavelengths.

8.4.2.2 Microbiological Studies

The first assays used *Leuconostoc mesenteroides* (ATCC 9135) and *Tetrahymena pyriformis*, but *Lactobacillus casei* (ATCC 7469) is the microorganism most commonly employed nowadays in these studies.

8.4.2.3 Biological Methods

These studies are based on the biological responses of rats and chickens to riboflavin. They have been displaced by the previously described methods.

8.5 NIACIN

8.5.1 Generalities

Niacin includes two vitamers, nicotinic acid and nicotinamide. Nicotinic acid is pyridine 3-carboxylic acid. It is slightly soluble in water and alcohol and easily soluble in alkalis, but is insoluble in both acetone and diethyl-ether. Nicotinamide is the active form, which functions as a constituent of two coenzymes: nicotinamide adenine dinucleotide (NAD) and nicotinamide adenine dinucleotide phosphate (NADP). In animal tissues, the reduced stages of these coenzymes have been found as the principal forms of niacin. Nicotinamide is very soluble in water, easily soluble in alcohol, soluble in glycerol, and slightly soluble in ether.

Niacin is categorized as a vitamin because its precursor, tryptophan, is an essential amino acid, so the human synthesis of niacin is dependent upon diets. Preformed niacin is widely distributed in plant and animal foods. The typical preformed niacin sources in diets are meat and meat products, cereals, dairy products, beverages, and eggs. However, cereals with esterified niacin in complexes have this vitamin unavailable for absorption, but its bioavailability can be increased by treatment with alkali to hydrolyze the esters. Coffee can be a source of niacin, as nicotinic acid is liberated in coffee by roasting.

8.5.2 Analytical Methods

8.5.2.1 Physicochemical Determination

- **Spectrophotometric determination**: These methods require several extraction and cleanup steps for removing interfering compounds.

  Final determination is based on the reaction between niacin and cyanogen bromide, which produces pyridinium compounds that finally react with aromatic amines, giving colored compounds. This method is hardly used nowadays because it does not distinguish between nicotinamide and nicotinic acid and it employs cyanogen bromide, which is a noxious and unstable reagent.
- **HPLC determination**: This technique has been used extensively for determining only one vitamer or both vitamers together. Simultaneous determination is difficult due to the differences in basicity and polarity of both vitamers and the interference problems. After water-phase extraction and sometimes cleanup steps as deproteinization, nicotinic acid and/or nicotinamide are analyzed by ion pair reversed-phase HPLC with UV detection. Most methods use different ion pair reagents for each vitamer, but a single ion pair reagent for nicotinamide and nicotinic acid can be also employed.

8.5.2.2 Microbiological Studies

Niacin microbiological determination uses some Lactobacilli such as *L. arabinosus* (ATCC 2112), *L. plantarum* (ATCC 8014), and *L. casei* (ATCC 7469), as well as *Leuconostoc mesenteroides* (ATCC 9135). *Leuconostoc mesenteroides* is not useful for nicotinamide determination. In general, growth is measured by turbidity, but acidimetric measurements can also be used. The latter determination requires longer incubation periods.

8.6 PYRIDOXINE

8.6.1 Generalities

Vitamin B_6 is the generic term for 3-hydroxy-2-methyl-pyridine derivatives with the biological activity of pyridoxine. It occurs in three vitamers: pyridoxine (pyridoxol), pyridoxal, and pyridoxamine, in which alcohol, aldehyde, and amine groups, respectively, are located at the 4-position of the pyridine ring. All three vitamers are comparably active. Pyridoxal-5′-phosphate and sometimes pyridoxamine-5′-phosphate are the coenzyme forms of vitamin B_6. Pyridoxine is stable in acid solutions. It is rapidly destroyed by light. Pyridoxine is more stable than either pyridoxal or pyridoxamine. Vitamin B_6 in plants which contain essentially pyridoxine is lost less than the vitamin in animal foods which contain mostly pyridoxal and pyridoxamine. Many vitamin B_6-dependent enzymes are known, and almost all of them act on amino acids in reactions which include transamination, decarboxylation, dehydration, desulfydration, racemization, cleavage, and synthesis, among other different processes.

No food is a particularly rich source of pyridoxine, as this vitamin is widely distributed in foods. Meats and cereals are the richest sources. In contrast, fruits are poor sources. In some vegetables, the form of vitamin B_6 is present as unavailable glucosides. In many foods the vitamin is bound to protein. If foods are stored for a long time (1 year), losses of vitamin B_6 may be important.

8.6.2 Analytical Methods

8.6.2.1 Physicochemical Determination

- **Fluorimetric determination**: Even though B_6 vitamers exhibit native fluorescence, a previous purification and separation of vitamers by column chromatography is required for the final fluorimetric determination. For this reason HPLC methods are the most commonly used.
- **HPLC determination**: In many methods all six vitamers are extracted mainly in acidic conditions (trichloroacetic acid, metaphosphoric acid, or sulfosalicylic acid). Other methods hydrolyze the phosphorylated forms of the vitamin by H_2SO_4 in conjunction with autoclaving or by enzymatic treatment, thus determining only the nonphosphorylated vitamers.

Several HPLC systems have been used for the final analysis, including ion-exchange HPLC, reversed-phase HPLC, and ion pair reversed-phase HPLC. Detectors used in all HPLC systems are mainly fluorescence ones, although UV detectors have been sometimes used.

8.6.2.2 Microbiological Studies

The main problem related to vitamin B_6 microbiological determination is the different growth response to different vitamers. This problem can be solved by separating the vitamers by cation-exchange chromatography on a Dowex AG 50 W-X8 column. *Saccharomyces uvarum* (ATCC 9080) and *Kloeckera brevis* (ATCC 9774) are the test microorganisms most frequently used. When *S. uvarum* is employed, an additional problem can be the inhibition of growth due to sample extracts with high salt concentrations.

8.7 BIOTIN

8.7.1 Generalities

Biotin is a sulfur-containing bicyclic compound in which tetrahydrotiophene and imidazolidone rings are fused and there is a valeric acid as the side chain. From the eight possible stereoisomers, only the dextrorotatory (D-(+)-biotin) is ordinarily found in nature and is the only one biologically active. This compound is sensitive to heat. Biotin is widely distributed in nature in the free form or covalently bound to proteins or peptides. Good sources of biotin are organ meats, egg yolks, and milk.

8.7.2 Analytical Methods

8.7.2.1 Physicochemical Determination

The proposed methods are based on spectrometric, chromatographic, and protein-binding techniques. As biotin occurs in food both in free form and covalently bound to proteins, in many methods it is necessary to break these bonds by acid or enzymatic hydrolysis. The former is more frequently used than the latter, since acid treatment converts D-biocytin into D-biotin and allows a total determination of biotin. Enzymatic treatment with papain leaves D-biocytin unchanged and only D-biotin is determined. Despite this, some methods propose enzymatic hydrolysis rather than acid hydrolysis because it does not induce any degradation of biotin.

- **Spectrophotometric determination**: The first spectrophotometric methods proposed for biotin quantification lacked adequate sensitivity and are hardly applied nowadays. The most interesting spectrophotometric method for biotin analysis is based on the reaction of biotin with periodate, resulting in iodate as the reaction product. This iodate reacts with iodide, forming triiodide which is finally measured at 350 to 352 nm. This method has a sensitivity down the microgram level.

- **Chromatographic determination**: Many chromatographic techniques have been proposed for biotin determination, including thin-layer chromatography, gas chromatography (after derivatization of biotin to sylil ester), and HPLC, which is the most common chromatographic technique in biotin analysis. Most proposed methods are reversed-phase HPLC on C_{18} columns, but anion-exchange HPLC and normal-phase HPLC have been already used. Direct detection of the vitamin by UV is not sensitive enough due to the lower content of biotin in foodstuffs, thus a pre- or postcolumn derivatization into fluorescent derivatives is necessary to obtain a satisfactory detection limit (10^{-8} or 10^{-9} g ml^{-1}).
- **Protein-binding assays**: These techniques allow a rapid and sensitive determination of biotin and, therefore, are commonly employed for routine analysis of a great number of samples. In all these methods the binding protein is the avidin or the streptoavidin, using radioactive isotopes, enzymes, or chemiluminescent substances as labels. The disadvantage of these assays is their lack of discrimination between biotin and its metabolites. Combining protein-binding assays and HPLC can solve this problem.

8.7.2.2 Microbiological Studies

These methods were the first developed for biotin analysis and are still accepted as official methods because of their high sensitivity (nanogram level). The species most commonly employed for biotin determination are some Lactobacilli such as *L. arabinosus* (ATCC 2112), *L. plantarum* (ATCC 8014), and *L. casei* (ATCC 7469); *Ochromonas danica*, *Neurospora crassa*, and *Saccharomyces cerevisae* (ATCC 4228) are also employed. *Kloeckera brevis* (ATCC 9774) is useful for *a* radiometric-microbiological determination based on the measurement of $^{14}CO_2$ formed by the test microorganism from ^{14}C-labeled L-methionine.

8.7.2.3 Biological Methods

These studies are based on the effect of biotin on the growth of rats or chickens, as well as on the determination of the activity of certain biotin-dependent enzymes such as pyruvate carboxylase.

8.8 FOLIC ACID

8.8.1 Generalities

The folic acid molecule consists of a pteroic acid nucleus (a pterine molecule coupled to *p*-aminobenzoic acid) conjugated with one to seven molecules of glutamic acid. Because of this structural component, folic acid is also called pteroylglutamic acid. The pyrazine ring can also present different reduced forms such as dihydropteroylglutamic acid and tetrahydropteroylglutamic acid. Different single carbon substituent groups such as formyl, methyl, and methylene can be also attached to N^5 and N^{10} positions of the pyrazine ring. This results in the existence of over 150 different forms of this vitamin in nature which may differ in their bioavailability. All biologically active forms of folic acid are known as folacin(s) or folate(s).

Folic acid is synthetized by plant and certain microorganisms and passed on to higher animals through the food chain. Good sources of folate include liver, dark-green leafy vegetables, legumes, wheat germ, egg yolks, and yeast. Although the monoglutamate form is present in foods, most of the folates are in the form of polyglutamate which is important for their analysis, since in both physicochemical and microbiological analyses enzymatic deconjugation treatment is necessary to the hydrolyze polyglutamate to mono- or diglutamate forms. Folates are labile to changes in pH, the presence of oxidizing agents, and exposure to heat and light.

8.8.2 Analytical Methods

8.8.2.1 Physicochemical Determination

- **HPLC determination**: The two main problems in the the development of successful HPLC methods for folates determination are the presence of many different forms of this vitamin and their low levels in foodstuffs. Additional problems are their instability to light, heat, and oxidation. For preventing oxidation losses of folate, certain antioxidants such as ascorbic acid or nitrogen flushing are employed during the extraction procedure.

  After extraction and enzymatic deconjugation, sample extracts are cleaned up by strong anion-exchange chromatography. Final analysis is performed by reversed-phase HPLC with fluorescence or UV detection.
- **Protein-binding assays**: Several enzyme protein-binding assays and radio-protein-binding assays have also been developed and commercialized as kits for rapid evaluation of this vitamin in blood, plasma, and different foodstuffs. Despite their rapidity, these methods lack adequate discrimination between different forms of folates.

8.8.2.2 Microbiological Studies

Due to the difficulties in the estimation of total folates in food by physicochemical methods, official methods are based on microbiological assays. As in the previously described methods, special care must be taken to protect labile folates from oxidation and photochemical degradation. For evaluation of free folic acid, *Enterococcus hirae* and *Streptococcus faecalis* (ATCC 7830) can be used, but they do not respond to other folate forms present in food. The most widely accepted procedure for total folates determination uses *Lactobacillus casei* ssp. *rhamnosus* (ATCC 7469) as the test microorganism, since this bacterium responds to most native folates present in food. The main problem involved with *Lactobacillus* is the presence of food-bound folates that do not promote the growth of these bacteria. It is known that the number of glutamyl residues linked to the pteroyl group is negatively related to the *L. casei* growing response. For this reason an enzymatic treatment of foods with folate conjugase is necessary to hydrolyze folylpolyglutamates to mono- or diglutamate forms. Recent studies have found that treatments with protease and α-amylase improve the folate determination in high protein and high carbohydrate foods.

8.9 VITAMIN B_{12}

8.9.1 Generalities

Vitamin B_{12} is the name given to a group of active vitamers called cobalamins which contain cobalt. The structure of this vitamin consists of a macrocyclic ring (usually called "corrin") formed by four reduced pyrrole rings linked together. Two of the four pyrrole rings are joined directly rather than through a single methylidyne carbon. Below the corrin ring system there is a 5,6-dimethyl-benzimidazole riboside that is linked with one end to the central cobalt atom. The form called cyanocobalamin contains a cyanide group coordinately bound to the cobalt atom of the vitamin B_{12} molecule. The cyanide group may be removed and substituted by a hydroxy, aqua, nitro, methyl, or 5′-adenosyl group called "hydroxycobalamin," "aquacobalamin," "nitrocobalamin," "methylcobalamin," and "5′-adenosylcobalamin," respectiviely.

Crystalline vitamin B_{12} is stable to heating at 100°C. Aqueous solutions of vitamin B_{12} at pH 4 to 7 can be autoclaved with very little loss. Cyanocobalamin is the most stable form. Nevertheless, the cobalamins are very sensitive in the presence of reducing agents, as well as in light (especially adenosyl- and methylcobalamins) and alkaline conditions. Photolysis results in two compounds stable only under anaerobic conditions.

In plant foods, vitamin B_{12} cannot be detected, unless it is contaminated with microorganisms (which may occur in legumes and root vegetables). For humans, the most reliable sources of this vitamin are animal foods such as meats and seafood. Algae can also contain vitamin B_{12}, but the bioavailability has been questioned.

8.9.2 Analytical Methods

8.9.2.1 Physicochemical Determination

- **Spectrophotometric determination**: Even though this technique has been used for vitamin B_{12} determination, it is not a suitable method for a complex sample matrix.
- **HPLC determination**: This technique has been widely used, since, in general, it is less sensitive than the microbiological method. The various isomers of vitamin B_{12} (cyanocobalamins) can be analyzed by reversed-phase HPLC. Pre- and postcolumn labeling, column switching, and solid-phase extraction (SPE) may be used (Iwase and Ono, 1997).
- **Protein-binding assays:** The **radioisotope dilution assay** (RIDA) has been employed for determining vitamin B_{12} in foods, using radiolabeled B_{12} and hog intrinsic factor (IF), the most specific B_{12}-binding protein. Several kits for the RIDA method are commercially available. A very good correlation coefficient has been found between the microbiological method and the RIDA method, but a radioisotope is used and radioisotope facilities and apparatuses are necessary for the RIDA method. The **direct**

competitive ELISA enzyme immunoassay ("Ridascreen®" test, Digen Ltd., U.K., 1998), developed in a microplate with a very specific vitamin B_{12} antibody, can be used for vitamin B_{12} quantification both in foods and in pharmaceuticals. This technique is very sensitive. In foods such as milk, 250 ng/l can be measured. The fat content of oily samples needs to be previously removed. Depending on the foodstuff analyzed, previous enzymatic treatment of the sample may be required. Vitamin B_{12} can also be analyzed in foods with a fully automated chemiluminescence B_{12} analyzer, with the acridinium ester-labeled B_{12} derivative and IF. Vitamin B_{12} is extracted from foods applying the method reported in the microbiological techniques paragraph, but, in the case of the chemiluminescence method, vitamin B_{12} is extracted with distilled water from pieces of filter paper after 1 day at 4°C in the dark. Standards are also used. The compounds extracted from the filter papers are determined spectrophotometrically by measuring absorbance at 361 nm. The vitamin B_{12} extracts are directly applied to the chemiluminescence B_{12} analyzer. If compared with the microbiological method, the chemiluminescence method appears to be 5 to 10 times less sensitive, but simpler, quicker, highly selective, and reproducible (Watanabe et al., 1998).

8.9.2.2 Microbiological Studies

Microbiological studies can be used for determining vitamin B_{12} in addition to radioisotopic techniques (Rougereau et al., 1997). The following strains are most often used: *Lactobacillus lactis* (ATCC 8000), *L. leischmannii* (ATCC 7830) (Schneider, 1987), and *Escherichia coli*, according to the classical methods of determination. The extraction of vitamin B_{12} can be carried out by homogenizing the food samples in acetate buffer, followed by extraction of total vitamin B_{12} from the homogenates by the method of boiling with KCN at acidic pH (Frenkel et al., 1980). After centrifuging, the supernatant is used for the assay. Then a paper chromatography of the extract is carried out. The filter paper is dried, cut into pieces, and used as samples for the microbiological assay. Samples usually need to be diluted for the microbiological method, and a B_{12} assay medium is required. When *L. leischmannii* (ATCC 7830) is employed, the turbidity (%T) of the *L. leischmannii* test culture is measured at 600 nm. Microbiological methods are very sensitive, but they are usually tedious and time consuming, for it takes long to prepare and incubate the samples and the tissue must be cultured and the strain preserved. A well-trained full-time technician is also necessary. In addition, when *L. leichmanii* is used, vitamin B_{12} analogs inactive for humans are also determined (Watanabe et al., 1998).

8.10 VITAMIN C

8.10.1 Generalities

L-Ascorbic acid (AA) and L-dehydroascorbic acid (DHAA) are the two biologically vitamin C active compounds; AA is a strong reducing agent and DHAA is the oxidized derivative of the AA. Interconversions of these compounds are carried out by glutathione dehydrogenase and ascorbate oxidase, among other enzymes. Vitamin C is a strong reducing agent and thus an important antioxidant. Vitamin C biochemical functions are based on its ability to donate one or two electrons.

Vitamin C is readily lost in cooking because of its water solubility. The availability of this vitamin is higher whenever vegetables are eaten raw, so fresh fruit and salad are the most reliable sources for vitamin C in the diet. Vitamin C can be depleted when fresh food is stored for a long time. Vitamin C can be oxidized to dehydroascorbate and then irreversibly degraded by hydrolytic opening of the lactone ring, which occurs in the presence of oxygen. Alkali, heat, light, and copper enhance the rate of oxidation.

Plant origin foods are the almost exclusive sources of vitamin C. In most developed countries this vitamin is used as a dietary supplement and chemical preservative, so soft drinks, cereals, cakes, confectionery, and fish and meat products might become an important dietary source of vitamin C. West Indian cherries, blackcurrants, oranges, lemons, strawberries, and kiwi fruit, among others, are good vitamin C sources, as well as most green leafy vegetables and also potatoes (the new ones in particular), because of the large amount generally eaten.

8.10.2 Analytical Methods

Vitamin C analysis must be performed at low pH and, if necessary, in the presence of a chelating agent because this vitamin can be easily oxidized, especially at high pH conditions. The presence of Cu^{2+} and Fe^{3+} ions also contributes to this process. The biologically active DHAA can be reconverted to AA by reduction.

8.10.2.1 Physicochemical Determination

- Chemical methods: These are based on the power of L-AA as a reducing agent. These techniques are more rapid, more reproducible, and less specific than bioanalysis and microbilogical assays for the complex biological relationship among the compounds that show vitamin C activity, as well as the similarities between these molecules and other that are not active. Chemical methods can be divided into two groups. The first one includes the determination of the reducing form of L-AA. The second one measures the total "vitamin C," represented by L-AA, DHAA and 2,3-diketogulonic acid.

  At present, for AA extraction, most methods use metaphosphoric acid. Sometimes, samples are cleaned up by solid-phase chromatography on a Sep-pak C_{18} cartridge.

  The first chemical methods were based on the oxidizing-reducing properties of L-AA or on its property of developing colored hydrazides by reaction with anyline diazotized derivatives. The extract is a reducing agent and can be measured by treatment with an oxidizing agent such as 2,6-dichlorophenolindophenol, proposed in 1927 by Tillmans. Other suitable agents are iodine and methylene-blue. About 50 years ago, "total" vitamin C was determined by reaction of 2,3-diketogulonic acid with 2,4-dinitrophenylhydrazine; after adding sulfuric acid, the oxazone formed gave a red color that could be read at 540 nm (Roe et al., 1948).

- Titration: The extract is dissolved with a mixture of metaphosphoric acid, glacial acetic acid, and water. After shaking and adding a starch solution, a titration with iodine solution is carried out.
- 3,6-Dichlorophenolindo titrimetric method (AOAC Int. 967.21, 45.1.14, 1995): L-AA is oxidized to L-DHAA by 2,6-dichloroindophenol. In the presence of significant amounts of iron or copper in the matrix to be analyzed, a chelating agent (such as ethylenediaminetetraacetic acid) should be included with the extraction. After the vitamin C extraction steps, sample extract is diluted. Three replicates of both standard and sample are titrated with 3,6-dichlorophenolindo solution. The red-color end point should last at least 10 sec to be valid. With colored samples, vitamin C needs to be determined by observing the change of transmittance at 545 nm. L-DHAA can be determined by first converting it to L-AA with a suitable reducing agent.
- Electrochemical methods: These methods for determining L-AA are simpler. Polarography developed by Lindquist and Ferroha in 1975 is very specific and requires a low amount of sample and a minimum preparation.
- Enzymatic determination: AA is transformed into DHAA using ascorbate oxidase.
- Colorimetric determination: AA is oxidized with 2,6-dichorophenolindophenol into DHAA and then it is converted into oxazone with 2,4-dinitrophenylhydrazine. After recoloring with sulfuric acid, it is finally measured at 520 nm.
- Microfluorometric method (AOAC Int. 967.22, 45.1.15, 1995): A fluorescent quinoxaline compound is formed upon reaction of DHAA with *o*-phenylenediamine and measured at Ex = 356 nm, Em = 440 nm. With this method, it is possible to measure both AA and DHAA, for the former is previously oxidized to DHAA. To compensate for the presence of interfering substances, blanks need to be run using boric acid prior to the addition of the diamine solution. In blanks, development of fluorescent quinoxaline is prevented by the formation of H_3BO_3-DHAA complex prior to additon of the phenylenediamine solution.
- **HPLC determination**: Nowadays, HPLC separations are the most used because of the high sensitivity and selectivity of this technique (Rougereau et al., 1997; Cserháti and Forgács, 1999). For determining "total vitamin C," both AA and DHAA are determined. However, in some procedures, AA is separated. Vitamin C extraction is carried out using several chemicals (HPO_3, TCA, EDTA, EDTA and citric acid, water, phosphate buffer), generally followed by dilution and injection of the samples.

Special preparation and extraction techniques have been described such as DHAA reduction, HPO_3/ethanol utilization, enzymatic oxidation and *o*-phenilendiamine derivatization, oxidation of AA followed by 2,4-dinitrophenylhydrazine (DNFH) derivatization, and $HClO_4$, among many others.

Normal-phase adsorption HPLC has been employed, as well as ion-exchange chromatography, bounded-NH_2 partition chromatography, reversed-phase chromatography, and ion pair reversed-phase chromatography. C_{18} columns have been the most employed. RP-8, NH_2, and silica columns have been also employed. UV detection has been the most used technique. However, electrochemical detection, as well as fluorescence after derivatization, can also be used.

When an NH_2 column is used, the retention time is strongly influenced by temperature (Gilpin and Sisco, 1980). A 35°C column temperature may be good for AA analysis, because a good retention time is obtained without affecting the stability of this vitamin. The problem of using reversed-phase HPLC on an NH_2 column is that after about 100 injections the column is worn out, probably due to the Schiffts' base formation from NH_2 groups. However, both C_8 and C_{18} columns do not show this problem.

8.10.2.2 Microbiological Studies

For qualitative estimations, they are quicker but less used than chemical methods. *Myrothecium verrucacia* can be used (Meyer et al., 1965).

8.10.2.3 Biological Methods

In 1922 Sherman et al., described a bioanalysis assay based on determining the minimum amount of vitamin C necessary for protecting the Guinea pigs from scurvy. Bioanalyses are expensive, less precise, and time consuming. They have the advantage of measuring all the compounds with vitamin C activity, and excluding those which do not show that activity. Nowadays, bioanalysis based on dental structure histological changes is the more specific method.

REFERENCES

General References

Association of Vitamin Chemists, Inc. *Methods of Vitamin Assay*, 3rd ed., Interscience Publishers, New York, 1966.

Basu, T.K. and Dickerson, J.W.T., *Vitamins in Human Health & Disease,* CAB International, U.K., 1996.

Christie, A.A. and Wiggins, R.A., Developments in vitamin analysis, in *Developments in Food Analysis Techniques*, Vol. I, King, R.D., Ed., Applied Science Publishers, London, 1978, pp. 1–42.

Combs, G., Fundamental aspects in nutrition & health, in *The Vitamins*, Academic Press, New York, 1992.

Cserháti, T. and Forgács, E., *Chromatography in Food Science & Technology,* Technomic Publishing Company, Lancaster, PA, 1999.

Cuellar-Rodríguez, S., Vitaminas, in Farmacología de las Enfermedades Endocrinas, Metabólicas e Inmunológicas, PNFC, Consejo General de Colegios Oficiales de Farmacéuticos de España, Ed., Acción Médica, Madrid, Barcelona, Spain, 2000.

Dolan, J.W., Improving an ion-pairing method, *LC-GC Int.*, October, 624–628, 1996.

Duggan, D.E., Bowman, R.L., Brodie, B.B., and Udenfriend S., A spectrophotofluorometric study of compounds of biological interest, *Arch. Biochem. Biophys.*, 68, 1–14, 1957.

Dyke, S.F., *The Chemistry of the Vitamins*, Interscience Publishers, New York, 1965, chap. 2.

Food and Agriculture Organization (FAO), Requirements of Vitamin A, Iron, Folate, & Vitamin B_{12}. Report of a Joint FAO/WHO Expert Consultation, FAO Food & Nutrition Series No. 23, Food & Agriculture Organization, Rome, 1988, p. 107.

Favell, D.J., Vitamin analysis. Microbiological methods, *J. Micronutr. Anal.*, 7, 237–246, 1990.

Friedrich, W., *Handbuch der Vitamine*, Urban & Schwarzenberg, Munich, 1987.

Garcia Almansa, A., Vitaminas (I and II), *Nutr. Clín.*, 3, 1983.

Gregory, J.F., Methods of vitamin assay for nutritional evaluation of food processing, *Food Technol.*, January, 75–80, 1983.

Hashmi, M., *Assay of Vitamins in Pharmaceutical Preparations*, John Wiley & Sons, New York, 1973.

Hedlund, B., Analysis of vitamins in food by HPLC, in *Recent Developments in Food Analysis*, Baltes W., Czedík-Eysenberg, P.B., and Pfannhauser, W., Eds., Verlag-Chemie, Weinheim, 1982, pp. 27–32.

Horwitt, M.K., Liebert, E., Kreisler, O., and Wittman, P., Investigations of human requirements for B-complex vitamins, Bulletin of the National Research Council No. 116, National Academy of Sciences, Washington, D.C., 1948, p. 106.

Hozova, B. and Sorman, L., *Combination of Irradiation & Thermal Processing Food Irradiation*, Elsevier Science Publishers, London, 1991, pp. 207–234.

Hozova, B. and Takacsova, M., The influence of combined storage procedures of foods on B vitamins content demonstrated at the example of heat sterilisation and irradiation, *Die Nahrung*, 38(4), 345–351, 1993.

Hytten, F.E., Nutrition, in *Clinical Physiology in Obstetrics*, Hytten, F. and Chamberlain, G., Eds., Blackwell, Oxford, 1980, pp. 163–192.

IUPAC, Commission on biochemical nomenclature, *Biochemistry*, 147(1), 1970.

Jansen, B.C.P., *The Vitamins*, 2nd ed., Vol. V, Sebrell, W.H. and Harris, R.S., Eds., Academic Press, New York, 1972, p. 99.

Leach, D., Stadalius, M., Berus, J., and Snyder, L., Reversed-phase HPLC of basic samples, *LC-GC Int.*, 1(5), 22–30, 1988.

Leenheer, A.P. and Lambert, W.E., *Modern Chromatographic Analysis of the Vitamins*, Marcel Dekker, New York, 1985.

Lehninger, A.L., *Biochemistry*, Worth Publishers, New York, 1983.

López, C. and Cos, A., Vitaminas, in *Manual de Alimentación y Nutrición para Educadores*, López, C. and Vázquez, C., Eds., Caja Madrid, Madrid, 1992.

Lyon, J., *Le Grand Livre des Vitamines*, Martínez, R., Ed., Barcelona, 1987.

Macrae, R., Nicolson, I.A., Richardson, D.P., and Scott, K.J., Chromatographic determination of vitamins in dairy products, RSC Proc. in Challenges in Dairy Analytical Techniques, Reading, MA, 1984, pp. 155–166.

Merck and Co., Inc., *Merck Index*, 11th ed., Merck & Co., Inc., Whitehouse Station, NJ, 1989.

Morawski, J., Analysis of dairy products by HPLC, *Food Technol.*, 38(4), 70–78, 1984.

National Research Council (NRC), *Recommended Dietary Allowances*, 10th ed., National Academy Press, Washington, D.C., 1989.

Ottaway, P., Vitamins as technical aids in food processing, *Food Technol. Int.*, 111–114, 1994.
Parris, N.A., Non-aqueous reversed-phase liquid chromatography. A neglected approach to the analysis of low polarity samples, *J. Chromatogr.*, 157, 161–170, 1978.
Passmore, R. and Eastwood, M.A., *Human Nutrition and Dietetics*, 8th ed., Davidson, S. and Passmore, R., Eds., Churchill Livingstone, New York, 1986.
Rougereau, A., Person, O., and Rougereau, G., Determination of vitamins, in *Analysis of Food Constituents,* Multon, J.-L., Ed. Wiley-VCH, New York, 1997.
Shah J., Riboflavin, in *Methods of Vitamin Assay,* 4th ed. Augustin, J., et al., Eds., John Wiley & Sons, New York, 1984.
Shortt, C., Micronutrients and health, I: vitamins, *Trends Food Sci. Technol.,* July, 17–19, 1990.
Snyder, L.R., Glajch, J.L., and Kirkland, J.J., *Practical HPLC Method Development*, John Wiley & Sons, New York, 1988.
Steiner, F., *Applications of Narrow-Bore Columns in HPLC*, Hewlett-Packard GmbH, Germany, 1992.
Steiner, I., Hruschka, G., Washüttl, J., and Kroyer, G., Selected B-vitamins in foodstuffs after conventional cooking and microwave treatment, *Ernährung/Nutrition,* 17(4), 221–225, 1993.
Strohecker, R. and Henning, H.M., *Vitamin Assay-Tested Methods,* Translated by D.D. Libman, Verlag-Chemie, Weinheim, 1966.
Thurnham, D.I., Bender, D.A., Scott, J., and Halsted, C.H., Water-soluble vitamins, in *Human Nutrition and Dietetics,* 10th ed., Garrow, J.S., James, W.P.T., and Ralph, A., Eds., Churchill Livingstone, London, 2000.
Toma, R. and Tabekhia, M., High performance liquid chromatographic analysis of B-vitamins in rice and rice products, *J. Food Sci.,* 44(1), 263–268, 1979.
Uherova, R., Fedorova, N., and Dubravicky, J., The study of vitamin interaction with the reactive constituents of smoking agent, *Die Nahrung,* 37(1), 85–87, 1993.
Uherova, R., Hozova, B., and Smirnov, V., The effect of microwave heating on retention of some B vitamins, *Food Chem.,* 46, 293–295, 1993.
Vandemark, F. and Schmidt, G., Determination of water soluble vitamins in pharmaceutical preparations using liquid chromatography, *J. Liq. Chromatogr.,* 4(7), 1157–1171, 1981.
Vivanco, F., Palacios, J.M., y Garcia Almansa, A., *Alimentación y Nutrición*, Dirección General de Sanidad, Ed., Sección de Nutrición, Instituto de Investigaciones Clínicas y Médicas, Madrid, Spain, 1990.
Weerdhof, T., Wiersum, M., and Reissenweber, H., Application of liquid chromatography in food analysis, *J. Chromatogr.,* 83, 455–460, 1972.
Welt, B. and Tong, C., Effect of microwave power and electromagnetic energy, *J. Microwave Power Electromagn. Energy,* 28(4), 187–195, 1993.
Wills, R., Shaw, C., and Day, W., Analysis of water soluble vitamins by high pressure liquid chromatography, *J. Chromatogr. Sci.,* 15(July), 262–266, 1977.
Woidich, H., Pfannahuser, W., and Blaicher, G., cited in *Modern Chromatographic Analysis of the Vitamins* (Leenheer, A.P. and Lambert, W.E., 1985), *Lebensm. Chem. Gericht. Chem.,* 32, 74, 1978.
Yost, R.W., Ettre, L.S., and Conlon, R.D., In *Practical Liquid Chromatography-An Introduction*, 1st ed., The Perkin-Elmer Corporation, Norwalk, CT, 1981.

Multivitamin Method References

Ang, C. and Moseley, F., Determination of thiamin and riboflavin in meat and meat products by high-pressure liquid chromatography, *J. Agric. Food Chem.,* 28(3), 483–486, 1980.

Barna, E. and Dworschak, E., Determination of thiamine (vitamin B_1) and riboflavin (vitamin B_2) in meat and liver by high-performance liquid chromatography, *J. Chromatogr.*, A 668, 359–363, 1994.

Barna, E., Comparison of data obtained by HPLC and microbiological determination of riboflavin in ready-to-eat foods, *Acta Aliment.*, 21(1), 3–9, 1991.

Bognar, A., Bestimmung von Riboflavin und Thiamin in Lebensmitteln mit Hilfe der Hochdruck-Flüssigkeits-Chromatographie (HPLC), *Dtsch. Lebensm. Rundsch.*, 77(12), 431–436, 1981.

Bötticher, B., Bötticher, D., and Kluthe, R., An HPLC method for the simultaneous determination of B_1, B_2 and B_6 vitamers in food, *Akt. Ernähr.*, 14, 117–118, 1989.

Chase, G., Landen, W., and Soliman, A., Method modification for liquid chromatographic determination of thiamine, riboflavin, and pyridoxine in medical foods, *J. Assoc. Off. Anal. Chem. Int.*, 76(6), 1276–1280, 1993.

Chase, G., Landen, W., Eitenmiller, R., and Soliman, A., Liquid chromatographic determination of thiamine, riboflavin, and pyridoxine in infant formula. *J. Assoc. Off. Anal. Chem. Int.*, 75(3), 561–565, 1992.

Fellman, J., Artz, W., Tassinari, P., Cole, C., and Augustin, J., Simultaneous determination of thiamin and riboflavin in selected foods by high-performance liquid chromatography, *J. Food Sci.*, 47, 2048–2050, 1982.

Fernando, S. and Murphy, P., HPLC determination of thiamin and riboflavin in soybeans and tofu, *J. Agric. Food Chem.*, 38, 163–167, 1990.

Finglas, P. and Faulks, R., The HPLC analysis of thiamin and riboflavin in potatoes, *Food Chem.*, 15, 37–44, 1984.

Gauch, R., Leuenberger, U., and Müller, U., Bestimmung der wasserlöslichen Vitamine B_1, B_2, B_6 und B_{12} in Milch durch HPLC, *Z. Lebensm. Unters. Forsch.*, 195, 312–315, 1992.

Hägg, M. and Kumpulainen, J., Thiamine and riboflavin contents in domestic and imported cereal products in Finland, *J. Food Comp. Anal.*, 6, 299–306, 1993.

Hägg, M. and Kumpulainen, J., Thiamine and riboflavin contents in finnish pig, heifer, and cow livers and in pork loin, *J. Food Comp. Anal.*, 7, 1–6, 1994a.

Hägg, M. and Kumpulainen, J., Thiamine and riboflavin contents of finnish breads and their corresponding flours, *J. Food Comp. Anal.*, 7, 94–101, 1994b.

Hägg, M., Effect of various commercially available enzymes in the liquid chromatographic determination with external standardization of thiamine and riboflavin in foods, *J. Assoc. Off. Anal. Chem. Int.*, 77(3), 681–686, 1994.

Hasselmann, C., Franck, D., Grimm, P., Diop, P., and Soules, C., High-performance liquid chromatographic analysis of thiamin and riboflavin in dietetic foods, *J. Micronutr. Anal.*, 5, 269–279, 1989.

Kamman, J., Labuza, T., and Warthesen, J., Thiamin and riboflavin analysis by high-performance liquid chromatography, *J. Food Sci.*, 45, 1499–1504, 1980.

Mauro, D. and Wetzel, D., Simultaneous determination of thiamine and riboflavin in enriched cereal based products by high-performance liquid chromatography using selective detection, *J. Chromatogr.*, 299, 281–287, 1984.

Nail, P.A., Thomas, M.R., and Eakin, B.S., The effect of thiamin and riboflavin supplementation on the level of those vitamins in human breast milk and urine, *Am. J. Clin. Nutr.*, 33, 198–204, 1980.

Reyes, E. and Subryan, L., An improved method of simultaneous HPLC assay of riboflavin and thiamin in selected cereal products, *J. Food Comp. Anal.*, 2, 41–47, 1989.

Sims, A. and Shoemaker, D., Simultaneous liquid chromatographic determination of thiamine and riboflavin in selected foods, *J. Assoc. Off. Anal. Chem. Int.*, 76(5), 1156–1160, 1993.

Skurray, G., A rapid method for selectively determining small amounts of niacin, riboflavin and thiamine in foods, *Food Chem.*, 7, 77–80, 1981.

Wehling, R. and Wetzel, D., Simultaneous determination of pyridoxine, riboflavin, and thiamin in fortified cereal products by high-performance liquid chromatography, *J. Agric. Food Chem.*, 32, 1326–1331, 1984.

Wimalasiri, P. and Wills, R., Simultaneous analysis of thiamin and riboflavin in foods by high-performance liquid chromatography, *J. Chromatogr.*, 318, 412–416, 1985.

Specific Vitamin B_1 References

Abdel-Kader, Z., Comparison of AOAC and high-performance liquid chromatographic methods for thiamin determination in foods, *Food Chem.*, 43, 393–397, 1992.

Association of Official Analytical Chemists (AOAC), Thiamine (vitamin B_1) in foods. Fluorometric method. Final action, No. 942.23, 15th ed., Arlington, VA, 1990.

Ayi, B., Yuhas, D., Moffett, K., Joyce, D., and Deangelis, N., Liquid chromatographic determination of thiamine in infant formula products by using ultraviolet detection, *J. Assoc. Off. Anal. Chem.*, 68(6), 1087–1092, 1985.

Bailey, A. and Finglas, P., A normal phase high performance liquid chromatographic method for the determination of thiamin in blood and tissue samples, *J. Micronutr. Anal.*, 7, 147–157, 1990.

Bauer, A., Steger, U., and Wallnöfer, P., Thiamingehalt von Grünkern in Abhängigkeit von der Art der küchentechnischen Verarbeitung, *Dtsch. Lebensm.-Rundsch.*, 87(6), 180–182, 1991.

Bettendorff, L., Grandfils, C., De Rycker, C., and Schoffeniels, E., Determination of thiamine and its phosphate esters in human blood serum at femtomole levels, *J. Chromatogr.*, 382, 297–302, 1986.

Bognar, A., Determination of vitamin B_1 in food by high-performance liquid chromatography and postcolumn derivatization, *Fresenius J. Anal. Chem.*, 343, 155–156, 1992.

Bötticher, B. and Bötticher, D., Simple rapid determination of thiamin by a HPLC method in foods, body fluids, urine and faeces, *Int. J. Vit. Nutr. Res.*, 56, 155–159, 1986.

Ellefson, W., Thiamine, in *Methods of Vitamin Assay,* 4th ed., Augustin, J. et al., Eds., John Wiley & Sons, New York, 1984.

Fox, J., Ackerman, S., and Thayer, D., Fluorometric determination of thiamine vitamers in chicken, *J. Assoc. Off. Anal. Chem. Int.*, 75(2), 346–354, 1992.

Gubler, C., Thiamine, in *Handbook of Vitamins,* 2nd ed., Machlin, L.J., Ed., Marcel Dekker, New York, 1991.

Herve, C., Beyne, P., and Delacoux, E., Determination of thiamine and its phosphate esters in human erythrocytes by high-performance liquid chromatography with isocratic elution, *J. Chromatogr., B,* 653, 217–220, 1994.

Hilker, D.M. and Clifford, A.J., Thiamin analysis and separation of thiamin phosphate esters by high-performance liquid chromatography, *J. Chromatogr. Biomed. Appl.*, 231, 433–438, 1982.

Kawasaki, C., Modified thiamine compounds, in *Vitamins and Hormones,* Vol. XXI, Harris, R.S., Ed., Academic Press, New York, 1963.

Laschi-loquerie, A., Vallas, S., Viollet, J., Leclercq, M., and Fayol, V., High performance liquid chromatographic determination of total thiamine in biological and food products, *Int. J. Vit. Nutr. Res.,* 62, 248–251, 1992.

McCormick, D.B., Thiamin, in *Modern Nutrition in Health and Disease*, 7th ed., Shils, M.E. and Young, V.R., Eds., Lea & Febiger, Philadelphia, 1988, pp. 355–361.

Nobile S., Savage V., and Huber U., A new procedure for the determination of thiamine in foods, *Int. J. Vit. Nutr. Res.*, 42, 444–450, 1972.

Ollilainen, V., Vahteristo, L., Uusi-Raura, A., Varo, P., Koivistoinen, P., and Huttunen, J., The HPLC determination of total thiamin (Vitamin B_1) in foods, *J. Food Comp. Anal.*, 6, 152–165, 1993.

Ryan, M. and Ingle, J., Fluorometric reaction rate method for the determination of thiamine, *Anal. Chem.*, 52, 2177–2184, 1980.

Sander, S., Hahn, A., Stein, J., and Rehner, G., Comparative studies on the high-performance liquid chromatographic determination of thiamine and its phosphate esters with chloroethylthiamine as an internal standard using pre- and postcolumn derivatization procedures, *J. Chromatogr.*, 558, 115–124, 1991.

Valls, F., Checa, M.A., Fernández Muiño, M.A., and Sancho, M.T., Determination of thiamin in cooked sausages, *J. Agric. Food Chem.*, 47, 170–173, 1999.

Vanderslice, J. and Huang, M., Liquid chromatographic analysis of thiamin and its phosphates in food products using amprolium as an internal standard, *J. Micronutr. Anal.*, 2, 189–199, 1986.

Vidal-Valverde, C. and Reche, A., An improved high performance liquid chromatographic method for thiamin analysis in foods, *Z. Lebensm. Unters. Forsch.*, 191, 313–318, 1990a.

Wielders, J. and Mink, C., Quantitative analysis of total thiamine in human blood, milk and cerebrospinal fluid by reversed-phase ion-pair high-performance liquid chromatograhpy, *J. Chromatogr.*, 277, 145–156, 1983.

Williams, R.D., Mason, H.L., Smith, B.F., and Wilder, R.M., Induced thiamin deficiency and the thiamine requirement of man: further observations, *Arch. Int. Med.*, 69, 721–738, 1942.

Specific Vitamin B_2 References

Association of Official Analytical Chemists (AOAC), Riboflavin (Vitamin B_2) in foods and vitamin preparations. Fluorometric Method. First action 1970. Final action 1971, No. 970.65, 15th ed., Arlington, VA, 1990.

Ashoor, S., Knox, M., Olsen, J., and Deger, D., Improved liquid chromatographic determination of riboflavin in milk and dairy products, *J. Assoc. Off. Anal. Chem.*, 68(4), 693–696, 1985.

Bilic, N. and Sieber, R., Determination of flavins in dairy products by high-performance liquid chromatography using sorboflavin as internal standard, *J. Chromatogr.*, 511, 359–366, 1990.

Bourquin, A. and Sherman, H.C., Quantitative determination of vitamin B_2, *J. Am. Chem. Soc.*, 53, 3501, 1931.

Cooperman, J.M. and López, R., Riboflavin, in *Handbook of Vitamins*, 2nd ed., Machlin, L.J., Ed., Marcel Dekker, New York, 1991.

Johnsson, H. and Branzell, C., High performance liquid chromatographic determination of riboflavin in food. A comparison with a microbiological method, *Int. J. Vit. Nutr. Res.*, 57, 53–58, 1987.

Kader, Z. and Ryley, J., The riboflavin content of blanched soya beans, a comparison of three methods of analysis, *J. Micronutr. Anal.*, 4, 169–175, 1988.

Lumley, I.D. and Wiggins, R.A., Determination of riboflavin and flavin mononucleotide in foodstuffs using high-performance liquid chromatography and a column-enrichment technique, *Analyst*, 106(10), 1103–1108, 1981.

Matts, F., Riboflavin, in *Vitamins in Medicine*, 4th ed. Vol. I. Barker, B. and Bender, D. Eds., William Heinemann Medical Books LTD., London, 1980.

Ollilainen, V., Mattila, P., Varo, P., Koivistoinen, P., and Huttunen, J., The HPLC determination of total riboflavin in foods, *J. Micronutr. Anal.*, 8, 199–207, 1990.

Reyes, E., Norris, K., Taylor, C., and Potts, D., Comparison of paired-ion liquid chromatographic method with AOAC fluorometric and microbiological method for riboflavin determination in selected foods, *J. Assoc. Off. Anal. Chem.*, 71(1), 16–19, 1988.

Russell, L. and Vanderslice, J., Comments on the standard fluorometric determination of riboflavin in foods and biological tissues, *Food Chem.*, 43, 79–82, 1992a.

Suhara, T., Torida, S., Ito, K., and Fukumoto, Y., Determination of vitamin B_2 in feed premixes by high-performance liquid chromatography, *Shiryo Kenkyu Hokoku (Tokyo Hishiryo Kensasho)*, 10, 152–163, 1985.

Valls, F., Sancho, M.T., Fernández Muiño, M.A., and Checa, M.A., Determination of total riboflavin in cooked sausages, *J. Agric. Food Chem.*, 47, 1067–1070, 1999.

Vidal-Valverde, C. and Reche, A., Reliable system for the analysis of riboflavin in foods by high performance liquid chromatography and UV detection, *J. Liq. Chromatogr.*, 13 (10), 2089–2101, 1990b.

Watada, A. and Tran, T., A sensitive high-performance liquid chromatography method for analyzing riboflavin in fresh fruits and vegetables, *J. Liq. Chromatogr.*, 8(9), 1651–1662, 1985.

Woodcock, E.A., Warthesen, J.J., and Labuza, T.P., Riboflavin photochemical degradation in pasta measured by high-performance liquid chromatography, *J. Food Sci.*, 47, 545–549, 555, 1982.

Specific Vitamin B_3 References

Balschukat, D. and Kress, E., Use of column switching for the determination of niacinamide in compound feed, *J. Chromatogr.*, 502, 79–85, 1990.

Cantoni, C., Cattaneo, P., and Lauri, S., Sulla nicotinamide nei prodotti carnei e dati sperimentali, *Ing. Alimentare,* 5, 42–43, 1992.

Dawson, K., Unklesbay, N., and Hedrick, H., HPLC determination of riboflavin, niacin, and thiamin in beef, pork, and lamb after alternate heat-processing methods, *J. Agric. Food Chem.*, 36, 1176–1179, 1988.

Horwitt, M.K., Harper, A.E., and Henderson, L.M., Niacin-tryptophan relationships for evaluating niacin equivalents, *Am. J. Clin. Nutr.*, 34, 423–427, 1981.

Fontaine, J. and Hörr, J., Determination of supplemented nicotinic acid or nicotinamide in compound feed by HPLC after cleaning of the feed extract with cationic sorbent extraction, *Agribiol. Res.*, 46(1), 10–19, 1993.

Fujita, T., Kanbe, K., and Sasaki, K., Studies on high-performance liquid chromatographic determination of nicotinic acid in foods, *Seikatsu Eisei*, 27, 30, 1983.

Gigliotti, C. and Daghetta, A., Indagine conoscitiva sul contenuto di acido nicotinico negli insaccati freschi, *Ind. Alimen.*, 32 (February), 113–120, 1993.

Hamano, T., Mitsuhashi, Y., Aoki, N., and Yamamoto, S., Simultaneous determination of niacin and niacinamide in meats by high-performance liquid chromatography, *J. Chromatogr.*, 457, 403–408, 1988.

Johnson, B.C., The microbiological determination of nicotinic acid, nicotinamide, and nicotinuric acid, *J. Biol. Chem.*, 159, 227, 1945.

Kitada, Y., Inone, M., Tamase, K., Imou, M., Hasuike, A., Sasaki, M., and Tanigawa, K., Analysis of nicotinic acid and nicotinamide in foods by ion-pair HPLC, *Eiyo Shokuryo,* 35, 121–124, 1982.

McKee, R.N., Kang-Lee, Y.A., Panaqua, M., and Swendseid, M.E., Determination of nicotinamide and metabolic products in urine by high performance liquid chromatography, *J. Chromatogr.*, 230, 309–317, 1982.

Sakai, H., Suzuki, A., Izawa, K., and Niimi, T., Determination of nicotinamide in premixes by high performance liquid chromatography, *Shiryo Kenkyu Hokoku (Tokyo Hishiryo Kensasho)*, 10, 122–129, 1985.

Snell, E.E. and Wright, L.D., A microbiological method for the determination of nicotinic acid, *J. Biol. Chem.*, 139, 675, 1941.

Takatsuki, K., Suzuki, S., Sato, M., Sakai, K., and Ushizawa, I., Liquid chromatographic determination of free and added niacin and niacinamide in beef and pork, *J. Assoc. Off. Anal. Chem.*, 70(4), 698–702, 1987.

Trugo, L., Macrae, R., and Trugo, N., Determination of nicotinic acid in instant coffee using high-performance liquid chromatography, *J. Micronutr. Anal.*, 1, 55–63, 1985.

Tyler, T. and Genzale, J., Liquid chromatographic determination of total niacin in beef, semolina, and cottage cheese, *J. Assoc. Off. Anal. Chem.*, 73(3), 467–469, 1990.

Valls, F., Sancho, M.T., Fernández Muiño, M.A., and Checa, M.A., Simultaneous determination of nicotinic acid and nicotinamide in cooked sausages, *J. Agric. Food Chem.* 48, 3392–3395, 2000.

Vidal-Valverde, C. and Reche, A., Determination of available niacin in legumes and meat by high-performance liquid chromatography, *J. Agric. Food Chem.*, 39, 116–121, 1991.

Williams, W.L., Yeast microbiological method for determination of nicotinic acid, *J. Biol. Chem.*, 166, 397, 1946.

Yoshida, K., Yamamoto, Y., and Fujiware, M., Simple analytical method for niacin (nicotinic acid) and nicotinamide in foods by high-performance liquid chromatography, *Shokuhin Eiseigaku Zasshi*, 23(6), 428–433, 1982.

Specific Vitamin B_6 References

Ang, C., Cenciarelli, M., and Eitenmiller, R., A simple liquid chromatographic method for determination of B_6 vitamers in raw and cooked chicken, *J. Food Sci.*, 53(2), 371–375, 1988.

Bessey, O.A., Adam, D.J., and Hansen, A.E., Intake of vitamin B_6 and infantile convulsions: a first approximation of requirements of pyridoxine in infants, *Pediatrics*, 20, 33–44, 1957.

Bitsch, R. and Möller, J., Evaluation of the vitamin B_6 content in foods by HPLC analysis, Dtsch. Forschungsgem., grant Bi 218/51, 77–79, 1988.

Bitsch, R., and Möller, J., Analysis of B_6 vitamers in foods using a modified high-performance liquid chromatographic method, *J. Chromatogr.*, 463, 207–211, 1989.

Bognar, A., Bestimmung von Vitamin B_6 in Lebensmitteln mit Hilfe der HochdruckFlüssig-Chromatographie (HPLC), *Z. Lebensm. Unters. Forsch.*, 181, 200–205, 1985.

Bognar, A., Studies on the influence of cooking on the vitamin B_6 content of food, *Bioavailability*, 93, 346–351, 1993.

Coburn, S.P. and Mahuren, J.D., A versatile cation-exchange procedure for measuring the seven major forms of vitamin B_6 in biological samples, *Anal. Biochem.*, 129, 310–317, 1983.

Driskell, J.A., Vitamin B_6, in *Handbook of Vitamins-Nutritional, Biochemical and Clinical Aspects*, Machlin, L.J., Ed., Marcel Dekker, New York, 1984, pp. 379–401.

Gregory, J.F. and Kirk, J.R., cited in *Determination of Vitamin B_6 in Animal Tissues by Reverse-Phase High-Performance Liquid Chromatography* (Gregory, J.F., Manley, D.B., and Kirk, J.R., 1981), *Am. J. Clin. Nutr.*, 32, 879, 1979.

Gregory, J.F. and Sartain, D.B., Improved chromatographic determination of free and glycosylated forms of vitamin B_6 in foods, *J. Agric. Food Chem.*, 39, 899–905, 1991.

Gregory, J.F., Methods for determination of vitamin B_6 in foods and other biological materials: a critical review, *J. Food Comp. Anal.*, 1, 105–123, 1988.

Gregory, J.F., Manley, D.B., and Kirk, J.R., Determination of vitamin B_6 in animal tissues by reverse-phase high-performance liquid chromatography, *J. Agric. Food Chem.*, 29, 921–927, 1981.

Hozova, B., Sorman L., Dudasova S., and Rajniakova A., Study of retention of vitamin B_6 in non-traditionally canned foods on using an experimental plan method, *Die Nahrung*, 32(4), 413–414, 1988.

Lim, K., Young, R., and Driskell, J., Separation of vitamin B_6 components by high-performance liquid chromatography, *J. Chromatogr.*, 188, 285–288, 1980.

Möller, J. and Bitsch, R., Einsatz der Hochleistungs-flüssigchromatographie (HPLC) zur simultanen Bestimmung der Pyridoxin-derivate aus Biologischen Materialien, *Ernährung. Umsch.*, 35(9), 324–328, 1988.

Reitzer-Bergaentzle, M., Marchioni, E., and Hasselmann, C., HPLC determination of vitamin B_6 in foods after precolumn derivatization of free and phosphorylated vitamers into pyridoxol, *Food Chem.*, 48, 321–324, 1993.

Schoonhoven, J., Schrijver, J., Berg, H., and Haenen, G., Reliable and sensitive high-performance liquid chromatographic method with fluorometric detection for the analysis of vitamin B_6 in foods and feeds, *J. Agric. Food Chem.*, 42, 1475–1480, 1994.

Valls, F., Sancho, M.T., Fernández Muiño, M.A., and Checa, M.A., Determination of vitamin B_6 in cooked sausages, *J. Agric. Food Chem.*, 49, 38–41, 2001.

Vanderslice, J., Maire, C., Doherty, R., and Beecher, G., Sulfosalicylic acid as an extraction agent for vitamin B_6 in food, *J. Agric. Food Chem.*, 28, 1145–1149, 1980.

Vanderslice, J., Brownlee, S., and Cortissoz, M., Liquid chromatographic determination of vitamin B_6 in foods, *J. Assoc. Off. Anal. Chem.*, 67(5), 999–1007, 1984.

Vanderslice, J.T., Stewart, K.K., and Yarmas, M.M., Liquid chromatographic separation and quantification of B_6 vitamers and their metabolite, pyridoxic acid, *J. Chromatogr.*, 176, 280–285, 1979.

Williams, A.K. and Cole, P.D., Vitamin B_6: ion exchange chromatography of pyridoxal, pyridoxol and pyridoxamine, *J. Agric. Food Chem.*, 23, 915–916, 1975.

Wong, F.F., Analysis of vitamin B_6 in extractives of food materials by high-performance liquid chromatography, *J. Agric. Food Chem.*, 26, 1444–1446, 1978.

Yasumoto, K., Tadera, K., Tsuji, H., and Mitsuda, H., Semi-automated system for analysis of vitamin B_6 complex by ion-exchange column chromatography, *J. Nutr. Sci. Vitaminol.*, 21, 117–127, 1975.

Specific Biotin References

Desvene, P.L., Coustal, S., and Frappier, F., Separation of biotin and its analogs by high-performance liquid chromatography: convenient labeling for ultraviolet or fluorimetric detection, *Anal Biochem.*, 128, 359–362, 1983.

Grosvenor, A.L., Feltus, A., Conover, R.C., Daunert, S., and Anderson, K.W., Development of binding assays in microfabricated picoliter vials: an assay for biotin, *Anal Chem.*, 72, 2590–2594, 2000.

Hayakawa, K. and Oizumi, J., Determination of free biotin in plasma by liquid chromatography with fluorimetric detection, *J. Chromatogr.*, 413, 247–250, 1987.

Hentz, N.G. and Bachas, L.G., Class-selective detection systems for liquid chromatography-based on the streptavidin-biotin interaction, *Anal. Chem.*, 67, 1014–1018, 1993.

Przyjazny, A., Hentz, N.G., and Bachas, L.G., Sensitive and selective liquid chromatographic postcolumn reaction detection system for biotin and biocytin using a homogeneus fluorophore linked assay, *J. Chromatogr.*, 654, 79–86, 1993.

Lahély, S., Ndaw, S., Arella, F., and Hasselmann, C., Determination of biotin in foods by high-performance liquid chromatography with postcolumn derivatization and fluorimetric detection, *Food Chem.*, 65, 253–258, 1999.

Livaniou, E., Costopoulou, D., Vassiliadou, I., Leondiadis, L., Nyalala, J.O., Ithakissios, D.S., and Evangelatos, G.P., Analytical techniques for determining biotin, *J. Chromatogr.*, A, 881, 331–343, 2000.

Stein, J., Hahn, A., Lembke, B., and Rehner, G., High-performance liquid chromatographic determination of biotin in biological materials after crown ether-catalized fluorescence derivatization with panacyl bromide, *Anal. Biochem.*, 200, 89–94, 1992.

Whitehead, C.C., Armstrong, J.A., and Waddington, D., The determination of the availability to chicks of biotin in feed ingredients by a bioassay based on the response of blood pyruvate carboxilase (EC 6.4.1.1.) activity, *Br. J. Nutr.*, 48, 81–88, 1982.

Yoshida, T., Uetake, A., Nakai, C., Nimura, M., and Kinoshita, T., Liquid chromatographic determination of biotin using 1pyrenyldiazomethane as a precolumn fluorescent labeling reagent, *J. Chromatogr.*, 456, 421–426, 1988.

Specific folate References

Day, B.P.F. and Gregory, J.F., III, Determination of folacin derivatives in selected foods by high performance liquid chromatography, *J. Agric. Food. Chem.*, 29, 374–377, 1981.

Desouza, S. and Eitenmiller, R., Effects of different enzyme treatments on extraction of total folate from various foods prior to microbiological assay and radioassay, *J. Micronutr. Anal.*, 7, 37–57, 1990.

Finglas, P.M., Faure, U., and Southgate, D.A.T., First BRC intercomparison on the determination of folates in food, *Food Chem.*, 46, 199–213, 1993.

Finglas, P.M., Wigertz, K., Vahteristo, L., Witthöft, C., Southon, S., and De Froidmont-Görtz, I., Standarisation of HPLC techniques for the determination of naturally-occurring folates in food, *Food Chem.*, 64, 245–255, 1999.

Flynn, L.M., Microbiological assays of folic acid activity in foods: changes in results with variation in experimental conditions and methods, *J. Assoc. Offic. Anal. Chem.*, 48, 1230–1235, 1965.

Gregory, J.F., III, Sartain, D.B., and Day, B.P.F., Fluorometric determination of folacin in biological materials using high performance liquid chromatography, *J. Nutr.*, 114, 341–353, 1984.

Holt, D.L., Wehling, R.L., and Zeece, M.G., Determination of native folates in milk and other dairy products by high-performance liquid chromatography, *J. Chromatogr.*, 449, 271–279, 1988.

Lucock, M.D., Green, M., Priestnall, M., Daskalakis, I., Levene, M.I., and Hartley, R., Optimisation of chromatographic conditions for the determination of folates in foods and biological tissues for nutritional and clinical work, *Food Chem.*, 53, 329–338, 1995.

Martin, J.I., Landen, W.I., Jr., and Soliman, A.G.M., Application of a trienzyme extraction for total folate determination in foods, *J. Assoc. Off. Anal. Chem.*, 73(5), 805–808, 1990.

Pfeiffer, C.M., Rogers, L.M., and Gregory, J.F., Determination of folate in cereal grain food products using trienzyme extraction and combined affinity and reverse-phase liquid chromatography, *J. Agric. Food Chem.*, 45, 407–413, 1997.

Phillips, D.R. and Wright, A.J.A., Studies on the response of *Lactobacillus casei* to different folate monoglutamates, *Br. J. Nutr.*, 47, 183–189, 1982.

Rader, J.I., Weaver, C.M., and Angyal, G., Use of a microbiological assay with tri-enzime extraction for measurement of pre-fortification levels of folates in enriched cereal-grain products, *Food Chem.*, 62(4), 451–465, 1998.

Ridascreen®-Folsäure, Direct competitive ELISA enzymoinmunoassay for folic acid (Ref. 16311), r-biopharm, ATOM, Digen Ltd., U.K., 1998.

Reingold, R.N., Perkins, E.G., and Picciano, M.F., Separation of folate derivatives by *in situ* paired-ion high pressure liquid chromatography, *J. Chromatogr.*, 190, 237–240, 1980.

Tamura, T., Mizuno, Y., Johnston, K.E., and Jacob R.A., Food folate assay with protease, α-amilase and folate conjugase treatments, *J. Agric. Food Chem.*, 45, 135–139, 1997.

Vatehristo, L., Ollilainen, V., Koivistoinen, P.E., and Varo, P., Improvements in the analysis of reduced folate monoglutamates and folic acid in food by high-performance liquid chromatography, *J. Agric. Food Chem.*, 44, 477–482, 1996a.

Vatehristo, L., Finglas, P.M., Witthöft, C.M., Wigertz, K., Seale, R., and de Froidmont-Görtz, I., Third EU-MAT intercomparison study on food folates analysis using HPLC procedures, *Food Chem.*, 57(1), 109–111, 1996b.

Wigertz, K. and Jägerstad, M., Comparison of a HPLC and radioprotein-binding assay for the determination of folates in milk samples, *Food Chem.*, 54(4), 429–436, 1995.

Specific Vitamin B_{12} References

Bennink, M.R. and Ono, K., Vitamin B_{12}, E and D contents of raw and cooked beef, *J. Food Sci.*, 47, 1786–1792, 1982.

Casey, P.J., Speckman, K.R., Ebert, F.J., and Hobbs, W.E., Radioisotope dilution technique for determination of vitamin B_{12} in foods, *J. Assoc. Off. Anal. Chem.*, 65, 85–88, 1982.

Eitenmiller, R.R., Landen, W.O., and Augustin, J., Jr., Vitamin analysis, in *Food Analysis,* 2nd ed., Nielsen, S.S., Ed., An Aspen Publication, Chapman & Hall Food Science Titles, Gaithersburg, MD, 1998.

Fie, M., Zee, J.A., and Amiot, J., Séparation et quantification des isomères de la vitamine B_{12} dans le lait et certains produits latiers par chromatographie liquide haute performance et par radio-essai, *Sci. Alim.*, 14, 763–775, 1994.

Ford, S., Gallery, J., Nichols, A., and Shambee, M., High-performance liquid chromatographic analysis of the (cyanoaquo) stereoisomers of several putative vitamin B_{12} precursors, *J. Chromatogr.,* 537, 235–247, 1991.

Frenkel, E.P., Prough, R., and Kitchens, R.L., Measurement of tissue vitamin B_{12} by radioisotopic competitive inhibition assay and quantitation of tissue cobalamin fractions, *Methods Enzymol.*, 67, 31–40, 1980.

Iwase, H. and Ono, I., Determination of cyanocobalamin in foods by high-performance liquid chromatography with visible detection after solid-phase extraction and membrane filtration for the precolumn separation of lipophilic species, *J. Chromatogr. A*, 771, 127–134, 1997.

Jacobsen, D.W., Green, R., Quadros, E.V., and Montejano, Y.D., Rapid analysis of cobalamin coenzymes and related corrinoid analogues by high-performance liquid chromatography, *Anal. Biochem.*, 120, 394–403, 1982.

Kralova, B., Rauch, P., and Cerna, J., Use of vitamin B_{12} radioassay in the analysis of biological materials, mainly foods, *Nahrung,* 26, 803–810, 1982.

Richardson, R.J., Favell, D.J., Gidley, G.C., and Jones, G.H., Application of a commercial radioassay test kit to the determination of vitamin B_{12} in food, *Analyst,* 103, 865–868, 1978.

Ridascreen®-Test, Direct competitive ELISA enzymoinmunoassay for vitamin B_{12} (Ref. 16309), r-biopharm, ATOM, Digen Ltd., U.K., 1998.

Schneider, Z., Purification and estimation of vitamin B_{12}, in *Comprehensive B_{12}*, Schneider, Z. and Stroiñski, A., Eds., De Gruyter, Berlin, 1987, pp. 111–155.

Watanabe, F., Takenaka, S., Abe, K., Tamura, Y., and Nakano, Y., Comparison of a microbiological assay and a fully automated chemiluminescent system for the determination of vitamin B_{12} in food, *J. Agric. Food Chem.*, 46(4), 1433–1436, 1998.

Specific Vitamin C References

Association of Official Analytical Chemists International., *Official Methods of Analysis,* 16th ed., AOAC International, Gaithersburg, MD, 1995.

Barja De Quiroga, G., López, M., Pérez, R., and Rojas, C., Simultaneous determination of two antioxidants, uric and ascorbic acid, in animal tissue by high-performance liquid chromatography, *Anal. Biochem.,* 199, 81–85, 1991.

Bessey, O.A. and King, C.G., The distribution of vitamin C in plant and animal tissues and its determination, *J. Biol. Chem.,* 103, 687, 1933.

Buslig, B.S., Wilson, C.W., III, and Shaw, P.E., High-performance liquid chromatographic separation of carboxylic acids with anion-exchange and reverse-phase columns, *J. Agric. Food Chem.,* 30, 342–345, 1982.

Carr, R.S., Bally, M.B., Thomas, P., and Neff, J.M., Comparison of methods for determination of ascorbic acid in animal tissues, *Anal. Chem.,* 55, 1229–1232, 1983.

Coustard, J.M. and Sudraud, G., Séparation des acides ascorbique et isoascorbique par chromatographie de paires d'ions sur phase inverse, *J. Chromatogr.,* 219, 338–342, 1981.

Dennison, D., Brawley, T., and Hunter, G., Rapid high-performance liquid chromatographic determination of ascorbic acid and combined ascorbic acid-dehydroascorbic acid in beverages, *J. Agric. Food Chem.,* 29, 927–929, 1981.

Finley, J.M. and Duang, D., Resolution of ascorbic, dehydroascorbic and diketogulonic acids by paired-ion reversed-phase chromatography, *J. Chromatogr.,* 207, 449–453, 1981.

Gilpin, R.K. and Sisco, W.R., cited in *Modern Chromatographic Analysis of the Vitamins* (Leenheer, A.P. and Lambert, W.E., 1985), *J. Chromatogr.,* 194, 285, 1980.

Harris, L.J., Mapson, L.W., and Wang, Y.L., Vitamin methods. A simple potentiometric method for determining ascorbic acid suitable for use with colored extracts, *Biochem. J.,* 36, 183, 1942.

Hedlund, B., Analysis of vitamins in food by HPLC, in *Recent Developments in Food Analysis,* Baltes, W., Czedík-Eysenberg, P.B., and Pfannhauser, W., Eds., Verlag Chemie, Weinheim, 1982, pp. 27–32.

Iwase, H. and Ono, I., Determination of ascorbic acid and dehydroascorbic acid in juices by high-performance liquid chromatography with electrochemical detection using L-cysteine as precolumn reductant, *J. Chromatogr.,* A 654, 215–220, 1993.

Kodaka, K., Inagaki, S., Ujie, T., Ueno, T., and Suda H., Determination of total vitamin C in foods by high-performance liquid chromatography, *Vitamin,* 59(5), 451–455, 1985.

Kutnink, M. and Omaye, S., Determination of ascorbic acid, erythorbic acid, and uric acid in cured meats by high performance liquid chromatography, *J. Food Sci.,* 52(1), 53–56, 1987.

Lindquist, J. and Ferroha, S.M., cited in *Modern Chromatographic Analysis of the Vitamins* (Leenheer, A.P. and Lambert, W.E., 1985), *Analyst (London),* 100, 377, 1975.

Meyer, B.J., Lagally, H., and Dodds, M.L., cited in *Modern Chromatographic Analysis of the Vitamins* (Leenheer, A.P. and Lambert, W.E., 1985), *Anal. Biochem.,* 4, 88, 1965.

Moledina, K.H. and Flink, J.M., Determination of ascorbic acid in plant food products by high performance liquid chromatography, *Lebensm. Wiss. Unters. Technol.,* 15, 351–358, 1982.

Parolari, G., L'acido ascorbico nelle carni, *Ind. Conserve,* 57, 19–21, 1982.

Pelletier, O., Vitamin C (L-ascorbic and dehydro-L-ascorbic acid), in *Methods of Vitamin Assay,* 4th ed., Augustin, J., Klein, B.P., Becker, D.A., and Venugopal, P.B., Eds., John Wiley & Sons, New York, 1985, pp. 334–336, 338, and 341.

Roe, J.H., Mills, M.B., Oesterling, M.J., and Dameron, C.M., cited in *Modern Chromatographic Analysis of the Vitamins* (Leenheer, A.P. and Lambert, W.E., 1985), *J. Biol. Chem.,* 174, 201, 1948.

Rizzolo, A., Forni, E., and Polesello, A., HPLC assay of ascorbic acid in fresh and processed fruit and vegetables, *Food Chem.,* 14, 189–199, 1984.

Sawamura, M., Ooishi, S., and Li, Z., Reduction of dehydroascorbic acid by sodium hydrosulfide and liquid chromatographic determination of vitamin C in citrus juices, *J. Sci. Food Agric.,* 53, 279–281, 1990.

Schmall, M., Pifer, C.W., and Wollish, E.G., Determination of ascorbic acid by a new colorimetric reaction, *Anal. Chem.,* 25, 1486, 1953.

Sherman, H.C., Lamer, V.K., and Campbell, H.L., The quantitative determination of the antiscorbutic vitamin, *J. Am. Chem. Soc.,* 44, 165, 1922.

Schüep, W. and Keck, E., Measurement of ascorbic acid and erythorbic acid in processed meat by HPLC, *Z. Lebensm. Unters. Forsch.,* 191, 290–292, 1990.

Sood, S., Sartori, L., Wittmer, D., and Haney, W., High-pressure liquid chromatographic determination of ascorbic acid in selected foods and multivitamin products, *Anal. Chem.,* 48(6), 796–798, 1976.

Speek, A., Schrijver, J., and Schreurs, W., Fluorometric determination of total vitamin C and total isovitamin C in foodstuffs and beverages by high-performance liquid chromatography with precolumn derivatization, *J. Agric. Food Chem.,* 32, 352–355, 1984.

Tillmans, J., cited in *Modern Chromatographic Analysis of the Vitamins* (Leenheer, A.P. and Lambert, W.E., 1985), *Z. Lebensm. Unters. Forsch.,* 54, 33, 1927.

Tsao, C.S. and Salimi, S.L., Differential determination of L-ascorbic acid and D-isoascorbic acid by reversed-phase high-performance liquid chromatography with electrochemical detection, *J. Chromatogr.,* 245(3), 355–358, 1982.

Valls-Garcia, F., Determinación de vitaminas en productos cárnicos cocidos mediante cromatografía líquida de alta eficacia. I. Optimización de los métodos de extracción y analíticos. II. Evaluación de las pérdidas vitamínicas durante el proceso industrial de fabricación, Tesis Doctoral, Universidad de Burgos, 1997.

Vanderslice, J. and Higgs, D., Chromatographic separation of ascorbic acid, isoascorbic acid, dehydroascorbic acid and dehydroisoascorbic acid and their quantitation in food products, *J. Micronutr. Anal.,* 4, 109–118, 1988.

Vanderslice, J. and Higgs, D., Separation of ascorbic acid, isoascorbic acid, dehydroascorbic acid and dehydroisoascorbic acid in food and animal tissue, *J. Micronutr. Anal.,* 7, 67–70, 1990.

9 Amino Acid Analysis

Larry R. Massom

CONTENTS

9.1 INTRODUCTION

Although there are over 200 amino acids found in nature, genes (DNA) only contain codons for 20 amino acids (see Table 9.1 for a list of these amino acids and selected properties). These 20 amino acids are the foundation for all proteins, including all enzymes and structural protein. Some of these amino acids are then modified during and/or after the protein is synthesized. Other nonprotein amino acids are synthesized in the body. Many of the above-mentioned 200 amino acids do not occur in mammalian tissue but can be present in plant tissue. Some are intermediates in various metabolic pathways; for example, ornithine is part of the urea cycle. Other nonprotein amino acids such as γ-aminobutyric acid (GABA) are neurotransmitters.

TABLE 9.1
Structures and Selected Properties of 20 Amino Acids

Type	Name 3-Letter Abbreviation Symbol	Structure	Molecular Weight	pKa	pI	Classification
	Glycine Gly G $C_2H_5NO_3$	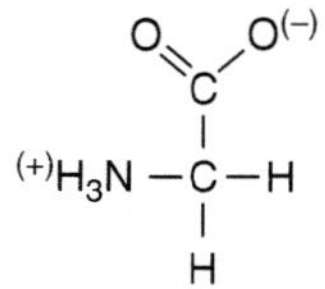	75.07	2.34 9.60	6.0	Nonessential
Acidic	L-Aspartic Asp D $C_4H_9NO_4$	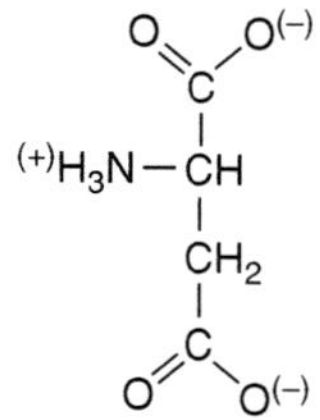	133.10	1.88 3.65 9.60	2.8	Nonessential
	L-Glutamic (Glutamate) Glu E $C_5H_8NO_4$	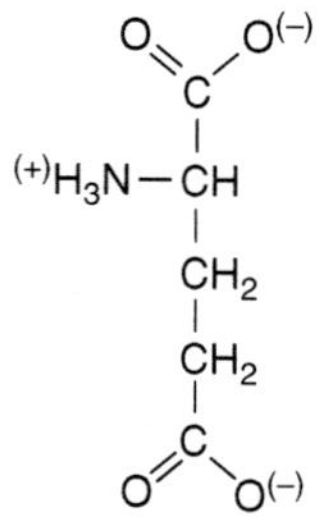	147.13	2.19 4.25 9.67	3.2	Nonessential

Imino acid	Proline Pro P $C_5H_9NO_2$	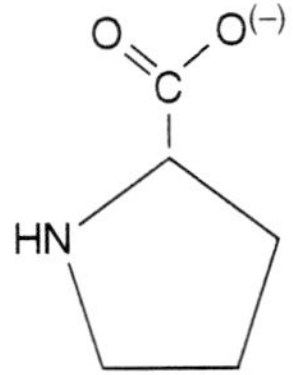	115.13	1.99 10.60	6.3	Nonessential
Hydroxyl	Threonine Thr T $C_4H_9NO_3$	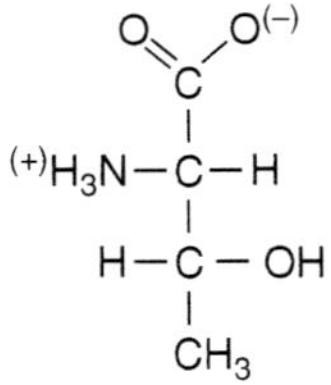	119.12	2.15 9.12	5.6	Essential
	Serine Ser S $C_3H_7NO_3$	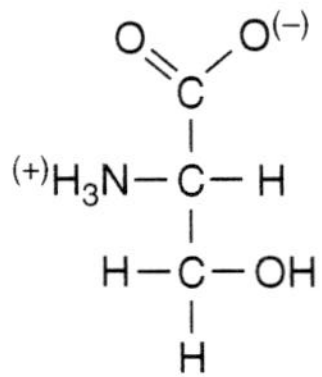	105.09	2.21 9.15	5.7	Nonessential

(*Continued*)

TABLE 9.1 (Continued)
Structures and Selected Properties of 20 Amino Acids

Type	Name 3-Letter Abbreviation Symbol	Structure	Molecular Weight	pKa	pI	Classification
Basic	L-Arginine Arg R $C_6H_{14}N_2O_2$	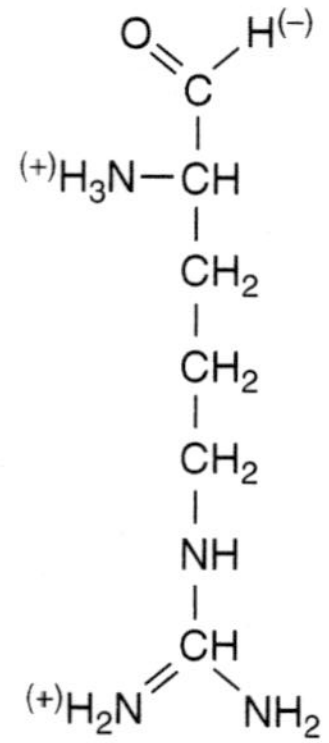	174.20	2.18 9.09 13.2	10.8	Semi-essential, required under certain metabolic conditions
	L-Lysine Lys K $C_6H_{14}N_2O_2$	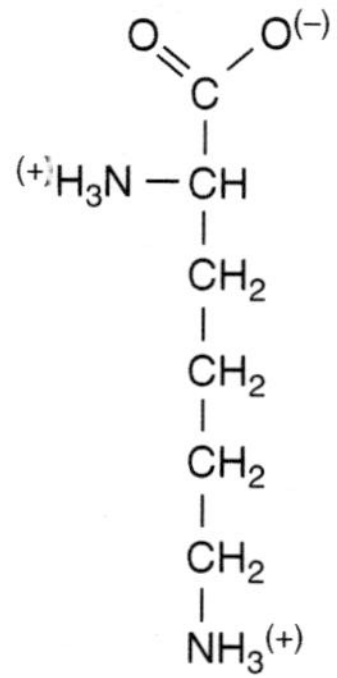	146.19	2.20 8.90 10.28	9.6	Essential

	L-Histidine His H $C_6H_9N_3O_2$	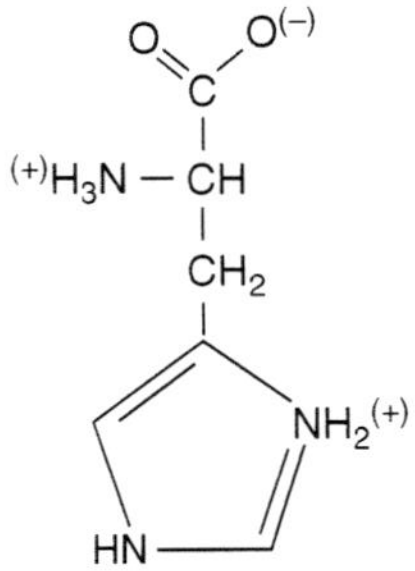	155.16	1.78 5.97 8.97	7.5	Essential for infants
Amide	L-Asparagine Asn N $C_4H_8N_2O_3$	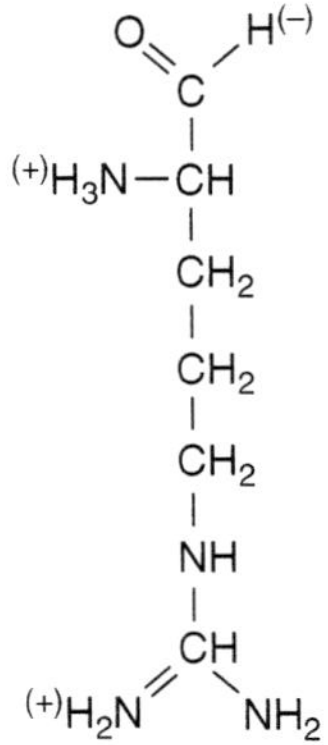	132.12	2.02 8.80	5.4	Nonessential

(Continued)

TABLE 9.1 (Continued)
Structures and Selected Properties of 20 Amino Acids

Type	Name 3-Letter Abbreviation Symbol	Structure	Molecular Weight	pKa	pI	Classification
	L-Glutamine Gln Q $C_5H_{10}N_2O_3$	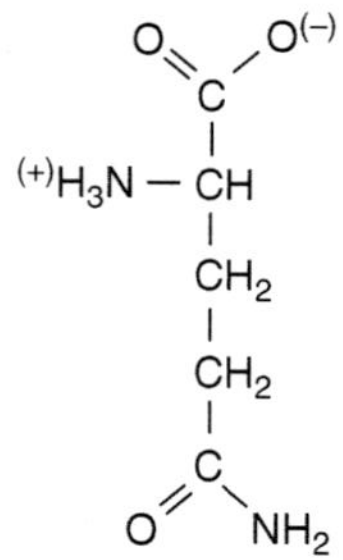	146.15	2.17 9.13	5.7	Nonessential
Aliphatic	L-Alanine Ala A $C_3H_7NO_2$	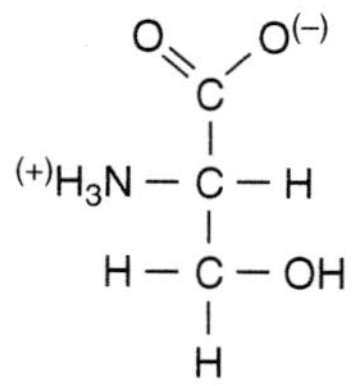	89.09	3.60 10.19	6.0	Nonessential

L-Isoleucine Ile I $C_6H_{13}NO_2$	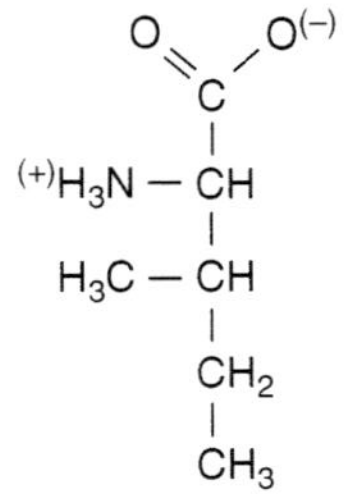	131.17	2.32 9.76	6.0	Essential
L-Leucine Leu L $C_6H_{13}NO_2$	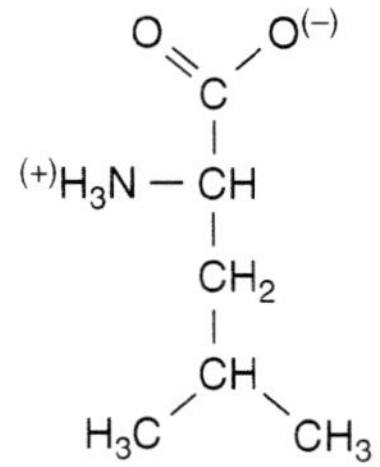	131.17	2.36 9.60	6.0	Essential
L-Valine Val V $C_5H_{11}NO_2$	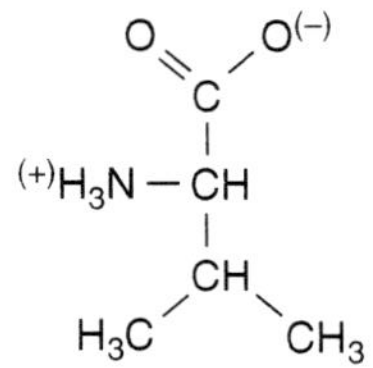	117.15	2.29 9.72	6.0	Essential

(*Continued*)

TABLE 9.1 (Continued)
Structures and Selected Properties of 20 Amino Acids

Type	Name 3-Letter Abbreviation Symbol	Structure	Molecular Weight	pKa	pI	Classification
Aromatic	L-Phenylalanine Phe F $C_9H_{11}NO_2$	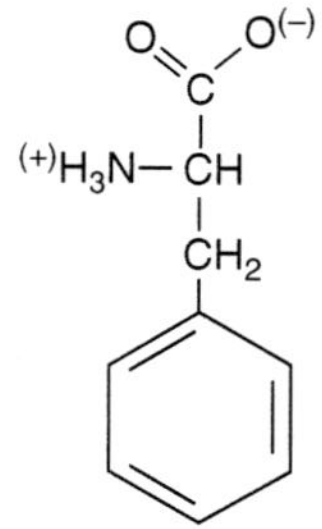	165.19	2.58 9.24	5.5	Essential, can replace tyrosine
	L-Tyrosine Tyr Y $C_9H_{11}NO_3$	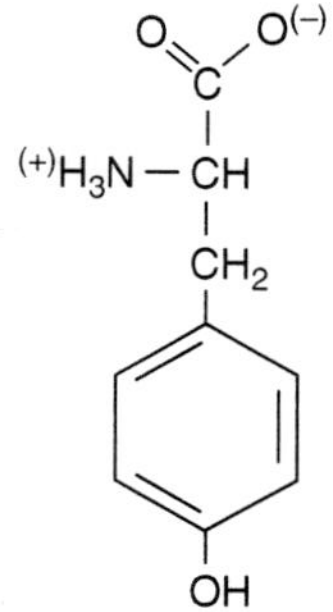	181.19	2.20 9.11 10.07	5.7	Nonessential, can be replaced by phenylalanine

	L-Tryptophan Try W $C_{11}H_{12}N_2O_2$	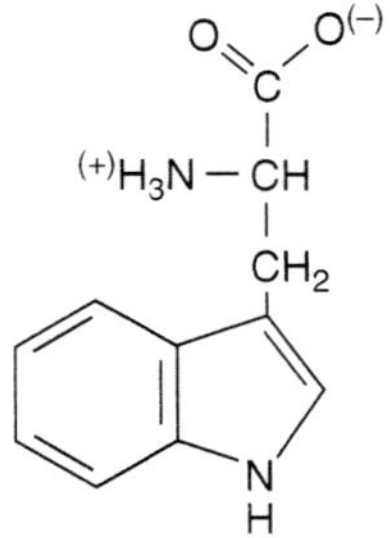	204.22	2.38 9.39	5.9	Essential
Sulfur containg	L-Cysteine Cys C $C_3H_7NO_2S$	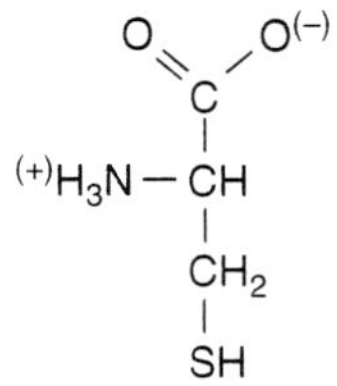	121.16	1.71 8.33 10.78	5.0	Nonessential, can partially replace the essential amino acid methionine
	L-Methionine Met M $C_5H_4NO_2S$	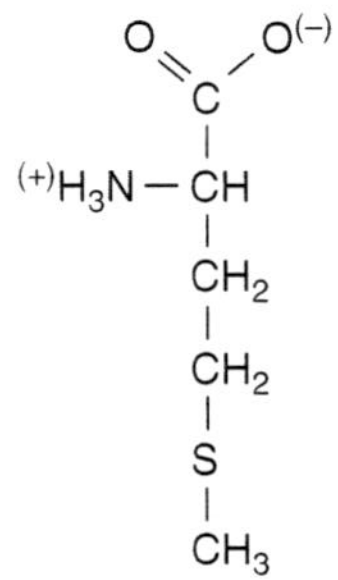	149.21	2.28 9.21	5.7	Essential

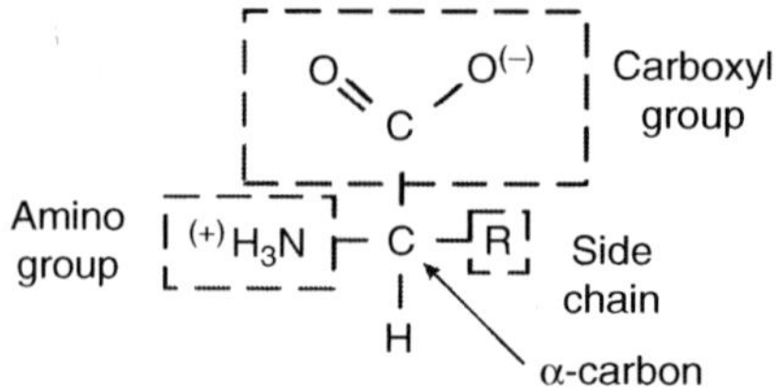

FIGURE 9.1 Generic structure of amino acids.

Structurally, the 20 amino acids found in proteins are α-amino carboxylic acids (see Figure 9.1), differing only in their side chains (R groups, see Table 9.1) with the exception of proline which is an α-imino, carboxylic acid. Some other nonprotein amino acids, however, only vary as to the location of the amine group, e.g., GABA, a γ-amine. Others may not even be a carboxylic acid; taurine, an amino acid important in the production of bile, is a β-amino sulfonic acid (see Figure 9.2)

Nutritionists have primarily focused on the 20 amino acids found in proteins, with special emphasis on the approximately 10 amino acids that are considered essential. Essential amino acids are those that must be present in a healthy diet because they cannot be synthesized by the human body. The exact list of amino acids that are considered essential varies slightly according to factors such as the age of the individual or the individual's state of health. For example, histidine is only essential in infants.

Since the focus has been on the ten essential amino acids, most nutritional laboratories have optimized their methods around these plus some of the other protein amino acids. Recently, nutritional supplements have begun to include some of the nonprotein amino acids such as ornithine in their formulations. These non-protein amino acids may cause interference when analysis is performed by methods optimized for standard nutritional work. There are amino acid analytical methods developed for physiological matrices, which include many of the nonprotein amino acids. These methods may be adapted for nutritional matrices. There is also some

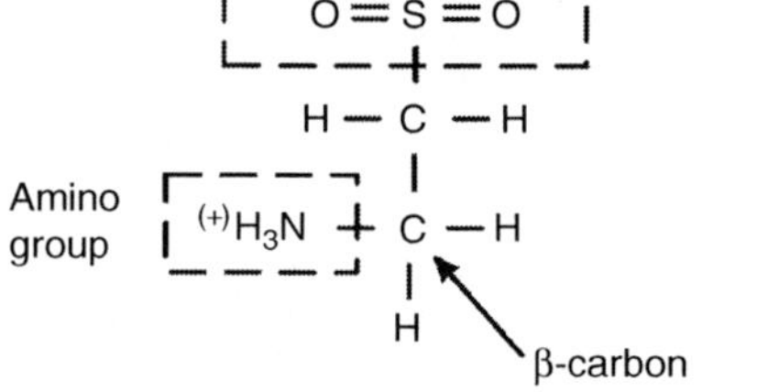

FIGURE 9.2 Taurine.

interest in non-physiological amino acids, e.g., 4-hydroxy-isoleucine, which are not found in mammalian tissue. The physiological methods will not include such amino acids, and specialized methods may need to be developed for these. It is important to know if a matrix contains these nonprotein amino acids; otherwise, they may either not be quantified by the method and/or interfere with the quantification of other amino acids.

Another important property of the protein amino acids is that they are all L-isomers, with the exception of glycine (glycine only has one configuration). The body cannot use the D-isomers of these amino acids, and, in fact, some D-isomers may be toxic. An exception is D-methionine, which the body can convert into L-methionine through transamination. Naturally produced amino acids are L-isomer; however, manufactured amino acids can be a racemic mixture with half of the molecules being D-isomers. These can then be separated and only the L-isomer used to fortify foods. The amino acid analysis methods normally employed in the nutritional laboratory will not differentiate the D- and L-isomers. There are, however, chiral chromatography and enzymatic methods that can be used if the product is suspected of containing D-isomers.

The amino acids in nutritional matrices may be present as free amino acids, peptides, proteins, and/or in polymers with nonpeptide bonds. Each of these forms may need to be prepared for analysis by different methods. Free amino acids can be directly extracted from the matrix and analyzed. Peptides and proteins must be hydrolyzed to free the amino acids before analysis. The analyses of the hydrolyzed matrices will give "total" amino acids, that is, those amino acids that were part of peptides/proteins as well as any free amino acids. The analysis of nonpeptide bonded polymers of amino acids such as aspartame often needs to be analyzed by specialized methods.

9.2 ACID HYDROLYSIS OF PROTEINS AND PEPTIDES

Acid hydrolysis is the primary method used to cleave the peptide bonds in proteins to release the amino acids for analysis. The conditions used in acid hydrolysis of foods for amino acid analysis are a set of compromises. These compromises are necessary because the stability of each amino acid varies, with threonine and serine being the most labile amino acids. The strength of the peptide bonds between the various amino acids also varies. Bonds involving valine, isoleucine, and leucine are the most difficult to hydrolyze. The compendium method (AOAC, 2000) using 6 *N* hydrochloric acid (HCl) at 110°C for 24 h is a method designed to minimize the degradation of the more labile amino acids while maximizing the release of all amino acids from the proteins. For most analytical work this will give reproducible, reliable results. There are, however, many pitfalls in this method that require experience and attention to detail. Many of these pitfalls are matrix and amino acid dependent. When more precise values are needed for the more difficult to hydrolyze or more labile amino acids, a time course of 6 *N* HCl hydrolysis can be conducted. Usually, the digestion times are 16, 24, and 72 h (Zumwalt et al., 1987). The values for the more labile amino acids such as threonine and serine are then determined by extrapolating back to the zero time line. The value for the difficult-to-hydrolyze amino acids is determined by their maximum values. Time line analysis is not routine for most nutritional labs because of the expense and the delay in acquiring results.

For matrices that are not pure protein, the hydrolysis is usually done "wet." The HCl and sample are together in the same vessel, usually a culture tube with a Teflon-lined cap or an ampule. A tight seal on the tube is important for two reasons. First, the presence of oxygen will degrade the amino acids and therefore needs to be minimized. With most food matrices, it is sufficient to either apply a vacuum or flush the tube with nitrogen and then quickly seal the tube. The competing hydrolysis of the carbohydrates and fats in the food matrix will quickly consume any remaining oxygen. Second, the tube needs to be sealed to prevent the loss of HCl. Pure HCl at room conditions is a gas. Concentrated HCl reagent (concentrated HCl) is the HCl gas dissolved in water to a concentration of approximately 37% or 12 *N*. The solubility of gases, unlike most solids, decreases as the temperature increases. This loss of HCl will result in incomplete hydrolysis, preferentially causing a low recovery in glutamic acid and serine. However, if the HCl loss is severe enough, all amino acid recoveries will suffer.

Most methods of acid hydrolysis agree that the acid should be HCl and that the concentration should be 6 *N*. Early researchers tried different concentrations and used different acids, but 6 *N* HCl gives the best recoveries. HCl is also easily removed from the hydrolysate by evaporation. This may be necessary for some of the precolumn derivatization methods. The amount of acid recommended is expressed in the molar ratio of acid to nitrogen content, with a ratio of 2800:1 being reported by Robel (1973) as being optimal. However, the matrix can affect the amount of acid required. A matrix high in soluble carbohydrates such as molasses may require a higher acid to nitrogen ratio (Roache and Gehrke, 1970).

Temperature and time are two interdependent variables that have been examined in detail by several researchers. Some of these researchers (Roache and Gehrke, 1970) have concluded that a temperature of 145°C for 6 h is a good alternative to the 110°C for 24 h used in compendial methods.

Protective agents such as phenol, 2-mercaptoethanol, and thioglycolic acid can be added to the hydrolysis step. It is believed that at least some of the tyrosine lost during acid hydrolysis may be due to its reaction with chlorine in the presence of any oxidizing agents. The addition of phenol, a halogen scavenger, may reduce this tyrosine loss. Bromine contamination will also result in the loss of tyrosine. The addition of reducing agents such as 2-mercaptoethanol and thioglycolic acid has been reported to also improve tyrosine recoveries. Inglis (1983) has reported that the addition of tryptamine increases the yield of tryptophan during acid hydrolysis.

Wet hydrolysis may not yield good results when using pure proteins, especially when the sample size is limited. Pure proteins can be hydrolyzed by "gas-phase hydrolysis" (Meltzer et al., 1987). With gas-phase hydrolysis, the HCl is placed in a separate container from the sample. The containers are then sealed into a secondary container. The secondary container is commonly a glass desiccator sealed with a Teflon gasket and placed in a desiccator cage. The desiccator is cycled between an inert gas, commonly nitrogen, and a vacuum, with the final cycle maintaining a vacuum in the desiccator. The desiccator is then placed in an oven at 110°C for the appropriate time.

On heating, the HCl gas is released from the 6 *N* HCl, and the gas then hydrolyzes the protein. This is a good technique for pure proteins of limited quantity, since most contaminants present in the 6 *N* HCl are usually nonvolatile and therefore cannot contaminate the samples. Another advantage is that the HCl and moisture that may condense on the samples can easily be removed with vacuum, leaving the hydrolyzed protein in a state ready for precolumn derivatization or to be dissolved in the appropriate solvent for chromatography. The disadvantage is that some of the protective agents that can be added to the 6 *N* HCl are not volatile and are therefore ineffective with gas-phase hydrolysis.

Not all amino acids can be analyzed using only acid hydrolysis. Tryptophan is completely destroyed by acid hydrolysis, and therefore, the protein must be hydrolyzed using alkaline conditions. Glutamine and asparagine are completely converted to glutamic acid and aspartic acid during acid hydrolysis. Therefore, it must be considered that the glutamic acid and aspartic acid values include glutamine and asparagine. This is often denoted as Glx instead of Gln or Glu and Asx instead of Asn or Asp. Glutamine can be quantified if it is first converted to a compound that is stable during acid hydrolysis. Kuhn et al. (1996) report a method for pure proteins where the glutamine is converted to diaminobutyric acid by reaction with bis(1, 1-trifluroacetoxy)iodobenze. The protein is then hydrolyzed and the diaminobutyric acid is quantified.

The sulfur-containing amino acids cysteine and methionine are partially degraded and/or oxidized during acid hydrolysis, especially in the presence of carbohydrate. Cysteine is partially converted (nonreproducibly) to cysteic acid, and methionine is partially converted (nonreproducibly) to methionine sulfoxide and methionine sulfone. These amino acids can be stabilized by completely oxidizing them to cysteic acid and methionine sulfone with performic acid before acid hydrolysis. This procedure will be discussed in more detail later.

9.3 ALKALINE HYDROLYSIS OF PROTEINS AND PEPTIDES FOR TRYPTOPHAN ANALYSIS

As has been stated, tryptophan is completely destroyed by acid hydrolysis. Some methods protect tryptophan with compounds such as 2-mercaptoethanol, but these methods do not always give reproducible results. Other methods for the liberation of tryptophan have included proteolysis by enzymes. An early method developed by Spies and Chambers (1949) did not require freed tryptophan. It is a colorimetric assay, reacting the tryptophan with *p*-dimethylaminobenzaldehyde (PDBA). This method has been useful in pure proteins, but many food matrices have interfering substances. In addition, tryptophan is one of only two amino acids with a strong extinction coefficient in a usable ultraviolet (UV) range (approximately 280 nm depending on solvent, etc.). However, the most reliable method of tryptophan analysis is to release the amino acid from protein with alkaline hydrolysis (Lucas et al., 1980) and then use chromatography for quantification.

Several different alkalis have been used for the hydrolysis of tryptophan, including lithium, sodium, potassium, calcium, and barium hydroxides. Barium hydroxide has been reported to be less effective than either sodium hydroxide or lithium hydroxide. Also, the barium needs to be completely removed from the sample before chromatography. Usually, this is achieved by precipitation, which often reduces the recovery of tryptophan. Barium is also more toxic than the other cations. Potassium hydroxide and sodium hydroxide attack glass more vigorously than lithium or calcium hydroxide, but calcium hydroxide has solubility problems. Therefore, sodium and lithium hydroxides have become the most commonly used alkalis with very little difference in performance.

As with acid hydrolysis, alkaline hydrolysis is a compromise of maximizing the release of tryptophan while minimizing its degradation. If sodium hydroxide is being used, some researchers have reported less degradation if it contains 5% stannous chloride. The concentration of either sodium hydroxide or lithium hydroxide is usually 4 or 4.2 M. It is important to remove as much oxygen as practical by sonicating, vacuum, nitrogen flushing, or a combination of these techniques. The reaction is commonly carried out at 110°C for 20 h. Higher temperatures for shorter times have also been shown effective, such as 145°C for 4 to 8 h depending on the matrix (Roache and Gehrke, 1970).

The alkaline hydrolysis of tryptophan degrades many of the other amino acids; therefore, it is usually used solely for the analysis of tryptophan. Because of this, tryptophan's strong absorbance at 280 nm can be used for quantification, thereby eliminating pre- or postcolumn derivatization. Reversed-phase chromatography using UV detection is an excellent quantification method for tryptophan.

9.4 PERFORMIC ACID OXIDATION OF CYSTEINE AND METHIONINE

As has been mentioned, the sulfur-containing amino acids cysteine and methionine are partially and irreproducibly converted to cysteic acid, methionine sulfoxide, and methionine sulfone, respectively, during acid hydrolysis. The most often used technique for overcoming this difficulty is to quantitatively convert these amino acids to derivatives that are stable during acid hydrolysis. Performic acid will oxidize cysteine to cysteic acid and methionine to methionine sulfone, both of which are stable during acid hydrolysis (Moore, 1963; Zumwalt et al., 1987). Cystine, the disulfide dimer of cysteine is also oxidized to cysteic acid. Performic acid oxidation cleaves the disulfide and oxidizes the thiol groups to sulfonate groups yielding two cysteic acids. The quantification of cysteine, even in the free state, is difficult. The thiol groups of two cysteines are easily oxidized by oxygen from the air to produce cystine. Acid hydrolysis promotes this reaction. This led to the early discovery of cystine as an amino acid and only later was it realized that cystine was simply a dimer of cysteine. Cysteine is the amino acid that has a DNA (gene) codon and not cystine. Also, the proportion of cysteine that exists in a protein as cystine is often small.

The performic acid used in oxidation of the sulfur-containing amino acids is produced in the laboratory by reacting formic acid with hydrogen peroxide. One part of 30% hydrogen peroxide is mixed with nine parts of 88% formic acid. This mixture is cooled (the hydrogen peroxide and formic acid may be cooled before mixing) and allowed to react for 1 h. The 1-h reaction time will maximize the concentration of performic acid. The cold performic acid is then added to the sample and allowed to react for 16 h while being kept cold, i.e., a cold water bath, a refrigerator, or in a −20°C freezer.

After oxidation, the sample is then acid hydrolyzed, but first the excess performic acid must be removed. This can be accomplished in several ways: (1) the oxidized sample may be evaporated to dryness by vacuum and heat or (2) a reducing agent such as hydrogen bromide or sodium bisulfite can be added. The hydrogen bromide must be completely removed before acid hydrolysis. This can be accomplished by a vacuum. Because of this, it is often used to ensure the complete removal of performic acid when the oxidized sample is evaporated to dryness for other purposes. The sodium bisulfite does not need to be removed. The sulfur dioxide evolved from the neutralization of the performic acid with sodium bisulfite should be removed by degassing and nitrogen purging.

9.5 MICROWAVE HYDROLYSIS

The major drawback to the compendium acid hydrolysis method is the long (24 h) digestion time. This offsets the advances made in the "quantification" portion of the analysis by improved high-performance liquid chromatography (HPLC) methods. In 1987, Chen et al. were the first to report the use of microwave ovens to hydrolyze proteins. They performed wet hydrolysis of 0.2- to 0.5-mg samples of ribonuclease A and insulin B chain in 6 *N* HCl and a 1:1 propionic acid–12 *N* HCl mixture. The reaction vessels were customized Teflon vials with silicon septa and Teflon caps. Earlier work in microwave digestion had shown that glass vessels could not withstand the pressures generated during hydrolysis. Each vial was flushed with nitrogen and heated in a household microwave oven for 3 to 5 min. The recoveries were in good agreement with the theoretical values and those obtained by conventional hydrolysis (24 h, 6 *N* HCl at 110°C). Chen et al. (1987) were unable to measure the acid temperature and pressures generated in the digestion vessels. Others have shown typical acid temperature to be in the range of 130° to 175°C with pressures greater than 100 psi.

Several researchers (Engelhart, 1997) have investigated the use of microwaves to hydrolyze proteins since Chen et al. (1987). Their findings have indicated that microwave hydrolysis is successful using the same reagents as conventional hydrolysis. Various researchers have found that the reaction time required varies with the protein or peptide. The difficult-to-hydrolyze peptide bonds such as Val-Val, Ile-Ile, etc. require more time, in some cases 60 min or more. However, this is still far shorter than the time required for conventional acid hydrolysis. However, the more labile amino acids threonine and serine can show greater losses at the higher temperatures generated during microwave acid hydrolysis.

Under strictly controlled conditions, amino acids have also been shown to be at least as stable under microwave hydrolysis as with conventional hydrolysis. Degradation of amino acids during microwave hydrolysis appears to be equivalent to that which occurs during conventional hydrolysis. The degree of racemization seems to be significantly less with microwave hydrolysis.

In wet hydrolysis, specialized reaction vessels are required for microwave hydrolysis. The material used to construct these vessels must (1) be transparent to microwaves, (2) be able to resist strong acids or alkalis, (3) be thermally stable at elevated temperatures (typically 100° to 180°C), (4) be able to withstand pressures of well over 100 psi, and (5) not interact with the microwaves. The materials commonly used are fluoropolymers such as Teflon. The vessels need to have a mechanism to allow them to be evacuated and flushed with nitrogen. For safety reasons, a pressure-release mechanism should also be incorporated. A number of reaction vessels have exploded during microwave digestion. Every precaution needs to be taken to prevent explosions for the protection of the laboratory personnel and the lab. For gas-phase hydrolysis, open autosampler vials are commonly placed into a reaction vessel that has the same properties as described for wet hydrolysis.

As stated earlier, the conditions of microwave acid hydrolysis must be tightly controlled for good results. Temperature seems to be one of the most critical parameters in using microwave acid hydrolysis. If the temperature is too high, there is increased loss of the more labile amino acids; however, if the temperature is not high enough, there is incomplete hydrolysis. Because of the temperature sensitivity of the hydrolysis, the inexpensive household microwave oven will not give consistent results. Microwave acid hydrolysis of proteins requires a laboratory-grade microwave oven, which has temperature probes with fully automated temperature feedback control. The expense of such a microwave oven, along with the expense of specialized reactor vessels, requires a relatively high initial capital investment as compared to the compendium acid hydrolysis.

Microwave acid hydrolysis is not a panacea for protein hydrolysis. It has the same problems and pitfalls as the compendium hydrolysis. Its chief advantage is a much shorter hydrolysis time, allowing more rapid turnover of samples.

9.6 FREE AMINO ACID EXTRACTION

Amino acid fortification is usually done with pure free amino acids. Since the concentration of free amino acids is extremely low in most "naturally" occurring food matrices, extracting the free amino acids will verify the fortification.

The high solubility in water of most amino acids makes their extraction straightforward; however, it is advisable to have the extractant the same or similar to the initial mobile phase used in the chromatography. If precolumn derivatization is used, the extractant should be either the solvent needed for the precolumn derivatization or a solvent easily exchanged to the derivatization solvent.

A weak HCl solution is a good extractant for free amino acids. If the chromatography is ion exchange with postcolumn derivatization, the extracted amino acids can be chromatographed directly. The weak HCl solution will protonate the amino acids so that they will have the correct charge for application to the ion-exchange column. If precolumn derivatization is used, the weak HCl can be removed by evaporation under vacuum and the sample redissolved in the solvent dictated by the precolumn derivatization.

Recently, there has been interest in fortification and supplementation with cystine. Research indicates that increased cystine intake is beneficial to people and animals under stress. Pure cystine, however, is not very soluble in aqueous extractants unless the pH is either below 2 or above 8. Cystine is soluble in a 1.0 *N* HCl solution, and as long as any dilutions keep the HCl concentration greater than 0.1 *N*, the pH should remain below 2. This pH range can be chromatographed directly with ion-exchange postcolumn reaction chromatography. Another difficulty with cystine is that it is not very reactive with *o*-phthalaldehyde (OPA, discussed later), a common postcolumn and precolumn derivative. However, ninhydrin (also, discussed later) will react well.

Monosodium glutamate (MSG) is the sodium salt of glutamic acid and therefore easily quantified as free glutamic acid. Of course, this yields total free glutamic acid and does not distinguish between glutamic acid added as MSG or in some other form.

9.7 INTERNAL STANDARDS AND CONTROLS

The processes of both acid and alkaline hydrolysis are prone to sample loss, not only because of the extreme conditions needed for hydrolysis, but also because of the number of manipulations required to prepare the sample for chromatography. Any method with this potential for sample loss should use a suitable internal standard.

An internal standard needs to be added as early as possible in the procedure so that the maximum number of steps can be monitored. For amino acid hydrolysis, this should be right after the sample is quantitatively aliquoted and definitely before hydrolysis or oxidation.

The internal standard should have chemical and physical properties as close to those of the analyte as possible. If these conditions are met, then anything that may affect the recovery of the analyte will be reflected by a proportional loss of the internal standard. Ideally, the recovery of the internal standard should be near 100%. If it is, the concentration of the analyte in the sample may be adjusted using the internal standard recovery data. Too great a loss of the internal standard should be an indication that the preparation of the sample was flawed and should be repeated.

Another factor that must be considered when selecting an internal standard is possible interference with the analyte. In the case of amino acid quantification with HPLC, the internal standard must not coelute with the other amino acids. The internal standard's relative retention time to the amino acids of interest is also important. If the amino acids being quantified are eluted near the beginning of the gradient, it is more efficient to have the internal standard elute also near the beginning. This can greatly reduce the length of the chromatography run, allowing more samples to be chromatographed and making the results available earlier.

For amino acid analysis, the internal standard is usually any number of amino acids that either do not occur in nature or are not present in the matrices being analyzed. For acid hydrolysis, norleucine is probably the most widely used internal standard, but other amino acids such as β-alanine are also used (see Table 9.2). An internal standard commonly used for the alkaline digest of tryptophan is 5-methyl tryptophan.

The use of an internal standard does not eliminate the need for control samples; it only supplements the control data. Because the effectiveness of the hydrolysis is dependent on the matrix being analyzed, the control should be either the same matrix or a matrix with very similar properties. This means a general nutrition lab will need

TABLE 9.2
Partial List of Internal Standards

L-Norleucine
β-Alanine
L-Homoarginine hydrochloride
L-Norvaline
5-β-(4-Pyridylethyl)-dl-penicillamine
L-Homoserine

more than one control. In a lab that analyzes many different matrices, it is impossible to track a control for each matrix. Therefore, the lab needs to select controls that best reflect the range of matrices and protein content. Not every control needs to be analyzed with each batch of samples as long as the similar samples are batched together with a similar control.

9.8 DETECTION OF AMINO ACIDS

Today, most amino acid analysis is conducted with HPLC. Gas chromatographic methods are less commonly used. The detection of amino acids, however, is not as straightforward as with other substances. Only tyrosine, tryptophan, and phenylalanine have significant UV absorption at useful wavelengths, and only tryptophan and tyrosine are fluorescent. This leaves either using a universal detection method, such as refractive index, light scattering, or very low UV wavelengths, or labeling the amino acids by derivatization with a substance that absorbs in the visible/UV wavelengths or fluoresces. The derivatization can be done either after elution from an HPLC column (postcolumn derivatization) or before HPLC separation (precolumn derivatization). Both of these systems have their advantages and disadvantages.

9.8.1 Postcolumn Derivatization

As stated, postcolumn derivatization labels the amino acids after elution from the HPLC column. Postcolumn derivatization involves introducing the derivatization reagents to the HPLC column eluant before the detector and providing the necessary conditions so that the reaction can proceed either to completion or reproducibly to a stage to allow adequate detection. The typical postcolumn reactor system (see Figure 9.3) consists of a pump(s) to deliver the derivatization reagent(s) into the column eluant, a reactor to provide the necessary mixing and time for the reaction to occur, and usually a heater for the reactor. The reactor in most cases is a length of tubing of the same internal diameter as the tubing leading from the column to the detector. The length (volume) of the tubing along with the flow rate controls the time the reaction has to occur. Since the postcolumn pressures are much less than precolumn pressures, the choice of material used is wider. Typically, the tubing is polypropylene, PEEK, or stainless steel. The material needs to be compatible with the reagents and thermally stable at the reactor temperature.

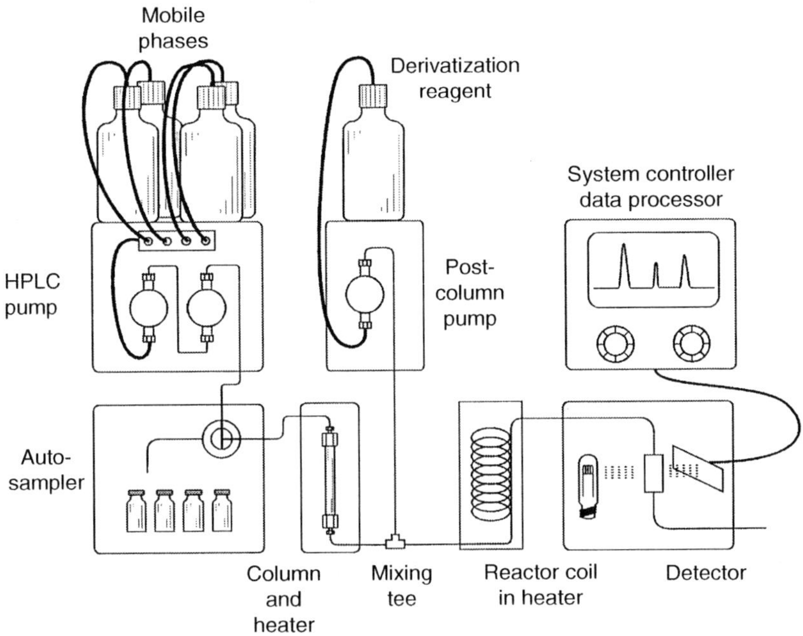

FIGURE 9.3 Chromatograph with a postcolumn reactor.

Commonly for postcolumn derivatization of amino acids, the separation is done by cation-exchange chromatography (discussed later). In this case, the separation of the individual amino acids is based solely on the properties of the amino acids. Another advantage of postcolumn derivatization is that there is much less sample preparation needed. Typically, the hydrolyte is only filtered and diluted with an appropriate solvent, either water or the initial mobile phase. The automation of the derivatization reduces the amount of analyst time. The disadvantages of postcolumn derivatization are (1) that the addition of reagents to the eluant stream can broaden the peaks and reduce resolution, (2) the high cost of the postcolumn reactor, and (3) that the choice of derivatization is restricted by the kinetics of the reaction and compatibility between the derivatization reagents and the mobile phases. For operations running relatively large numbers of samples, postcolumn derivatization is a good choice. The savings in analyst time can easily offset the cost of the post-column reactor.

The two most commonly used postcolumn detection reagents are ninhydrin (Spackman et al., 1958) and OPA (Roth, 1971), with ninhydrin as the more commonly used (see Table 9.3 for a summary of these derivatizations).

9.8.1.1 Postcolumn Derivatization with Ninhydrin

The ninhydrin reagent contains two active components: ninhydrin and hydrindantin (reduced ninhydrin). Since ninhydrin, hydrindantin, and the reaction product Ruhemann's Purple are insoluble in water, an organic solvent that is miscible with water

TABLE 9.3
Partial List of Amino Acid Derivatization Methods

Method	Chromatography	Detection	Remarks
Postcolumn			
Ninhydin	Ion-exchange column Mobile phase: pH and NaCl gradient, usually citrate either sodium or lithium	Visible: 570 nm for primary amines; 440 nm for secondary amines	Detects primary and secondary amines
OPA (*o*-phthalaldehyde)	Ion-exchange column Mobile phase: pH and NaCl gradient, usually citrate either sodium or lithium	Fluorescence: Ex λ 340 nm Em λ 455 nm	Will only detect secondary amines if first oxidized with sodium hypochlorite
Precolumn			
OPA (*o*-phthalaldehyde)	Column: reversed-phase C_8 or C_{18}	Fluorescence: Ex λ 340 nm Em λ 455 nm	Will only detect secondary amines if first oxidized with sodium hypochlorite
PITC (phenylisothiocyanate) (Pico-Tag, Waters Corp.)	Column: reversed-phase C_8 or C_{18}	UV: 254 nm (λmax = 269 nm)	Detects primary and secondary amines
FMOC (9-fluornylethyl chloroformate)	Column: reversed-phase C_8 or C_{18}	Fluorescence: Ex λ 263 nm Em λ 313 nm	Detects primary and secondary amines
Dansyl chloride	Column: reversed-phase C_8 or C_{18}	Fluorescence: Ex λ 360 nm Em λ 470 nm	Detects primary and secondary amines
Dabsyl chloride	Column: reversed-phase C_8 or C_{18}	Visible: 436 nm	Detects primary and secondary amines

must also be present. Many ninhydrin formulations use DMSO (dimethyl sulfoxide). DMSO presents a hazard in the laboratory since it is readily absorbed through the skin and will act as a carrier for almost anything dissolved in it. Another solvent used is methyl cellosolve. Methyl cellosolve is also absorbed through the skin and can produce severe headaches. Pickering Laboratories (Mountain View, CA) in its ninhydrin reagent TRIONE uses Sulfolane in place of DMSO.

The ninhydrin oxidizes the amino acid, releasing the amine group as ammonium, and is thereby reduced, forming hydrindantin. The hydrindantin then reacts with the liberated amine to form Ruhemann's Purple. The ninhydrin reagent contains additional hydrindantin to insure adequate concentrations to drive this reaction to completion. Ruhemann's Purple absorbs light at 570 nm. Secondary amino acids react differently. Again, the amino acid is oxidized to release the amine, which then reacts with the ninhydrin to form a yellow chromophore, which has a maximum absorbance at 440 nm. This allows the quantification of both primary and secondary amino acids, but requires the detector to either monitor two (2) wavelengths simultaneously or be able to switch between the two (2) wavelengths at the appropriate times.

The ninhydrin derivatization occurs very slowly at temperatures much below 120°C; therefore, the reactor must be heated. Typically, the reactor temperature is set between 125° and 135°C. The upper temperature approaches the boiling point of the reagents at typical reactor pressures. This may lead to precipitation of some of the components and blockage of tubing, including the detector flow cell. If the tubing or flow cell is completely blocked, it often cannot be unblocked and will usually need to be replaced.

Since the amino acid is destroyed during ninhydrin derivatization and forms the same reaction product regardless of the original amino acid (except for the secondary amino acids), ninhydrin derivatization will only work as a postcolumn method.

9.8.1.2 Postcolumn Derivatization with *o*-Phthalaldehyde

A very sensitive and specific postcolumn derivatization reagent is OPA. OPA reacts with primary amines in the presence of a reducing reagent such as 2-mercaptoethanol, producing a fluorescent substance with an excitation wavelength of 340 nm and an emission wavelength of 455 nm. The reaction occurs in an alkaline borate buffer in the range of pH 9 to 11. The most commonly used form of borate is the potassium salt; however, the sodium salt has also been used.

The postcolumn reactor flow rate is equal to the column mobile phase flow rate. The volume of the reaction coil is configured to give a reaction time of 20 to 30 sec. The reaction will proceed adequately at room temperature, but with some systems, a higher response may be obtained with an elevated reaction coil temperature. A temperature of 50° to 60°C will usually maximize the response. This temperature is far lower than the 120° to 130°C required for the ninhydrin derivatization and therefore should not cause problems. The reaction coil can be made of stainless steel, polypropylene, Teflon, or PEEK.

Two amino acids, cysteine and proline, give poor or no fluorescence with OPA derivatization. The very low fluorescence of cysteine-OPA will only be a problem with free cysteine or its dimer, cystine. Since cysteine can be converted to cysteic acid using performic acid oxidation, this drawback can be eliminated because cysteic acid-OPA has a strong fluorescence.

Since proline is a secondary amine, it cannot react with OPA unless its pyrrolidyl ring is first opened, producing a primary amine. This can be accomplished by oxidizing proline with sodium hypochlorite (Dong et al., 1985). The sodium hypochlorite oxidation is done postcolumn using a second postcolumn reagent pump, mixing tee, and a heated reaction coil placed upstream of the OPA mixing tee (see Figure 9.4). A 0.0075% sodium hypochlorite solution in the borate buffer is pumped with a flow rate of one half of the column mobile phase and OPA reagent flow rate. The hypochlorite oxidation of proline occurs very slowly at room temperature. Therefore, the reaction coil must be heated to between 50 and 60°C. The volume of the reaction coil should allow a reaction time of 20 to 25 sec.

Lysine also yields lower fluorescence than the other amino acids. However, the addition of a detergent, typically Brij or sodium dodecyl sulfate (SDS), to the reaction will increase the signal.

Unlike ninhydrin, the reaction of amino acids with OPA does not destroy the amino acids, but adds the fluorescent tag to the primary amine. This allows OPA to also be used as a precolumn derivative, since the derivatization product is unique for each amino acid.

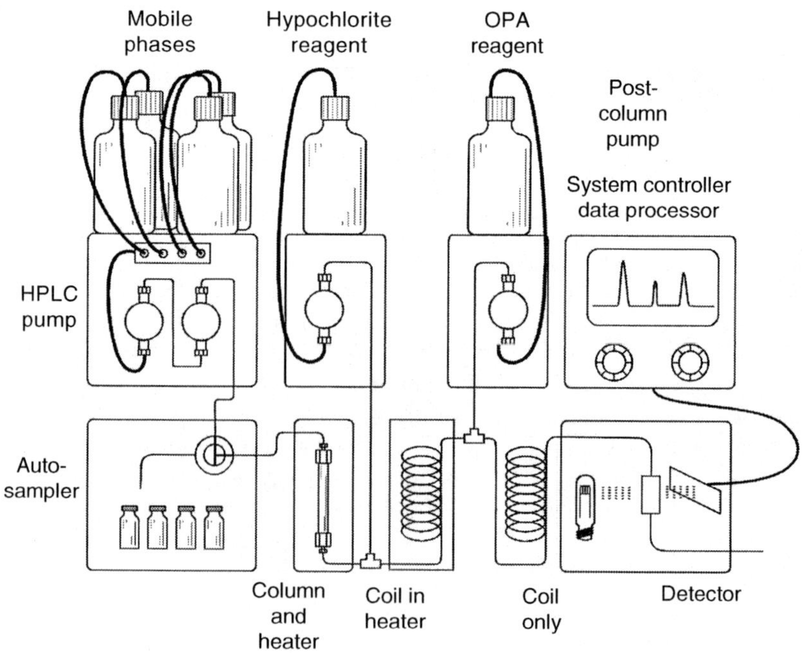

FIGURE 9.4 Chromatograph with a postcolumn reactor and postcolumn hypochlorite oxidation.

9.8.2 Precolumn Derivatization

Precolumn derivatization has advantages and disadvantages as compared to postcolumn derivatization. The chromatography for most precolumn derivatization methods is commonly reversed phase using either C8 or C18 on silica. Silica supports can withstand higher pressures (usually up to 5000 psi) than the polymeric supports needed for ion-exchange chromatography used with postcolumn derivatization. This translates to higher flow rates and shorter run times.

The disadvantage of precolumn derivatization is increased manipulation of the sample before it can be chromatographed. Most autosamplers today can automate the derivatization reaction, thus freeing lab personnel for other activities. However, the derivatization reaction will usually not occur on the still very acidic hydrolyte. This necessitates the prior removal of the acid and water from the sample. This is commonly done by a vacuum. To completely remove the acid from the sample, the sample often has to be dried, redissolved (usually in water), and then dried a second time. Once dried, the sample is redissolved using a solvent compatible with the derivatization reaction. The additional time required by these steps can offset the savings achieved by using reversed-phase chromatography. Gas-phase hydrolysis of pure proteins eliminates the need to remove the acid from the sample.

With some precolumn derivatization methods, the derivatization reagents must also be removed before the sample can be chromatographed. This eliminates the automation of the reaction with the autosampler and increases the amount of sample preparation time.

9.8.2.1 *o*-Phthalaldehyde Precolumn Derivatization

OPA precolumn derivatization is probably the most commonly used precolumn method. OPA precolumn derivatized amino acids are detected with fluorescent detectors using the same wavelengths used for postcolumn OPA. The derivatized sample can also be loaded directly on the reversed-phase column, allowing automation of the derivatization reaction with an autosampler. The sample, however, does need to be evaporated to dryness after hydrolysis for removal of the acid. The removal of the acid needs to be complete, usually requiring redissolving and redrying or strongly buffered to an alkaline pH.

The OPA derivatization reaction is rapid at room temperature, but, as with most chemical reactions, the time to "completion" can be long. This means that the derivatization reaction must be stopped at a carefully controlled time to get repeatable, accurate results. The reaction can be stopped either by injecting the sample onto the column or adding a stopping reagent to the reaction. Either of these means is applicable to autosampler automation (Carpino and Han, 1972).

The precolumn derivatization with OPA is conducted under similar conditions to the postcolumn derivatization condition. The reaction is carried out in a borate buffer with a pH of 9 to 11, containing Brij or SDS, and 2-mercaptaethanol. The reaction occurs at room temperature and can be stopped either by injecting the mixture onto the column, or by lowering the pH of the reaction by the addition of an appropriate buffer such as potassium dihydrogen phosphate with a pH around 4.3. This latter method is preferable since the results are easier to reproduce.

Precolumn derivatization with OPA has the same drawbacks as postcolumn derivatization. Secondary amino acids such as proline will not react unless first oxidized. Cysteine and cystine will not react, but can first be oxidized to cysteic acid. The addition of Brij or SDS to the reaction increases the fluorescence of lysine.

9.8.2.2 9-Fluorenylmethyl Chloroformate Derivatization

9-Fluorenylmethyl chloroformate (FMOC) (Carpino et al., 1972) is another popular precolumn derivatization reagent. FMOC derivatization produces a very stable fluorescent product with both primary and secondary amines. The reaction is very straightforward. The sample is first mixed with a pH 7.7 borate buffer. Next the FMOC dissolved in acetone is added. The reaction is rapid, taking about 40 s.

The major drawback to the FMOC derivatization is that unlike OPA, FMOC is itself fluorescent and reacts with water to give a fluorescent product. If excess FMOC is not removed, it elutes in a broad band in the middle of the profile interfering with amino acid quantification. The excess reagent can be dealt with in two ways. A very hydrophobic amine such as 1-amino-adamantane can be added. This will then elute after lysine, the last amino acid to elute with reversed-phase chromatography. This, of course, lengthens the chromatography run time. The FMOC can also be extracted from the reaction by liquid–liquid extraction using pentane. This is usually done twice. The liquid–liquid extraction complicates automation of the reaction.

9.8.2.3 Other Precolumn Derivatization Methods

There have been many other precolumn derivatization methods developed, each with its own list of pros and cons. Table 9.3 lists some of the other derivatization methods. Some of these methods are available in commercial kits such as Waters Corporation's (Millford, MA) Pico-Tag and AccQ-Tag. These kits ensure quality reagents and very reproducible results. Commercial kits can be of benefit to the laboratory that does not conduct routine analysis of amino acids. However, these kits are not foolproof, nor do they guarantee good results. These kits may not work for all matrices, or the conditions may need modification with different matrices. No kit can completely replace experience.

9.9 CHROMATOGRAPHY OF AMINO ACIDS

The HPLC methods for amino acid analysis fall predominately into two types: ion exchange and reverse phase. Ion-exchange chromatography (IEC) is predominately used with postcolumn derivatization, while reversed-phase chromatography (RPC) is used with precolumn derivatization.

9.9.1 Ion-Exchange Chromatography

The IEC used for amino acid analysis is strong cation exchange (scx), that is, the stationary phase has a negative charge. This negative charge attracts positively charged cations and repels negatively charge anions. The carboxyl and amine groups on amino acids are a weak acid and a weak base, respectively. Under acidic conditions (pH very much below 7), both the carboxyl group and the amine group will be protonated. This means that the carboxyl group will have a neutral charge (—COOH) and the amine will have a positive charge ($-NH_3^+$). If there are no ionizable side chains, the net charge of the amino acid will be positive and retained by the scx support.

The pH at which ionizable functional groups are completely protonated can be predicted from each group's pKa. The pKa is the pH at which, statistically, one half of the groups will be protonated. The pKa is not just a property of the ion alone, but is affected by neighboring structures. This means the pKa of the alpha carboxyl and amino group will vary from amino acid to amino acid (see Table 9.1 for the pKa's of the 20 amino acids). In general, the pKa of the alpha carboxyl is around 2.3, and for the alpha-carbon amine about 9.4. For practical purposes, an ion can be considered completely protonated when the pH is 1.5 units below its pKa and deprotonated at 1.5 pH units above its pKa. At a pH around 3, most of the α-carbon carboxyls and α-amines will be protonated. This will result in the amino acid having a net positive charge, which in turn will cause the amino acid to charge pair with the scx support. As the pH of the mobile phase is increased, however, the carboxyl and amines will lose protons. The carboxyl groups will become negative ($—COO^-$) and the amines will become less positive, giving the amino acid a net negative charge. This will result in the scx support repelling (eluting) the amino acids.

As can be seen in Table 9.1, however, some amino acids have ionizable side chains with their own pKa. This obviously will affect the net charge of the amino acids. As the mobile phase goes from very acidic to more basic, the acidic amino acids (carboxyl group as part of their side chain) will gain a net negative charge at low pH and will elute first. The basic amino acids will retain a net positive charge until a much higher pH is obtained and will elute last. The elution order of amino acids can be predicted (with a few exceptions) by the pH at which its net charge is neutral (see Figure 9.5). This pH of the net neutral charge is known as the isoelectric point and is quantified as pI (see Table 9.1).

As stated, the pIs of the amino acids only give an estimate of the elution order. The other factor which affects elution order is the hydrophobicity of the amino acids. The scx support used in amino acid analysis is usually sulfonated styrene. The use of styrene instead of silica is necessary because of the pH range of the mobile phases needed for amino acid separation. The pH range can be from pH 2.7 to over pH 11 for the column-regenerating mobile phases used to strip any ionic compounds from the column. Silica supports are not stable when the pH of the mobile phases fall much below 3 or above 7.5. When the pH is near 2, the covalent bond attaching the bonded phase will begin to be hydrolyzed, resulting in the degradation of the support. When the pH is above 7.5, the alkaline condition will begin dissolving the silica. Styrene is hydrophobic and therefore will interact with the hydrophobic portion of the amino acid structure. Having two modes of separation is referred to as being "mixed mode." Under most circumstances, mixed mode chromatography is considered undesirable since it complicates method development. For amino acid chromatography, the mixed mode aids in resolving the more hydrophobic amino acids that elute in the center of the profile.

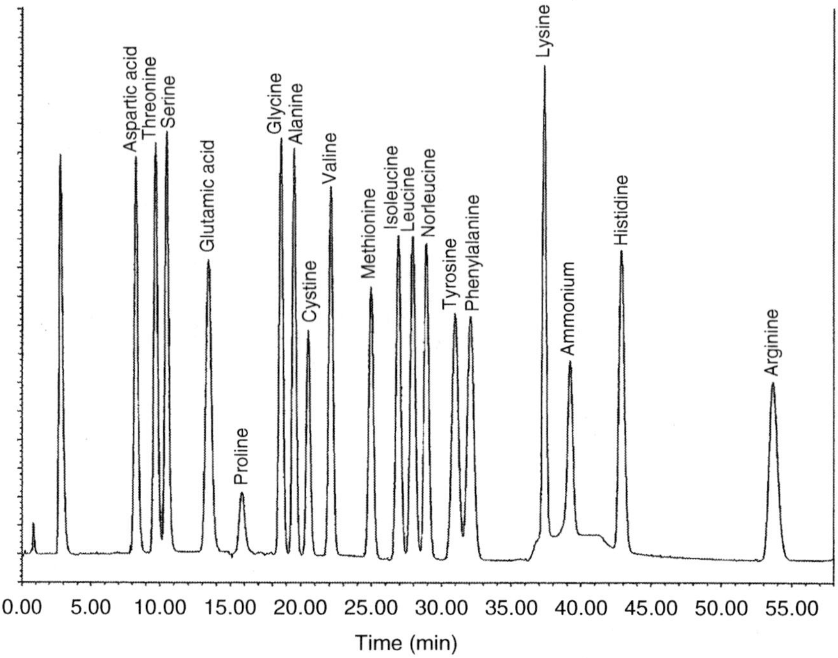

FIGURE 9.5 Chromatogram of ion-exchange separation of amino acids with postcolumn ninhydrin derivatization.

Most IEC methods for amino acids use citrate as a buffering agent at a concentration of 0.2 *M*. Citrate has pKa's of 3.14, 4.77, and 6.39; therefore, citrate works well, having a buffering capacity down to about pH 2.7 and up to about pH 7.5. pH 7.5 is still below the pIs of several amino acids (see Table 9.1). If a pH gradient is the only means of elution, the elution time of basic amino acids will be extremely long. To reduce the elution times of the basic amino acids, a salt gradient occurs simultaneously. The initial concentration of the salt cation is 0.2 *N*, with the final concentration reaching 1.2 *N*. The cation competes with the amino acids for the negatively charged sites on the scx support. As the cation concentration increases, it becomes more competitive, eluting the amino acids. Sodium and lithium are used as the cation, with chlorine being the salt's anion. Research has shown that sodium works best for most matrices; however, lithium works better for physiological fluids. The lithium cation competes more strongly than the sodium cation. This provides better resolution of some amino acids that are impossible to resolve with the sodium cations. The scx needs to be equilibrated with whichever cation is being used, so the columns are often sold for either sodium mobile phases or lithium mobile phases. A note of caution: the chloride ion will react with stainless steel. Since most HPLCs contain stainless steel fittings and tubing, this can cause some problems with corrosion.

Since citrate loses its buffering action above pH 7.5, some mobile phase systems use the borate ion as the final mobile phase buffering agent. Borate has pKas of 9.14,

12.74, and 13.8, making it an excellent buffer at very high pHs, but has no buffering action at neutral or acidic pHs. The change in ions from citrate to borate can cause baseline problems. Pierce Chemical Co. (Rockford, IL) manufactures a proprietary buffer (Buffelute) system which is citrate free.

Many of the gradients used, especially early on, were step gradients. A step gradient occurs when the switch from one mobile phase to the next is abrupt. Step gradients are easier to automate, requiring simple switching at the appropriate time. A drawback of step gradients is that a series of mobile phases with small increments in pH and salt concentration must be made. With a step gradient, a minimum number of distinct mobile phases used is three, with six or more being typical. With the advent of cheap computers, continuous gradients can be accurately and repeatedly mixed by modern HPLCs. This allows reducing the number of different mobile phase down to as few as two.

Other substances are often retained on the scx support that cannot be eluted with 1.2 *N* sodium (or lithium) chloride at pHs of 7.5 or 8. Divalent cations such as calcium can be especially hard to remove and will "poison" the support, making it unusable. Therefore, it is common practice to "regenerate" the column either after every sample or on a routine basis. The most common regeneration mobile phase is 0.2 *N* sodium or lithium hydroxide. Many regeneration mobile phases will also contain a divalent chelater, usually EDTA [(ethylenedinitrilo)-tetraacetic acid].

The scx support for amino acid chromatography is a polymer, usually styrene. Polymers cannot withstand pressures as high as silica-based supports. Most polymers should not be subject to pressures over 2000 psi. If the support is subjected to too high of a pressure, it will be "crushed," forming fine particles. The fine particles will migrate down the column, with the mobile phase becoming trapped in the frit at the end of the column. A column frit is a porous disk of stainless steel or a polymer used to retain the support in the column, but allowing the mobile phase to pass through. The pore size of the frit is typically 0.2 to 0.5 μm. The intact support cannot enter these pores, but the crushed support fines can and will usually become lodged. This will restrict the flow of the mobile phase, causing high pressures. A good diagnostic for this is to reverse the mobile phase flow through the column. The pressure will then drop for a while, but when the fines have migrated to the frit at the other end of the column, the frit will plug up and the pressure will increase again. The crushed support also has less volume than the intact support. This creates voids in the column, which will broaden the peaks and thus reduce resolution.

This lower pressure limit for polymeric supports translates into lower flow rates, typically below 0.5 ml/min, causing long run times. Grunau et al. (1992) show an elution of 99 amino acids using the Pickering lithium buffers and column. A 135-min gradient was required to resolve all 99 amino acids. This is extreme, and for most nutritional applications the gradient with a regenerating cycle will take a little over 1 h. However, the Grunau et al. (1992) paper is a good reference for the relative elution time of many amino acids.

The column temperature also affects the retention time. The shift (shortening) in retention time due to increasing column temperature is significantly different for each amino acid. This makes column temperature an effective tool in improving resolution. The difference in effect of temperature on retention time can be used to shift the order of elution with some amino acid pairs. In addition, the column pressure is also inversely affected by temperature. The shorter retention time with the decrease in pressure favors using an elevated temperature. Typically, for IEC of amino acids the column temperature will range from 40° to 70°C.

Systems vary in important characteristics such as dead volume, which can greatly affect how closely the true gradient follows the program gradient. Because of the (dead) volume of the system, the true gradient will always lag behind the program gradient. Some of the older systems had a large dead volume due to the need for large pulse dampeners. The lag between the true and program gradient at the low flow rates needed for IEC could greatly extend the run time of the gradient. Also, column heaters are not all alike. Even though the true temperature of two column heaters may be identical, the configuration will affect the internal temperature of the column. Factors such as these are the reason that various parameters such as column temperature and gradient must be optimized for each system.

9.9.2 Reversed-Phase Chromatography

Reversed-phase chromatography has a hydrophobic stationary phase and a hydrophilic mobile phase. The stationary phase most often is a porous silica bead with an alkyl-bonded phase. The diameters of the silica beads typically are 5 or 10 μm; however, recently, smaller bead diameters of 2 and 3 μm and less are gaining in use. The pore size typically ranges from 60 to 300 Å. The silica support can withstand much higher pressures than the polymer supports, typically up to 5000 psi. This tolerance of higher pressure allows faster flow rates and, therefore, shorter retention times, speeding up the chromatography. But, as stated earlier, a silica support can only tolerate a pH range of 2 to 7.5.

The principle behind reversed-phase chromatography is the partitioning of the sample components between the hydrophobic stationary phase and the hydrophilic mobile phase. This is the same principle as liquid–liquid extraction. Whether or not the sample components are more attracted to the stationary phase or mobile phase depends on the components' relative solubility in hydrophobic and hydrophilic solvents. The relative amount of time the component spends in each phase determines the component's retention time. The hydrophobic components will have longer retention times.

The proportion of time that a component spends in each phase can be adjusted by changing the hydrophobic/hydrophilic characteristics of the two phases. The selections and proportions of its constituents adjust the hydrophobicity of the mobile phase. Decreasing or eliminating water from the mobile phase makes it more hydrophobic. Using increasingly more hydrophobic organic solvents in the mobile phase will also increase its hydrophobicity. The most typical organic solvents used for amino acid chromatography are methanol and/or acetonitrile, which are fairly hydrophilic. Tetrahydrofuran (THF) is used less often and usually only as a portion of the organic constituent of the mobile phase.

The hydrophobicity of the stationary phase can be controlled by the selection of the alkyl chain bonded to the silica bead. The longer the alkyl chain is, the more hydrophobic the stationary phase. Octadecylsilane (ODS, C18) is the most commonly used support, but C8 is also used. A bead diameter of 5 μm is popular, but the use of 2- or 3- μm beads is increasing. The pore size for amino acids is not as critical as with larger molecules, and 60 Å or so will work fine.

The coating of the silica with the alkane is never complete, leaving silanol groups on the bead surface. Silanols are acidic (negatively charged) and will interact with protonated bases (positively charged), resulting in mixed mode chromatography as discussed earlier. This means that the pH of the mobile phase will affect the elutions of the amino acids. The mobile phase is therefore buffered. The most common buffer is acetate, and the pH is generally slightly acid or neutral. Remember, a high pH mobile phase (greater than pH 7.5) will dissolve the silica support and must be avoided.

In general, the order of elution for the amino acids is the following: the acidic and polar amino acids elute first, followed by the amino acids with short alkyl side chains, and finally the more hydrophobic amino acids (see Figure 9.6). Varying the mobile phase pH, the amount and type of organic solvent, and its ionic strength will affect the relative retention times of the amino acids. Adjustments of these parameters are necessary to optimize the chromatography. Jarrett et al. (1986) studied the affects of these parameters for OPA precolumn derivatized amino acids on C8 silica. This study is an excellent guide to predict which parameter modification will

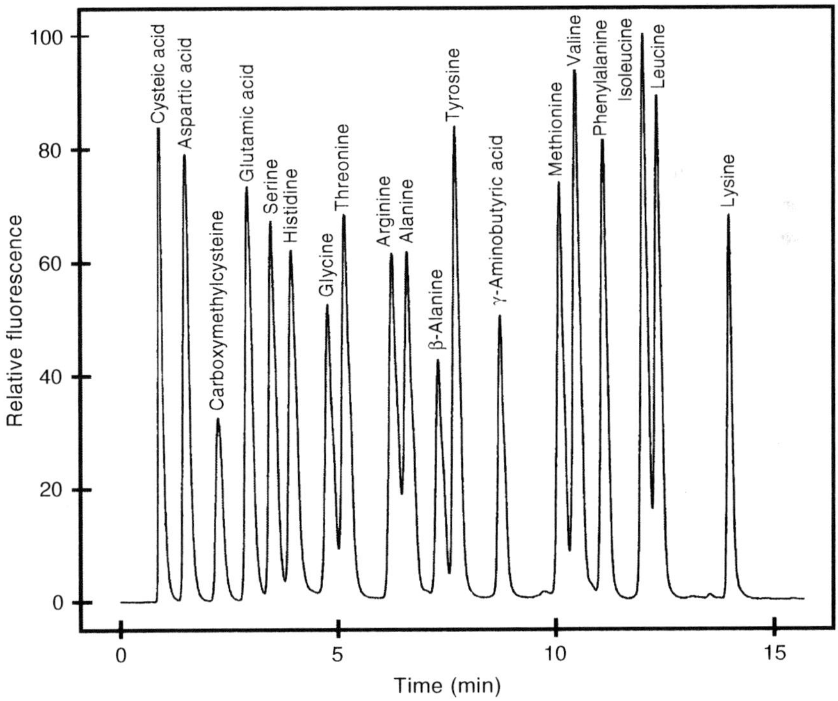

FIGURE 9.6 Chromatogram of reversed-phase separation of amino acids with precolumn OPA derivatization.

improve the resolution of the various amino acids. Because of the small size of an amino acid molecule, the derivatization reagent will have an affect on the chromatography. Therefore, changing the various parameters will have different effects depending on which derivatization method is employed.

9.10 SUMMARY

Our understanding of the nutritional role of amino acids is rapidly expanding. Not only must the essential amino acids be supplied in the diet for good health, but it is being discovered that dietary intake of nonessential amino acids can improve our health and our ability to deal with and recover from stress. The sources of stress not only include illness and injury, but also exercise and fast-paced, high-pressure modern living. Today's health consciousness demands foods and supplements that not only contain the dietary requirements of the essential amino acids, but other amino acids as well. Because of this rapid expansion in understanding the importance of amino acids, health conscious consumers are demanding new products containing these amino acids. To meet this demand, producers and suppliers are rapidly providing these products. The amino acids composition of foods not only varies from food to food, but also within any single food due to different plant varieties, soils, weather, agricultural practices, processing, etc. Fortification with pure amino acids also has variables. Therefore, laboratory analysis of foods and supplements is essential.

The analysis of amino acids is a complex issue. The first step in analyzing amino acids in proteins (which are most of the amino acids found in nature) is to release the amino acids to the free state by hydrolysis. Unless great expense and effort is expended, the conditions of hydrolysis must be a compromise to maximize amino acid release while minimizing amino acid degradation. Different matrices require different conditions. Time, temperature, amount of acid, presence of protective agents, etc. must be adjusted to meet the needs of the matrix.

Once the amino acids are free, they must be separated from each other and any other components of the sample. Today, HPLC is the most commonly used method of separation. The chromatography is usually ion-exchange chromatography (IEC) or reversed-phase chromatography (RPC). Amino acids IEC is almost exclusive on polymeric scx due to the use of the alkaline mobile phase. The buffering agent is predominately citrate. The RPC of amino acids is predominately done with a C8 or C18 silica stationary phase, with mobile phases buffered with acetate and containing either methanol or acetonitrile as the organic portion. Even though this appears to be standardized, there are still many variables which need to be optimized for each system to achieve adequate resolution of the amino acids included in the profile.

Once separated, the amino acids must then be quantified. The physical and chemical properties of amino acids, with the exception of tyrosine, tryptophan, and phenylalanine, limit detection to universal detectors such as refractive index (RI) or light scattering. The drawback of these detectors is that they are universal and any non-amino acid components will be detected, often interfering with the detection of the amino acids themselves. Derivatization coupled with either UV/visible or fluorescent detection is much more specific, eliminating the detection of most potentially interfering substances. For postcolumn derivatization, ninhydrin and OPA are used. There are a number of precolumn methods, including several commercially available kits. OPA is probably the most used precolumn derivatization method.

All the various methods for hydrolysis, chromatography, and derivatization, combined with the need to optimize these methods for the different matrices make amino acid analysis somewhat daunting. It is hoped that this chapter has served as an introduction and will aid in either the selection of a laboratory to conduct the analysis or as an aid to those who are beginning or conducting the analysis.

ACKNOWLEDGMENTS

I would like to thank Jane Sabbatini and Casey Rollins for proofreading the manuscript. I would also like to thank Harry Jarrett for proofreading the manuscipt and providing the chromatogram shown in Figure 9.6.

REFERENCES

AOAC Official Method of Analysis, sec 982.30, Association of Official Analytical Chemists, Washington, D.C., 2000.

Carpino, L.A. and Han, G.Y., The 9-fluorenylmethoxycarbonyl amino-protecting group, J. Org. Chem., 37, 3404, 1972.

Chen, S.T., Chu, S.H., and Wang, K.T., Rapid hydrolysis of proteins and peptides by means of microwave technology and its application to amino acid analysis, Int. J. Pept. Protein Res., 30(4), 572, 1987.

Dong, M.W. and Gant, J.R., High-speed liquid chromatographic analysis of amino acids by post-column sodium hypochlorite-*o*-phthalaldehyde reaction, J. Chromatogr., 327, 17, 1985.

Engelhart, W.G., Micorwave hydrolysis of proteins and peptides for amino acid analysis, in Microwave-Enhanced Chemistry, Kingston, H.M. and Haswell, S.J., Eds., American Chemical Society, Washington, D.C., 1997.

Grunau, J.A. and Swiader, J.M., Chromatography of 99 amino acids and other ninhydrin-reactive compounds in the Pickering lithium gradient system, J. Chromatogr., 594, 165, 1992.

Kuhn, K.S., Stehle, P., and Furst, P., Quantitative analyses of glutamine in peptides and proteins, J. Argic. Chem., 44, 1808, 1996.

Inglis, A.S., Single hydrolysis method for all amino acids, including cysteine and tryptophan, Methods Enzymol., 91, 27, 1983.

Jarret, H.W., Cooksy, K.D., Ellis, B., and Anderson, J.M., The separation of *o*-phthalaldehyde derivatives of amino acids by reversed-phase chromatography on octylsilica columns, Anal. Biochem., 153. 189, 1986.

Lucas, B. and Sotelo, A., Effect of different alkalies, temperature, and hydrolysis times on tryptophan determination of pure proteins and of foods, Anal. Biochem., 109, 192, 1980.

Meltzer, N.M., Tous, G.I., Gruber, S., and Stein, S., Gas-phase hydrolysis of proteins and peptides, Anal. Biochem., 160, 236, 1987.

Moore, S., On the determination of cystine as cysteic acid, J. Biol. Chem., 238(1), 235, 1963.

Roach, D. and Gehrke, C.W., The hydrolysis of proteins, J. Chromatogr., 52, 393, 1970.

Robel, E., Effect of the volume of HCl in minimizing amino acid losses when carbohydrate-containing samples are hydrolyzed, Poultry Sci., 52, 604, 1973.

Roth, M., Fluorescence reaction for amino acids, Anal. Chem., 43(7), 880, 1971.

Spackman, D.H., Stein, W.H., and Moore, S., Anal. Chem., 30, 1190, 1958.

Spies, J.R. and Chamber, D.C., Chemical determination of tryptophan in proteins, Anal. Chem., 221, 1249, 1949.

Zumwalt, R.W., Absheer, J.S., Kaiser, F.E., and Gehrke, C.W., Acid hydrolysis of proteins for chromatographic analysis of amino acids, J. Assoc. Off. Anal. Chem., 70(1), 147, 1987.

10 Carbohydrates and Other Electrochemically Active Compounds in Functional Foods

William R. LaCourse

CONTENTS

10.1 FUNCTIONAL FOODS

Functional foods are a new concept in the United States. They are a rapidly emerging concept in the area of nutrition. They have become one of the most interesting areas of study capturing the interest of not only the consumers but also of scientists (Clydesdale, 1997; Milner, 1999). The reason for this emerging interest is the widespread belief that certain foods and selected food components are associated with enhanced physical and mental performance along with reduced risk of disease (Craig, 1997; Messina and Messina, 1996). In this regard, the importance of functional foods can be attributed to Hippocrates and his now famous tenet, "Let food be thy medicine and medicine be thy food" a concept that goes back to ca. 400 BC (Hasler, 1996; Milner and Craig, 2000). Now almost 2500 years later as we mark the beginning of a new millennium, the concept of food as medicine has taken on philosophical importance evolving into the concept of nutritional essentiality—certain foods and specific food components are essential for the health and survival of higher animal—thus forming the basis of functional foods. More recently, the relationships between diet and disease have become major areas of investigation in nutrition especially involving these functional foods (Harper, 1999).

Although functional foods have been around for sometime, they seemed to have little impact on the consuming public at large. The consumer acceptance of functional foods is growing rapidly and the markets for them have assumed global proportions. Functional foods are one of the largest and fastest growing consumer markets in Asia, Europe, and here in the United States. Japan, the birthplace of functional foods, has had for sometime now the most developed and sophisticated consumer market for functional foods (Jago, 2000). To our knowledge, Japan is the only country with a formal regulatory approval process for functional foods. Officially designated as Foods for Specified Health Use (FOSHU), these functional foods have now come to bear the official seal of approval from the Japanese Ministry of Health and Welfare (Arai, 1996). With hundreds of products currently licensed as FOSHU foods, Japan is a good example of a successful and thriving functional foods industry and continues to provide leading-edge nutritional products. Generally speaking, all foods are functional foods in that they provide flavor, taste, aroma, and nutritional value. Nevertheless, more recently, the term functional food has come to mean one of providing a specific physiological benefit in addition to providing the basic nutritional requirements.

10.1.1 What is a Functional Food?

Despite their ubiquitous nature, functional foods have never been precisely defined. The term functional food was first introduced in Japan in the mid-1980s. Functional foods refer to processed foods containing significant levels of biologically active components that specifically impart health benefits or desirable physiological functions in addition to being nutritious. Other organizations have proposed definitions for functional foods. The Institute of Medicine's Food and Nutrition Board (IOM/FNB) defined functional foods as "any food or food ingredient that may provide a health benefit beyond the traditional nutrients it contains" (IOM/NAS, 1994). Functional foods have received favorable reviews by the Functional Foods for Health (FFH) program, a joint program of the University of Illinois at Chicago and the University of Illinois at Urbana-Champaign. In the United States, there is no legal recognition of the functional foods category. Health conscious consumers, especially the baby boomer generation, have provided the boost making functional foods the leading trend in the U.S. food industry (Meyer, 1998). Other terms such as dietary supplements, nutraceuticals, designer foods, phytochemicals, and food additives have been presented to the consuming public. Here are some definitions. Dietary supplements, according to the Dietary Supplement Health and Education Act (DSHEA) of 1994, are defined as "products intended to supplement the diet to enhance health." These include vitamins, minerals, amino acids, herbs, and other botanicals. A dietary supplement is "not represented as a conventional food or a sole item of a meal or the diet." A nutraceutical was recently defined as a "diet supplement that delivers a concentrated form of a biologically active component of food in a nonfood matrix in order to enhance health" (Zeisel, 1999). Dietary supplements and nutraceuticals are distinct from functional foods, which deliver an active ingredient within the food matrix, and food additives, which enhance the flavor or aroma without enhancing the nutritional value of a food.

One class of foods that falls under the broad category of functional foods is designer foods. A designer food is created by modifying the composition of a basic food product to incorporate certain nutrients or other bioactive compounds for a specific health purpose. A designer food may be modified by genetic engineering, hybridization techniques, or feeding specific diets to livestock or poultry. A phytochemical is a physiologically active, naturally occurring compound produced by plants that is not classified as a nutrient but nevertheless provides a health benefit when consumed. Varieties of phytochemicals have been discovered and are now available to the consuming public. Some designer foods have been modified to incorporate phytochemicals and some phytochemicals are available in supplement form.

Another class of functional foods is prebiotics and probiotics. A prebiotic is defined as a nondigestible food ingredient that beneficially affects the host by selectively stimulating the growth or activity of one or a limited number of bacteria in the colon. An example of a prebiotic is the inulin-type fructans. Despite the limited work done with prebiotics, studies with these inulin-type fructans have generated sufficient data for their consideration as putative functional food components. A probiotic on the other hand is defined as a viable microbial dietary supplement that beneficially affects the host through its effects in the intestinal tract. Present examples of probiotics include fermented dairy products such as yogurt or freeze-dried cultures. Future examples of probiotics will include fermented vegetables and meats (Roberfroid, 2000). This brings one to the definition of nutrition.

According to the Webster's New World Dictionary, nutrition is defined as, "a series of processes by which an organism takes in and assimilates food for promoting growth and replacing worn or injured tissues." It follows that a nutrient is a substance in food that is essential for maintaining health. These include carbohydrates, proteins, lipids, vitamins, and minerals. Therefore, functional foods contain nutrients in the proper balance to not only help maintain health, but they also contain other compounds that promote and protect health.

There have been major conferences on functional foods as part of a new global awareness on the subject. The first international conference on East–West Perspectives on Functional Foods was held in September 1995 in Singapore (*Nutr. Rev.*, 1996). The issues addressed were historical and cultural perspectives on functional foods, their role in disease prevention and health promotion, scientific and technological advances within different regions and countries, scientific data on the effects of various bioactive and nutritive food components on human physiological systems, selected case studies of specific foods or classes of foods, the safety of functional foods and substantiation of their putative health benefits, and a discussion of an appropriate definition of functional foods and regulatory issues. A conference was held in Copenhagen, Denmark in October 1998. This conference provided a global analysis of the scientific, commercial, and regulatory activities in North America, Western Europe, and the Pacific region in the area of functional foods and nutraceuticals. At the American Association of Cereal Chemists (AACC)-sponsored conference on *Functional Foods: Strategies for the Food Industry*, held in June 1999 in Newport Beach, California, the meeting focused on the development and production of functional foods tailored specifically to the food industry. At the Annual Experimental Biology 2000 meetings, the third annual symposium on *Functional Foods for Health Promotion* was held in April 2000 in San Diego, California. This symposium, sponsored by International Life Sciences Institute of North America, focused on factors that affect the physiological responses to biologically active components in foods. More recently, an international conference and exhibition on Nutraceuticals and Functional Foods was held in September 2000 in Houston, Texas. This conference was sponsored by the International Union of Food Science and Technology.

The Institute of Food Technologist (IFT) recently published a Scientific Status Summary on functional foods. This Scientific Status Summary reviewed the research to date for the primary plant and animal foods associated with physiological benefits. Although much has been written about bioactive compounds in this regard (Kuhn, 1998), the publication of the IFT focused on whole foods rather than specific components isolated from foods. The functional foods from plant sources included oat products, soy and soy products, flaxseed, tomatoes, garlic, broccoli and other cruciferous vegetables, citrus fruits, cranberries, green tea, wines, and grapes. The functional foods from animal sources included fish, dairy products, and beef.

The functional foods mentioned above received notoriety for sorts because of the rapid growth of consumer interest in the importance of their bioactive compounds. Lycopene, for example, the primary phytochemical found in tomatoes, has significant antioxidant potential in vitro and has been linked with preventing prostate cancer and cardiovascular disease in humans. Isoflavones in soy and lignans in flaxseed oil, acting as weak estrogens, may play a role in reduced risk of breast cancer. Garlic, containing organosulfur compounds, may have a role in lower blood pressure, reduced cancer risk, and other functions. Limonoids in citrus fruits, glucosinolates in cruciferous vegetables, and catechins in green tea appear to have protective roles in a variety of human cancers. Probiotics, in fermented dairy products such as yogurt, are associated with reduced risk of colon cancer, low serum cholesterol, and other effects resulting from altered gut flora.

The rapidly growing consumer and public interest in functional foods has produced a great demand for information. In the United States, this interest in functional foods is fueled by the following factors: rising healthcare costs, rising interest in wellness programs through diet, an aging population, changes in federal food laws affecting product and label claims, and recent advances in science and technology. Attitudinal research done by the International Food Information Council (IFIC) Foundation shows that consumers today are much more interested in the concepts and principles of functional foods than many health professionals realize. To date, the Federal Food, Drug and Cosmetic Act (FFDCA) has not provided a formal federal statutory definition of functional foods. Therefore, the Food and Drug Administration (FDA) has no federal authority to formally regulate functional foods under the category of foods, including the expanded category of dietary supplements. Since FDA regulates foods based on their intended use, this is determined largely by the label and label claims accompanying the product. Scientific research in food sciences suggests many potential health benefits from food components. These health benefits could expand the health claims now allowed to be considered and identified by the FDA.

10.1.2 Scientific Criteria for Functional Foods

There is increasing scientific evidence suggesting that physiologically active components in functional foods, of either plant or animal origin, may have health benefits. This is supported by research efforts at a number of academic, scientific, and regulatory organizations with efforts underway to establish scientific criteria to help support the health claims for functional food components or the foods themselves. From a regulatory standpoint, two types of claims are allowed on foods and dietary supplements in compliance with FDA regulations: (a) structure/function claims describing effects on the normal function of the body and (b) health (or reduced disease risk) claims implying a relationship between diet or diet components and a health (or disease) condition as approved by the FDA and supported by significant scientific agreement. To narrow the gap between foods and dietary supplements, the FDA recently announced the creation of a new Office of Nutritional

Products, Labeling and Dietary Supplements. This replaces the Office of Special Nutritionals and the Office of Food Labeling. Nevertheless, it should be emphasized that functional foods are not a universal substitute for poor eating habits. Emphasis, however, must be on a balanced overall dietary model based on the U.S. Dietary Recommendations: one that is plant-based, low in animal fat, high in fiber, and consuming 5–9 servings of fruits and vegetables daily.

Consumers are becoming increasingly health conscious, seeking out and incorporating functional foods into their diets in pursuit of good health maintenance. The area of functional foods is in its infancy and gaining popularity. Any health claims associated with functional foods must be based on sound scientific criteria (Clydesdale, 1997). In pursuit of this sound scientific principle, a large body of solid scientific research is needed to validate the potential benefits of any particular functional food or its component. In this regard, one must consider several factors such as the complexity of the functional food matrix, effects on the functional food, compensatory metabolic changes accompanying dietary changes, and the need for biomarkers of disease development.

And last, but not the least, the consuming public must have a clear understanding of, and a strong confidence in, the sound scientific criteria used to substantiate the health effects and claims regarding functional foods. The scientific community is only now beginning to understand functional foods and their ability to deliver potential health benefits. Until this knowledge is complete, consumers must be aware that functional foods are not magic bullets or a total panacea. Examples of functional food components include carbohydrates, carotenoids, fatty acids, flavonoids, glucosinolates, phenols, plant sterols, prebiotics, probiotics, saponins, soy proteins, sulfides, thiols, and tannins. This review will focus on carbohydrates in functional foods.

10.1.3 Carbohydrates in Nutrition

As mentioned above, functional components are those ingredients that impart functionality to the functional food. This section will focus on reviewing the information about carbohydrates as specific food components. Before discussing the carbohydrates in functional foods, let us take a look at carbohydrates in food. The goal here is to explore the scientific literature on food carbohydrates in general with nutritional properties.

Carbohydrates as a food group contribute significantly to nutrition. Scientists have been studying for sometime now how carbohydrates work individually and in concert with other food components to regulate physical and mental stamina and modulate stress. Since the physical and chemical structures of foods vary, calories available from some foods are available sooner than from others. Food is fuel, and depending on the activity planned for the day, the food you eat should suit the occasion. A diet planned to include certain foods will help you obtain and sustain the kind of specific energy you need for peak performance.

As energy yielding compounds, carbohydrates together with proteins, fats, and a few minerals represent the sine qua non of a nutritionally essential diet. In the United States, most people are waking up to the fact that a diet rich in plant foods is healthy and protective. The modern nutritional wisdom of a diet rich in carbohydrates is reflected in the new MyPyramid food guide introduced by the United States Department of Agriculture (USDA). In addition to exercise, the steps to a healthier you encourages the consumption of grains, vegetables, fruits, and beans, a long way from the four basic food groups. The role of carbohydrates in foods and their importance in nutrition has been discussed and debated in a recent article on healthy eating (Herman, 1997). The proponents maintain that a diet rich in high-fiber and carbohydrate (along with low fat) is nutritionally healthy with high-fiber, complex carbohydrates comprising 60%–75% of the diet (Herman, 1997). Opponents contend that fear of fat has led to an excessive, over consumption of simple carbohydrates (refined sugar and white flour foods) rather than complex carbohydrates (whole grains, vegetables, and beans) (Herman, 1997). The debate goes on to mention that Americans have not reduced their fat intake and are loading up on the wrong kinds of carbohydrates. This leads to the next most important question. Are there good and bad carbohydrates?

A recent commentary in the *Lancet* suggests that there are (Katan, 1999). In one case study, the diets of over 65,000 women were examined over a six-year period (Herman, 1997). Results from these studies showed that women who consumed a low-fiber diet along with plenty of soft drinks were two and a half times more likely to get diabetes than those who consumed a high-fiber diet with very few soft drinks. This study showed that sugar was not the only culprit. Other dietary culprits included white bread, white rice, and potatoes. This finding led to the controversial glycemic index (GI) concept, which was developed as a tool to help people monitor and stabilize their blood sugar level.

Most processed foods contain considerable amounts of refined sugar or sweeteners. How has this impacted the nutritional quality of foods? A recent report entitled "Changes in Nutritional Quality of Food Product Offerings and Purchases by USDA's Economic Research Service" indicates that the nutritional quality of foods has not improved (Chapman, 2000). This study analyzed the nutritional index of five food categories. The specific nutrient content of certain foods showed significant changes. Processed meats and bacon had significant changes in carbohydrates, while in cookies, the level of fiber improved. Increases in desirable nutrients were generally offset by increases in undesirable nutrients and vice versa.

Carbohydrates can be both a boon and a bane from a nutritional standpoint. For example, carbohydrates as mood-enhancing foods, has long been recognized. Specifically, the carbohydrate connection to levels of serotonin, a brain chemical messenger that causes upbeat moods (positive changes in attitude and behavior) is well known (Turner, 1999). Research studies on the effects of serotonin levels in brains show that both simple sugars (candy, cookies, and honey) and complex carbohydrates (whole grains and most vegetables) cause upbeat moods in different ways. This is how it works: when carbohydrates are consumed, they trigger the release of insulin, which in turn increases the level of tryptophan, an amino acid in the brain. Tryptophan is then converted to serotonin, responsible for the upbeat mood of feeling well. There is one difference, however. Simple sugars produce positive mood shifts, mental function, and overall energy by providing empty calories in the short run, which is an apparent bane of the carbohydrate food. Complex carbohydrates on the other hand produce the healthy serotonin during high stress times that provide the required nutrients instead of the empty calories, which is an apparent boon of the carbohydrate food.

The science of carbohydrates never tasted sweeter. It helps a consumer to understand that sugar is sugar. This is true for refined white sugar, maple syrup, or barley malt. Even though refined white sugar has come to be regarded as the culprit, sugar consumption in any form may not be good for anyone. For example, sugars help promote tooth decay and undermine the consumption of more nutritious foods. Excessive sugar consumption can be especially bad for those at risk for diabetes, obesity, hypoglycemia, or coronary heart disease. Consumer misconception about the so-called natural sweeteners is well known. Just because it says "natural," does not mean it is better for you. Some sweeteners are known to contain complex carbohydrates fortified with vitamins and minerals. With the exception of molasses, nutritionally speaking there is no advantage of one sweetener over the other (Cheney, 1995).

It is important to understand how the body processes sugar. Sugars are simple carbohydrates. They occur as monosaccharides (glucose, fructose, or galactose) or disaccharides (sucrose, lactose, or maltose). Simple carbohydrates are found naturally in milk, fruits, and vegetables as well as in processed foods such as table sugar, candy, and soft drinks. Polysaccharides are complex carbohydrates composed of long glucose chains plus fiber and other nutrients. They are found in grains (cereal, rice bread, and pasta), legumes (lentils, peas, and beans), and starchy vegetables (corn, potatoes, and peas). Regardless of whether it is a simple or complex carbohydrate, the primary dietary function of a carbohydrate is to provide energy. Enzymes of the digestive system hydrolyze all carbohydrates to the simple monosaccharide glucose, commonly referred to as blood sugar, for energy (Cheney, 1995).

From the foregoing discussion, it is clearly evident that consuming large amounts of simple sugars on a regular basis puts a strain on our physiological systems. Because they are packed with fiber and nutrients, complex carbohydrates are considered by many nutrition and health experts to be the foundation of a healthy diet (Quagliani, 1997). They are also considered the building blocks of food because of their broad functionality (Ohr, 1998). For example, research on starches and hydrocolloids have improved the quality of cereals, candy, and low-fat foods. Starches and hydrocolloids are two of the more widely known fat mimetics used in food. The Whistler Center for Carbohydrate Research at Purdue University in West LaFayette, Indiana is well known for their research studies on complex carbohydrates.

Nevertheless, the American Dietetic Association recommends that 55%–60% of the total daily calories come from both simple and complex carbohydrates (Quagliani, 1997). This recommendation calls for a diet rich in carbohydrates found in nutrition-packed milk, vegetables, and fruits and allowing for higher-fat and higher-calorie sweets as well. The recommendation goes on to mention that the source for most of the carbohydrates should be the complex type.

The structure of carbohydrates has a major impact on their function in different applications. The variability of carbohydrate structures can be seen in their chain length, the type of carbohydrate units in the chain, and the position of glycosidic linkages that affect physicochemical properties. A Joint FAO/WHO (Food and

Agriculture Organization/World Health Organization) Expert Consultation on Carbohydrates in Nutrition was held in Rome, Italy, April 1997 (FAO, 1998). This consultation was wide ranging in its scope and published the report essentially as a reference document. This report reviewed and outlined the state of the knowledge on the important roles of carbohydrates in human nutrition and health. The Consultation discussions focused on the following: (1) the great strides made in carbohydrate chemistry in the development of a variety of new food products, many of which were based on nutritional considerations; (2) a growing understanding of the diverse physiological roles of carbohydrates in digestion and fermentation in the gut; (3) the influence of carbohydrates on physical performance through glycogen loading; (4) issues with important implications for agricultural production, the food industry, and public health policy; (5) formulation and implementation of science-based dietary guidelines for the nutritional sciences; and (6) role of carbohydrates in noncommunicable diseases. This consultation reached several conclusions on each of the topics discussed and made a number of scientifically sound, up-to-date, pragmatic recommendations on carbohydrates in nutrition.

10.1.4 Carbohydrates in Food

An understanding of the carbohydrate functionality requires an understanding of the structure, chemical, and physical properties of carbohydrates. Carbohydrates in food comprise anything from a simple carbohydrate (simple sugar or monosaccharide) to highly complex carbohydrates (polysaccharides). Because each of these carbohydrates has certain properties, the structural diversity found in food carbohydrates offer different functional properties. The functional properties are generally derived from the carbohydrate content and are directly related to the carbohydrate structure. Let us take a look at the various carbohydrates found in food.

All carbohydrates are generated via the process known as photosynthesis from the sun's energy combined with carbon dioxide and water in the plant. Carbohydrates are important organic compounds composed of carbon, hydrogen, and oxygen. Chemically, they are polyhydroxy aldehydes, ketones, alcohols, or acids. They have important functional roles both in plants as well as in the human body. Food carbohydrates come in two principal groups: simple carbohydrates and complex carbohydrates. Simple carbohydrates consist of any soluble, digestible, and absorbable sugars such as monosaccharides, disaccharides, and sugar alcohols or polyols. Complex carbohydrates comprise the oligosaccharides and polysaccharides (starches and dietary fiber).

Monosaccharides are the simplest sugars containing a single sugar unit. These include glucose, fructose, and galactose. Monosaccharides are the building blocks from which all the other carbohydrates are made. Two monosaccharide units link together to form disaccharides. The most common disaccharides are sucrose (glucose + fructose), lactose (glucose + galactose), and maltose (glucose + glucose). These monosaccharides and disaccharides are available as pure crystalline compounds for use as food ingredients. As a result, their functional properties follow a relatively predictable pattern. Sugar alcohols or polyols are produced by hydrogenating monosaccharides such as glucose, thereby transforming the reducing end of the glucose molecule to a hydroxyl (−OH) group. The most common sugar alcohols are sorbitol, mannitol, and xylitol.

Oligosaccharides contain anywhere from three to nine monosaccharide units. Oligosaccharides formed mainly from the hydrolysis of starch are known as malto-oligosaccharides. Other oligosaccharides include raffinose (a trisaccharide), stachyose (a tetrasaccharide), and fructo-oligosaccharides (inulins).

Polysaccharides are made from linking multiple monosaccharide units. These long chain molecules are known as starch in the plant system and glycogen in the animal system. Starch polysaccharides primarily contain α-glucans. These include amylose, amylopectin, modified starches, and resistant starches. Fiber, also known as nonstarch polysaccharides (NSP), primarily contains components derived from the plant cell wall such as cellulose, hemicellulose, and pectin. Other polysaccharides include gums and hydrocolloids.

Now that we have defined and identified the important food carbohydrates, the next step is to classify these food carbohydrates as important dietary carbohydrates combining chemistry and nutrition. Efforts trying to reconcile the chemistry and nutritional aspects that best reflect the physiological properties have led to a terminology of dietary carbohydrates describing these carbohydrates and their various fractions and subfractions (FAO, 1998).

10.1.5 Dietary Carbohydrate Terminology

10.1.5.1 Sugars

Sugars generally refer to simple carbohydrates. These include the monosaccharides and disaccharides. On the other hand, sugar generally refers to table sugar (purified sucrose) and is also known as refined sugar. These terms, "sugars" and "sugar," are used to inform the consumer between those that are considered healthy carbohydrates and those that are not.

10.1.5.2 Complex Carbohydrates

First described in a report entitled "Dietary Goals for the United States" published in 1977 (FAO, 1998; U.S. Senate Select Committee on Nutrition and Human Needs, 1977), complex carbohydrates were used to mean whole grains, fruits, and vegetables and to distinguish them from sugars. More recently, complex carbohydrates have come to mean starch, dietary fiber, or nondigestible oligosaccharides. It is now realized that starch, a complex carbohydrate by any definition, occurs in various forms. These variable forms of starch are metabolized differently, with some forms of starch being readily absorbed and having a high GI and other forms being resistant to digestion. Complex carbohydrates are found plentiful in grains (cereal, rice, pasta, and bread), starchy vegetables (corn, peas, and potatoes) and legumes (lentils, peas, and dried beans).

Dietary carbohydrates were further classified into available and unavailable carbohydrates (FAO, 1998). Available carbohydrate was defined as soluble sugars and starches, whereas unavailable carbohydrate was defined as mainly hemicellulose and cellulose (fiber). The concept of dietary carbohydrates described as available and unavailable was based on the two important properties of carbohydrates: digestibility and fermentability.

10.1.5.3 Resistant Starch

Resistant starch (RS) was defined as "starch and starch degradation products not absorbed in the small intestine of healthy humans" by Englyst and Cummings (1990). The finding of RS has added to our understanding of the important role carbohydrates play in human health. The different nutritional characteristics of starch, whether digested and absorbed readily or slowly or undigested, led to their nutritional classification as readily digestible, slowly digestible, or RS (Englyst, Kingman, and Cummings, 1992). The principal forms of RS are RS1 (physically trapped starch), RS2 (RS granules), and RS3 (retrograded starch) (Cho, Prosky, and Dreher, 1999; Englyst and Cummings, 1990; Englyst, Kingman, and Cummings, 1992). RS is formed when starch or starch-containing foods are heated, for example during bread baking. RS also has many characteristics of the fermentable dietary fiber, and is important to both the consumer and the food processor in that it may provide a nutritional boost to the dietary fiber content of the food in question.

10.1.5.4 Modified Starch

Starch is the main carbohydrate component of many food products. Furthermore, starch and starch degradation products provide a wide range of functional properties. Starch is primarily composed of amylose and amylopectin. The proportions of amylose and amylopectin in a starch-containing food can vary. These variations can be brought about by food processing and genetic engineering techniques. For example, corn starch containing high amylose and high amylopectin have been available for sometime now with different functional and nutritional properties. High amylose starches gelatinize only at higher temperatures, are more susceptible to retrogradation, and have a tendency to form amylose–lipid complexes. These properties can be fully utilized in formulating foods with high RS content.

Today, stringent demands are being placed on this important complex carbohydrate, starch, by the food industry. Not long ago, starch was seen as an unattractive collection of complex macromolecules lacking the fine challenging structural features. It is not only expected for it to behave as a super molecule with great physical and chemical attributes, but also to display unique functional as well as nutritional properties. Unfortunately, native starches do not measure up to such requirements and must be modified to improve their functional properties. Starches can be modified physically, chemically, or genetically (Eliasson and Gudmundsson, 1996). Physical modifications involve changes in starch structure or phase behavior, interaction between starch and protein or lipid components, and do not involve covalent linkages. Specifically, these modifications include partial hydrolysis leading to the formation of dextrins and pregelatinization. Chemical modifications involve changes through covalent linkages. Specifically, these include cross-linking or oxidation and introduction of side groups (FAO, 1998). Genetic modifications involve changes in the composition of amylose and amylopectin (Eliasson and Gudmundsson, 1996). Modification of starch using tools of biotechnology is gaining importance worldwide. Today, carbohydrate scientists, using advanced analytical technology and biotechnology for modifying starches both in vivo and in vitro, are gaining insight into the fine structure of starch molecules. Ultimately, these modifications have resulted in improved functional properties such as reduced viscosity, improved gel stability, appearance, texture, mouthfeel, and resistance to heat (FAO, 1998). Modified starches as green emulsifiers (fat replacers), is another growing area of interest. Some modified starches may be partially resistant to digestion in the small intestine, thereby contributing to RS. Soluble and insoluble starches play many important roles in the human intestine.

10.1.5.5 Dietary Fiber

Fiber is an NSP derived from plants. The term "dietary fiber" was originally described as "that portion of food which is derived from cellular walls of plants which is digested very poorly by human beings" (Trowell, 1972). This description of dietary fiber had more to do with a physiological concept rather than any mention of a food carbohydrate. Nevertheless, the implication of dietary fiber in health and disease was made based on the etiology of a number of diseases seen in the West (Burkitt and Trowell, 1975).

The components of dietary fiber are derived mainly from the cell walls of plant material in the diet. These include cellulose, hemicellulose (β-glucans, arabinoxylans, etc.), and pectin, collectively referred to as the NSP. Food gums, also called hydrocolloids, are polysaccharides and are considered dietary fiber. Lignin, a non-carbohydrate component of the cell wall and a nonpolysaccharide, is also included as fiber that is very tough. Most edible fruits and vegetables, with the exception of carrots, contain low levels of lignin.

Dietary fiber is classified as soluble fiber (gums/hydrocolloids, pectins) or insoluble fiber (cellulose, hemicellulose). Dietary fiber is neither broken down completely nor absorbed. This is because there is no digestive enzyme capable of hydrolyzing the dietary polysaccharide containing glucose units. Dietary fiber is not digested in the human upper intestine, thus contributing directly very little to the nutritional value. Nevertheless, they can provide some energy after fermentation in the colon. They do contribute, however, by primarily modulating the digestion and metabolism of other nutrients. For example, the physiological properties of soluble fiber can be seen in its principal effects on glucose and lipid absorption from the small intestine, whereas those of insoluble fiber can be seen in its pronounced effects on bowel behavior.

10.1.6 Carbohydrates in Functional Foods

The foregoing discussion took a look at food carbohydrates in general. Here we will focus on the important carbohydrates in functional foods. In addition to being nutritious, these dietary carbohydrates should also be acceptable and functional. Food carbohydrates with important functional properties include complex carbohydrates, RS, and dietary fibers (Cho, Prosky, and Dreher, 1999).

Complex carbohydrates are a large food group and include starch and NSP. They have widely differing chemical structures and physiological effects. The nutritional and physiological importance of complex carbohydrates in foods is increasingly being recognized by both consumers and scientists (Agus et al., 2000; Axelsen et al., 2000; Brown, 1996; Cho, et al., 1999; Daly et al., 2000; Liu et al., 2000; Meyer et al., 2000; Ohr, 1999; Parks and Hellerstein, 2000; Smith et al., 2000; Thompson et al., 2000). Dietary fibers as food carbohydrate have been studied for their functional properties (Jenkins et al., 2000; Lu et al., 2000; Ohr, 1999; Walqvist, 1994). The functional properties of oligosaccharides with beneficial health effects have been studied (Ohr, 1999; Oku, 1996). Oligosaccharides, as prebiotics in the maintenance of gut health, have been investigated (Coussement, 1997; Ohr, 1999). Inulin derived from chicory (Roberfroid, 2000; Sheehy and Morrissey, 1998), fructo-oligosaccharides derived from inulin (Coussement, 1997; Lu et al., 2000; Ohr, 1999; Roberfroid, 1996), and human milk oligosaccharides (Engfer, 2000) have also been investigated for their functional properties with beneficial health effects.

10.1.7 Analytical Aspects of Functional Foods

Like all biochemical sciences, advances in food and nutrition sciences can be empowered by advances in analytical sciences. The analysis of foods in general and functional foods in particular, is not easy because the food matrix is a multicomponent system comprised of carbohydrates, amino acids, proteins, lipids, flavonoids, flavor, aroma, and taste compounds. Food analysis remains the absolute that determines, adds to, and advances the knowledge of nutrition science as an evolving process. It is therefore important to establish the necessary analytical methodologies with standard reference materials. Once the proper food analyses are performed, this analytical information must then be presented in a manner that is both useful and informative to the consuming public. The importance of food analysis and the information gained from such analytical studies is being recognized by institutes such as the National Institutes of Health (NIH), the FDA, the USDA, and the National Institute of Standards and Technology (NIST) (Dupont, 1999).

Spectroscopy was the early technique and tool of choice in food analysis. Specifically, near-infrared (NIR) spectroscopy was used with many applications for the rapid analysis of foods. NIR was used in the systematic analysis of food components such as moisture, protein, and fat, and also sugars and organic acids. The advantages of this technique were its simplicity and versatility. It is still used along with other spectroscopic techniques such as Fourier transform infrared (FTIR) and nuclear magnetic resonance (NMR) spectroscopies.

One of the challenges in the area of functional foods remains the analysis of the particular food component(s), its physicochemical properties and their role in different applications, and ultimately their relevance to nutrition. Efforts to establish analytical methods using standard extracts for botanicals have been the subject of a number of presentations offered at the Texas A&M short course on Nutraceuticals and Functional Foods (Mannie, 2000). The identification of food components present in a food sample can be rapidly established by chromatographic methods. Chromatography as a tool and technique is rapidly gaining importance in food analysis. Commercial growth and success have led to increased availability and reliability of analytical techniques and tools. Current state-of-the-art technology includes chromatographic methods such as high-performance liquid chromatography

(HPLC) and gas chromatography (GC). Both HPLC and GC, independently and in tandem with mass spectrometry (MS), combine to make them the most powerful analytical techniques available to the food scientist.

HPLC is emerging as the technique of choice for the routine analysis of functional foods. This can be seen in HPLC applications to botanicals and herbals (Mannie, 2000). HPLC methods are being validated and promoted by several organizations involved in the analysis of functional foods. For example, the Institute for Nutraceutical Advancement's Methods Validation Program (INAMVP), an industry-wide consortium, selects and validates methods for quantifying compounds used in dietary supplements. The United States Pharmacopoeia (USP) has created an independent herbal division that will offer methods of analysis for bulk herbs, standardized herbal extracts, and finished herbal products which will be published in the USP National Formulary. Efforts are underway at the American Herbal Pharmacopoeia (AHP) to publish complete monographs on HPLC protocols for a variety of herbal products. The FDA has designed a clear framework for labeling, allowable structure/function claims, and scientific substantiation for dietary supplement products.

One of the challenges in these analyses is to scientifically substantiate the amount and efficacy of the active ingredient(s) in functional foods. Standardized testing for botanical products is an ongoing effort in the industry today. Establishing standardized marker compounds aids in the quantitative assessment of the quality of the herbal extract. Even though GC and spectrophotometry have been used in the analysis of herbal products, HPLC, with inherent features such as speed, precision, and specificity, is the current method of choice (Ohr, 1999).

HPLC has also been used to analyze herbal tonic products to ensure their therapeutic potency. HPLC profiles of raw botanical materials have been developed with active standards. Testing the finished herbal product is a real challenge. The herbal food matrix is a complex one consisting of not only herbs, but also other components such as flavor compounds, grains, and other ingredients. Therefore, isolating the herbal components is not an option. On the other hand, the raw herbal material is routinely analyzed to meet the standard specified by the HPLC profile(s) as quality assurance.

10.1.8 Analytical Aspects of Carbohydrates in Functional Foods

The nutritional value of a dietary food carbohydrate(s) depends to a large extent on the amount and type of carbohydrates that is provided to elicit a function. This underlying concept of nutrition forms the basis for all analytical methods used to determine the carbohydrate nutritive value of a food. One of the most commonly performed analyses is the determination of carbohydrates. As mentioned above, these carbohydrates include simple sugars, sugar alcohols, oligosaccharides, and polysaccharides.

HPLC, GC, and enzymatic methods have been traditionally used for the specific analysis of monosaccharides, disaccharides, and sugar alcohols (polyols) (Asp, 1995; Greenfield and Southgate, 1992; Southgate, 1991). HPLC with refractive index detection (RID) and GC have been the techniques of choice for the simultaneous determination of different monosaccharides. The enzymatic methods, while being very specific, require highly purified enzymes for the determination of specific individual sugars in a mixture without the need for a large investment in analytical instrumentation. This can be seen in the analysis of a single sugar such as glucose, as the end point in the analysis of starch. While there are available HPLC methods for the determination of sugar alcohols or polyols, they have usually been determined by GC methods as alditol acetate derivatives.

HPLC and GC have been the techniques of choice for analyzing oligosaccharides. These techniques have worked well for small oligosaccharides or purified preparations. However, these techniques have not been successful for the analysis of oligosaccharides in complex foods. Here, the carbohydrates in complex foods have been initially hydrolyzed using enzymes to free the monosaccharides followed by their determination using HPLC. For example, the malto-oligosaccharides have to be extracted before their analysis of starch. Otherwise, they are recovered as starch.

Oligosaccharides, by definition, comprise anywhere from three to nine monosaccharide units. On the other hand, polysaccharides contain 10 or more monosaccharide units. From an analysis standpoint, oligosaccharides can be separated from polysaccharides based on their solubility in 80% (v/v) aqueous ethanol. This alcohol solubility of carbohydrates is dependent not only on their degree of polymerization (DP), but also on their molecular structure. For example, a highly branched complex carbohydrate may exhibit greater solubility in 80% ethanol despite a considerably higher DP ($DP > 10$). Nevertheless, analytically speaking, the separation of oligosaccharides from polysaccharides is one that cannot be based on their DP (Asp et al., 1992).

Starch in foods has been analyzed mainly by enzymatic hydrolysis to free the glucose followed by its specific determination. From a nutritional standpoint, starch is present in foods as digestible and RS. RS is that which is not absorbed in the small intestine of healthy humans. From an analytical standpoint, the distinction between digestible and resistant is unclear (Björk, 1996). Nevertheless, RS is virtually insoluble in water and techniques for analyzing total starch in foods have employed either an alkaline (KOH) or organic solvent (DMSO) treatment to successfully disperse the crystalline starch fractions and keep them solubilized. Quantitative analysis of RS in foods has utilized enzymes for their determination (Björk, 1996). Nevertheless, techniques for analyzing RS in foods need to be established and tested in formal collaborative studies.

NSP have been determined using a combination of procedures. Starch is first solubilized and subjected to enzyme hydrolysis. After removal of low molecular weight carbohydrates including starch hydrolysis products, the NSP is hydrolyzed to release their monomeric components followed by their quantitative analysis. The critical step is the acid hydrolysis of NSP that must be optimized to achieve complete hydrolysis without degradation of the free monomers (Cho, Prosky, and Dreher, 1999; FAO, 1998). The free monomeric units obtained from NSP have specifically been analyzed as alditol acetate derivatives using GC. The preferred method for analyzing uronic acids derived from pectic substances remains the colorimetric method. There is also a colorimetric method available for determination of total NSP. NSP can be fractionated into cellulose and noncellulose polysaccharides using a combination of sequential extraction and hydrolysis. For example, cellulose can first be treated with concentrated acid to disperse the polymer followed by hydrolysis in dilute sulfuric acid. HPLC analysis of NSP has been gaining in popularity.

Dietary fiber in foods has been analyzed by three different approaches, which have been subjected to comprehensive testing in formal collaborative studies, receiving approval as official methods from the Association of Official Analytical Chemists (AOAC) and the Bureau Communautaire de Reference (BCR) of the European Community (Cho, Prosky, and Dreher, 1999; FAO, 1998). Brief descriptions are as follows:

- *Enzymatic-gravimetric methods of Prosky et al. and Lee et al.*—These official AOAC methods were established from procedures directed at simulating the digestion in the small intestine of humans to obtain an undigested residue as a measure of dietary fiber. This residue is then corrected for any associated protein and ash. In the absence of KOH or DMSO, starch that resists the enzyme amylase in the assay remains as fiber component. The sample is milled and subjected to enzymatic hydrolysis using a heat-stable amylase at 100°C. Two of the principal forms of RS, RS1 (physically enclosed starch within intact cell structures) and RS2 (raw starch granules) will not be included. RS3 (retrograded amylose) is included and is the principal form of RS in processed foods. Lignin, a noncarbohydrate component of the cell wall and a nonpolysaccharide, is also included along with some tannins. These components comprise a very small proportion of most foods, but can be found in substantial amounts in some unconventional raw materials or special fiber preparations (Cho, Prosky, and Dreher, 1999; FAO, 1998).
- *Enzymatic-chemical methods of Englyst et al.*—These methods measure the NSP specifically, either as total NSP (reducing substances) using the colorimetric method or as individual monomeric components using techniques such as GC or HPLC. DMSO is used in the initial step to ensure complete removal of starch and lignin is not analyzed. The difference between the AOAC methods and this method is mainly due to exclusion of RS and lignin (Cho, Prosky, and Dreher, 1999; FAO, 1998).
- *Enzymatic-chemical method of Theander et al. (also known as the Uppsala method)*—The hydrolysis conditions and monomeric determination using GC bear a close semblance to the Englyst method. The difference is that the DMSO step is eliminated and a gravimetric estimate of lignin is added to obtain the dietary fiber. Results obtained using this method and the AOAC methods are in close agreement (Cho, Prosky, and Dreher, 1999; FAO, 1998).

10.2 PULSED ELECTROCHEMICAL DETECTION OF CARBOHYDRATES IN FUNCTIONAL FOODS

As mentioned in the foregoing section, carbohydrates in foods and functional foods can be analyzed using the techniques of HPLC or GC. HPLC techniques utilizing sensitive electrochemical detection are becoming increasingly popular compared to

GC, for their direct detection of carbohydrates, with high sensitivity and specificity and without the need for sample derivatization. Specifically, the technique of high-performance anion-exchange chromatography with pulsed electrochemical detection (HPAEC-PED) is fast becoming the technique of choice for this. HPLC techniques with RI detection are greatly limited in their ability to analyze complex carbohydrates in foods.

Since the inception of PED, the food industry has been both a pioneer and benefactor of this technology. The success and acceptance of PED for the analysis of foods and beverages are indicated by the large number of PED-related publications and officially approved methods by regulatory agencies. Officially approved methods include the following:

- ISO 11292 (International Standards Organization)—Instant Coffee: Determination of free and total carbohydrates using HPAEC.
- ICUMSA (International Commission for Uniform Methods of Sugar Analysis)—Determination of molasses.

The following brief review of electrocatalytic detection mechanisms at noble metal electrodes and fundamental theory of PED has been designed to help the reader understand and explore the breadth and analytical utility of PED techniques for carbohydrates in food and beverage analysis.

10.2.1 Electrochemical Detection in LC

Electrochemical detection (ED) in HPLC has proven to be a powerful analytical technique for the determination of compounds containing electroactive groups. The high sensitivity and selectivity of ED is ideally suited for complex samples, as evinced by its application to the determination of neurotransmitters in complex biological samples (i.e., brain extracts). Neurotransmitters are typically aromatic compounds (e.g., phenols, aminophenols, catecholamines, and other metabolic amines), which are detected easily by anodic reactions at a constant (direct current (dc)) applied potential at inert electrodes (Adams, 1969; Kissinger, 1984). The most common electrode materials are Au, Pt, and carbon. Electronic resonance in aromatic molecules stabilizes free-radical intermediate products of anodic oxidations, and, as a consequence, the activation barrier for the electrochemical reaction is lowered significantly. In contrast, absence of π-resonance for aliphatic compounds results in very low oxidation rates even though the reactions may be favored thermodynamically (Adams, 1969). Stabilization of free-radical products from aliphatic compounds can be achieved alternatively via their adsorption to the surface of noble metal electrodes. Unfortunately, adsorption of organic molecules and free-radicals also has the consequence of fouling of the electrode and loss of its activity (Breiter, 1963; Gilman, 1963, 1967; Giner, 1964). The historical perspective of nonreactivity for aliphatic compounds at noble metal electrodes can be attributed to surface fouling as a result of high, but transient, catalytic activity.

Even for reversible redox couples that are considered to be well behaved, dc amperometry is often accompanied by the practice of disassembling the electrochemical cell and mechanically polishing the working electrode. In this manner, fouling from nonspecific adsorption processes or mechanistic consequences is physically removed from the electrode surface. An alternate approach is to combine electrochemical detection with online cleaning. Hence, to maintain uniform and reproducible electrode activity at noble metal electrodes for polar aliphatic compounds, PED was developed (Hughes, Meschi, and Johnson, 1981).

10.2.2 Historical Perspective

Virtually, every publication concerning voltammetric data obtained on noble metal electrodes describes a procedure for electrode pretreatment to clean and reactivate electrodes that have become fouled by adsorption of solution impurities. Preparation and reactivation of noble metal electrodes are commonly achieved by the repetitive application of cyclic potential scans or alternated positive and negative potential pulses. In fact, the use of alternating positive and negative potential pulses to reactivate noble metal electrodes has been employed since early in the twentieth century (Armstrong, Himsworth, and Butler, 1934; Hammett, 1924). In the 1950s and 1960s, the development of hydrocarbon fuel cells inspired research on pulsed waveforms to maintain the activity of their noble metal electrodes (Breiter, 1963; Gilman, 1963, 1967; Giner, 1964). In the 1970s, the advent of electrochemical detection in flow-injection and LC systems intensified the interest in pulsed waveforms to facilitate electroanalytical detection at noble metal electrodes (Clark, Fleishman, and Fletcher, 1972; MacDonald and Duke, 1973; Stulik and Hora, 1976).

Pulsed amperometric detection (PAD) was first introduced in 1981 for the detection of aliphatic alcohols at Pt electrodes (Hughes and Johnson, 1981; Hughes, Meschi, and Johnson, 1981). The Johnson group at Iowa State University pioneered research on pulsed waveforms at noble metal electrodes for the detection of polar aliphatic compounds. In conjunction with Dionex Corporation (Sunnyvale, California), the first commercial detector dedicated to PAD was joined with HPAEC for the direct determination of carbohydrates (Edwards and Haak, 1983; Rocklin and Pohl, 1983). The application of triple-step potential-time waveforms in electrochemical detection at noble metal electrodes was known as pulsed amperometric detection (PAD) (Edwards and Haak, 1983; Hughes and Johnson, 1982; Johnson and LaCourse, 1990; Rocklin and Pohl, 1983). PAD was soon followed by pulsed coulometric detection (PCD) (Neuberger and Johnson, 1988, 1990), which integrated the amperometric signal. Regardless of the specific form of signal measurement, all of these techniques are now grouped under the title of PAD. Potential sweep-pulsed coulometric detection (PS-PCD) (Neuberger and Johnson, 1988; Welch et al., 1989), which incorporated a triangular potential scan to the detection step of PCD has become known as integrated pulsed amperometric detection (IPAD) (LaCourse and Johnson, 1990) and less commonly as integrated voltammetric detection (IVD) (LaCourse, 1997). The "integrated" description refers to the coulometric rejection of the oxide background. All detection strategies based on the application of multistep potential-time waveforms at noble metal electrodes for electrochemical detection in HPLC fall under the term "pulsed electrochemical detection" (PED) (Johnson and LaCourse, 1990, 1992; LaCourse, 1993, 1997).

Although increased sensitivity and reproducibility has also been reported for pulsed potential cleaning at carbon electrodes by several researchers (Berger, 1985; Ewing, Dayton, and Wightman, 1981; Fleet and Little, 1974; Tengyl, 1984; Van Rooijan and Poppe, 1991), those electrodes have not generally been successful for the detection of polar aliphatic compounds. This effect is attributable to the absence of appropriate electrocatalytic properties of carbon surfaces to support the anodic oxygen-transfer reaction mechanisms of polar aliphatic compounds.

10.2.3 Electrocatalysis at Noble Metal Electrodes

The majority of easily detected compounds at solid anodes under constant applied potentials are self-stabilized via π-resonance. Therefore, a desirable characteristic of electrodes in dc amperometry is inert. The electrode serves as a sink to provide and remove electrons with no direct involvement in the reaction mechanism. Since π-resonance does not exist in polar aliphatic compounds (e.g., carbohydrates), stabilization of reaction intermediates is actively achieved via adsorption at "clean" noble metal electrodes. Faradaic processes that benefit from electrode surface interactions are described as electrocatalytic. Unfortunately, an undesirable consequence of this approach is the accumulation of adsorbed carbonaceous materials, which eventually foul the electrode surface.

Noble metals (i.e., Pt and Au) are commonly considered to be inert, but under electrochemical conditions these electrodes are quite active. Figure 10.1A shows the current–potential (*i–E*) plot for an Au rotating disc electrode (RDE) in 0.1 M NaOH with (·····) and without (—) dissolved O_2. The anodic signals for the positive scan correspond to the formation of surface oxide (wave a) and the breakdown of water to O_2 (wave b). The cathodic peak on the following reverse scan corresponds to the cathodic dissolution of the surface oxide (wave c). If dissolved O_2 is present, a cathodic wave (wave d) is observed during the positive and negative scans. Except for hydrogen adsorption, reduction, and oxidation at potentials < -500 mV, all other features of Pt electrodes are similar to that of Au electrodes in 0.1 M NaOH, see Figure 10.1B. Note that wave d for O_2 reduction is well resolved from wave a for an Au electrode, whereas there is significant overlap of O_2 reduction and oxide formation for a Pt electrode. This difference accounts for the general preference of Au over Pt electrodes for the majority of PED applications.

From the voltammetric data, it is apparent that surface oxide is reversibly formed and dissolved by the application of alternating positive and negative potentials, respectively. Oxide-free, or clean, noble metal surfaces have an affinity to adsorb organic compounds. Upon changing to a more positive potential, electrocatalytic oxidation of adsorbed compounds is promoted via anodic oxygen-transfer from H_2O by transient, intermediate products in the surface-oxide formation mechanism (i.e., AuOH and PtOH). Any fouling, which results as a consequence of the catalytic detection process or adsorbed compounds is oxidatively desorbed from the electrode surface by application of a large positive potential causing the formation of a stable surface oxide (i.e., AuO and PtO). These surface oxides are quite inert and must

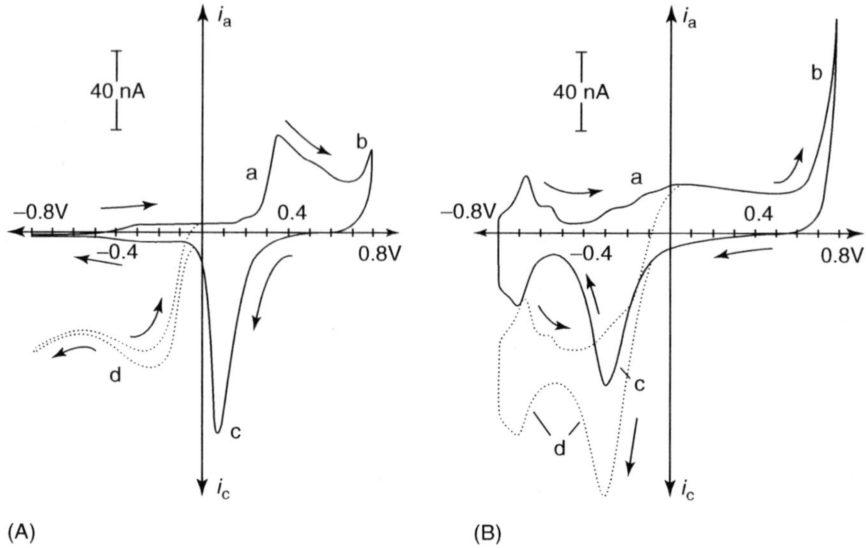

FIGURE 10.1 Cyclic voltammetric response (*i–E*) for (A) Au and (B) Pt rotating disc electrode (RDE). Conditions: Rotation speed, 900 rpm; scan rate, 200 mV/s; Ag/AgCl reference electrode. Solutions: (—) 0.1 M NaOH, deaerated; (·····) 0.1 M NaOH.

be cathodically dissolved by a negative potential excursion to restore the native reactivity of the oxide-free metal surface.

The three steps (1) electrocatalytic oxidation for detection, (2) oxide formation to clean the electrode surface, and (3) oxide removal to reactivate the electrode and preadsorb analyte for the next cycle form the basis of all PED techniques. Three modes of anodic electrocatalytic detection can occur at noble metal electrodes.

Mode I: Direct detection at oxide-free surfaces—At potentials $<$ ca. $+200$ mV (Figure 10.1A) oxidation of the compound can occur with little or no concurrent formation of surface oxide. The surface-stabilized oxidation results in a product which may leave the diffusion layer, readsorb for further oxidation, or foul the electrode surface. The baseline signal originates primarily from double-layer charging, which decays quickly to a virtual zero value. All alcohol-containing compounds (e.g., carbohydrates) are detected by Mode I at Au electrodes in alkaline solutions and Pt electrodes in acidic solutions (Austin et al., 1984; Edwards and Haak, 1983; Edwards, Pohl, and Rubin, 1987; Hughes and Johnson, 1981, 1982, 1983; Hughes, Meschi, and Johnson, 1981; Neuburger and Johnson, 1987; Olechno et al., 1987; Rocklin and Pohl, 1983).

Mode II: Direct oxide-catalyzed detection—In contrast to Mode I detections, Mode II detections require the concurrent formation of surface oxide. Hence, Mode II detections occur at potentials $>$ ca. $+150$ mV (Figure 10.1A). Oxidation of preadsorbed analyte is the primary contributor to the analytical signal, however, simultaneous catalytic oxidation of analyte in the diffusion layer is not excluded. The oxidation products may leave the diffusion layer or foul the electrode surface.

Readsorption of analyte and detection products is attenuated by the surface oxide. The background signal, resulting from anodic formation of surface oxide, is large and has a deleterious effect on quantitation in HPLC-PAD. Aliphatic amines and amino acids (Johnson et al., 1986; Polta and Johnson, 1983; Welch et al., 1989) are detected by Mode II at Au and Pt electrodes in alkaline solutions. Numerous sulfur compounds (Johnson and Polta, 1986; LaCourse and Owens, 1995; Ngoviatchai and Johnson, 1988; Polta and Johnson, 1986; Polta, Johnson, and Luecke, 1986) are also detected by Mode II at Au and Pt electrodes in both alkaline and acidic solutions.

Mode III: Indirect detections at oxide surfaces—Essential to Mode I and Mode II detections is the preadsorption of the analyte at oxide-free surfaces at negative potentials before electrocatalytic oxidation of the analyte itself. Species which adsorb strongly to the electrode surface and are electroinactive interfere with the oxide formation process. Preadsorbed species reduce the effective surface area of the electrode surface, and the analyte signal originates from a suppression of oxide formation. Since the baseline signal results from anodic currents from surface-oxide formation at a clean electrode surface, a negative peak results. Detection as a result of the suppression of surface-oxide formation is known as Mode III detection. Sulfur-containing and inorganic compounds have been detected by Mode III (Johnson and LaCourse, 1992; LaCourse, 1993, 1997).

Most frequently, PED is applied at Au electrodes under alkaline conditions ($pH > 12$), see Figure 10.1A. All aldehydes, including reducing sugars, are anodically detected (Mode I) during the positive potential excursion at the oxide-free surface in the region of ca. -0.6 to $+0.2$ V (Mode I). Large anodic signals are obtained for alcohols, polyalcohols, and nonreducing sugars in the region of ca. -0.3 to $+0.2$ V (Mode I) with an attenuated signal for most compounds from ca. $+0.2$ to $+0.6$ V. Nitrogen and sulfur-containing compounds, for which a nonbonded electron pair is present, are adsorbed at oxide-free Au surfaces for $E <$ ca. $+0.1$ V and can be anodically detected by oxide-catalyzed reactions during the positive scan or $E >$ ca. $+0.1$ V (Mode II). Detections at $E >$ ca. $+0.8$ V are not recommended because of the deleterious effects of evolution of O_2. Although there is little or no individual compound selectivity, functional group selectivity is clearly evident. Electroinactive surface-adsorbing species can be detected by suppression of the oxide formation process (Mode III) at potentials $>$ ca. $+0.2$ V.

Voltammetric resolution of complex mixtures is futile, since electrocatalytic-based detection of various members within a class of compounds is controlled primarily by the dependence of the catalytic surface state on the electrode potential rather than by the redox potentials (E°) of the reactants. Therefore, general selectivity is achieved via chromatographic separation before electrocatalytic detection. This conclusion does not preclude limited selectivity from control of detection parameters.

10.2.4 Detection Strategies in PED

The detection of carbohydrates, which represents the vast majority of food applications, occurs at electrode surfaces that are virtually free of oxides (Mode I). Mode I detections are best implemented using PAD waveforms. The electrochemical nature of carbohydrates is best elicited via studying the voltammetric response of glucose.

10.2.4.1 Mode I: Aldehyde- and Alcohol-Containing Compounds

The current–potential (*i*–*E*) response is shown in Figure 10.2 for an Au RDE in 0.1 M NaOH with (—) and without (----) glucose in the absence of dissolved O_2. With the presence of glucose (—), a reducing sugar, an anodic wave is observed on the positive scan beginning at ca. −0.6 V (wave e). This wave corresponds to oxidation of the aldehyde group to the carboxylate anion in this alkaline media. A much larger anodic signal is obtained for the combined oxidations of the alcohol and aldehyde groups in the region of ca. −0.2 to +0.4 V (wave f). The anodic signal is attenuated abruptly during the positive scan with the onset of oxide formation (wave a). The signal for ca. +0.4 to +0.6 V (wave g) in addition to that for oxide formation results from the anodic desorption of adsorbed glucose or intermediate products simultaneously with the formation of surface oxide on the Au electrode. The absence of signal on the negative scan in the region of ca. +0.8 to +0.2 V indicates the absence of activity by the oxide-covered electrode surface. Following cathodic dissolution of the oxide on the negative scan to produce wave c, the surface activity for glucose oxidation is immediately returned and an anodic peak (wave h) is observed for oxidation of alcohol and aldehyde groups on glucose. Anodic waves e, f, g, and h are all observed to increase in signal intensity with increases in glucose concentration.

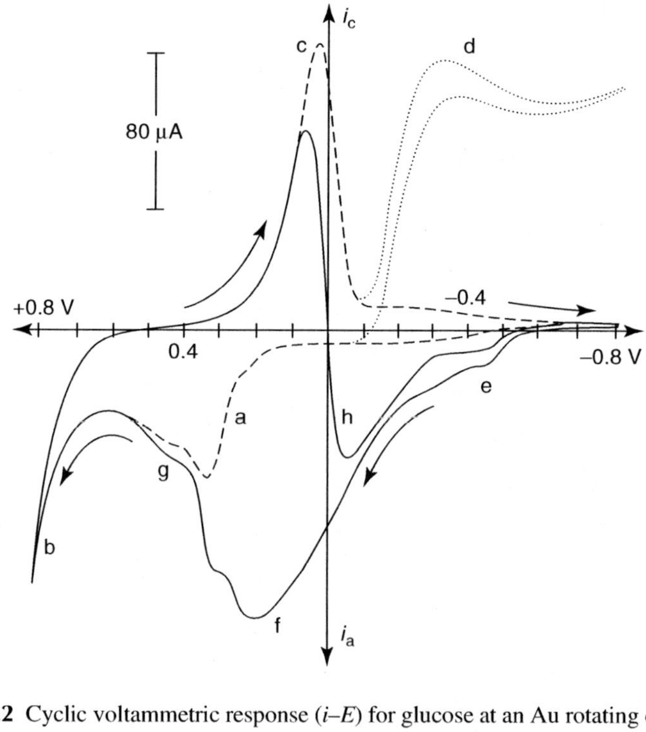

FIGURE 10.2 Cyclic voltammetric response (*i*–*E*) for glucose at an Au rotating disc electrode (RDE). Conditions: Rotation speed, 1000 rpm; potential scan, 200 mV/s; and Ag/AgCl reference electrode. Solutions: (·····) 0.1 M NaOH with dissolved O_2; (----) 0.1 M NaOH, deaerated; and (—) 0.2 mM glucose, deaerated.

Although PAD for carbohydrates can be performed at either Au or Pt electrodes in alkaline media, the use of Au electrodes has the distinct advantage that detection can be achieved without simultaneous reduction of dissolved O_2 (wave d, Figure 10.2). The residual (·····) response characteristics correspond to those discussed in Figure 10.1A. Similar voltammetric response curves are obtained for all other carbohydrates, except that nonreducing carbohydrates will not show a wave commencing at −0.6 V corresponding to the oxidation of the aldehyde group.

Although similar electrode processes are involved in the oxidations of all carbohydrates, significant differences exist in their reaction dynamics (Van Rooijan and Poppe, 1991). These differences are reflected in the shape of current versus bulk concentration (i–C^b) plots. Calibration plots for sorbitol and glucose show an extended linear range as compared to calibration plots for sucrose and maltose. Fructose's response is intermediate to those of glucose and sucrose. Two tentative explanations are offered for the nonlinear i–C^b plots for reactions characterized as being under surface control. First, the reactant molecules can be strongly adsorbed and, therefore, response is controlled by the adsorption isotherm for that reactant. Second, if detection products are strongly adsorbed with the result of surface fouling, the current response will be attenuated more abruptly during the positive scan for large values of C^b for which full coverage of the surface is achieved more quickly. These effects necessitate the use of calibration curves to determine each carbohydrate's range of linearity.

It is evident from Figure 10.2 that satisfactory results will be obtained for a rather wide range of detection potentials (E_{det}). Since uniformly low levels of dissolved O_2 are difficult to maintain in HPLC systems, a choice of $E_{det} = +0.0$ to $+0.2$ V for the PAD waveform is a reasonable first estimate for universal detection of carbohydrates without interference from dissolved O_2.

10.2.4.1.1 Pulsed Amperometric Detection

Oxide-free detections can and were first implemented with a three-step potential-time waveform (Table 10.1) at a frequency of ca. 2 to 0.5 Hz, which is appropriate for HPLC applications to maintain chromatographic peak integrity. The detection potential (E_{det}) is chosen to be appropriate for the desired functional group, and the faradaic signal can be sampled during a short time (e.g., 16.7 ms) after a delay of t_{del}. Typical values of t_{det} are in the range of 100–600 ms. Following the detection process, the electrode surface is oxidatively cleaned by a positive potential step to E_{oxd} (+0.6 to +0.8 V) for t_{oxd} (50–200 ms) and then cathodically reactivated by a large negative step to E_{red} (−0.8 to −0.2 V) for t_{red} (100–600 ms) before the next detection cycle.

Although the simple three-step waveform gives the highest sensitivity, a four-step waveform known, or quadruple-potential waveform (Table 10.1), gives better long-term reproducibility. In the quad pulse waveform, a large negative potential step E_{red} is applied for a brief period t_{red} after the detection step to reduce any partially solvated Au species back to metallic Au. In addition, this step invokes cathodic cleaning of the electrode surface. The negative potential step is following a brief positive potential step (E_{oxd}, t_{oxd}) to activate the electrode surface, which is followed by a potential pulse to affect adsorption and preconcentration of the anlayte on the electrode surface (E_{ads}, t_{ads}). Amperometric detection under the control of any step potential-time waveform is known as pulsed amperometric detection (PAD).

TABLE 10.1
Pulsed Amperometric Detection (PAD) Waveform and Table of Waveform Parameters for Carbohydrates in 0.1 M NaOH at an Au Electrode

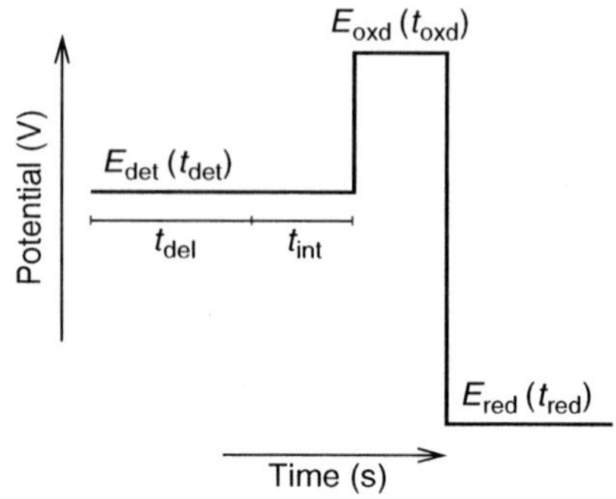

(A) Triple-Step PAD Waveform

Potential (mV versus Ag/AgCl)			Time (ms)		
Parameter	**General**	**Optimized**	**Parameter**	**General**	**Optimized**
E_{det}	−200 to +400	+200	t_{det}	>40	440
			t_{del}	>20	240
			t_{int}	>20	200
E_{oxd}	+300 to +800	+800	t_{oxd}	>60	180
E_{red}	−800 to +100	−300/−800[a]	t_{red}	>60	360

(B) Quadruple-Step PAD Waveform

Parameter	Time (ms)	Potential (mV versus Ag/AgCl)	Integration Period
E_{detect}	0	+100	
	200	+100	Begin
	400	+100	End
$E_{reduce\ and\ clean}$	410	−2000	
	420	−2000	
$E_{activate}$	430	+600	
$E_{adsorption}$	440	−100	
	500	−100	

[a] E_{red}, −300 mV gives better relative standard deviations (RSDs), but it should only be used for relatively clean samples; E_{red}, −800 mV provides superior reductive cleaning.

Anodic detection of the aldehyde and alcohol functionalities in carbohydrates occurs in a potential region where there is only a very small background signal for the concurrent formation of surface oxide, Mode I. The development of positive peaks for carbohydrates and polyalcohols by Mode I in LC-PAD is illustrated in Figure 10.3A by the chronoamperometric (*i–t*) response curves generated following the potential step from E_{red} to E_{det} in the PAD waveform. The residual current (curve a)

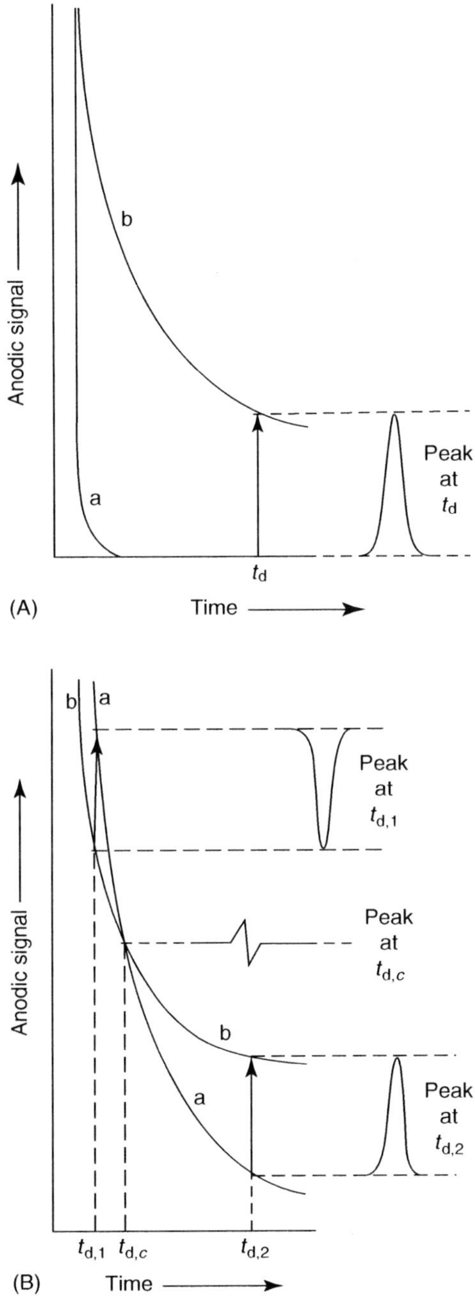

FIGURE 10.3 Chronoamperometric response (i–t) following a potential step from E_{red} to E_{det} in the pulsed amperometric detection (PAD) waveform without (a) and with (b) analyte present to illustrate the origins of chromatographic baseline and peak signals in liquid chromatography–pulsed amperometric detection (LC-PAD) for (A) Mode I and (B) Mode II detection. The delay before current sampling is denoted by t_d (i.e., t_{del}).

decays quickly, and the baseline signal in HPLC-PAD is minimal for $t_{del} >$ ca. 100 ms. Curve b in Figure 10.3A represents the transient *i–t* response for the presence of the reactant and the peak shown is representative of the corresponding anodic signal expected in HPLC-PAD for the value of t_{del} indicated.

As with any detection system, an important consideration is the minimum detectable signal that can be recovered. The problem of differentiating analytical signal from noise becomes increasingly difficult as the signal source becomes weaker (i.e., trace level concentrations). The ability of an instrument to discriminate between signal and noise is expressed in the signal-to-noise ratio (S/N) of the detection system. Any reduction in noise represents an increase in the S/N, and an enhancement of the detection system.

The S/N for measurements of transient amperometric signals is influenced by the instrumental strategy used for sampling of the electrode current. A major noise component of the chronoamperometric signal is sinusoidal and correlated with 60-Hz line frequency. Hence a common strategy for current sampling in PAD involves some form of signal averaging over the period of one 60-Hz oscillation. Since there is no contribution to signal strength from the 60-Hz noise, the time integral of the 60-Hz sinusoidal noise signal (16.7 ms) is zero. Extension of this strategy to the integration of an integral number (m) of 16.7-ms periods results in a significant increase in analytical signal strength while maintaining a 60-Hz noise signal of zero. Typically, the integration period (t_{int}) is 200 ms, which is $m = 12$. Since 200 ms is an integral multiple of one period of 50-Hz oscillation (i.e., 20 ms), 50-Hz line noise is also minimized.

HPLC-PAD and waveform optimization—The selection of potential values for the PAD waveform in HPLC-PAD has been discerned traditionally from the (*i–E*) response obtained for the analyte(s) of interest using cyclic voltammetry (CV). Cyclic voltammograms are generated using triangular potential–time (*E–t*) waveforms. LaCourse and Johnson (1993) have shown that the analyte response using CV, in which potential and time are coupled, is biased by the electrode surface at previous potential values in the scan. This effect is more noticeable for Mode I detections of strongly fouling compounds. In addition, the contributions of currents from oxide formation and double-layer charging are significantly different in CV versus PAD. Aside from inappropriate potential selection of PAD waveforms from CV data, CV is virtually useless for determining optimum time parameters. PAD waveform optimization is best performed using a voltammetric response obtained when a PAD waveform is applied at a hydrodynamically controlled electrode (i.e., RDE) with small incremental changes in one of the parameters of the waveform (e.g., E_{det}) for each cycle of a multistep waveform. Such an approach is known as pulsed voltammetry (PV). PV has proven to be the definitive method for the optimization of PAD waveforms, especially for Mode I detections. Figure 10.4 shows the "background-corrected" i–E_{det} response (positive scan direction) at an Au RDE for equimolar concentration of several carbohydrates in 0.1 M NaOH. As expected, the wave for the aldehyde group (glucose and maltose) begins at ca. −600 mV and quickly plateaus. The peak signal in the region of −200 to +400 mV corresponds to the oxidation of the aldehyde group (glucose and maltose) and the alcohol groups (all carbohydrates). The response for all carbohydrates is inhibited by the formation of surface oxide (not shown) at ca.

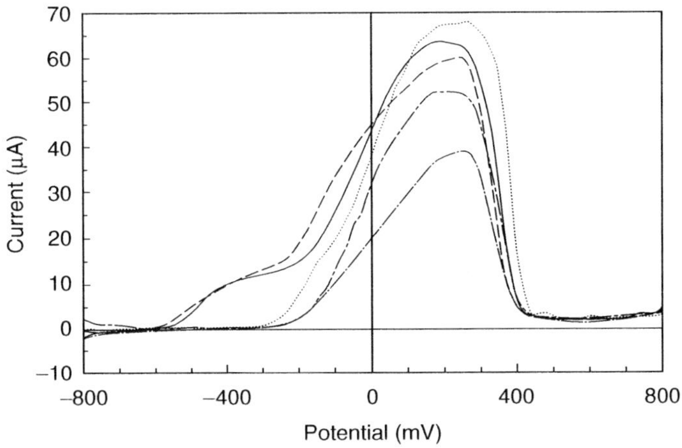

FIGURE 10.4 Pulsed voltammetry (PV) response for five carbohydrates at the Au rotating disc electrode (RDE) in 0.1 M NaOH. Solutions (0.1 mM): (·····) sorbitol; (—) glucose; (— - —) fructose; (— · —) sucrose; (----) maltose.

+400 mV; hence, carbohydrates represent oxide-free detections. A maximum response is obtained at $E_{det} = +180$ mV for all carbohydrates and the application of this value is universal and results in the highly sensitive detection of virtually all carbohydrates. The advantage of PV is clearly evident when one compares the PV responses in Figure 10.4 with the CV plot of glucose in Figure 10.2. Table 10.1 lists the acceptable ranges and optimized value of E_{det} for the PAD waveform of glucose at Au in 0.1 M NaOH. In addition to peak amplitudes, S/N plots can be obtained using PV. A detailed description of PV has been published (LaCourse and Johnson, 1993).

Effect of pH and organic modifiers—Since carbohydrates are readily separated based upon anion formation in alkaline media on anion-exchange columns (Paskach, Lieker, and Thielecke, 1991), chromatographic separations are controllable via changes in pH, ionic strength, and organic modifiers. The consideration of pH-gradient elution must recognize the effect of pH change on the background signal as well as the choice of E_{det} for maximum sensitivity in HPLC-PAD (LaCourse, Mead, and Johnson, 1990; Mead, Larew, and Johnson, 1989). The potential for onset of oxide formation at Au electrodes shifts to more negative values with increases in pH at a rate of ca. -60 mV pH^{-1}. The effect of pH on the oxide formation process is attributable to the pH-dependent nature of Au oxide formation, i.e., $Au(H_2O)_{ads} \rightarrow Au\text{–}OH + H^+ + e^-$. Because the optimal choice of E_{det} corresponds approximately to the value for onset of oxide formation, values of E_{det} should be adjusted by the amount -60 mV pH^{-1} from the value of $+200$ mV recommended for 0.1 M NaOH. The negative shift in oxide formation with increasing pH can be reflected by a large baseline change in LC-PAD under pH-gradient elution when E_{det} remains constant throughout the gradient (LaCourse, Jackson, and Johnson, 1989). This effect can be alleviated to a great extent by substitution of a pH-sensitive glass-membrane electrode for the Ag/AgCl reference electrode in the PAD cell. Because the response of the glass-membrane electrode is ca. -60 mV pH^{-1}, the value of E_{det} is automatically adjusted during execution of pH gradients. Under ionic-strength conditions suitable for electrochemical detection (i.e., $\mu > 50$ mM), the effect of changing ionic strength is reflected as minor perturbations in the background signal from oxide formation. This effect is not noticeable under isocratic HPLC conditions. Under gradient conditions (e.g., increasing acetate concentration), both positive and negative baseline drifts have been observed. In comparison to ionic-strength effects, changes in the concentration of organic modifiers can have a much greater effect on the baseline signal in HPLC-PAD (LaCourse and Johnson, 1990). This can occur, even for electroinactive organic additives, because the modifiers are frequently adsorbed at the electrode surface with a resulting suppression of the oxide formation process (Mode III). In addition to alteration of the HPLC-PAD baseline, adsorbed organic modifiers can severely attenuate the analytical signal for carbohydrates by interfering with access to specific adsorption sites on the electrode needed for the reaction to occur.

10.2.4.2 Mode II: Amine- and Sulfur-Containing Compounds

In addition to carbohydrates and alcohols, compounds which also have amine or sulfur moieties (e.g., aminoalcohols, aminosugars, thiosugars) rely on the detection of the hydroxyl groups to take advantage of the simplicities of PAD waveforms for Mode I detections. The amine or sulfur moieties are exploited to increase the adsorption of the analyte to the electrode surface. Strong adsorption of reacting molecules is considered very beneficial because the residence time of the molecule on the electrode surface is increased substantially, thereby increasing the probability for a successful faradaic reaction. In the case of compounds containing only amine or sulfur groups, only Mode II detections are often available.

Voltammetry—Although amine compounds can be detected directly at Pt (Johnson, 1986) and Au (Dobberpuhl and Johnson, 1995; Welch et al., 1989), electrodes only under alkaline conditions, numerous organic, and inorganic sulfur-containing compounds can be detected across the entire pH range at both electrode materials. Both functional groups are detected by the oxide-catalyzed mechanism of Mode II. The Au electrode is preferred over the Pt electrode to minimize interference from dissolved O_2.

The current–potential (*i–E*) response is shown in Figure 10.5A for an Au RDE in 0.1 M NaOH with (—) and without (----) lysine in the absence of dissolved O_2. With the presence of 10 μM lysine (—), an anodic wave is observed on the positive scan beginning at ca. +0.1 V (wave e). This wave corresponds to oxidation of the amine moiety. It is highly significant that lysine is detected only during the positive potential scan when surface oxide is being generated. No signal is observed for the negative scan when oxide growth has ceased. Hence, dc amperometric detection in the range of ca. +0.2 to +0.9 V is not expected to produce useful analytical signals at an Au electrode covered by inert oxide. Anodic wave e is observed to increase in signal intensity with increases in lysine concentration. All other amines and amino acids display similar voltammetric behavior with that of lysine. Reduction of dissolved O_2 occurs for $E <$ ca. −0.1 V (wave D) during both the positive and negative scans. Figure 10.5B shows pulsed voltammograms of lysine at an Au electrode in 0.1 M NaOH, which is in general agreement with the CV of lysine

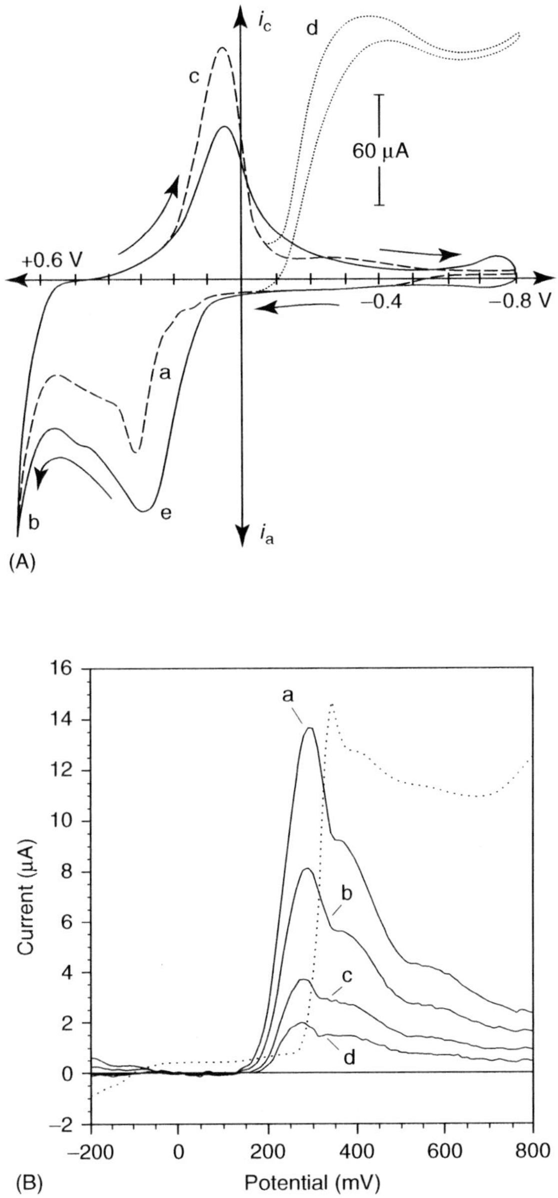

FIGURE 10.5 Current–potential (*i–E*) response curves for lysine at an Au rotating disc electrode (RDE) by (A) cyclic and (B) pulsed voltammetry (PV). Conditions: Rotation speed, 1000 rpm; potential scan, 200 $mV \cdot s^{-1}$; and Ag/AgCl reference electrode. Solutions: (⋯⋯) 0.1 M NaOH with dissolved O_2; (----) 0.1 M NaOH, deaerated; and (—) 0.2 mM lysine, deaerated. PV responses of lysine are background-corrected. Conditions: Rotation speed, 900 rpm. Waveform: E_{det}, variable; t_{det}, 450 ms; t_{del}, 240 ms; t_{int}, 200 ms; E_{oxd}, +800 mV; t_{oxd}, 180 ms; E_{red}, −800 mV, t_{red}, 360 ms. Solutions: (—) lysine; a, 100 μM; b, 50 μM; c, 20 μM; d, 10 μM; and (⋯⋯) 0.1 M NaOH, deaerated.

(Figure 10.5A). Note that the majority of the analytical signal is in the region of ca. +180 to +500 mV, which corresponds to the oxidation of the amine group. Although the signal is present over the range of ca. +500 to +800 mV, it quickly decays, and the inert oxide layer on the electrode surface predominates the total current. Voltammetrically, amines behave similarly to that of amino acids. Therefore, the guidelines which correspond to amino acid detection apply equally well to that of aliphatic amines.

Numerous organic and inorganic sulfur compounds are adsorbed at the oxide-free surfaces of Au and Pt electrodes and can be detected by Mode II (Johnson and Polta, 1986; Ngoviatchai and Johnson, 1988; Polta, Johnson, and Leucke, 1986; Vandeberg, Kowagoe, and Johnson, 1992). These compounds include thioalcohols, thioethers, thiophenes, thiocarbamates, organic thiophosphates, and numerous inorganic compounds. Adsorption is a prerequisite to detection and therefore at least one nonbonded electron pair must reside on the S-atom. The kinetics for detection of adsorbed S-compounds are quite favorable at pHs from 0 to 14. Since alcohol and amine groups are detected only under highly acidic or alkaline conditions, the detection of sulfur compounds under mildly acidic conditions is highly selective. In addition, these pH conditions are compatible with reversed-phase separations.

The PV response is shown in Figure 10.6 for an Au RDE in 250 mM acetate buffer (pH = 3.75) with (—) and without (----) penicillamine. With the presence of (a) 50, (b) 20, and (c) 10 mM penicillamine, an anodic wave is observed on the

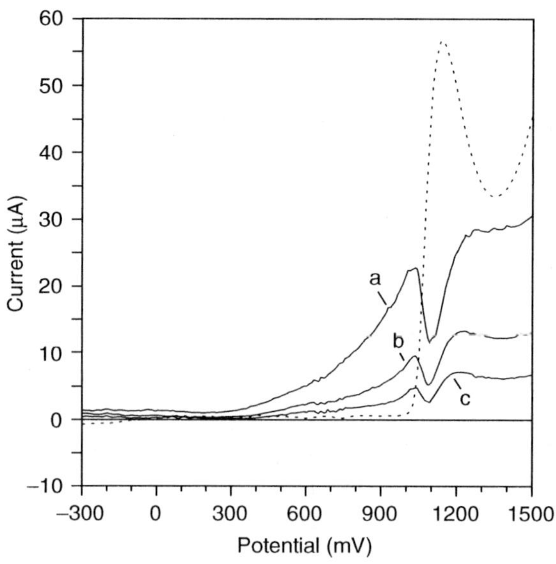

FIGURE 10.6 Pulsed voltammetric (PV) response of penicillamine at an Au rotating disc electrode (RDE) in 250 mM acetate buffer, pH 3.75. This plot is background-corrected. Conditions: Rotation speed, 900 rpm. Waveform: E_{det}, variable; t_{det}, 450 ms; t_{del}, 240 ms; t_{int}, 200 ms; E_{oxd}, +1500 mV; t_{oxd}, 180 ms; E_{red}, −300 mV, t_{red}, 360 ms. Solutions: (—) penicillamine; a, 50 μM; b, 20 μM; c, 10 μM; and (· · ···) residual, deaerated.

positive scan at ca. +300 to ca. + 1500 mV. This wave corresponds to the oxidation of the thiol and, to a limited degree, the amine moiety. The lack of response of the amine moiety is educed from the facts that the amine moiety is protonated at this pH, and valine shows little or no response under these same conditions. The anodic wave is observed to increase in signal intensity with increases in penicillamine concentration. The adsorption of sulfur-containing compounds, including penicillamine, is so strong that the onset of oxide formation is pushed to higher potentials. An artifact of the subtraction of the unperturbed residual response from the analyte response with an altered background results in the negative dip at ca. +1050 mV. This perturbation of the oxide in the presence of analyte leads to deleterious consequences (e.g., negative peaks) when using PAD waveforms. In general, all sulfur compounds give very similar voltammetric profiles, except for the degree of perturbation of the residual response.

10.2.4.2.1 Integrated Pulsed Amperometric Detection

Anodic detection of amine- and sulfur-containing compounds occurs in a potential region where there is a significant signal for the concurrent formation of surface oxide. Oxide-catalyzed detections are denoted as Mode II. The origins of detection peaks in LC-PAD based on Mode II are illustrated in Figure 10.3B. Here, the residual *i–t* response (curve a) corresponds to double-layer charging and to the formation of surface oxide. The current from surface-oxide formation decays much more slowly than the current from double-layer charging. Baseline signals for Mode II typically have a nonzero value over a large range of t_{del} values (see Figure 10.3B, curve a). The *i–t* response corresponding to the presence of adsorbed analyte is represented as curve b. For small $t_{d,1}$ (i.e., $t_{del,1}$), negative peaks can be obtained in HPLC-PAD because of initial inhibition of the oxide formation process by the adsorbed analyte. For larger values of $t_{d,2}$ (i.e., $t_{del,2}$), positive peaks are obtained when sufficient oxide has been produced to catalyze the anodic reaction of the adsorbate. In the case of an unfortunate choice of an intermediate value of $t_{d,c}$ at the crossover point, a detection peak might not be obtained. Choice of $t_{del} >$ ca. 150 ms of a PAD waveform usually is sufficient to assure positive HPLC-PAD peaks based on Mode II.

As discussed for Figure 10.3B, a large baseline signal is encountered in HPLC-PAD for the oxide-catalyzed detections of amino acids and sulfur compounds (Mode II). Furthermore, the large baseline current is frequently observed to drift to large anodic values, especially for new or freshly polished electrodes. This drift is the consequence of a slow growth in the true electrode surface area as a result of surface reconstruction caused by the oxide on–off cycles in the applied multistep waveforms. As listed in Table 10.2, Mode II detections performed with PAD are subject to a number of disadvantages because of the formation of surface oxide, which is required and concomitant with the detection of amine- and sulfur-based compounds.

The disadvantages listed in Table 10.2 are either alleviated or greatly diminished by the use of the waveform in Table 10.3. Here, the electrode current is integrated electronically throughout a rapid cyclic scan of the detection potential (E_{det}) within the pulsed waveform. The potential scan proceeds into (positive scan) and backs out of (negative scan) the region of the oxide-catalyzed reaction for detection by Mode II.

TABLE 10.2
Disadvantages of Pulsed Amperometric Detection (PAD) for Oxide-Catalyzed Detections

Baseline/Background Sensitivities (*B*)	
pH Organic modifiers Ionic strength (*I*) Temperature (*T*)	Any changes or gradients in any of these variables will lead to baseline drifting. The baseline drift is attributable to variations in the extent of surface-oxide formation.
Sample-Related Effects	
Poor signal-to-noise ratio (*P*)	The sample current is only a fraction of the total signal. The majority of signal for Mode II detections is derived from surface-oxide formation. Oxide-formation signal tends to be noisy.
Postpeak dips	The presence of the analyte at the electrode surface interferes with surface-oxide formation. Hence, the background is different in the presence and absence of the analyte, which often results in a dip after the chromatographic peak.

The anodic charge for oxide formed on the positive sweep tends to be compensated by the corresponding cathodic charge (opposite polarity) for dissolution of the oxide on the negative sweep. Hence, the background signal on the electronic integrator at the end of the detection period can be virtually zero and is relatively unaffected by the gradual change of electrode area. IPAD combines CV with potential pulse cleaning to maintain uniform electrode activity.

Waveform optimization—There are several requirements pertaining to the cyclic sweep of E_{det} in IPAD which must be satisfied to achieve maximum success for applications to LC:

(i) Cyclic scan of E_{det} must begin (E_{dst}) and end (E_{dnd}) at a value for which the electrode is free of surface oxide, i.e., $E <$ ca. $+0.1$ V versus SCE for Au in 0.1 M NaOH (see Figure 10.1A).
(ii) Value of E_{dnd} should not extend into the region for cathodic detection of dissolved O_2, i.e., $E < -0.1$ V versus SCE (see Figure 10.1A), if dissolved O_2 is present.
(iii) Positive scan maximum (E_{dmx}) must not extend beyond the value for anodic solvent breakdown, i.e., ca. $+0.7$ V versus SCE (see Figure 10.1A). When the majority of signal in an oxide-catalyzed (Mode II) detection can be found at the onset of oxide formation where analyte coverage versus rate of oxide formation is most optimal, a greater S/N can be achieved by not scanning to the possible limit of this criterion.

It is to be noted from the residual *i–E* curve for Au in Figure 10.1A that only a small potential region centered at ca. -0.1 V versus SCE is appropriate to satisfy criteria (i) and (ii) in 0.1 M NaOH containing dissolved O_2. The purging of solvents with He and use of O_2-impermeable tubing can result in virtually O_2-free conditions which greatly relaxes constraints resulting from (i) and (ii).

TABLE 10.3
Optimized Integrated Pulsed Amperometric Detection (IPAD) Waveform for Thiocompounds and Summary of Parameter Functions

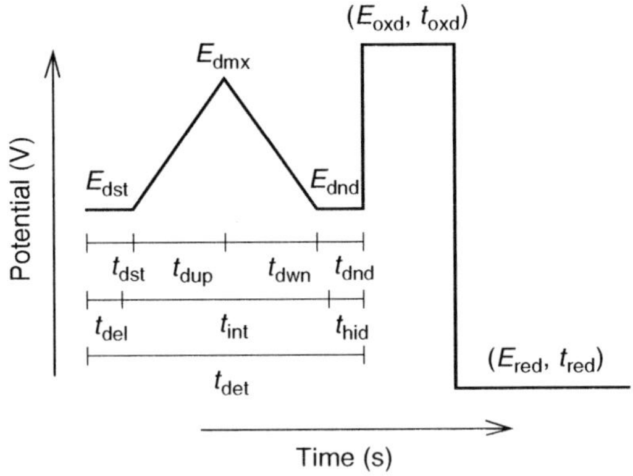

Parameter	Description	Acetate Buffer, pH 5.5, *E* (mV), *t* (ms)
E_{dst}	Starting potential of scan, before oxide formation	−350
E_{dmx}	Maximum potential of scan for analyte oxidation	1500
E_{dnd}	Ending potential of scan, more negative than oxide dissolution	−350
E_{oxd}	Oxidation potential to initiate formation of oxide	1700
E_{red}	Reduction potential to initiate dissolution of inert oxide	−650
t_{dst}	Time at E_{dst}	20
t_{dup}	Time for scan up from E_{dst} to E_{dmx}	218
t_{dwn}	Time for scan down from E_{dmx} to E_{dnd}	218
t_{dnd}	Time at E_{dnd}	0
t_{det}	Total time of detection step	456
t_{del}	Delay time needed to overcome double-layer charging	20
t_{int}	Integration time: signal collection/background rejection	436
t_{hld}	Hold time to complete oxide dissolution at E_{red}	0
t_{oxd}	Time at E_{oxd} to achieve complete formation of oxide	100
t_{red}	Time at E_{red} to achieve complete dissolution of inert surface oxide	200

At present, the IPAD waveform should be optimized utilizing the criteria presented above in relation to CV data and using PV, to optimize the pulsed potential steps as described for optimizing PAD waveforms. Table 10.4 lists both the general and accepted IPAD waveform parameters under several pH conditions. Presently, work is underway to develop an IPAD optimization program similar to that of PV.

It is stressed that the advantage of IPAD, in comparison to PAD, relates to minimization of baseline drift for oxide-catalyzed detections (Mode II). Comparisons of IPAD with PAD for carbohydrates at the oxide-free surfaces (Mode I) have indicated virtually no significant difference between the two techniques, and the continued use of PAD for carbohydrate and alcohol detections is recommended.

TABLE 10.4
Summary of the Analytical Figures of Merit Achievable with High-Performance Liquid Chromatography with Integrated Pulsed Amperometric Detection (HPLC-IPAD). These Data were Collected by Injecting Each Compound Separately with a Mobile Phase Chosen to Yield Similar k' Values for Each Compound

Compound	Target Level (ppb)	LOD[a] (ppb)	LOL[a] (ppb)	CV[a] at 50 ppb (%)
Amoxicillin	10	1	50	1.4
Ampicillin	10	2	500	1.9
Cephapirin	20	1	200	1.7
Penicillin G	5	5	500	3.6
Cloxacillin	10	5	500	2.4

[a] LOD, limit of detection ($S/N = 3$); LOL, limit of linearity; CV, coefficient of variation.

HPLC-IPAD—A typical gradient system for the separation of a complex mixture often requires the combination of pH, organic modifier, and ionic-strength gradients. The current produced by analyte detection based on Mode II is accompanied by current from surface-oxide formation. Consequently, variables (e.g., pH, organic modifier, and ionic strength) which affect the rates of oxide formation and dissolution will be reflected as drift in the baseline in HPLC-PAD. For these reasons, IPAD was designed to apply a waveform which coulometrically rejects the oxide background by summing the charges owing to oxide formation and oxide dissolution which are expected to be of equivalent magnitude but opposite polarity. IPAD can virtually eliminate drift and changes associated with variations in composition (e.g., ionic strength and organic modifier) of the mobile phase, as well as changes in the total surface area of the noble metal electrode surface. Changes in pH may require the simultaneous use of a pH reference electrode (LaCourse and Johnson, 1990; LaCourse, Mead, and Johnson, 1990).

10.3 REVIEW OF FOOD APPLICATIONS

HPAEC-PAD is now well established for the direct detection of monosaccharides, disaccharides, oligosaccharides, polysaccharides, and sugar alcohols. This technique is proving to be more than capable of analyzing the various components obtained from the hydrolysis of complex carbohydrates and dietary fiber. The technique of HPAEC-PAD has recently been reviewed along with its application to the analysis of the components derived from the hydrolysis of complex carbohydrates and dietary fiber (Henshall, 1999). Human milk oligosaccharides in one study were analyzed by HPAEC-PED using stachyose as the internal standard (Engfer et al., 2000). Oligosaccharides produced by the action of an enzyme, lichenase [(1,3)(1,4)-β-D-glucan-4-glucanhydrolase, E.C. 3.2.1.73] on oat β-glucans were analyzed by HPAEC-PAD. These oligosaccharides obtained from lichenase-digested oat β-glucans were identified as tri- and tetrasaccharides (Weisz et al., 1999). Maltodextrins, dextrans, inulin, and other oligosaccharides have been analyzed by HPAEC-PAD (Dionex Corp.).

The following summary of food applications covers the history, breadth, and strong analytical utility of PED techniques for food and beverage analysis.

10.3.1 Carbohydrate Applications

10.3.1.1 Beverages

Hughes and Johnson (1982) describe the earliest carbohydrate analysis of beverages using PAD detection in which they quantified five components in different brands of beer. This work demonstrated the superiority of PAD combined with AEC for carbohydrate analysis over RID using real samples. Samples were prepared by passing them through a solid-phase extraction (SPE) cartridge with a mixed-bed resin before analysis by cation-exchange chromatography-PAD with postcolumn addition of NaOH. In Figure 10.7, chromatograms showing the separation and detection of a light beer and a regular beer are shown. To demonstrate the versatility

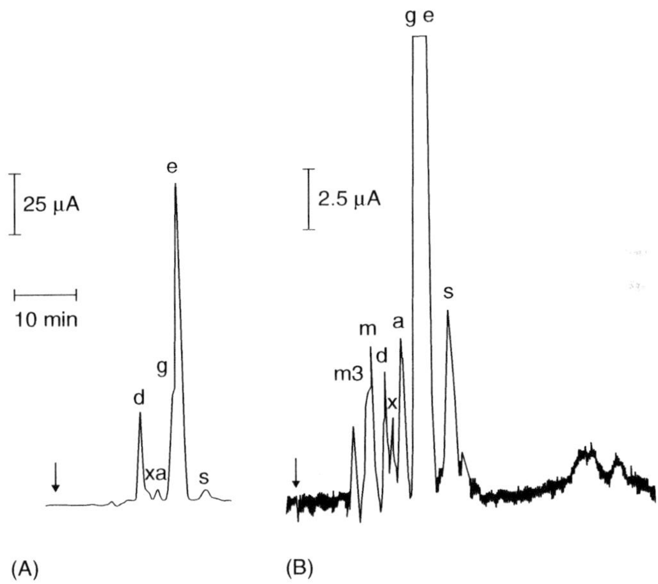

FIGURE 10.7 Chromatographic analysis of (A) Miller Lite beer and (B) Coors beer. Peak identities: m3, maltotriose; m, maltose; d, dextrose; x, xylose; a, arabinose; g, glycerol; e, ethanol; and s, sorbitol. Chromatographic conditions: Ca(II) loaded cation-exchange column (Hamilton) maintained at 85°C, with a water mobile phase at a flow rate of 0.5 mL/min. Postcolumn addition of base was necessary so 6.1 M NaOH was delivered at 0.01 mL/min. The detector was house constructed with a Pt wire working electrode, a Pt auxillary electrode, and a saturated calomel reference electrode. (Reprinted from Hughes, S. and Johnson, D.C., *J. Agric. Food. Chem.*, 30(4), 712–714, 1982.)

of this technique, carbohydrates in ginger ale and coconut milk were also determined. In each case, PAD provided a fast, direct analysis with few matrix peaks discernable.

Beer was also analyzed by Larew, Mead, and Johnson (1998) for the purpose of determining the total carbohydrate content in beer. In a flow injection scheme, an enzyme reactor is added before the PAD detection step facilitates the detection of oligosaccharides which do not have the high response factors that monosaccharides have. The glucoamylase reactor converts the disaccharides and oligosaccharides to glucose and the total carbohydrate content is calculated from the glucose peak. This method also simplifies the quantitation because only one calibration curve is required. Light beer and regular beer are compared to show the application of the enzyme reactor. The percent carbohydrate present in the light beer is known and compared favorably with the experimentally determined value.

Soga, Inoue, and Yamaguchi (1992) analyzed the carbohydrate and sugar alcohol content of beer using hydrophillic interaction chromatography with postcolumn addition of NaOH to facilitate PAD detection. For the separation, an acetonitrile (ACN)/water mobile phase was used. Determination of the monosaccharide and oligosaccharide profiles of beer during various stages in the brewing process was performed to show the applicability of HPAEC-PAD to the quality control of beer at various stages in the brewing process, from the starting materials to the final product (Dionex Technical Manual, 1995).

Wine was analyzed by Hughes and Johnson (1983) and carbohydrates and alcohols were determined. Cation-exchange chromatography with a water mobile phase was used for the separation, and postcolumn addition of NaOH was necessary for PAD detection. Both dry and sweet wines were analyzed, and as expected, the dry wine had much less fructose and dextrose than the sweet wine. A peak for sorbitol was not observed for any of the wines tested. Because sorbitol is not naturally present in grapes, its presence in wine would indicate adulteration of the wine with other fruit juices.

Bernal et al. (1996) analyzed wines made in different regions by HPAEC-PAD to detect differences in the grapes used and in wine-making techniques. Sample cleanup consisted of elution through a C-18 SPE cartridge before analysis to remove the pigments present in red wine. White wine was not subjected to this process. Seven monosaccharides were determined in both red and white wines. Wines which had been fermented for 6 and 12 months were used in this study and the results show the disappearance of the sugars with fermentation time as expected. Differences with respect to the winemaker were also evident presumably because of varied wine-making techniques employed in the process.

Wine coolers and brandy were analyzed for aliphatic alcohols using ion-exclusion chromatography-PAD with minimal sample preparation (LaCourse et al., 1991). In each case, glycerol and ethanol were found with methanol (MeOH) present in trace amounts.

Bernal et al. (1996) also determined seven monosaccharides present in instant coffee. A C-18 SPE step was again used before analysis to protect the chromatography system. Two of the coffees used in this study were adulterated. This was easily detectable because the levels of glucose and fructose were elevated compared to the unadulterated coffees. Free carbohydrates and total carbohydrate content of soluble coffee was determined in a multiple laboratory study (Prodolliet, Bugner, and Feinberg, 1995). Eleven different laboratories analyzed six different coffee samples and the results were compared to determine the precision of the HPAEC-PAD method in the different laboratories. When the carbohydrate content in the coffee was greater than 0.3%, the precision of the HPAEC-PAD method was excellent. In another study, carbohydrate content in instant coffee was determined by HPAEC-PAD (Dionex Technical Manual, 1995).

10.3.1.2 Fruit Juices

Determination of carbohydrates in beverages has received some attention because they can be used for quality control of the juice, to identify cases of adulteration of the juice, and to aid in the determination of total carbohydrate content for labeling purposes (White and Widmer, 1990). Since PAD is selective for carbohydrates, the only sample preparation necessary is dilution and filtration. The primary sources of carbohydrates present in the juices analyzed are glucose, fructose, and sucrose. Corradini, Cristalli, and Corradini (1993) compared juices from apples, grapefruit, and oranges from different commercial sources and show that the carbohydrate content is similar in each case. This work was followed up with a later paper in which fruit concentrates, juices, and nectars from apple, peach, and pear were analyzed (Corradini, Cristalli, and Corradini, 1994). This work showed that the differences in the carbohydrate profiles between the juices is easily discerned by HPAEC-PAD and would be quite useful in the detection of fruit juice adulteration. White and Widmer compare the performance of PAD with RID (1990) for carbohydrate analysis. PAD compares favorably with RID because it is rapid, simple, and sensitive with detection limits 100 times more sensitive than that for RID. Juices from a number of different fruits were analyzed by Zook and LaCourse (1995) to demonstrate the application of PAD for determination of carbohydrate content in foods and beverages. Figure 10.8 shows a separation of the carbohydrates in orange juice. Carbohydrate content of fruit juices was shown to be an easy, routine analysis (Dionex Technical Manual, 1995).

Carbohydrate analysis of foods is an effective method of determining adulteration. Products such as juices can be manufactured more cheaply through the addition of inexpensive sweeteners, high fructose corn syrup, and invert sugar are examples. Since PAD is selective and sensitive for carbohydrates, it is ideally suited for such a determination. Many adulterants used are quite similar to the fruit juice to which it is being added to avoid detection of the adulteration by regulators (Swallow, Low, and Petrus, 1991). Swallow, Low, and Petrus (1991) and White and Cancalon (1992) present HPAEC-PAD methods for the determination of orange juice adulteration with beet medium invert sugar (BMIS). The oligosaccharide profile of the invert sugar is compared with that of the pure juice and the questionable juice. Even at levels as low as 5% BMIS added, the presence of adulteration is evident. White and Cancalon (1992) improved the method of Swallow et al. (1991) by reducing the run time from 71 to 31 min. These results are further supported by a similar study in which orange juice adulteration is determined by examination of the

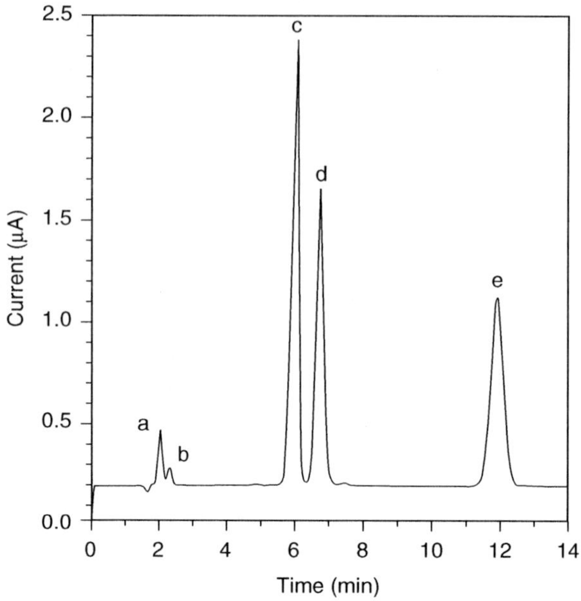

FIGURE 10.8 Chromatogram of carbohydrates in orange juice. Peak identities: a, xylitol; b, sorbitol; c, glucose; d, fructose; e, sucrose. Chromatographic conditions: CarboPac PA1 anion-exchange column (Dionex), 100 mM NaOH mobile phase with a 1 mL/min flow rate; PAD at an Au electrode. (Reprinted from Zook, C.M. and LaCourse, W.R., *Curr. Sep.*, 14(2), 48–52, 1995. With permission.)

oligosaccharide profile (Dionex Technical Manual, 1995). Low and Wudrich (1993) also show that adulteration of fruit juice is readily determined by HPLC-PAD. All of these papers show that PAD is a simple, rapid, and direct method for the determination of carbohydrate profiles in juices and sweeteners.

Wang and Zopf (1989) determined oligosaccharides present in milk using ion-exchange chromatography followed by PAD. The eluted oligosaccharides were collected individually and subjected to acid hydrolysis. The monosaccharides were then determined using the same chromatographic system.

Lactose and sucrose were determined in a chocolate milk sample using HPAEC-PAD (Peschet and Giacalone, 1991). The sample was prepared by simply diluting 100 mg chocolate milk in 1 L water. The resulting chromatogram was very clean with the analyte peaks clearly detectable. No background peaks were observed. These papers demonstrate the ease of application of PAD to a very complex milk matrix.

10.3.1.3 Food Additives and Nutriments

Ginseng saponins are the major components of ginseng, which is gaining importance as an herbal medicine. Park et al. (1994) determined ginseng saponins from ginseng root to compare the performance of HPAEC-PAD with HPLC-UV. With the PAD system, nanogram level detection limits were achieved which were two orders of magnitude lower than HPLC-UV analysis.

As discussed above, carbohydrate analysis of foods is an effective means of identifying cases of adulteration. Low (1996) describes several projects in which adulteration was easily detected using the carbohydrate profile from both the pure food and the sweeteners used in the adulteration process. Grapefruit juice and honey adulteration can be determined by comparing the fingerprint obtained for these products using HPAEC-PAD with the fingerprint obtained for sweeteners such as high fructose corn syrup. Using these profiles, adulteration can be easily identified.

Analysis of honey for carbohydrate content is a challenging problem. Glucose and fructose are the major components, but in addition, honey contains at least 11disaccharides, 11 trisaccharides, and other higher oligosaccharides. Swallow and Low (1990) analyzed honey from different sources to evaluate the applicability of HPAEC-PAD for this analysis. Glucose and fructose were determined by simply diluting the honey with water and analyzing by HPAEC-PAD. Analysis of the purified oligosaccharide samples revealed a unique fingerprint for honeys from different natural sources. Figure 10.9 shows a chromatogram of the carbohydrate profile obtained for a pure honey. Fingerprints of this type can be used to detect adulteration of honey with inexpensive sweeteners. In a later paper, Swallow and Low demonstrated the use of HPAEC-PAD for the determination of honey adulteration (1994). In this work, samples from 44 different honey sources were analyzed as above (Swallow and Low, 1994) and although the oligosaccharide

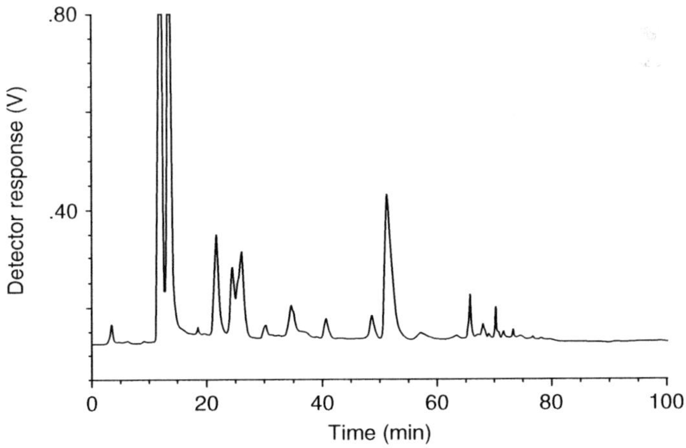

FIGURE 10.9 Chromatogram of the oligosaccharide profile of a pure honey sample. Chromatographic conditions: Two CarboPac PA1 columns (Dionex) connected in series; gradient elution was used with NaOH as the primary eluent. Postcolumn addition of NaOH was used to minimize baseline drift. PAD at an Au electrode was used. (Reprinted from Swallow, K.W. and Low, N.H., *J. Assoc. Off. Anal. Chem.*, 77(3), 695–702, 1994. With permission.)

profiles were similar, a unique fingerprint was discerned for each one. The sources of the variation in the honey are numerous, ranging from the botanical source to the storage temperature. Honey samples were then purposely adulterated with high fructose corn syrup and invert syrup to determine changes in the profile. In both cases, the presence of the adultering substance was easily detected. Determination of adulteration of maple syrup by HPAEC-PAD was also shown by Stuckel and Low (1995).

Monosaccharides and disaccharides in molasses were determined by HPAEC-PAD, which is the ICUMSA official method (Dionex Technical Manual, 1995). This was an interlaboratory study which involved 11 laboratories. The reproducibility of this method was excellent for the different laboratories and the results agreed with a GC method.

Larew and Johnson (1988) quantified the percentage of malto-oligosaccharides found in corn syrup using HPAEC-PAD. This work is distinguished from other corn syrup determinations because they added a glucoamylase reactor between the separation and the detection step. The enzyme reactor served to increase the sensitivity for malto-oligosaccharides and, since the analytes are converted to glucose during the enzyme reaction, only a glucose calibration curve is required for calibration. This system worked well for the corn syrup samples, and it did decrease the work required in the quantitation step. Stefansson and Lu (1993) also determined malto-oligomers in corn syrup directly using reversed-phase ion-pair chromatography-PAD to determine the compatibility of the ion-pair chromatography with PAD. This system proved compatible with PAD and more versatile than ion-exchange chromatography because the ion-pair agent could be changed to affect different chromatographic selectivity.

10.3.1.4 Solid Foods

HPAEC-PAD is applicable for complex samples. One of the earliest papers using HPAEC-PAD showed the analysis of high fructose corn syrup and potato chips (Edwards and Haak, 1983). The corn syrup assay was of immediate interest to manufacturers, because the presence of higher saccharides affects the taste of the syrup by imparting a bitter taste. The use of HPAEC-PAD for quality control of product was demonstrated with the analysis of glucose, sucrose, and lactose to determine the level of seasoning present in the finished product. Rocklin and Pohl (1983) published a more detailed study of determining glucose, fructose, and lactose from extracts of flavored potato chips in the presence of high concentrations of salt.

Van Riel and Olieman (1991) took advantage of the sensitivity of HPAEC-PAD in their sample preparation of dairy products for mono- and oligosaccharide analysis using ultrafiltration. For dairy products containing 0.5% individual saccharides, 1 g was diluted to 100 and 2 mL of the resulting solution filtered. Recovery of galactose, fructose, saccharose, lactulose, glucose, maltose, maltotriose, and maltopentaose were determined by spiking 50 mg of each into 10 g skim milk. Recoveries ranged from 98.2% for maltopentaose to 101.8% for saccharose. Repeatabilities were less than 5.3%. Sample matrices included fruit yogurt, candy, and infant formula. Figure 10.10 shows the HPAEC-PAD traces for each of these samples.

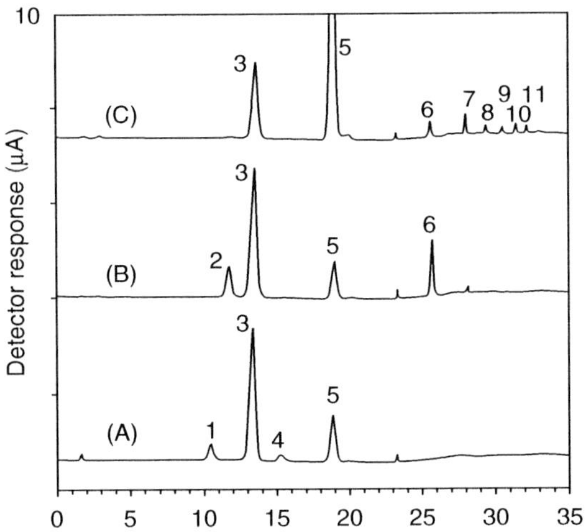

FIGURE 10.10 Chromatograms of saccharides in (A) fruit yogurt, (B) candy, and (C) infant formula. Peak identification: 1, D-galactose; 2, D-glucose; 3, saccharose; 4, D-fructose; 5, lactose; 6, maltose; 7, maltotriose; 8, maltotetraose; 9, maltopentaose; 10, maltohexaose; 11, maltoheptaose. Conditions: Linear gradient from 0 to 0.25 M sodium acetate in 0.1 M sodium hydroxide on CarboPAC PA1 column (Dionex). Detection: Pulsed amperometric detection (PAD). (Reprinted from Van Riel, J. and Olieman, C., *Carbohydr. Res.*, 215, 39–46, 1991.)

Common sugars, such as glucose, fructose, and sucrose, in a wide variety of foods can be determined after dilution and filtration before HPAEC-PAD analysis. Such foods include tomato ketchup, butterscotch candy, and chocolate (Dionex Technical Manual, 1995). Sample preparation for the high-fat chocolate include supercritical fluid extraction to remove the fat. Impurities in sweeteners can be detected, such as glucose and fructose in sucrose.

Artificial sweeteners include sucrose derivatives as well as sugar alcohols. Sucralose is a sweetener which is 400–800 times as sweet as sucrose. It is a selectively chlorinated sucrose. Ichiki and coworkers (1996) determined sucralose in foods, and trace impurities 4-Cl-galactose and 1,6-Cl-fructose were detected in a sucralose sample (Dionex Technical Manual, 1995). Other sugar substitutes such as ketose, maltrin, and inulin are also separated and detected using HPAEC-PAD. Sugar alcohols including glycerol, sorbitol, and mannitol have been determined in hard candies and chewing gum. Sample preparation involves either dissolution or sonication followed by dilution.

The determination of lactose, galactose, and dextrose in grated cheeses was improved by using AEC with PAD (Pollman, 1989). Previous LC methods suffered from a lack of sensitivity, failure to resolve galactose and dextrose, and interference from the presence of salt in the cheese in the determination of dextrose.

Monosaccharide analysis following acid hydrolysis is a common use of HPAEC-PAD. Garleb, Bourquin, and Fahey (1989, 1991) found that HPAEC-PAD technology was superior to previous HPLC techniques, and that the hydrolysis method was the limiting factor in neutral monosaccharide analysis. Their aim was to compare three hydrolytic methods, one using trifluoroacetic acid (TFA) and two using sulfuric acid (H_2SO_4) under different temperature and time parameters. The substrates hydrolyzed included xylan, microcrystalline cellulose, wheat bran, apples, wheat straw, and alfalfa. Monosaccharides quantified were arabinose, xylose, glucose, galactose, mannose, and rhamnose. They found that using TFA for hydrolysis led to less monosaccharide degradation, but that the procedures using H_2SO_4 yielded greater recovery of monosaccharides. Also, the temperature, rather than duration of hydrolysis, yielded higher recoveries of monosaccharides for certain substrates.

Quigley and Englyst (1992) analyzed NSP, which are measured as an index of dietary fiber. Their analysis required enzymatic release of starch followed by its acid hydrolysis. Neutral and amino sugars are released using 12 M H_2SO_4 at 35°C for 1 h, followed by 2 M H_2SO_4 at 100°C for 1 h. Sample preparation following hydrolysis consisted of addition of internal standard for both neutral and amino sugars, and addition of NaOH for amino sugars to adjust the pH. Removal of sulfate ions is accomplished online using a precolumn guard and column switching. Samples hydrolyzed and analyzed for neutral sugars were guar gum, mucin, chin, and cellulose, and the NSP residue from haricot bean, oat bran, mycoprotein, wheat bran, soya bran, wholemeal flour, carrot, cabbage, garden pea, sugar beet, mushroom, and ileostomy effluent. The other components of NSP are uronic acids. Quigley and Englyst (1994) released uronic acids from pectin, in which values for uronic acids are known, to understand the time course of release. These values were used to determine hydrolysis, either by acid or enzymes, times for release of uronic acids from NSP from food sources. The samples analyzed include linseed, carrot, potato, haricot bean, Brussels sprout, cabbage, parsnip, and wholemeal bread. Determination of glycouronic acids by HPAEC-PAD was also accomplished by Martens and Frankenberger (1990a, b). Samples of orange peel, acacia powder, straw, animal waste, and soil were acid hydrolyzed followed by enzyme catalysis to extract the galacturoic and glucoronic acids. The HPAEC-PAD method was superior to an existing HPLC-UV method because of the sensitivity and precision achievable.

Pectin is an important natural polymer because it is responsible for determining cell wall strength and flexibility. Oligogalacturonic acids are released from polygalacturonic acid in pectin and are regulators of physiological responses in plants. Hotchkiss and Hicks (1990, 1993) reported on the analysis for oligogalacturonic acids in plant tissues by HPAEC-PAD. HPAEC-PAD analysis yielded a direct method in which a complete oligosaccharide profile was obtained. In a later paper, Hotchkiss, El-Bahtimy, and Fishman (1996) investigated the pectin structure in peaches to gain insight into the ripening process. Peach tissues were extracted and enzyme digested with endo-polygalacturonase with analysis performed at 0, 15, and 45 min to observe the progression of the enzymatic digestion. Results were compared with a high-performance size exclusion analysis to provide a more complete picture of the peach ripening process.

Galacturonate in pectin was analyzed by HPAEC-PAD and compared to an existing colorimetric procedure (Garleb, Bourquin, and Fahey, 1991). Samples of apple, cucumber, celery, grapefruit, and radish were lyophilized and the pectin was isolated from the total dietary fiber fraction. No statistically significant difference between the HPAEC-PAD and colorimetric method was detected; however, the colorimetric procedure could not differentiate between glucuronate and galacturonate. In addition, the HPAEC-PAD procedure was used to determine the chemical composition of the pectin isolated from the samples. In this case, the HPAEC-PAD method was superior to a gas–liquid chromatography method primarily because PAD eliminates the need for derivitization.

During times of stress, plants will accumulate low molecular weight metabolites within their tissues. Adams et al. (1993) analyzed tissues from several types of plants for evidence of stress through the presence of sugar alcohols, specifically inositol. Samples of flower petals were extracted with a liquid–liquid extraction, desalted, vacuum-dried, reconstituted in water, and passed though a polystyrene-divinylbenzene SPE cartridge, and analyzed by HPAEC-PAD. HPAEC-PAD proved to be excellent for the complex samples which resulted from the extraction process with no discernable interferences for the peaks of interest.

A specialty pectin, which is used as a fat substitute, is a partially methylated polygalacturonic acid. Pectins from different sources have different oligogalacturonic acid profiles. Separation of pectins of various DPs is possible using AEC with an acetate gradient.

Koizumi and coworkers (1989) used HPAEC-PAD to determine D-gluco-oligosaccharides and D-gluco-polysaccharides with DP $\geq$ 50. They also determined that sensitivity is not independent of molecular weight, rather it increases for each HCOH group present, which makes PAD a good detection choice for oligosaccharides. In addition, Koizumi and coworkers demonstrated the separation of depolymerization products of amylopectin portion of starch from different crops including rice, corn, sweet potato, and edible canna. Baseline separation of amylopectins from DP = 6 to 60 is accomplished in 40 min (Koizumi, Fukuda, and Hizukuri, 1991). Under gradient conditions, acetate ion is typically the preferred pusher ion, in that it offers rapid equilibration, is PAD-inactive, and allows for near maximal resolution of carbohydrate moieties. Recently, Wong and Jane (1995) compared the effects of acetate and nitrate as pushing agents in HPAEC-PAD for the separation and detection of debranched amylopectin. They found that in comparison with the commonly used pushing agent, nitrate offered greater reproducibility, accuracy, and lower limits of detection. Figure 10.11 shows the HPAEC-PAD chromatographic profiles of the enzymatically debranched tapioca and wheat amylopectins using a nitrate gradient. Impressively, baseline resolution of peaks up to a DP of 66 was achieved. The separations were designed to be completed within 100 min.

Slaughter and Livingston III (1994) used HPAEC-PAD to analyze cereal grasses for fructan isomers. They found the HPAEC-PAD method was far superior to existing methods of analysis for isomeric fructans, which were a combination of reverse- and normal-phase separation. The HPAEC-PAD method was complete within 20 min. Resolution of closely related oligosaccharides clearly highlights the power of high-pH AEC, especially when combined with direct detection via PAD.

Determination of trace levels of glycosides in grape musts was undertaken by Pastore, Lavagnini, and Versini (1993). They hydrolyzed the glycosides and analyzed the monosaccharides derived, thus circumventing analysis of the great variety of aglyconic constituents. Analysis of the monosaccharides included separation on a

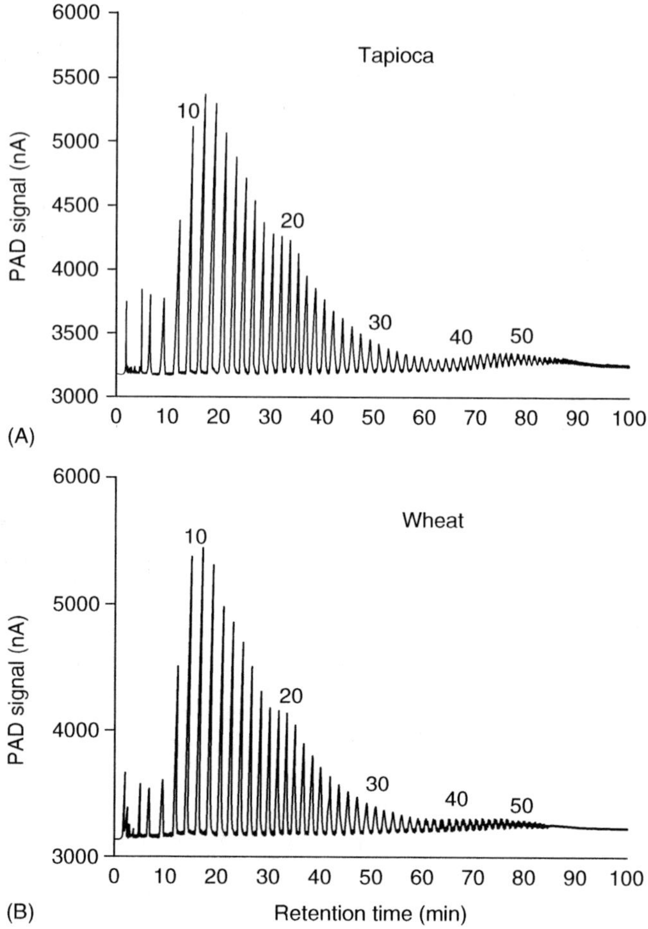

FIGURE 10.11 Chromatographic profiles of the enzymatically debranched (A) tapioca amylopectin and (B) wheat amylopectin using high-performance anion-exchange chromatography with pulsed amperometric detection (HPAEC-PAD). Peak numbers indicate the degree of polymerization (DP). (Reprinted from Wong, K.S. and Jane, J., *J. Liq. Chromatogr.*, 18, 63, 1995. With permission.)

CarboPAC PA1 column using 20 mM NaOH at 15°C. Regeneration of the column using 0.4 M acetic acid was performed daily. Decreasing retention times, and increasing peak heights, were accounted for by frequently injecting standards. Detection using PAD took place at an Au microband electrode fabricated by the authors. The authors determined that the microband electrode did not exhibit lower limits of detection, but that stability of the baseline was more rapidly attained. Reproducibility for the microband electrode was 3% for 10 repeated injections of 20 μL of 2 mg/L glucose. The sugars detected in the grape must included rhamnose, arabinose, glucose, xylose, fructose, and apiose. The fructose is conjectured to come from the must and not from a hydrolyzed glycoside. This is the first report of xylose as a component of grape must glycosides.

Frias and coworkers (1996) undertook the analysis of the raffinose family of oligosaccharides (RFO). Determination of RFO is important because these compounds have been associated with flatus in animals and humans, and have therefore limited the consumption of foods with high RFO content. Extraction of RFOs from pea seed flour is accomplished using 80% ethanol, and the solution is further purified using C-18 SPE cartridges. Analysis of the eluent was performed using capillary zone electrophoresis (CZE)-UV_{220} and HPAEC-PAD. Sugars quantitated were sucrose, raffinose, stachyose, and verbascose. Lactose was used as an internal standard for CZE. Results were similar using each detection scheme, although the HPAEC-PAD was ca. 1000X more sensitive than the CZE-UV.

α-Tomatine in various types of tomatoes was determined using HPLC-PAD by Friedman, Levin, and McDonald (1994). α-Tomatine is a glycoalkaloid found in ripening tomatoes. It is reported to be toxic and have antifungal activity, and its content in tomatoes decreases as the fruit ripens. Separation of glycoalkaloids extracted from fresh tomato was accomplished using reversed-phase chromatography with phosphate/citrate buffer (pH 3.0)/ACN/MeOH, followed by PAD at an Au electrode at a detection potential of 600 mV. Figure 10.12 shows chromatograms of ripe red (A) and unripe green (B) field-grown tomatoes. The α-tomatine in the red ripe tomato is much less than in the unripe green tomato, as expected.

Using the extraction procedure and HPLC-PAD as described in their 1994 paper, Friedman and Levin (1995) determined the α-tomatine in different parts of the tomato plant as well as in various tomato-base foods. Ripe, red tomatoes contained the least α-tomatine. The content of α-tomatine was greater in unripe-green tomatoes, and greatest in leaves, small stems, and flowers of the tomato plant. The authors conjecture that the presence of α-tomatine helps to protect against insects and fugal pathogens. They found no α-tomatine in eggplants, as predicted; but did find small

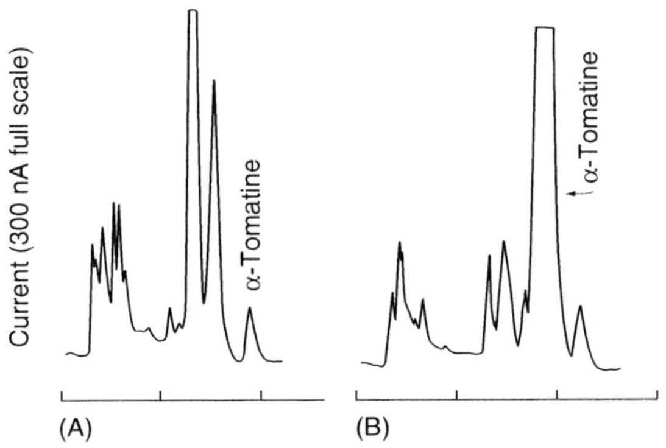

FIGURE 10.12 High-performance liquid chromatograms of α-tomatine in field-grown (A) ripe red tomatoes and (B) unripe green tomatoes. Conditions: Column, Supelcosil LC-ABZ; mobile phase 25% ACN, 15% MeOH, 100 mM sodium phosphate (monobasic) pH 3. (Reprinted from Friedman, M., Levin, C.E., and McDonald, G.M., *J. Agric. Food Chem.*, 42, 1959–1964, 1994.)

amounts in tomatillos, which are not closely related to tomatoes. The authors were able to quantitate the amount of α-tomatine in prepared tomato products such as stewed red tomatoes, pickled green tomatoes, microwaved green tomatoes, and fried green tomatoes.

10.4 REVIEW OF AGRICULTURAL APPLICATIONS

The classification of tobacco products as cigarette or cigar by the Bureau of Alcohol, Tobacco and Firearms is critical for the purpose of application of the correct taxation rate. Since the monosaccharide profile of cigarette and cigar tobacco differs because of the differences in processing, Zook et al. (1996) present a method in which the monosaccharide profile is elucidated and used for this purpose. Samples were prepared by extraction with water, C-18 SPE cleanup, and HPAEC-PAD analysis. Using this method, cigarettes are easily differentiated from cigars because cigarettes contain three to ten times more glucose and fructose and the cigars tested lacked sucrose in all cases.

Fructan is known to accumulate in the tissues of temperate plants during times of cool weather. Since the analysis for fructan is difficult, Chatterton et al. (1989) present a method for the analysis of the synthetic, intermediate fructosylsucrose. Purified sample extracts were analyzed by HPAEC-PAD with a limit of detection (LOD) of 1 mg/mL.

Wilson, Cataldo, and Andersen (1995) studied stress indicators to determine stress in trees. In this work, the presence of sugar alcohols and saccharides was monitored. Samples were prepared by liquid–liquid extraction, followed by drying, reconstitution in water, and analysis by HPAEC-PAD. The PAD procedure was compared with an existing colormetric assay and an HPLC-RID method. The PAD was superior to both methods. Starch was also analyzed by first extracting from the tissue with base and the extract was digested with amyloglucosidase. The resulting samples were analyzed by HPAEC-PAD for glucose. The analysis took only 4 min, so large numbers of samples can be reliably and quickly analyzed with the presented method.

Glucooligosaccharides and polysaccharides were analyzed by hydrophilic interaction chromatography with PAD to evaluate the technique for investigation of polysaccharide structure. Feste and Kahn (1992) were successful in fractionating glucooligosaccharides with a run time of 26 min. Oligosaccharide structure was also the focus of the work done by Ammeraal and coworkers (1991). In this case, the authors were using HPAEC-PAD to determine the structure of linear and branched glucose oligosaccharides. The authors were successful in separating oligosaccharide chains with two or three branch isomers which are singly branched.

10.5 REVIEW OF INDUSTRIAL APPLICATIONS

Sodium carboxymethylcellulose was analyzed by HPAEC-PAD for the purpose of quality control (Kragten, Kamerling, and Vliegenthart, 1992). This analyte is important in the food and coating industry. The samples were prepared by hot acid hydrolysis. Because the HPAEC-PAD method is direct, it has advantages over existing methods which require derivatization.

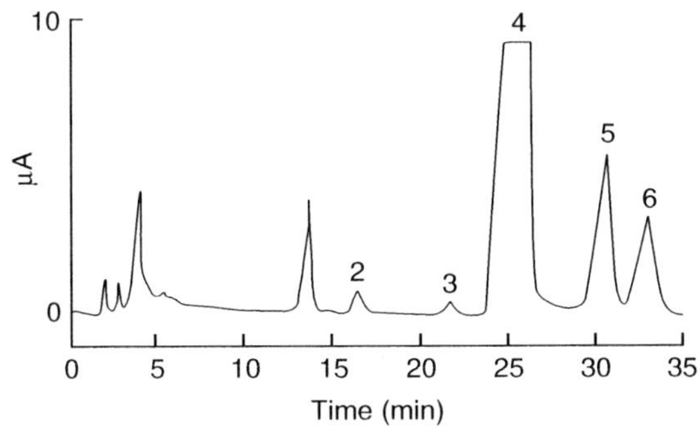

FIGURE 10.13 Chromatogram of an unbleached kraft pulp hydrolyzate. Peaks: 1, 2-deoxyglucose; 2, arabinose; 3, galactose; 4, glucose; 5, xylose; 6, mannose. Chromatographic conditions: CarboPac PA-1 + guard (Dionex Corp.). Eluent: 3 mM NaOH; PAD-1 detector (Dionex) with an Au working electrode and Ag/AgCl reference. (Reprinted from Sullivan, J. and Douek, M., *J. Chromatogr.*, 671(1–2), 339–350, 1994. With permission.)

The paper industry is notorious for discharging toxic chemicals into natural receiving waters. Ozonolysis is being investigated as an alternative to the bleaching process because it would eliminate the use of chlorinated organic compounds currently used for this purpose. However, with current methods of ozonolysis, the ozone modifies the cellulose such that the pulp properties are lost. Van Nifterik et al. (1993) use HPAEC-PAD to study the action of ozone on cellulose to minimize the modification process. Samples were ozonated then passed through a polyvinylpyrrolidone filter cartridge before analysis.

Carbohydrate analysis of samples in various stages of the paper-making process is important to the optimization of the process. Breakdown of polysaccharides to monosaccharides is undesirable. To monitor paper processing, Sullivan and Douek (1994) developed an HPAEC-PAD method for determining monosaccharides in wood, pulp, and process liquor samples. This method yielded an easy, fast method of determining monosaccharides in a quite complex matrix. Figure 10.13 shows a chromatogram using the HPAEC-PAD method with an SPE cleanup step of an unbleached kraft pulp hydrolyzate. The chromatogram is quite clean with minimal matrix peaks.

Free glycerol content in biofuels is an important indicator of the performance of a particular fuel. Lozano et al. (1996) used HPLC-PAD to determine glycerol in fuel after an extraction of the glycerol with a water/alcohol mixture. The HPLC-PAD method was compared to an established enzymatic method. The HPLC-PAD method compared favorably with the established method because although there was no statistical difference in the results from the two methods, the HPLC-PAD method is easier to perform.

10.6 FOOD-RELATED, NONCARBOHYDRATE APPLICATIONS

Several studies have focused on the detection of aliphatic amines by HPLC-PAD (Dobberpuhl, Hoekstra, and Johnson, 1996; Dobberpuhl and Johnson, 1995), but amine detection are oxide-catalyzed and are best performed using IPAD. A fine

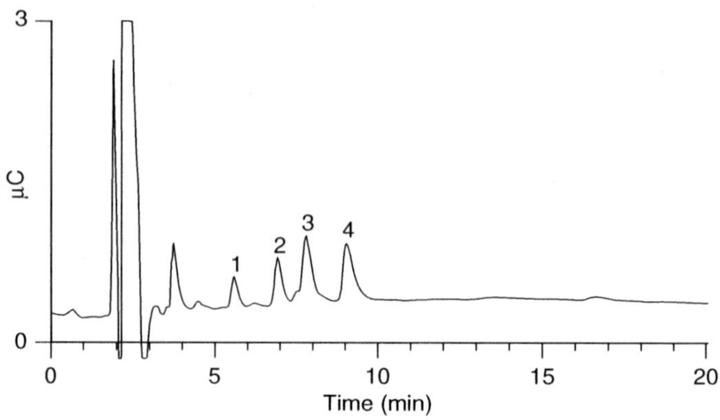

FIGURE 10.14 High-performance anion-exchange chromatography with pulsed amperometric detection (HPLC-PAD) analysis of biogenic amines in fish using HPLC-PAD. Peaks: 1, putrescine; 2, histidine; 3, cadarverine; and 4, histamine. (Reprinted from Draisci, R. et. al., *Chromatographia*, 35, 584–590, 1993. With permission.)

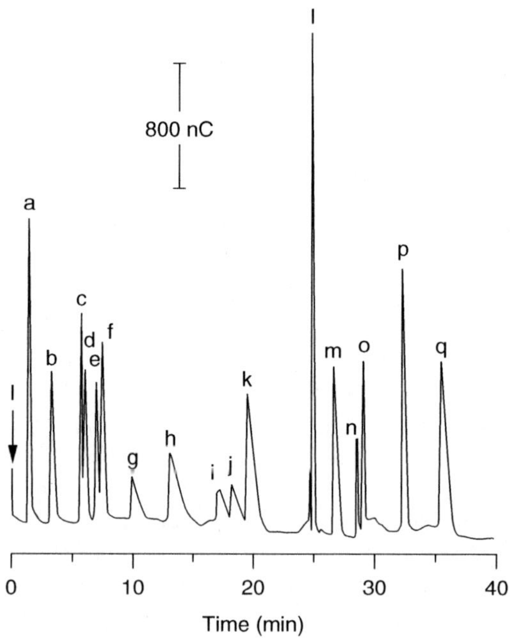

FIGURE 10.15 Separation of 17-component amino acid hydrolyzate. Conditions: Column, Dionex AS-8; gradient. Peaks (25 nmol each, except 12.5 nmol for cystine): a, arginine; b, lysine; c, threonine; d, alanine; e, glycine; f, serine; g, valine; h, proline; i, isoleucine; j, leucine; k, methionine; l, histidine; m, phenylalanine; n, glutamic acid; o, aspartic acid; p, cystine; and q, tyrosine. (Reprinted from Welch, L.E., LaCourse, W.R., Mead, D.A., Jr., and Johnson, D.C., *Anal. Chem.*, 61, 555, 1989.)

example involves the use of HPLC-IPAD to directly determine biogenic amines as indicators of seafood spoilage (Draisci et al., 1993). They separated putrescine, histidine, cadaverine, and histamine using a cation-exchange column with an ACN gradient. Figure 10.14 shows the chromatograms of amines extracted from spoiled canned herrings. Detection was performed by IPAD following postcolumn addition of NaOH.

Simple mixtures of amino acids can be separated isocratically using AEC (Welch et al., 1990). For complex mixtures, it is essential to perform separations with gradient-elution chromatography. Figure 10.15 shows the chromatogram of a protein hydrolyzate containing 17 amino acid residues using HPLC-IPAD with a pH reference electrode (Welch et al., 1989). The amino acid mixture was separated using an anion-exchange column with a quaternary gradient, which incorporated both a pH and an organic modifier gradient. An improved separation of 21 amino acids was performed by Martens and Frankenberger (1992). Presently, detection limits (i.e., $S/N = 3$) for HPLC-IPAD for amino acids are typically 1–50 pmol injected with comparable sensitivities for primary and secondary amino acids.

LaCourse and Owens (1995) have demonstrated the superiority of IPAD over PAD for the determination of thiocompounds using standard reversed-phase conditions. These results were as expected since sulfur detections are Mode II. Figure 10.16 shows response of biotin (160 pmol injected) using microbore chromatography.

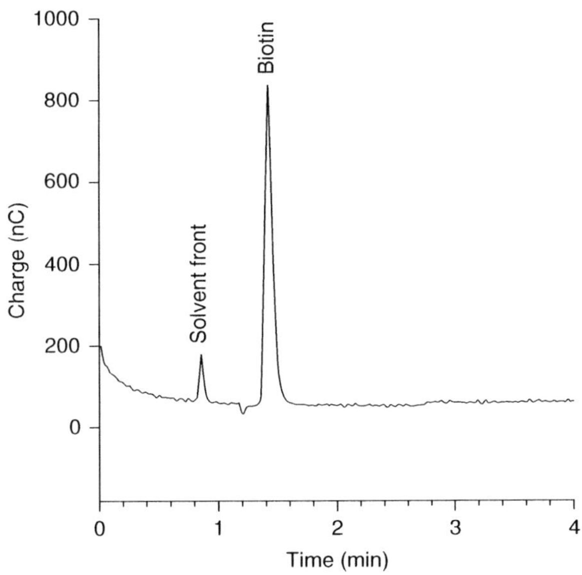

FIGURE 10.16 Chromatogram of biotin (165 pmol injected) using High-performance anion-exchange chromatography with integrated pulsed amperometric detection (HPLC-IPAD). Conditions: C_{18} 1-mm SepStik column; mobile phase, 90/10 vol% pH 7.2 phosphate buffer, 100 mM ACN; 0.5 μL injection volume; flow rate, 70 μL/min; and detection, IPAD. (Reprinted from LaCourse, W.R., Dasenbrock, C.O., and Zook, C.M., *Sem. Food Anal.*, 2, 5–41, 1997. With permission.)

Biotin, a thioether, was detected sensitively without derivatization. In addition, IPAD enables the direct determination of thio-redox couples (i.e., –SH/–S–S–) and numerous other sulfur moieties at a single Au electrode. The high selectivity of PED for thiocompounds under mildly acidic conditions reduces sample preparation and produces simpler chromatograms of complex mixtures. Figure 10.17 shows the determination of thiocompounds in (A) grapefruit juice, (B) watermelon, and (C) chicken liver. In general, the IPAD waveform gives lowers LODs, more stable baselines, and eliminates oxide-induced artifacts.

More recently, the analysis of milk for sulfur-containing antibiotic residues was performed using a C-8 column with a mobile phase of 20% 500 mM acetate buffer

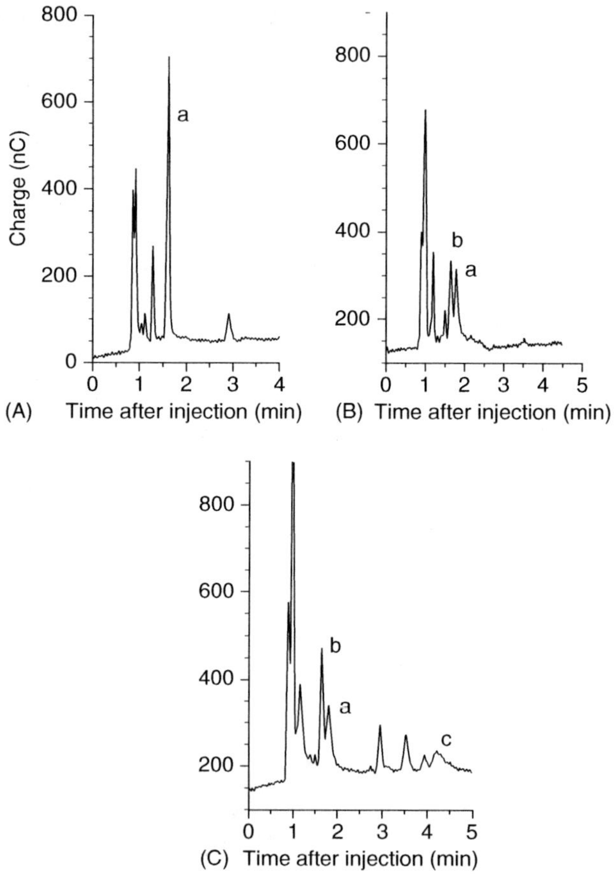

FIGURE 10.17 Detection of thiocompounds in various samples by High-performance anion-exchange chromatography with integrated pulsed amperometric detection (HPLC-IPAD). Samples were (A) grapefruit juice, (B) watermelon, and (C) chicken liver. Condition: C_{18}, 1-mm SepStik column; 99.6% 100 mM phosphate buffer (pH 3)/0.4% acetonitrile; 65 μL/min; 0.5 μL injection volume; and detection, IPAD waveform in Table 10.3. Peaks: a, GSH; b, methionine; and c, GSSG. (Reprinted from LaCourse, W.R. and Owens, G.S., *Anal. Chim. Acta*, 307, 301, 1995. With permission.)

(pH 3.75)/5% ACN/75% water (Dasenbrock, Zook, and LaCourse, 1996). Table 10.4 shows that detection limits of 1–5 ppb were achieved with single injections of each compound. All of these detection limits are at or below the target levels for antibiotic residues in milk set by the FDA. The IPAD analysis was also compared with a UV detection scheme by adding a UV detector in line. In all cases, the IPAD detection was more sensitive and selective for the compounds of interest than UV detection.

Many of the early efforts to discern the mechanism of anodic oxygen transfer utilized inorganic species as test species. It only follows that these compounds are also applicable to PED following ion chromatography. Wagner and McGarrity (1992) used ion chromatography followed by PAD to determine sulfite in beer. Correlation coefficients of .997 or better were obtained with excellent spike recoveries. HPAEC-PAD results were in good agreement with the standard pararosaniline method. In fact, this method was subsequently automated resulting in excellent precision and accuracy (Wagner, 1995).

10.7 CONCLUSIONS

The significance of PED within chemical and biochemical analysis can be best appreciated in view of the commonly held impression, based on attempted detections at constant (dc) applied potential, that polar aliphatic compounds are generally not electroactive. Furthermore, these compounds often do not possess chromophoric or fluorophoric groups and, as a consequence, direct and sensitive detection by photometric techniques is not possible. PED has been applied to the direct, sensitive, and reproducible detection of a large variety of polar aliphatic compounds in foods. These compounds include carbohydrates, amines, amino acids, and thiocompounds. This technique is simple, in that it requires no derivatization.

PED is ideally suited to the detection of virtually all carbohydrates and carbohydrate analogs. In addition, the optimized waveform for any one compound with a class is applicable to all others in that class. IPAD facilitates Mode II detections and increases gradient compatibility. PED is compatible with all aqueous-based separations, and sulfur-based compounds are selectively detected under typical reversed-phase conditions.

PED offers many advantages over alternate detection schemes for LC and capillary electrophoresis. Because electrochemical detection relies on reaction at the electrode surface, detector cells can be miniaturized without sacrificing sensitivity. This advantage makes them especially suited for microbore techniques. As microbore techniques gain prominence, low-volume detection schemes will become more important. Pulsed potential cleaning eliminates the need for daily polishing of the electrode which renders PED more convenient experimentally than dc amperometry. PED is sensitive and selective for specific functional groups on the analyte, excluding components which do not contain those functional groups. In this manner, some potential interferents are not detected and the data interpretation is simplified. When analyzing complex sample matrices, such as food matrices, this advantage becomes quite significant.

ACKNOWLEDGMENTS

The author wishes to thank Suresh Ralapati for his contributions to this Chapter. The author also acknowledges the support of Dionex Corporation.

REFERENCES

Adams, P., Zegeer, A., Bohnert, H.J., and Jensen, R.G., *Anal. Biochem.*, 214(1), 321, 1993.

Adams, R.N., *Electrochemistry at Solid Electrodes*, Marcel Dekker, New York, 1969.

Agus, M.S.D., Swain, J.F., Larson, C.L., Eckert, D.S., and Ludwig, D.S., *Am. J. Clin. Nutr.*, 71, 901, 2000.

Ammeraal, R.N., Delgado, G.A., Tenbarge, F.L., and Friedman, R.B., *Carbohydr. Res.*, 215 (1), 179, 1991.

Arai, S., *Biosci. Biotechnol. Biochem.*, 60, 9, 1996.

Armstrong, G., Himsworth, F.R., and Butler, J.A.V., *Proc. R. Soc. London (A)*, 143, 89, 1934.

Asp, N.-G., *Am. J. Clin. Nutr.*, 61(4), 930S, 1995.

Asp, N.-G., Schweizer, T.F., Southgate, D.A.T., and Theander, O., Dietary fibre analysis, *Dietary Fibre. A Component of Food. Nutritional Function in Health and Disease*, Schweizer, T.F. and Edwards, C.A. (Eds.), Springer, London, p. 57, 1992.

Austin, D.S., Polta, J.A., Polta, T.Z., Tang, A.P.C., Cabelka, T.D., and Johnson, D.C., *J. Electroanal. Chem.*, 108, 227, 1984.

Axelsen, M. et al., *Am. J. Clin. Nutr.*, 71, 1108, 2000.

Berger, T.A., U.S. Patent No. 4,496,454, Hewlett-Packard Corp., Avondale, PA, 1985.

Bernal, J.L., Del Nozal, M.J., Toribio, L., and Del Alamo, M., *J. Agric. Food Chem.*, 44(2), 507, 1996.

Björk, I., Starch: Nutritional aspects, *Carbohydrates in Food*, Eliasson, A.-C. (Ed.), Marcel Dekker, Inc., New York, p. 505, 1996.

Breiter, M.W., *Electrochim. Acta*, 8, 973, 1963.

Brown, I., *Nutr. Rev.*, 54(11), Part II, S115, 1996.

Burkitt, D.P. and Trowell, H.S., *Refined Carbohydrate Foods and Disease: Some Implications of Dietary Fibre*, Academic Press, London, United Kingdom, 1975.

Chapman, N., *Prepared Foods*, February, 2000.

Chatterton, J.N., Harrison, P.A., Thornley, W.R., and Bennett, J.H., *Plant Physiol. Biochem.*, 27(2), 289, 1989.

Cheney, S.J., *Vegetarian Times*, June, 1995.

Cho, S.S., Prosky, L., and Dreher, M. (Eds.), *Complex Carbohydrates in Foods*, Marcel Dekker, Inc., New York, 1999.

Clark, D., Fleishman, M., and Fletcher, D., *J. Electroanal. Chem.*, 36, 137, 1972.

Clydesdale, F.M., *Nutr. Rev.*, 55, 413, 1997.

Corradini, C., Cristalli, A., and Corradini, D., *J. Liq. Chromatogr.*, 16(16), 3471, 1993.

Corradini, C., Cristalli, A., and Corradini, D., *Ital. J. Food Sci.*, 6(1), 103, 1994.

Coussement, P., Non-digestible oligosaccharides for functional foods, *Proc. First Vitafoods Int. Conf.*, Blenford, D.E. (Ed.), Food Tech Europe, Copenhagen, Denmark, 1997.

Craig, W.J., *J. Am. Diet. Assoc.*, 97, S199, 1997.

Daly, M.E. et al., *Am. J. Clin. Nutr.*, 71, 1516, 2000.

Dasenbrock, C.O., Zook, C.M., and LaCourse, W.R., "Analysis of Antibiotics in Milk using HPLC-IPAD" paper presented at Ohio Valley Chromatography Symposium, Oxford, OH, June, 1996.

Dionex Application Note 67, Determination of Plant Derived Neutral Oligo and Polysaccharides, Dionex Corporation, Sunnyvale, CA.

Dionex Technical Manual, *Better Solutions for Food and Beverage Analysis*, Dionex Corporation, Sunnyvale, CA, 1995.
Dobberpuhl, D.A., Hoekstra, J.C., and Johnson, D.C., *Anal. Chim. Acta*, 322, 55, 1996.
Dobberpuhl, D.A. and Johnson, D.C., *Anal. Chem.*, 67, 1254, 1995.
Draisci, R., Cavalli, S., Lucentini, L., and Stacchini, A., *Chromatographia*, 35, 9, 1993.
DuPont, J., *Nutr. Today*, 34(6), 234, 1999.
Edwards, P. and Haak, K.K., *Am. Lab.*, April, 78, 1983.
Edwards, W.T., Pohl, C.A., and Rubin, R., *Tappi J.*, 70, 138, 1987.
Eliasson, A.-C. and Gudmundsson, M., Starch: Physicochemical and functional aspects, *Carbohydrates in Food*, Eliasson, A.-C. (Ed.), Marcel Dekker, Inc., New York, p. 431, 1996.
Engfer, M.B. et al., *Am. J. Clin. Nutr.*, 71, 1589, 2000.
Englyst, H.N. and Cummings, J.H., Non-starch polysaccharides (dietary fiber) and resistant starch, *New Developments in Dietary Fiber*, Furda, I. and Brine, J. (Eds.), Plenum Press, New York, p. 205, 1990.
Englyst, H.N., Kingman, S.M., and Cummings, J.H., *Eur. J. Clin. Nutr.*, 46, S33, 1992.
Ewing, A.G., Dayton, M.A., and Wightman, R.M., *Anal. Chem.*, 53, 1842, 1981.
FAO, Carbohydrates in nutrition, Report of a Joint FAO/WHO Expert Consultation, *FAO Food and Nutrition Paper 66*, Rome, Italy, 1998.
Feste, A.S. and Khan, I., *J. Chromatogr.*, 630(1–2), 129, 1992.
Fleet, B. and Little, C.J., *J. Chromatogr. Sci.*, 12, 747, 1974.
Frias, J., Price, K.R., Fenwick, G.R., Hedley, C.L., Sorensen, H., and Vidal-Valverde, C., *J. Chromatogr.*, 719, 213, 1996.
Friedman, M. and Levin, C.E., *J. Agric. Food Chem.*, 43, 1507, 1995.
Friedman, M., Levin, C.E., and McDonald, G.M., *J. Agric. Food Chem.*, 42, 1959, 1994.
Garleb, K.A., Bourquin, L.D., and Fahey, G.C., Jr., *J. Agric. Food Chem.*, 37, 1287, 1989.
Garleb, K.A., Bourquin, L.D.. and Fahey, G.C., Jr., *J. Food Sci.*, 56(2), 423, 1991.
Gilman, S., *J. Phys. Chem.*, 67, 78, 1963.
Gilman, S., , *Electroanalytical Chemistry*, Bard, A.J. (Ed.), Vol. 2, Marcel Dekker, New York, 1967.
Giner, J., *Electrochim. Acta*, 9, 63, 1964.
Greenfield, H. and Southgate, D.A.T., *Food Composition Data. Production, Management and Use*, Elsevier Applied Science, London, United Kingdom, 1992.
Hammett, L.P., *J. Am. Chem. Soc.*, 46, 7, 1924.
Harper, A.E., *Nutr. Today*, 34(6), 216, 1999.
Hasler, C.M., *Nutr. Rev.*, 54(11), Part II, S6, 1996.
Henshall, A., High-performance anion exchange chromatography with pulsed amperometric detection (HPAE-PAD): A powerful tool for the analysis of dietary fiber and complex carbohydrates, *Complex Carbohydrates in Foods*, Cho, S.S., Prosky, L., and Dreher, M. (Eds.), Marcel Dekker, Inc., New York, p. 267, 1999.
Herman, P., *Vegetarian Times*, August, 1997.
Hotchkiss, A.T., Jr., El-Bahtimy, K., and Fishman, M.L., *Mod. Methods Plant Anal.*, 17, 129, 1996.
Hotchkiss, A.T., Jr. and Hicks, K.B., *Anal. Biochem.*, 184(2), 200, 1990.
Hotchkiss, A.T., Jr. and Hicks, K.B., *Carbohyr. Res.*, 247, 1, 1993.
Hughes, S. and Johnson, D.C., *Anal. Chim. Acta*, 132, 11, 1981.
Hughes, S. and Johnson, D.C., *J. Agric. Food. Chem.*, 30(4), 712, 1982.
Hughes, S. and Johnson, D.C., *Anal. Chim. Acta*, 149, 1, 1983.
Hughes, S., Meschi, P.L., and Johnson, D.C., *Anal. Chim. Acta*, 132, 11, 1981.
Ichiki, H., Semma, M., Sekiguchi, Y., Nakamura, M., and Ito, Y., *Nippon Shokuhin Kagaku Gakkaishi*, 2(2), 119, 1996.

IOM/NAS, *Institute of Medicine/National Academy of Sciences*, Thomas, P.R. and Earl, R. (Eds.), National Academy Press, Washington DC, WA, p. 109, 1994.
Jago, D., *Prepared Foods*, April, 2000.
Jenkins, D.J.A. et al., *Am. J. Clin. Nutr.*, 71, 401, 2000.
Johnson, D.C., *Nature*, 321, 451, 1986.
Johnson, D.C. and LaCourse, W.R., *Anal. Chem.*, 62, 589A, 1990.
Johnson, D.C. and LaCourse, W.R., *Electroanalysis*, 4, 367, 1992.
Johnson, D.C., Polta, J.A., Polta, T.Z., Neuburger, G.G., Johnson, J., Tang, A.P.C., Yeo, I.H., and Baur, J., *J. Chem. Doc. Faraday Trans.*, 1(82), 108, 1986.
Johnson, D.C. and Polta, T.Z., *Chromatogr. Forum*, 1, 37, 1986.
Katan, M.B., *Lancet*, March 27, 1999.
Kissinger, P.T., *Laboratory Techniques in Electroanalytical Chemistry*, Kissinger, P.T and Heineman, W.R (Eds.), Marcel Dekker, New York, 1984.
Koizumi, K., Fukuda, M., Hizukuri, S., *J Chromatogr*, 585(2), 233, 1991.
Koizumi, K., Kubota, Y., Tanimoto, T., and Okada, Y., *J. Chromatogr.*, 464(2), 365, 1989.
Kragten, E.A., Kamerling, J.P., and Vliegenthart, J.F.G., *J. Chromatogr.*, 623(1), 49, 1992.
Kuhn, M.E., *Food Process.*, 59(5), 21, 1998.
LaCourse, W.R., *Analusis*, 21, 181, 1993.
LaCourse, W.R., *Pulsed Electrochemical Detection in High Performance Liquid Chromatography*, John Wiley and Sons, New York, 1997.
LaCourse, W.R., Dasenbrock, C.O., and Zook, C.M., *Sem. Food Anal.*, 2, 5, 1997.
LaCourse, W.R., Jackson, W.A., and Johnson D.C., *Anal. Chem.*, 61(22), 2466, 1989.
LaCourse, W.R. and Johnson, D.C., *Advances in Ion Chromatography*, Jandik, P. and Cassidy, R.M. (Eds.), Vol. 2, Century International Inc., Medfield, MA, 1990.
LaCourse, W.R. and Johnson, D.C., *Carbohydr. Res.*, 215, 159, 1991.
LaCourse, W.R. and Johnson, D.C., *Anal. Chem.*, 65, 50, 1993.
LaCourse, W.R., Johnson, D.C., Rey, M.A., and Slingsby, R.W., *Anal. Chem.*, 63(2), 134, 1991.
LaCourse, W.R., Mead, D.A., Jr., and Johnson, D.C., *Anal. Chem.*, 62(2), 220, 1990.
LaCourse, W.R. and Owens, G.S., *Anal. Chim. Acta*, 307, 301, 1995.
Larew, L.A. and Johnson, D.C., *Anal. Chem.*, 60(18), 1867, 1988.
Larew, L.A, Mead, D.A., Jr., and Johnson, D.C., *Anal. Chim. Acta*, 204(1–2), 43, 1988.
Liu, S. et al., *Am. J. Clin. Nutr.*, 71, 1455, 2000.
Low, N.H., *Am. Lab.*, 28(6), 35M, 1996.
Low, N.H. and Wudrich, G.G., *J. Agric. Food Chem.*, 41(6), 902, 1993.
Lozano, P., Chirat, N., Graille, J., and Pioch, D., *Fresenius' J. Anal. Chem.*, 354(3), 319, 1996.
Lu, Z.X. et al., *Am. J. Clin. Nutr.*, 71, 1123, 2000.
MacDonald, A. and Duke, P.D., *J. Chromatogr.*, 83, 331, 1973.
Mannie, E., *Prepared Foods*, June, 2000.
Martens, D.A. and Frankenberger, W.T., Jr., *Chromatographia*, 30(5–6), 249, 1990a.
Martens, D.A. and Frankenberger, W.T., Jr., *Chromatographia*, 30(11–12), 651, 1990b.
Martens, D.A. and Frankenberger, W.T., Jr., *J. Liq. Chromatogr.*, 15(3), 423, 1992.
Mead, D.A, Jr., Larew, L.A., and Johnson, D.C., *Advances in Ion Chromatography*, Jandik, P. and Cassidy, R.M. (Eds.), Vol. 1, Century International Inc., Medfield, MA, 1989.
Messina, M. and Messina, V., *Adv. Exp. Med. Biol.*, 401, 207, 1996.
Meyer, A., *Food Process.*, 58(8), 32, 1998.
Meyer, K.A. et al., *Am. J. Clin. Nutr.*, 71, 921, 2000.
Milner, J.A., *J. Nutr.*, 129, 1395S, 1999.
Milner, J.A. and Craig, L., *Am. J. Clin. Nutr.*, 71 (Suppl.), 1653S, 2000.
Neuberger, G.G. and Johnson, D.C., *Anal. Chem.*, 60, 2288, 1988.

Neuberger, G.G. and Johnson, D.C., *Anal. Chim. Acta*, 192, 205, 1990.

Neuburger, G.G. and Johnson, D.C., *Anal. Chem.*, 59, 203, 1987.

Ngoviatchai, A. and Johnson, D.C., *Anal. Chim. Acta*, 215, 1, 1988.

Ohr, L.M., *Prepared Foods*, October, 1998.

Ohr, L.M., *Prepared Foods*, May, 1999.

Oku, T., *Nutr. Rev.*, 54(11), Part II, S59, 1996.

Olechno, J., Carter, S.R., Edwards, W.T., and Gillen, D.G., *Am. Biotechnol. Lab.*, 5, 38, 1987.

Park, M.K., Park, J.H., Lee, M.Y., Kim, S.J., and Park, I.J., *J. Liq. Chromatogr.*, 17(5), 1171, 1994.

Parks, E.J. and Hellerstein, M.K., *Am. J. Clin. Nutr.*, 71, 412, 2000.

Paskach, T.J., Lieker, P.J., and Thielecke, K., *Carbohydr. Res.*, 215, 1, 1991.

Pastore, P., Lavagnini, I., and Versini, G., *J. Chromatogr.*, 634, 47, 1993.

Peschet, J.L. and Giacalone, A., *Ind. Aliment Agric.*, 108(7–8), 583, 1991.

Pollman, R.M., *J. AOAC*, 72(3), 425, 1989.

Polta, J.A. and Johnson, D.C., *J. Liq. Chromatogr.*, 6, 1727, 1983.

Polta, T.Z. and Johnson, D.C., *J. Electroanal. Chem.*, 209, 159, 1986.

Polta, T.Z., Johnson, D.C., and Luecke, G.R., *J. Electroanal. Chem.*, 209, 171, 1986.

Prodolliet, J., Bugner, E., and Feinberg, M., *J. AOAC*, 78(3), 768, 1995.

Quagliani, D., *Better Homes Gardens*, May, 1997.

Quigley, M.E. and Englyst, H.N., *Analyst*, 117, 1715, 1992.

Quigley, M.E. and Englyst, H.N., *Analyst*, 119, 1511, 1994.

Roberfroid, M.B., *Nutr. Rev.*, 54(11), Part II, S38, 1996.

Roberfroid, M.B., *Am. J. Clin. Nutr.*, 71 (Suppl.), 1682S, 2000.

Rocklin, R.D. and Pohl, C.A., *J. Liq. Chromatogr.*, 6(9), 1577, 1983.

Sheehy, P.J.A. and Morrissey, P.A., Functional foods: Prospects and perspectives, *Nutritional Aspects of Food Processing and Ingredients*, Henry, C.J.K. and Heppell, N.J. (Eds.), Aspen Publishers, Gaithersburg, MD, p. 45, 1998.

Slaughter, L.H. and Livingston, D.P., III, *Carbohydr. Res.*, 253, 287, 1994.

Smith, S.R. et al., *Am. J. Clin. Nutr.*, 71, 450, 2000.

Soga, T., Inoue, Y., and Yamaguchi, K., *J. Chromatogr.*, 625, 151, 1992.

Southgate, D.A.T., *Determination of Food Carbohydrates*. Elsevier Science Publishers, Ltd., Barking, United Kingdom, 1991.

Stefansson, M. and Lu, B., *Chromatographia*, 35(1–2), 61, 1993.

Stuckel, J.G. and Low, N.H., *J. Agric. Food Chem.*, 43(2), 3046, 1995.

Stulik, W. and Hora, V., *J. Electroanal. Chem.*, 70, 253, 1976.

Sullivan, J. and Douek, M., *J. Chromatogr.*, 671(1–2), 339, 1994.

Swallow, K.W. and Low, N.H., *J. Agric. Food Chem.*, 38(9), 1828, 1990.

Swallow, K.W. and Low, N.H., *J. Assoc. Off. Anal. Chem.*, 77(3), 695–702, 1994.

Swallow, K.W., Low, N.H., and Petrus, D.R., *J. Assoc. Off. Anal. Chem.*, 74(2), 341, 1991.

Tengyl, J., *Electrochemical Detectors*, Ryan, T.H. (Ed.), Plenum Press, New York and London, 1984.

Thompson, F.E. et al., *Am. J. Clin. Nutr.*, 71, 1503, 2000.

Trowell, H., *Rev. Eur. Etud. Clin. Biol.*, 17(4), 345, 1972.

Turner, L., *Vegetarian Times*, February, 1999.

U.S. Senate Select Committee on Nutrition and Human Needs, *Dietary Goals for the United States*, 2nd edition, U.S. Government Printing Office, Washington DC, WA, 1977.

Van Nifterik, L., Xu, J., Laurent, J.L., Mathieu, J., and Rakoto, C., *J. Chromatogr.*, 640(1–2), 335, 1993.

Van Riel, J. and Olieman, C., *Carbohydr.Res.*, 215, 39–46, 1991.

Van Rooijan, H.W. and Poppe, H., *Anal. Chim. Acta*, 130, 91, 1991.

Vandeberg, P.G., Kowagoe, J.L., and Johnson, D.C., *Anal. Chim. Acta*, 260(1), 1, 1992.

Wagner, H.P., *J. Am. Soc. Brew. Chem.*, 53, 82, 1995.

Wagner, H.P. and McGarrity, M.J., *J. Am. Soc. Brew. Chem.*, 50, 1, 1992.

Walqvist, M.L., Functional foods in the control of obesity, *Functional Foods: Designer Foods, Pharmafoods, Nutraceuticals*, Goldberg, I. (Ed.), Chapman and Hall, London, United Kingdom, p. 71, 1994.

Wang, W.T. and Zopf, D., *Carbohydr. Res.*, 189, 1, 1989.

Weisz, J. et al., Poster No. AAFC 052, *Food Science at Work: Technical Communications from the Guelph Food Research Community*, September 16, Guelph, Ontario, Canada, 1999.

Welch, L.E., LaCourse, W.R., Mead, D.A., Jr., and Johnson, D.C., *Anal. Chem.*, 61, 555, 1989.

Welch, L.E., LaCourse, W.R., Mead, D.A., Jr., and Johnson, D.C., *Talanta*, 37, 377, 1990.

White, D.R., Jr. and Cancalon, P.F., *J. Assoc. Off. Anal. Chem.*, 75(3), 584, 1992.

White, D.R., Jr. and Widmer, W.W., *J. Agric. Food Chem.*, 38(10), 1918, 1990.

Wilson, R., Cataldo, A., and Andersen, C.P., *Can. J. For. Res.*, 25(12), 2022, 1995.

Williamson, D.F., *Nutr. Rev.*, 54(11), Part II, S1, 1996.

Wong, K.S. and Jane, J., *J. Liq. Chromatogr.*, 18, 63, 1995.

Zeisel, S.H., *Science*, 285, 1853, 1999.

Zook, C.M. and LaCourse, W.R., *Curr. Sep.*, 14(2), 48, 1995.

Zook, C.M., Patel, P.M., LaCourse, W.R., and Ralapati, S., *J. Agric. Food Chem.*, 44(7), 1773, 1996.

Index

A

B

C

D

G

H

I

K

L

M

N

O

P

Q

R

S

W